Introduction to fiber optics

Recent advances in the development of low-loss optical fibers have revolutionized the field of telecommunications, and fiber-based networks form a key part of international communication systems. This comprehensive book provides an introduction to the physical principles of optical fibers and discusses in detail their use in modern optical communication systems and sensor technology.

The authors begin by setting out the basic propagation characteristics of single-mode and multimode optical fibers. In later chapters they cover optical sources (such as laser diodes), optical detectors, and fiber optic communication system design. They also treat a wide variety of related topics such as doped fiber amplifiers, soliton propagation, dispersion compensation, fiber Bragg gratings and fiber sensors, as well as measurement techniques for the characterization of optical fibers.

Throughout the book, physical and engineering aspects of the subject are interwoven, and many worked examples and exercises are included. It will be an ideal textbook for undergraduate or graduate students taking courses in optical fiber communications, photonics, or optoelectronics.

Ajoy Ghatak is a professor of physics at the Indian Institute of Technology, New Delhi. He obtained his M.Sc. from Delhi University and Ph.D. from Cornell University. An accomplished author and physicist, he has been a Fellow of the Optical Society of America since 1995.

K. Thyagarajan is a professor of physics at the Indian Institute of Technology, New Delhi. He received his B.Sc. and M.Sc. from Delhi University and his Ph.D. from the Indian Institute of Technology, Delhi. He is a member of the Executive Committee of the Optical Society of India. *Introduction to fiber optics* is Professors Ghatak and Thyagarajan's fourth book together.

Introduction to fiber optics

AJOY GHATAK   K. THYAGARAJAN

PUBLISHED BY THE PRESS SYNDICATE OF THE UNIVERSITY OF CAMBRIDGE
The Pitt Building, Trumpington Street, Cambridge CB2 1RP, United Kingdom

CAMBRIDGE UNIVERSITY PRESS
The Edinburgh Building, Cambridge CB2 2RU, UK http://www.cup.cam.ac.uk
40 West 20th Street, New York, NY 10011-4211, USA http://www.cup.org
10 Stamford Road, Oakleigh, Melbourne 3166, Australia

First published 1998

Printed in the United States of America

Typeset in Times Roman

Library of Congress Cataloging-in-Publication Data
Ghatak, A. K. (Ajoy K.). 1939–
Introduction to fiber optics / Ajoy Ghatak and K. Thyagarajan.
p. cm.
ISBN 0-521-57120-0 (hb). – ISBN 0-521-57785-3 (pb)
1. Fiber optics. I. Thyagarajan, K.
TA1800.G48 1998
621.36'92–DC21 97-36652
CIP

A catalog record for this book is available from the British Library.

ISBN 0 521 57120 0 hardback
ISBN 0 521 57785 3 paperback

Contents

Preface

The dramatic reduction in transmission loss of optical fibers coupled with equally important developments in the area of light sources and detectors have brought about a phenomenal growth of the fiber optic industry during the past two decades. Indeed, the birth of optical fiber communications coincided with the fabrication of low-loss optical fibers and operation of room temperature semiconductor lasers in 1970. Since then, the scientific and technological progress in the field has been so phenomenal that optical fiber communication systems find themselves already in the fifth generation within a span of about 25 years. Broadband optical fiber amplifiers coupled with wavelength division multiplexing techniques and soliton communication systems are some of the very important developments that have taken place in the past few years, which are already revolutionizing the field of fiber optics. Although the major application of optical fibers has been in the area of telecommunications, many new related areas such as fiber optic sensors, fiber optic devices and components, and integrated optics have witnessed considerable growth. In addition, optical fibers allow us to perform many interesting and simple experiments permitting us to understand basic physical principles.

With the all-pervading applications of optical fibers, many educational institutions have started courses on fiber optics. At our Institute, we have a three-semester M.Tech. program on Optoelectronics and Optical Communications (jointly run by the Physics and Electrical Engineering Departments) in which we have an extensive coverage of the theory of optical fibers and optical fiber communications and also many experiments and projects associated with it. We also have an elective paper on fiber optics for our M.Sc. (Physics) students. The present book is an outgrowth of the lectures that have been delivered both to our M.Sc. as well as to our M.Tech. students during the past fifteen years. Many of the experiments described in the book have also evolved as simple and elegant demonstration of optics principles to our undergraduate engineering students taking a course on Optics. The material presented here and also the associated experiments have been very successfully used in various summer and winter schools in the area of fiber optics conducted by our Institute. It was felt that there is a need today of a textbook at the undergraduate level covering the field from the basic concepts to the very recent advances, including various applications of this exciting field.

The book aims to cover the field of fiber optics and its many applications at an undergraduate level. The book also contains many solved and unsolved problems, some of which will give the reader a greater feel for numbers while the others are expected to help in a greater understanding of the concepts developed in the book. We would greatly appreciate receiving suggestions for further improvement of the book. We would also be very grateful if any errors in the book are pointed out to us.

New Delhi
March 1997

Ajoy Ghatak
K. Thyagarajan

Acknowledgments

It is our most pleasant task to acknowledge the help that we have received from numerous individuals in the writing of this book. We have been working in the general area of guided wave optics for over twenty years and the writing of this book has itself taken over ten years. During this period we have had close interactions with many individuals. In particular, we are deeply indebted to Professor M. S. Sodha (who introduced us to this field) and to our colleagues Professor Ishwar Goyal, Dr. Banshidhar Gupta, Dr. Ajit Kumar, Professor Arun Kumar, Professor Bishnu Pal, Professor Anurag Sharma, Professor Enakshi Sharma, Dr. Raj Shenoy, Dr. Ramanand Tewari and Dr. Ravi Varshney for many enlightening discussions and stimulating collaborations which have enriched our understanding of this exciting field. One of us (AG) used a part of this book in presenting a course of lectures (during the summer sessions of 1987 and 1996) at University of Karlsruhe, Germany. The many stimulating discussions with Professor Gerhard Grau, Professor Wolfgang Freude and Professor Elmer Sauter are gratefully acknowledged.

A major portion of Chapter 11 (on sources for optical fiber communications) was very kindly written by Dr. Raj Shenoy; for this we are very grateful to him. We are also indebted to Mr. Prem Bhavnani, Mr. Rajiv Jindal, Ms. Jagneet Kaur, Mr. Parthasarathi Palai, Mr. K. V. Prakash and Mr. Vipul Rastogi for their help in carrying out many of the detailed calculations and also in the drawing of many figures in the book. Special thanks are due to Mrs. Lata Mathur and Mrs. Bhawna Bharadwaj and also the Continuing Education Program of IIT Delhi for their help and support in the preparation of this book. We are also grateful to Professor V. S. Raju, Director IIT Delhi, for his encouragement and support of this work.

We would like to thank the many authors and publishers for granting us permissions to use their work in this book. In particular, we would like to thank Professor R. R. Alfano and Professor I. Bennion for very kindly providing us with the glossy prints of figures 16.6 and 21.13.

The writing of this book used up many of our weekends and vacations, which we would normally have spent with our families and it is indeed extremely difficult to acknowledge their sacrifice. Our very special gratitude to Gopa, Raji, Arjun, Divyasmita, Amitabh, Kalyani and Krishnan for their patience and understanding.

Abbreviations

Å	Angstrom
APD	avalanche photo diode
ASE	amplified spontaneous emission
AT&T	American telegraph & telephone
BER	bit error rate
BH	buried heterostructure
BW	bandwidth
CCITT	Comite consultatif internationale telegraphique et telephonique
CSF	conventional single mode fiber
cw	continuous wave
dB	decibel
DBR	distributed Bragg reflector
DCF	dispersion compensating fiber
DFB	distributed feedback
DH	double heterostructure
DMA	differential mode attenuation
DSF	dispersion shifted fiber
EDFA	erbium doped fiber amplifier
EMD	equilibrium mode dispersion
ESA	excited-state absorption
ESI	equivalent step index
eV	electron volt
e-h	electron-hole
FP	Fabry-Perot
fs	femtosecond
FTIR	frustrated total internal reflection
FWHM	full width at half maximum
FWM	four wave mixing
GHz	gigahertz
GSA	ground state absorption
GVD	group velocity dispersion
GW	gigawatt
HWP	half wave plate
kHz	kilohertz
kV	kilovolt
kW	kilowatt
LCP	left circularly polarized wave
LD	laser diode
LED	light emitting diode
LHS	left hand side
LP	linearly polarized
LPS	limited phase space
Mb/s	mega (million) bits per second
MFD	mode field diameter

MHz	megahertz
MJ	megajoule
MLM	multilongitudinal mode
μm	micron (micrometer)
μs	microsecond
ms	millisecond
MW	megawatt
mW	milliwatt
MZ	Mach-Zehnder
NA	numerical aperture
NA	numerical aperture
NEC	Nippon electric company
NLSE	nonlinear Schrodinger equation
nm	nanometer
ns	nanosecond
PAM	pulse amplitude modulation
PCM	pulse code modulation
PIN	p (doped) intrinsic n (doped)
ps	picosecond
QWP	quarter wave plate
rad	radian
RCP	right circularly polarized wave
RHS	right hand side
rms	root mean square
RNF	refracted near field
RZ	return to zero
SBC	Soleil Babinet compensator
SIF	step index fiber
SNR	signal to noise ratio
SOP	state of polarization
SPM	self-phase modulation
Tb/s	tera (trillion) bits per second
TDM	time division multiplexing
TE	transverse electric
TEM	transverse electromagnetic
TM	transverse magnetic
TNF	transmitted near field
UV	ultra violet
WDM	wavelength division multiplexing
WKB	Wentzel Kramers Brillouin
ZMDW	zero material dispersion wavelength

›n: The fiber optics revolution

en a demand for increased capacity of transmission of infor-
ts and engineers continuously pursue technological routes
oal. The technological advances ever since the invention
have indeed revolutionized the area of telecommunication
e availability of the laser, which is a coherent source of
ed communication engineers with a suitable carrier wave
enormously large amounts of information compared with
...waves and microwaves. Although the dream of carrying millions of telephone (audio) or video channels through a single light beam is yet to be realized, the technology is slowly edging toward making this dream a reality.

A typical lightwave communication system consists of a lightwave transmitter, which is usually a semiconductor laser diode (emitting in the invisible infrared region of the optical spectrum) with associated electronics for modulating it with the signals; a transmission channel – namely, the optical fiber to carry the modulated light beam; and finally, a receiver, which consists of an optical detector and associated electronics for retrieving the signal (see Figure 1.1). The information – that is, the signal to be transmitted – is usually coded into a digital stream of light pulses by modulating the laser diode. These optical pulses then travel through the optical fiber in the form of guided waves and are received by the optical detector from which the signal is then decoded and retrieved.

At the heart of a lightwave communication system is the optical fiber, which acts as the transmission channel carrying the light beam loaded with information. It consists of a dielectric core (usually doped silica) of high refractive index surrounded by a lower refractive index cladding (see Figure 1.2). Incidentally, silica is the primary constituent of sand, which is found in so much abundance on our earth. Guidance of light through the optical fiber takes place by the phenomenon of total internal reflection. Sending the information-loaded light beams through optical fibers instead of through the open atmosphere protects the light beam from atmospheric uncertainties such as rain, fog, pollution, and so forth.

One of the key elements in the fiber optics revolution has been the dramatic improvement in the transmission characteristics of optical fibers. These include the attenuation of the light beam as well as the distortion in the optical signals as they race through the optical fiber. Figure 1.3 shows the dramatic reduction in the propagation loss of optical radiation through glass from ancient times to present. The steep fall in loss beginning in 1970 as the technology advanced rapidly is very apparent. It was indeed the development of low-loss optical fibers (20 dB/km at the He–Ne laser wavelength of 633 nm) in 1970 by Corning Glass Works in the United States that made practical the use of optical fibers as a viable transmission medium in lightwave communication systems. Figure 1.4 shows the wavelength variation of loss of a typical silica optical fiber showing the low-loss operating wavelength windows of 1300 nm and 1550 nm.

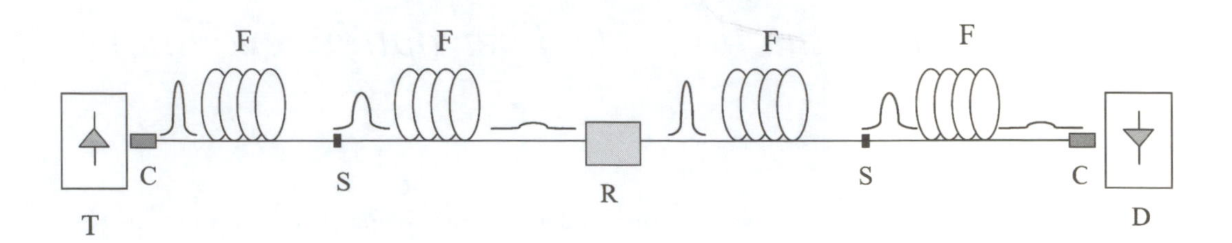

T Transmitter
C Connector
F Fiber
S Splice
R Repeater
D Detector

Fig. 1.1: A typical fiber optic communication system consisting of an optical transmitter (laser diode or LED), the transmission medium (optical fiber), and the optical receiver (photodetector). Information is sent in the form of optical pulses through the link.

Although a variety of optical fibers are available, the fibers in most use today are the so-called single-mode fibers with a core diameter of about 10 μm and an overall diameter of 125 μm. Optical fibers with typical losses in the range of 0.2 dB/km at 1550 nm and capable of transmission at 2–10 Gbit/s (Gb/s) are now commercially available. (A loss figure of 0.2 dB/km would imply a 50% power loss after propagating through a distance of about 15 km; the corresponding power loss for the best glass available in 1966 was about 1000 dB/km, which implies a 50% power loss in a distance of about 3 m!) Most currently installed systems are based on communication at a 1300-nm optical window of transmission. The choice of this wavelength was dictated by the fact that around an operating wavelength of 1300 nm the optical pulses propagate through a conventional single-mode fiber with almost no pulse broadening. Because silica has the lowest loss in the 1550-nm wavelength band, special fibers known as dispersion-shifted fibers have been developed to have negligible dispersion in the 1550-nm band, thus providing us with fibers having lowest loss and almost negligible dispersion.

In the lightwave communication systems in operation today, the signals have to be regenerated every 30–60 km to ensure that information is intelligibly retrieved at the receiving end. This is necessary either because the light pulses have become attenuated, and hence the signal levels have fallen below the detectable level, or because the spreading of the pulses has resulted in an overlapping of adjacent pulses leading to a loss of information. Until now this regeneration had to be achieved by first converting the optical signals into electrical signals, regenerating the signals electrically, and then once again converting the electrical signals into optical signals by modulating another semiconductor laser; such devices are called regenerators. Recent developments in optical amplifiers based on erbium- (a rare earth element) doped silica optical fibers have opened up possibilities of amplifying optical signals directly in the optical domain without the need of conversion to electrical signals. Because of amplification in the optical domain itself, such systems are not limited by the speed of the electronic circuitry and indeed can amplify multiple signals transmitted via different wavelengths simultaneously. For example, Figure 1.5 shows a typical gain spectrum of an erbium-doped fiber amplifier where one can note the flat gain over a wavelength band as large as 30 nm. Fortuitously, the gain band of

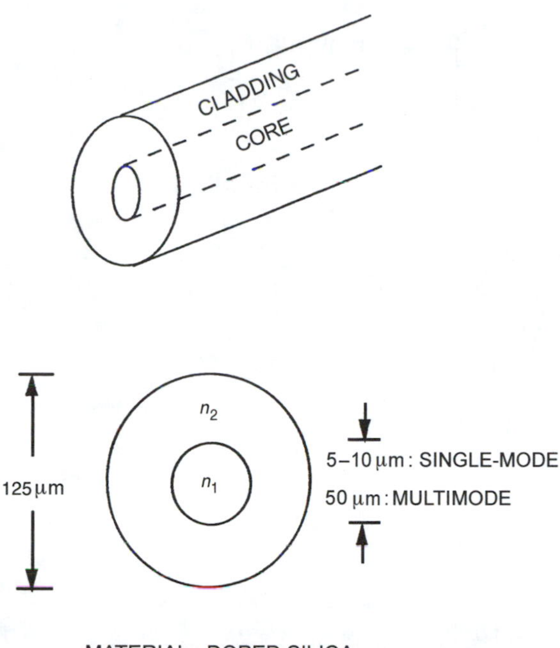

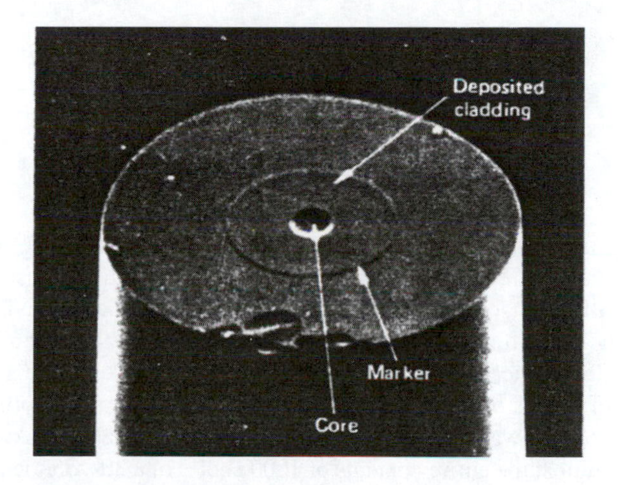

Fig. 1.2: (a) A typical optical fiber consisting of a doped silica core surrounded by a pure silica cladding of slightly lower refractive index. Light guidance takes place through the phenomenon of total internal reflection. (b) A scanning electron micrograph of an etched fiber showing clearly the core and the cladding. [After Miya et al. (1979).]

such optical amplifiers falls exactly on the low-loss transmission window of silica-based fibers. Indeed, the wavelength band of 100 nm around 1550 nm of the low-loss window of silica-based optical fibers (from 1500 to 1600 nm) corresponds to 12,500 GHz of bandwidth. This may be compared with the total radio bandwidth of only 25 GHz. Although these give the total accessible bandwidth figures, utilizing even a fraction of this available bandwidth gives us an enormous potential.

The coincidence of the low-loss window and the wide-bandwidth erbium-doped optical amplifiers has opened up possibilities of having wavelength division multiplexed communication systems (i.e., systems in which multiple wavelengths are used to carry independent signals, thus multiplying the

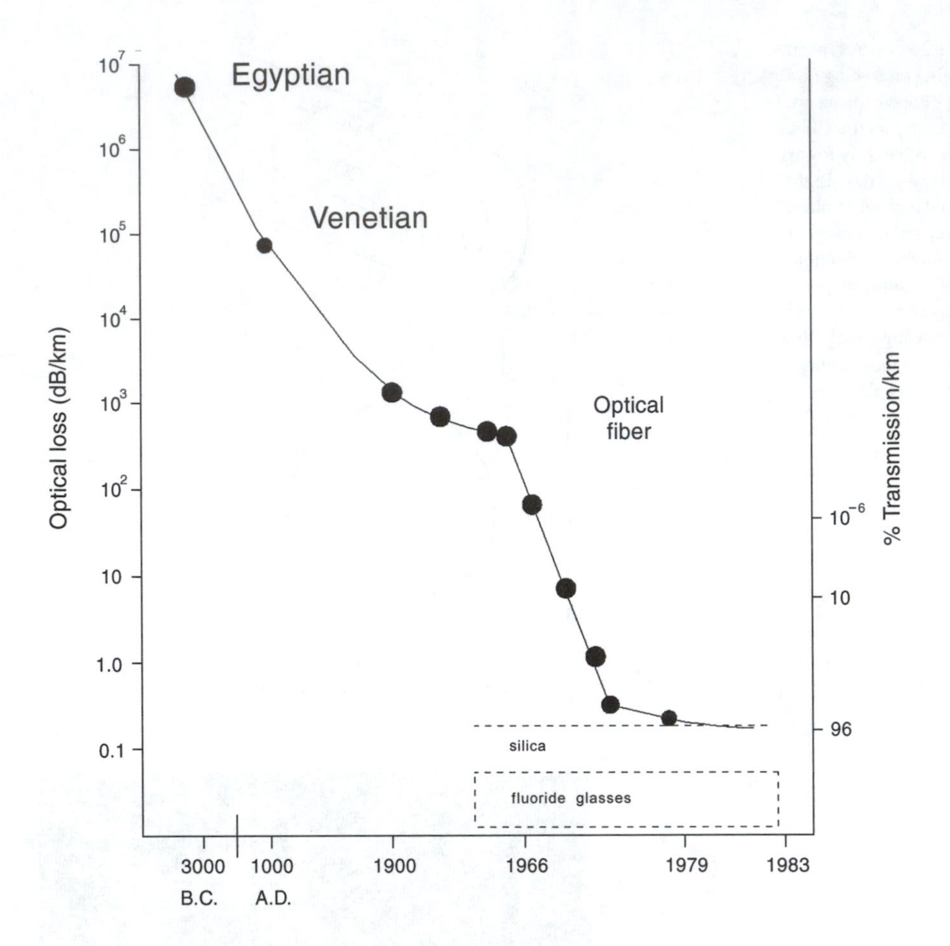

Fig. 1.3: Figure shows the dramatic reduction in transmission loss in optical glass from ancient times to present. [After Nagel (1989).]

capacity of an individual fiber) capable of carrying enormous rates of information. Indeed, recent reports have shown successful transmission at the rate of 1.1 trillion bits per second (1.1 Tb/s) over 150 km and 2.6 Tb/s over 120 km using 132 different wavelengths in the interval 1529.03–1563.86 nm (European Conference on Optical Communication, 1996). Figure 1.6 shows the setup and results of the 1.1-Tb/s experiment that were accomplished with 55 different optical wavelengths carrying independent signals. This corresponds to sending almost the entire contents of 1000 copies of a 30-volume encyclopedia in 1 s!

There is also a lot of research activity on special kinds of fibers – namely, dispersion-compensating fibers (DCFs). This has arisen because the existing underground network already contains more than 70 million km of fibers optimized for operation at 1300 nm. Because today's optical amplifiers operate only in the 1550-nm region, the question that arises is whether it is possible to use the existing network of fibers to send signals at 1550 nm. Since they are not optimized for 1550-nm operation, such fibers exhibit a significant amount of dispersion at 1550 nm, leading to distortion of signals. The newly developed DCFs have very large dispersions but have a sign that is opposite those of the 1300-nm fibers. Hence, by appropriately choosing the lengths of these fibers, one can indeed compensate for the distortion and thus use the

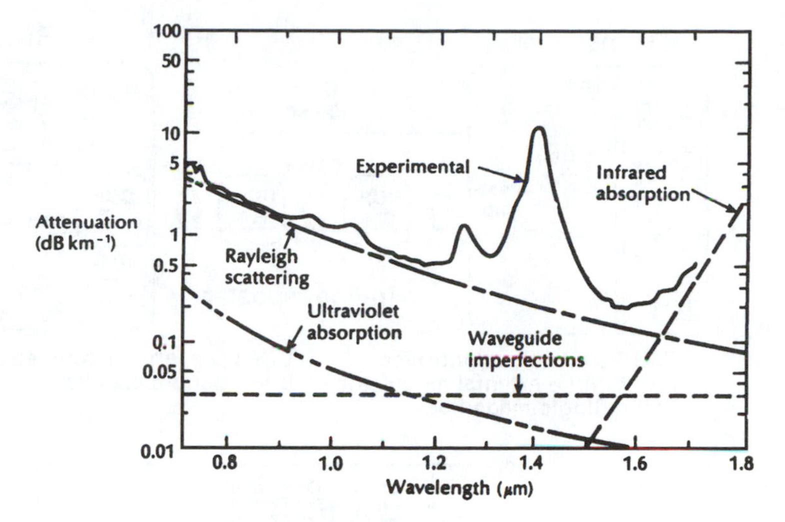

Fig. 1.4: Loss spectrum of a typical low-loss optical fiber. [After Miya et al. (1979).]

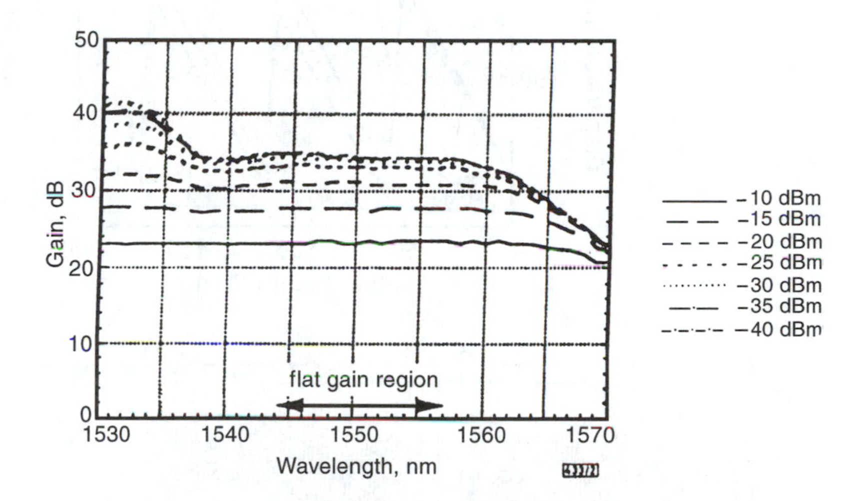

Fig. 1.5: A typical gain spectrum of a flat gain erbium-doped optical fiber amplifier. [After Yoshida, Kuwano, and Iwashita (1995).]

existing network, resulting in significant cost savings. Indeed, the experiment shown in Figure 1.6 used the 1300-nm optimized fibers and used DCFs in the link.

Another very significant development involves use of optical nonlinear effects in optical fibers to compensate for any distortion of signal due to dispersion in the fiber. Such light pulses, in which nonlinear effects cancel dispersion effects completely, are termed solitons. The nonlinear effects become very predominant because even low powers can lead to very large intensities (light power per unit area) of light beams since the cross-sectional areas of the beams that are guided by the fiber can be as small as 50 μm^2. Figure 1.7 shows a dramatic experiment demonstrating soliton propagation through 180 million km of fibers at a data rate of 10 billion pulses per second. Although such extremely large capacities have been demonstrated in experiments using soliton propagation, their implementation in real systems is still to come.

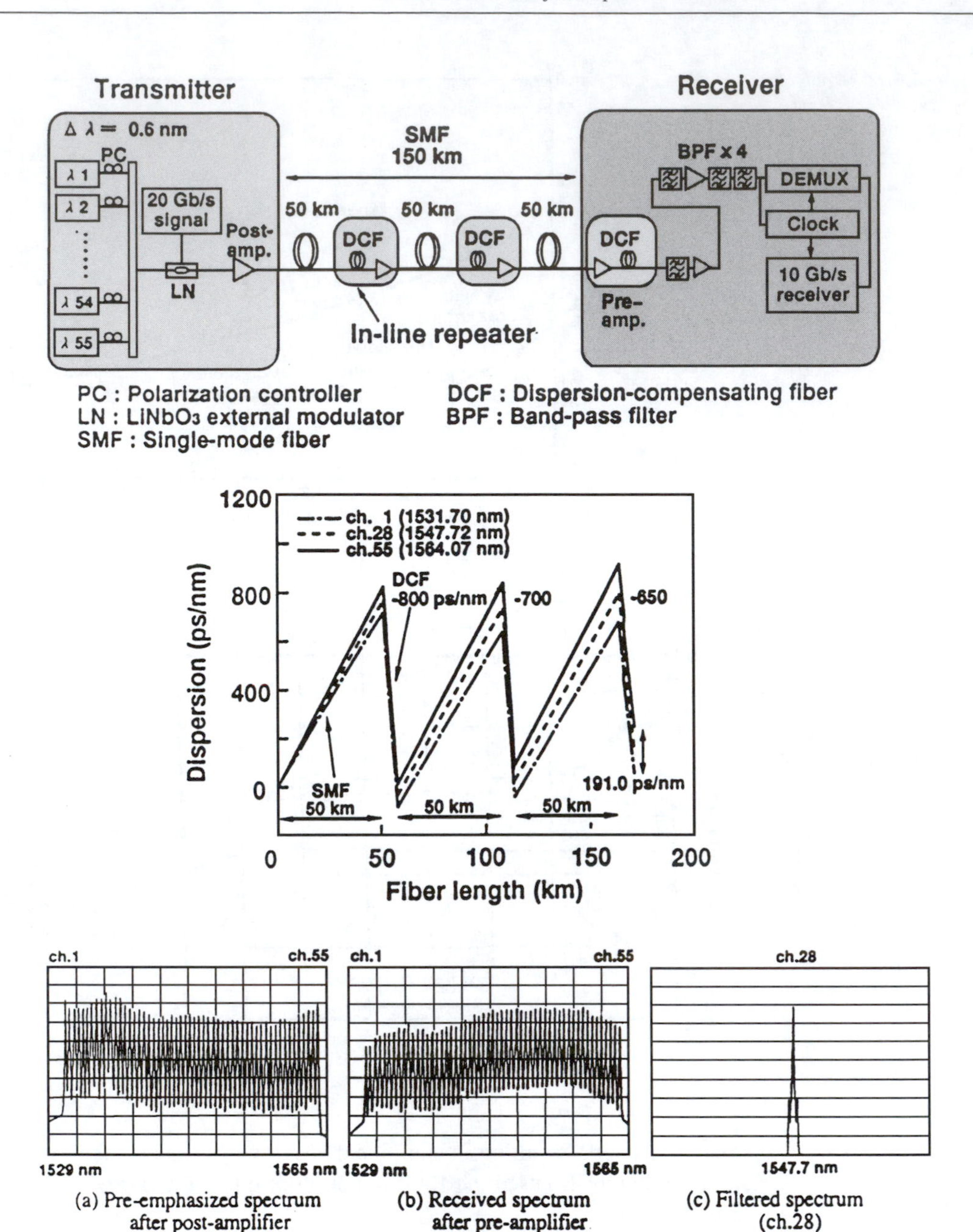

Fig. 1.6: (a) Experimental setup for 55 wavelength WDM (wavelength division multiplexing) transmission. (b) The corresponding dispersion map of the fiber link. (c) Spectra of the 55 wavelengths used in the experiment (H: 3.6 nm/div, V: 5 Db/div, Res: 0.1 nm). [After Onaka et al. (1996).]

Optical fibers have indeed revolutionized the field of telecommunication and are the backbone of today's global communication networks. Indeed, the growth rate has been so phenomenal that the performance of lightwave systems has been doubling every 1.34 years [Midwinter (1994)]. This may be compared with the growth rate of silicon chips, which has a doubling time of 2.2 years.

Lightwave communication systems employing rare earth-doped fibers and solitons are expected to lead to near "zero loss" and near "infinite bandwidth" systems, thereby providing us with a network capable of handling almost all our information needs and resulting in a true information-based society.

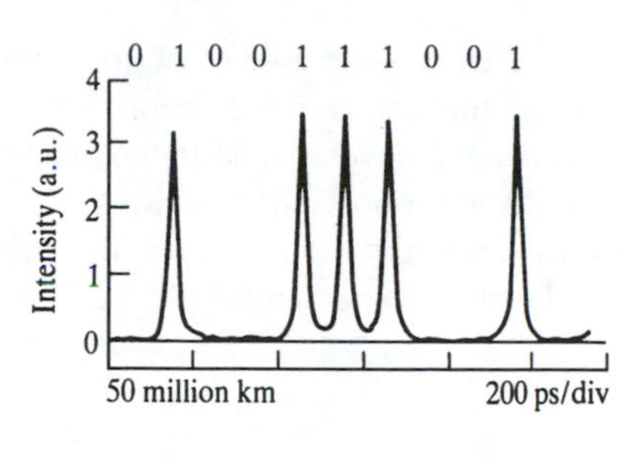

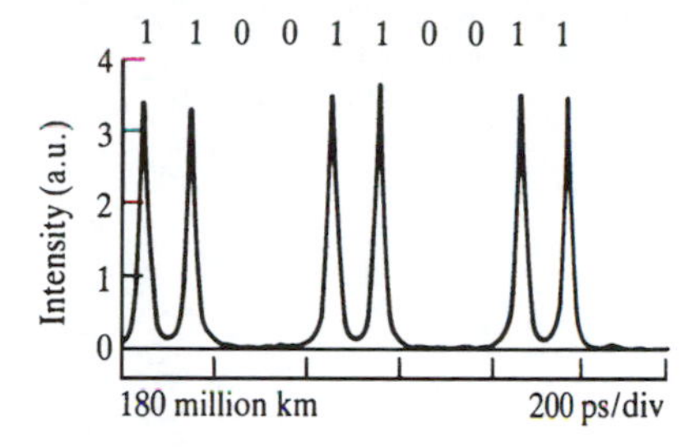

Fig. 1.7: Transmitted data pattern of (1100110011) as received after 180 million km of propagation in a recirculating fiber loop experiment. [After Nakazawa (1994).]

In the following chapters of the book we aim to develop the field of fiber optics from the basic principles to the present-day technology involving erbium-doped fiber amplifiers, DCFs, and solitons. In Chapter 2 we briefly discuss some basic optical effects such as interference, diffraction, and polarization. This is followed by a general discussion on the simplest optical fiber – namely, the step index optical fiber in Chapter 3. Chapters 4 and 5 discuss the analysis of optical waveguides in terms of a ray picture that is valid for multimode fibers. Chapter 6 discusses the important concept of material dispersion; it was the observation of zero material dispersion around 1300 nm in silica-based fibers that shifted operation into the 1300-nm wavelength window. Chapters 7–10 discuss the characteristics of optical waveguides in terms of the more rigorous modal picture and show the enormous information-carrying capacities of single-mode fiber. Design issues of single-mode fibers are discussed in Chapter 10.

Chapters 11 and 12 give brief discussions on sources and detectors used in fiber optic communication systems with special emphasis on the specific characteristics relevant to communication systems only. Chapter 13 gives a brief exposure to the issues involved in the design of a simple optical fiber communication system.

Chapters 14–16 deal with the very recent developments in optical fiber systems – namely, optical fiber amplifiers, dispersion compensation principles, and soliton propagation through optical fibers. The chapters discuss the basic concepts and aim to illustrate the tremendous impact that these have in the area of fiber optics.

Chapters 17 and 18 discuss an important area of fiber optics – namely, fiber optic components and fiber optic sensors, which are finding widespread applications in many diverse areas. The discussion is restricted to components and sensors based only on single-mode fibers since these are expected to find greater application in view of their greater sensitivities and better performance characteristics.

Chapters 19 and 20 discuss some of the important methods for characterizing multimode and single-mode fibers.

Chapter 21 discusses the interesting area of periodic interactions in waveguides. Fiber Bragg gratings that use periodic interaction seem to be one of the most active areas of research and development in the fiber optics field today in view of both demonstrated and proposed applications.

Chapters 22–24 discuss some advanced topics in fiber optics. These chapters are expected to further illustrate the beautiful physics that fiber optics offers us.

2

Basic optics

2.1 Introduction

This chapter gives an elementary introduction to polarization, interference, and diffraction characteristics of a light wave. The principles developed in this chapter are used in understanding the basic concepts in fiber optics.

2.2 Plane polarized waves

A linearly polarized plane wave is the simplest electromagnetic wave, and if we assume the plane wave to be propagating in the $+z$-direction, the corresponding electric field can be written in the form

$$\mathbf{E} = \hat{\mathbf{x}} E_0 \cos(\omega t - kz + \theta) \tag{2.1}$$

where the electric field vector is assumed to oscillate in the x direction. In equation (2.1), $\omega(=2\pi\nu)$ is the angular frequency and k is the propagation constant. For propagation in free space

$$k = \frac{\omega}{c}$$

where c represents the velocity of light in free space; when propagating in a medium characterized by refractive index n, we have

$$k = \frac{\omega}{v}$$

where

$$v = \frac{c}{n}$$

represents the velocity of the electromagnetic wave in that medium. Equation (2.1) describes an x-polarized wave. It is also known as a *linearly polarized* wave because the electric vector is oscillating along a specific axis (see Figure 2.1). A linearly polarized wave is also known as a *plane polarized* wave because the electric field is always confined to a particular plane; for the x-polarized wave

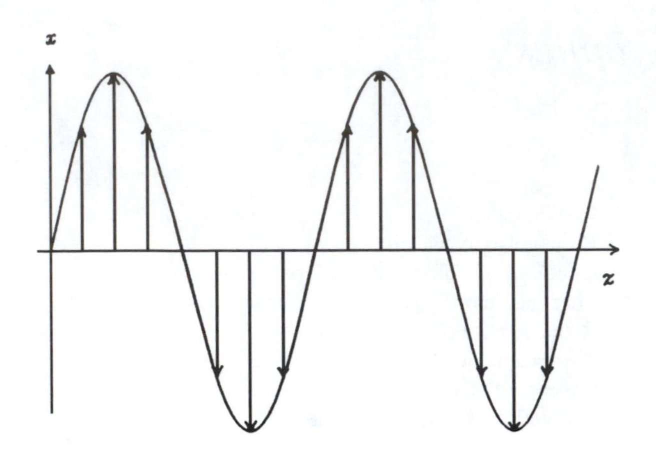

Fig. 2.1: An x-polarized plane electromagnetic wave propagating along the z direction. The arrow represents the direction and magnitude of the electric field vector at a particular instant of time.

propagating along the z-direction, the electric field is confined to the x–z plane (see Figure 2.1).

In the complex representation we may write

$$\mathbf{E} = \hat{\mathbf{x}} E_0 e^{i(\omega t - kz + \theta)} \tag{2.2}$$

where the actual field is the real part of the above equation. In general, a plane wave propagating in the direction of $\mathbf{k}$ is described by the equation

$$\mathbf{E} = \hat{\mathbf{e}} E_0 e^{i(\omega t - \mathbf{k}\cdot\mathbf{r} + \theta)} \tag{2.3}$$

where $\hat{\mathbf{e}}$ represents the unit vector along the direction of polarization, and for the wave to be transverse we must have

$$\mathbf{k} \cdot \hat{\mathbf{e}} = 0$$

The corresponding magnetic field is given by

$$\mathbf{H} = \frac{1}{\omega \mu_0} [\mathbf{k} \times \mathbf{E}] \tag{2.4}$$

The Poynting vector $\mathbf{S}$ is defined by

$$\mathbf{S} = \mathbf{E} \times \mathbf{H} \tag{2.5}$$

and points along the direction of propagation of the wave in an isotropic medium. We may interpret $\mathbf{S} \cdot \mathbf{da}$ as the electromagnetic energy crossing the area $\mathbf{da}$ per unit time. For an x-polarized plane wave propagating along the z-direction, the electric field is given by equation (2.1). Using equation (2.4), the corresponding magnetic field is given by

$$\mathbf{H} = \frac{k}{\omega \mu_0} \hat{\mathbf{y}} E_0 \cos(\omega t - kz + \theta) \tag{2.6}$$

Thus

$$\mathbf{S} = \frac{k}{\omega \mu_0} E_0^2 \cos^2(\omega t - kz + \theta) \tag{2.7}$$

For light waves $\omega \sim 10^{15}\,\mathrm{s}^{-1}$ and the $\cos^2$ term in equation (2.7) oscillates very rapidly so any detector would record only a time average. Since

$$\begin{aligned}\langle\cos^2(\omega t - kz + \theta)\rangle &= \underset{T\to\infty}{\mathrm{Lt}}\ \frac{1}{2T}\int_{-T}^{T}\cos^2(\omega t - kz + \theta)\,dt \\ &= \frac{1}{2}\end{aligned}$$

we obtain

$$\langle S\rangle = \frac{k}{2\omega\mu_0}E_0^2 = \frac{n}{2c\mu_0}E_0^2 \tag{2.8}$$

The time-averaged quantity $\langle S\rangle$ is also referred to as the intensity of the electromagnetic wave and represents the average energy crossing a unit area (perpendicular to the direction of propagation) per unit time.

Example 2.1: Consider a laser beam at 0.633 μm with a power of 1 mW and a cross-sectional area of 3 mm^2. Assuming a beam of uniform intensity

$$I = \frac{10^{-3}}{3\times 10^{-6}} = \frac{1}{3}\times 10^3\ \mathrm{W/m^2} = \frac{n}{2c\mu_0}E_0^2$$

For air $n \simeq 1$ and using $c = 3\times 10^8$ m/s, $\mu_0 = 4\pi\times 10^{-7}$ SI units, we have the corresponding electric field as

$$E_0 \simeq 501\ \mathrm{V/m}$$

Example 2.2: Consider a bulb emitting 10 W of optical power. Since the emission is uniform along all directions, the corresponding intensity at a distance of 10 m is

$$I = \frac{10}{4\pi\times 10^2} \simeq 7.96\times 10^{-3}\ \mathrm{W/m^2}$$

The corresponding electric field is given by

$$E_0 = \left(\frac{2c\mu_0 I}{n}\right)^{1/2} \simeq 2.4\ \mathrm{V/m}$$

Example 2.3: Since a laser beam is almost perfectly parallel, it can be focused to a spot of a radius of a few wavelengths. If we consider a beam of 1 mW at 0.633 focused by a lens to a spot of radius 6 μm, then the resulting intensity is

$$I = \frac{1\,\mathrm{mW}}{\pi\times(6\,\mu\mathrm{m})^2} \simeq 8.8\times 10^6\ \mathrm{W/m^2}$$

The corresponding electric field is

$$E_0 \simeq 8.1\times 10^4\ \mathrm{V/m}$$

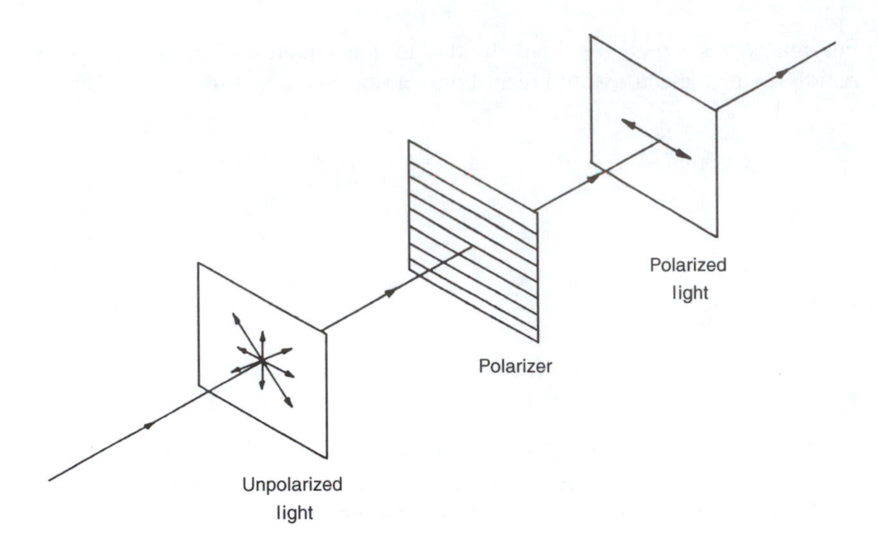

Fig. 2.2: Production of a plane polarized light from an unpolarized light source using a polarizer. The horizontal lines in the polarizer represent the pass axis.

2.2.1 *Production of plane polarized waves*

The ordinary light wave, like the one coming from a sodium lamp or from the sun, is unpolarized – that is, the electric vector (on a plane transverse to the direction of propagation) keeps changing its direction in a random manner as shown in Figure 2.2. If we allow the unpolarized wave to fall on a polarizer (such as a polaroid), then the wave emerging from the polaroid will be linearly polarized. In Figure 2.2 the lines shown on the polaroid represent what is usually referred to as the "pass axis" of the polaroid – that is, the electric field perpendicular to the pass axis gets absorbed. Polaroid sheets are extensively used for producing linearly polarized light waves. As an interesting corollary, we note that if a second polaroid (whose pass axis is at right angles to the pass axis of the first polaroid) is put immediately after the first polaroid, then no light will come through; the polaroids are said to be in a "crossed-position."

2.3 Circularly and elliptically polarized waves

Let us consider the superposition of two plane waves, one polarized in the x-direction and the other polarized in the y-direction, with a phase difference of $\pi/2$ between them.

$$\begin{aligned} \mathbf{E}_1 &= E_0\hat{\mathbf{x}}\cos(\omega t - kz) \\ \mathbf{E}_2 &= E_0\hat{\mathbf{y}}\cos(\omega t - kz - \pi/2) \end{aligned} \tag{2.9}$$

where we have assumed the amplitude of the two waves to be the same ($=E_0$). The resultant electric field is given by

$$\mathbf{E} = E_0[\hat{\mathbf{x}}\cos(\omega t - kz) + \hat{\mathbf{y}}\sin(\omega t - kz)] \tag{2.10}$$

which describes a *right circularly polarized* (RCP) wave. At any particular value of z, the tip of the E-vector, with increasing time t, can easily be shown to rotate on the circumference of a circle. For example, at $z = 0$, the x and y

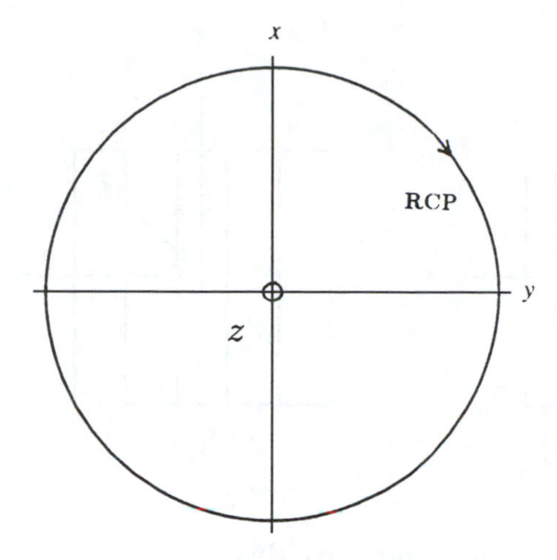

Fig. 2.3: Representation of a RCP beam propagating along the z-direction.

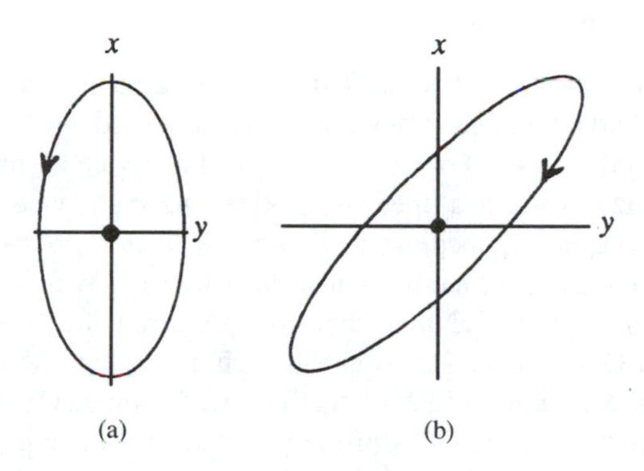

Fig. 2.4: Two different elliptical states of polarization represented by equation (2.12) with (a) $E_1 = \frac{1}{2}E_0$ and $\phi = -\pi/2$, and (b) $E_1 = E_0$ and $\phi = \pi/3$.

components of the electric vector are given by

$$E_x = E_0 \cos \omega t, \quad E_y = E_0 \sin \omega t; \tag{2.11}$$

thus the tip of the electric vector rotates on a circle in the clockwise direction (see Figure 2.3). For advancing a right-handed corkscrew along the z-direction, we would have to rotate it along the clockwise direction, and therefore it is said to represent an RCP wave. When propagating in air or in any isotropic medium, the state of polarization (SOP) is maintained – that is, a linearly polarized wave will remain linearly polarized; similarly, an RCP wave will remain RCP.

In general, the superposition of two waves with

$$E_x = E_0 \cos(\omega t - kz) \quad \text{and} \quad E_y = E_1 \cos(\omega t - kz - \phi) \tag{2.12}$$

will represent an elliptically polarized wave. Figure 2.4 shows two different elliptically polarized waves. Figure 2.4(a) corresponds to

$$E_1 = \frac{1}{2}E_0 \quad \text{and} \quad \phi = -\pi/2, \tag{2.13}$$

representing a left elliptically polarized wave, and Figure 2.4(b) corresponds to

$$E_1 = E_0 \quad \text{and} \quad \phi = \pi/3. \tag{2.14}$$

representing a right elliptically polarized wave.

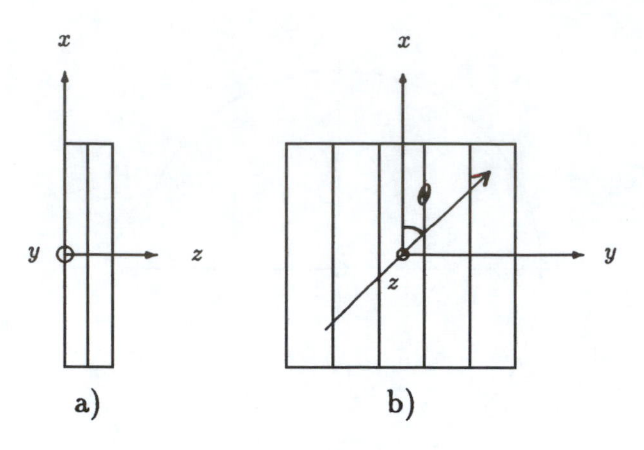

Fig. 2.5: A plane polarized wave is incident normally on a quarter-wave plate with the plane of polarization making an angle θ with the optic axis (x-axis in the figure). (a) Side view, (b) front view.

2.4 Propagation of a light wave through a quarter-wave plate

A quarter-wave plate (QWP) is a device that is used in many experiments involving fiber and integrated optics and is basically used to change the SOP of a propagating optical wave. For example, using this device we may transform a circularly polarized wave to a linearly polarized wave and vice versa. A QWP is made of an *anisotropic* medium like calcite or quartz. We will not go into the details of an anisotropic medium; it suffices here to say that inside a crystal like that of calcite, quartz, $LiNbO_3$, there is a preferred direction known as the *optic axis*. In a QWP the crystal is cut in such a way that the optic axis is at right angles to the thickness of the plate. Figure 2.5 shows a QWP with its optic along the x-direction. The modes of the QWP are x- and y-polarized; the x-polarized mode, which is polarized along the optic axis, is usually referred to as the extraordinary wave, and the y-polarized mode is referred to as the ordinary wave. By modes we imply that if the incident wave is x-polarized, then it will propagate as an x-polarized wave; similarly, if the incident wave is y-polarized, then it will propagate as a y-polarized wave without any change in the SOP but with a velocity slightly different from the velocity of the x-polarized wave. On the other hand, if a circularly polarized wave is incident, we must express this as a linear combination of x- and y-polarized waves and then consider the independent propagation of the two waves.

As an example, we consider the incidence of a wave polarized at an angle of 45° to the x-axis (see Figure 2.5). At $z = 0$, it can be expressed as

$$E_x = E_0 \cos\theta \cos\omega t, \quad E_y = E_0 \sin\theta \cos\omega t \tag{2.15}$$

where $\theta = 45°$. The refractive indices "seen" by the x- and y-polarized waves are different and are usually denoted by n_e and n_o, respectively. These are known as the extraordinary and ordinary refractive indices and are known constants of the crystal (see Table 2.1). Thus, the x- and y-polarized waves propagate with velocities c/n_e and c/n_o, respectively, and therefore inside the crystal we will have

$$E_x = \frac{E_0}{\sqrt{2}} \cos(\omega t - k_e z) \tag{2.16}$$

Table 2.1. *Ordinary and extraordinary indices for calcite, quartz, and rutile*

Crystal	n_o	n_e	λ (μm)
Calcite	1.658	1.486	0.589
Quartz	1.544	1.553	0.589
Rutile	2.621	2.919	0.579

and

$$E_y = \frac{E_0}{\sqrt{2}} \cos(\omega t - k_o z) \tag{2.17}$$

where

$$k_o = \frac{\omega}{c} n_o, \quad \text{and} \quad k_e = \frac{\omega}{c} n_e \tag{2.18}$$

Thus, as the wave propagates, a phase difference is set up between the two components. We assume $n_o > n_e$ (which is true for negative crystals like calcite). Let the thickness d of the crystal be such that

$$(k_o - k_e) d = \pi/2$$

or

$$d = \frac{\lambda_0}{4(n_o - n_e)} \tag{2.19}$$

where $\lambda_0 (= \omega/2\pi c)$ represents the free space wavelength. Then at $z = d$, the x and y components of the polarization are given by

$$E_x = \frac{E_0}{\sqrt{2}} \cos(\omega t - \phi) \tag{2.20}$$

and

$$\begin{aligned} E_y &= \frac{E_0}{\sqrt{2}} \cos\{\omega t - [(\phi + \pi/2)]\} \\ &= \frac{E_0}{\sqrt{2}} \sin(\omega t - \phi) \end{aligned} \tag{2.21}$$

where $\phi = k_e d$, equations (2.20) and (2.21) describe an RCP wave.

Similarly, a *half-wave plate* (HWP) is a device that introduces a phase difference of π between the two electric field components; thus, its thickness will be twice the thickness of a QWP. A linearly polarized lightwave remains linearly polarized after passing through an HWP; however, the orientation of the plane of polarization can be changed by a desired amount using an HWP.

As a simple example, starting with an unpolarized lightwave, we can produce a circularly polarized wave by passing it first through a polaroid and then through an appropriately oriented QWP (see Figure 2.6).

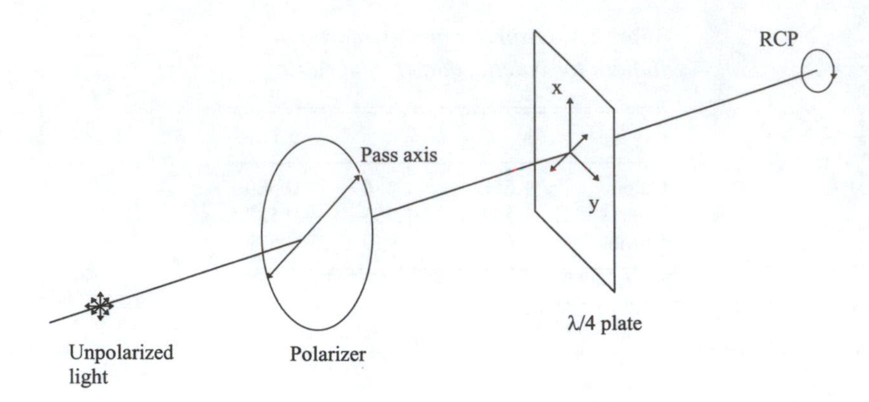

Fig. 2.6: Production of RCP light from unpolarized light using a combination of a polarizer and a QWP.

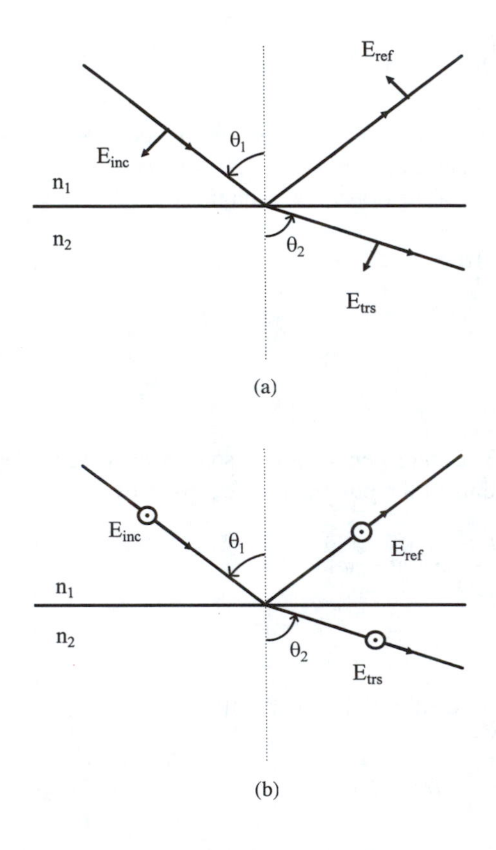

Fig. 2.7: Reflection and transmission of a plane wave incident at an interface between media of refractive indices n_1 and n_2. (a) p polarization, (b) s polarization.

2.5 Reflection at a plane interface

Consider a plane interface formed between two media of refractive indices n_1 and n_2 (see Figure 2.7). When an electromagnetic wave is incident on such an interface, it will, in general, give rise to a reflected wave and a transmitted wave. The amplitudes of the reflected and transmitted waves can be obtained by using the electromagnetic boundary conditions at the interface. The amplitude reflection and transmission coefficients defined as the ratio of the electric field of the reflected and transmitted waves to the electric field of the incident wave are given by [see, e.g., Ghatak and Thyagarajan (1989), Chapter 2]

$$r_p = \frac{n_1 \cos\theta_2 - n_2 \cos\theta_1}{n_1 \cos\theta_2 + n_2 \cos\theta_1} \tag{2.22}$$

$$t_p = \frac{2n_1 \cos\theta_1}{n_2 \cos\theta_1 + n_1 \cos\theta_2} \tag{2.23}$$

$$r_s = \frac{n_1 \cos\theta_1 - n_2 \cos\theta_2}{n_1 \cos\theta_1 + n_2 \cos\theta_2} \tag{2.24}$$

$$t_s = \frac{2n_1 \cos\theta_1}{n_1 \cos\theta_1 + n_2 \cos\theta_2} \tag{2.25}$$

where the subscripts p and s correspond to the incident wave polarized in the plane of incidence and perpendicular to it, respectively; θ_1 and θ_2 are the angles of incidence and refraction, respectively (see Figure 2.7), and are related through Snell's law:

$$n_1 \sin\theta_1 = n_2 \sin\theta_2 \tag{2.26}$$

Example 2.4: Let us consider normal incidence on an air–silica interface with the refractive index of silica of 1.45. The corresponding energy reflection coefficient is

$$R = |r|^2 = \left(\frac{n_1 - n_2}{n_1 + n_2}\right)^2 \simeq 0.03$$

For incidence on a GaAs–air interface, using the refractive index of GaAs as 3.6, we have

$$R = 0.32$$

thus showing a strong reflection. This large reflection coefficient at a semiconductor–air interface is used for feedback in a semiconductor laser in which the end facets are just cleaved semiconductor substrates.

2.5.1 *Brewster angle*

We note from equation (2.22) that $r_p = 0$ when

$$n_1 \cos\theta_2 = n_2 \cos\theta_1 \tag{2.27}$$

Using equation (2.26), we have

$$\begin{aligned}\frac{n_2^2}{n_1^2}\cos^2\theta_1 &= (1 - \sin^2\theta_2)\\ &= 1 - \frac{n_1^2}{n_2^2}\sin^2\theta_1\end{aligned}$$

or

$$\frac{n_2^2}{n_1^2} = \sec^2\theta_1 - \frac{n_1^2}{n_2^2}\tan^2\theta_1 = 1 + \left(1 - \frac{n_1^2}{n_2^2}\right)\tan^2\theta_1$$

Thus

$$\tan\theta_1 = \frac{n_2}{n_1} \tag{2.28}$$

This angle is referred to as the Brewster angle and is such that at this angle of incidence the reflection coefficient is zero for the polarization parallel to the plane of incidence. No such angle exists for the s polarization. Thus, if an unpolarized beam is incident at the Brewster angle, the reflected light will be linearly polarized – in fact, s-polarized.

2.5.2 *Total internal reflection*

When a wave is incident from a denser medium on an interface separating two media, then, since $n_1 > n_2$, we have from equation (2.26), $\theta_2 > \theta_1$. For an angle of incidence $\theta_1 = \theta_c$ such that

$$n_1 \sin\theta_c = n_2 \tag{2.29}$$

the refracted wave just grazes the interface. This angle is referred to as the critical angle, and for angles of incidence greater than the critical angle there is no refracted wave and the wave is said to suffer total internal reflection.

When $n_1 \sin\theta_1 > n_2$, then $\sin\theta_2 > 1$ and $\cos\theta_2$ becomes a pure imaginary quantity

$$\cos\theta_2 = (1 - \sin^2\theta_2)^{1/2} = -\frac{i}{n_2}\left(n_1^2\sin^2\theta_1 - n_2^2\right)^{1/2}$$

Thus, the reflection coefficients become complex. For example, we have

$$\begin{aligned} r_s &= \frac{n_1\cos\theta_1 + i\left(n_1^2\sin^2\theta_1 - n_2^2\right)^{1/2}}{n_1\cos\theta_1 - i\left(n_1^2\sin^2\theta_1 - n_2^2\right)^{1/2}} \\ &= \exp\left[i2\tan^{-1}\left\{\frac{\sqrt{n_1^2\sin^2\theta_1 - n_2^2}}{n_1\cos\theta_1}\right\}\right] \end{aligned} \tag{2.30}$$

Thus, the corresponding energy reflection coefficient is $R_s = |r_s|^2 = 1$ showing that the reflection is total.

One can similarly show that for $\theta_1 > \theta_c$, $R_p = |r_p|^2 = 1$.

The fact that even under total internal condition, t_s and t_p are nonzero [see equations (2.23) and (2.25)] is interpreted from the fact that, even under conditions of total internal reflection, there is a finite value of electric field even in the rarer medium. This is called an evanescent wave. One can show that the field in the rarer medium decays exponentially away from the interface and there is no net energy flow into the rarer medium. The presence of such an evanescent wave is very important from the point of view of many fiber optic components such as directional couplers (see Chapter 17).

> **Example 2.5:** For an interface between silica ($n_1 = 1.45$) and air ($n_2 = 1.0$), the critical angle is $\theta_c = \sin^{-1}(n_2/n_1) = 43.6°$.

> **Example 2.6:** For an interface between doped silica (n_1=1.46) and pure silica ($n_2 = 1.45$), the critical angle is $\theta_c = 83.3°$.

2.6 Two-beam interference

Whenever two waves superpose, one obtains an intensity distribution that is known as the interference pattern. We will consider here the interference pattern produced by waves emanating from two point sources. As is well known, a stationary interference pattern is observed when the two interfering waves maintain a constant phase difference. For light waves, because of the very process of emission, one cannot observe a stationary interference pattern between the waves emanating from two independent sources, although interference does take place. Thus, one tries to derive the interfering waves from a single wave so that a definite phase relationship is maintained.

Let S_1 and S_2 represent two coherent point sources emitting waves of wavelength λ (see Figure 2.8(a)). We wish to determine the interference pattern on photographic plates P_1 and P_2. The intensity distribution is given by

$$I = 4I_0 \cos^2 \delta/2 \tag{2.31}$$

where, I_0 is the intensity produced by either of the waves independently and

$$\delta = \frac{2\pi}{\lambda}\Delta \tag{2.32}$$

where

$$\Delta = S_1Q \sim S_2Q \tag{2.33}$$

represents the path difference between the two interfering waves. Thus, when

$$\delta = 2m\pi$$
$$\Rightarrow \Delta = S_1Q \sim S_2Q = m\lambda; \quad m = 0, 1, 2, \ldots \text{(bright fringe)} \tag{2.34}$$

we will have a bright fringe and when

$$\delta = (2m+1)\pi$$
$$\Rightarrow \Delta = S_1Q \sim S_2Q = \left(m + \frac{1}{2}\right)\lambda; \quad m = 0, 1, 2, \ldots \text{(dark fringe)} \tag{2.35}$$

we will have a dark fringe. Using simple geometry one can show that the locus of the points (on the plane P_1) so that $S_1Q \sim S_2Q = \Delta$ is a hyperbola given

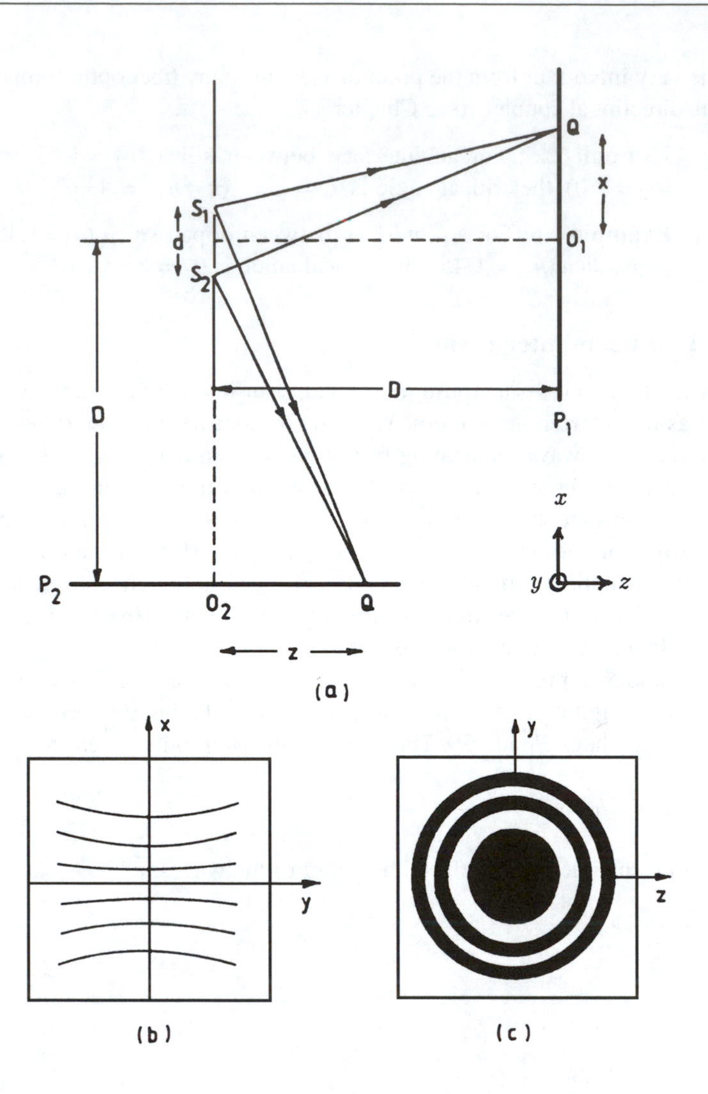

Fig. 2.8: (a) Schematic of the arrangement to observe two-beam interference, (b) fringe pattern on plane P_1, and (c) fringe pattern on plane P_2.

by [see e.g., Ghatak (1992), Chapter 11]

$$(d^2 - \Delta^2)x^2 - \Delta^2 y^2 = \Delta^2 \left[D^2 + \frac{1}{4}(d^2 - \Delta^2) \right] \tag{2.36}$$

Now,

$$\Delta = 0 \Rightarrow x = 0$$

which represents the central bright fringe and coincides with the y-axis. Equation (2.36) can be written in the form

$$x = \pm\sqrt{\frac{\Delta^2}{d^2 - \Delta^2}} \left[y^2 + D^2 + \frac{1}{4}(d^2 - \Delta^2) \right]^{1/2} \tag{2.37}$$

For values of y such that

$$y^2 \ll D^2$$

(i.e., close to the x-axis) the loci are almost straight lines parallel to the y-axis and one obtains almost straight line fringes as shown in Figure 2.8(b).

For $\Delta \ll d$ and $y \ll D$, we may write equation (2.37) as

$$x = \pm \frac{\Delta D}{d}$$

The positions of maxima and minima are determined by the values of Δ [see equations (2.34) and (2.35)]. The corresponding fringe width would be

$$\beta = \frac{\lambda D}{d} \tag{2.38}$$

Thus, for $D = 50$ cm, $d = 0.05$ cm, and $\lambda = 6000$ Å , $\beta = 0.06$ cm.

We next consider the interference pattern produced on the plane P_2. As an example, let us assume for simplicity $\lambda = 6 \times 10^{-5}$ cm and $d = 6 \times 10^{-3}$ cm, so that d is an integral multiple of λ. Thus

$$S_1 O_2 - S_2 O_2 = d = 100\lambda = p\lambda$$

and the point O_2 will be a bright spot. Obviously, the fringe pattern on the plane P_2 will be circular (see Figure 2.8(c)) because the locus of point Q so that $S_1 Q - S_2 Q$ is a constant will be a circle on the y–z plane whose center is at O_2. The point Q will be bright if

$$S_1 Q - S_2 Q = (S_1 O_2 - S_2 O_2) - m\lambda = (p - m)\lambda$$

where $m = 0, 1, 2, \ldots$. Elementary algebra shows that the radius z of the fringes is given by

$$z = \left[\left\{ \frac{Dd}{(p - m)\lambda} - \frac{1}{2}(p - m)\lambda \right\}^2 - \left(D - \frac{1}{2}d \right)^2 \right]^{1/2} \tag{2.39}$$

and it will correspond to the $(p - m)$th bright fringe. For example, for $d = 0.6$ mm, $\lambda = 6 \times 10^{-5}$ cm, $D = 50$ cm, we have $p = 1000$, and the fringes corresponding to $m = 1$, 2, and 3 (which will correspond to the 999th, 998th, and 997th fringe) will have radii 2.24 cm, 3.17 cm, and 3.88 cm, respectively.

It can easily be seen that if the distance d is increased, the central fringe will become a higher order fringe and thus the fringes will emerge from the center. On the other hand, if the distance d is decreased, the central fringe will become a lower order fringe and thus the fringes will collapse to the center.

2.6.1 Fiber optic Mach–Zehnder interferometer

One of the important applications of optical fibers is in the sensing of var- break ious external parameters such as pressure, temperature, magnetic field, etc. (see Chapter 18). A fiber optic Mach–Zehnder interferometer is one of the basic configurations used in high-sensitivity measurements. In such an

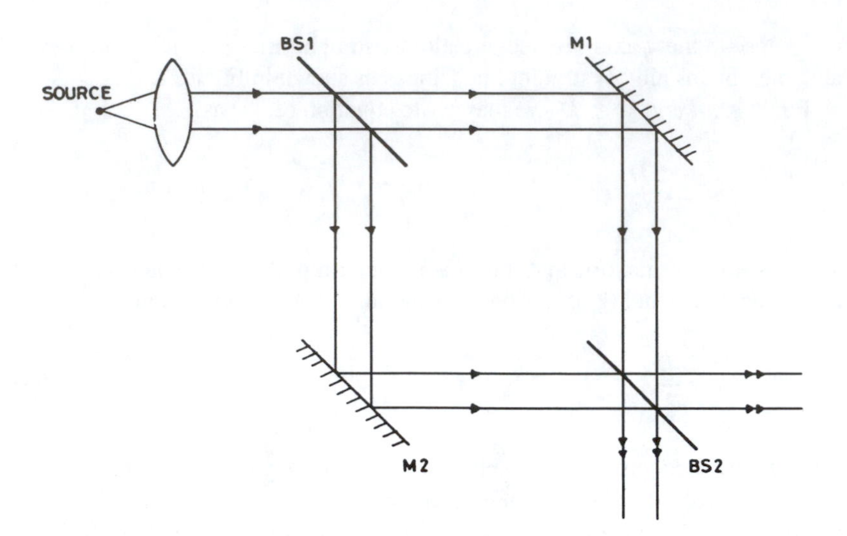

Fig. 2.9: Configuration of a conventional Mach–Zehnder interferometer.

arrangement, changes in the phase of light propagating through a single-mode fiber caused by an external parameter are very sensitively measured by an interference effect. Fiber optic Mach–Zehnder interferometers are one of the most sensitive sensors with predicted capabilities exceeding those of many conventional sensors. In this chapter, we discuss the basic configuration of a fiber optic Mach–Zehnder interferometer.

Figure 2.9 shows a conventional Mach–Zehnder interferometer using bulk optical elements in which light from a source is first split into two arms with a beam splitter and then recombined by another beam splitter after propagating through two different arms. Depending on the phase difference between the two arms, at the output of the interferometer one obtains an interference pattern much like in any other interferometer.

The corresponding fiber optic version of the Mach–Zehnder interferometer[1] is shown in Figure 2.10, in which the two arms of the interferometer are replaced by two single-mode fibers; the beams emerging from the two fibers will almost be diverging spherical waves and are superposed with a beam splitter. If the output ends of the two fibers are adjusted as shown in Figure 2.10, then the two diverging spherical waves will have their centers of curvature collinear with the fiber axis. In such a case, circular fringes will be formed on a screen placed as shown in Figure 2.8(c).

This arrangement is very similar to the interference pattern formed by two point sources on a screen placed perpendicular to the line joining the sources (see Section 2.6). In contrast to a point source that emits uniformly along all directions, in the present case, the interfering beams have almost a Gaussian amplitude distribution. Thus, the circular fringe pattern will have an overall intensity pattern as in a Gaussian beam. As discussed in Section 2.6, when the output ends of the fiber are brought closer to each other, the circular fringes expand, and when they are moved away from each other, they contract (where is the zero-order fringe in this system?).

[1] If the two beam splitters were replaced by fiber optic directional couplers, we would then have an all-fiber Mach–Zehnder interferometer.

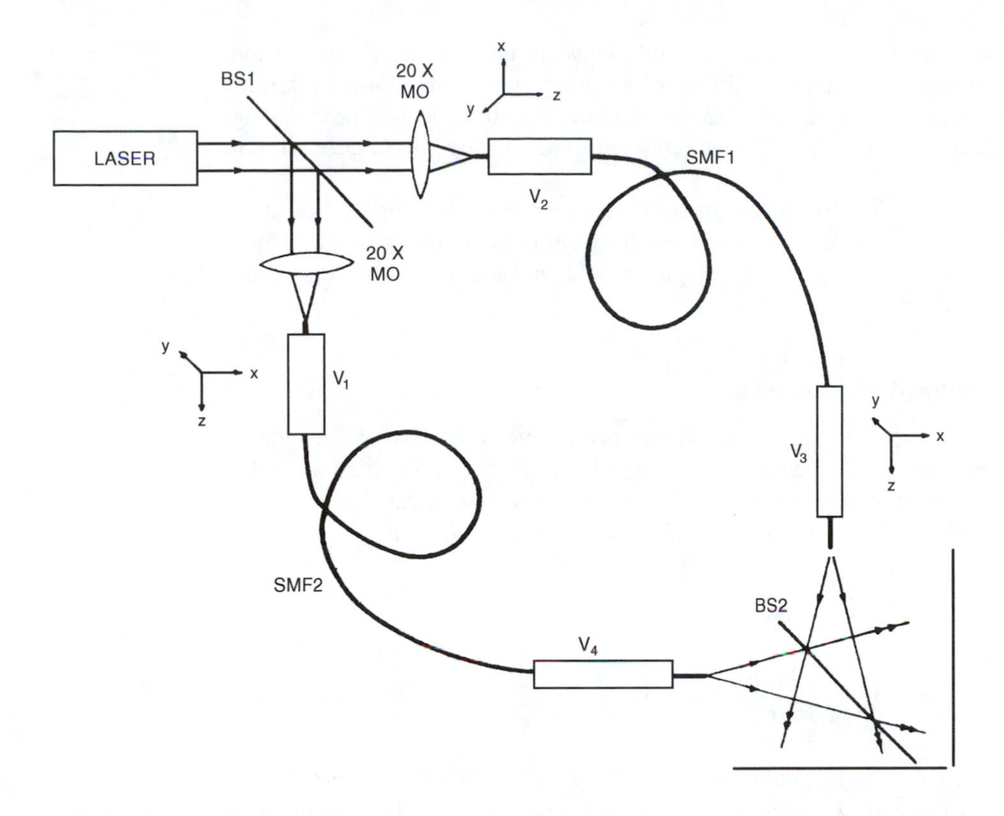

Fig. 2.10: Schematic of a fiber optic Mach–Zehnder interferometer.

The total phase difference between the interfering beams at any point Q on the screen, which is labelled as P_2 in Figure 2.8(a), is given by

$$\begin{aligned}\phi &= \frac{2\pi}{\lambda_0}(S_2Q - S_1Q) + (\phi_2 - \phi_1)\\ &= \frac{2\pi}{\lambda_0}\left(\sqrt{R_2^2 + r^2} - \sqrt{R_1^2 + r^2}\right) + (\phi_2 - \phi_1)\end{aligned} \tag{2.40}$$

where $r(=O_2Q)$ is the distance of the observation point from the center of the screen, R_1 and R_2 are the distances of the fiber ends from the screen, and ϕ_1 and ϕ_2 are the phase shifts suffered by the beam in propagating through the two fiber arms; these include any constant phase difference between the two arms due to differences in fiber lengths, and so forth as well as any difference due to external perturbation. In Figure 2.8(a), $R_1 = D + d/2$ and $R_2 = D - d/2$.

If each of the beam splitters is a 50:50 beam splitter, then the time-averaged intensity detected by an optical detector is

$$I = I_0 \cos^2\left(\frac{\phi}{2}\right) \tag{2.41}$$

where I_0 is the input intensity. It is seen from equation (2.41) that the intensity at any point depends on the phase difference between the two arms. Thus, if the phase difference ϕ is an integral multiple of 2π, then the intensity is maximum, whereas if it is an odd integral multiple of π, then the intensity zero (what happened to energy conservation?). Thus, the intensity at any point on the screen depends on the phase difference between the two arms of the interferometer. Any external perturbation differentially affecting the light propagating through the two arms will change $(\phi_2 - \phi_1)$ and thus the interference pattern itself.

A fiber optic Mach–Zehnder interferometer as shown in Figure 2.10 can be used as a sensor for detecting and measuring various external parameters such as temperature, pressure, strain, magnetic field, and so forth. This is discussed in greater detail in Chapter 18.

2.7 Concept of coherence

In Section 2.6 we discussed the interference of two waves that are generated from one wave by the method of division of wavefront (see Figure 2.8). If the source emits only one wavelength λ_1, then the fringes are formed everywhere on the screen. The positions of maxima and minima are given by

$$\begin{aligned} x &= m\frac{\lambda_1 D}{d} \quad \text{maxima} \\ &= \left(m + \frac{1}{2}\right)\frac{\lambda_1 D}{d} \quad \text{minima} \end{aligned} \tag{2.42}$$

where we have used equations (2.37), (2.34), and (2.35). If the source emits another wavelength λ_2 along with λ_1, then the positions of maxima and minima of λ_2 are given by

$$\begin{aligned} x &= n\frac{\lambda_2 D}{d} \quad \text{maxima} \\ &= \left(n + \frac{1}{2}\right)\frac{\lambda_2 D}{d} \quad \text{minima} \end{aligned} \tag{2.43}$$

At the center of the screen P_1 in Fig. 2.8(a), $x = 0$ and the maxima of both wavelengths coincide exactly. As we move away from the center, since the fringe spacings of λ_1 (which is equal to $\lambda_1 D/d$) and of λ_2 (equal to $\lambda_2 D/d$) are unequal, at some value of x, the maxima of λ_1 and minima of λ_2 will start to overlap. In such a region, there would be no fringes observable. For this to happen at a value of x, the optical path difference (OPD) of λ_1 must be $m\lambda_1$, whereas that of λ_2 should be $(m + \frac{1}{2})\lambda_2$. Thus

$$\text{OPD} = m\lambda_1 = \left(m + \frac{1}{2}\right)\lambda_2 \tag{2.44}$$

Hence

$$m = \frac{\text{OPD}}{\lambda_1} = \frac{\text{OPD}}{\lambda_2} - \frac{1}{2}$$

giving

$$\text{OPD} = \frac{\lambda_1\lambda_2}{2(\lambda_1 - \lambda_2)} \simeq \frac{\lambda^2}{2\Delta\lambda} \tag{2.45}$$

where $\Delta\lambda = \lambda_1 - \lambda_2$ and we have assumed $\lambda_1\lambda_2 \simeq \lambda^2$, where λ is the average wavelength.

The quantity $\lambda^2/2\Delta\lambda$ represents the maximum OPD so as to obtain an observable interference pattern in the presence of two wavelengths spaced by $\Delta\lambda$. If the source emits a continuous spectrum of width $\Delta\lambda$, then one can show that for good contrast the path difference should be less than

$$l_c = \frac{\lambda^2}{\Delta\lambda} \tag{2.46}$$

The above quantity is referred to as the coherence length. Equation (2.46) can also be written in terms of frequency as

$$l_c = \frac{c}{\Delta\nu} \tag{2.47}$$

The coherence time τ_c is defined as

$$\tau_c = \frac{l_c}{c} = \frac{1}{\Delta\nu} \tag{2.48}$$

Thus, the more monochromatic a wave is, the larger is the coherence length or coherence time.

Example 2.7: An LED at 850 nm has a typical $\Delta\lambda$ of 30 nm. The corresponding coherence length is

$$l_c \simeq 24\,\mu\text{m}$$

On the other hand, a laser diode at the same wavelength has a spectral width of 2nm, giving a coherence length of

$$l_c \simeq 0.36\,\text{mm}$$

Example 2.8: A He–Ne laser has a $\Delta\nu = 1.5$ GHz. The corresponding coherence length is

$$l_c = 20\,\text{cm}$$

2.8 Diffraction of a Gaussian beam

An infinitely entended uniform plane wave such as the one described by equation (2.1) propagates as a plane wave, and the transverse amplitude and phase distribution do not change as the wave propagates. On the other hand, if the beam is characterized by any transverse amplitude/phase distribution, then as the beam propagates, the transverse amplitude and phase distributions change. This is due to the phenomenon of diffraction.

The diffraction phenomenon can be easily understood by considering the diffraction of a Gaussian beam. Indeed, when a laser oscillates in its fundamental mode, the transverse amplitude distribution of the output beam is Gaussian; similarly, the transverse amplitude distribution of the fundamental mode of an optical fiber is very nearly Gaussian. Therefore, the study of the diffraction of a Gaussian beam is of considerable importance in fiber optics.

We consider the propagation along the z-direction of a Gaussian beam whose amplitude distribution on the plane $z = 0$ is given by

$$u(x, y, 0) = A \exp\left[-\frac{x^2 + y^2}{w_0^2}\right] \tag{2.49}$$

where w_0 is usually referred to as the "spot size" of the beam; it represents the radial distance at which the intensity falls off by a factor of $1/e^2$. The above equation implies that the phase front is plane at $z = 0$. As the beam propagates along the z-axis, diffraction occurs and one obtains [see, e.g., Ghatak and Thyagarajan (1989), Chapter 2]

$$u(x, y, z) = \frac{iA\pi}{\lambda} \frac{2w_0^2}{2z + ikw_0^2} \exp\left[-ik\left(z + \frac{x^2 + y^2}{2R(z)}\right)\right] e^{-\frac{(x^2+y^2)}{w^2(z)}} \tag{2.50}$$

where

$$R(z) = z\left[1 + \frac{\pi^2 w_0^4}{\lambda^2 z^2}\right] \tag{2.51}$$

represents the radius of curvature of the wavefront and

$$w(z) = w_0\left(1 + \frac{\lambda^2 z^2}{\pi^2 w_0^4}\right)^{1/2} \tag{2.52}$$

represents the z-dependent spot size of the beam. For large values of z,

$$w(z) \simeq \frac{\lambda z}{\pi w_0} \tag{2.53}$$

implying

$$\tan\theta \simeq \frac{w(z)}{z} \simeq \frac{\lambda}{\pi w_0} \tag{2.54}$$

where θ is the semiangle of the cone defining the diffraction divergence of the beam. Thus, for $\lambda \simeq 0.6\,\mu$m [using equations (2.54) and (2.52)]

$$\theta \simeq 0.011^\circ \text{ for } w_0 = 1\,\text{mm} \Rightarrow w \simeq 2.16\,\text{mm at } z = 10\,\text{m}$$

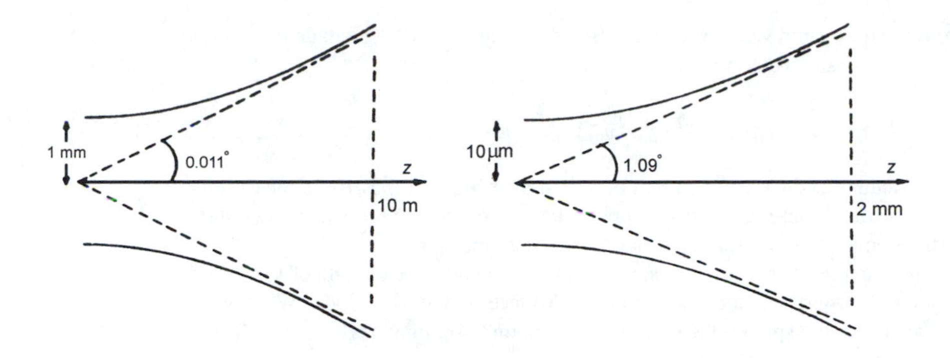

Fig. 2.11: Diffraction of a Gaussian beam. Note that the diffraction divergence (θ) increases when the waist of the beam is reduced.

and

$$\theta \simeq 1.09^\circ \text{ for } w_0 = 10\,\mu\text{m} \Rightarrow w \simeq 39.5\,\mu\text{m at } z = 2\,\text{mm}$$

(see Figure 2.11). Notice that θ increases with a decrease in w_0 (the smaller the size of the aperture, the greater the diffraction). Further, for a given w_0, θ decreases with a decrease in λ; indeed, as $\lambda \to 0$ there is no diffraction that is the geometrical optics limit.

Example 2.9: The output from a single-mode fiber operating at 1300 nm is approximately a Gaussian beam with $w_0 = 5\,\mu$m. Thus, the corresponding divergence is

$$\theta \simeq \tan^{-1}\left(\frac{\lambda_0}{\pi w_0}\right) \simeq 4.73^\circ$$

Thus, if a screen is placed a distance of 10 cm in front of the fiber, the radius of the beam is approximately 8.3 mm.

Problems

2.1 Obtain the electric field produced on the retina when a person standing at a distance of 10 m from a bulb emitting 20 W of optical energy looks at it. You may assume $\lambda = 0.5\ \mu$m, pupil radius = 2 mm, and the eye lens to retina distance is 25 mm.

2.2 For the example 2.1, obtain the corresponding magnetic field.

2.3 (a) The energy density of the Sun on the Earth's surface is approximately 1.35 kW/m^2. Assume that if one looks directly into the sun, the radius of the image of the sun (on the retina) is approximately 0.2 mm. Calculate the power density in the image assuming the pupil radius to be 1 mm.

(b) A diffraction limited laser beam (of diameter 2 mm) is incident on the pupil of the eye. Assume that the wavelength of the laser is 633 nm and that the focal length of eye lens is 25 mm. Calculate the power density on the retina.

[ANSWER: (a) 30 kW/m^2, (b) 10 MW/m^2.]

2.4 Write down a complete expression for a plane wave traveling with its **k** in the y–z plane making an angle of 30° with the z-axis.

2.5 The Fraunhofer diffraction pattern of a Gaussian beam coming out of a laser operating at 1 μm is measured and it is found that the intensity drops by 50%

from the maximum value when the diffraction angle is $0.1°$. Calculate the waist size of the beam. You may use

$$w^2(z) = w_0^2\left(1 + \lambda^2 z^2/\pi^2 w_0^4\right)$$

2.6 The Fraunhofer diffraction pattern of a Gaussian beam at $0.5\ \mu$m is measured and it is found that the intensity drops to 10% of the maximum value when the diffraction angle is $1°$. Calculate the waist size of the beam.

2.7 A Gaussian laser beam with a beam width of 10 cm and a wavelength of $1.06\ \mu$m is pointed toward the moon, which is at a distance of 3.76×10^5 km. What will be the size of the spot on the surface of the moon? You may neglect any effects due to atmospheric turbulence.

2.8 Obtain the reflection coefficients from an air–glass ($n = 1.5$) interface when the angle of incidence is $0°$, $30°$, $60°$, and $85°$ for s and p polarizations.

2.9 Two coherent laser beams at a wavelength of $0.6328\ \mu$m making angles of $+45°$ and $-45°$ with the normal to a screen are incident on it. What is the spacing between the fringes that will be observed on the screen?

2.10 A single-frequency laser at 1300 nm has a spectral width of 100 MHz. What is the corresponding coherence length?

3

Basic characteristics of the optical fiber

3.1 Introduction

An optical waveguide is a structure that can guide a light beam from one place to another. The most extensively used optical waveguide is the step index optical fiber that consists of a cylindrical central dielectric core, clad by a dielectric material of a slightly lower refractive index (see Figure 3.1(a)). The corresponding refractive index distribution (in the transverse direction) is given by

$$\begin{aligned} n(r) &= n_1, \quad 0 < r < a \quad \text{core} \\ &= n_2, \quad r > a \quad \text{cladding} \end{aligned} \tag{3.1}$$

where r represents the cylindrical radial coordinate and a represents the radius of the core. Actually, the core extends only to a finite distance b (see Figure 3.1(b)); however, for all practical purposes, we will assume the cladding to extend to infinity. Typically, for a step index (multimode) silica fiber,

$$n_1 \simeq 1.48, \quad n_2 \simeq 1.46, \quad a \simeq 25\ \mu\text{m}, \quad b \simeq 62.5\ \mu\text{m} \tag{3.2}$$

In this chapter, we discuss the various characteristics of the optical fiber – namely, its light-gathering power and its loss- and pulse-broadening characteristics. Throughout the chapter we use ray optics, which is valid for highly multimoded waveguides.

To understand light guidance in an optical fiber, we consider a ray entering the fiber as shown in Figure 3.1(a). If the angle of incidence (at the core–cladding interface) ϕ is greater than the critical angle

$$\phi_c = \sin^{-1}\left(\frac{n_2}{n_1}\right) \tag{3.3}$$

then the ray will undergo total internal reflection at that interface. Further, because of the cylindrical symmetry in the fiber structure, this ray will suffer total internal reflection at the lower interface also and therefore will get guided through the core by repeated total internal reflections. Even for a bent fiber, light guidance can take place through multiple total internal reflections (see Figure 3.2). Figure 3.3 shows the actual guidance of a light beam as it propagates through a long optical fiber.

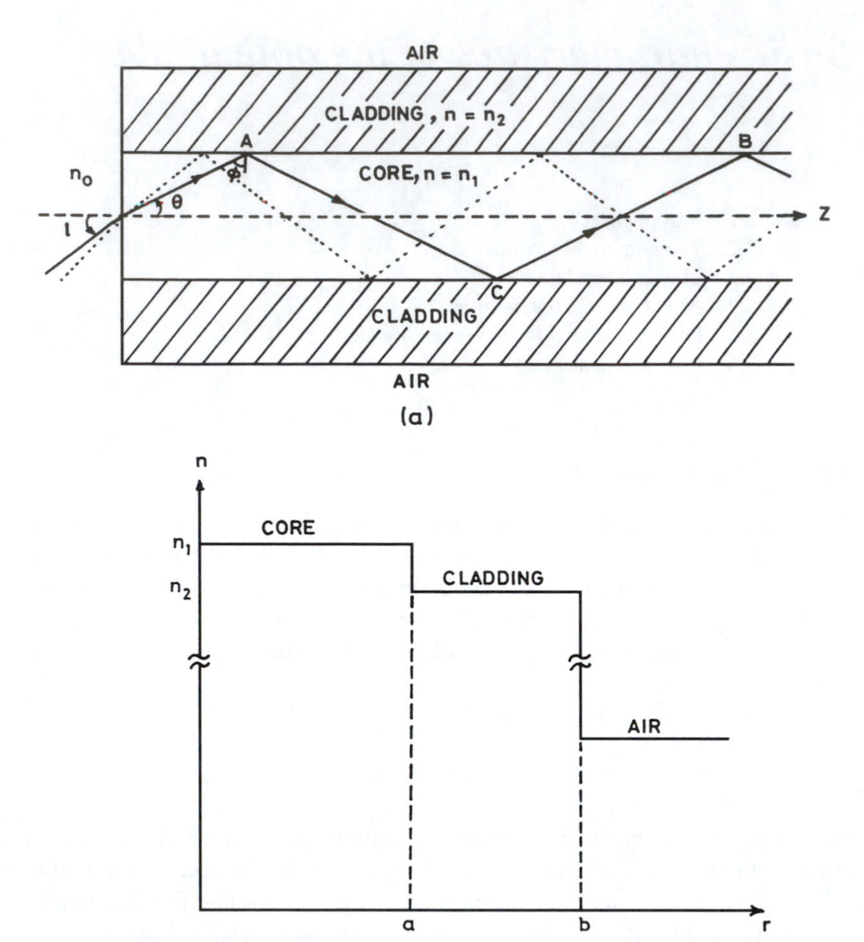

Fig. 3.1: (a) A glass fiber that consists of a cylindrical central core, clad by a material of slightly lower refractive index. Light rays impinging on the core–cladding interface at an angle greater than the critical angle are trapped inside the core of the waveguide. Rays making larger angles with the axis take a longer to traverse the length of the fiber. (b) Refractive index distribution of a cladded optical fiber that consists of a cylindrical glass structure surrounded by a material of slightly lower refractive index. In a typical (multimode) fiber, we may have the core refractive index $n_1 \simeq 1.5$, $\Delta = 0.01$, core radius $a = 25\,\mu$m, $2b = 125\,\mu$m.

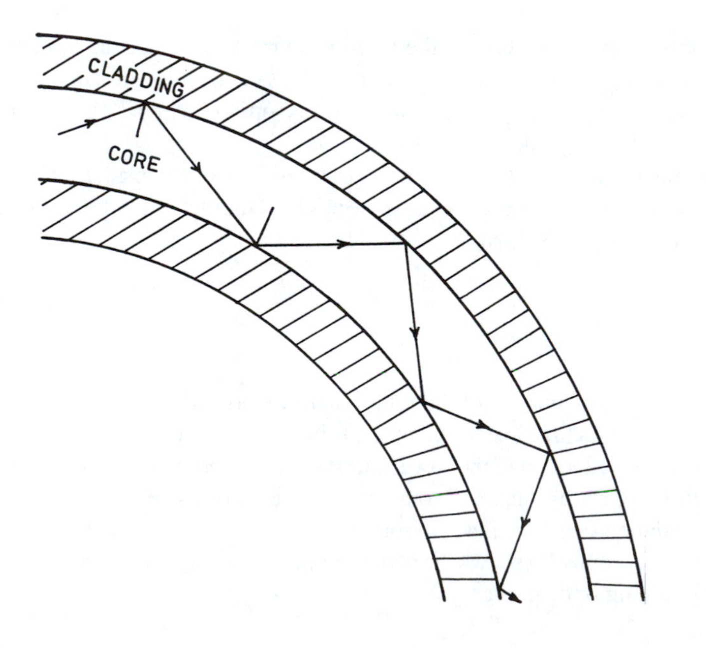

Fig. 3.2: Rays propagating through a bent fiber. Notice that whereas the angle of incidence (at the core–cladding interface) remains constant in a straight fiber [see Figure 3.1(a)], it changes in a bent fiber. Thus, a ray may eventually hit the core–cladding interface at an angle less than the critical angle and be refracted away.

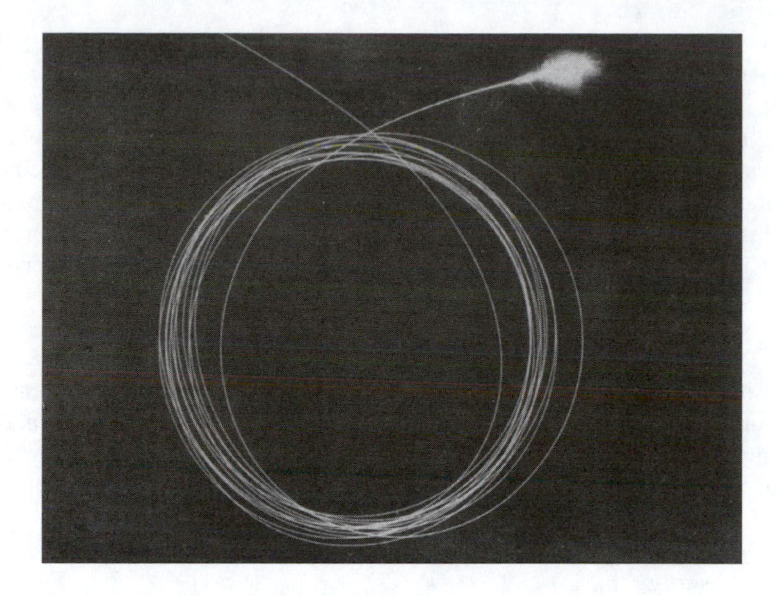

Fig. 3.3: A long optical fiber carrying a light beam.

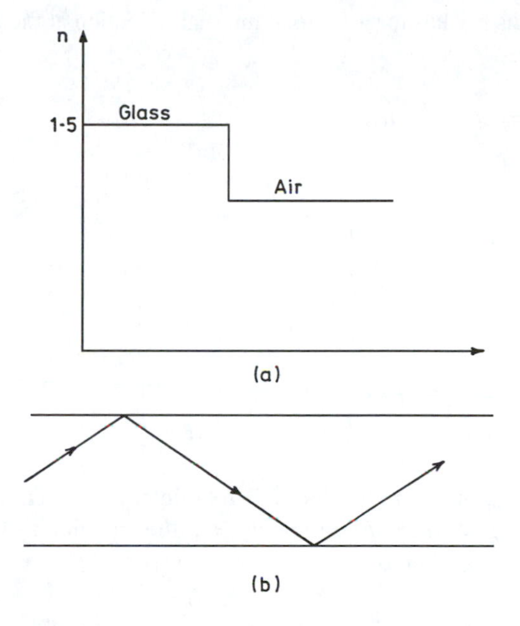

Fig. 3.4: (a) Refractive index distribution of an unclad fiber that consists of a cylindrical glass structure made of homogeneous material. (b) Light rays impinging on the glass–air interface at an angle greater than the critical angle are trapped inside the core of the waveguide through total internal reflections.

The phenomenon of guidance by multiple total internal reflections was in fact demonstrated by John Tyndall as early as 1854. In this demonstration, Tyndall showed that light travels along the curved path of water emanating from an illuminated vessel. However, fiber optics really developed in the 1950s with the works of Hopkins and Kapany in the United Kingdom and of Van Heel in Holland; these works led to use of the optical fiber in many optical devices.

The necessity of a cladded fiber (Figure 3.1) rather than a bare fiber (Figure 3.4) was felt because of the fact that, for transmission of light from one place to another, the fiber must be supported, and supporting structures may considerably distort the fiber, thereby affecting the guidance of the lightwave (see also Problems 3.1–3.3). This can be avoided by choosing a sufficiently thick cladding. Further, in a fiber bundle, in the absence of the cladding, light can leak through from one fiber to another.[1]

It is of interest to mention that the retina of the human eye consists of a large number of rods and cones that have the same kind of structure as the optical fiber – that is, they consist of dielectric cylindrical rods surrounded by another dielectric of slightly lower refractive index and the core diameters are in the range of a few microns. The light absorbed in these *light guides* generates electrical signals, which are then transmitted to the brain through various nerves.

3.2 The numerical aperture

We return to Figure 3.1(a) and consider a ray that is incident on the entrance aperture of the fiber making an angle i with the axis. Let the refracted ray make an angle θ with the fiber axis. Assuming the outside medium to have a refractive index n_0 (which for most practical cases is unity), we get

$$\frac{\sin i}{\sin\theta} = \frac{n_1}{n_0}$$

Obviously, if this ray has to suffer total internal reflection at the core–cladding interface,

$$\sin\phi (= \cos\theta) > n_2/n_1 \tag{3.4}$$

Thus

$$\sin\theta < \left[1 - \left(\frac{n_2}{n_1}\right)^2\right]^{1/2}$$

and we must have

$$\sin i < \frac{n_1}{n_0}\left[1 - \left(\frac{n_2}{n_1}\right)^2\right]^{1/2} = \left[\frac{n_1^2 - n_2^2}{n_0^2}\right]^{1/2} \tag{3.5}$$

If $(n_1^2 - n_2^2) \geq n_0^2$, then for all values of i total internal reflection will occur at the core–cladding interface. Assuming $n_0 = 1$, the maximum value of $\sin i$ for a ray to be guided is given by

$$\begin{aligned} \sin i_m &= \left(n_1^2 - n_2^2\right)^{1/2} \quad && \text{when } n_1^2 < n_2^2 + 1 \\ &= 1 && \text{when } n_1^2 > n_2^2 + 1 \end{aligned} \tag{3.6}$$

Thus, if a cone of light is incident on one end of the fiber, it will be guided through the fiber provided the semiangle of the cone is less than i_m. This angle is a measure of the light-gathering power of the fiber and, as such, one defines

[1] This leakage is due to the fact that when a wave undergoes total internal reflection, it actually penetrates a small region of the rarer medium [see, e.g., Ghatak and Thyagarajan (1978) Chapter 11]. The wave in the rarer medium is known as the evanescent wave, which can couple light from one fiber to another. Thus, in the absence of the cladding, the light may leak away to an adjacent fiber (see Chapter 17).

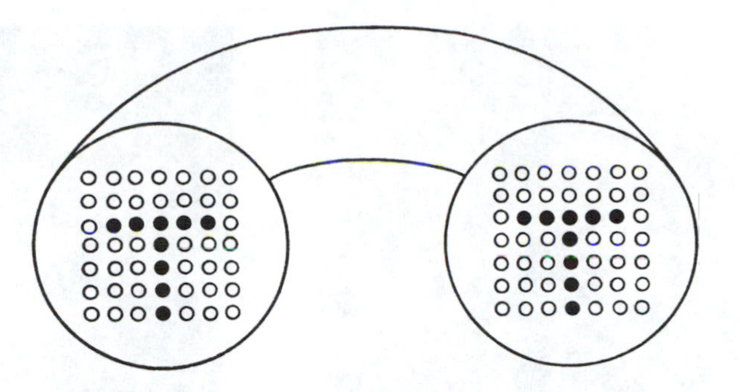

Fig. 3.5: A bundle of aligned fibers. A bright (or dark) spot at the input end of the fiber produces a bright (or dark) spot at the output end. Thus, an image will be transmitted (in the form of bright and dark spots) through a bundle of aligned fibers.

the numerical aperture (NA) of the fiber by the following equation

$$\mathrm{NA} = \left(n_1^2 - n_2^2\right)^{1/2} \tag{3.7}$$

where we have assumed that $n_1^2 < n_2^2 + 1$, which is true for all practical fibers. To get a numerical appreciation we note that for a typical fiber $n_1 = 1.48$, $n_2 = 1.46$ giving

$$\mathrm{NA} = 0.242$$

which implies $i_m \simeq 14°$. The NA of a fiber is a very important property and essentially determines the efficiency of coupling from a source to the fiber as well as losses across a misaligned joint in a splice. In Chapter 19 we discuss briefly an experimental procedure to measure NA of a fiber.

3.3 The coherent bundle

If a large number of fibers are put together, this forms what is known as a bundle. If the fibers are not aligned – that is, they are all jumbled up – the bundle is said to form an incoherent bundle. However, if the fibers are aligned properly – that is, if the relative positions of the fibers in the input and output ends are the same – the bundle is said to form a coherent bundle. Now, if a fiber is illuminated at one of its ends, then there will be a bright spot at the other end of the fiber; thus, a coherent bundle can transmit an image from one end to another (see Figure 3.5). On the other hand, in an incoherent bundle the output image will be scrambled. Because of this property, an incoherent bundle can be used as a coder; the transmitted image can be decoded by using a similar bundle in the reverse direction at the output end. In a bundle, because there can be hundreds of thousands of fibers, decoding without the original bundle configuration would be extremely difficult.

Perhaps the most important application of a coherent bundle is in a fiber optic endoscope, which can be put inside a human body and the interior of the body can be viewed from outside; for illuminating the portion that is to be seen, the bundle is enclosed in a sheath of fibers that carry light from outside to the interior of the body (see Figure 3.6). Each fiber transmits light from a small portion of the object and therefore the resolution is directly related to the packing density. A state-of-art fiberscope can have about 10,000 fibers, which would form a bundle of about 1 mm in diameter capable of resolving objects

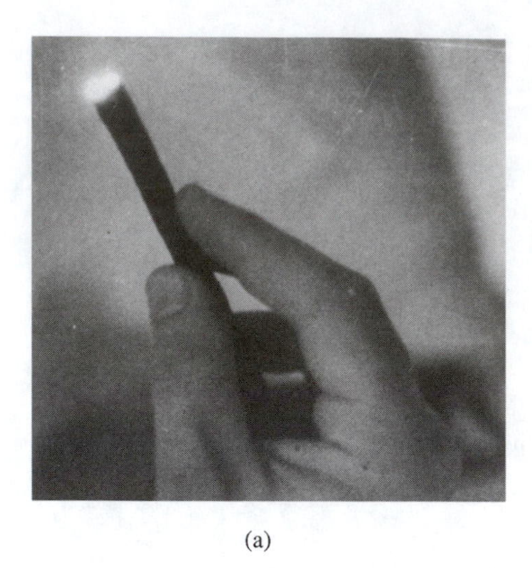

(a)

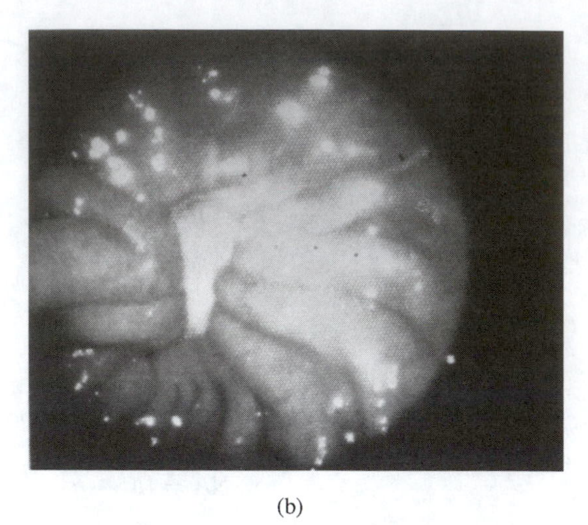

(b)

Fig. 3.6: (a) An optical fiber medical probe called an endoscope enables doctors to examine the inner parts of the human body; (b) a stomach ulcer as seen through an endoscope. [Photographs courtesy United States Information Service, New Delhi.]

70 μm across. Fiber optic bundles can also be used for viewing otherwise inaccessible parts of a machine.

3.4 Attenuation in optical fibers

Attenuation and dispersion represent the two most important characteristics of an optical fiber that determine repeater spacings in a fiber optic communication system (see Chapter 13). Obviously, the lower the attenuation (and similarly lower the dispersion) the greater will be the required repeater spacings and therefore the lower will be the cost of the communication system. Pulse dispersion will be discussed in the next section; in this section we briefly discuss the various attenuation mechanisms in an optical fiber.

The attenuation of an optical beam is usually measured in decibels (dB). If an input power P_1 results in an output power P_2, then the loss in decibels is given by

$$\alpha = 10 \log_{10} \frac{P_1}{P_2} \tag{3.8}$$

Thus, if the output power is only half the input power, then the loss is 10 log $2 \simeq 3$ dB. Similarly, a loss of 30 dB corresponds to

$$\log \frac{P_1}{P_2} = 3 \Rightarrow P_2 = \frac{1}{1000} P_1$$

Figure 3.7 shows the evolution of losses in glasses from ancient times. As can be seen until about mid-1960s, the losses in "pure" glass had been about over 1000 dB/km. These were primarily due to traces of impurities present in it. In 1966, Kao and Hockam suggested the use of optical fibers in communication systems and mentioned that for optical fiber communication to be a viable proposition, the losses should be less than 20 dB/km. Around 1966, the loss of the best available glass, which was about 1000 dB/km, implied a 50% loss in power after propagating through a 3-m length. Kao and Hockam's suggestion led to

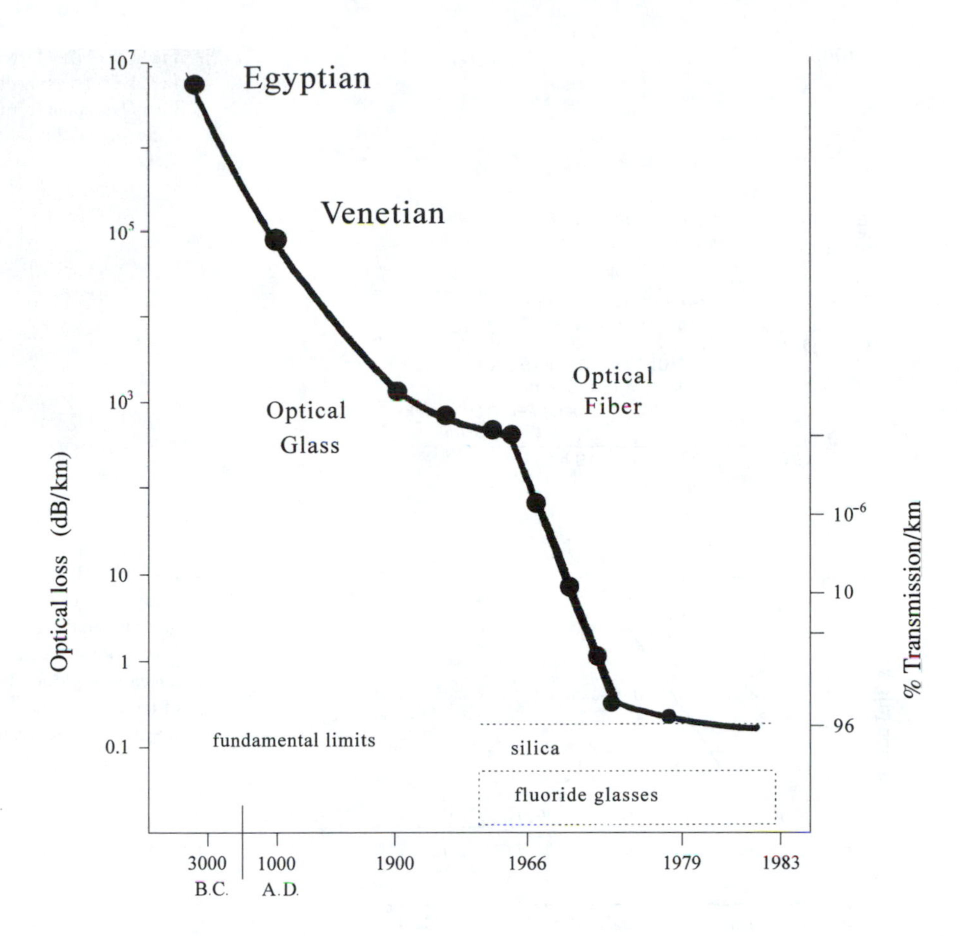

Fig. 3.7: Reduction of optical loss from the ancient times of Egyptian glasses to today's silica fibers. [After Nagel (1989).]

immense activity on the purification of fused silica, and in 1970 Corning Glass Works in the United States announced the fabrication of silica fibers having a loss of about 17 dB/km (at $\lambda_0 \simeq 0.6328$ μm). Since then the technology has been continuously improving and the current state-of-art fabricated fibers have losses ≤ 0.2 dB/km at 1.55 μm (see Figure 3.8); a loss of 0.2 dB/km implies 95.5% transmission after propagating through a 1-km length of the fiber. In Figure 3.9, a comparison is made of typical attenuation curves for various guiding media along with the frequency range at which they operate. The loss curve for the optical fiber appears very sharp because of the logarithmic frequency scale.

3.5 Pulse dispersion in step index optical fibers

Pulse dispersion represents one of the most important characteristics of an optical fiber that determines the information-carrying capacity of a fiber optic communication system.

As shown in Figure 3.1, the simplest type of optical fiber consists of a thin cylindrical structure of transparent glassy material of uniform refractive index n_1 surrounded by a cladding of another material of uniform but slightly lower refractive index n_2. These fibers are referred to as step index fibers because of the step discontinuity of the index profile at the core–cladding interface.

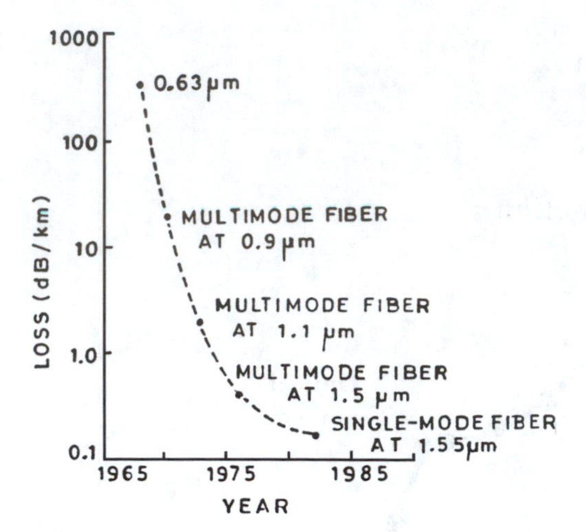

Fig. 3.8: The decrease in loss over the years of silica fibers [Adapted from Schwartz (1984).]

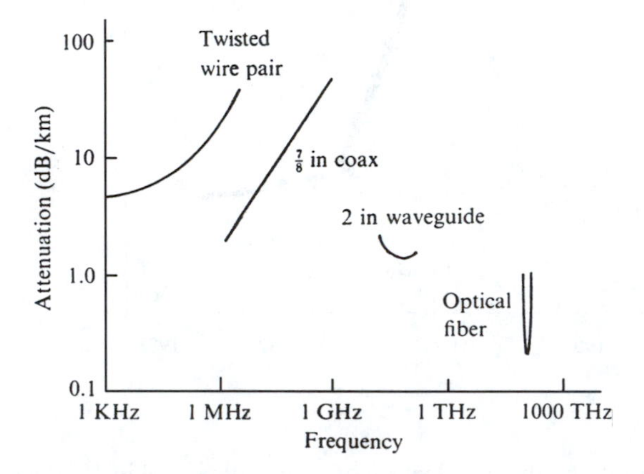

Fig. 3.9: Typical attenuation of various guiding media. The loss curve for the optical fiber appears very sharp because of the logarithmic frequency scale. [Adapted from Henry (1984).]

In digital communication systems, information to be sent is first coded in the form of pulses and then these pulses of light are transmitted from the transmitter to the receiver where the information is decoded (see Chapter 13). The larger the number of pulses that can be sent per unit time and still be resolvable at the receiver end, the larger will be the transmission capacity of the system. A pulse of light sent into a fiber broadens in time as it propagates through the fiber; this phenomenon is known as pulse dispersion and happens primarily for two reasons:

(1) Different rays take different times to propagate through a given length of the fiber (this is also known as intermodal dispersion) and
(2) Any given source emits over a range of wavelengths and, because of the intrinsic property of the material, different wavelengths take different amounts of time to propagate along the same path (also referred to as material dispersion)[2].

[2]There is also a third mechanism called waveguide dispersion that is important only in single-mode fibers (see Chapter 10). Both waveguide dispersion and material dispersion form part of what is known as intramodal dispersion.

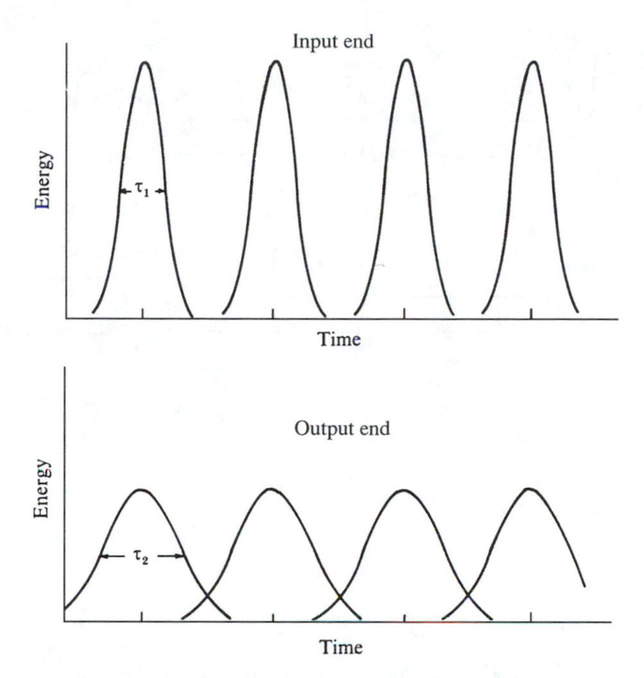

Fig. 3.10: A series of pulses, each of width τ_1 (at the input end of the fiber), after transmission through the fiber emerges as a series of pulses of width τ_2 ($> \tau_1$). If the broadening of the pulses is large, then adjacent pulses will overlap at the output end and may not be resolvable. Thus, pulse broadening determines the minimum separation between adjacent pulses, which in turn determines the maximum information-carrying capacity of the optical fiber.

To understand the first mechanism causing pulse dispersion, we note that in the fiber shown in Figure 3.1, the rays making larger angles with the axis have to traverse a longer optical path length and they take a longer time to reach the output end. Consequently, the pulse broadens as it propagates through the fiber (see Figure 3.10). Hence, even though two pulses may be well resolved at the input end, because of broadening of the pulses they may not be so at the output end. When the output pulses are not resolvable, no information can be retrieved. Thus, the smaller the pulse dispersion, the greater the information-carrying capacity of the system.

We next calculate the amount of dispersion in a step index fiber. Referring to Figure 3.1 for a ray making an angle θ with the axis, the distance AB is traversed in time

$$t = \frac{AC + CB}{c/n_1} = \frac{n_1(AB)}{c\cos\theta} \tag{3.9}$$

where c/n_1 represents the speed of light in a medium of refractive index n_1, with c being the speed of light in free space. Because the ray path will repeat itself, the time taken by a ray to traverse length L of the fiber will be

$$t = \frac{n_1 L}{c\cos\theta} \tag{3.10}$$

The above expression shows that the time taken by a ray is a function of the angle θ made by the ray with the z-axis, which leads to pulse dispersion. If we assume that all rays lying between 0 and θ_c are present, then the time taken by rays corresponding, respectively, to $\theta = 0$ and $\theta = \theta_c = \cos^{-1}(n_2/n_1)$ will be

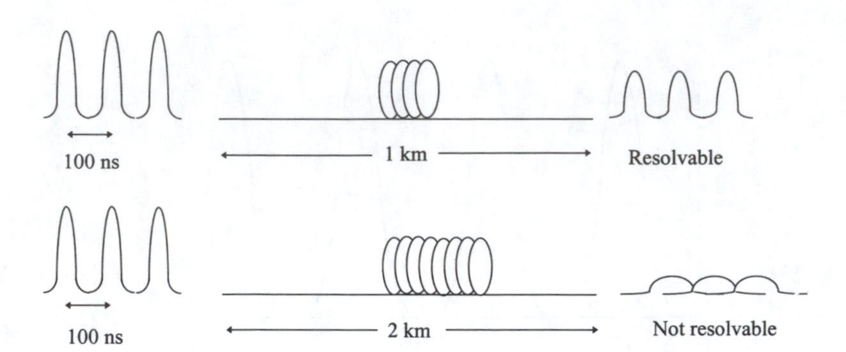

Fig. 3.11: Pulses separated by 100 ns, at the input end would be quite resolvable at the end of 1 km of the fiber. The same pulses would not be resolvable at the end of 2 km of the fiber.

given by

$$t_{\min} = \frac{n_1 L}{c} \tag{3.11}$$

$$t_{\max} = \frac{n_1^2 L}{c n_2} \tag{3.12}$$

Hence, if all the input rays were excited simultaneously, at the output end the rays would occupy a time interval of duration

$$T = t_{\max} - t_{\min} = \frac{n_1 L}{c}\left(\frac{n_1}{n_2} - 1\right) \tag{3.13}$$

For a typical fiber, if we assume

$$n_1 = 1.5, \quad \frac{n_1 - n_2}{n_2} \simeq 0.01, \quad L = 1\,\text{km}$$

one obtains

$$T \simeq 50\ \text{ns/km} \tag{3.14}$$

that is, an impulse after traversing through the fiber of length 1 km broadens to a pulse of about 50 ns duration. Thus, two pulses separated by say 100 ns at the input end would be quite resolvable at the end of 1 km of the fiber; however, they would be unresolvable at the end of 2 km (see Figure 3.11). Hence, in a 1-megabit-per-second (1 Mb/s) fiber optic system, where we have one pulse every 10^{-6} s, 50-ns/km dispersion would require repeaters to be placed every 3–4 km. On the other hand, in a 1000-Mb/s fiber optic communication system, where we require transmitting one pulse every 10^{-9} s, a dispersion of 50 ns/km would result in intolerable broadening even within 50 m or so, which would be highly inefficient and uneconomical from a system point of view.

As mentioned earlier, material dispersion that is due to the dependence of the refractive index of the fiber material on wavelength also leads to pulse dispersion that is due to the different traversal times taken by different wavelength components of the source (see Chapter 6). For optical fibers based on silica, material dispersion causes a pulse dispersion of around 90 ps/km for a source of spectral width 1 nm and operating at a wavelength of 0.85 μm. Thus, for an LED operating at a wavelength of 0.85 μm and having a spectral width of 30 nm, the contribution from material dispersion would be around 2.7 ns/km. Thus, for step index multimode fibers, the contribution from material dispersion is rather small and can be neglected. To achieve systems with very high information-carrying capacity, it is necessary to reduce pulse dispersion; two alternative solutions exist: one involving the use of graded index fibers and the other involving single-mode fibers. In Chapter 4 we discuss ray paths and pulse dispersion in graded index optical waveguides and in Chapter 10 we consider dispersion in single-mode fibers. We will see later that in graded index fibers, where the pulse dispersion that is due to different times taken by different rays can be minimized, material dispersion can play an important role. In single-mode fibers, where the first form of dispersion is absent, material dispersion plays a dominant role along with waveguide dispersion. These aspects are discussed in detail in later chapters.

We should mention here that a rigorous analysis of the propagation in fiber would involve the solution of Maxwell's equations, which are discussed in Chapter 8. Ray optics is valid when the waveguide parameter

$$V = \frac{2\pi}{\lambda_0} a \left(n_1^2 - n_1^2\right)^{1/2} \tag{3.15}$$

is greater than about 10; in the above equation λ_0 represents the wavelength of light in free space. For the parameters given in equation (3.2) and operating at $\lambda_0 = 0.8\,\mu$m

$$V \simeq 48$$

and the ray analysis should be very accurate. However, with the same values of n_1 and n_2 if the core radius was 2 μm, the value of V would have been 3.8 and the ray analysis would not be applicable at all. For such fibers the wave theory has to be used (see Chapter 8).

3.6 Loss mechanisms

The principal sources of attenuation in an optical fiber can be broadly classified into two groups: absorptive and radiative.

3.6.1 Absorptive losses

Absorptive losses can be further subdivided into intrinsic and extrinsic losses. Intrinsic absorption is caused by interaction of the propagating lightwave with one or more major components of glass that constitute the fiber's material composition. An example of such an interaction is the infrared absorption band of SiO_2. However, in the wavelength regions of interest to optical communication (0.8–0.9 μm and 1.2–1.5 μm), infrared absorption tails make negligible contributions (see Figure 3.12).

Table 3.1. *Absorption loss in silica glass due to presence of 1 ppm of different metals and OH^- ions as impurities*

Impurities	Loss due to 1 ppm of impurity (dB/km)	Absorption peak (μm)
Fe^{2+}	0.68	1.1
Fe^{3+}	0.15	0.4
Cu^{2+}	1.1	0.85
Cr^{3+}	1.6	0.625
V^{4+}	2.7	0.725
OH^-	1.0	0.95
OH^-	2.0	1.24
OH^-	4.0	1.38

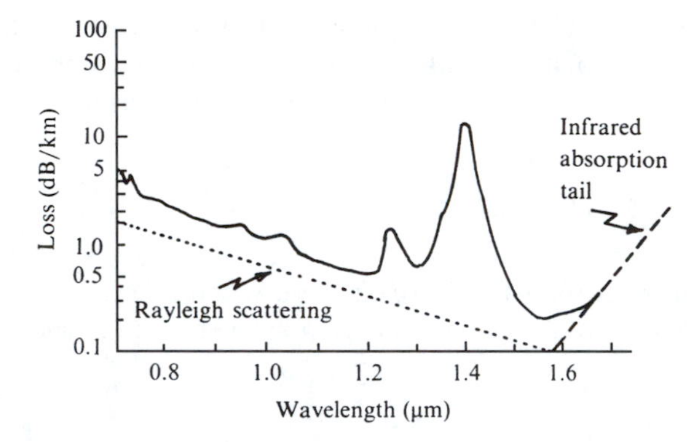

Fig. 3.12: The solid curve represents a typical loss spectrum of a silica fiber fabricated around 1979 and has been adapted from Miya et al. (1979). The dotted curve represents the Rayleigh scattering loss, which varies as λ^{-4}, and the dashed curve represents the infrared absorption tail.

On the other hand, extrinsic absorption is caused by the presence of minute quantities of materials like transition metal ions (e.g., Fe^{2+}, Cu^{2+}, Cr^{3+}, and so forth) and is also due to OH^- ions dissolved in glass (see Table 3.1). For example, the presence of 1 part per million (ppm) of Fe^{2+} would lead to a loss of 0.68 dB/km at 1.1 μm – thus the necessity of ultrapure fibers. The presence of OH^- ions leads to absorption peaks at 0.72, 0.88, 0.95, 1.13, 1.24, and 1.38 μm. The broad peaks at 1.24 and 1.38 μm in Figure 3.12 are due to the presence of OH^- ions. Fortunately, the absorption bands are narrow enough that ultrapure fibers exhibit losses ≤ 0.2 dB/km at $\lambda_0 \simeq 1.55$ μm. Various technologies have been developed for fabrication of extremely low-loss optical fibers [see, e.g., Croft et al. (1985), French et al. (1979), Pal (1979)]. Using the vapor-phase axial deposition (VAD) technique, it has been possible to reduce the residual OH^- content to less than 1 part per billion (ppb); the corresponding loss curve is shown in Figure 3.13. Such fibers provide a wider low-loss window in silica-based fibers, permitting the use of wavelength division multiplexing in fiber optic systems.

3.6.2 *Radiative losses*

Radiative losses occur when a guided light beam gets coupled to radiation propagating in the cladding. Rayleigh scattering is predominantly responsible

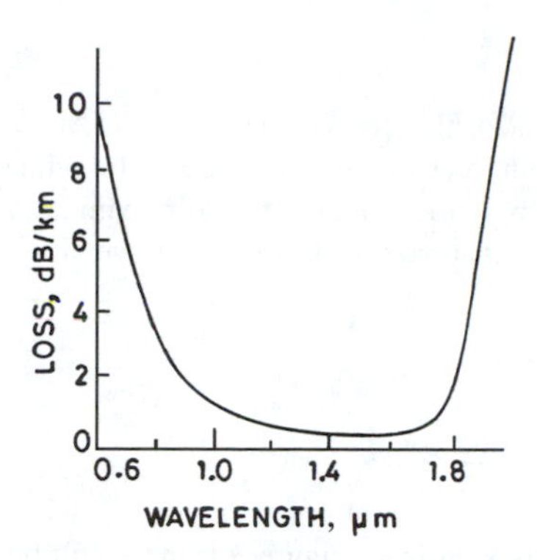

Fig. 3.13: Loss spectrum of an ultimately low OH^- content VAD optical fiber. [Adapted from Moriyama et al. (1980).] Such fibers provide a wide low-loss window.

for such coupling and is a fundamental mechanism that is caused by small-scale (small compared with the wavelength of the lightwave) inhomogeneities that are frozen into the fiber. These inhomogeneities are produced during the fabrication of the fiber and result in composition and density fluctuations. The loss due to Rayleigh scattering is proportional to λ_0^{-4} and obviously decreases rapidly with increase in wavelength (see the dotted curve in Figure 3.12). At 1.55-μm wavelength, theory predicts a Rayleigh scattering loss of ~0.15 dB/km in fused silica, which represents the ultimate loss limit of the optical fiber (at 1.55 μm). Indeed, the attenuation coefficient due to Rayleigh scattering in (pure) fused silica is given by the following approximate formula

$$\alpha(\lambda) = \alpha_0 \left(\frac{\lambda_0}{\lambda} \right)^4 \tag{3.16}$$

where

$$\alpha_0 = 1.7 \text{ dB/km} \quad \text{at } \lambda_0 = 0.85\,\mu\text{m} \tag{3.17}$$

The above equation predicts the Rayleigh scattering loss to be 0.31 dB/km and 0.15 dB/km at 1.3-μm and 1.55-μm wavelengths, respectively.

Dopants like GeO_2, P_2O_5, etc. (which are used so that the fiber has a specific refractive index profile), also lead to an increase in the Rayleigh scattering loss. Thus, larger NA fibers that are produced by larger levels of doping generally have higher Rayleigh scattering loss.

Radiative losses may also be caused by fiber bending or by imperfections in the fiber such as core–cladding interface irregularities, diameter fluctuations, and so forth.

A possible futuristic system is based on infrared fibers ($\lambda_0 \geq 2\,\mu$m) where, because of the larger wavelength, the Rayleigh scattering loss is extremely small. Obviously, silica-based fibers cannot be used because of the occurrence of the infrared absorption band (see Figures 3.12 and 3.13). On the other hand, for fluoride glasses, the absorption peak lies in the far infrared (~50 μm), the tail of which has negligible value at 2 μm. Thus, infrared fibers ($\lambda_0 \geq 2\,\mu$m) would have extremely low loss ($\leq$0.01 dB/km), implying repeater spacings $\geq$1000 km, which would indeed be a fantastic technological achievement. However, at present there are problems not only in the fabrication of fibers with such low losses but also in the fabrication of reliable sources and detectors at these wavelengths.

Problems

3.1 (a) Consider a step index fiber for which $n_1 = 1.475$, $n_2 = 1.460$, and $a = 25\ \mu$m. What is the maximum value of θ (see Figure 3.1) for which the rays will be guided through the fiber? (b) Corresponding to the maximum value of θ, calculate the number of reflections that would take place in traversing a kilometer length of the fiber.

Solution: (a)

$$\theta_m = \cos^{-1}\frac{n_2}{n_1} \simeq 8.2^\circ$$

(b) As can be readily seen from Figure 3.1, there will be one reflection for a ray traversing a distance $2a/\tan\theta$ along the length of the fiber. Thus, the number of reflections (for a ray propagating with $\theta \simeq \theta_m$) that would take place in traversing a distance L along the length of the fiber would be given by

$$\frac{\tan\theta_m}{2a}L$$

implying there will be approximately 2.88 million reflections in traversing a kilometer length of the fiber.

3.2 In the above problem assume a loss of only 0.01% of power at each reflection at the core–cladding interface. Calculate the corresponding loss in dB/km.

Solution: Loss in dB/reflection $= 10\log\frac{1}{0.9999} \simeq 4.3 \times 10^{-4}$. Thus, there will be a loss of about 1234 dB after traversing a kilometer length of the fiber. Thus, the core–cladding interface should be extremely smooth.

3.3 Repeat the calculations in the above two problems for a bare silica fiber for which $n_1 = 1.46$, $n_2 = 1.0$, and $a = 25\ \mu$m.

[ANSWER: $\theta_m \simeq 46.8^\circ$; ~21 million reflections per kilometer; ~9150 dB/km.]

3.4 Consider a bare fiber consisting of a core of refractive index 1.48 and having air ($n_2 = 1$) as cladding. What is its NA? What is the maximum incident angle up to which light can be guided by the fiber?

[ANSWER: NA $= 1.09$, $i_m = \pi/2$.]

3.5 Consider a fiber with $n_1 = 1.48$, $n_2 = 1.46$, and with its end placed in water ($n_0 = 1.33$). What is the maximum angle of incidence for guidance?

[ANSWER: $i_m = 10.5^\circ$.]

3.6 Polymer optical fibers with high-purity polymethyl methacrylate core and fluorinated polymer cladding are commercially available with a NA of 0.50. What is the corresponding maximum angle of acceptance?

[ANSWER: $i_m = 60^\circ$.]

3.7 Consider a fiber from which the cladding is removed over a short length as shown in Figure 3.14. Assume that the core and cladding refractive indices are 1.5 and 1.4, respectively. What will happen to the output power if the bare portion of the fiber is covered by a liquid whose refractive index is varied from 1.4 to 1.5?

3.8 Consider an optical fiber consisting of a core and a cladding made of different materials with widely differing dispersion characteristics. Figure 3.15 shows a typical wavelength variation of the refractive index of three glasses: ADF10, BaF7,

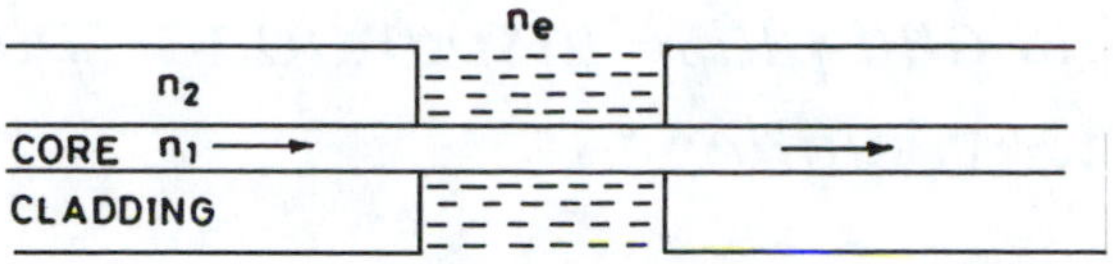

Fig. 3.14: For Problem 3.7.

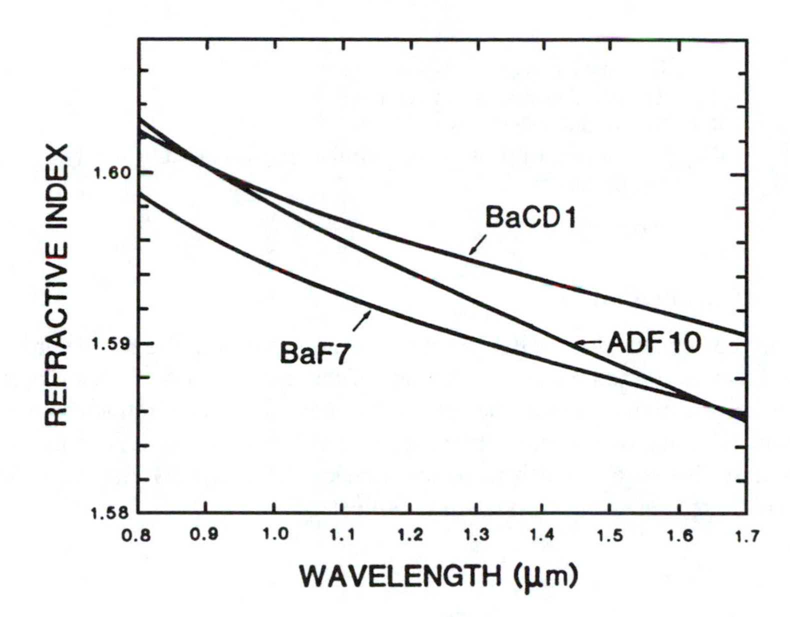

Fig. 3.15: Wavelength variation of the refractive index of three glasses. [After Morishita (1989).]

and BaCD1 from Hoya Optical Glass technical data [Morishita (1989)]. Consider two fibers formed with

(a) BaCD1 core and ADF10 cladding
(b) ADF10 core and BaF7 cladding

In what wavelength regions would the two fibers act as guiding structures?

4

Ray paths and pulse dispersion in planar optical waveguides

4.1 Introduction

In the previous chapter we discussed the phenomenon of pulse broadening in step index fibers and mentioned that one of the techniques of reducing pulse broadening is to use graded index fibers. To study the effect of refractive index gradient on pulse dispersion, in this chapter we obtain the ray paths and calculate the pulse dispersion in slab waveguides characterized by a refractive index variation that depends only on the x coordinate:

$$n = n(x) \tag{4.1}$$

Using Snell's law, we first derive the one-dimensional ray equation, the solution of which would give the ray paths. From the ray paths, we calculate the ray transit times – that is, the time taken by a specific ray to propagate through a certain distance of the waveguide. We consider a special class of graded index waveguides – namely, the power law profile – and obtain an expression for the ray transit time. In Chapter 5 we obtain the optimum profile shape that corresponds to minimum pulse dispersion. Although the transit time calculations correspond to slab waveguides, the final results obtained for pulse dispersion are rigorously valid for the corresponding situation in optical fibers characterized by a power law profile.

4.2 The one-dimensional ray equation

We consider a medium in which the refractive index depends only on the x coordinate as given by equation (4.1). A medium with continuously varying refractive index given by equation (4.1) can be thought of as a limiting case of a medium consisting of a set of thin layers, each characterized by a specific value of the refractive index (see Figure 4.1(a)). To trace out the rays through such a stack of thin layers we can use Snell's law according to which

$$n_1 \sin\phi_1 = n_2 \sin\phi_2 = n_3 \sin\phi_3 = \ldots \quad \text{constant}$$

where $\phi_1, \phi_2, \ldots$ are the angles of incidence at various interfaces as shown in Figure 4.1(a). If $\theta_1, \theta_2, \ldots$ are the corresponding angles that the rays make with

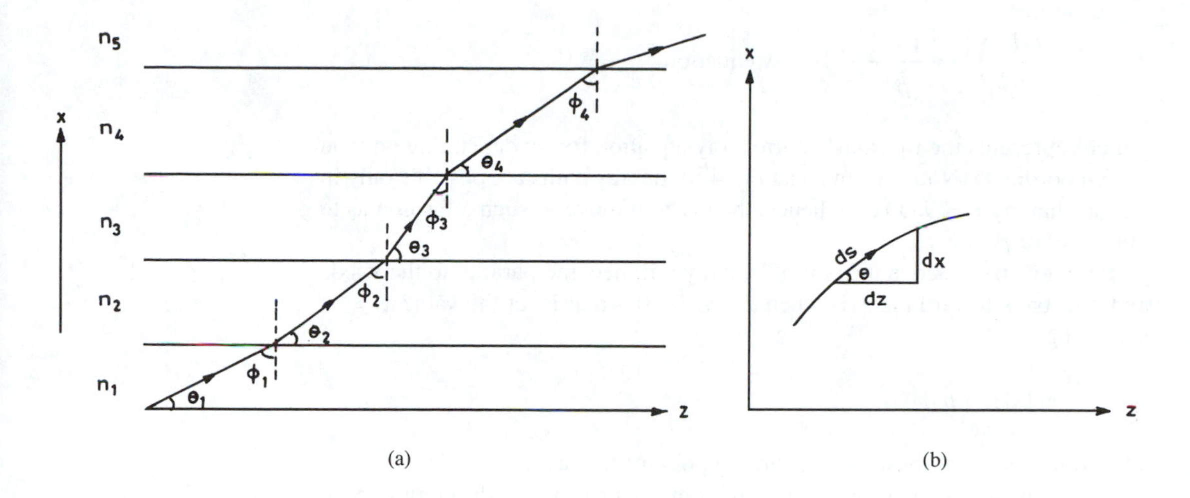

Fig. 4.1: (a) A continuously varying refractive index distribution can be approximated by a layered structure, and ray paths in such a layered structure can be obtained by using Snell's law. (b) As the number of layers increases and the thickness of each layer reduces to zero, the ray path becomes a continuous curve.

the z-axis, then

$$n_1 \cos\theta_1 = n_2 \cos\theta_2 = n_3 \cos\theta_3 = \ldots \text{ constant, say } \tilde{\beta} \tag{4.2}$$

(we have chosen the z-axis in such a way that the ray lies in the x–z plane).

When the refractive index variation is continuous, the thickness of each layer becomes infinitesimally small and the piecewise straight lines shown in Figure 4.1(a) form a continuous curve as shown in Figure 4.1(b). From equation (4.2) we can infer that the rays bend in such a way that the product $n(x)\cos\theta(x)$ remains constant, which we denote by $\tilde{\beta}$. Thus

$$n(x)\cos\theta(x) = \tilde{\beta} \quad \text{(invariant of the ray path)} \tag{4.3}$$

Now if ds represents the infinitesimal arc length along the ray path, then from Figure 4.1(b) we have

$$(ds)^2 = (dx)^2 + (dz)^2$$

or

$$\left(\frac{ds}{dz}\right)^2 = \left(\frac{dx}{dz}\right)^2 + 1 \tag{4.4}$$

Since

$$\cos\theta = \frac{dz}{ds}$$

we obtain

$$\frac{ds}{dz} = \frac{1}{\cos\theta(x)} = \frac{n(x)}{\tilde{\beta}}$$

Substituting in equation (4.4) we get

$$\left(\frac{dx}{dz}\right)^2 = \frac{n^2(x)}{\tilde{\beta}^2} - 1 \quad \text{ray equation} \tag{4.5}$$

which represents the rigorously correct ray equation for n^2 depending only on the x-coordinate. Notice from equation (4.5) that ray paths are possible only in regions having $\tilde{\beta} \leq n(x)$ and, hence, the ray will move in such a fashion as to always have $\tilde{\beta} \leq n(x)$.

For a given refractive index profile, a ray will become parallel to the z-axis and turn back toward the axis when $dx/dz = 0$ – that is, at the value $x = x_t$ satisfying

$$n(x_t) = \tilde{\beta} \tag{4.6}$$

The point $x = x_t$ is known as the turning point of the ray.

Equation (4.5) can be put in a more convenient form by differentiating it with respect to z

$$2\frac{dx}{dz}\frac{d^2x}{dz^2} = \frac{1}{\tilde{\beta}^2}\frac{dn^2}{dx}\frac{dx}{dz}$$

or

$$\frac{d^2x}{dz^2} = \frac{1}{2\tilde{\beta}^2}\frac{dn^2}{dx} \quad \text{ray equation} \tag{4.7}$$

which is another form of the ray equation. We consider some simple examples.

4.2.1 Ray paths in a homogeneous medium

In a homogeneous medium, n^2 is a constant and equation (4.7) simplifies to

$$\frac{d^2x}{dz^2} = 0$$

the solution of which is

$$x(z) = A + Bz$$

which represents a straight line. Thus, we get the obvious result that the ray paths in a homogeneous medium are straight lines.

4.2.2 Ray paths in square law media

A square law medium is characterized by the following refractive index distribution (see Figure 4.2(a)).

$$\begin{aligned} n^2(x) &= n_1^2\left[1 - 2\Delta\left(\frac{x}{a}\right)^2\right], \quad |x| < a \quad \text{core} \\ &= n_1^2[1 - 2\Delta] = n_2^2, \quad |x| > a \quad \text{cladding} \end{aligned} \tag{4.8}$$

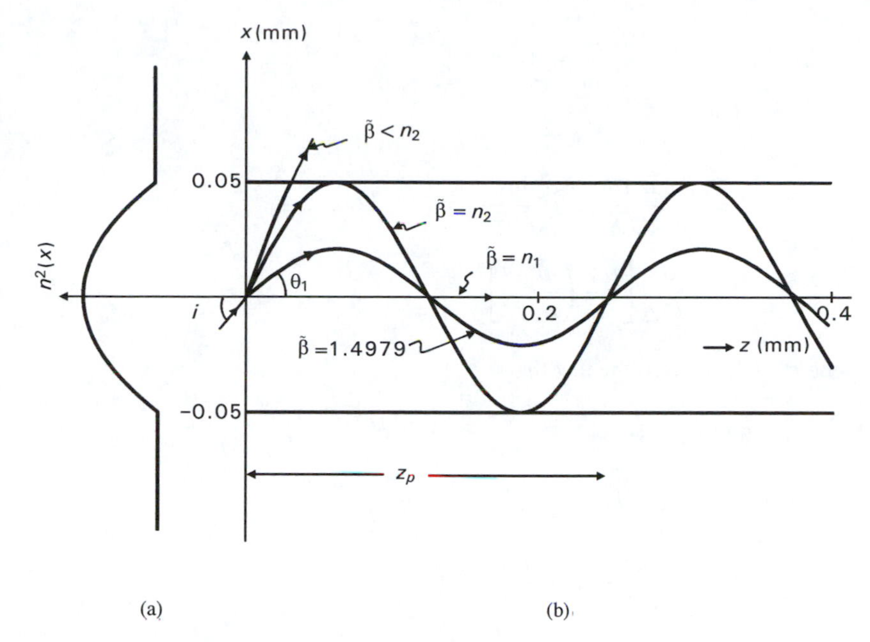

Fig. 4.2: (a) Parabolic variation of refractive index; $n_1 = 1.5$, $\Delta = 0.01$, and $a = 50\,\mu$m. (b) Exact ray paths in a parabolic index slab waveguide for different launch angles. The ray that propagates along the axis has $\tilde{\beta} = n_1$ and the ray that turns back at $x = a$ has $\tilde{\beta} = n_2$. All guided rays have $n_2 < \tilde{\beta} < n_1$. A ray launched with $\tilde{\beta} < n_2$ will be refracted away and corresponds to a refracting ray.

Ray paths in square law media are of great importance as they readily lead to very important results for parabolic index fibers, which are extensively used in fiber optic communication systems.

We consider ray paths in the core of the waveguide. Substituting for $n^2(x)$ from equation (4.8) in equation (4.7) we obtain

$$\frac{d^2x}{dz^2} + \Gamma^2 x(z) = 0 \tag{4.9}$$

where

$$\Gamma = \frac{n_1\sqrt{2\Delta}}{\tilde{\beta}a} \tag{4.10}$$

The general solution of equation (4.9) is given by

$$x(z) = A\sin\Gamma z + B\cos\Gamma z \quad \text{rigorously correct ray paths in square law media} \tag{4.11}$$

which represents the general ray path. In a parabolic index fiber the meridional ray paths are rigorously given by equation (4.11) and, without any loss of generality, we may assume

$$x(0) = 0$$

which implies $B = 0$. Thus

$$x(z) = A\sin\Gamma z$$

If the ray makes an angle θ_1 with the z-axis at $z = 0$, then

$$\tan\theta_1 = \left.\frac{dx}{dz}\right|_{z=0} = A\Gamma$$

or

$$A = \frac{\tilde{\beta}\, a \tan\theta_1}{n_1\sqrt{2\Delta}} = \frac{a\sin\theta_1}{\sqrt{2\Delta}}$$

$$= \frac{a}{\sqrt{2\Delta}}\left[1 - \left(\frac{\tilde{\beta}}{n_1}\right)^2\right]^{\frac{1}{2}} \tag{4.12}$$

where we have used the fact that

$$\tilde{\beta} = n_1 \cos\theta_1$$

We therefore obtain

$$x(z) = \frac{a\sin\theta_1}{\sqrt{2\Delta}}\sin\left(\frac{n_1\sqrt{2\Delta}}{a\tilde{\beta}}\,z\right) \tag{4.13}$$

The periodic length of the sinusoidal ray path is given by

$$z_p = \frac{2\pi}{\Gamma} = \frac{2\pi a\tilde{\beta}}{n_1\sqrt{2\Delta}} \tag{4.14}$$

Typical ray paths are shown in Figure 4.2; each ray corresponds to a specific value of $\tilde{\beta}$ and bends in such a way that the product $n(x)\cos\theta(x)$ remains constant. The figure corresponds to $n_1 = 1.5$, $\Delta = 0.01$, and $a = 50\,\mu\text{m}$ (see Problem 4.1). For the axial ray, $\tilde{\beta} = n_1(=1.5)$. The ray with $\tilde{\beta} = n_2 = 1.4849(\theta_1 = 8.13°)$ is tangential at the core–cladding interface. The NA of the waveguide is discussed in Problem 4.2.

The ray paths shown in Figure 4.2 correspond to $x = 0$ at $z = 0$. In general, the ray path is completely determined if the values of $\tilde{\beta}$ and the x coordinate of the ray at $z = 0$ are known (see Problem 4.4).

The ray paths given by equation (4.13) are valid only for $|x| < a$ since the refractive index profile is parabolic only in the region $|x| < a$. Now if $\tilde{\beta}$ is such that the turning point $x = x_t$ lies between $x = +a$ and $x = -a$, then the ray will propagate by periodically oscillating around the z-axis as shown in Figure 4.2. Such rays will form guided rays. For this to happen, $\tilde{\beta}$ must be $> n_2$ since the turning point is determined by $n(x_t) = \tilde{\beta}$ and the refractive index reduces monotonically from the axis. Also from equation (4.3) and the fact that the maximum value of n is n_1, we must have $\tilde{\beta} \leq n_1$. Hence, for guided rays we must have

$$n_2 < \tilde{\beta} < n_1 \quad \text{guided rays} \tag{4.15}$$

If $\tilde{\beta} < n_2$, then the ray will intersect the core–cladding interface at a finite angle and will be transmitted into the cladding (see Figure 4.2). Further, since the cladding has uniform refractive index, the ray will travel in a straight line. Hence, such rays are not guided and are called *refracting rays*.

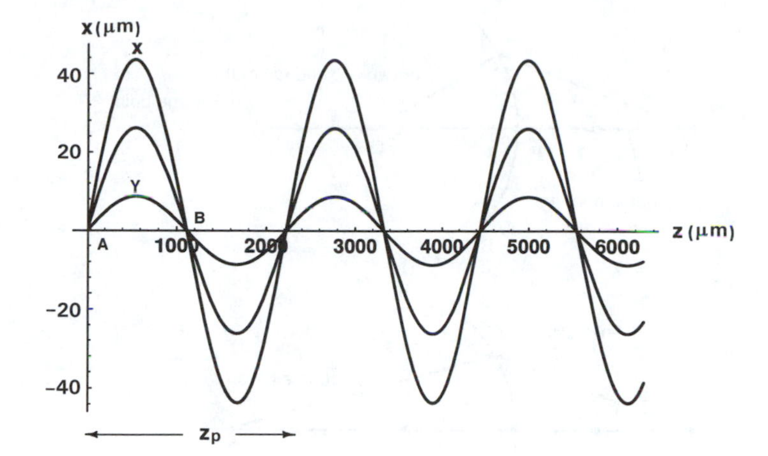

Fig. 4.3: Ray paths in the paraxial approximation in a parabolic index waveguide. Under this approximation, all rays have the same period. The figure corresponds to $n_1 = 1.50$, $\Delta = 0.01$, and $a = 50\,\mu$m (see Problem 4.1).

$$0 < \tilde{\beta} < n_2 \quad \text{refracting rays} \tag{4.16}$$

The rays shown in Figure 4.2(b) have slightly different periods (see Problem 4.1). However, if θ_1 is restricted to small values so that $\cos\theta_1 \simeq 1$ (as is indeed the case for usual launch conditions), $\tilde{\beta} \simeq n_1$ and all rays have the same periodic length given by

$$z_p \simeq \frac{2\pi a}{\sqrt{2\Delta}} \tag{4.17}$$

which is independent of the launch angle θ_1 (see Figure 4.3). Figure 4.3 corresponds to $n_1 = 1.5$, $\Delta = 0.01$, and $a = 50\,\mu$m (see Problem 4.1), the paraxial period being 2.22 mm. Thus, connecting the two points A and B, there are an infinite number of nearby paths, implying that the optical path length $\int n\,ds$ must be stationary and all rays would take the same amount of time.

Physically, although the ray AXB traverses a larger path in comparison to AYB, it does so in a medium of lower "average" refractive index – thus, the greater path length is compensated by a greater "average speed" and, hence, all rays take the same amount of time (in the language of wave optics, all modes have the same group velocity; see Chapter 7).

We may pause here for a moment and discuss some elementary paraxial optics. Let APB denote a curved surface separating two media of refractive indices n_1 and n_2 (see Figure 4.4). Let I be the paraxial image point of the object point O. Let us consider a point Q on the left of I. If we ask what rays connect the points O and Q, the answer is the straight line path OPQ [see Figure 4.4(a)]. Only this will be the allowed ray path connecting O and Q. This path corresponds to L_{op} (=optical path length) being a minimum – that is, the time taken by the ray OPQ is a minimum – and all nearby paths like OAQ take a longer time

$$n_1 \cdot OA + n_2 \cdot AQ > n_1 \cdot OP + n_2 \cdot PQ$$

Similarly, for a point R on the right of the point I (see Figure 4.4(b)), the only allowed ray path (connecting O and R) is the straight line path OPR, which corresponds to L_{op} being maximum and all nearby paths like OAR take less

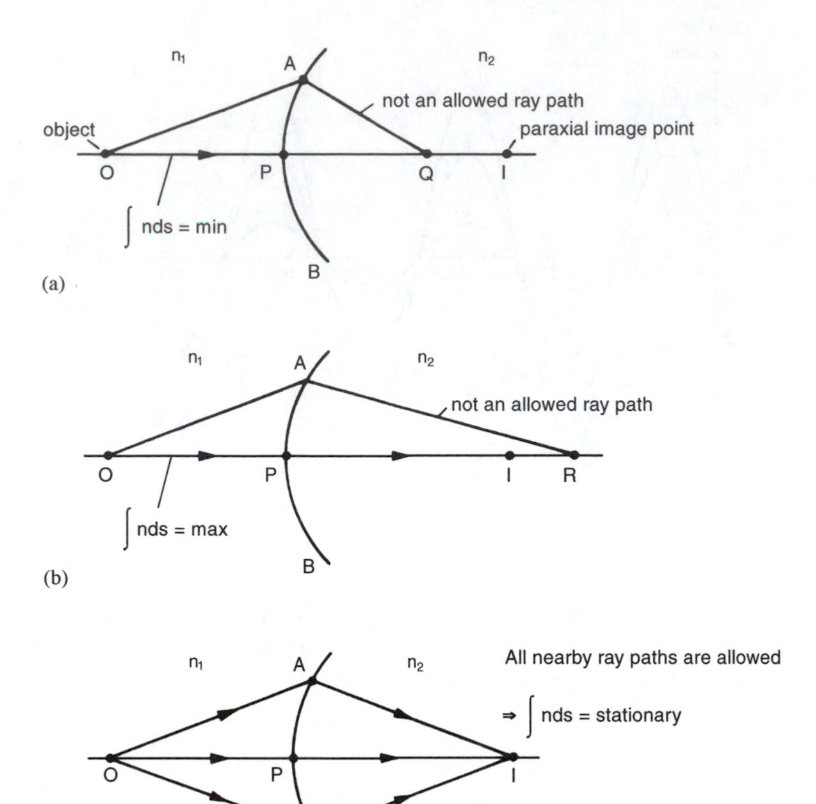

Fig. 4.4: APB represents a refracting surface separating media of refractive indices n_1 and n_2 and I is the paraxial image point of O. (a) OPQ is an allowed ray path and is such that its optical path length is a minimum – that is, it is less than the optical path lengths of all nearby paths such as OAQ. (b) Again, OPR is an allowed ray path and its optical path length corresponds to a maximum – that is, all nearby paths such as OAR have an optical path length less than OPR. (c) All nearby ray paths such as OPI, OAI, OBI, have equal optical path lengths and thus in the present case the optical path length assumes a stationary value. [Adapted from Ghatak and Sauter (1989)].

time

$$n_1 \cdot OA + n_2 \cdot AR < n_1 \cdot OP + n_2 \cdot PR$$

Finally, for the paraxial image point I, all nearby ray paths are allowed and L_{op} will be stationary so that (see Figure 4.4(c))

$$n_1 OA + n_2 AI = n_1 OP + n_2 PI = n_1 \text{OB} + n_2 \text{BI} \tag{4.18}$$

and all such paraxial rays take the same amount of time.

Returning to Figure 4.3, rays connecting A and B correspond to

$$\int n\,ds = \text{stationary} \tag{4.19}$$

and all paraxial rays take the same amount of time. This would lead to extremely small pulse dispersion. We give below an exact analysis of transit time calculations leading to the calculation of pulse dispersion.

4.3 Transit time calculations

One of the important characteristics of an optical waveguide is pulse dispersion, which is the temporal spreading of a pulse of light launched into the waveguide. One of the mechanisms that leads to pulse dispersion is the difference in time taken by different rays. Thus, if all rays are launched simultaneously in a waveguide, at the output the rays will arrive at different times and, hence, will correspond to a temporal dispersion of the pulse. To calculate the pulse dispersion, we calculate the time taken by a ray to traverse a given length of the waveguide.

Let τ_p represent the time taken for the ray to traverse the length z_p (see Figure 4.2(b)). Now the time taken to travel an arc length ds along the ray is

$$d\tau = \frac{ds}{v(x)} = \frac{1}{c} n(x)\, ds \tag{4.20}$$

where $v(x) = c/n(x)$ is the velocity of the ray along the ray path. Hence, the time taken for one period is

$$\tau_p = \int \frac{n(x)\, ds}{c} \tag{4.21}$$

where the integral is over one period. Now

$$ds = [(dx)^2 + (dz)^2]^{1/2} = \left[1 + \left(\frac{dz}{dx}\right)^2\right]^{1/2} dx$$

which, using the ray equation (equation (4.5)), can be written as

$$ds = \frac{n(x)}{[n^2(x) - \tilde{\beta}^2]^{1/2}} dx \tag{4.22}$$

Hence, the time taken to travel over a quarter of a period (i.e., from the axis to the turning point) is $\tau_p/4$, which is given by

$$\frac{1}{4}\tau_p = \frac{1}{c} \int_0^{x_t} \frac{n^2(x)}{[n^2(x) - \tilde{\beta}^2]^{1/2}} dx \tag{4.23}$$

where x_t is the x coordinate at the turning point. The above equation is valid for an arbitrary profile. Some specific profiles are considered below.

4.3.1 Pulse dispersion in a parabolic index medium

We evaluate equation (4.23) for a parabolic index medium characterized by equation (4.8). Since the turning point is defined by $n(x_t) = \tilde{\beta}$, using equation (4.8) we have

$$n_1^2\left[1 - 2\Delta\left(\frac{x_t}{a}\right)^2\right] = \tilde{\beta}^2$$

or

$$x_t = \frac{a}{n_1\sqrt{2\Delta}}[n_1^2 - \tilde{\beta}^2]^{1/2} \tag{4.24}$$

Thus

$$\tau_p = \frac{4}{c}\int_0^{x_t} \frac{n_1^2[1-2\Delta(\frac{x}{a})^2]}{\{n_1^2[1-2\Delta(\frac{x}{a})^2]-\tilde{\beta}^2\}^{1/2}}dx$$
$$= \frac{4}{c}\int_0^{x_t} \frac{n_1^2-2\Delta n_1^2(\frac{x}{a})^2-\tilde{\beta}^2+\tilde{\beta}^2}{[n_1^2-2\Delta n_1^2(\frac{x}{a})^2-\tilde{\beta}^2]^{1/2}}dx$$

or

$$\tau_p = \frac{4}{c}\left[\frac{n_1\sqrt{2\Delta}}{a}\int_0^{x_t}\sqrt{x_t^2-x^2}\,dx + \frac{\tilde{\beta}^2 a}{n_1\sqrt{2\Delta}}\int_0^{x_t}\frac{dx}{\sqrt{x_t^2-x^2}}\right] \tag{4.25}$$

where x_t is defined through equation (4.24). Carrying out the elementary integration, we get

$$\tau_p = \frac{\pi}{c}\frac{a}{n_1\sqrt{2\Delta}}\left[n_1^2+\tilde{\beta}^2\right] \tag{4.26}$$

Now, from equation (4.13) it readily follows that the turning point (see Figure 4.2(b)) corresponds to $z = z_p/4$ and

$$\frac{1}{4}\Gamma z_p = \frac{\pi}{2}$$

implying (see equation (4.14))

$$z_p = \frac{2\pi}{\Gamma} = \frac{2\pi a\tilde{\beta}}{n_1\sqrt{2\Delta}} \tag{4.27}$$

Thus, if $\tau(z)$ represents the time taken by the ray to traverse the distance z (which covers many periods) then

$$\frac{\tau(z)}{z} = \frac{\tau_p}{z_p} = \frac{1}{2c}\left[\tilde{\beta}+\frac{n_1^2}{\tilde{\beta}}\right] \tag{4.28}$$

Since

$$n_2 < n(x) < n_1 \tag{4.29}$$

we have for guided rays

$$n_2 < \tilde{\beta} < n_1 \tag{4.30}$$

A ray with $\tilde{\beta} = n_1$ takes the minimum time given by

$$\tau_{\min}(z) = \frac{1}{2c}\left[n_1+\frac{n_1^2}{n_1}\right]z = \frac{n_1}{c}z \tag{4.31}$$

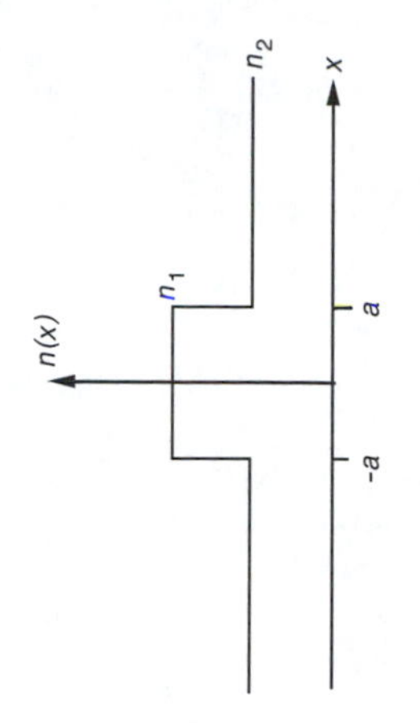

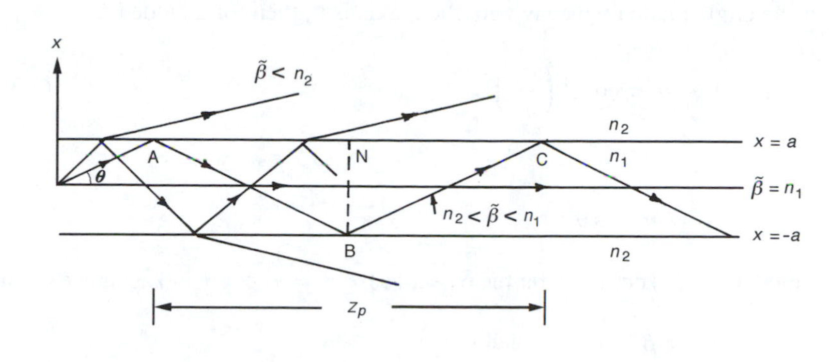

Fig. 4.5: Guided and refracting rays in a step index planar waveguide. For guided rays, $n_2 < \tilde{\beta} < n_1$, and for refracting rays, $\tilde{\beta} < n_2$.

whereas a ray with $\tilde{\beta} = n_2$ takes the maximum time given by

$$\tau_{\max}(z) = \frac{1}{2c}\left[n_2 + \frac{n_1^2}{n_2}\right] z \tag{4.32}$$

Obviously, $\tau_{\min}$ corresponds to the axial ray for which $\theta = 0$ and $\tau_{\max}$ to a ray so that it becomes parallel to the z-axis at $x = a$ (see the curve corresponding to $\tilde{\beta} = n_2$ in Figure 4.2). Thus, the pulse dispersion is given by

$$\begin{aligned} \Delta\tau &= \tau_{\max} - \tau_{\min} \\ &= \frac{1}{2cn_2}(n_1 - n_2)^2 z \\ &\simeq \frac{n_1}{2c}\Delta^2 z \end{aligned} \tag{4.33}$$

where in the last step we have assumed $n_1 \simeq n_2$. Thus, for $n_1 = 1.46$, $\Delta = 0.01$ we have

$$\Delta\tau \simeq 250 \text{ ps/km}$$

Note that $\Delta\tau$ is proportional to Δ^2; thus, for reducing pulse dispersion, one must have waveguides with small NA.

4.3.2 *Pulse dispersion in a planar step index waveguide*

To appreciate the small pulse dispersion given by equation (4.33) in a square law medium, we now obtain the pulse dispersion in a step index planar waveguide characterized by (see Figure 4.5)

$$\begin{aligned} n^2(x) &= n_1^2 \quad |x| < a \quad \text{(core)} \\ &= n_2^2 = n_1^2(1 - 2\Delta) \quad |x| > a \quad \text{(cladding)} \end{aligned} \tag{4.34}$$

Because the core and cladding are homogeneous media, ray paths in both media are straight lines (see Figure 4.5). For a ray to be guided in the core, it must suffer total internal reflection at the core–cladding interfaces at $x = \pm a$. Hence, if the angle made by the ray with the z-axis is θ, then for a guided ray

$$0 < \theta < \cos^{-1}\left(\frac{n_2}{n_1}\right) \tag{4.35}$$

or

$$n_1 > n_1 \cos\theta > n_2$$

Since $\tilde{\beta} = n(x)\cos\theta(x)$, for the present case $\tilde{\beta} = n_1 \cos\theta$ and again we obtain

$$\begin{aligned} n_2 < \tilde{\beta} < n_1 \quad & \text{guided rays} \\ \tilde{\beta} < n_2 \quad & \text{refracted rays} \end{aligned} \tag{4.36}$$

To obtain the transit time we must obtain the ray period z_p and the time taken τ_p to cover one period. Now, from Figure 4.5 we have

$$z_p = AC = 2AN = \frac{4a}{\tan\theta} \tag{4.37}$$

Also

$$\tau_p = \frac{AB + BC}{c/n_1} = \frac{2n_1}{c}AB = \frac{4an_1}{c\sin\theta} \tag{4.38}$$

Thus, the time taken to cover one period is

$$\tau_p = \frac{n_1 z_p}{c\cos\theta} = \frac{n_1^2}{c\tilde{\beta}} z_p \tag{4.39}$$

where we have used $\tilde{\beta} = n_1 \cos\theta$. Thus, the time taken to cover distance z is

$$\tau = \frac{\tau_p}{z_p} z = \frac{n_1^2}{c\tilde{\beta}} z \tag{4.40}$$

Thus

$$\tau_{\text{min}} = \frac{n_1 z}{c} \tag{4.41}$$

$$\tau_{\text{max}} = \frac{n_1^2 z}{c n_2} \tag{4.42}$$

The pulse dispersion is given by

$$\begin{aligned} \Delta\tau &= \tau_{\text{max}} - \tau_{\text{min}} \\ &= \frac{n_1}{cn_2}(n_1 - n_2) z \\ &\simeq \frac{n_1 \Delta}{c} z \end{aligned} \tag{4.43}$$

where we have used $n_1 - n_2 \simeq n_1\Delta$ and $n_1 \simeq n_2$. For $n_1 \simeq 1.46$, $\Delta = 0.01$, we have

$$\Delta\tau \simeq 50\,\text{ns/km}$$

Comparing equation (4.43) with equation (4.33), we see that the pulse dispersion in a square law medium is reduced by a factor $\Delta/2$ compared with a step index waveguide. For $\Delta = 0.01$, this corresponds to a reduction by a factor of 200.

Although the above results are derived for a slab waveguide, it so happens that equations (4.33) and (4.43) are rigorously valid for parabolic index and step index optical *fibers*, respectively. Because of this small pulse dispersion, the first- and second-generation optical communication systems used near-parabolic index (multimode) optical fibers.

4.4 Transit time calculations in a medium characterized by a power law profile

In this section we calculate the pulse dispersion for rays propagating through a graded index medium characterized by the following refractive index profile (see Figure 4.6)

$$\begin{aligned} n^2(x) &= n_1^2\left[1 - 2\Delta\left|\frac{x}{a}\right|^q\right] = n_1^2 - \gamma\,|x|^q\ :\ |x| < a \\ &= n_1^2(1 - 2\Delta) = n_2^2\ :\ |x| > a \end{aligned} \tag{4.44}$$

where

$$\gamma = \frac{n_1^2 2\Delta}{a^q} \tag{4.45}$$

The profile described by the above equation is usually referred to as the "power law profile."[1] Obviously, $q = 2$ represents the parabolic index profile (see Section 4.3.1) and $q = \infty$ represents the step index profile (see Section 4.3.2).

In Section 4.2.2 we showed that ray paths for $q = 2$ media are sinusoidal and in Problem 4.6 we will obtain the ray paths for $q = 1$. In general, it is not possible to determine the ray path analytically for an arbitrary value of q; however, we can still make some general observations:

(1) For a ray launched in the core if $\tilde{\beta} < n_2$, the ray will hit the core–cladding interface and will be refracted away.
(2) Rays for which $n_2 < \tilde{\beta} < n_1$ will be guided through the waveguide.
(3) Guided rays will be periodic – somewhat similar to those shown in Figure 4.2.
(4) The guided rays will become parallel to the z-axis at $x = x_t$ where

$$\tilde{\beta} = n(x = x_t) \tag{4.46}$$

At $x = x_t$, $dx/dz = 0$ (see equation (4.5)).

[1] The reader may note that for a graded index fiber characterized by a power law profile (see Chapter 5), x is to be replaced by the cylindrical radial coordinate r and the modulus sign is not required. For a planar waveguide, the modulus sign is necessary except when q is an even integer.

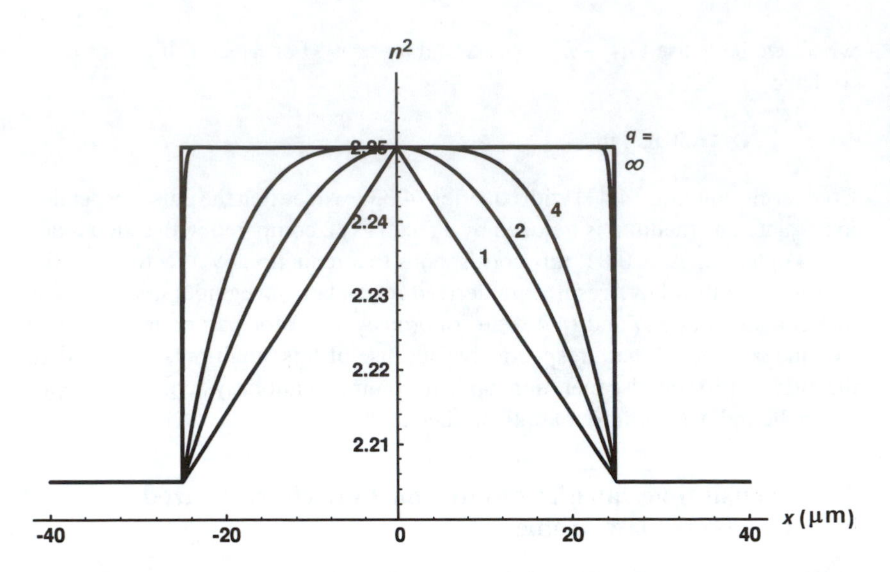

Fig. 4.6: Refractive index variation corresponding to a power law profile given by equation (4.44). Parabolic profile corresponds to $q = 2$ and step index profile corresponds to $q = \infty$.

Thus, if τ_p represents the time taken for the ray to traverse the length z_p, then (see equation (4.23))

$$\frac{1}{4}\tau_p = \frac{1}{c}\int_0^{x_t} \frac{n^2(x)}{[n^2(x) - \tilde{\beta}^2]^{1/2}}dx \tag{4.47}$$

Now, the periodic length of the ray path is given by (using equation 4.5)

$$\frac{1}{4}z_p = \int_0^{x_t} \frac{dx}{dx/dz} = \tilde{\beta}\int_0^{x_t} \frac{dx}{[n^2(x) - \tilde{\beta}^2]^{1/2}} \tag{4.48}$$

Substituting equation (4.44) in equation (4.47) we get

$$\begin{aligned}\frac{1}{4}\tau_p &= \frac{1}{c}\left[n_1^2\int_0^{x_t} \frac{dx}{[n^2(x) - \tilde{\beta}^2]^{1/2}} - \gamma\int_0^{x_t} \frac{x^q}{[n^2(x) - \tilde{\beta}^2]^{1/2}}dx\right] \\ &= \frac{1}{c}\left[\frac{z_p n_1^2}{4\tilde{\beta}} - I_1\right] \end{aligned} \tag{4.49}$$

where we have used equation (4.48) and

$$I_1 = \gamma\int_0^{x_t} \frac{x^q}{\sqrt{f(x)}}dx$$

with

$$f(x) = n^2(x) - \tilde{\beta}^2 = \left(n_1^2 - \tilde{\beta}^2\right) - \gamma x^q$$

We may rewrite

$$I_1 = \gamma\int_0^{x_t} \frac{x^{q+1}}{\sqrt{x^2 f(x)}}dx = \gamma\int_0^{x_t} \frac{x^{q+1}}{\sqrt{g(x)}}dx \tag{4.50}$$

where

$$g(x) = x^2 f(x) = (n_1^2 - \tilde{\beta}^2)x^2 - \gamma x^{q+2}$$

Thus

$$x^{q+1} = \frac{1}{\gamma(q+2)}\left[(n_1^2 - \tilde{\beta}^2)2x - \frac{dg}{dx}\right] \tag{4.51}$$

Substituting in equation (4.50) we get

$$\begin{aligned} I_1 &= \frac{1}{q+2}\left[2(n_1^2 - \tilde{\beta}^2)\int_0^{x_t} \frac{dx}{\sqrt{f(x)}} - \int_0^{x_t} \frac{dg/dx}{\sqrt{g(x)}}\, dx\right] \\ &= \frac{2(n_1^2 - \tilde{\beta}^2)}{q+2}\frac{z_p}{4\tilde{\beta}} - \frac{2}{q+2}[g^{1/2}]_0^{x_t} \end{aligned}$$

But $g^{1/2} = x\sqrt{n^2(x) - \tilde{\beta}^2}$ vanishes at $x = 0$ and at $x = x_t$. Thus

$$I_1 = \frac{z_p}{2\tilde{\beta}}\frac{(n_1^2 - \tilde{\beta}^2)}{q+2} \tag{4.52}$$

Substituting in equation (4.49) and simplifying we get

$$\frac{\tau_p}{z_p} = A_1\tilde{\beta} + \frac{B_1}{\tilde{\beta}} \tag{4.53}$$

where

$$A_1 = \frac{2}{c\,(2+q)}, \qquad B_1 = \frac{qn_1^2}{c\,(2+q)} \tag{4.54}$$

If the ray covers many periods, then we may write for the time taken to propagate through a distance z,

$$\tau = \left[A_1\tilde{\beta} + \frac{B_1}{\tilde{\beta}}\right] z \tag{4.55}$$

Although the above result has been derived for a slab waveguide, it is rigorously valid even for fibers. The above equation is used in Chapter 5 to obtain the optimum refractive index profile for minimum pulse dispersion in graded index optical fibers.

Problems

4.1 Consider a square law medium with $n_1 = 1.5$, $\Delta = 0.01$, and $a = 50\ \mu$m. Assuming that the ray is launched as shown in Figure 4.2, calculate the amplitude A and the period z_p of the ray paths for $\theta_1 = 1°, 4°$, and $\cos^{-1}(\frac{n_2}{n_1}) = 8.13°$. Show that for $\theta_1 > 8.13°$, the ray will be refracted away.

Solution: For $\theta_1 = 1°$

$$\tilde{\beta} = n_1 \cos\theta_1 = 1.49977$$

Similarly for $\theta_1 = 4°$ and $8.13°$, $\tilde{\beta} = 1.49635$ and 1.48492, respectively. Thus,

$$z_p \simeq 2.221, 2.216 \quad \text{and} \quad 2.199 \text{ mm}$$

for $\theta_1 = 1°, 4°$, and $8.13°$, respectively. For the ray not to hit the core–cladding interface, we must have $\tilde{\beta} > n_2$. On the other hand, for the ray to be refracted away, we must have $\tilde{\beta} < n_2$ or

$$\cos\theta_1 < \frac{n_2}{n_1} = 0.98995$$

$$\Longrightarrow \quad \theta_1 > 8.13°$$

Thus, for $\theta_1 > 8.13°$, the ray will be refracted away.

4.2 (a) Consider rays launched at an axial point of a parabolic index waveguide. Show that all rays launched in such a way that (see Figure 4.2)

$$i < i_m = \sin^{-1}\sqrt{n_1^2 - n_2^2} \tag{4.56}$$

will be guided through the waveguide. Show also that for $i > i_m$, the rays will be refracted away at the core–cladding interface.
(b) Similarly show that for off-axis launching at point x_1, guidance will take place if the launching is such that

$$i < i_0 = \sin^{-1}\sqrt{n^2(x_1) - n_2^2} \tag{4.57}$$

The above results are valid for an arbitrary variation of $n^2(x)$. Note that the acceptance angle of graded index waveguide depends on the point of incidence of the rays on the front face of the waveguide. This is the basic principle behind the refracted near-field technique for index profiling of optical fibers (see Section 19.4).

Solution: (a) Applying Snell's law at the incident point we have (see Figure 4.2)

$$\sin i = n_1 \sin\theta_1$$

$$\Longrightarrow \tilde{\beta} = n_1 \cos\theta_1 = \sqrt{n_1^2 - \sin^2 i} \tag{4.58}$$

For guidance to take place we must have $\tilde{\beta} > n_2$, implying $\sqrt{n_1^2 - \sin^2 i} > n_2$ or

$$i < i_m = \sin^{-1}\sqrt{n_1^2 - n_2^2} \tag{4.59}$$

The quantity $\sin i_m$ is known as the axial NA of the waveguide.
(b) Similarly for the off-axis point,

$$\sin i = n(x_1) \sin\theta_1$$

$$\Longrightarrow \tilde{\beta} = n(x_1) \cos\theta_1 = \sqrt{n^2(x_1) - \sin^2 i} \tag{4.60}$$

For guidance we must have

$$\tilde{\beta} > n_2 \Longrightarrow \sqrt{n^2(x_1) - \sin^2 i} > n_2$$

or

$$i < i_0 = \sin^{-1} \sqrt{n^2(x_1) - n_2^2}$$

4.3 For a ray launched parallel to the z-axis at $x = x_0$, show that the ray path in a square law medium is given by

$$x = x_0 \cos \Gamma z \tag{4.61}$$

where Γ is given by equation (4.10) with

$$\tilde{\beta} = n(x_0) = n_1 \left[1 - 2\Delta \left(\frac{x_0}{a}\right)^2\right]^{1/2}$$

4.4 Consider a ray launched on a square law medium at $x = x_0$ making an angle θ_0 (inside the square law medium) with the z-axis. Show that the ray path is given by

$$x(z) = \frac{\tan \theta_0}{\Gamma} \sin \Gamma z + x_0 \cos \Gamma z \tag{4.62}$$

where Γ is given by equation (4.10) with

$$\tilde{\beta} = n_1 \left[1 - 2\Delta \left(\frac{x_0}{a}\right)^2\right]^{1/2} \cos \theta_0$$

Hence, show that the ray paths are identical to the ones obtained in Problem 4.1 provided the latter are displaced along the z-axis.

4.5 In a parabolic index medium given by equation (4.8) with $n_1 = 1.5$, $\Delta = 0.01$, and $a = 50\ \mu$m, a ray is launched parallel to the z-axis at $x = x_0 = 15\ \mu$m. What is the angle that the ray would make with the z-axis when it crosses it?

[HINT: Use the fact that $n(x) \cos \theta(x)$ remains invariant.]

[ANSWER: 2.43°.]

4.6 Consider a medium with a triangular refractive index profile (see $q = 1$ profile in Figure 4.6)

$$\begin{aligned} n^2(x) &= n_1^2 \left(1 - 2\Delta \frac{|x|}{a}\right); \quad |x| < a \\ &= n_1^2(1 - 2\Delta) = n_2^2; \quad |x| > a \end{aligned} \tag{4.63}$$

Obtain the ray paths in the medium.

Solution: Guided rays will again correspond to $n_2 < \tilde{\beta} < n_1$ and such rays will be confined to the region $|x| < a$. For $a > x > 0$, the ray equation (equation (4.7)) gives us

$$\frac{d^2x}{dz^2} = -\frac{n_1^2 \Delta}{a\tilde{\beta}} \tag{4.64}$$

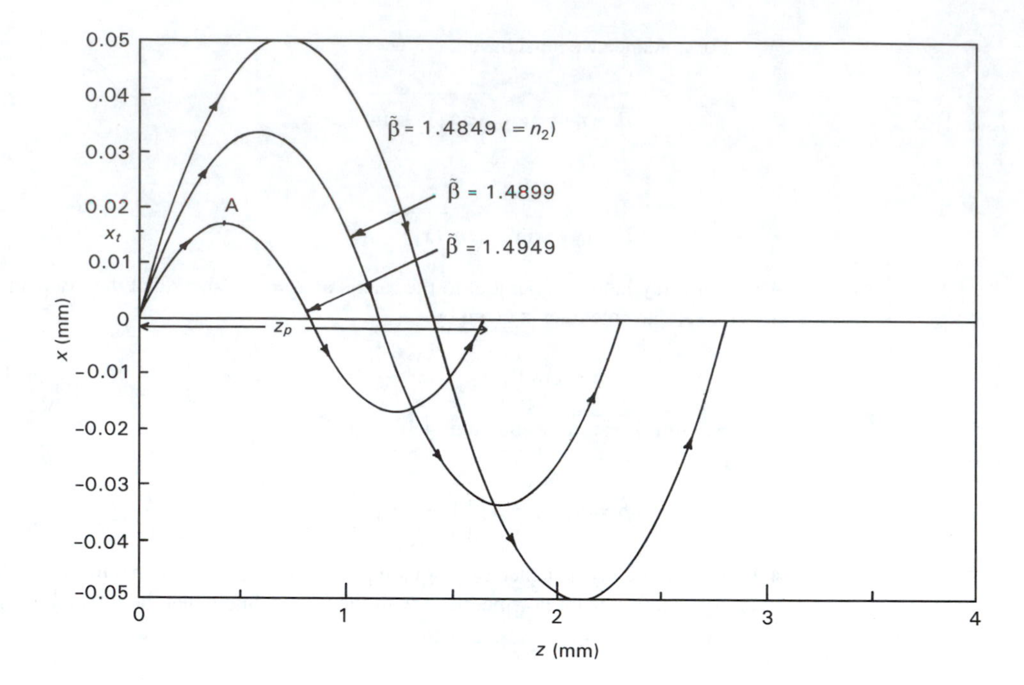

Fig. 4.7: Ray paths in a triangular refactive index profile waveguide.

whose general solution is

$$x(z) = -\frac{n_1^2 \Delta}{2a\tilde{\beta}^2} z^2 + Az + B, \quad x > 0 \tag{4.65}$$

where A and B are constants. Similarly for $x < 0$, the ray path is

$$x(z) = \frac{n_1^2 \Delta}{2a\tilde{\beta}^2} z^2 + A'z + B', \quad x > 0 \tag{4.66}$$

where the constants A' and B' have to be chosen in such a way that $x(z)$ and dx/dz are continuous at $x = 0$. (If dx/dz was not continuous at $x = 0$, d^2x/dz^2 would become infinite, which would contradict equation (4.64); similarly $x(z)$ should also be continuous at $x = 0$.) Thus, the ray paths in a linearly varying refractive index profile are parabolas. If the ray is launched at $x = 0$, $z = 0$, and making an angle θ_0 with the z-axis, then we have

$$B = 0 \quad \text{and} \quad A = \left.\frac{dx}{dz}\right|_{z=z_0} = \tan\theta_0 \tag{4.67}$$

Thus, the ray path would be (see Figure 4.7).

$$\begin{aligned} x(z) &= -\frac{n_1^2 \Delta}{2a\tilde{\beta}^2} z^2 + \frac{\left(n_1^2 - \tilde{\beta}^2\right)^{1/2}}{\tilde{\beta}} z; \quad x > 0, 0 < z < \frac{1}{2} z_p \\ &= \frac{n_1^2 \Delta}{2a\tilde{\beta}^2}\left(z - \frac{1}{2} z_p\right)^2 - \frac{\left(n_1^2 - \tilde{\beta}^2\right)}{\tilde{\beta}}\left(z - \frac{1}{2} z_p\right); \\ &\qquad x < 0, \frac{1}{2} z_p < z < z_p \end{aligned} \tag{4.68}$$

where

$$\frac{1}{2} z_p = \frac{2a\tilde{\beta}\left(n_1^2 - \tilde{\beta}^2\right)^{1/2}}{n_1^2 \Delta} \tag{4.69}$$

Beyond $z = z_p$, the ray path will repeat itself.

4.7 Consider a parabolic index waveguide (see equation (4.8)) with

$$n_1 = 1.75, \quad n_2 = 1.677, \quad a = 25\ \mu\text{m}$$

Calculate the NA at the axis and at a point 20 μm from the axis.

[ANSWER: $\sin\alpha_m \simeq 0.5$, $\sin\alpha_0 \simeq 0.3$, implying $\alpha_m \simeq 30°$ and $\alpha_0 \simeq 17.5°$.]

4.8 Consider a square law medium with $n_1 = 1.5$, $\Delta = 0.01$, and $a = 50\ \mu$m. Assuming $x(0) = 0$, plot the ray paths $x(z)$ for $\tilde{\beta} = 1.5$, 1.49, 1.4849, and 1.47.

4.9 Calculate the maximum angle of acceptance of rays for an axial point for a waveguide of Problem 4.7 when the medium in front of the waveguide is air and when it is water ($n = 1.33$).

[ANSWER: 30° and 22.1°.]

4.10 Give the range of $\tilde{\beta}$ values for guided rays for refractive index profiles shown in Figure 4.8 with $n_1 > n_2 > n_3$.

4.11 A ray of light is incident on a core–cladding interface of a waveguide as shown in Figure 4.9. Can this ray excite a guided ray in the waveguide? Give reasons.

4.12 Consider a square law medium described by a refractive index profile given by equation (4.8) with $n_1 = 1.5$, $\Delta = 0.01$, and $a = 30\ \mu$m and of length L. A point source is placed 10 μm from the axis at the entrance face.

(a) Under paraxial approximation, what should be the minimum length L so that the rays emerging from the medium are parallel?
(b) At what angle will the rays be emerging from the exit face?

You may neglect refraction effects at the input and output faces.

[ANSWER: (a) $L \simeq 333\ \mu$m, (b) $\theta \simeq 2.7°$.]

4.13 Consider a slab waveguide characterized by the following refractive index distribution

$$n^2(x) = n_1^2 \operatorname{sech}^2 gx \tag{4.70}$$

Obtain the ray paths in such a medium. Obtain the time taken by a ray to propagate through a length z of the medium and show that all rays take the same amount of time.

Solution: Substituting from equation (4.70) in equation (4.5) we obtain

$$\left(\frac{dx}{dz}\right)^2 = \frac{n_1^2 \operatorname{sech}^2 gx - \tilde{\beta}^2}{\tilde{\beta}^2}$$

$$= \frac{\left(n_1^2 - \tilde{\beta}^2\right) - \tilde{\beta}^2 \sinh^2 gx}{\tilde{\beta}^2 \cosh^2 gx}$$

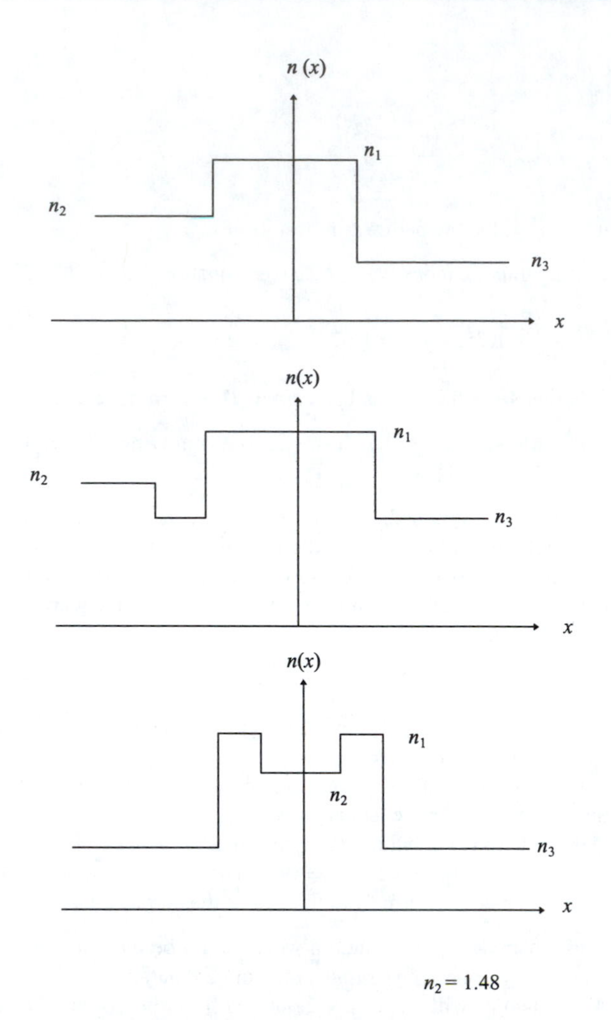

Fig. 4.8: Figure for Problem 4.10.

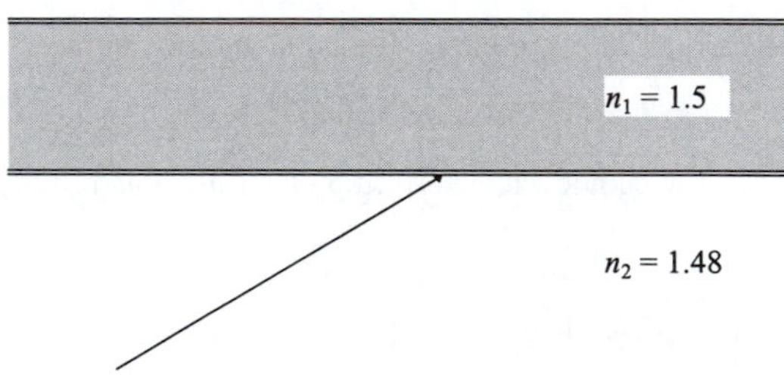

Fig. 4.9: Figure for Problem 4.11.

or

$$\int dz = \frac{1}{g} \int \frac{1}{\sqrt{1-\xi^2}}\, d\xi \tag{4.71}$$

where

$$\xi = \frac{\tilde{\beta}}{\sqrt{n_1^2 - \tilde{\beta}^2}} \sinh gx \tag{4.72}$$

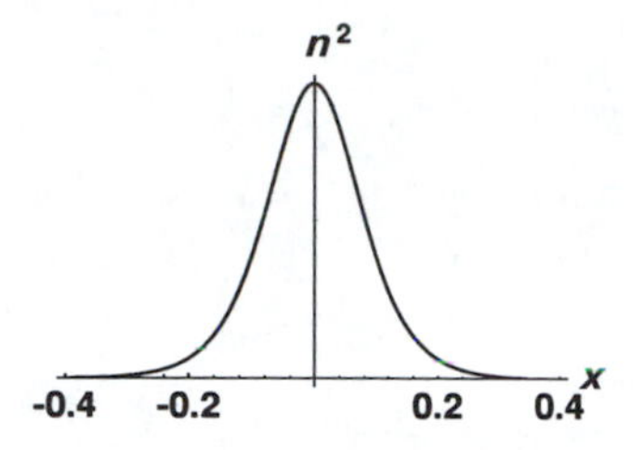

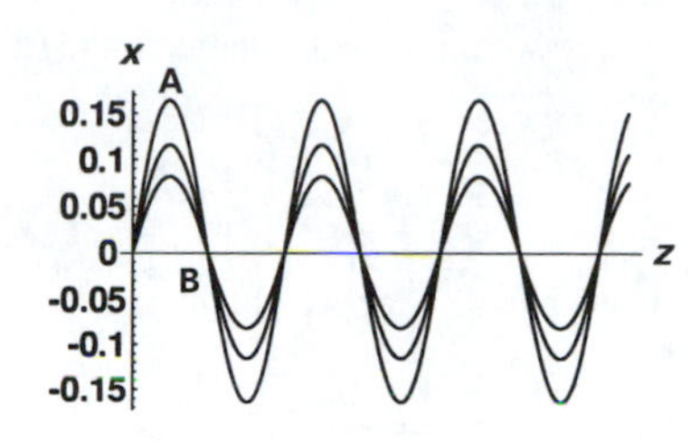

Fig. 4.10: Exact ray paths in a medium characterized by equation (4.70). All rays have exactly the same period and thus they take the same amount of time to propagate over a given length of the medium.

Thus

$$z = \frac{1}{g} \sin^{-1} \xi + \text{constant}$$

or

$$\xi = \sin(gz + \phi)$$

implying that

$$x(z) = \frac{1}{g} \sinh^{-1} \left[\frac{\sqrt{n_1^2 - \tilde{\beta}^2}}{\tilde{\beta}} \sin gz \right] \tag{4.73}$$

where in the last step we have assumed that at $z = 0$, $x = 0$ – that is, on-axis launching. Typical ray paths are shown in Figure 4.10 and the period

$$z_p = \frac{2\pi}{g} \tag{4.74}$$

is independent of the launching angle. Thus *all* rays originating from point O pass through point B, and therefore from Fermat's principle it readily follows that *all* rays must take the same amount of time. This can also be seen by calculating the time that it takes the ray to go from O to A. Now the time taken by a ray to go from O to A is

$$\begin{aligned} \frac{1}{4}\tau_p &= \int \frac{ds}{c/n(x)} \\ &= \frac{1}{c} \int_0^{z_p/4} n(x) \left[1 + \left(\frac{dx}{dz} \right)^2 \right]^{1/2} dz \end{aligned} \tag{4.75}$$

Using equations (4.5) and (4.70) we obtain

$$\tau_p = \frac{4n_1^2}{c\tilde{\beta}} \int_0^{z_p/4} \text{sech}^2[gx(z)]\, dz$$

Now

$$\begin{aligned} \text{sech}^2[gx(z)] &= \frac{1}{1 + \sinh^2 gx(z)} \\ &= \frac{1}{1 + \gamma^2 \sin^2 gz} \end{aligned}$$

where we have used equation (4.73) and

$$\gamma^2 = \frac{n_1^2 - \tilde{\beta}^2}{\tilde{\beta}^2}$$

If we put $\psi = gz$ we obtain

$$\tau_p = \frac{4n_1^2}{c\tilde{\beta}g}\int_0^{\pi/2}\frac{d\psi}{1+\gamma^2\sin^2\psi}$$

$$= \frac{4n_1^2}{c\tilde{\beta}g}\frac{1}{\sqrt{1+\gamma^2}}\tan^{-1}\left\{\sqrt{1+\gamma^2}\tan\psi\right\}\Bigg|_0^{\pi/2} = \frac{2\pi n_1}{cg} \qquad (4.76)$$

which is indeed independent of $\tilde{\beta}$ – that is, the launching angle of the ray. Thus, for a refractive index variation given by equation (4.70), all rays lying in the x–z plane and emanating from a point will again meet at one point.

5

Pulse dispersion in graded index optical fibers

5.1 Introduction

As mentioned in Chapter 3, dispersion along with attenuation determines the information-carrying capacity of the fiber optic system. As such, a study of the dispersion characteristics of an optical fiber is a subject of considerable importance. In this chapter we discuss dispersion characteristics of graded index fibers characterized by the following refractive index distribution

$$\begin{aligned} n^2(r) &= n_1^2\left[1 - 2\Delta\left(\frac{r}{a}\right)^q\right]; \quad 0 < r < a \\ &= n_2^2 = n_1^2(1 - 2\Delta); \quad r > a \end{aligned} \tag{5.1}$$

where r corresponds to a cylindrical radial coordinate, n_1 represents the value of the refractive index on the axis (i.e., at $r = 0$), and n_2 represents the refractive index of the cladding; $q = 1$, $q = 2$, and $q = \infty$ correspond to the linear, parabolic, and step index profiles, respectively (see Figure 5.1). Equation (5.1) describes what is usually referred to as a power law profile, which gives an accurate description of the refractive index variation in most multimode fibers. One of the main advantages of the power law profile is that by choosing different values of q one can describe a variety of profiles and in addition we can get rigorously correct expressions for the time taken by different rays to propagate through a certain distance of the fiber; our objective is to determine the optimum value of q that would lead to minimum pulse dispersion.

We use ray optics to study the dispersion characteristics. We should mention that ray optics is applicable to highly multimoded fibers – that is, when

$$V \geq 10$$

where

$$V = \frac{2\pi}{\lambda_0} a\sqrt{n_1^2 - n_2^2} \tag{5.2}$$

λ_0 being the free space wavelength of the propagating light beam; the quantity V is known as the waveguide parameter of the fiber. Modes in optical waveguides are discussed in Chapters 7–9. In Section 9.4 we show that the total number of modes in a highly multimoded graded index optical fiber (characterized by

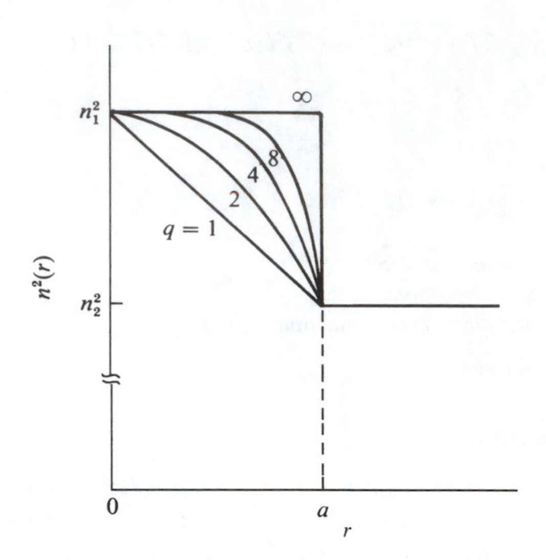

Fig. 5.1: The power law profile as given by equation (5.1).

equation (5.1)) are approximately given by (see Section 9.4)

$$N \approx \frac{q}{2(2+q)} V^2$$

Thus, a parabolic index ($q = 2$) fiber with $V = 10$ will support approximately 25 modes. Similarly, a step index ($q = \infty$) fiber with V=10 will support approximately 50 modes. When the fiber supports such a large number of modes, then the continuum (ray) description should give very accurate results.

5.2 Pulse dispersion

In Chapter 4 we carried out a detailed analysis of ray paths in planar optical waveguides. In a cylindrically symmetric structure (where $n = n(r)$ – as in an optical fiber), the ray equations and their solutions are much more involved; we discuss these in Chapter 23, where we show that because of the rotational and translational invariance of the optical fiber, a ray is characterized by two invariants $\tilde{\beta}$ and $\tilde{l}$. The invariant $\tilde{\beta}$ is defined by the following equation (see equation (23.7))

$$\tilde{\beta} = n(r) \cos\theta(r) \tag{5.3}$$

where $\theta(r)$ is the angle that the ray makes with the z-axis. The invariant $\tilde{\beta}$ is the same as that discussed in the previous chapter and is a consequence of the fact that the refractive index does not depend on the z-coordinate (translational invariance). The cylindrical symmetry (ϕ independence) leads to the invariant $\tilde{l}$, which is usually referred to as the skewness parameter. Meridional rays (which are confined to a particular plane) are characterized by $\tilde{l} = 0$; thus, *all* rays in a planar waveguide have zero skewness parameter.

Using the ray equations derived in Section 23.2, we have, in Section 5.3, derived the following expression for the time taken by a ray to traverse a certain distance z though the optical fiber (characterized by equation (5.1))

$$\tau(z) = \left[A\tilde{\beta} + \frac{B}{\tilde{\beta}} \right] z \tag{5.4}$$

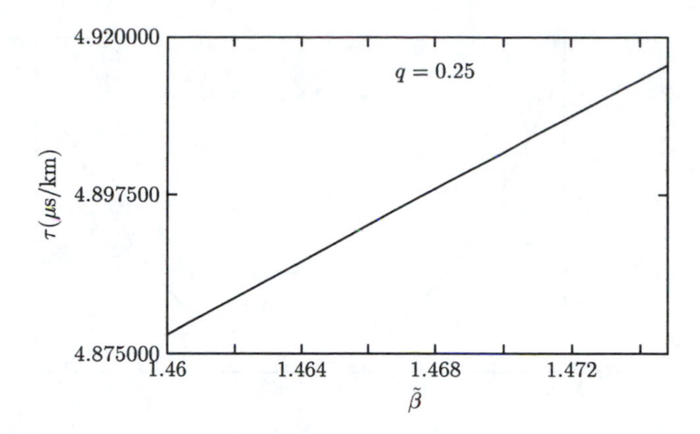

Fig. 5.2: Variation of $\tau(\tilde{\beta})$ with $\tilde{\beta}$ for $q = 0.25$.

where

$$A = \frac{2}{c(2+q)}, \quad B = \frac{qn_1^2}{c(2+q)} \tag{5.5}$$

c being the velocity of light in free space. We may note the following points:

(i) The time taken is independent of the skewness parameter $\tilde{l}$; this is a characteristic of the power law profile.

(ii) The time taken depends only on the invariant $\tilde{\beta}$ and for guided rays

$$n_2 < \tilde{\beta} < n_1 \tag{5.6}$$

(iii) Since $\tau(z)$ is independent of $\tilde{l}$, equation (5.4) is identical to the corresponding result derived for a planar waveguide (see equation (4.55)).

(iv) We have neglected material dispersion – that is, we have neglected the dependence of the refractive index on the wavelength; a detailed discussion (on material dispersion) is given in Chapter 6. Thus, the temporal broadening of a pulse (as given by equation (5.4)) represents the broadening due to the fact that different rays take different amounts of time to propagate through the fiber. This is also referred to as intermodal dispersion (see also Chapter 9).

In Figures 5.2–5.5 we have plotted $\tau(\tilde{\beta})$ as a function of $\tilde{\beta}$ for

$$n_2 = 1.46, \quad \Delta = 0.01 \, (\Rightarrow n_1 \simeq 1.4748) \tag{5.7}$$

corresponding to

$$q = 0.25, 1.98, 2.00, \text{ and } \infty$$

respectively, with $n_2 < \tilde{\beta} < n_1$. The vertical axis gives the time taken by the ray to traverse a kilometer length of the fiber. We consider some specific cases:

Case 1: $q = \infty$ (step index fibers). For $q = \infty$, $A = 0$, and $B = n_1^2/c$. Thus,

$$\tau(\tilde{\beta}) = \frac{n_1^2}{c\tilde{\beta}} z \tag{5.8}$$

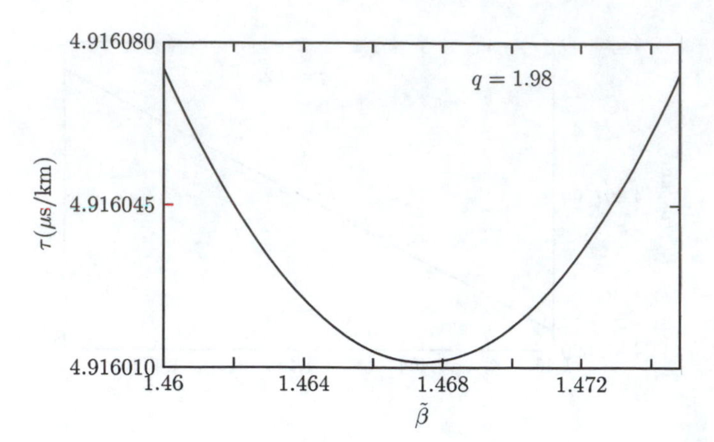

Fig. 5.3: Variation of $\tau(\tilde{\beta})$ with $\tilde{\beta}$ for $q = 1.98$.

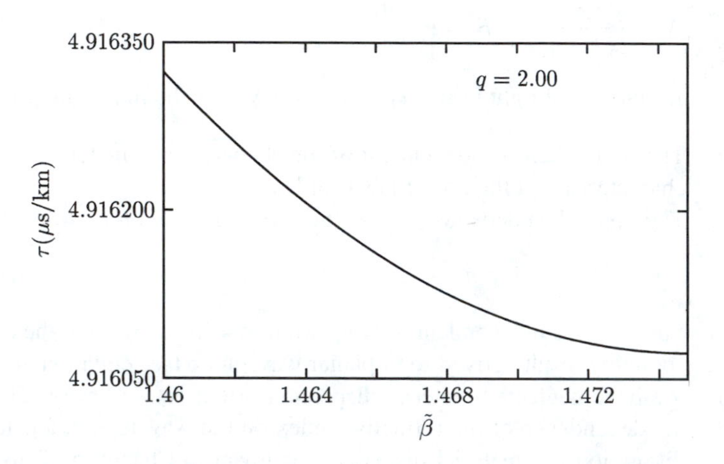

Fig. 5.4: Variation of $\tau(\tilde{\beta})$ with $\tilde{\beta}$ for $q = 2.00$.

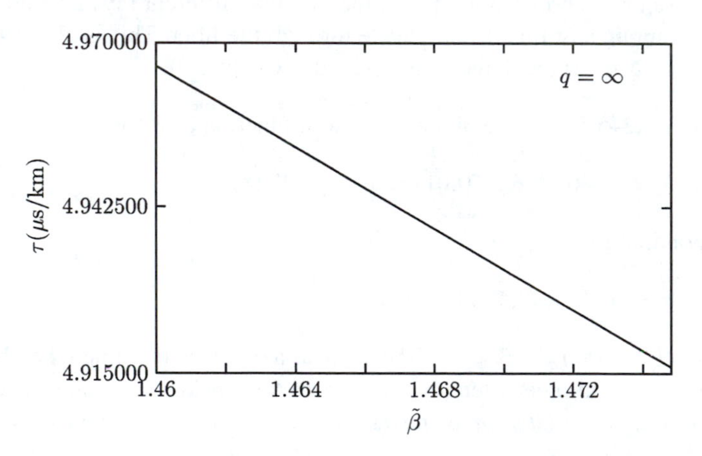

Fig. 5.5: Variation of $\tau(\tilde{\beta})$ with $\tilde{\beta}$ for $q = \infty$.

represents a monotonically decreasing function of $\tilde{\beta}$ (see Figure 5.5). Since $n_2 < \tilde{\beta} < n_1$, we get

$$\tau_{\max} = \tau(\tilde{\beta} = n_2) = \frac{n_1^2}{cn_2} z$$

which corresponds to the ray that is incident at the core–cladding interface at the critical angle. Further

$$\tau_{\min} = \tau(\tilde{\beta} = n_1) = \frac{z}{c/n_1}$$

which corresponds to the ray propagating parallel to the z-axis. Thus, the pulse dispersion is given by

$$\begin{aligned} \Delta\tau = \tau_{\max} - \tau_{\min} &= \frac{n_1}{c}\frac{(n_1 - n_2)}{n_2} L \\ &\simeq \frac{n_1 \Delta}{c} L \end{aligned} \tag{5.9}$$

consistent with the result obtained in Chapter 3. Thus, for $n_1 \simeq 1.46$, $\Delta \simeq 0.01$ we obtain

$$\Delta\tau \simeq 50 \text{ ns/km}$$

Case 2: $q = 2$ (parabolic index fibers). For $q = 2$, $A = 1/2c$, and $B = n_1^2/2c$. Thus

$$\tau(\tilde{\beta}) = \frac{1}{2c}\left[\tilde{\beta} + \frac{n_1^2}{\tilde{\beta}}\right] z \tag{5.10}$$

which is again a monotonically decreasing function of $\tilde{\beta}$ for $n_2 < \tilde{\beta} < n_1$ (see Figure 5.4). Indeed

$$\tau_{\max} = \tau(\tilde{\beta} = n_2) = \frac{1}{2c}\left[n_2 + \frac{n_1^2}{n_2}\right] z$$

and

$$\tau_{\min} = \tau(\tilde{\beta} = n_1) = \frac{z}{c/n_1}$$

giving the following expression for pulse dispersion

$$\Delta\tau = \tau_{\max} - \tau_{\min} \simeq \frac{n_2 \Delta^2}{2c} z \tag{5.11}$$

For $n_1 \simeq 1.46$, $\Delta \simeq 0.01$ we obtain

$$\Delta\tau \simeq \frac{1}{4} \text{ ns/km}$$

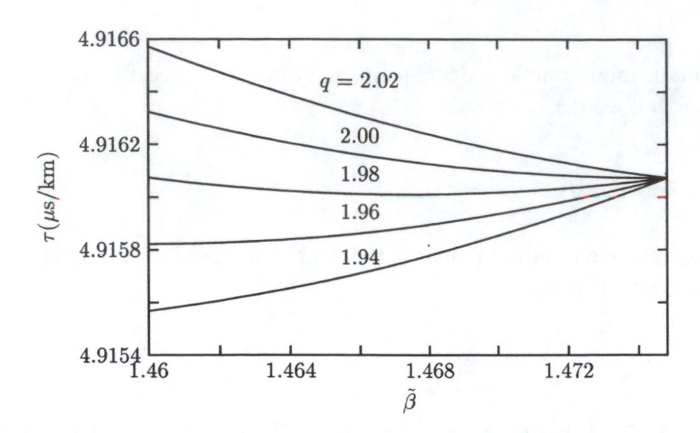

Fig. 5.6: Variation of the ray transit time τ given by equation (5.4) with $\tilde{\beta}$ for different power law profiles. Note that for $q > 2$ and $q < 2 - 4\Delta$, τ varies monotonically with $\tilde{\beta}$ and for $2 - 4\Delta < q < 2$, τ has a minimum in the range $n_2 < \tilde{\beta} < n_1$, the range for guided rays. Calculations correspond to $\Delta = 0.01$.

Case 3: The optimum profile. From analysis of a parabolic index fiber, we find that there is a tremendous decrease in the pulse dispersion as we go from a step index fiber to a parabolic index fiber. Typical variations of τ with $\tilde{\beta}$ (for q values in the vicinity of 2) are shown in Figure 5.6. To obtain the optimum value of q corresponding to minimum pulse dispersion, we first note that for $n_2 < \tilde{\beta} < n_1$, $\tau(\tilde{\beta})$ is a monotonically increasing or decreasing function except for the values of q given by

$$2 - 4\Delta < q < 2$$

This can be easily shown by finding the value of $\tilde{\beta}$ for which $d\tau/d\tilde{\beta} = 0$. The condition immediately gives

$$A - \frac{B}{\tilde{\beta}_m^2} = 0 \Rightarrow \tilde{\beta}_m = \sqrt{\frac{B}{A}} = n_1\sqrt{\frac{q}{2}} \tag{5.12}$$

Since for guided rays $n_2 < \tilde{\beta} < n_1$ and $n_2 = n_1\sqrt{1 - 2\Delta}$, the value of $\tilde{\beta}$ given by equation (5.12) lies in the range $n_2 < \tilde{\beta} < n_1$ only for $2 - 4\Delta < q < 2$. For $q > 2$, $\tau(\tilde{\beta})$ decreases monotonically as $\tilde{\beta}$ increases from n_2 to n_1 and the pulse dispersion is given by

$$\begin{aligned}\Delta\tau(q > 2) &= \tau(\tilde{\beta} = n_2) - \tau(\tilde{\beta} = n_1) \\ &= \frac{(n_1 - n_2)}{c(q + 2)}\left(\frac{qn_1}{n_2} - 2\right)z \end{aligned} \tag{5.13}$$

For $q < 2 - 4\Delta$, $\tau(\tilde{\beta})$ increases monotonically as $\tilde{\beta}$ increases from n_2 to n_1 and the corresponding pulse dispersion is given by

$$\begin{aligned}\Delta\tau(q < 2 - 4\Delta) &= \tau(\tilde{\beta} = n_1) - \tau(\tilde{\beta} = n_2) \\ &= \frac{(n_1 - n_2)}{c(q + 2)}\left(2 - \frac{qn_1}{n_2}\right)z \end{aligned} \tag{5.14}$$

Finally

$$\Delta\tau(2 - 4\Delta < q < 2) = \mathrm{Max}(\Delta\tau_1, \Delta\tau_2) \tag{5.15}$$

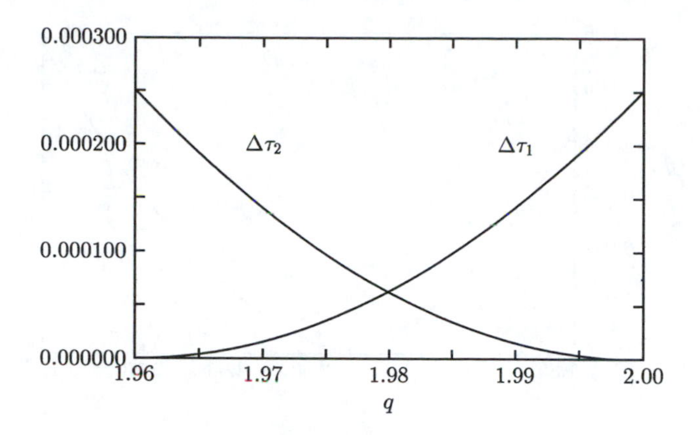

Fig. 5.7: Variation of $\Delta\tau_1$ and $\Delta\tau_2$ [given by equations (5.16) and (5.17)] with q in the range $2-4\Delta < q < 2$. Optimum value of q corresponds to the case when $\Delta\tau_1 = \Delta\tau_2$.

where

$$\Delta\tau_1 = \tau(\tilde{\beta} = n_2) - \tau(\tilde{\beta} = \tilde{\beta}_m) \tag{5.16}$$

$$\Delta\tau_2 = \tau(\tilde{\beta} = n_1) - \tau(\tilde{\beta} = \tilde{\beta}_m) \tag{5.17}$$

Figure 5.7 shows the plots of $\Delta\tau_1$ and $\Delta\tau_2$ as a function of q in the domain $2-4\Delta < q < 2$. Obviously, the point of intersection of the two curves corresponds to minimum pulse dispersion. This point of intersection corresponds to $\Delta\tau_1 = \Delta\tau_2$ – that is, $\tau(\tilde{\beta} = n_2) = \tau(\tilde{\beta} = n_1)$ – which implies

$$q = \frac{2n_2}{n_1} = 2\sqrt{1-2\Delta} \simeq 2 - 2\Delta \quad \text{optimum profile} \tag{5.18}$$

The corresponding pulse dispersion is given by

$$\Delta\tau \simeq \frac{n_1}{8c}\Delta^2 z \tag{5.19}$$

Thus, for $n_1 \simeq 1.46$ and $\Delta = 0.01$ we get

$$\left.\begin{aligned} \Delta\tau &\simeq 50 \text{ ns/km} && \text{for } q = \infty \\ &\simeq \tfrac{1}{4} \text{ ns/km} && \text{for } q = 2 \\ &\simeq \tfrac{1}{16} \text{ ns/km} && \text{for } q = 1.98 \end{aligned}\right\} \tag{5.20}$$

We mention here the following points:

(i) Equation (5.20) tells us that for a parabolic index fiber, the pulse dispersion is reduced by a factor of about 200 in comparison to the step index optical fiber. This is because the first- and second-generation optical communication system used near-parabolic index fibers. Since graded index multimode fibers have large core diameters (in comparison to the single-mode fibers), they are much easier to splice and are still used in optical communication systems. We have a more detailed discussion on design issues in Chapter 13.

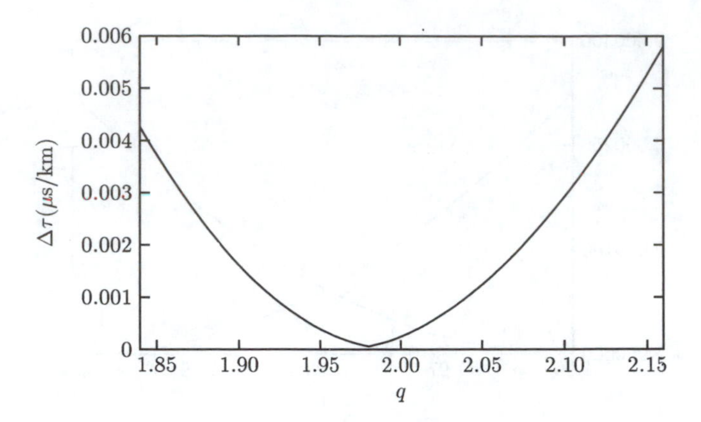

Fig. 5.8: Variation of the pulse dispersion $\Delta\tau$ with q. The optimum corresponds to $q \simeq 2 - 2\Delta$.

(ii) In Section 9.4 we show that if one carries out the modal analysis of graded index fibers, then in the WKB approximation one obtains the same expression for the pulse dispersion and for the optimum profile.

(iii) In the analysis given above we have neglected material dispersion (see Chapter 6), which must be included in calculating the total dispersion and the information-carrying capacity (see Chapter 13).

(iv) From equation (5.20) we find that the pulse dispersion is very sensitive to the value of q. Thus, as q increases from 1.98 to 2.00, the pulse dispersion increases by a factor of 4. Figure 5.8 shows the variation of $\Delta\tau$ as a function of q for fiber parameters given by equation (5.7). For an actual fiber, the q parameter usually depends on the wavelength, implying that a finite spectral width of the source may result in significant pulse dispersion.

5.2.1 Effect of material dispersion on the optimum profile

In Chapter 6 we show that the group velocity of a pulse is given by

$$\frac{1}{v_g} = \frac{d}{d\omega}\left[\frac{\omega}{c}\, n(\omega)\right] \tag{5.21}$$

Thus

$$v_g = \frac{c}{n + \omega\frac{dn}{d\omega}} \tag{5.22}$$

In Section 5.4 we used the fact that the velocity of the ray is given by c/n; since a light pulse contains a spectrum of frequencies, we must replace n by the group refractive index [see equation (6.6)]:

$$n \to n + \omega\frac{dn}{d\omega} = n - \lambda_0\frac{dn}{d\lambda_0}$$

The expression for $\tau(z)$ will be the same as given by equation (5.4) except that A and B are now given by [Ankiewicz and Pask (1977)]

$$A = \frac{2N_1}{n_1} \frac{\left(1 + \frac{1}{4}y\right)}{c(q+2)} \tag{5.23}$$

$$B = \frac{1}{2} q n_1^2 A - \frac{1}{4c} N_1 n_1 y \tag{5.24}$$

where

$$y = \frac{2n_1}{N_1} \frac{\omega}{\Delta} \frac{d\Delta}{d\omega} \tag{5.25}$$

and

$$N_1 = n_1 + \omega \frac{dn_1}{d\omega} \tag{5.26}$$

One can again carry out an analysis similar to the one given above to obtain the following expression for the optimum value of q.

$$q_{\text{opt}} \simeq 2 + y - \Delta\left(2 + \frac{y}{2}\right) \tag{5.27}$$

The corresponding temporal dispersion is given by

$$(\Delta\tau)_{\text{opt}} \simeq \frac{N_1 z}{c} \frac{\Delta^2}{4(2-\Delta)} \tag{5.28}$$

Notice from equation (5.27) that q_{opt} depends on λ_0 due to the wavelength dependence of Δ and y. Thus, a fiber with an optimum profile at 1300 nm will not correspond to an optimum profile at another operating wavelength.

5.3 Derivation of the expression for $\tau(\tilde{\beta})$

In Section 23.2 we show that for a cylindrically symmetric profile – that is, for

$$n = n(r) \tag{5.29}$$

the ray path $r(z)$ is obtained by solving the equation

$$\frac{dr}{dz} = \pm \frac{\sqrt{f(r)}}{\tilde{\beta}} \tag{5.30}$$

or

$$\int dz = \pm \tilde{\beta} \int \frac{dr}{\sqrt{f(r)}} \tag{5.31}$$

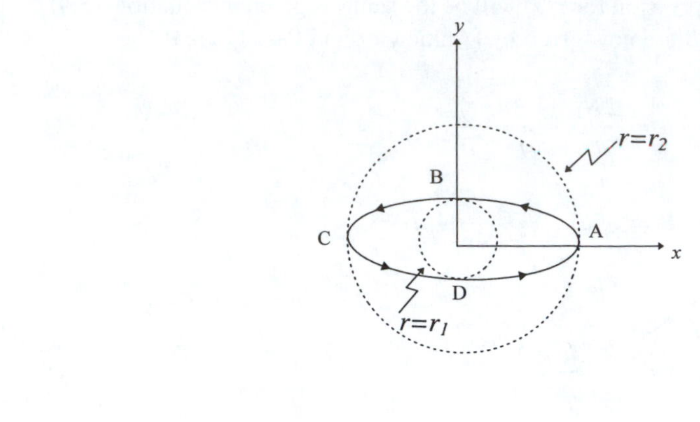

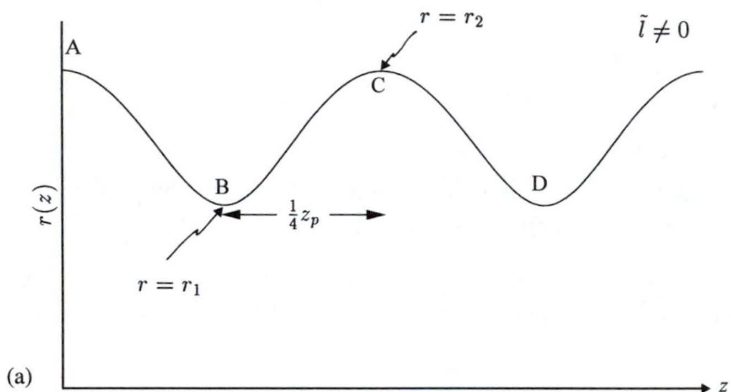

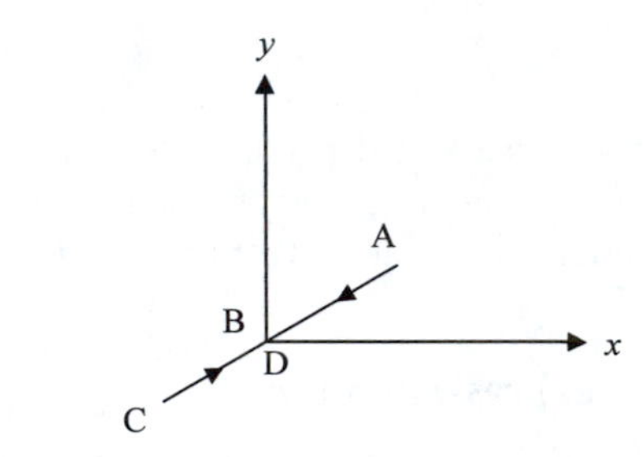

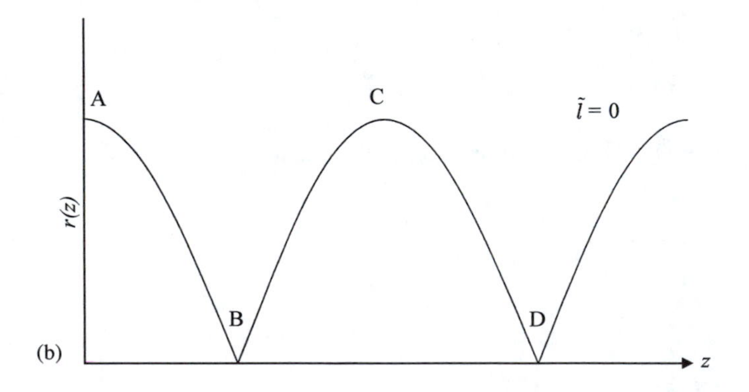

Fig. 5.9: Typical variations of $r(z)$ with z for $\tilde{l} \neq 0$ (skew rays) (a) and $\tilde{l} = 0$ (meridional rays) (b) for the power law profile. Note that these are not the actual ray paths; the curves represent the variation of the magnitude of the distance from the axis. (For $\tilde{l} \neq 0$, the rays are not confined to a plane.) Insets show projections of the ray paths in the transverse plane.

where

$$f(r) = n^2(r) - \tilde{\beta}^2 - \frac{\tilde{l}^2}{(r/a)^2} \tag{5.32}$$

We also show there that for bound rays in a fiber characterized by a power law profile, the variation of $r(z)$ is of the form shown in Figure 5.9(a) with the periodic distance z_p, given by

$$z_p = 4\tilde{\beta} \int_{r_1}^{r_2} \frac{dr}{\sqrt{f(r)}} \tag{5.33}$$

where $r = r_1$ and $r = r_2$ are the inner and outer caustics where $f(r)$ vanishes – that is

$$f(r) = 0 \text{ at } r = r_1 \text{ and at } r = r_2 \tag{5.34}$$

It may be mentioned that for $\tilde{l} = 0$, $r_1 = 0$; however, $f(0) \neq 0$ – see Figure 5.9(b).

We now calculate the time taken by a ray to traverse a certain distance of the fiber; we follow the analysis given by Ankiewicz and Pask (1977). Now the time taken to traverse the distance $\frac{1}{4}z_p$ is given by

$$\frac{1}{4}\tau_p = \int_{\mathrm{A}\to\mathrm{B}} \frac{ds}{c/n(r)} \tag{5.35}$$

where the integration is carried on the ray path. Since

$$n\frac{dz}{ds} = \tilde{\beta} \tag{5.36}$$

we can write

$$\begin{aligned} \frac{1}{4}\,\tau_p &= \frac{1}{c\tilde{\beta}} \int_{z_0}^{z_0+\frac{1}{4}z_p} n^2(r)\,dz \\ &= \frac{n_1^2}{c\tilde{\beta}} \left[\int_{z_0}^{z_0+\frac{1}{4}z_p} dz - 2\Delta \int_{z_0}^{z_0+\frac{1}{4}z_p} \left(\frac{r}{a}\right)^q dz \right] \\ &= \frac{n_1^2}{4c\tilde{\beta}} z_p - \frac{n_1^2 2\Delta}{c\tilde{\beta}} I_1 \end{aligned} \tag{5.37}$$

where

$$\begin{aligned} I_1 &= \int_{z_0}^{z_0+\frac{1}{4}z_p} (r/a)^q\,dz = \int_{r_1}^{r_2} \frac{(r/a)^q}{dr/dz}\,dr \\ &= \tilde{\beta} \int_{r_1}^{r_2} \frac{(r/a)^q}{\sqrt{f(r)}} dr = \frac{\tilde{\beta}}{a^q} \int_{r_1}^{r_2} \frac{r^{q+1}}{\sqrt{g(r)}} dr \end{aligned} \tag{5.38}$$

where

$$g(r) = r^2 f(r) = r^2 \left[n_1^2 \left\{ 1 - 2\Delta \left(\frac{r}{a} \right)^q \right\} - \frac{\tilde{l}^2}{(r/a)^2} - \tilde{\beta}^2 \right]$$

$$= (n_1^2 - \tilde{\beta}^2) r^2 - 2 n_1^2 \Delta \frac{r^{q+2}}{a^q} - \tilde{l}^2 a^2 \tag{5.39}$$

Thus

$$\frac{dg}{dr} = 2\left(n_1^2 - \tilde{\beta}^2\right) r - 2\Delta\,(q+2)\, n_1^2 \frac{r^{q+1}}{a^q} \tag{5.40}$$

or

$$\frac{r^{q+1}}{a^q} = \frac{1}{2\Delta\,(q+2)\, n_1^2} \left[2\left(n_1^2 - \tilde{\beta}^2\right) r - \frac{dg}{dr} \right] \tag{5.41}$$

Substituting in equation (5.38) we get

$$I_1 = \frac{\tilde{\beta}}{2\Delta(q+2) n_1^2} \left[2\left(n_1^2 - \tilde{\beta}^2\right) \int_{r_1}^{r_2} \frac{r dr}{\sqrt{g(r)}} - \int_{r_1}^{r_2} \frac{dg/dr}{\sqrt{g(r)}} dr \right]$$

$$= \frac{\tilde{\beta}}{2\Delta(q+2) n_1^2} \left[2\left(n_1^2 - \tilde{\beta}^2\right) \int_{r_1}^{r_2} \frac{dr}{f(r)} - 2 g^{1/2} \Big|_{r_1}^{r_2} \right]$$

$$= \frac{\tilde{\beta}}{2\Delta(q+2) n_1^2} \left[2\left(n_1^2 - \tilde{\beta}^2\right) \frac{z_p}{4\tilde{\beta}} - 2\sqrt{r^2 f(r)} \Big|_{r_1}^{r_2} \right] \tag{5.42}$$

where we have used equation (5.33). Since dr/dz (and therefore $f(r)$) vanishes at $r = r_1$ and $r = r_2$, the second term is zero. We therefore obtain

$$I_1 = \frac{(n_1^2 - \tilde{\beta}^2) z_p}{4\Delta\,(q+2)\, n_1^2}$$

substituting in equation (5.37), we get

$$\frac{1}{4}\tau_p = \left[\frac{n_1^2}{c\tilde{\beta}} - \frac{2\left(n_1^2 - \tilde{\beta}^2\right)}{c\tilde{\beta}(q+2)} \right] \frac{1}{4} z_p \tag{5.43}$$

Propagating over a distance that contains many periods, we may write

$$\tau(z) = \left(A\tilde{\beta} + \frac{B}{\tilde{\beta}} \right) z$$

where

$$A = \frac{2}{c(2+q)}, \quad B = \frac{q n_1^2}{c(2+q)}$$

Problems

5.1 Evaluate the integral in equation (5.33) for $q = 2$ and show that

$$z_p = \frac{\pi a \tilde{\beta}}{n_1 \sqrt{2\Delta}} \tag{5.44}$$

Notice the independence on $\tilde{l}$.

5.2 Evaluate the integral in equation (5.33) for arbitrary q but $\tilde{l} = 0$ and show that

$$z_p = \frac{4\sqrt{\pi} a \tilde{\beta}}{q[n_1\sqrt{2\Delta}]^{2/q}} \frac{\Gamma(1/q)}{\Gamma\left(\frac{1}{q} + \frac{1}{2}\right)} \left[n_1^2 - \tilde{\beta}^2\right]^{\frac{2-q}{2q}} \tag{5.45}$$

For $q = 2$, the above equation simplifies to equation (5.44).

6

Material dispersion

6.1 Introduction

Until now we have considered the broadening of an optical pulse due to different rays taking different amounts of time to propagate through a certain length of a fiber. However, because any source of light would have a certain spectral width $\Delta\lambda_0$ and each spectral component would, in general, travel with a different group velocity, we would always have dispersion. This is referred to as material dispersion and is a characteristic of the material only; material dispersion plays a very important role in design of a fiber optic communication system.

6.2 Calculation of material dispersion

In Section 6.3 we show that when a temporal pulse propagates through a homogeneous medium, it propagates with a group velocity v_g given by the following equation

$$v_g = \frac{1}{(dk/d\omega)} \tag{6.1}$$

where

$$k(\omega) = \frac{\omega}{c} n(\omega) \tag{6.2}$$

represents the propagation constant and $n(\omega)$ represents the frequency-dependent refractive index. Thus

$$\frac{1}{v_g} = \frac{dk}{d\omega}$$

$$= \frac{d}{d\omega}\left[\frac{\omega}{c} n(\omega)\right]$$

or

$$\frac{1}{v_g} = \frac{1}{c}\left[n(\omega) + \omega\frac{dn}{d\omega}\right] \tag{6.3}$$

Usually one expresses the group velocity in terms of the free space wavelength λ_0, which is related to the frequency through the following relation

$$\lambda_0 = \frac{2\pi c}{\omega}$$

Thus

$$\omega \frac{dn}{d\omega} = \frac{2\pi c}{\lambda_0}\left[\frac{dn}{d\lambda_0}\left(-\frac{2\pi c}{\omega^2}\right)\right]$$
$$= -\lambda_0 \frac{dn}{d\lambda_0}$$

Equation (6.3) can therefore be written as

$$\frac{1}{v_g} = \frac{1}{c}\left[n(\lambda_0) - \lambda_0 \frac{dn}{d\lambda_0}\right] \tag{6.4}$$

Thus, the time taken by a pulse to traverse length L of fiber is given by

$$\tau = \tau(\lambda_0) = \frac{L}{v_g} = \frac{L}{c}\left[n(\lambda_0) - \lambda_0 \frac{dn}{d\lambda_0}\right] \tag{6.5}$$

which is dependent on the wavelength λ_0. The quantity

$$N(\lambda_0) = n(\lambda_0) - \lambda_0 \frac{dn}{d\lambda_0} \tag{6.6}$$

is also referred to as the group refractive index since $c/N(\lambda_0)$ gives the group velocity.

If the source is characterized by spectral width $\Delta\lambda_0$, then each wavelength component will traverse with a different group velocity, resulting in temporal broadening of the pulse. This broadening is given by

$$\Delta\tau = \frac{d\tau}{d\lambda_0}\Delta\lambda_0 = -\frac{L}{c}\lambda_0 \frac{d^2 n}{d\lambda_0^2}\Delta\lambda_0$$

or

$$\Delta\tau = -\frac{L}{c}\left(\lambda_0^2 \frac{d^2 n}{d\lambda_0^2}\right)\left(\frac{\Delta\lambda_0}{\lambda_0}\right) \tag{6.7}$$

The quantity $(\lambda_0^2 d^2 n/d\lambda_0^2)$ is a dimensionless quantity. The above broadening is referred to as material dispersion and occurs when a pulse propagates through any dispersive medium. Since material dispersion as given by equation (6.7) is proportional to the spectral width $\Delta\lambda_0$ and also to the length L traversed in the medium, it is usually specified in units of picoseconds per kilometer (length of

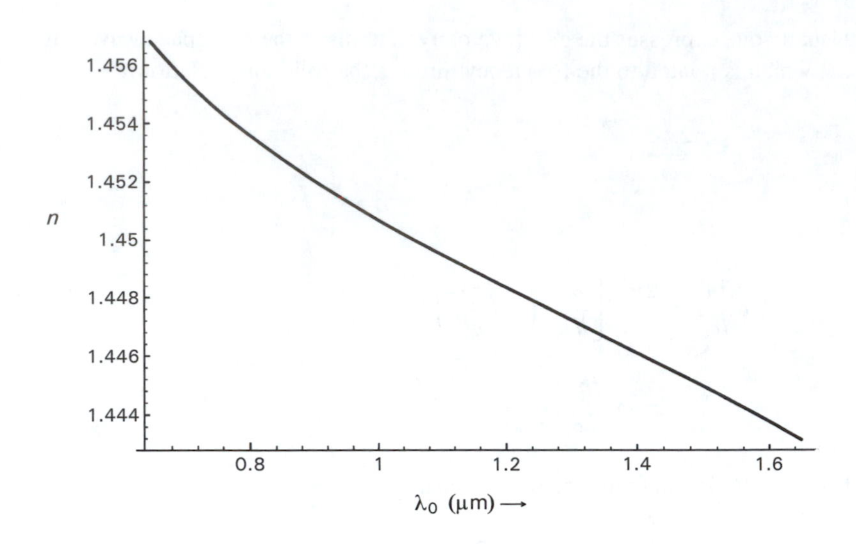

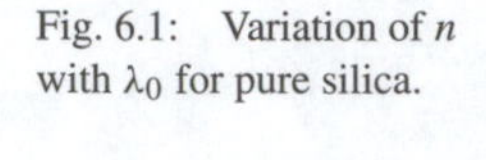
Fig. 6.1: Variation of n with λ_0 for pure silica.

the fiber) per nanometer (spectral width of the source).

$$D_m = \frac{\Delta\tau}{L\Delta\lambda_0} = -\frac{1}{\lambda_0 c}\left(\lambda_0^2 \frac{d^2 n}{d\lambda_0^2}\right) \times 10^9 \text{ (ps/km·nm)} \tag{6.8}$$

where λ_0 is measured in micrometers and $c = 3 \times 10^5$ km/s.

6.2.1 Material dispersion in pure and doped silica

As an example, we first consider fused silica, which forms the basic component of optical fibers. An accurate (empirical) expression for the refractive index variation for fused silica is given by [Paek et al. 1981]

$$n(\lambda_0) = C_0 + C_1\lambda_0^2 + C_2\lambda_0^4 + \frac{C_3}{(\lambda_0^2 - l)} + \frac{C_4}{(\lambda_0^2 - l)^2} + \frac{C_5}{(\lambda_0^2 - l)^3} \tag{6.9}$$

where

$$\begin{aligned} C_0 &= 1.4508554, \quad & C_1 &= -0.0031268 \\ C_2 &= -0.0000381, \quad & C_3 &= 0.0030270 \\ C_4 &= -0.0000779, \quad & C_5 &= 0.0000018 \\ l &= 0.035 \end{aligned} \tag{6.10}$$

and λ_0 is measured in micrometers. Table 6.1 shows the values of $n(\lambda_0)$, $dn/d\lambda_0$, and $d^2n/d\lambda_0^2$ for pure silica as obtained from equation (6.9) for some wavelengths and Figures 6.1–6.3 show the corresponding wavelength variations.

To have some numerical appreciation, we see from Table 6.1 that at $\lambda_0 = 0.8\,\mu$m

$$\frac{d^2n}{d\lambda_0^2} \simeq 0.04\,\mu\text{m}^{-2}$$

Table 6.1. *Variation of* n, $dn/d\lambda_0$, *and* $d^2n/d\lambda_0^2$ *with* λ_0 *for pure silica*

$\lambda_0(\mu m)$	$n(\lambda_0)$	$\frac{dn}{d\lambda_0}(\mu m^{-1})$	$\frac{d^2n}{d\lambda_0^2}(\mu m^{-2})$
0.65	1.45685128	−0.02714381	0.10309425
0.70	1.45560969	−0.02276059	0.07410101
0.75	1.45455521	−0.01958463	0.05412515
0.80	1.45363725	−0.01725159	0.03997833
0.85	1.45282003	−0.01552236	0.02972035
0.90	1.45207767	−0.01423535	0.02212732
0.95	1.45139101	−0.01327862	0.01640339
1.00	1.45074564	−0.01257282	0.01201730
1.05	1.45013051	−0.01206070	0.00860611
1.10	1.44953705	−0.01170022	0.00591684
1.15	1.44895849	−0.01146001	0.00376986
1.20	1.44838944	−0.01131637	0.00203553
1.25	1.44782555	−0.01125123	0.00061889
1.30	1.44726325	−0.01125037	−0.00055057
1.35	1.44669962	−0.01130300	−0.00152585
1.40	1.44613221	−0.01140040	−0.00234725
1.45	1.44555895	−0.01153568	−0.00304575
1.50	1.44497810	−0.01170333	−0.00364537
1.55	1.44438815	−0.01189888	−0.00416491
1.60	1.44378781	−0.01211873	−0.00461922
1.65	1.44317592	−0.01235992	−0.00502012

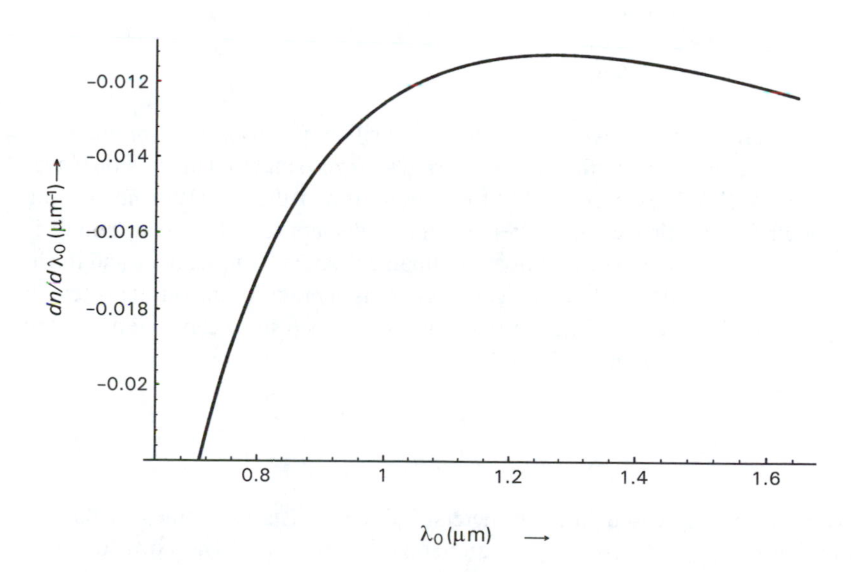

Fig. 6.2: Variation of $dn/d\lambda_0$ with λ_0 for pure silica as obtained from equation (6.9).

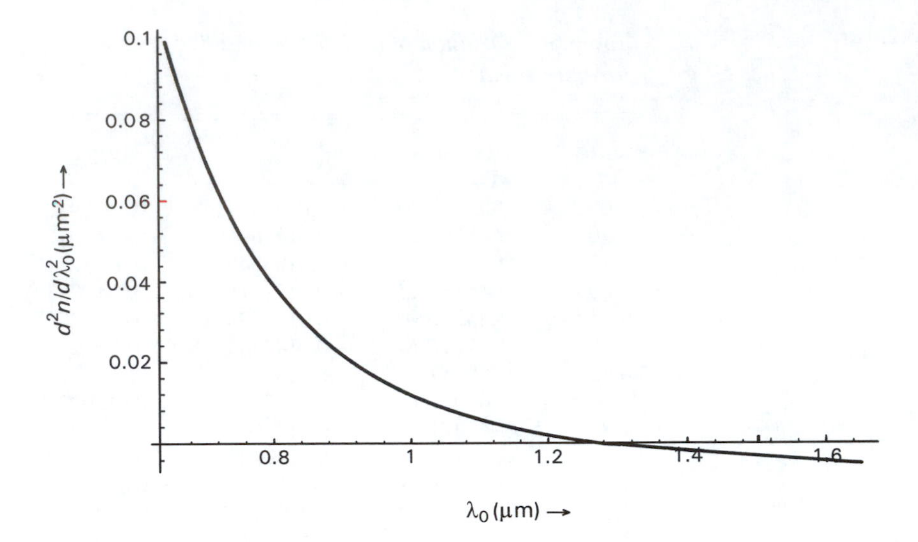

Fig. 6.3: Variation of $d^2n/d\lambda_0^2$ with λ_0 for pure silica as obtained from equation (6.9). Note that at $\lambda_0 \simeq 1.27\,\mu$m, $d^2n/d\lambda_0^2 \simeq 0$; this wavelength is referred to as the ZMDW.

implying a dispersion of 110 ps/km·nm. Thus, for an LED source at 0.8 μm with $\Delta\lambda_0 = 25$ nm (which represents a typical value for a GaAs LED), we obtain

$$\Delta\tau \simeq 2.7\ \text{ns/km}$$

Similarly at $\lambda_0 \simeq 1.55\,\mu$m,

$$\frac{d^2n}{d\lambda_0^2} \simeq 0.004\,\mu\text{m}^{-2}$$

implying a dispersion of 20 ps/km·nm. Thus, for a laser diode at 1.55 μm with $\Delta\lambda_0 \simeq 2$ nm we obtain

$$\Delta\tau \simeq 40\ \text{ps/km}$$

Now, from Figure 6.3 we see that for fused silica $d^2n/d\lambda_0^2 \simeq 0$ around $\lambda_0 \simeq 1.27\,\mu$m. This wavelength is hence referred to as zero material dispersion wavelength (ZMDW). A pulse of light centered around the ZMDW and passing through fused silica would suffer negligible dispersion.[1] This wavelength is of great importance in optical fiber communication, and the second- and third-generation optical fiber communication systems operate around this wavelength region (see Chapter 13). The refractive index of doped silica can be represented by the following empirical formula

$$n^2(\lambda_0) - 1 = \frac{b_1\lambda_0^2}{\lambda_0^2 - a_1} + \frac{b_2\lambda_0^2}{\lambda_0^2 - a_2} + \frac{b_3\lambda_0^2}{\lambda_0^2 - a_3} \tag{6.11}$$

where λ_0 is expressed in micrometers. Table 6.2 lists the values of the coefficients $a_1, a_2, \ldots, b_3$ for pure and some doped silica. The corresponding

[1] For operation around ZMDW, the dispersion is determined by the next higher order term – namely, $d^3k/d\omega^3$ (see Problem 6.5).

Table 6.2. *Values of coefficients in Sellemeier's formula [equation (6.11)] for pure and doped silica*

Sample	Dopant (mole %)	a_1	a_2	a_3	b_1	b_2	b_3
A	Pure SiO_2	0.004679148	0.01351206	97.93400	0.6961663	0.4079426	0.8974794
B	GeO_2 (6.3)	0.007290464	0.01050294	97.93428	0.7083952	0.4203993	0.8663412
C	GeO_2 (19.3)	0.005847345	0.01552717	97.93484	0.7347008	0.4461191	0.8081698
D	B_2O_3 (5.2)	0.004981838	0.01375664	97.93353	0.6910021	0.4022430	0.9439644
E	P_2O_5 (10.5)	0.005202431	0.01287730	97.93401	0.7058489	0.4176021	0.8952753

Note: Adapted from Kimura (1986).

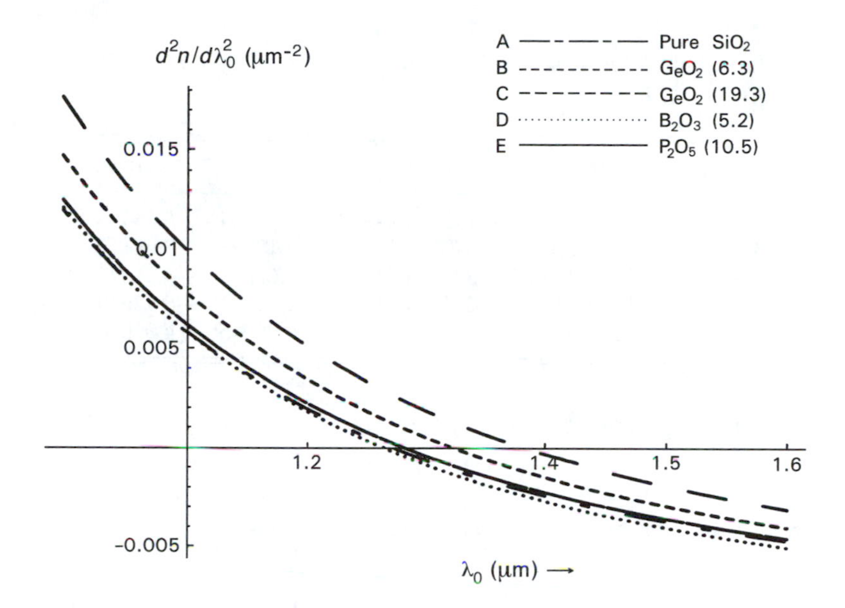

Fig. 6.4: Variation of $d^2n/d\lambda_0^2$ for pure and doped silica. The labels A–E correspond to various samples given in Table 6.2. Observe that doping changes the ZMDW slightly.

variation of $d^2n/d\lambda_0^2$ with λ_0 is shown in Figure 6.4. As can be seen, material dispersion changes with doping.

6.2.2 *Material dispersion in fluoride glasses*

Present day optical fiber systems are based on silica glasses. In the last few years, fluoride-based fibers have been investigated in detail for operation in the mid-infrared (2–5 μm) wavelength region because of their predicted ultralow loss of 10^{-3} dB/km. Such fibers may find applications in longwave repeaterless telecommunication links and intercontinental submarine links. Table 6.3 gives the composition of various fluoride-based glasses, and Table 6.4 lists the values of the Sellemeier coefficients *A* to *E* that determine the refractive index through the relation

$$n(\lambda_0) = A\lambda_0^{-4} + B\lambda_0^{-2} + C + D\lambda_0^2 + E\lambda_0^4 \tag{6.12}$$

Table 6.3. *Various fluoride glasses for applications in infrared fiber optic communications*

	Concentration (mole %)								
Material	ZrF_4	BaF_2	LaF_3	NaF	HfF_4	AlF_3	CaF_2	GdF_3	YF_3
ABCY		22				40	22		16
HBL		33	5		62				
ZBG	63	33						4	
ZBLAN	53	20	4	20		3			

Note: Adapted from Mendez and Sunak (1986).

Table 6.4. *Sellemeier coefficients of various fluoride glasses given in Table 6.3*

Material	$A \times 10^6$	$B \times 10^3$	C	$D \times 10^3$	$E \times 10^6$
ABCY	7.67742	2.16195	1.42969	−1.28304	−5.35487
HBL[a]	−28.61020	3.11470	1.50294	−1.17821	−2.64123
ZBG	93.67070	2.94329	1.51236	−1.25045	−4.01026
ZBLA[a]	−300.80370	4.03214	1.51272	−1.21921	−6.77630
ZBLAN	93.67070	2.94329	1.49136	−1.25045	−4.01026

[a] Fitted data.
Note: Adapted from Mendez and Sunak (1986).

where λ_0 is in micrometers. From equation (6.12) we have

$$\frac{dn}{d\lambda_0} = -4A\lambda_0^{-5} - 2B\lambda_0^{-3} + 2D\lambda_0 + 4E\lambda_0^3 \tag{6.13}$$

$$\frac{d^2n}{d\lambda_0^2} = 20A\lambda_0^{-6} + 6B\lambda_0^{-4} + 2D + 12E\lambda_0^2 \tag{6.14}$$

Figure 6.5 shows the variation of n with λ_0 for the various glasses, and Figure 6.6 shows the variation of $d^2n/d\lambda_0^2$ with λ_0.

6.3 Derivation of group velocity

A plane monochromatic wave propagating along the z-axis through an infinitely extended homogeneous medium is described by

$$\psi = Ae^{i(\omega t - kz)} \tag{6.15}$$

where $k(\omega)$ represents the propagation constant. This monochromatic wave

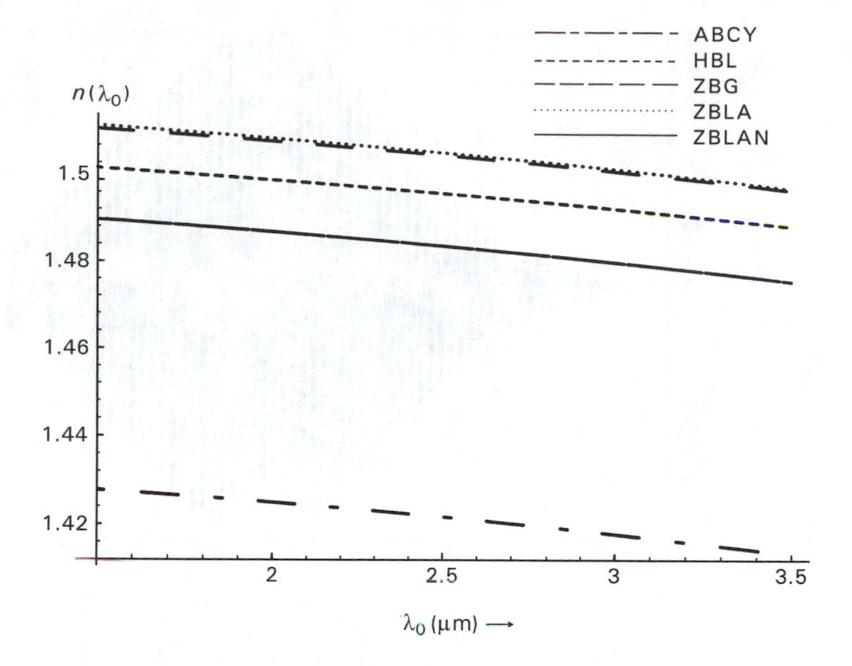

Fig. 6.5: Variation of n with λ_0 for some fluoride glasses. The composition of the various glasses is given in Table 6.3. [Adapted from Mendez and Sunak (1986).]

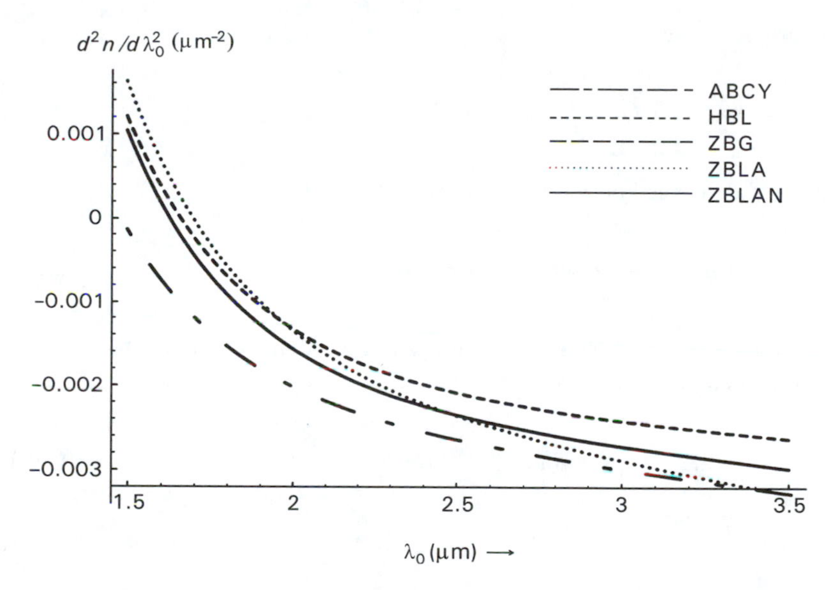

Fig. 6.6: Variation of $d^2n/d\lambda_0^2$ with λ_0 for typical fluoride glasses listed in Table 6.3.

travels with a velocity given by

$$v_p = \frac{\omega}{k} \tag{6.16}$$

which is referred to as the phase velocity and represents the velocity at which a surface of constant phase advances in the medium. A monochromatic wave described by equation (6.15) extends in the entire time domain, $-\infty < t < \infty$, which is a practical impossibility.

We next consider a temporal pulse described by the function $\Psi(z = 0, t)$ at $z = 0$ (see Figure 6.7). To study its evolution we represent it as a superposition

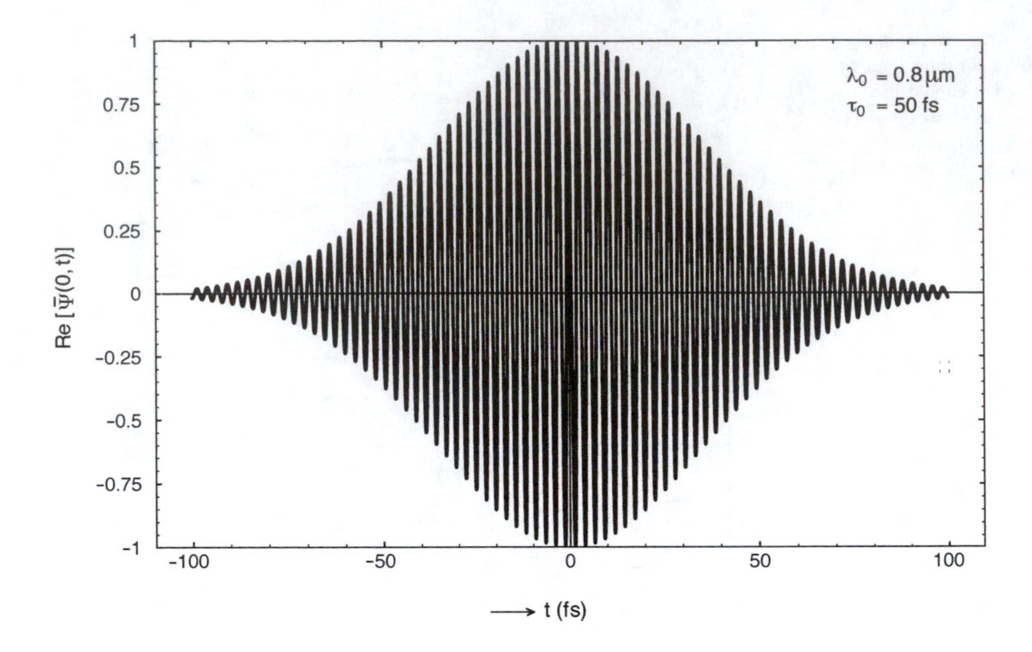

Fig. 6.7: A Gaussian pulse represented by equation (6.22). For clarity, $\omega_0 \tau$ has been assumed to be small so that only a few oscillations occur within the pulse; in actual practice, there would be a large number of oscillations within the pulse.

of harmonic waves

$$\Psi(z = 0, t) = \int_{-\infty}^{+\infty} A(\omega) e^{i\omega t} d\omega \tag{6.17}$$

Taking the inverse Fourier transform we get

$$A(\omega) = \frac{1}{2\pi} \int \Psi(z = 0, t) e^{-i\omega t} dt \tag{6.18}$$

Each frequency component would propagate according to equation (6.15); thus, the total field at z would be given by

$$\Psi(z, t) = \int_{-\infty}^{\infty} A(\omega) e^{i(\omega t - kz)} d\omega \tag{6.19}$$

We write

$$A(\omega) = |A(\omega)| e^{i\phi(\omega)} \tag{6.20}$$

to obtain

$$\Psi(z, t) = \int |A(\omega)| e^{i[\omega t - k(\omega) z + \phi(\omega)]} d\omega \tag{6.21}$$

In most cases, an optical pulse propagating through a fiber can be approximated by a Gaussian temporal distribution.

$$\Psi(z = 0, t) = C e^{-t^2/\tau_0^2} e^{i\omega_0 t} \tag{6.22}$$

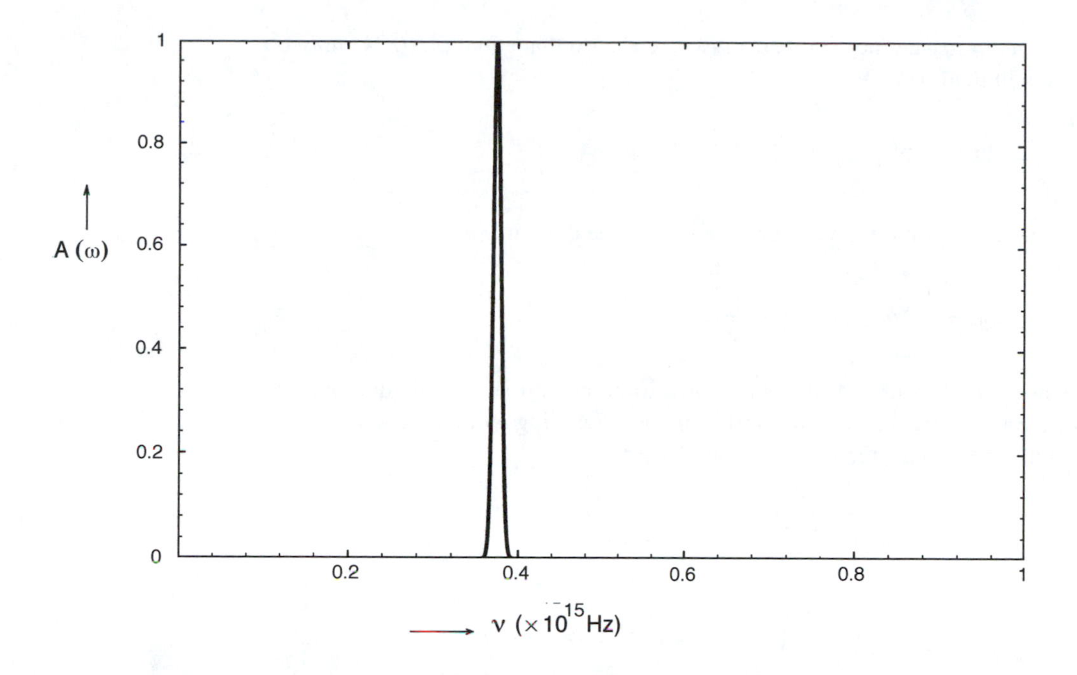

Fig. 6.8: The Fourier spectrum corresponding to the pulse shown in Fig. 6.7. The width of the Fourier spectrum is inversely proportional to τ_0.

which represents a Gaussian pulse of temporal width $2\tau_0$. Substituting in equation (6.18), we have

$$A(\omega) = \frac{C}{2\pi}\int_{-\infty}^{\infty} e^{-t^2/\tau_0^2} e^{i(\omega_0-\omega)t}\, dt$$
$$= \frac{C\tau_0}{2\sqrt{\pi}} \exp\left[-\frac{\tau_0^2}{4}(\omega-\omega_0)^2\right] \tag{6.23}$$

Typically for a 1-ns pulse at $\lambda_0 \simeq 0.8\,\mu$m

$$\tau_0 = 10^{-9}\,\text{s} \quad \text{and} \quad \omega_0 \simeq 2.36 \times 10^{15}\,\text{s}^{-1}$$

The function $A(\omega)$ has a spectral width given by

$$\Delta\omega \sim \frac{2}{\tau_0}$$

giving

$$\frac{\Delta\omega}{\omega_0} \simeq \frac{2}{\tau_0\omega_0} \simeq 1.0 \times 10^{-6}$$

The quantity $\Delta\omega/\omega_0$ is usually referred to as the spectral purity of the pulse and for such a small value of $\Delta\omega/\omega_0$, the frequency spectrum $A(\omega)$ is an extremely sharply peaked function around $\omega = \omega_0$. Thus, $A(\omega)$ is very sharply peaked at $\omega = \omega_0$ (see Figure 6.8). The longer the pulse, the greater will be the sharpness of the frequency spectrum $A(\omega)$. In general, in almost all cases of practical interest, $|A(\omega)|$ is very sharply peaked when ω lies in a small interval $\Delta\omega$

around $\omega = \omega_0$ and negligible everywhere else so that equation (6.21) may be written in the form

$$\Psi(z,t) \simeq \int_{\Delta\omega} |A(\omega)| e^{i[\omega t - kz + \phi(\omega)]} d\omega \tag{6.24}$$

where the integration now extends only over the domain

$$\omega_0 - \frac{1}{2}\Delta\omega \lesssim \omega \lesssim \omega_0 + \frac{1}{2}\Delta\omega \tag{6.25}$$

Outside this domain, $|A(\omega)|$ is assumed to be negligible. In this domain, we assume $n(\omega)$, and hence $k(\omega)$, to be smoothly varying so that we can make a Taylor series expansion of $k(\omega)$ around $\omega = \omega_0$:

$$k(\omega) = k(\omega_0) + \left(\frac{dk}{d\omega}\right)_{\omega=\omega_0} (\omega - \omega_0) + \frac{1}{2}\left(\frac{d^2k}{d\omega^2}\right)_{\omega=\omega_0} (\omega - \omega_0)^2 + \cdots \tag{6.26}$$

Similarly, we may write

$$\phi(\omega) = \phi(\omega_0) + \left(\frac{d\phi}{d\omega}\right)_{\omega=\omega_0} (\omega - \omega_0) + \frac{1}{2}\left(\frac{d^2\phi}{d\omega^2}\right)_{\omega=\omega_0} (\omega - \omega_0)^2 + \cdots \tag{6.27}$$

We substitute the above expansions in equation (6.24) and assume that in the domain of integration $\Delta\omega$, $k(\omega)$ and $\phi(\omega)$ do not vary appreciably so that we may neglect terms that are proportional to $(\omega - \omega_0)^2$ and higher powers[2] of $(\omega - \omega_0)$; under this approximation we get

$$\Psi(z,t) \simeq e^{i(\omega_0 t - k_0 z + \phi_0)} f(z,t) \tag{6.28}$$

where

$$f(z,t) = \int_{\Delta\omega} |A(\omega)| \exp\left\{ i(\omega - \omega_0)\left[t - \left(\frac{dk}{d\omega}\right)_{\omega=\omega_0} z + \left(\frac{d\phi}{d\omega}\right)_{\omega=\omega_0} \right]\right\} d\omega$$

represents the envelope of the wave packet. Thus, we have

$$f(z,t) = \int_{\Delta\omega} |A(\omega)| \exp\left[-i\frac{(\omega - \omega_0)}{v_g}(z - z_0 - v_g t)\right] d\omega \tag{6.29}$$

[2]The effect of the term proportional to $(\omega - \omega_0)^2$ is discussed in Section 6.4.

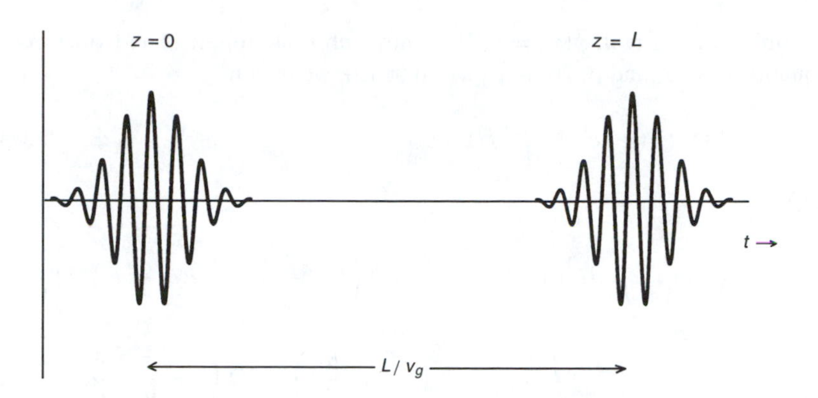

Fig. 6.9: Distortionless propagation of a wave packet with group velocity v_g, which occurs when k is strictly linearly related to ω.

where

$$v_g \equiv \left[\left(\frac{dk}{d\omega} \right)_{\omega=\omega_0} \right]^{-1}, \quad z_0 = v_g \left(\frac{d\phi}{d\omega} \right)_{\omega=\omega_0} \tag{6.30}$$

We may note that

$$f(z,0) = \int_{\Delta\omega} |A(\omega)| \exp\left[-i \frac{(\omega - \omega_0)}{v_g}(z - z_0) \right] d\omega$$

It is immediately seen that at $z = z_0$, the integrand is everywhere positive and the value of the integral is maximum. For $|z - z_0| \geq v_g/\Delta\omega$, the exponential function (in the integrand) oscillates rapidly in the domain of integration and the value of the integral is very small. This immediately leads to the (uncertainty) relation

$$\Delta\omega \frac{\Delta z}{v_g} \geq 1$$

where Δz represents the spatial extent of the pulse.

Since the function $f(z,t)$ depends on z and t only through the combination $z - v_g t$, this implies that the pulse propagates without any distortion with velocity v_g; this velocity is known as the group velocity of the pulse (see Figure 6.9). The pulse remains undistorted as long as the neglect of second-order and higher terms in equations (6.26) and (6.27) is justified. However, if the $k - \omega$ relation is strictly linear (as is indeed the case for electromagnetic waves in free space), the pulse will never undergo any distortion.

Thus, at $t = 0$ the pulse is sharply peaked at $z = z_0$ and as t increases the center of the pulse, $z_c(t)$, moves according to the equation (see Figure 6.9)

$$z_c(t) = z_0 + v_g t \tag{6.31}$$

6.4 Broadening of a Gaussian pulse

We now make an explicit calculation taking into account the second-order terms in equations (6.26) and (6.27) for a Gaussian pulse (see equation 6.22) for which $A(\omega)$ is given by equation (6.23).

Since $A(\omega)$ is real, $\phi(\omega) = 0$. If we now substitute for $A(\omega)$ and $k(\omega)$ from equations (6.23) and (6.26) in equation (6.19), we obtain

$$\Psi(z, t) = Ce^{i(\omega_0 t - k_0 z)} f(z, t) \tag{6.32}$$

where

$$\begin{aligned} f(z, t) &= \int A(\omega) \exp\left\{i\left[(\omega - \omega_0)\left(t - \frac{z}{v_g}\right) - \frac{1}{2}\alpha(\omega - \omega_0)^2 z\right]\right\} d\omega \\ &\simeq \frac{\tau_0}{2\sqrt{\pi}} \int \exp\left[-\Omega^2\left(\frac{\tau_0^2}{4} + i\frac{\alpha}{2}z\right) + i\Omega\left(t - \frac{z}{v_g}\right)\right] d\Omega \end{aligned} \tag{6.33}$$

where $\Omega = \omega - \omega_0$ and

$$\alpha = \left.\frac{d^2 k}{d\omega^2}\right|_{\omega = \omega_0} \tag{6.34}$$

Thus

$$f(z, t) = \left(1 + i\frac{2\alpha z}{\tau_0^2}\right)^{-1/2} \exp\left[\frac{-\left(t - \frac{z}{v_g}\right)^2}{\left(\tau_0^2 + 2i\alpha z\right)}\right] \tag{6.35}$$

and $\Psi(z, t)$ can be written as

$$\Psi(z, t) = \frac{C}{(\tau(z)/\tau_0)^{1/2}} \exp\left[-\frac{\left(t - \frac{z}{v_g}\right)^2}{\tau^2(z)}\right] \exp[i(\Phi(z, t) - k_0 z)] \tag{6.36}$$

where[3]

$$\Phi(z, t) = \omega_0 t + \kappa\left(t - \frac{z}{v_g}\right)^2 - \frac{1}{2}\tan^{-1}\left(\frac{2\alpha z}{\tau_0^2}\right) \tag{6.37}$$

$$\kappa = \frac{2\alpha z}{\tau_0^4}\left(1 + \frac{4\alpha^2 z^2}{\tau_0^4}\right)^{-1} \tag{6.38}$$

and

$$\tau^2(z) = \tau_0^2\left(1 + \frac{4\alpha^2 z^2}{\tau_0^4}\right) \tag{6.39}$$

[3]The second term in equation (6.37) leads to the phenomenon of chirping, which is discussed in Chapter 15.

Fig. 6.10: The temporal broadening of a Gaussian pulse. (a) The electric field distribution and (b) the corresponding intensity variation.

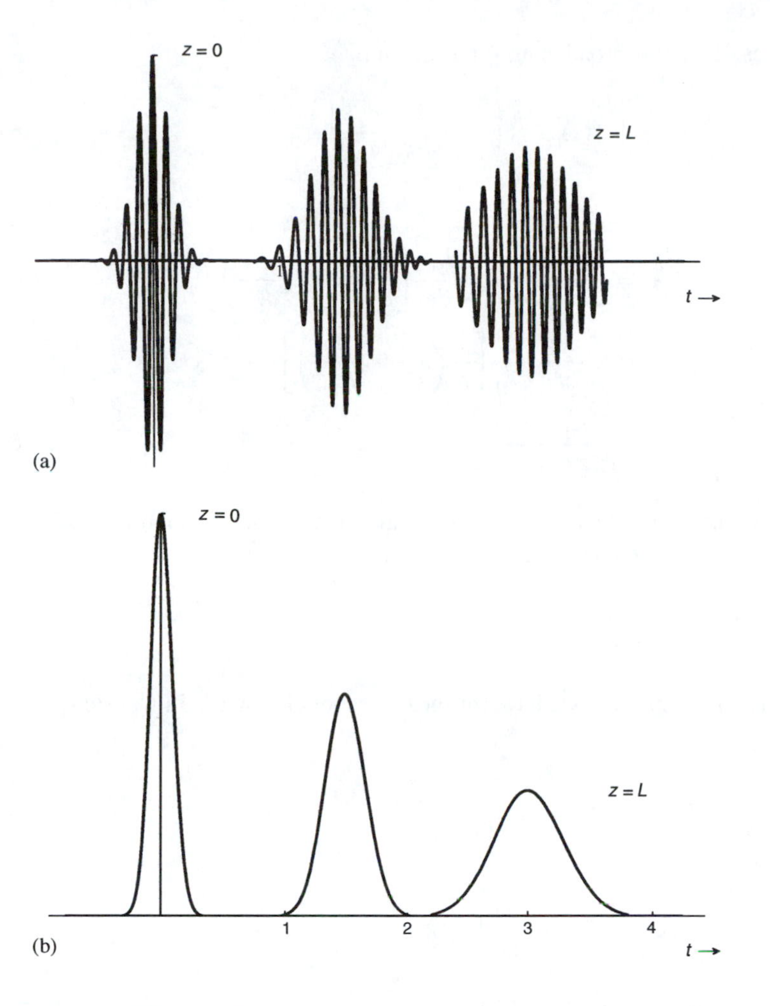

The corresponding intensity distribution is given by

$$I(z, t) = \frac{I_0}{(\tau(z)/\tau_0)} \exp\left[\frac{-2\left(t - \frac{z}{v_g}\right)^2}{\tau^2(z)}\right] \tag{6.40}$$

which is plotted in Figure 6.10(b) for different values of z.

Figure 6.10 clearly shows the broadening of the pulse and shows that the peak of the pulse moves with the group velocity given by equation (6.1). We note that

$$\int_{-\infty}^{+\infty} I(z, t)\, dt$$

is independent of z, showing that the total energy contained in the pulse is conserved. From equation (6.40) we may note the following:

The pulse width at any value of z is given by

$$\tau(z) = \tau_0 \left[1 + \frac{4z^2}{\tau_0^4}\left(\frac{d^2k}{d\omega^2}\right)^2\right]^{1/2} \tag{6.41}$$

Thus, the pulse broadening $\Delta\tau$ is given by

$$\Delta\tau \equiv \left[\tau^2(z) - \tau_0^2\right]^{1/2} = \frac{2z}{\tau_0}\left|\frac{d^2k}{d\omega^2}\right| \tag{6.42}$$

Now

$$\begin{aligned} \frac{d^2k}{d\omega^2} &= \frac{d}{d\omega}\left(\frac{dk}{d\omega}\right) = \frac{d}{d\omega}\left[\frac{1}{c}\left(n - \lambda_0\frac{dn}{d\lambda_0}\right)\right] \\ &= -\frac{\lambda_0^2}{2\pi c}\frac{d}{d\lambda_0}\left[\frac{1}{c}\left(n - \lambda_0\frac{dn}{d\lambda_0}\right)\right] \\ &= \frac{\lambda_0^3}{2\pi c^2}\frac{d^2n}{d\lambda_0^2} \end{aligned} \tag{6.43}$$

Also, the spectral width of the Gaussian pulse given by equation (6.22) can be obtained from equation (6.23) as

$$\Delta\omega \simeq \frac{2}{\tau_0} \tag{6.44}$$

Since $\omega = 2\pi c/\lambda_0$, we have for the corresponding width in wavelength

$$\Delta\lambda_0 \simeq \frac{\lambda_0^2}{\pi c \tau_0} \tag{6.45}$$

Thus

$$|\Delta\tau| \simeq \frac{z}{\lambda_0 c}\left|\lambda_0^2\frac{d^2n}{d\lambda_0^2}\right|\Delta\lambda_0 \tag{6.46}$$

consistent with equation (6.7).

Problems

6.1 The refractive index for fused silica in the wavelength region $0.5\,\mu\text{m} < \lambda_0 < 1.6\,\mu\text{m}$ is approximately given by the following empirical formula (also referred to as Sellemeier's equation)

$$n(\lambda_0) = C_0 - a\lambda_0^2 + \frac{b}{\lambda_0^2} \tag{6.47}$$

where

$$C_0 = 1.451; \quad a = b = 0.003$$

and λ_0 is measured in micrometers. Simple differentiation gives us (see equation (6.6))

$$N(\lambda_0) = C_0 + a\lambda_0^2 + 3b/\lambda_0^2 \tag{6.48}$$

At $\lambda_0 = 0.8\,\mu\text{m}$, $n(\lambda_0) \simeq 1.4538$, $N(\lambda_0) \simeq 1.4670$, indicating a difference of 0.9% between phase and group velocities.

Using the less accurate (but more convenient to use) empirical formula given by equation (6.47), calculate the zero material dispersion wavelength and also the pulse dispersion at 0.8 μm and 1.55 μm and compare with the results given earlier.

[ANSWER: 1.316 μm, 101 ps/km·nm at 0.8 μm and 14.9 ps/km·nm at 1.55 μm.]

6.2 Consider a 10-ps pulse at $\lambda_0 = 0.85\,\mu$m propagating through pure silica. Given that $d^2k/d\omega^2 \simeq 3.258 \times 10^{-26}$ s^2/m at $\lambda_0 = 0.85\,\mu$m, obtain the maximum distance over which the pulse would propagate almost undistorted.

Solution: For a 10-ps pulse, $\tau_0 = 10^{-11}$ s and

$$\Delta\omega \sim \frac{2}{\tau_0} = 2 \times 10^{11}\,\text{s}^{-1}$$

Thus, for distortionless propagation (see Problem 6.6)

$$z \ll z_d = \frac{2}{\alpha(\Delta\omega)^2} \simeq 1.5\,\text{km}$$

The distance z_d in this case is large because of the extremely small spectral width of the source; indeed, for the given source

$$\Delta\lambda_0 = \frac{\lambda_0^2}{2\pi c}\Delta\omega \simeq 0.08\,\text{nm}$$

6.3 In Problem 6.2, the spectral width of the pulse was assumed to be that determined due to the finite temporal width τ_0 of the pulse. This is true if the pulse is produced by a monochromatic source. If the source spectral width is much larger than $1/\tau_0$, then the dispersion is much higher. Thus, consider an LED source at $\lambda_0 \simeq 0.85\,\mu$m with a spectral width of 30 nm. Calculate the broadening in 1 km due to material dispersion.

Solution: At $\lambda_0 = 0.85\ \mu$m, $d^2n/d\lambda_0^2 \simeq 3 \times 10^{10}$ m^{-2} and thus from equation (6.8) we have

$$\Delta\tau \simeq 2.6\ \text{ns/km}$$

6.4 If in the above problem the source is a laser at 0.85 μm with a spectral width of 1 nm, obtain the pulse broadening.

Solution: Since broadening is proportional to $\Delta\lambda_0$, the corresponding broadening is

$$\Delta\tau \simeq 90\ \text{ps/km}$$

Note the significant reduction in broadening due to the smaller spectral width.

6.5 Around ZMDW, pulse distortion would be caused by $d^3n/d\lambda_0^3$. Estimate the dispersion that is due to this term.

Solution: We consider two closely spaced wavelengths λ_0 and $\lambda_0 + \Delta\lambda_0$ and let $\tau(\lambda_0)$ and $\tau(\lambda_0 + \Delta\lambda_0)$ represent the respective time delays. Then we may write

$$\tau(\lambda_0 + \Delta\lambda_0) = \tau(\lambda_0) + \Delta\lambda_0\frac{d\tau}{d\lambda_0} + \frac{(\Delta\lambda_0)^2}{2}\frac{d^2\tau}{d\lambda_0^2} \tag{6.49}$$

Now $d\tau/d\lambda_0 = -L(\lambda_0/c)d^2n/d\lambda_0^2$ and it would vanish at the ZMDW and we would have

$$\Delta\tau = \tau(\lambda_0 + \Delta\lambda_0) - \tau(\lambda_0) = \frac{(\Delta\lambda_0)^2}{2}\frac{d^2\tau}{d\lambda_0^2} \tag{6.50}$$

Also

$$\frac{d^2\tau}{d\lambda_0^2} = \frac{d}{d\lambda_0}\left(\frac{d\tau}{d\lambda_0}\right) = \frac{d}{d\lambda_0}\left(-\frac{L\lambda_0}{c}\frac{d^2n}{d\lambda_0^2}\right)$$
$$= -\frac{L\lambda_0}{c}\frac{d^3n}{d\lambda_0^3} \tag{6.51}$$

since $d^2n/d\lambda_0^2 = 0$. Thus

$$\Delta\tau = -L(\Delta\lambda_0)^2\left(\frac{\lambda_0}{2c}\frac{d^3n}{d\lambda_0^3}\right) \tag{6.52}$$

which is the broadening due to material dispersion. Note that the broadening is proportional to $(\Delta\lambda_0)^2$ and to L.

6.6 For the wave packet given by equation (6.24) show that the packet remains undistorted for $z \ll z_d$ where

$$z_d = \frac{2}{\alpha(\Delta\omega)^2} \tag{6.53}$$

Assume $\phi(\omega) = 0$.

Solution: If we substitute for $k(\omega)$ from equation (6.26) in equation (6.24) we will obtain

$$\Psi(z,t) \simeq e^{i(\omega_0 t - k_0 z)}\int_{\Delta\omega}|A(\omega)|$$
$$\times \exp\left[-i\frac{\omega-\omega_0}{v_g}(z - v_g t) - i\frac{\alpha}{2}(\omega-\omega_0)^2 z\right]d\omega \tag{6.54}$$

where α is given by equation (6.34). Since $|A(\omega)|$ is assumed to have a negligible value for $|\omega - \omega_0| \geq \Delta\omega$, the term involving α will make a negligible contribution for

$$z \ll z_d \tag{6.55}$$

where

$$z_d = \frac{2}{\alpha(\Delta\omega)^2}$$

Indeed, the pulse will appreciably broaden for $z \geq z_d$.

6.7 Consider a source at frequency ω_0 (and corresponding wavelength λ_0) being sinusoidally modulated at frequency $\Omega \ll \omega_0$ – that is, with a field

$$\psi(t, z=0) = (A + B\cos\Omega t)e^{i\omega_0 t} \tag{6.56}$$

If such a beam passes through a medium, calculate the time variation of the field at distance L.

Solution: The field variation of the source at $z = 0$ can be written as

$$\psi(t, z = 0) = Ae^{i\omega_0 t} + \frac{B}{2}e^{i(\omega_0+\Omega)t} + \frac{B}{2}e^{i(\omega_0-\Omega)t} \tag{6.57}$$

that is, the input consists of frequencies ω_0, $\omega_0 + \Omega$, and $\omega_0 - \Omega$. If k_0, k_+, and k_- represent the propagation constant of the medium at frequencies ω_0, $\omega_0 + \Omega$, and $\omega_0 - \Omega$, respectively, then we have

$$\psi(t, z) = Ae^{i(\omega_0 t - k_0 z)} + \frac{B}{2}e^{i[(\omega_0+\Omega)t - k_+ z]} + \frac{B}{2}e^{i[(\omega_0-\Omega)t - k_- z]} \tag{6.58}$$

Since $\Omega \ll \omega_0$, we have

$$\begin{aligned} k_+ &\simeq k_0 + \left(\frac{dk}{d\omega}\right)_{\omega=\omega_0} \Omega \\ &= k_0 + \frac{\Omega}{v_g} \end{aligned} \tag{6.59}$$

where $v_g = (dk/d\omega)^{-1}$ is the group velocity at ω_0. Similarly

$$k_- \simeq k_0 - \frac{\Omega}{v_g} \tag{6.60}$$

Substituting for k_+ and k_- in equation (6.58) we have, after some simplifications

$$\psi(t, L) = \left\{A + B\cos\left[\Omega\left(t - \frac{L}{v_g}\right)\right]\right\} e^{i(\omega_0 t - k_0 z)} \tag{6.61}$$

Thus, modulation of the beam propagates at the group velocity v_g. From equation (6.61) we can see that if we measure the phase of the modulation as a function of the wavelength λ_0 of the source, we essentially obtain the variation of v_g with λ_0. Such a technique is indeed used in measurement of dispersion characteristics of a single-mode fiber.

6.8 In Problem 6.7 we assumed the source to be a monochromatic source at λ_0. If the source contains two wavelengths, λ_1 and λ_2, the modulation at the two wavelengths propagates with different velocities. For what minimum value of Ω (the modulation frequency) will one obtain zero modulation at $z = L$?

Solution: If v_{g1} and v_{g2} represent the group velocities at λ_1 and λ_2, respectively, then it follows that when the delay between the two modulated beams becomes equal to one-half period of modulation, the output will essentially have no modulation. Thus, for a modulation frequency Ω_0 such that

$$\left|\frac{L}{v_{g1}} - \frac{L}{v_{g2}}\right| = \frac{\pi}{\Omega_0} \tag{6.62}$$

there is no modulation at the output. If λ_1 and λ_2 are close, then we have

$$\frac{1}{v_{g2}} - \frac{1}{v_{g1}} = \left(\frac{dk}{d\omega}\right)_{\lambda_2} - \left(\frac{dk}{d\omega}\right)_{\lambda_1}$$
$$\simeq -\frac{\lambda_c}{c}\Delta\lambda\frac{d^2n}{d\lambda_0^2} \tag{6.63}$$

where $\Delta\lambda = \lambda_2 - \lambda_1$ and $\lambda_c = (\lambda_1 + \lambda_2)/2$. In deriving equation (6.63) we have used equation (6.43). Thus

$$\frac{1}{\Omega_0} = \frac{\lambda_c L}{\pi c}\Delta\lambda\left|\frac{d^2n}{d\lambda_0^2}\right| \tag{6.64}$$

or in terms of frequency $f_0 = \Omega_0/2\pi$,

$$f_0 = \frac{1}{2\left(\frac{\lambda_c L}{c}\Delta\lambda\left|\frac{d^2n}{d\lambda_0^2}\right|\right)}$$
$$= \frac{1}{2\Delta\tau} \tag{6.65}$$

where $\Delta\tau$ is given by equation (6.46). The frequency f_0 corresponds approximately to the 3-dB bandwidth of the medium. Equation (6.65) gives an approximate relationship between pulse dispersion $\Delta\tau$ and the 3-dB bandwidth f_0.

6.9 Assume

$$\Psi(z = 0, t) = Ce^{-\frac{t^2}{\tau^2}(1+ig)}e^{i\omega_0 t}$$

(Such a pulse is known as a chirped pulse – see Chapter 15.) Show that

$$A(\omega) = \frac{E_0\tau}{2\sqrt{\pi}(1+ig)^{1/2}}\exp\left[-\frac{(\omega-\omega_0)^2\tau^2}{4(1+ig)}\right]$$

Substitute for $A(\omega)$ in equation (6.19) and show using equation (6.26) that $\Psi(z, t)$ is given by equation (6.32) with

$$f(z,t) = (1 - \sigma g + ig)^{-1/2}$$
$$\times \exp\left[-\frac{(t - z/v_g)^2\{1 + i[g - \sigma(g^2+1)]\}}{\tau^2[1 - 2\sigma g + \sigma^2(1+g^2)]}\right]$$

where

$$\sigma = \frac{2\alpha z}{\tau^2}$$

Obviously for $\alpha g > 0$, the pulse will first undergo compression and attain its minimum temporal width at

$$z = \frac{\tau^2 g}{2\alpha(1+g^2)}$$

7

Modes in planar waveguides

7.1 Introduction

In this chapter we carry out a detailed modal analysis of planar waveguides characterized by refractive index distribution depending only on the x coordinate – that is

$$n^2 = n^2(x)$$

For such waveguides, Maxwell's equations reduce to two independent sets of equations: the first set corresponding to what are known as TE (transverse electric) modes, where the electric field does not have a longitudinal component, and the second set corresponding to what are known as TM (transverse magnetic) modes, where the magnetic field does not have a longitudinal component. While carrying out the modal analysis, we try to illustrate almost all the salient points associated with the modes of a waveguide, therefore making it easier to understand the physical principles of more complicated guiding structures.

In Section 7.2 we discuss Maxwell's equations in inhomogeneous media and study the classification of TE and TM modes in planar waveguides. In Sections 7.3 and 7.4 we have detailed discussions on TE modes of a symmetric step index and parabolic index waveguides, respectively. We try to illustrate the relationship between the ray and modal analyses.

7.2 Maxwell's equations in inhomogeneous media: TE and TM modes in planar waveguides

In this section we derive the equations that are the starting points for modal analysis. We start with Maxwell's equations, which for an isotropic, linear, nonconducting, and nonmagnetic medium take the form

$$\nabla \times \boldsymbol{\mathcal{E}} = -\partial \boldsymbol{\mathcal{B}}/\partial t = -\mu_0 \partial \boldsymbol{\mathcal{H}}/\partial t \tag{7.1}$$

$$\nabla \times \boldsymbol{\mathcal{H}} = \partial \boldsymbol{\mathcal{D}}/\partial t = \epsilon_0 n^2 \partial \boldsymbol{\mathcal{E}}/\partial t \tag{7.2}$$

$$\nabla \cdot \mathcal{D} = 0 \tag{7.3}$$

$$\nabla \cdot \mathcal{B} = 0 \tag{7.4}$$

where we have used the constitutive relations

$$\mathcal{B} = \mu_0 \mathcal{H} \tag{7.5}$$

$$\mathcal{D} = \epsilon \mathcal{E} = \epsilon_0 n^2 \mathcal{E} \tag{7.6}$$

in which $\mathcal{E}$, $\mathcal{D}$, $\mathcal{B}$, and $\mathcal{H}$ represent the electric field, electric displacement, magnetic induction, and magnetic intensity, respectively, $\mu_0(=4\pi \times 10^{-7}\ \mathrm{Ns^2/C^2})$ represents the free space magnetic permeability, $\epsilon(=\epsilon_0 K = \epsilon_0 n^2)$ represents the dielectric permittivity of the medium, K and n are, respectively, the dielectric constant and the refractive index, and $\epsilon_0(=8.854 \times 10^{-12}\ \mathrm{C^2/\,N\,m^2})$ is the permittivity of free space. Now taking the curl of equation (7.1) and using equation (7.2) we get

$$\nabla \times (\nabla \times \mathcal{E}) = -\mu_0 \frac{\partial}{\partial t}(\nabla \times \mathcal{H}) = -\mu_0 \epsilon_0 n^2 \frac{\partial^2 \mathcal{E}}{\partial t^2}$$

or

$$\nabla(\nabla \cdot \mathcal{E}) - \nabla^2 \mathcal{E} = -\epsilon_0 \mu_0 n^2 \frac{\partial^2 \mathcal{E}}{\partial t^2} \tag{7.7}$$

Further

$$0 = \nabla \cdot \mathcal{D} = \epsilon_0 \nabla \cdot (n^2 \mathcal{E}) = \epsilon_0 [\nabla n^2 \cdot \mathcal{E} + n^2 \nabla \cdot \mathcal{E}]$$

Thus

$$\nabla \cdot \mathcal{E} = -(1/n^2) \nabla n^2 \cdot \mathcal{E} \tag{7.8}$$

Substituting in equation (7.7) we obtain

$$\nabla^2 \mathcal{E} + \nabla \left(\frac{1}{n^2} \nabla n^2 \cdot \mathcal{E} \right) - \epsilon_0 \mu_0 n^2 \frac{\partial^2 \mathcal{E}}{\partial t^2} = 0 \tag{7.9}$$

The above equation shows that for an inhomogeneous medium the equations for $\mathcal{E}_x$, $\mathcal{E}_y$, and $\mathcal{E}_z$ are coupled. For a homogeneous medium, the second term on the left hand side (LHS) vanishes and each Cartesian component of the electric vector satisfies the scalar wave equation.

In a similar manner, taking the curl of equation (7.2) and using equations (7.1) and (7.4) we get

$$\nabla^2 \mathcal{H} + \frac{1}{n^2} \nabla n^2 \times (\nabla \times \mathcal{H}) - \epsilon_0 \mu_0 n^2 \frac{\partial^2 \mathcal{H}}{\partial t^2} = 0 \tag{7.10}$$

If the refractive index varies only in the transverse direction – that is

$$n^2 = n^2(x, y) \tag{7.11}$$

then writing each Cartesian component of equations (7.9) and (7.10) one can easily see that the time and z part can be separated out. Thus, if the refractive index is independent of the z coordinate, then the solutions of equations (7.9) and (7.10) can be written in the form

$$\mathcal{E} = \mathbf{E}(x, y)e^{i(\omega t - \beta z)} \tag{7.12}$$

$$\mathcal{H} = \mathbf{H}(x, y)e^{i(\omega t - \beta z)} \tag{7.13}$$

where β is known as the propagation constant. Equations (7.12) and (7.13) define the modes of the system. We note from equations (7.12) and (7.13) that modes represent special field distributions that suffer a phase change only as they propagate through the waveguide along z; the transverse field distributions described by $\mathbf{E}(x, y)$ and $\mathbf{H}(x, y)$ do not change as the field propagates through the waveguide. The quantity β represents the propagation constant of the mode.

We next assume that the refractive index depends only on the x coordinate – that is

$$n^2 = n^2(x) \tag{7.14}$$

Then even the y part can be separated out, implying that the y and z dependences of the fields will be of the form $e^{-i(\gamma y + \beta z)}$. However, we can always choose the z-axis along the direction of propagation of the wave and we may, without any loss of generality, put $\gamma = 0$. Thus we may write

$$\mathcal{E}_j = E_j(x)e^{i(\omega t - \beta z)}; \quad j = x, y, z \tag{7.15}$$

$$\mathcal{H}_j = H_j(x)e^{i(\omega t - \beta z)}; \quad j = x, y, z \tag{7.16}$$

Substituting the above expressions for the electric and magnetic fields in equations (7.1) and (7.2) and taking their x, y, and z components we obtain

$$i\beta E_y = -i\omega\mu_0 H_x \tag{7.17}$$

$$\partial E_y/\partial x = -i\omega\mu_0 H_z \tag{7.18}$$

$$-i\beta H_x - \partial H_z/\partial x = i\omega\epsilon_0 n^2(x) E_y \tag{7.19}$$

$$i\beta H_y = i\omega\epsilon_0 n^2(x) E_x \tag{7.20}$$

$$\partial H_y/\partial x = i\omega\epsilon_0 n^2(x) E_z \tag{7.21}$$

$$-i\beta E_x - \partial E_z/\partial x = -i\omega\mu_0 H_y \tag{7.22}$$

As can be seen, the first three equations involve only E_y, H_x, and H_z and the last three equations involve only E_x, E_z, and H_y. Thus, for such a waveguide configuration, Maxwell's equations reduce to two independent sets of equations. The first set corresponds to nonvanishing values of E_y, H_x, and H_z with E_x, E_z, and H_y vanishing, giving rise to what are known as TE modes because the electric field has only a transverse component. The second set corresponds to nonvanishing values of E_x, E_z, and H_y with E_y, H_x, and H_z vanishing, giving

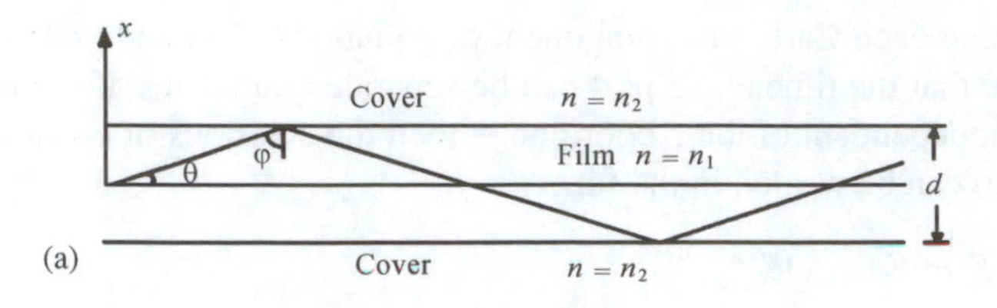

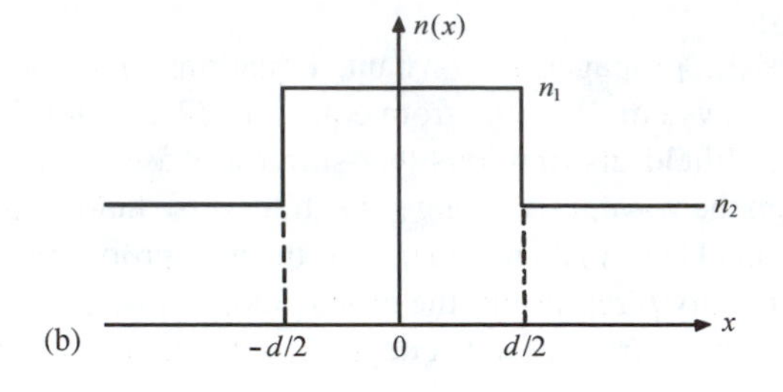

Fig. 7.1: (a) The simplest planar optical waveguide consists of a planar film (of refractive index n_1) sandwiched between two materials of lower refractive indices. Light guidance takes place by the phenomenon of total internal reflection. (b) The refractive index distribution for a symmetric planar waveguide.

rise to what are known as TM modes because the magnetic field now has only a transverse component. The propagation of waves in such planar waveguides may thus be described in terms of TE and TM modes. In the next two sections we discuss the TE and TM modes of a symmetric step index planar waveguide.

7.3 TE modes of a symmetric step index planar waveguide

In this and the following section we carry out a detailed modal analysis of a symmetric step index planar waveguide. We first consider TE modes: we substitute for H_x and H_z from equations (7.17) and (7.18) in equation (7.19) to obtain

$$d^2E_y/dx^2 + \left[k_0^2 n^2(x) - \beta^2\right] E_y = 0 \tag{7.23}$$

where

$$k_0 = \omega(\epsilon_0\mu_0)^{1/2} = \omega/c \tag{7.24}$$

is the free space wave number and $c\ (=1/(\epsilon_0\mu_0)^{1/2})$ is the speed of light in free space.

Until now our analysis has been valid for an arbitrary x-dependent profile. We now assume a specific profile given by (see Figure 7.1)

$$n(x) = \begin{cases} n_1; & |x| < d/2 \\ n_2; & |x| > d/2 \end{cases} \tag{7.25}$$

with $n_1 > n_2$. Using the above equations we will solve equation (7.23) subject to the appropriate boundary conditions at the discontinuities. Since E_y and H_z represent tangential components on the planes $x = \pm d/2$, they must be continuous at $x = \pm d/2$ and since H_z is proportional to dE_y/dx [see equation

(7.18)], we must have

$$E_y \text{ and } dE_y/dx \text{ continuous at } x = \pm d/2 \tag{7.26}$$

The above represents the boundary conditions that have to be satisfied.[1] Substituting for $n(x)$ in equation (7.23) we obtain

$$d^2E_y/dx^2 + \left(k_0^2 n_1^2 - \beta^2\right) E_y = 0; \quad |x| < d/2 \quad \text{film} \tag{7.27}$$

$$d^2E_y/dx^2 + \left(k_0^2 n_2^2 - \beta^2\right) E_y = 0; \quad |x| > d/2 \quad \text{cover} \tag{7.28}$$

Guided modes are those modes that are mainly confined to the film and hence their field should decay in the cover (i.e., in the region $|x| > d/2$) so that most of the energy associated with the mode lies inside the film. Thus, we must have

$$\beta^2 > k_0^2 n_2^2 \tag{7.29}$$

When $\beta^2 < k_0^2 n_2^2$, the solutions are oscillatory in the region $|x| > d/2$ and they correspond to what are known as radiation modes of the waveguide. These modes correspond to rays that undergo refraction (rather than total internal reflection) at the film–cover interface and when these are excited, they quickly leak away from the core of the waveguide. Some aspects of radiation modes are discussed in Chapter 24.

Furthermore, we must also have $\beta^2 < k_0^2 n_1^2$, otherwise the boundary conditions cannot be satisfied[2] at $x = \pm d/2$. Thus, for guided modes we must have

$$n_2^2 < \frac{\beta^2}{k_0^2} < n_1^2 \tag{7.30}$$

At this point it may be worthwhile to recall our discussion in Chapter 4 where we said that (for slab waveguides) guided rays correspond to

$$n_2 < \tilde{\beta} < n_1 \tag{7.31}$$

and refracting rays correspond to $\tilde{\beta} < n_2$; further, there cannot be any ray with $\tilde{\beta} > n_1$. Thus, $\tilde{\beta}$ (in ray optics) corresponds to β/k_0 in wave optics:

$$\tilde{\beta} \Leftrightarrow \frac{\beta}{k_0} \tag{7.32}$$

We write equations (7.27) and (7.28) in the form

$$d^2E_y/dx^2 + \kappa^2 E_y = 0; \quad |x| < d/2 \quad \text{film} \tag{7.33}$$

$$d^2E_y/dx^2 - \gamma^2 E_y = 0; \quad |x| > d/2 \quad \text{cover} \tag{7.34}$$

---

[1] The very fact that E_y satisfies equation (7.23) also implies that E_y and dE_y/dx are continuous unless $n^2(x)$ has an infinite discontinuity. This follows from the fact that if E_y' is discontinuous, then E_y'' will be a delta function and equation (7.23) will lead to an inconsistent equation. Thus, the continuity conditions are imbedded in Maxwell's equations. Here primes represent differentiation with respect to x.

[2] It is left as an exercise for the reader to show that if we assume $\beta^2 > k_0^2 n_1^2$ and also assume decaying fields in the region $|x| > d/2$, then the boundary conditions at $x = +d/2$ and at $x = -d/2$ can never be satisfied simultaneously.

where

$$\kappa^2 = k_0^2 n_1^2 - \beta^2 \tag{7.35}$$

and

$$\gamma^2 = \beta^2 - k_0^2 n_2^2 \tag{7.36}$$

The solution of equation (7.33) can be written in the form

$$E_y(x) = A \cos \kappa x + B \sin \kappa x; \quad |x| < d/2 \tag{7.37}$$

where A and B are constants. In the regions $x > d/2$ and $x < -d/2$, the solutions are $e^{\pm\gamma x}$ and, if we neglect the exponentially amplifying solution, we obtain

$$E_y(x) = \begin{cases} Ce^{\gamma x}; & x < -d/2 \\ De^{-\gamma x}; & x > d/2 \end{cases} \tag{7.38}$$

If we now apply the boundary conditions (namely, continuity of E_y and dE_y/dx at $x = \pm d/2$), we get four equations from which we can get the transcendental equation, which will determine the allowed values of β. This is indeed the general procedure for determining the propagation constants in asymmetric waveguides [see, e.g., Ghatak and Thyagarajan (1989), Section 14.2]; however, when the refractive index distribution is symmetric about $x = 0$ – that is, when

$$n^2(-x) = n^2(x) \tag{7.39}$$

the solutions are either symmetric or antisymmetric functions of x; thus, we must have

$$E_y(-x) = E_y(x) \quad \text{symmetric modes} \tag{7.40}$$

$$E_y(-x) = -E_y(x) \quad \text{antisymmetric modes} \tag{7.41}$$

(The proof of this theorem is discussed in Problem 7.5.) For the symmetric mode, we must have

$$E_y(x) = \begin{cases} A \cos \kappa x; & |x| < -d/2 \\ Ce^{-\gamma|x|}; & |x| > d/2 \end{cases} \tag{7.42}$$

Continuity of $E_y(x)$ and dE_y/dx at $x = \pm d/2$ gives us

$$A \cos(\kappa d/2) = Ce^{-\gamma d/2} \tag{7.43}$$

and

$$-\kappa A \sin(\kappa d/2) = -\gamma Ce^{-\gamma d/2} \tag{7.44}$$

respectively. Dividing equation (7.44) by equation (7.43) we get

$$\xi \tan \xi = \frac{\gamma d}{2} \tag{7.45}$$

where

$$\xi = \frac{\kappa d}{2} = \left(k_0^2 n_1^2 - \beta^2\right)^{1/2} \frac{d}{2} \tag{7.46}$$

Now

$$\frac{\gamma d}{2} = \left(\frac{1}{4}V^2 - \xi^2\right)^{1/2} \tag{7.47}$$

where

$$V = k_0 d \left(n_1^2 - n_2^2\right)^{1/2} \tag{7.48}$$

is known as the dimensionless waveguide parameter. Thus, equation (7.45) can be put in the form

$$\xi \tan \xi = \left(\frac{1}{4}V^2 - \xi^2\right)^{1/2} \tag{7.49}$$

Similarly, for the antisymmetric mode we have

$$E_y(x) = \begin{cases} B \sin \kappa x; & |x| < -d/2 \\ \frac{x}{|x|} D e^{-\gamma |x|}; & |x| > d/2 \end{cases} \tag{7.50}$$

and following an exactly similar procedure we get

$$-\xi \cot \xi = \left(\frac{1}{4}V^2 - \xi^2\right)^{1/2}$$

Thus, we have

$$\xi \tan \xi = \left[\left(\frac{V}{2}\right)^2 - \xi^2\right]^{1/2} \quad \text{for symmetric modes} \tag{7.51}$$

and

$$-\xi \cot \xi = \left[\left(\frac{V}{2}\right)^2 - \xi^2\right]^{1/2} \quad \text{for antisymmetric modes} \tag{7.52}$$

Since the equation

$$\eta = \sqrt{\left(\frac{V}{2}\right)^2 - \xi^2}$$

(for positive values of ξ) represents a circle (of radius $V/2$) in the first quadrant of the $\xi - \eta$ plane, the numerical evaluation of the allowed values of ξ (and

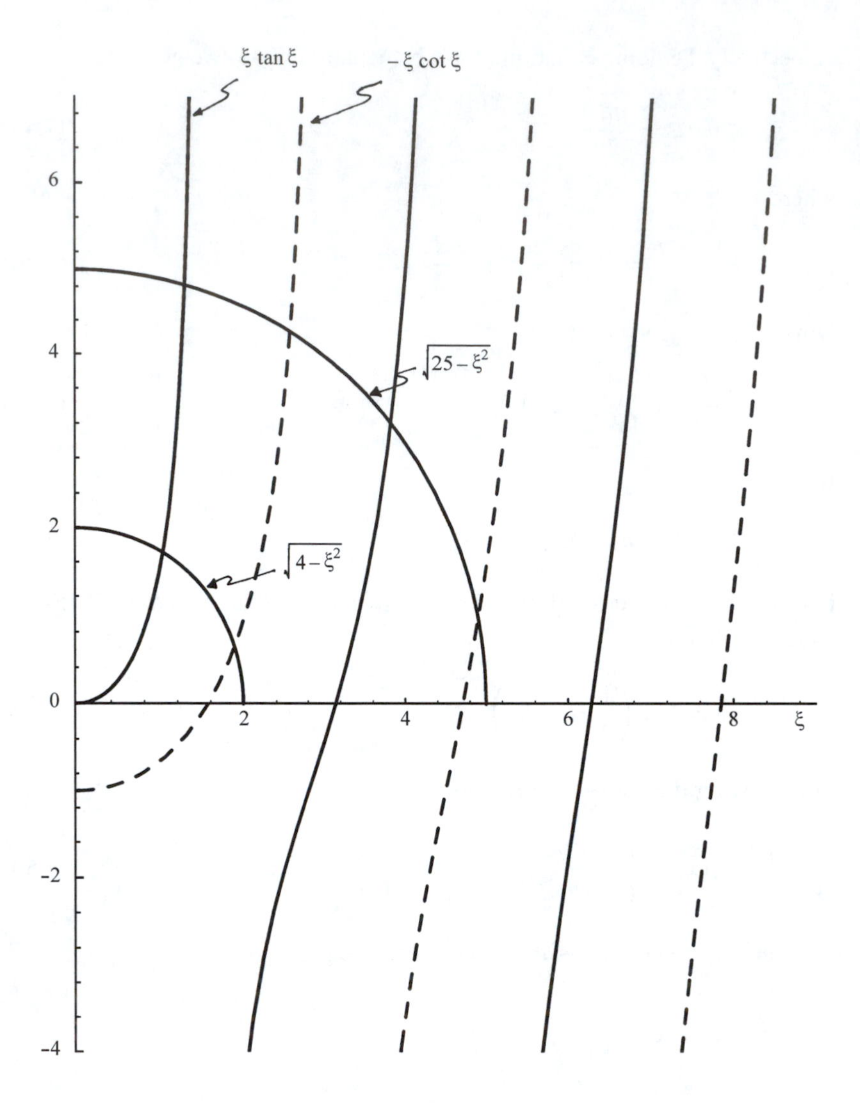

Fig. 7.2: The variation of $\xi \tan \xi$ (solid curves) and $-\xi \cot \xi$ (dashed curves) as a function of ξ. The points of intersection of the solid and dashed curves with the quadrant of a circle of radius $V/2$ determine the propagation constants of the optical waveguide corresponding to symmetric and antisymmetric modes, respectively.

hence of the propagation constants) is quite simple. In Figure 7.2 we have plotted the functions $\xi \tan \xi$ (solid curve) and $-\xi \cot \xi$ (dashed curve) as a function of ξ. Their points of intersection with the quadrant of the circle determine the allowed values of ξ and if we use equation (7.46) we can determine the corresponding values of β. Note that for guided modes the propagation constant β can assume only a discrete set of values (as determined by the transcendental equation).

The two circles in Figure 7.2 correspond to $V/2 = 2$ and $V/2 = 5$. Obviously, as can be seen from the figure, for $V = 4$ we will have one symmetric and one antisymmetric mode and for $V = 10$ we will have two symmetric and two antisymmetric modes.

It is often very convenient to define the dimensionless propagation constant

$$b \equiv \frac{\beta^2/k_0^2 - n_2^2}{n_1^2 - n_2^2} = 1 - \frac{\xi^2}{V^2/4} \tag{7.53}$$

Obviously, because of equation (7.30), we will have for guided modes

$$0 < b < 1 \tag{7.54}$$

Example 7.1: As an example, we consider a step index symmetric waveguide with

$$n_1 = 1.503 \quad n_2 = 1.500 \quad d = 4\ \mu\text{m} \tag{7.55}$$

For

$$\lambda_0 = 1\ \mu\text{m}$$

we will have

$$V = 2.385$$

and the waveguide will support only one symmetric TE mode with

$$\xi \simeq 0.81664 \Rightarrow b \simeq 0.531223, \quad \frac{\beta}{k_0} \simeq 1.50159 \tag{7.56}$$

For the same waveguide, if the operating wavelength were 0.5 μm, we would have

$$V = 4.771$$

and the waveguide would support one symmetric TE mode and one antisymmetric TE mode with

$$\xi \simeq 1.09426 \text{ and } 2.08132$$

implying

$$b \simeq 0.789584 \text{ and } 0.238762$$

The corresponding values of β/k_0 are

$$\frac{\beta}{k_0} \simeq 1.502369 \text{ and } 1.500717 \tag{7.57}$$

respectively. As the V value becomes larger, the waveguide supports more modes.

7.3.1 Physical understanding of modes

To have a physical understanding of modes, we consider the electric field pattern inside the film ($-d/2 < x < d/2$). For example, for a symmetric TE mode, this is given by (see equation (7.42))

$$E_y(x) = A \cos \kappa x$$

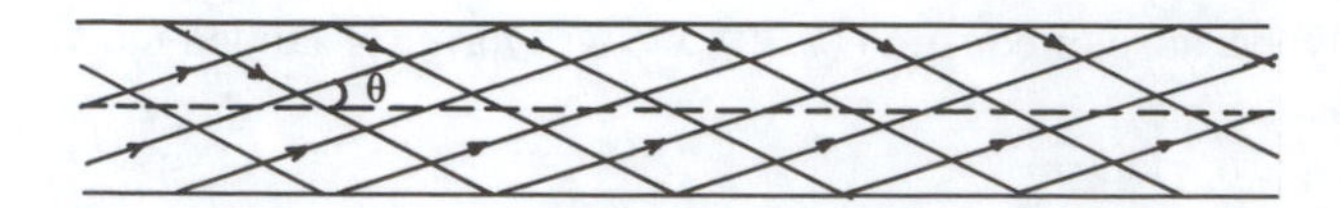

Fig. 7.3: A guided mode in a step index waveguide corresponds to the superposition of two plane waves (inside the film) propagating at discrete angles $\pm\theta$ $[=\cos^{-1}(\beta/k_0 n_1)]$ with the z-axis. Different modes will correspond to different (discrete) values of θ.

Thus, the complete field inside the film is given by

$$\mathcal{E}_y = A\cos\kappa x e^{i(\omega t-\beta z)}$$

$$= \frac{1}{2}Ae^{i(\omega t-\beta z-\kappa x)} + \frac{1}{2}Ae^{i(\omega t-\beta z+\kappa x)} \tag{7.58}$$

Now

$$e^{i(\omega t-k_x x-k_y y-k_z z)}$$

represents a wave propagating along the direction of **k** whose x, y, and z components are k_x, k_y, and k_z, respectively. Thus, for the two terms on the right-hand side (RHS) of equation (7.58) we have

$$k_x = \pm\kappa, \quad k_y = 0, \quad \text{and} \quad k_z = \beta \tag{7.59}$$

which represent plane waves with propagation vectors parallel to the x–z plane making angles $\pm\theta$ with the z-axis where

$$\tan\theta = k_x/k_z = \kappa/\beta$$

or

$$\cos\theta = \beta/(\beta^2+\kappa^2)^{1/2} = \beta/k_0 n_1 \tag{7.60}$$

Thus, a guided mode can be considered to be a superposition of a pair of plane waves that are propagating at angles $\pm\theta(=\pm\cos^{-1}(\beta/k_0 n_1))$ with the z-axis (see Figure 7.3). Since only discrete values of β are allowed (which we designate as β_m), only discrete angles of propagation of waves (or of the rays) are allowed. Each mode is characterized by a discrete angle of propagation θ_m; this is the basic principle of the prism film-coupling technique for determining the (discrete) propagation constants of an optical waveguide (see Chapter 19).

Referring to the waveguide discussed in Example 7.1, at $\lambda_0 = 1\ \mu$m we will have

$$\theta \simeq 2.48°$$

for the mode whose propagation constant is given by equation (7.56). For the same waveguide, at $\lambda_0 = 0.5\ \mu$m we will have

$$\theta \simeq 1.66° \quad \text{and} \quad 3.16°$$

corresponding to the symmetric and the antisymmetric modes, respectively (see equation (7.57)).

The concept of the cutoff of a mode can also be easily understood from the above discussion. Since guided waves correspond to

$$n_2 < \frac{\beta}{k_0} < n_1$$

we have

$$\frac{n_2}{n_1} < \cos\theta < 1 \tag{7.61}$$

The condition that β cannot be less than $k_0 n_2$ implies that $\cos\theta$ should be greater than n_2/n_1, which is nothing but the condition for total internal reflection at the core–cladding interface (see equation (4.35)). Thus, beyond cutoff – that is, for $V < V_c$ – the component waves no longer undergo total internal reflections at the boundaries.

From Figure 7.2 we can derive the following conclusions about TE modes (similar discussion can be made for TM modes, which are discussed in the next section):

(a) If $0 < V/2 < \pi/2$ – that is, when

$$0 < V < \pi \tag{7.62}$$

we have only one discrete (TE) mode of the waveguide and this mode is symmetric in x. When this happens, we refer to the waveguide as a *single-moded waveguide*. For example, for

$$n_1 = 1.50, \quad n_2 = 1.48, \quad \text{and} \quad d = 3\ \mu\text{m} \tag{7.63}$$

the waveguide will support only one TE mode for $\lambda_0 > 1.46\,\mu$m; actually the waveguide will support one TE and one TM mode – see discussion at the end of Section 7.4. For $\lambda_0 < 1.46\,\mu$m the same waveguide will support more than one mode.

(b) From Figure 7.2 it is easy to see that if $\pi/2 < V/2 < \pi$ (or, $\pi < V < 2\pi$) we will have one symmetric and one antisymmetric mode. In general, if

$$2m\pi < V < (2m+1)\pi \tag{7.64}$$

we will have $(m+1)$ symmetric modes and m antisymmetric modes, and if

$$(2m+1)\pi < V < (2m+2)\pi \tag{7.65}$$

we will have $(m+1)$ symmetric modes and $(m+1)$ antisymmetric modes where $m = 0, 1, 2, \ldots$. Thus, the total number of modes will be the integer closest to (and greater than) V/π. Thus, for the waveguide considered above, if the operating wavelength is made 0.6 μm, then $V = 2.44\pi$ and therefore we will have three modes (two symmetric and one antisymmetric).

(c) When the waveguide supports many modes (i.e., when $V \gg 1$), the points of intersection (in Figure 7.2) will be very close to $\xi = \pi/2, \pi$, $3\pi/2$, and so forth; thus, the propagation constants corresponding to the first few modes will be approximately given by the following equation:

$$\xi = \xi_m = \left(k_0^2 n_1^2 - \beta_m^2\right)^{1/2} d/2 \approx (m+1)\pi/2; \quad V \gg 1 \tag{7.66}$$

where

$m = 0, 2, 4, \ldots$ correspond to symmetric modes

and

$m = 1, 3, 5, \ldots$ correspond to antisymmetric modes

(d) It is obvious from Figure 7.2 that for the fundamental mode (which we will refer to as the zero-order mode), $\xi(=\kappa d/2)$ will always lie between 0 and $\pi/2$ and the corresponding field variation $E_y(x)$ will have no zeroes. For the next mode (which will be antisymmetric in x) $\xi(=\kappa d/2)$ will always lie between $\pi/2$ and π and therefore the corresponding $E_y(x)$ will have only one zero (at $x = 0$). It is easy to extend the analysis and prove that

the mth mode will have m zeroes in the corresponding field distribution (7.67)

The above statement is valid for an arbitrary waveguiding structure. The actual plot of the modal pattern for the first few modes is shown in Figure 7.4. It may be noted that the field spreads out more as the wavelength increases or as V number decreases.

(e) The variations of the normalized propagation constant b (defined in equation (7.53)) with V are plotted in Figure 7.5 for the first few modes. The solid curves (corresponding to TE modes) are universal – that is, for a given waveguide and a given operating wavelength we first have to determine V and then "read off" from the curves the exact values of b from which we can determine the values of β by using the following equation.

$$\beta^2 = k_0^2 \left[n_2^2 + b\left(n_1^2 - n_2^2\right)\right] \tag{7.68}$$

Since for guided modes β cannot be less than $k_0 n_2$, when (for a particular mode) β reaches the value equal to $k_0 n_2$ (i.e., when b becomes equal to zero), the mode is said to have reached "cutoff." Thus, at cutoff

$$\beta = k_0 n_2, \quad \gamma = 0, \quad \text{and} \quad b = 0 \tag{7.69}$$

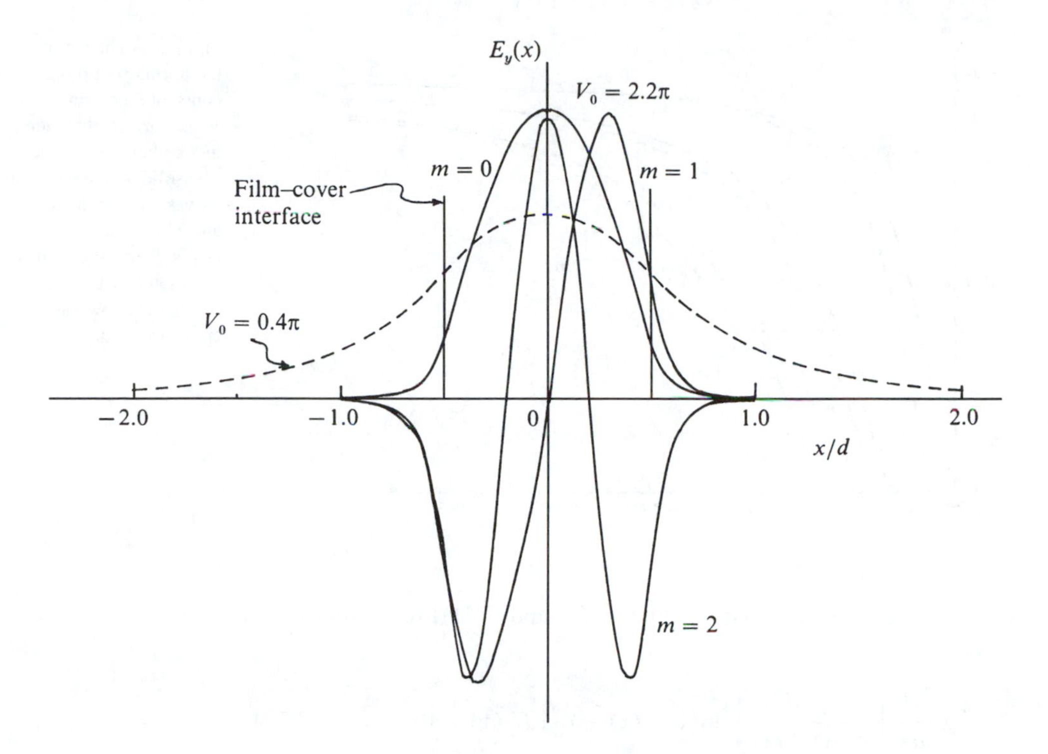

Fig. 7.4: The solid curves represent the modal fields for the symmetric step index planar waveguide with $V = 4.4\pi$; even and odd values of m correspond to symmetric and antisymmetric modes, respectively. The dashed curve represents the fundamental mode for $V = 0.8\pi$. All modes have been normalized to carry the same power.

For the symmetric waveguide that we have been discussing, at cutoff $\xi = V/2$ and hence the cutoff of TE modes is determined by

$$\frac{V}{2}\tan\left(\frac{V}{2}\right) = 0 \quad \text{symmetric modes} \tag{7.70}$$

$$\frac{V}{2}\cot\left(\frac{V}{2}\right) = 0 \quad \text{antisymmetric modes} \tag{7.71}$$

The above equation implies that the cutoff V values for various modes are given by

$$V_c = m\pi\,; \quad m = 0, 1, 2, 3, \ldots \tag{7.72}$$

where even and odd values of m correspond to symmetric and antisymmetric modes, respectively. Note that the fundamental mode has no cutoff and therefore there will always be at least one guided mode.

7.4 TM modes of a symmetric step index planar waveguide

In the above discussion we considered the TE modes of the waveguide. An exactly similar analysis can also be performed for the TM modes, which are characterized by field components E_x, E_z, and H_y (see equations (7.20)–(7.22)).

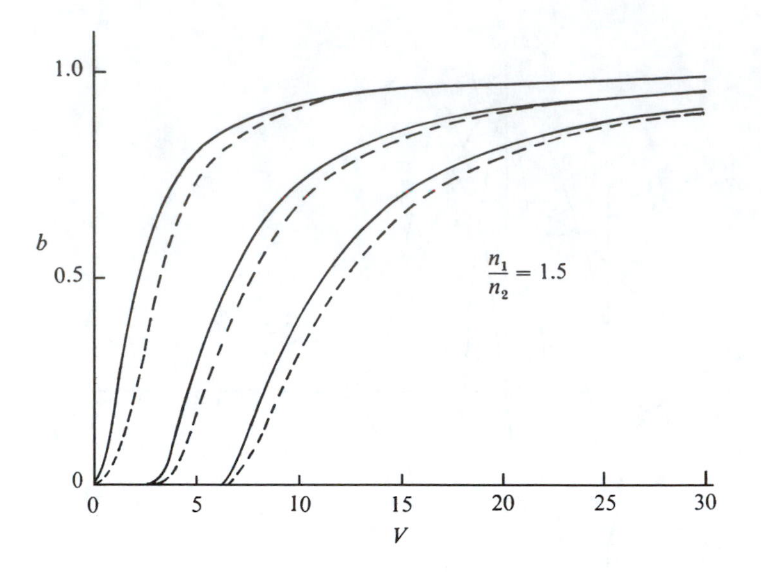

Fig. 7.5: The variation of the normalized propagation constant b as a function of V for a symmetric step index planar waveguide. The solid and the dashed curves correspond to TE and TM modes, respectively. The value of V at which $b = 0$ corresponds to what is known as the cutoff frequency.

If we substitute for E_x and E_z from equations (7.20) and (7.21) in equation (7.22) we will get

$$n^2(x)\frac{d}{dx}\left[\frac{1}{n^2(x)}\frac{dH_y}{dx}\right] + \left(k_0^2 n^2(x) - \beta^2\right) H_y(x) = 0 \tag{7.73}$$

which can be rewritten as

$$\frac{d^2 H_y}{dx^2} - \left[\frac{1}{n^2(x)}\frac{dn^2}{dx}\right]\frac{dH_y}{dx} + \left(k_0^2 n^2(x) - \beta^2\right) H_y(x) = 0 \tag{7.74}$$

The above equation is of a form that is somewhat different from the equation satisfied by E_y for TE modes (see equation (7.23)); however, for the step index waveguide shown in Figure 7.1, the refractive index is constant in each region and therefore we have

$$d^2 H_y/dx^2 + \left(k_0^2 n_1^2 - \beta^2\right) H_y(x) = 0; \quad |x| < d/2 \tag{7.75}$$

and

$$d^2 H_y/dx^2 - \left(\beta^2 - k_0^2 n_2^2\right) H_y(x) = 0; \quad |x| > d/2 \tag{7.76}$$

We must be careful about the boundary conditions. Since H_y and E_z are tangential components on the planes $x = \pm d/2$, we must have (see equation (7.21))

$$H_y \text{ and } \frac{1}{n^2}\frac{dH_y}{dx} \text{ continuous at } x = \pm\frac{d}{2} \tag{7.77}$$

This is also obvious from equation (7.73).[3] The solutions of equations (7.75) and (7.76) can be written immediately. Considering first the symmetric modes,

[3]Once again the condition that H_y and $(1/n^2)dH_y/dx$ should be continuous at $x = \pm d/2$ follows from equation (7.73) because if $(1/n^2)dH_y/dx$ were discontinuous $d/dx[(1/n^2)H_y']$ would be a delta function and equation (7.73) would lead to an inconsistent equation. Thus, the continuity of H_y and $(1/n^2)H_y'$ are contained in equation (7.73).

we have

$$H_y(x) = \begin{cases} A\cos\kappa x; & |x| < d/2 \\ Be^{-\gamma|x|}; & |x| > d/2 \end{cases} \tag{7.78}$$

where the symbols κ and γ are the same as given by equations (7.35) and (7.36). The boundary conditions given by equation (7.77) give us

$$A\cos(\kappa d/2) = Be^{-\gamma d/2}$$

$$\frac{1}{n_1^2}\left(-A\kappa\sin\frac{\kappa d}{2}\right) = \frac{1}{n_2^2}(-B\gamma e^{-\gamma d/2})$$

Dividing we get

$$\kappa\tan(\kappa d/2) = \left(n_1^2/n_2^2\right)\gamma$$

which can be rewritten in the form

$$\xi\tan\xi = \left(\frac{n_1}{n_2}\right)^2\left[\left(\frac{V}{2}\right)^2 - \xi^2\right]^{1/2} \quad \text{symmetric TM modes} \tag{7.79}$$

A similar derivation gives us

$$-\xi\cot\xi = \left(\frac{n_1}{n_2}\right)^2\left[\left(\frac{V}{2}\right)^2 - \xi^2\right]^{1/2} \quad \text{antisymmetric TM modes} \tag{7.80}$$

where ξ and V have been defined earlier [see equations (7.46) and (7.48)].

The numerical solutions of equations (7.79) and (7.80) can be discussed in a manner exactly similar to the TE case with the difference that the RHS of equations (7.79) and (7.80) now represents an ellipse whose semimajor axis (along the η direction) is of magnitude $(n_1^2/n_2^2)(V/2)$ and whose semiminor axis (along the ξ direction) is of magnitude $V/2$. All qualitative conclusions discussed in Section 7.3 for TE modes (namely, the cutoff frequencies and number of zeroes of various modes, the physical interpretation of modal fields etc.) will remain valid. We should also mention the following three points:

(a) Since $n_1 > n_2$, the point of intersection of the ellipse [of semimajor axis $(n_1^2/n_2^2)V/2$ and semiminor axis $(V/2)$] with the curves $\xi\tan\xi$ and $-\xi\cot\xi$ will have values of ξ greater than those corresponding to the TE mode. Thus from equation (7.46) we note that for a given V, the β values of TM_m modes are smaller than those of the corresponding TE_m modes.

(b) Although a waveguide for which $0 < V < \pi$ is referred to as a single moded waveguide, we actually have two modes (one TE and one TM) characterized by slightly different propagation constants. However, the incident field is usually linearly polarized and if **E** is along the

y-axis, the TE mode is excited and if $\mathbf{E}$ is along the x-axis, the TM mode is excited. *This result is quite general and is valid for all planar waveguides*. On the other hand, if the incident field has a polarization that makes an angle with the x-axis (or, if the field is elliptically polarized), then both TE and TM modes will be excited and, because they have slightly different propagation constants, they will superpose with different phases at different values of z, changing the state of the resultant polarization. As an example, we consider the incidence of a linearly polarized wave with the electric vector making an angle of 45° with the x- and y-axes. Thus, at $z = 0$ we have

$$\left.\begin{aligned} \mathcal{E}_x &= E_0 \cos \pi/4 \cos \omega t \\ \mathcal{E}_y &= E_0 \cos \pi/4 \cos \omega t \end{aligned}\right\} \quad \begin{aligned}&\text{field distribution}\\ &\text{at } z = 0\end{aligned} \tag{7.81}$$

where E_0 represents the transverse variation of the modal field that we have assumed to be the same for the TE and TM modes (see equations (7.42) and (7.78)) the values of β are assumed to be nearly equal. If the propagation constants for the TE and TM modes are denoted by β_0 and $(\beta_0 - \Delta\beta_0)$, respectively, then the field distributions for $z > 0$ will be

$$\left.\begin{aligned} \text{TE: } \mathcal{E}_y &= \tfrac{E_0}{\sqrt{2}} \cos[\omega t - \beta_0 z] \\ \text{TM: } \mathcal{E}_x &= \tfrac{E_0}{\sqrt{2}} \cos[\omega t - \beta_0 z + \Delta\beta_0 z] \end{aligned}\right\} \quad \begin{aligned}&\text{field distribution}\\ &\text{at } z > 0\end{aligned} \tag{7.82}$$

It can be readily seen that at $z = \pi/(2\Delta\beta_0)$ the beam will be circularly polarized and at $z = \pi/(\Delta\beta_0)$ the beam will be linearly polarized (with the electric vector now at right angles to the original direction) – for intermediate values of z, the beam will be elliptically polarized. At $z = L_b = 2\pi/\Delta\beta_0$, the original polarization state is restored and this characteristic length is referred to as the beat length.

(c) Similarly, for $\pi < V < 2\pi$, although the waveguide is referred to as a "two-moded waveguide," there are actually four modes (two TE and two TM), and so forth.

(d) For most practical waveguides, $n_1 \approx n_2$ and the propagation constants (and the field patterns) for the TE and TM modes are very nearly equal.

Example 7.2: We consider a planar waveguide with $n_1 = 1.5$, $n_2 = 1.0$, and $d = 0.555\ \mu$m. At $\lambda_0 = 1.3\ \mu$m, $V \simeq 3$ and one obtains

$$b(\text{TE}) \simeq 0.6280 \quad \text{and} \quad b(\text{TM}) \simeq 0.4491$$

The corresponding values of β/k_0 are given by

$$\left(\frac{\beta}{k_0}\right)_{\text{TE}} \simeq 1.336 \quad \text{and} \quad \left(\frac{\beta}{k_0}\right)_{\text{TM}} \simeq 1.2495$$

with $L_b \approx 15\ \mu$m.

7.5 The relative magnitude of the longitudinal components of the *E* and *H* fields

We first consider the TE modes. Using equations (7.17) and (7.18) we get

$$\left|\frac{H_z}{H_x}\right| = \frac{1}{\beta}\left|\frac{\partial E_y/\partial x}{E_y}\right|$$

Now from equation (7.42) we have (inside the film)

$$\left|\frac{E_y}{\partial E_y/\partial x}\right| \sim \frac{1}{\kappa}$$

which represents the characteristic distance for the spatial variation in the x direction. Thus

$$|H_z/H_x| \sim \kappa/\beta \tag{7.83}$$

Since

$$\kappa^2 = k_0^2 n_1^2 - \beta^2$$

we readily have

$$\kappa/\beta = \left(k_0^2 n_1^2/\beta^2 - 1\right)^{1/2}$$

Now, guided modes correspond to

$$n_2^2 < \beta^2/k_0^2 < n_1^2$$

therefore

$$0 < \frac{\kappa}{\beta} < \left(\frac{n_1^2 - n_2^2}{n_2^2}\right)^{1/2}$$

Thus

$$\left|\frac{H_z}{H_x}\right| \le \left(\frac{n_1^2 - n_2^2}{n_2^2}\right)^{1/2} \tag{7.84}$$

For $n_1 \approx 1.50$ and $n_2 \approx 1.49$, the RHS of the above equation is about 0.1, which shows that the longitudinal component is very weak in comparison to the transverse component. Thus, as long as $n_1 \approx n_2$, the mode can be approximately assumed to be a transverse electromagnetic mode. The same is also true for the TM modes.

The above discussion is valid, in general, as long as $n_1 \approx n_2$. In Chapters 8 and 9 we discuss round optical waveguides with cylindrical symmetry – even there, as long as the core and cladding refractive indices are nearly equal (which is true for all practical fibers), the longitudinal components of the electric and magnetic fields are usually negligible in comparison to the corresponding transverse components. When this is the case, the waveguide is referred to as *weakly guiding*.

7.6 Power associated with a mode

In this section we calculate the power associated with the TE mode. The power flow is given by

$$\langle \mathcal{S} \rangle = \langle \mathcal{E} \times \mathcal{H} \rangle \tag{7.85}$$

where $\mathcal{S}(= \mathcal{E} \times \mathcal{H})$ represents the Poynting vector and $\langle \ldots \rangle$ represents the time average of the quantity inside the angular brackets.

$$\langle f(t) \rangle = \frac{1}{T} \int_0^T f(t)\, dt \tag{7.86}$$

where $T(= 2\pi/\omega)$ represents the time period. Obviously, while calculating the Poynting vector we must take the real parts of $\mathcal{E}$ and $\mathcal{H}$. Therefore, for a TE mode, we write the electric field as

$$\mathcal{E}_y = E_y(x) \cos(\omega t - \beta z) \tag{7.87}$$

which is essentially the real part of equation (7.15). Now

$$-\mu_0 \frac{\partial \mathcal{H}}{\partial t} = \nabla \times \mathcal{E} = \begin{vmatrix} \hat{\mathbf{x}} & \hat{\mathbf{y}} & \hat{\mathbf{z}} \\ \frac{\partial}{\partial x} & \frac{\partial}{\partial y} & \frac{\partial}{\partial z} \\ 0 & \mathcal{E}_y & 0 \end{vmatrix} \tag{7.88}$$

Thus

$$-\mu_0 \frac{\partial \mathcal{H}_x}{\partial t} = -\frac{\partial \mathcal{E}_y}{\partial z} = -\beta E_y(x) \sin(\omega t - \beta z)$$

or

$$\mathcal{H}_x = -\frac{\beta}{\omega \mu_0} E_y(x) \cos(\omega t - \beta z) \tag{7.89}$$

Similarly

$$\mathcal{H}_z = -\frac{1}{\omega \mu_0} \frac{dE_y}{dx} \sin(\omega t - \beta z) \tag{7.90}$$

Now

$$\langle \mathcal{S}_x \rangle = \langle \mathcal{E}_y \mathcal{H}_z \rangle = 0 \tag{7.91}$$

and

$$\begin{aligned} \langle \mathcal{S}_z \rangle &= -\langle \mathcal{E}_y \mathcal{H}_x \rangle \\ &= \frac{\beta}{2\omega \mu_0} E_y^2(x) \end{aligned} \tag{7.92}$$

Although the above expression is rigorously valid only for the TE mode in a slab waveguide, it is approximately valid for all waveguides in the weakly guiding approximation.

The power associated with the mode (per unit length in the y direction) is given by

$$P = \frac{1}{2}\frac{\beta}{\omega\mu_0}\int_{-\infty}^{+\infty} E_y^2 dx \tag{7.93}$$

We consider the symmetric mode (see equation (7.42)) for which

$$P = \frac{1}{2}\frac{\beta}{\omega\mu_0}2\left(A^2\int_0^{d/2}\cos^2\kappa x dx + C^2\int_{d/2}^{\infty} e^{-2\gamma x}dx\right) \tag{7.94}$$

or

$$P = \frac{\beta}{2\omega\mu_0}A^2\left(\frac{d}{2} + \frac{1}{2\kappa}\sin\kappa d + \frac{C^2}{A^2}\frac{1}{\gamma}e^{-\gamma d}\right)$$

If we now use equation (7.43) for C/A we get

$$\begin{aligned} P &= \frac{\beta A^2}{4\omega\mu_0}\left\{d + \frac{2\sin(\kappa d/2)\cos(\kappa d/2)}{\kappa} + \frac{2}{\gamma}[1 - \sin^2(\kappa d/2)]\right\} \\ &= \frac{\beta A^2}{4\omega\mu_0}\left\{d + \frac{2}{\gamma} + \frac{2\sin(\kappa d/2)\cos(\kappa d/2)}{\gamma\kappa}[\gamma - \kappa\tan(\kappa d/2)]\right\} \end{aligned}$$

$$P = \frac{\beta A^2}{4\omega\mu_0}\left(d + \frac{2}{\gamma}\right) \tag{7.95}$$

where we have used equation (7.45). It may be mentioned that even for the antisymmetric TE mode, the power associated (per unit length in the y direction) is given by equation (7.95). Further, if we carry out a similar analysis for TM modes we obtain the following expression for the power flow

$$P = \frac{A^2\beta}{2\omega\epsilon_0 n_1^2}\left[\frac{d}{2} + \frac{(n_1 n_2)^2}{\gamma}\frac{k_0^2(n_1^2 - n_2^2)}{(n_2^4\kappa^2 + n_1^4\gamma^2)}\right] \tag{7.96}$$

for symmetric as well as antisymmetric modes.

7.7 Radiation modes

Until now we have considered the guided modes of the waveguide for which

$$n_2^2 < \beta^2/k_0^2 < n_1^2$$

There exists another class of modes for which[4]

$$\beta^2/k_0^2 < n_2^2 \tag{7.97}$$

[4]It is impossible to have $\beta/k_0 > n_1$ (see Problem 7.3).

These are referred to as the *radiation modes* of the waveguide. It can be immediately seen that for $(\beta^2/k_0^2) < n_2^2$ the wave equation (say for the TE modes) in the region $|x| > d/2$ takes the form

$$d^2 E_y/dx^2 + \delta^2 E_y = 0 \tag{7.98}$$

where

$$\delta^2 = k_0^2 n_2^2 - \beta^2 \tag{7.99}$$

which is now a positive quantity; thus the solutions in the region $|x| > d/2$ will be wavelike of the form

$$e^{\pm i\delta x} \tag{7.100}$$

We may recall that in the region $|x| > d/2$, the field associated with guided modes decayed exponentially in the x direction (see equation (7.42)). On the other hand, equation (7.100) tells us that the radiation modes correspond to oscillatory solutions in the cover.

7.8 Excitation of guided modes

We start with the wave equation satisfied by TE modes (see equation (7.23))

$$d^2\psi_m/dx^2 + \left[k_0^2 n^2(x) - \beta_m^2\right]\psi_m(x) = 0; \quad m = 0, 1, 2, \ldots \tag{7.101}$$

where $\psi_m(x)$ represents the field pattern corresponding to the propagation constant β_m; we have used the symbol $\psi_m(x)$ instead of the more complicated symbol $E_y^{(m)}(x)$. Using the condition that for guided modes $\psi_m(x)$ will go to zero as $x \to \pm\infty$, we can readily show that (see Problem 7.2)

$$\int_{-\infty}^{+\infty} \psi_m^*(x)\psi_k(x)\,dx = 0 \quad \text{for } m \neq k \tag{7.102}$$

which is known as the *orthogonality condition*. Since equation (7.101) is a linear equation, a constant multiple of $\psi_m(x)$ is also a solution and we can always choose the constant so that

$$\int_{-\infty}^{+\infty} |\psi_m(x)|^2\,dx = 1 \tag{7.103}$$

which is known as the *normalization condition*. Combining equations (7.102) and (7.103) we get the *orthonormality condition*

$$\int_{-\infty}^{+\infty} \psi_m^*(x)\psi_k(x)\,dx = \delta_{mk} \tag{7.104}$$

where δ_{mk} is the Kronecker delta function defined by the following equation.

$$\delta_{mk} = \begin{cases} 0 & \text{for } m \neq k \\ 1 & \text{for } m = k \end{cases} \tag{7.105}$$

Equation (7.104) represents the orthonormality condition satisfied by the discrete (guided) modes. The radiation modes (which form a continuum) also form an orthogonal set in the sense that

$$\int_{-\infty}^{+\infty} \psi^*_{\beta'}(x)\psi_\beta(x)\,dx = 0, \quad \text{for } \beta \neq \beta' \tag{7.106}$$

However, the integral for $\beta = \beta'$ is not defined and the orthonormality condition is in terms of the Dirac delta function.

The important point is that the finite number of guided modes along with the continuum of radiation modes form a complete set of functions in the sense that any "well-behaved" function of x can be expanded in terms of these functions – that is

$$\phi(x) = \sum_m c_m \psi_m(x) + \int c(\beta)\psi_\beta(x)\,d\beta \tag{7.107}$$

where the first term on the RHS represents a sum over discrete (guided) modes and the second term represents an integral over the continuum (radiation) modes. If we multiply equation (7.107) by $\psi^*_k(x)$ and integrate we will readily obtain

$$c_k = \int_{-\infty}^{+\infty} \psi^*_k(x)\phi(x)\,dx \tag{7.108}$$

where we have used equation (7.104). Now, let $E_y(x, z = 0)$ represent the actual incident field (polarized in the y-direction) at the entrance aperture of the waveguide ($z = 0$). The power launched in the mth mode will be (see equation (7.93))

$$\begin{aligned} P_m &= (1/2\omega\mu_0)\beta_m|c_m|^2 \int_{-\infty}^{+\infty} |\psi_m(x)|^2\,dx \\ &= (1/2\omega\mu_0)\beta_m\left|\int \psi^*_m(x)E_y(x, z = 0)\,dx\right|^2 \end{aligned} \tag{7.109}$$

As the beam propagates through the waveguide, the field in the region ($z > 0$) will be given by

$$E_y(x, z) = \sum_m c_m \psi_m(x)e^{-i\beta_m z} + \int c(\beta)\psi_\beta(x)e^{-i\beta z}\,d\beta \tag{7.110}$$

One can easily see that at an arbitrary value of z, the power in the mth mode will be proportional to

$$|c_m e^{-i\beta_m z}|^2 = |c_m|^2 \tag{7.111}$$

which remains constant with z. Different modes superpose with different phases at different values of z as a consequence of which – considering guided modes only – the transverse intensity distribution will vary with z (see Example 7.3 and Figure 21.2).

7.9 The parabolic index waveguide

In this section we discuss the TE modes of a parabolic index profile characterized by the following refractive index distribution (see also Section 4.2.2):[5]

$$n^2(x) = n_1^2\left[1 - 2\Delta\left(\frac{x}{a}\right)^2\right] \tag{7.112}$$

where n_1, Δ, and a are constants. For such an infinitely extended profile, we have only discrete guided modes and we will be able to obtain an analytical expression for the propagation constants. We will also be able to study the propagation of a beam through such a waveguide. Finally, the analysis enables us to understand the modal characteristics of a parabolic index fiber (see Chapter 9).

If we substitute equation (7.112) in equation (7.23) we obtain

$$\frac{d^2\psi}{dx^2} + \left\{k_0^2 n_1^2\left[1 - 2\Delta\left(\frac{x}{a}\right)^2\right] - \beta^2\right\}\psi(x) = 0 \tag{7.113}$$

where we have used the symbol ψ instead of E_y. The above equation can be written in the form

$$\frac{d^2\psi}{d\xi^2} + (\Lambda - \xi^2)\psi(\xi) = 0 \tag{7.114}$$

where

$$\xi = \gamma x; \qquad \gamma = \left[\frac{n_1 k_0 \sqrt{2\Delta}}{a}\right]^{1/2} \tag{7.115}$$

and

$$\Lambda = \frac{k_0^2 n_1^2 - \beta^2}{\gamma^2} = \left[k_0^2 n_1^2 - \beta^2\right]\frac{a}{n_1 k_0 \sqrt{2\Delta}} \tag{7.116}$$

Equation (7.114) is of the same form as the one-dimensional Schrodinger equation corresponding to the linear harmonic oscillator problem [see, e.g., Ghatak (1996), Chapter 7]. If we apply the boundary condition that

$$\psi(x) \to 0 \quad \text{as} \quad x \to \pm\infty$$

then we will find that Λ can take discrete values given by

$$\Lambda = (2m + 1); \quad m = 0, 1, 2, \ldots \tag{7.117}$$

the corresponding eigen functions being the Hermite–Gauss functions (see Appendix A). Thus, the normalized modal patterns (satisfying the condition

[5] We may mention here that equation (7.112) is rather hypothetical in the sense that for large values of $|x|$, equation (7.112) predicts negative values of n^2 – a realistic profile is truncated – that is, equation (7.112) is valid for $|x| < a$ beyond which the refractive index is constant. However, for a highly multimoded waveguide, when a large number of low-order modes are excited, the analysis of the infinitely extended profile gives results that are close to the results obtained by using the truncated profile.

given by equation (7.103)) and the corresponding propagation constants are given by

$$\psi_m(x) = N_m H_m(\xi) e^{-\frac{1}{2}\xi^2} \tag{7.118}$$

$$\beta_m = n_1 k_0 \left[1 - (2m+1) \frac{1}{n_1 k_0} \frac{\sqrt{2\Delta}}{a} \right]^{1/2} \tag{7.119}$$

where $m = 0, 1, 2, \ldots$ and

$$N_m = \left[\frac{\gamma}{2^m m! \sqrt{\pi}} \right]^{1/2} \tag{7.120}$$

represents the normalization constant so that

$$\int_{-\infty}^{+\infty} \psi_m(x) \psi_n(x) \, dx = \delta_{mn} \tag{7.121}$$

Note that the integration is over x and *not* over ξ. Further, the functions $H_m(\xi)$ are the well-known Hermite polynomials given by

$$\begin{aligned} &H_0(\xi) = 1, \quad H_1(\xi) = 2\xi \\ &H_2(\xi) = 4\xi^2 - 2, \quad H_3(\xi) = 8\xi^3 - 12\xi, \ldots \end{aligned} \tag{7.122}$$

and

$$H_{m+1}(\xi) = 2\xi H_m(\xi) - 2m H_{m-1}(\xi)$$

A few interesting points may be noted:

(a) Equations (7.117) and (7.118) represent *exact* solutions of equation (7.113), which can be verified by direct substitution.
(b) Since the profile is symmetric in x (i.e., $n^2(-x) = n^2(x)$), the modal fields are either symmetric in $x(m = 0, 2, 4, \ldots)$ or antisymmetric in $x(m = 1, 3, 5, \ldots)$.
(c) The function $H_0(\xi)$ (and hence $\psi_0(\xi)$) has no zeroes; the function $H_1(\xi)$ (and hence $\psi_1(\xi)$) has only one zero (at $\xi = 0$); the function $H_2(\xi)$ (and hence $\psi_2(\xi)$) has two zeroes (at $\xi = \pm\sqrt{\frac{1}{2}}$); the function $H_3(\xi)$ (and hence $\psi_3(\xi)$) has three zeroes (at $\xi = 0, \pm\sqrt{\frac{3}{2}}$) etc. In general, $H_m(\xi)$ (and hence $\psi_m(\xi)$) will have m zeroes – consistent with the remarks made in Section 7.3.
(d) Equation (7.119) tells us that for large values of m, β_m becomes imaginary. Thus, if we write $\beta_m = -i\gamma_m$, then the z dependence is $e^{-\gamma_m z}$, which represents the attenuating modes.

Example 7.3: As an example, we consider the launching of an off-axis y-polarized Gaussian beam whose spot size is the same as that of the fundamental mode:

$$\mathcal{E}_y(x, z = 0, t) = \frac{E_0}{\pi^{1/4}} e^{-\frac{1}{2}(\xi - \xi_0)^2} \cos \omega t \tag{7.123}$$

where $\xi_0 = \gamma x_0$ with x_0 being the x-coordinate of the center of the incident Gaussian beam at $z = 0$.

Expanding the Gaussian into waveguide modes,

$$\frac{E_0}{\pi^{1/4}} e^{-\frac{1}{2}(\xi-\xi_0)^2} = \sum_{m=0,\ldots}^{\infty} c_m \psi_m(x) \tag{7.124}$$

we get

$$\begin{aligned} c_m &= \frac{E_0}{\pi^{1/4}} \int_{-\infty}^{+\infty} \psi_m(x)\, e^{-\frac{1}{2}(\xi-\xi_0)^2} dx \\ &= \frac{E_0}{\sqrt{\gamma}} \frac{1}{\sqrt{m!}} \left(\frac{1}{2}\xi_0^2\right)^{m/2} e^{-\frac{1}{4}\xi_0^2} \end{aligned} \tag{7.125}$$

where (in evaluating the integral) we have used the generating function for Hermite polynomials[6] [see, e.g., Ghatak et al. (1995), Chapter 6]

$$G(\xi, t) = e^{-t^2+2\xi t} = \sum_{m=0,1,\ldots}^{\infty} \frac{1}{m!} H_m(\xi) t^m \tag{7.126}$$

Thus

$$\mathcal{E}_y(x, z, t) = \sum_{m=0,1,\ldots}^{\infty} c_m \psi_m(x) \cos(\omega t - \beta_m z) \tag{7.127}$$

The fractional power in the mth mode is given by

$$p_m = \frac{1}{m!} \left(\frac{1}{2}\xi_0^2\right)^m e^{-\frac{1}{2}\xi_0^2} \tag{7.128}$$

Obviously

$$\sum_{m=0,1,\ldots}^{\infty} p_m = 1 \tag{7.129}$$

In Figures 7.6 and 7.7, we have plotted p_m (as a function of m) for $\xi_0^2 = 20$ and $\xi_0^2 = 200$, respectively. The figures show the excitation efficiencies of various modes; a large value of ξ_0 implies launching at a point far from the axis.

Carrying out an analysis similar to that given in Section 7.6, we obtain

$$\langle S_x \rangle = -\frac{1}{2\omega\mu_0} \sum_m \sum_n c_m c_n \psi_m(x) \frac{d\psi_n}{dx} \sin[(\beta_m - \beta_n) z] \tag{7.130}$$

[6] If we multiply equation (7.126) by $\exp[-\frac{1}{2}(\xi - \xi_0)^2 - \frac{1}{2}\xi^2]$, carry out the integration, and compare the coefficients of t^m on both sides, we get equation (7.125).

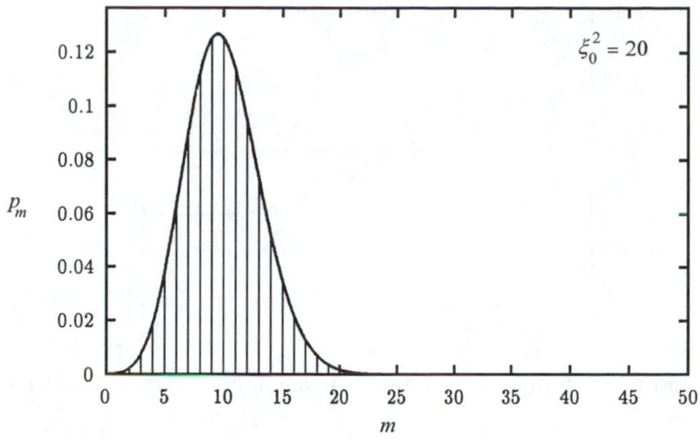

Fig. 7.6: Probabilities of excitation of various modes for $\xi_0^2 = 20$; the first 25 modes get primarily excited.

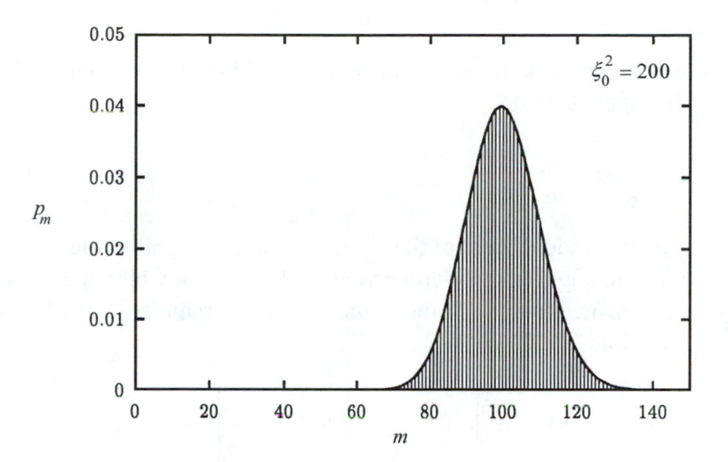

Fig. 7.7: Probabilities of excitation of various modes for $\xi_0^2 = 200$; only about 80 modes for m lying between 60 and 140 are excited.

and

$$\langle \mathcal{S}_z \rangle = \frac{1}{2\omega\mu_0} \sum_m \sum_n c_m c_n \psi_m(x) \psi_n(x) \beta_m \cos\left[(\beta_m - \beta_n) z\right] \tag{7.131}$$

Figure 7.8 shows the (magnitude and direction) of the Poynting vector for the following values of various parameters

$$a = 50\,\mu\text{m}, \quad k_0 = 10\,(\mu\text{m})^{-1} (\Rightarrow \lambda_0 \simeq 0.6328\,\mu\text{m})$$

$$n_1 = 1.5, \quad \Delta = 0.01, \quad \xi_0^2 = 20\,(\Rightarrow x_0 \simeq 21.7\,\mu\text{m})$$

The double sum has been carried out for m, n going from 0 to 25. The solid line in the figure shows the paraxial ray path.

Example 7.4: In most practical waveguides

$$\frac{1}{n_1 k_0} \frac{\sqrt{2\Delta}}{a} \ll 1 \tag{7.132}$$

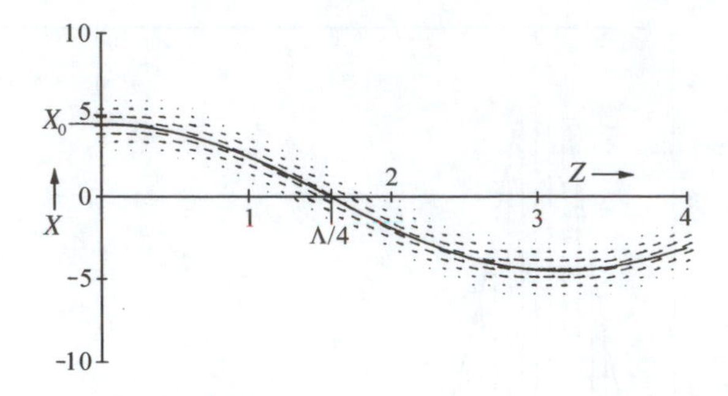

Fig. 7.8: The evolution of the Poynting vector as an off-axis Gaussian beam propagates through a parabolic index medium ($\xi_0^2 = 20$) [After Freude, Ghatak, and Grau (1997).]

and we may make a binomial expansion in equation (7.119) to obtain

$$\beta_m \simeq n_1 \frac{\omega}{c} - \left(m + \frac{1}{2}\right) \frac{\sqrt{2\Delta}}{a} \tag{7.133}$$

where we have neglected the second and higher order terms. Under this approximation

$$\frac{1}{v_g} = \frac{d\beta_m}{d\omega} \simeq \frac{n_1}{c} \tag{7.134}$$

which is independent of the mode number. Thus, all modes have (approximately) the same group velocity because of which the pulse dispersion in a parabolic index medium is extremely small. We write equation (7.127) as

$$\mathcal{E}_y(x, z, t) = \text{Re}\left[\sum_m c_m \psi_m(x) e^{i(\omega t - \beta_m z)}\right] \tag{7.135}$$

Substituting for c_m, $\psi_m(x)$ and β_m from equations (7.125), (7.118), and (7.133), respectively, we obtain

$$\begin{aligned} \mathcal{E}_y(x, z, t) = \text{Re}\Bigg[& e^{i\omega t} \frac{E_0}{\sqrt{\gamma}} \sum_m \left[\frac{1}{\sqrt{m!}} \left(\frac{1}{2}\xi_0^2\right)^{m/2} e^{-\frac{1}{4}\xi_0^2}\right]\Bigg] \\ & \times \left[\left(\frac{\gamma}{2^m m! \sqrt{\pi}}\right)^{1/2} H_m(\xi) e^{-\frac{1}{2}\xi^2}\right] \\ & \times \exp\left[-i\left(\frac{n_1 \omega}{c} - \frac{1}{2}\delta\right) z\right] [e^{-i\delta z}]^m \end{aligned} \tag{7.136}$$

where

$$\delta = \frac{\sqrt{2\Delta}}{a} \tag{7.137}$$

Thus

$$\mathcal{E}_y(x, z, t) = \text{Re}\left[e^{i\Phi} \frac{E_0}{\pi^{1/4}} e^{-\frac{1}{2}\xi^2 - \frac{1}{4}\xi_0^2} \sum_m \frac{1}{m!} H_m(\xi) g^m\right] \tag{7.138}$$

where

$$\Phi = \omega t - \left(n_1 \frac{\omega}{c} - \frac{1}{2} \frac{\sqrt{2\Delta}}{a} \right) z \tag{7.139}$$

and

$$g = \frac{1}{2} \xi_0 e^{-i\delta z} \tag{7.140}$$

If we use equation (7.126) and carry out elementary manipulations we obtain

$$\mathcal{E}_y(x, z, t) = \text{Re} \left[\frac{E_0}{\pi^{1/4}} \, e^{i\chi} e^{-\frac{1}{2}[\xi - \xi_0 \cos \delta z]^2} \right] \tag{7.141}$$

where $\chi = \Phi + \xi \xi_0 \sin \delta z - \frac{1}{4} \xi_0^2 \sin 2\delta z$. Thus

$$\langle |\mathcal{E}_y(x, z, t)|^2 \rangle = \frac{E_0^2}{2\sqrt{\pi}} \, e^{-(\xi - \xi_0 \cos \delta z)^2} \tag{7.142}$$

showing that the peak of the intensity distribution follows the (paraxial) ray trajectory (see equation (4.17))

$$x = x_0 \cos \delta z \tag{7.143}$$

Thus, the center of the Gaussian beam executes a sinusoidal path, which is nothing but the paraxial[7] ray path. From the above analysis we may note the following:

(i) As the Gaussian beam propagates, the power in each mode remains constant (there is no mode coupling). However, since β_m depends on the mode number, they, at different values of z, superpose in different phases to produce the intensity distribution shown in Figure 7.8.

(ii) The transverse width of the beam remains the same; this is because the spot size of the incident Gaussian beam was the same as the spot size of the fundamental mode.

(iii) The above analysis will be accurate as long as the launching point ($x = x_0$) is far away from the core–cladding interface.

Example 7.5: We next briefly consider the on-axis launching of a Gaussian beam; thus we assume

$$\Psi(x, z = 0) = E_0 \, e^{-x^2/w_i^2} \tag{7.144}$$

[7]Neglecting higher order terms in equation (7.133) is equivalent to the paraxial approximation.

Using equation (7.108) and properties of the Hermite polynomials[8] we obtain

$$c_m = \begin{cases} E_0 N_m w_0 \sqrt{\pi} \dfrac{m!}{(m/2)!} \left(\dfrac{w_0^2}{w_i^2} + 1 \right)^{-1/2} \left(\dfrac{1 - \frac{w_0^2}{w_i^2}}{1 + \frac{w_0^2}{w_i^2}} \right)^{m/2}; & m = 0, 2, 4, \ldots \\ 0; \quad m = 1, 3, 5, \ldots \end{cases} \tag{7.145}$$

where

$$w_0 = \frac{\sqrt{2}}{\gamma} = \sqrt{2} \left[\frac{a^2}{2\Delta\,(n_1 k_0)^2} \right]^{1/4} \tag{7.146}$$

If we substitute for c_m in equation (7.135) and sum the series (a more direct approach is discussed in Problem 7.13) we get

$$|\Psi(x, z)|^2 = E_0^2 \frac{w_i}{w(z)} \exp\left[-\frac{2x^2}{w^2(z)} \right] \tag{7.147}$$

where

$$w^2(z) = w_i^2 \left[\left(\frac{w_0}{w_i} \right)^4 \sin^2 \delta z + \cos^2 \delta z \right] \tag{7.148}$$

showing the modulating spot size of the propagating Gaussian beam. The following points may be noted

(i) When $w_i = w_0$, $w(z) = w_i$ (independent of z) and the beam propagates as the fundamental mode.

(ii) As $\Delta \to 0$, δ also tends to zero and we get

$$\begin{aligned} w^2(z) &= \frac{w_0^4}{w_i^2} (\delta z)^2 + w_i^2 \\ &= w_i^2 + \frac{\lambda_0^2}{n_1^2 \pi^2 w_i^2} z^2 \end{aligned} \tag{7.149}$$

which represents the spreading of a Gaussian beam as it propagates through a homogeneous medium of refractive index n_1 (see Section 2.8).

(iii) For given values of w_i, Δ and a as $\lambda_0 \to 0$

$$w^2(z) \to w_i^2 \cos^2 \delta z$$

which represents the geometrical optics limit.

[8] $\int_{-\infty}^{+\infty} e^{-\xi^2/a^2} H_{2m}(\xi)\, d\xi = \sqrt{\pi} a \frac{(2m)!}{m!} (a^2 - 1)^m; m = 0, 1, 2, \ldots$

Problems

7.1 Consider a symmetric planar waveguide with the following parameters:

$$n_1 = 1.50, \quad n_2 = 1.48, \quad d = 3.912\ \mu\text{m}$$

At $\lambda_0 = 1\ \mu$m, (a) show that there will be only two TE modes, the corresponding propagation constants being $\beta_0 = 9.4058\ \mu\text{m}^{-1}$ and $\beta_1 = 9.3525\ \mu\text{m}^{-1}$; calculate the discrete angles that the component waves make with the z-axis. (b) At $z = 0$, assume that the field in the core is given by

$$E_y(x) = 1.375 \times 10^4 \cos \kappa_0 x e^{i\omega t} + 1.309 \times 10^4 \sin \kappa_1 x e^{i\omega t}\ \text{V/m}$$

Show that equal power of 1 W is carried by the two modes. Calculate the transverse intensity distribution at

$$z = 0, \quad \pi/\Delta\beta, \quad 2\pi/\Delta\beta$$

where $\Delta\beta = \beta_0 - \beta_1$. Interpret the results physically.

7.2 We rewrite the wave equation determining the TE modes [equation (7.101)] as an eigenvalue equation

$$d^2\psi_m/dx^2 + k_0^2 n^2(x)\psi_m = \lambda_m \psi_m(x) \tag{7.150}$$

where $\lambda_m = \beta_m^2$ represents the eigenvalues of the operator $[(d^2/dx^2 + k_0^2 n^2(x))]$. Prove that all values of λ_m are real and that if $\lambda_m \neq \lambda_k$, the corresponding modal fields are orthogonal.

Solution: We rewrite the complex conjugate of the eigenvalue equation corresponding to the eigenvalue λ_k

$$d^2\psi_k^*/dx^2 + k_0^2 n^2(x)\psi_k^* = \lambda_k^* \psi_k^*(x) \tag{7.151}$$

We multiply equation (7.150) by ψ_k^* and equation (7.151) by ψ_m and subtract to obtain

$$\psi_k^* d^2\psi_m/dx^2 - \psi_m d^2\psi_k^*/dx^2 = (\lambda_m - \lambda_k^*)\psi_k^*(x)\psi_m(x)$$

The LHS is simply

$$\frac{d}{dx}\left(\psi_k^* \frac{d\psi_m}{dx} - \psi_m \frac{d\psi_k^*}{dx}\right)$$

therefore if we integrate from $-\infty$ to $+\infty$ we obtain

$$(\lambda_m - \lambda_k^*) \int_{-\infty}^{+\infty} \psi_k^*(x)\psi_m(x)\,dx = \left[\psi_k^* \frac{d\psi_m}{dx} - \psi_m \frac{d\psi_k^*}{dx}\right]_{-\infty}^{+\infty} \tag{7.152}$$

The RHS vanishes because for guided modes the fields vanish at $x = \pm\infty$. Thus, we get

$$(\lambda_m - \lambda_k^*) \int_{-\infty}^{+\infty} \psi_k^*(x)\psi_m(x)\,dx = 0 \tag{7.153}$$

For $k = m$, the integral $\int_{-\infty}^{+\infty} |\psi_m(x)|^2\, dx$ is positive definite and therefore we must have

$$\lambda_m = \lambda_m^* \tag{7.154}$$

proving that all eigenvalues β_m^2 must be real. Further, for $\lambda_m \neq \lambda_k$, we must have

$$\int_{-\infty}^{+\infty} \psi_k^*(x)\psi_m(x)dx = 0; \quad (\lambda_m \neq \lambda_k) \tag{7.155}$$

which represents the orthogonality condition.[9]

7.3 In the eigenvalue equation for TE modes (equation (7.23)) prove that β^2 cannot be greater than the maximum value of $k_0^2 n^2(x)$.

Solution: We rewrite equation (7.23) in the form

$$d^2 E_y/dx^2 = \alpha(x) E_y(x)$$

where

$$\alpha(x) = \beta^2 - k_0^2 n^2(x)$$

Now, if β^2 is greater than the maximum value of $k_0^2 n^2(x)$, then $\alpha(x)$ is positive *everywhere* and $d^2 E_y/dx^2$ has the same sign as $E_y(x)$ *everywhere*. Thus, if E_y is positive at some value of x, then $d^2 E_y/dx^2$ will also be positive. Hence, if $E_y' > 0$, then $E_y \to \infty$ as $x \to \infty$ because E_y' will keep on increasing. On the other hand, if $E_y' < 0$, then $E_y \to -\infty$ as $x \to -\infty$. Therefore, there must be some region where $\beta^2 < k_0^2 n^2(x)$.

7.4 In the specific profile shown in Figure 7.1 show explicitly that for $\beta^2 > k_0^2 n_1^2$, the boundary conditions cannot be satisfied.

7.5 Show that if $n^2(-x) = n^2(x)$, then the modal field patterns are either symmetric or antisymmetric functions of x.

Solution: We write the wave equation determining the TE modes (equation (7.23)) in the form

$$d^2 E_y(x)/dx^2 + k_0^2 n^2(x) E_y(x) = \beta^2 E_y(x) \tag{7.156}$$

Making the transformation $x \to -x$ we get

$$\frac{d^2}{dx^2} E_y(-x) + k_0^2 n^2(x) E_y(-x) = \beta^2 E_y(-x) \tag{7.157}$$

where we have used the fact that $n^2(-x) = n^2(x)$. Comparing equations (7.156) and (7.157) we see that $E_y(x)$ and $E_y(-x)$ satisfy the same equation and therefore

[9] If $\lambda_m = \lambda_k$, ψ_m and ψ_k are then two independent modal fields belonging to the same value of the propagation constant; the sets of modes are said to be *degenerate* and ψ_m and ψ_k are not necessarily orthogonal. It can easily be seen that any linear combination $C_1\psi_m + C_2\psi_k$ is also a possible mode belonging to the same propagation constant and it is always possible to construct modes that are mutually orthogonal to each other; the details are given in most textbooks on quantum mechanics [see, e.g., Ghatak (1996)].

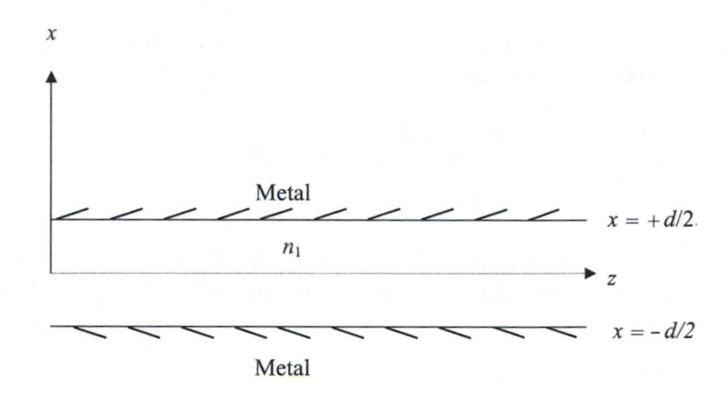

Fig. 7.9: A dielectric slab sandwiched between two perfectly conducting surfaces.

they are eigenfunctions belonging to the *same* value of β^2. Thus, if the mode is nondegenerate,[10] then $E_y(-x)$ must be a multiple of x – that is

$$E_y(-x) = \lambda E_y(x)$$

Making the transformation $x \to -x$ again, we get

$$E_y(x) = \lambda E_y(-x) = \lambda^2 E_y(x)$$

so that $\lambda^2 = 1$ or $\lambda = \pm 1$. Hence

$$E_y(-x) = \pm E_y(x) \tag{7.158}$$

proving the theorem. Modal fields belonging to the class $\lambda = +1$ and $\lambda = -1$ are symmetric and antisymmetric functions of x, respectively.

7.6 Show by direct substitution that with n^2 given by equation (7.112)

$$e^{-\frac{1}{2}\xi^2}, \quad \xi e^{-\frac{1}{2}\xi^2} \quad \text{and} \quad (4\xi^2 - 2)e^{-\frac{1}{2}\xi^2}$$

satisfy equation (7.113); the corresponding value of β^2 being given by equation (7.119) with $m = 0$, 1, and 2, respectively.

In general, show that $H_m(\xi)e^{-\frac{1}{2}\xi^2}$ satisfies equation (7.113), the corresponding value of β^2 being given by equation (7.119).

[HINTS: Hermite functions are polynomial solutions of $d^2y/d\xi^2 - 2\xi dy/d\xi + 2ny(\xi) = 0$.]

7.7 Consider a waveguide whose bounding surfaces are made of perfect conductors (see Figure 7.9) so that E_y may be assumed to vanish at $x = \pm d/2$. Consider the TE modes and show that

$$\beta_m^2 = k_0^2 n_1^2 - (m\pi/d)^2 \tag{7.159}$$

where $m = 1, 3, 5, \ldots$ correspond to the symmetric TE modes and $m = 2, 4, 6, \ldots$ correspond to the antisymmetric TE modes (two points should be noted: first, $d\psi/dx$ is discontinuous at $x = \pm d/2$ – this is because $n^2(x)$ has an infinite discontinuity at $x = \pm d/2$; second, for $m > k_0 n_1 d/\pi$, β_m becomes imaginary – these are the attenuating modes; there is, however, no absorption of energy).

[10] This theorem is therefore strictly true for nondegenerate modes only (see footnote in the solution to Problem 7.2). For degenerate modes the field patterns need not be symmetric or antisymmetric functions of x. However, even for degenerate modes, one can always construct appropriate linear combinations that are either symmetric or antisymmetric functions of x.

7.8 (a) Consider TE modes in a planar waveguide one of whose bounding surface (at $x = 0$) is a perfect conductor; further

$$
\begin{aligned}
n(x) &= n_1 \quad \text{for } 0 < x < h \\
&= n_2 \quad \text{for } x > h
\end{aligned}
\tag{7.160}
$$

Thus, one of the boundary conditions would be that $E_y(x)$ would vanish at $x = 0$. Solve equation (7.23) to obtain the following transcendental equation determining the propagation constants.

$$
-\xi \cot \xi = \sqrt{\alpha^2 - \xi^2}
\tag{7.161}
$$

where

$$
\xi = \left(k_0^2 n_1^2 - \beta^2\right)^{1/2} h \quad \text{and} \quad \alpha = k_0 h \left(n_1^2 - n_2^2\right)^{1/2}
\tag{7.162}
$$

Indeed, equation (7.161) gives the antisymmetric TE modes of the structure shown in Figure 7.1(a).
(b) For $\alpha = 3\pi$, determine the discrete values of ξ.
(c) For $n_1 = 1.5$ and $k_0 h = 10$ calculate n_2 and the corresponding values of β/k_0.

[ANSWER: (b) 2.8360, 5.6415, and 8.3388.]

7.9 (a) We next consider the structure for which the refractive index is n_2 for $|x| < a$ and n_1 for $a < |x| < a + h$ with a metal boundary at $|x| = a + h$ (see Figure 7.10a). For TE modes the field distribution is given by

$$
\begin{aligned}
E_y &= \begin{cases} A_s \cosh \gamma x; & \text{symmetric} \\ & \qquad |x| < a \\ A_a \sinh \gamma x; & \text{antisymmetric} \end{cases} \\
&= \begin{cases} B \sin\left[\kappa\left(a + h - |x|\right)\right]; & \text{symmetric} \\ & \qquad a < |x| < a + h \\ \frac{x}{|x|} B \sin\left[\kappa\left(a + h - |x|\right)\right]; & \text{antisymmetric} \end{cases}
\end{aligned}
\tag{7.163}
$$

where we have assumed $n_2 < \beta/k_0 < n_1; \gamma$ and κ are the same as in Section 7.3. Show that the propagation constants are determined from the following transcendental equation

$$
-\frac{(\alpha^2 - \xi^2)^{1/2}}{\xi} \tan \xi = (\coth \gamma a)^{\pm 1}
\tag{7.164}
$$

where ξ and α have been defined in equation (7.162); the + and − signs corresspond to symmetric and antisymmetric modes, respectively. Thus, for $\gamma a \gg 1$ (i.e., when the two waveguides are well separated) the symmetric and antisymmetric modes are almost degenerate and are given by equation (7.161).
(b) For $\alpha = 3\pi$ and $h/a = 0.2$, obtain the discrete values of ξ [see Figure 7.10b]; the corresponding modal fields are shown in Figures 7.11 and 7.12.

[ANSWER: 2.82154 (s), 2.84990 (a), 5.60021 (s), 5.68284 (a), 8.23678 (s), 8.47649 (a). s: symmetric modes; a: antisymmetric modes.]

7.10 Figure 7.10 can be assumed approximately to represent two identical waveguides separated by a certain distance. This is essentially a directional coupler and the

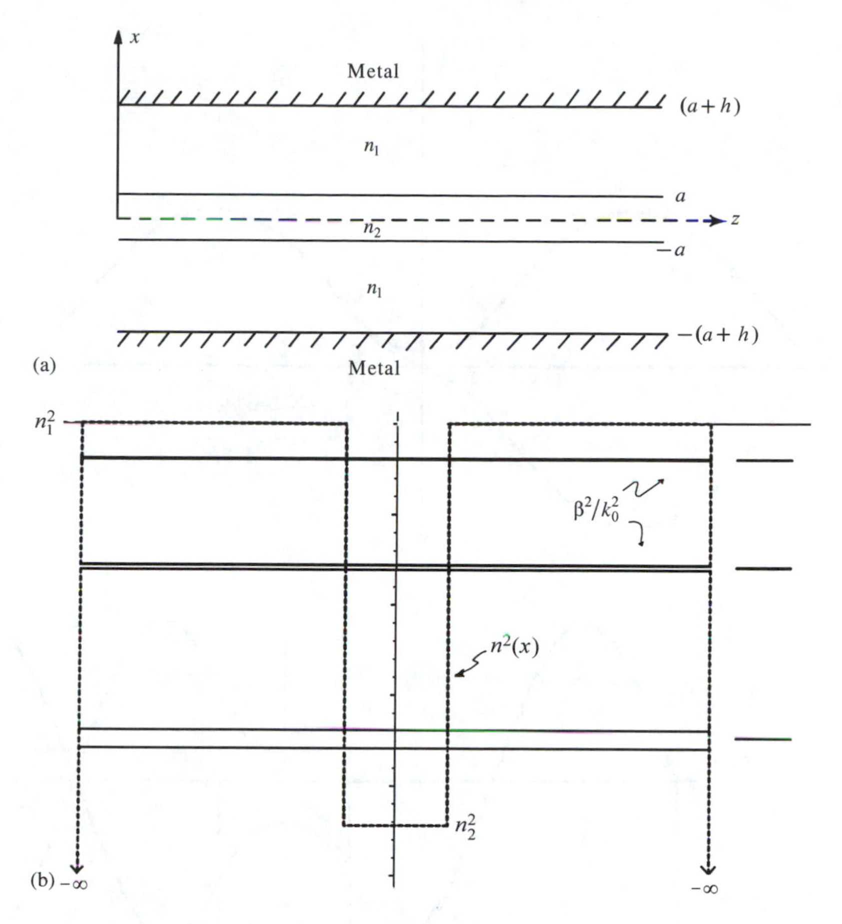

Fig. 7.10: The refractive index distribution corresponding to Problem 7.9 and the corrsponding discrete eigenvalues $(\beta/k_0)^2$ (shown as horizontal lines) for $\alpha = 3\pi$ and $h/a = 0.2$. The short horizontal lines on the extreme right represent the (degenerate) value of $(\beta/k_0)^2$ corresponding to two isolated structures.

previous problem describes the "supermodes" of the composite structure. Considering the two lowest order modes we may write as a solution of the wave equation

$$\Psi = \left[\psi_s(x)e^{i(\omega t-\beta_s z)} + \psi_a(x)e^{i(\omega t-\beta_a z)}\right]$$

$$= \left[\psi_s(x) + \psi_a(x)e^{i(\Delta\beta)z}\right]e^{i(\omega t-\beta_s z)} \tag{7.165}$$

where $\psi_s(x)$ and $\psi_a(x)$ represent the modal pattern for the symmetric and antisymmetric modes, respectively (see Figure 7.11), and $\Delta\beta = (\beta_s - \beta_a)$. Show that at $z = 0, 2\pi/\Delta\beta, 4\pi/\Delta\beta, \ldots$ most of the energy is in the first waveguide and at $z = \pi/\Delta\beta, 3\pi/\Delta\beta, \ldots$ there is almost complete transfer of power to the second waveguide.

7.11 Consider an interface of a dielectric and a metal, – that is

$$n^2 = \begin{cases} n_1^2; & x > 0 \\ n_m^2; & x < 0 \end{cases} \tag{7.166}$$

where n_m^2 is complex having a large negative part. Show that such a structure can support a TM mode whose propagation constant is given by

$$\beta^2\big/k_0^2 = n_1^2 n_m^2\big/\left(n_1^2 + n_m^2\right) \tag{7.167}$$

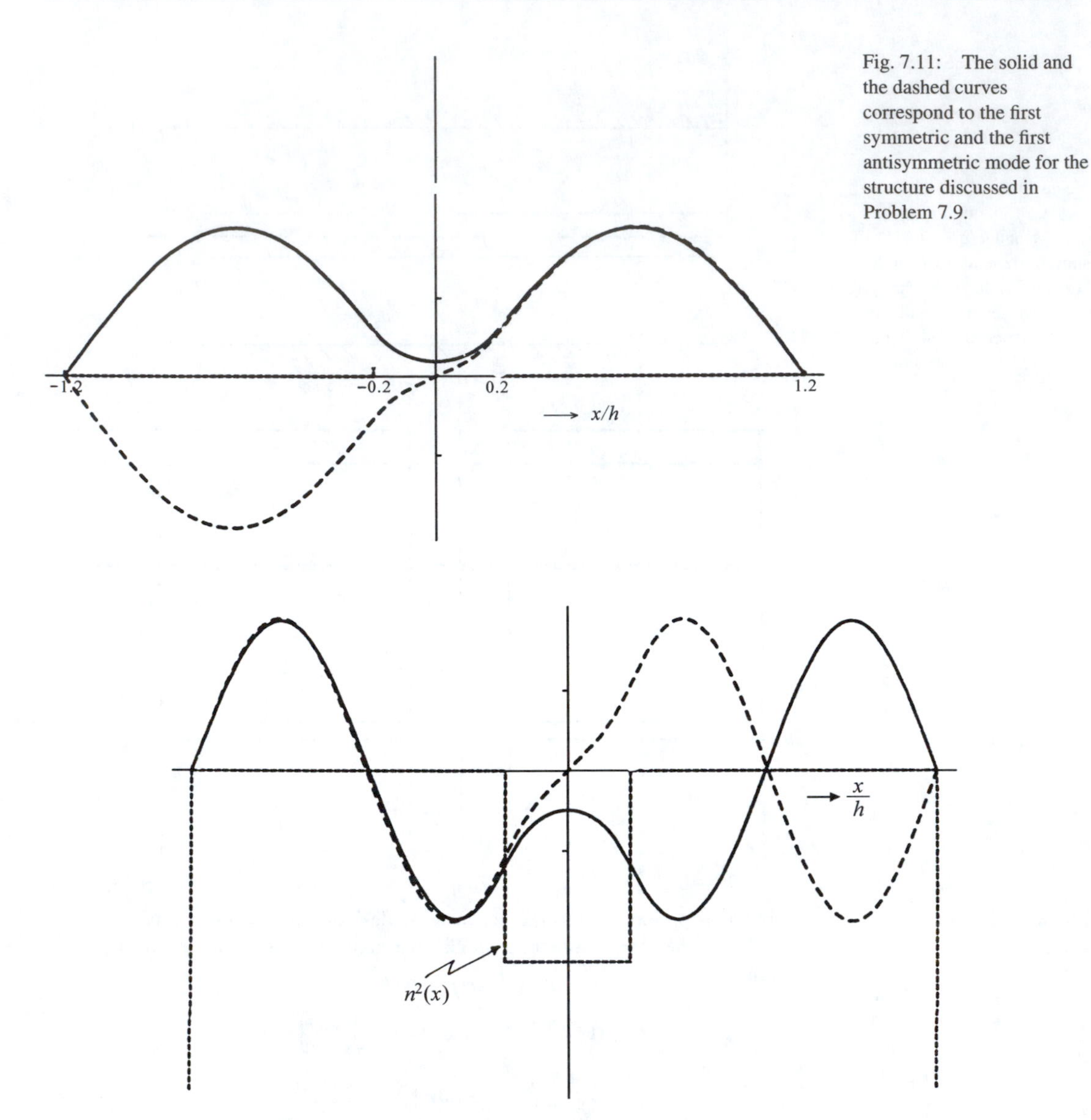

Fig. 7.11: The solid and the dashed curves correspond to the first symmetric and the first antisymmetric mode for the structure discussed in Problem 7.9.

Fig. 7.12: The solid and the dashed curves correspond to the second symmetric and the second antisymmetric mode for the structure discussed in Problem 7.9.

This mode is known as the surface plasmon mode and is of considerable interest in integrated optics. Show that the corresponding field decays exponentially in the $x > 0$ and $x < 0$ regions. Show also that for such a structure, TE modes can never exist.

7.12 Starting from the equation

$$\Psi(x, z) = \sum_m c_m \psi_m(x) e^{-i\beta_m z} \tag{7.168}$$

determine c_m and then show that

$$E_y(x, z) = \int_{-\infty}^{+\infty} K(x, x', z) \Psi(x', z = 0)\, dx' \tag{7.169}$$

where

$$K(x, x', z) = \sum_m \psi_m(x')\psi_m(x)e^{-i\beta_m z} \tag{7.170}$$

7.13 For the infinitely extended parabolic profile (equation (7.112)), evaluate $K(x, x', z)$ with β_m given by equation (7.133).

[HINTS: Use Mehler's formula [see, e.g., Ghatak, Goyal, and Chua (1995), Section 6.7.]

$$\sum_{n=0}^{\infty} \frac{H_n(x)H_n(y)}{n!}\left(\frac{z}{2}\right)^n = \frac{1}{(1-z^2)^{1/2}} \exp\left[\frac{2xyz - (x^2+y^2)z^2}{1-z^2}\right] \tag{7.171}$$

[ANSWER:
$K(x, x', z) = (\frac{i\gamma^2}{2\pi \sin \delta z})^{1/2} e^{-in_1k_0z} \exp[\frac{i\gamma^2 xx'}{\sin \delta z} - \frac{i\gamma^2}{2}(x^2 + x'^2)\cot \delta z]]$

7.14 Using the results of the previous two problems, evaluate $\Psi(x, z)$ when

(a) $\Psi(x, z = 0) = \frac{E_0}{\pi^{1/4}}\, e^{-1/2(\xi-\xi_0)^2}$
[see Example 7.4]

(b) $\Psi(x, z = 0) = E_0\, e^{-x^2/w_i^2}$
(see Example 7.5).

8

Propagation characteristics of a step index fiber

8.1 Introduction

The propagation characteristics, splice loss, etc. of the optical fiber, play an extremely important role in the design of a fiber optic communication system. In this chapter we carry out a detailed modal analysis of the step index fiber. This is followed by the extensively used Gaussian approximation for the fundamental mode of a step index fiber; in the Gaussian approximation the splice loss calculations, which are very straightforward, are discussed in Section 8.5. The Petermann-2 spot size and the far-field spot size are also discussed.

8.2 Scalar modes in the weakly guiding approximation

The step index fiber is characterized by the following refractive index distribution

$$\begin{aligned} n(r) &= n_1 \quad 0 < r < a \qquad \text{core} \\ &= n_2 \quad r > a \qquad \text{cladding} \end{aligned} \tag{8.1}$$

In actual fibers

$$n_1 \approx n_2 \tag{8.2}$$

and this allows use of the so-called scalar wave approximation (also known as the weakly guiding approximation). In this approximation, the modes are assumed to be nearly transverse and can have an arbitrary state of polarization. Thus, the two independent sets of modes can be assumed to be x-polarized and y-polarized, and in the scalar approximation they have the same propagation constants. These linearly polarized modes are usually referred to as LP modes. We may compare this with the discussion in Section 7.5 where we mentioned that when $n_1 \approx n_2$, the modes are nearly transverse and the propagation constants of the TE and TM modes are almost equal.

In the weakly guiding approximation, the transverse component of the electric field (E_x or E_y) satisfies the scalar wave equation

$$\nabla^2 \Psi = \epsilon_0 \mu_0 n^2 \frac{\partial^2 \Psi}{\partial t^2} \tag{8.3}$$

For n^2 depending only on the transverse coordinates (r, ϕ), we may write

$$\Psi(r, \phi, z, t) = \psi(r, \phi)e^{i(\omega t - \beta z)} \tag{8.4}$$

where ω is the angular frequency and β is known as the propagation constant. The above equation represents the modes of the system. Substituting in equation (8.3), we readily obtain

$$\left(\nabla^2 - \frac{\partial^2}{\partial z^2}\right)\psi + \left[\frac{\omega^2}{c^2} n^2(r, \phi) - \beta^2\right] \psi = 0 \tag{8.5}$$

In most practical fibers n^2 depends only on the cylindrical coordinate r and therefore it is convenient to use the cylindrical system of coordinates to obtain[1]

$$\frac{\partial^2 \psi}{\partial r^2} + \frac{1}{r}\frac{\partial \psi}{\partial r} + \frac{1}{r^2}\frac{\partial^2 \psi}{\partial \phi^2} + \left[k_0^2 n^2(r) - \beta^2\right] \psi = 0 \tag{8.6}$$

where

$$k_0 = \omega / c = 2\pi / \lambda_0 \tag{8.7}$$

is the free space wave number. Because the medium has cylindrical symmetry, we can solve equation (8.6) by the method of separation of variables:

$$\psi(r, \phi) = R(r)\Phi(\phi) \tag{8.8}$$

On substituting and dividing by $\psi(r, \phi)/r^2$, we obtain

$$\frac{r^2}{R}\left(\frac{d^2 R}{dr^2} + \frac{1}{r}\frac{dR}{dr}\right) + r^2\left[n^2(r)k_0^2 - \beta^2\right] = -\frac{1}{\Phi}\frac{d^2\Phi}{d\phi^2} = +l^2 \tag{8.9}$$

where l is a constant. The ϕ dependence will be of the form $\cos l\phi$ or $\sin l\phi$ and for the function to be single valued (i.e., $\Phi(\phi + 2\pi) = \Phi(\phi)$) we must have

$$l = 0, 1, 2, \ldots, \text{etc.} \tag{8.10}$$

(Negative values of l correspond to the same field distribution.) Since for each value of l there can be two independent states of polarization, modes with $l \geq 1$ are four-fold degenerate (corresponding to two orthogonal polarization states and to the ϕ dependence being $\cos l\phi$ or $\sin l\phi$); modes with $l = 0$ are ϕ independent and have two-fold degeneracy. The radial part of the equation gives us

$$r^2 \frac{d^2 R}{dr^2} + r\frac{dR}{dr} + \left\{\left[k_0^2 n^2(r) - \beta^2\right]r^2 - l^2\right\} R = 0 \tag{8.11}$$

The solution of the above equation for a step index profile is given in the next section. However, we can make some general comments about the solutions of equation (8.11) for an arbitrary cylindrically symmetric profile having a

[1] We should mention here that for an infinitely extended parabolic profile $n^2(r) = n_1^2 - \alpha(x^2 + y^2)$ it is equally convenient to use Cartesian coordinates (see Chapter 9).

refractive index that decreases monotonically from a value n_1 on the axis to a constant value n_2 beyond the core–cladding interface $r = a$ (see Figure 5.1). The solutions of equation (8.11) can be divided into two distinct classes (cf. Section 7.3):

(a) $$k_0^2 n_1^2 > \beta^2 > k_0^2 n_2^2 \tag{8.12}$$

For β^2 lying in the above range, the fields $R(r)$ are oscillatory in the core and decay in the cladding and β^2 assumes only discrete values; these are known as the *guided modes* of the system. For a given value of l, there will be several guided modes, which are designated LP_{lm} modes ($m = 1, 2, 3, \ldots$); LP stands for linearly polarized.[2] Further, since the modes are solutions of the scalar wave equation, they can be assumed to satisfy the orthonormality condition

$$\int_0^\infty \int_0^{2\pi} \psi_{lm}^*(r, \phi)\psi_{l'm'}(r, \phi) r \, dr \, d\phi = \delta_{ll'} \, \delta_{mm'} \tag{8.13}$$

(b) $$\beta^2 < k_0^2 n_2^2 \tag{8.14}$$

For such β values, the fields are oscillatory even in the cladding and β can assume a continuum of values. These are known as the *radiation modes*.

The guided and radiation modes form a complete set of modes in the sense that an arbitrary field distribution can be expanded in terms of these modes – that is

$$\psi(r, \phi, z) = \sum_{l,m} c_{lm} \psi_{lm}(r, \phi) e^{-i\beta_{lm} z} + \int c(\beta) \psi_\beta(r, \phi) e^{-i\beta z} d\beta \tag{8.15}$$

where the first term represents a sum over discrete modes and the second term an integral over the continuum of modes.[3] The quantity $|c_{lm}|^2$ is proportional to the power carried by the (l, m)th mode; the constants c_{lm} can be determined by knowing the incident field at $z = 0$ and using the orthonormality condition. Similarly $|c(\beta)|^2 \, d\beta$ is proportional to the power carried by radiation modes with β values lying between β and $\beta + d\beta$.

The calculation of the modal field distributions and the corresponding propagation constants are of extreme importance in the study of waveguides. For example, knowing the frequency dependence of the propagation constant one can calculate the temporal broadening of a pulse (see Chapter 10), which determines

[2] If one solves the vector wave equation, the modes are classified as HE_{lm}, EH_{lm}, TE_{0m}, and TM_{0m} modes, the correspondence is $\mathrm{LP}_{0m} = \mathrm{HE}_{1m}$; $\mathrm{LP}_{1m} = \mathrm{HE}_{2m}$, TM_{0m}, and $\mathrm{LP}_{lm} = \mathrm{HE}_{l+1,m}$, $\mathrm{EH}_{l-1,m} (l \geq 2)$ [see, e.g., Ghatak and Thyagarajan (1989), Section 13.14].

[3] The radiation modes can be further classified into propagating radiation modes $0 < \beta^2 < k_0^2 n_2^2$ and decaying radiation modes $-\infty < \beta^2 < 0$; the latter correspond to modes that decay exponentially along the z-axis.

the information-carrying capacity. Knowledge of the modal field distribution is essential for the calculation of excitation efficiencies and splice lossess at joints and the development of new fiber optic devices like directional couplers, etc. We now present a detailed modal analysis for step index optical fibers.

8.3 Modal analysis for a step index fiber

In this section, we obtain the modal fields and the corresponding propagation constants for a step index fiber for which the refractive index variation is given by equation (8.1). For such a fiber it is possible to obtain rigorous solutions of the vector equations [see, e.g., Snyder and Love (1988), Sodha and Ghatak (1977), Chapter 4]. However, most practical fibers used in communication are weakly guiding – that is, relative refractive index difference $(n_1 - n_2)/n_1 \ll 1$ and in such a case the radial part of the transverse component of the electric field satisfies the following equation (see equation (8.11)):

$$r^2\frac{d^2R}{dr^2} + r\frac{dR}{dr} + \left\{\left[k_0^2 n^2(r) - \beta^2\right]r^2 - l^2\right\}R = 0 \tag{8.16}$$

and the complete transverse field is given by

$$\Psi(r, \phi, z, t) = R(r)e^{i(\omega t - \beta z)}\begin{Bmatrix}\cos l\phi \\ \sin l\phi\end{Bmatrix} \tag{8.17}$$

If we use equation (8.1) for $n^2(r)$ in equation (8.16), we obtain

$$r^2\frac{d^2R}{dr^2} + r\frac{dR}{dr} + \left(U^2\frac{r^2}{a^2} - l^2\right)R = 0; \quad 0 < r < a \tag{8.18}$$

and

$$r^2\frac{d^2R}{dr^2} + r\frac{dR}{dr} - \left(W^2\frac{r^2}{a^2} + l^2\right)R = 0; \quad r > a \tag{8.19}$$

where

$$U = a\left(k_0^2 n_1^2 - \beta^2\right)^{1/2} \tag{8.20}$$

$$W = a\left(\beta^2 - k_0^2 n_2^2\right)^{1/2} \tag{8.21}$$

and the normalized waveguide parameter V is defined by

$$V = (U^2 + W^2)^{1/2} = k_0 a\left(n_1^2 - n_2^2\right)^{1/2} \tag{8.22}$$

Guided modes correspond to $n_2^2 k_0^2 < \beta^2 < n_1^2 k_0^2$ and therefore for guided modes both U and W are real.

Equations (8.18) and (8.19) are of the standard form of Bessel's equation [see, e.g., Irving and Mullineux (1959), Ghatak et al. (1995)]. The solutions of equation (8.18) are $J_l(x)$ and $Y_l(x)$ where $x = Ur/a$. The solution $Y_l(x)$ has to be rejected since it diverges as $x \to 0$. The solutions of equation (8.19) are the modified Bessel functions $K_l(\tilde{x})$ and $I_l(\tilde{x})$ with the asymptotic forms

$$K_l(\tilde{x}) \underset{\tilde{x}\to\infty}{\longrightarrow} \left(\frac{\pi}{2\tilde{x}}\right)^{\frac{1}{2}} e^{-\tilde{x}} \tag{8.23}$$

$$I_l(\tilde{x}) \underset{\tilde{x}\to\infty}{\longrightarrow} \frac{1}{(2\pi\tilde{x})^{\frac{1}{2}}} e^{\tilde{x}} \tag{8.24}$$

where $\tilde{x} = Wr/a$. Obviously, the solution $I_l(\tilde{x})$, which diverges as $\tilde{x} \to \infty$, has to be rejected. Thus, the transverse dependence of the modal field is given by

$$\psi(r, \phi) = \begin{cases} \dfrac{A}{J_l(U)} J_l\left(\dfrac{Ur}{a}\right) \begin{bmatrix} \cos l\phi \\ \sin l\phi \end{bmatrix}; & r < a \\ \dfrac{A}{K_l(W)} K_l\left(\dfrac{Wr}{a}\right) \begin{bmatrix} \cos l\phi \\ \sin l\phi \end{bmatrix}; & r > a \end{cases} \tag{8.25}$$

where we have assumed the continuity of ψ at the core–cladding interface. Continuity of $\partial\psi/\partial r$ at $r = a$ leads to[4]

$$\frac{U J_l'(U)}{J_l(U)} = \frac{W K_l'(W)}{K_l(W)} \tag{8.26}$$

Using the identities

$$\pm U J_l'(U) = l J_l(U) - U J_{l\pm 1}(U) \tag{8.27}$$

$$\pm W K_l'(W) = l K_l(W) \mp W K_{l\pm 1}(W) \tag{8.28}$$

$$J_{l+1}(U) = (2l/U) J_l(U) - J_{l-1}(U) \tag{8.29}$$

and

$$K_{l+1}(W) = (2l/W) K_l(W) + K_{l-1}(W) \tag{8.30}$$

equation (8.26) can be written in either of the following two forms

$$U \frac{J_{l+1}(U)}{J_l(U)} = W \frac{K_{l+1}(W)}{K_l(W)} \tag{8.31}$$

or

$$U \frac{J_{l-1}(U)}{J_l(U)} = -W \frac{K_{l-1}(W)}{K_l(W)} \tag{8.32}$$

[4]It should be mentioned that as long as ψ is assumed to satisfy the scalar wave equation (equation (8.6)), both ψ and $d\psi/dr$ have to be continuous at any refractive index discontinuity. This follows from the fact that if $d\psi/dr$ does not happen to be continuous, then $d^2\psi/dr^2$ will be a Dirac delta function, which would therefore be inconsistent with equation (8.6) unless, of course, there is an infinite discontinuity in n^2, which indeed happens at the interface between a dielectric and a perfect conductor.

However, using the proper limiting forms of $K_l(W)$ as $W \to 0$, one can show that[5]

$$\lim_{W\to 0} W\frac{K_{l-1}(W)}{K_l(W)} \to 0; \quad l = 0, 1, 2, \ldots \tag{8.33}$$

and therefore we use equation (8.32) for studying the modes. For $l = 0$ we get

$$U\frac{J_1(U)}{J_0(U)} = W\frac{K_1(W)}{K_0(W)} \tag{8.34}$$

where we have used the relations $J_{-1}(U) = -J_1(U)$ and $K_{-1}(W) = K_1(W)$.

We should mention here that the boundary conditions used in deriving the eigenvalue equation (equation (8.32)) are consistent with the approximation involved in using the scalar wave equation. For example, if ψ is assumed to represent E_y then, rigorously speaking, E_y and $\partial E_y/\partial r$ are *not* continuous at $r = a$ for all ϕ; indeed, one must make E_ϕ, E_z, and $n^2 E_r$ continuous at the interface $r = a$. However, if $n_1 \approx n_2$ the error involved is negligible [see, e.g., Sodha and Ghatak (1977)].

It is convenient to define the normalized propagation constant

$$b = \frac{\frac{\beta^2}{k_0^2} - n_2^2}{n_1^2 - n_2^2} = \frac{W^2}{V^2} \tag{8.35}$$

Thus

$$W = V\sqrt{b} \tag{8.36}$$

and

$$U = \sqrt{V^2 - W^2} = V\sqrt{1-b} \tag{8.37}$$

Since for guided modes

$$n_2^2 < \frac{\beta^2}{k_0^2} < n_1^2 \tag{8.38}$$

we will have

$$0 < b < 1 \tag{8.39}$$

[5] The limiting forms are

$$K_0(W) \underset{W\to 0}{\longrightarrow} -\ln(W/2)$$

and

$$K_l(W) \underset{W\to 0}{\longrightarrow} \frac{1}{2}\Gamma(l)(2/W)^l; \quad l > 0$$

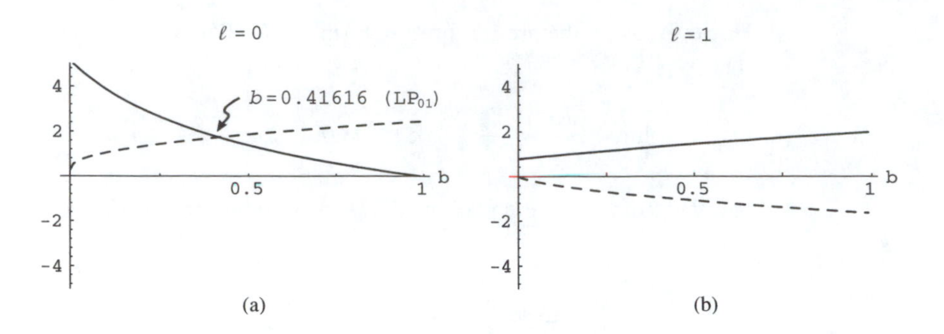

Fig. 8.1: (a) Variation of LHS (solid curve) and RHS (dashed curve) of equation (8.41) for $V = 2$. The point of intersection corresponds to $b = 0.41616$. (b) Variations of LHS (solid curve) and RHS (dashed curve) of equation (8.40) for $l = 1$ and $V = 2$. Since there are no points of intersection, there are no guided modes corresponding to $l = 1$.

Thus, equations (8.32) and (8.34) can be written in the form

$$V(1-b)^{1/2}\frac{J_{l-1}[V(1-b)^{1/2}]}{J_l[V(1-b)^{1/2}]} = -Vb^{1/2}\frac{K_{l-1}[Vb^{1/2}]}{K_l[Vb^{1/2}]}; \quad l \geq 1 \tag{8.40}$$

and

$$V(1-b)^{1/2}\frac{J_1[V(1-b)^{1/2}]}{J_0[V(1-b)^{1/2}]} = Vb^{1/2}\frac{K_1[Vb^{1/2}]}{K_0[Vb^{1/2}]}; \quad l = 0 \tag{8.41}$$

The solution of the above transcendental equations will give us universal curves describing the dependence of b (and therefore of U and W) on V. For a given value of l, there will be a finite number of solutions and the mth solution ($m = 1, 2, 3, \ldots$) is referred to as the LP_{lm} mode.

In Table 8.1 we have tabulated the values of b, U, and W for $1.5 < V < 2.5$ for the fundamental mode of a step index fiber. A convenient empirical formula for $b(V)$ is given by

$$b(V) = \left(A - \frac{B}{V}\right)^2; \quad 1.5 \lesssim V \lesssim 2.5 \tag{8.42}$$

where

$$A = 1.1428 \quad \text{and} \quad B = 0.996$$

In Figure 8.1 we have plotted the LHS and RHS of equations (8.41) and (8.40) corresponding to $l = 0$ and $l = 1$ for $V = 2$. It can be seen that there is only one mode corresponding to $l = 0$. The point of intersection gives the allowed value of b and we obtain

$$b = 0.41616 \text{ (for } V = 2) \Rightarrow \mathrm{LP}_{01} \text{ mode}$$

The simple empirical formula (equation (8.42)) gives $b \simeq 0.4158$. For $l = 1$ (and similarly for $l = 2, 3, \ldots$) (see Figure 8.1(b)), there are no points of intersection and therefore there are no modes.

Example 8.1: As an example, we consider a step index fiber characterized by

$$n_2 = 1.45, \quad \Delta = 0.0064, \quad a = 3.0\,\mu\text{m} \tag{8.43}$$

Table 8.1. *Variations of b, U, W, w_P/a, and w/a for the fundamental mode of the step index fiber*

V	b	U	W	w_P/a	w/a
1.500	0.22925	1.31689	0.71819	1.69342	1.78402
1.525	0.23955	1.32985	0.74639	1.65638	1.73858
1.550	0.24980	1.34252	0.77468	1.62172	1.69659
1.575	0.25997	1.35489	0.80305	1.58923	1.65769
1.600	0.27006	1.36698	0.83148	1.55872	1.62156
1.625	0.28007	1.37880	0.85997	1.53001	1.58793
1.650	0.28997	1.39034	0.88850	1.50295	1.55654
1.675	0.29977	1.40163	0.91709	1.47741	1.52720
1.700	0.30947	1.41267	0.94570	1.45326	1.49970
1.725	0.31905	1.42347	0.97435	1.43040	1.47387
1.750	0.32851	1.43403	1.00303	1.40872	1.44958
1.775	0.33785	1.44436	1.03172	1.38814	1.42668
1.800	0.34707	1.45448	1.06043	1.36858	1.40505
1.825	0.35616	1.46437	1.08914	1.34996	1.38460
1.850	0.36512	1.47406	1.11787	1.33223	1.36523
1.875	0.37396	1.48355	1.14660	1.31531	1.34684
1.900	0.38266	1.49285	1.17533	1.29915	1.32938
1.925	0.39123	1.50195	1.20406	1.28371	1.31276
1.950	0.39967	1.51087	1.23279	1.26893	1.29692
1.975	0.40798	1.51962	1.26150	1.25478	1.28182
2.000	0.41616	1.52818	1.29021	1.24122	1.26739
2.025	0.42421	1.53658	1.31891	1.22820	1.25359
2.050	0.43213	1.54482	1.34760	1.21571	1.24038
2.075	0.43992	1.55290	1.37627	1.20370	1.22772
2.100	0.44758	1.56082	1.40493	1.19216	1.21558
2.125	0.45512	1.56860	1.43357	1.18105	1.20391
2.150	0.46252	1.57622	1.46220	1.17035	1.19271
2.175	0.46981	1.58371	1.49080	1.16003	1.18192
2.200	0.47697	1.59106	1.51938	1.15009	1.17154
2.225	0.48401	1.59827	1.54795	1.14049	1.16154
2.250	0.49093	1.60536	1.57649	1.13122	1.15189
2.275	0.49773	1.61231	1.60501	1.12227	1.14258
2.300	0.50442	1.61915	1.63351	1.11361	1.13359
2.325	0.51099	1.62586	1.66199	1.10523	1.12491
2.350	0.51744	1.63246	1.69044	1.09713	1.11651
2.375	0.52379	1.63894	1.71887	1.08928	1.10838
2.400	0.53003	1.64531	1.74727	1.08168	1.10051
2.425	0.53616	1.65157	1.77565	1.07431	1.09288
2.450	0.54218	1.65773	1.80400	1.06716	1.08549
2.475	0.54810	1.66379	1.83233	1.06022	1.07832
2.500	0.55392	1.66974	1.86064	1.05349	1.07137

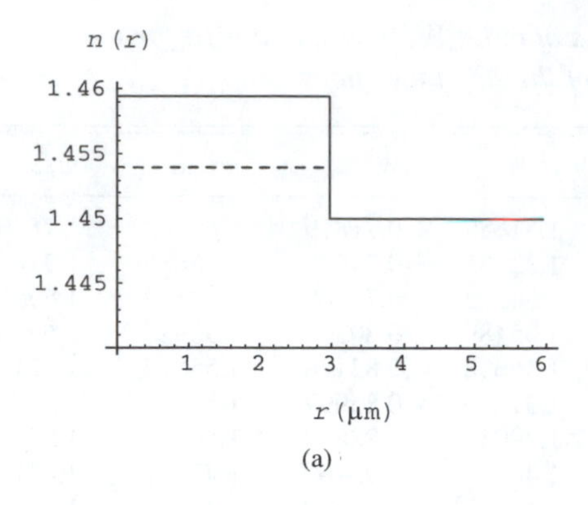

(a)

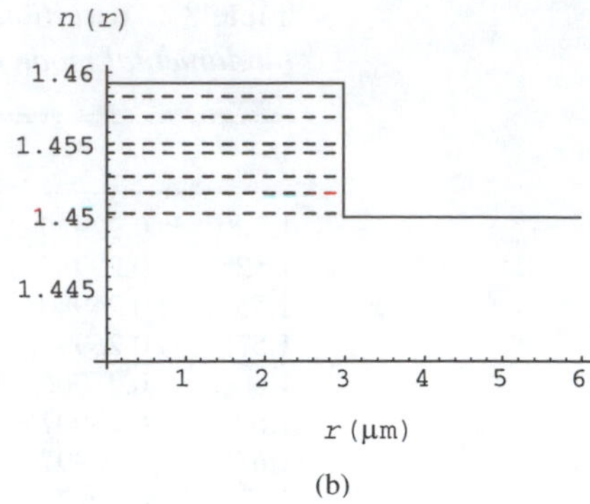

(b)

Fig. 8.2: The dashed horizontal lines correspond to the discrete values of β/k_0 of the guided modes corresponding to a step index fiber whose parameters are given by equation (8.43); (a) and (b) correspond to $\lambda_0 = 1.546\,\mu$m ($V = 2$) and 0.4757 μm ($V = 6.5$), respectively.

so that $V = 2.0$ at $\lambda_0 = 1.546\,\mu$m. Thus

$$\frac{\beta}{k_0} = \left[n_2^2 + b\left(n_1^2 - n_2^2\right)\right]^{1/2}$$

$$\simeq 1.4539$$

(see Figure 8.2(a)). Another fiber with

$$n_2 = 1.45, \quad \Delta = 0.010, \quad a = 2.0\,\mu\text{m} \tag{8.44}$$

will have $V = 2.0$ at $\lambda_0 = 1.288\,\mu$m. The corresponding value of β/k_0 will be given by

$$\frac{\beta}{k_0} \simeq 1.4560$$

(see Figure 8.3(a)).

Example 8.2: We next consider a higher value of V, say $V = 6.5$. In Figure 8.4 we have plotted the LHS and RHS of equation (8.41) [$l = 0$] and of equation (8.40) for $l = 1, 2, 3, 4$, and 5. One finds that there are two modes corresponding to $l = 0$ (which are referred to as LP_{01} and LP_{02} modes), two modes corresponding to $l = 1$ (which are referred to as LP_{11} and LP_{12} modes), and one mode each corresponding to $l = 2, l = 3$, and $l = 4$ (which are referred to as the LP_{21}, LP_{31}, and LP_{41} modes, respectively). The corresponding values of b are given by

$$\begin{aligned} b = {}& 0.897699\,(\text{LP}_{01}), \quad 0.475182(\text{LP}_{02});\\ & 0.742163\,(\text{LP}_{11}), \quad 0.179216(\text{LP}_{12});\\ & 0.541097\,(\text{LP}_{21});\\ & 0.300334\,(\text{LP}_{31});\\ & 0.027816\,(\text{LP}_{41}); \end{aligned} \tag{8.45}$$

The last value of b is very close to cutoff. Now, the fibers characterized by equations (8.43) and (8.44) have $V = 6.5$ for $\lambda_0 \simeq 0.4757\,\mu$m and 0.3963 μm, respectively. The corresponding values of β/k_0 are shown as dashed lines in Figures 8.2(b) and 8.3(b).

Table 8.2. *Cutoff frequencies of various LP_{lm} modes in a step index fiber*

$l = 0$ modes $(J_1(V_c) = 0)$		$l = 1$ modes $(J_0(V_c) = 0)$	
Mode	V_c	Mode	V_c
LP_{01}	0	LP_{11}	2.4048
LP_{02}	3.8317	LP_{12}	5.5201
LP_{03}	7.0156	LP_{13}	8.6537
LP_{04}	10.1735	LP_{14}	11.7915
$l = 2$ modes $(J_1(V_c) = 0\,; V_c \neq 0)$		$l = 3$ modes $(J_0(V_c) = 0; V_c \neq 0)$	
Mode	V_c	Mode	V_c
LP_{21}	3.8317	LP_{31}	5.1356
LP_{22}	7.0156	LP_{32}	8.4172
LP_{23}	10.1735	LP_{33}	11.6198
LP_{24}	13.3237	LP_{34}	14.7960

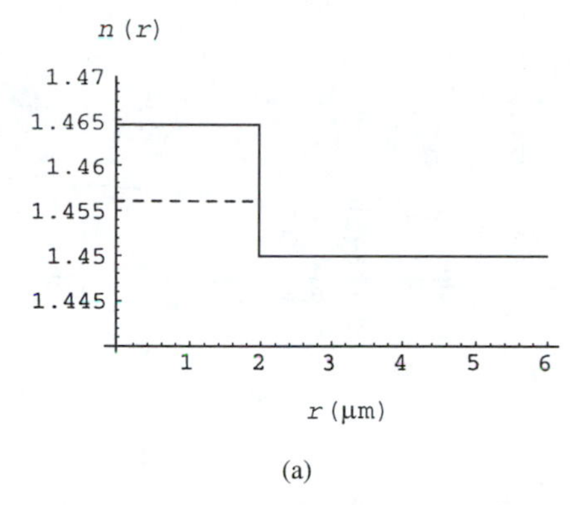

(a)

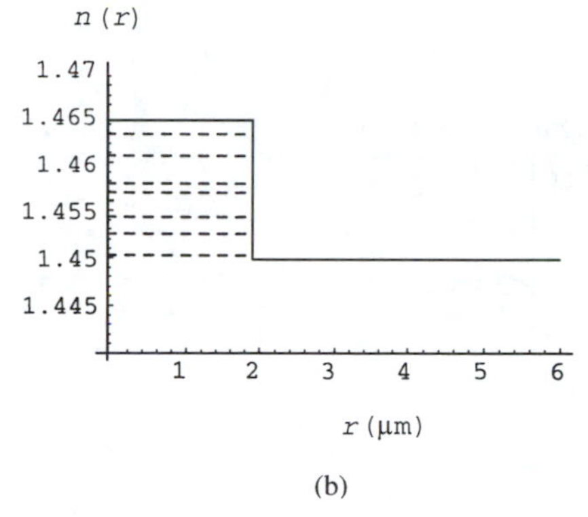

(b)

Fig. 8.3: The dashed horizontal lines correspond to the values of β/k_0 of the guided modes corresponding to a step index fiber whose parameters are given by equation (8.44); (a) and (b) correspond to $\lambda_0 = 1.288\ \mu$m ($V = 2$) and $0.3963\ \mu$m ($V = 6.5$), respectively.

By studying the zeros of Bessel functions (see Table 8.2) one can immediately see that, since $0 < b < 1$, there will be only a finite number of guided modes. For example, consider the LHS of equation (8.41).

$$f(b) = V(1-b)^{1/2}\frac{J_1[V(1-b)^{1/2}]}{J_0[V(1-b)^{1/2}]} \tag{8.46}$$

Now

$$f(b) = 0, \quad \text{when} \quad V(1-b)^{1/2} = 0, 3.8317, 7.0156, \ldots$$

and

$$f(b) = \infty, \quad \text{when} \quad V(1-b)^{1/2} = 2.4048, 5.5201, \ldots$$

For $V = 6.5$

$$f(b) = 0, \quad \text{when} \quad b = 1, 0.6525$$

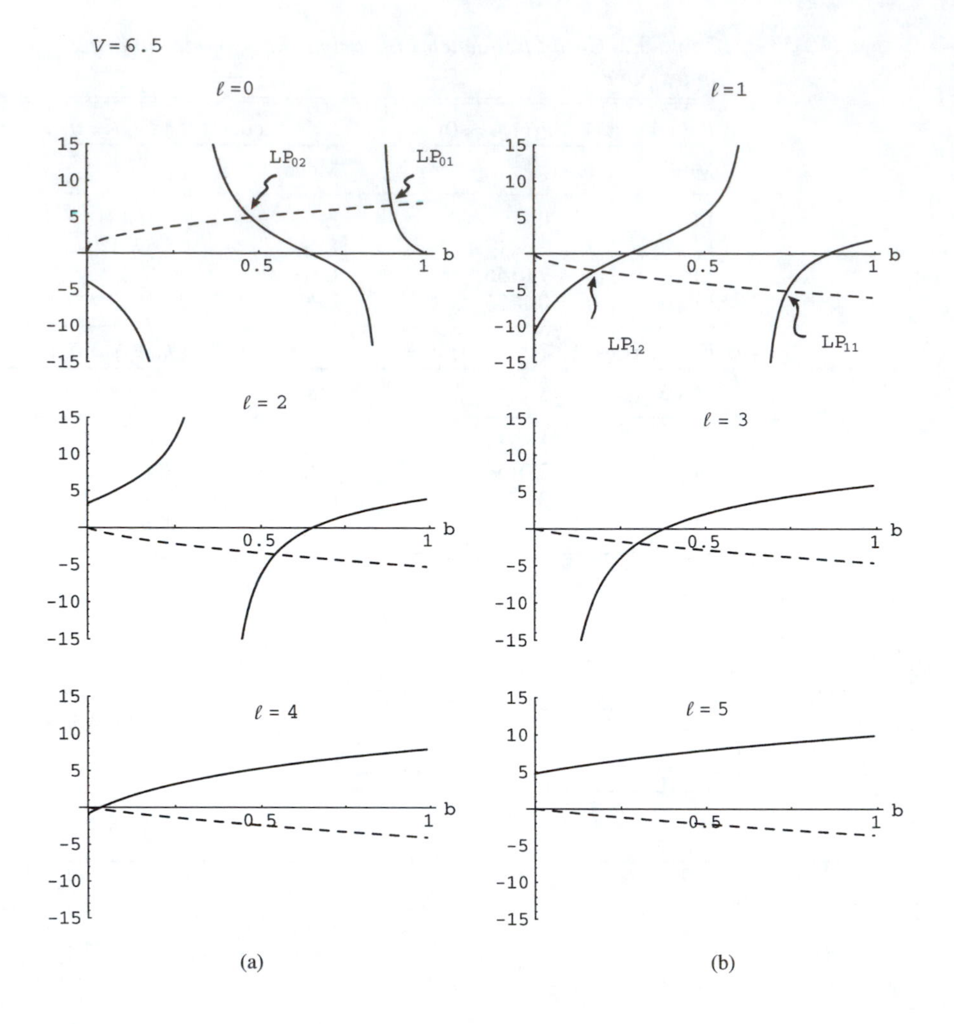

(a) (b)

Fig. 8.4: Variations of LHS (solid curves) and RHS (dashed curves) of equations (8.40) and (8.41) for $V = 6.5$ corresponding to $l = 0, 1, 2, 3, 4$ and 5. Points of intersection correspond to guided modes of the step index fiber.

and

$$f(b) = \infty, \quad \text{when} \quad b = 0.8631, 0.2788$$

(see Figure 8.4). By knowing the positions of zeros and infinities and the fact that $f(b)$ is positive in the vicinity of $b = 1$, one can easily make a qualitative plot and determine the number of modes and also the approximate values of b. One can make a similar analysis for $l \geq 1$.

The guided modes that are given by the points of intersection in Figure 8.4 are designated in decreasing values of b as LP_{l1}, LP_{l2}, LP_{l3}, etc. The variation of b with V forms a set of universal curves, which are plotted in Figure 8.5. As can be seen, at a particular V value there are only a finite number of modes.

The condition $b = 0$ (i.e., $\beta^2 = k_0^2 n_2^2$) corresponds to what is known as the *cutoff* of the mode. For $b < 0$, $\beta^2 < k_0^2 n_2^2$ and the fields are oscillatory even in the cladding and we have what are known as *radiation modes*. Obviously, at cutoff $\beta = k_0 n_2$ implying

$$b = 0, \quad W = 0, \quad U = V = V_c \tag{8.47}$$

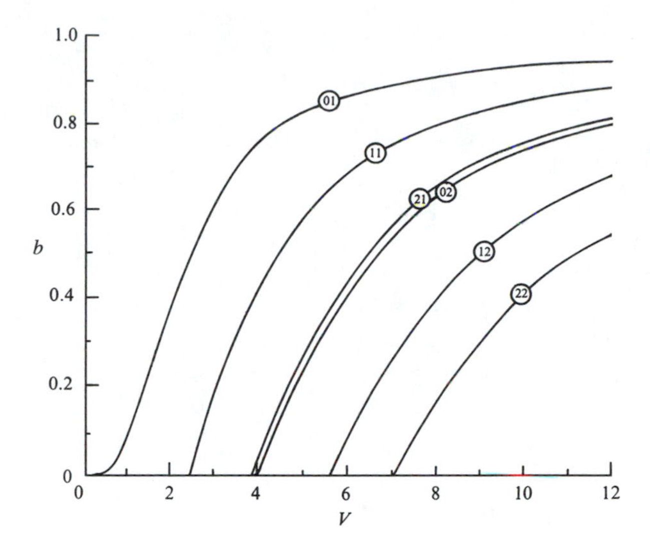

Fig. 8.5: Variation of the normalized propagation constant b with normalized frequency V for a step index fiber corresponding to some low-order modes. The cutoff frequencies of LP_{2m} and $LP_{0,m+1}$ modes are the same. [Adapted from Gloge (1971).]

The cutoffs of various modes are determined from the following equations

$$\left.\begin{array}{lll} l = 0 \text{ modes}: & J_1(V_c) = 0 & \\ l = 1 \text{ modes}: & J_0(V_c) = 0 & \\ l \geq 2 \text{ modes}: & J_{l-1}(V_c) = 0; & V_c \neq 0 \end{array}\right\} \tag{8.48}$$

Note that for $l \geq 2$, the root $V_c = 0$ must not be included since

$$\lim_{V \to 0} V \frac{J_{l-1}(V)}{J_l(V)} \neq 0 \quad \text{for } l \geq 2 \tag{8.49}$$

Thus, the cutoff V values (also known as normalized cutoff frequencies) occur at the zeroes of Bessel functions and are tabulated in Table 8.2.

As is obvious from the above analysis and also from Figure 8.5 for a step index fiber with

$$0 < V < 2.4048 \tag{8.50}$$

we will have only one guided mode – namely, the LP_{01} mode. Such a fiber is referred to as a single-mode fiber and is of tremendous importance in optical communication systems. As discussed in Chapter 10, the dispersion curves can be used to calculate the group velocity and the intramodal dispersion of various modes of a step index fiber. Similarly

for $2.4048 < V < 3.8317$, only LP_{01} and LP_{11} modes;
for $3.8317 < V < 5.1356$, only LP_{01}, LP_{11}, LP_{21}, and LP_{02} modes;
for $5.1356 < V < 5.5201$, only LP_{01}, LP_{11}, LP_{21}, LP_{02}, and LP_{31} modes, and so forth

will exist as guided modes. The numbers limiting the V value correspond to the zeroes of the Bessel functions. The curves in Figure 8.5 are universal – that is, for a given step index fiber and a given operating wavelength we have to first calculate the value of V and then "read off" the exact values of b from the

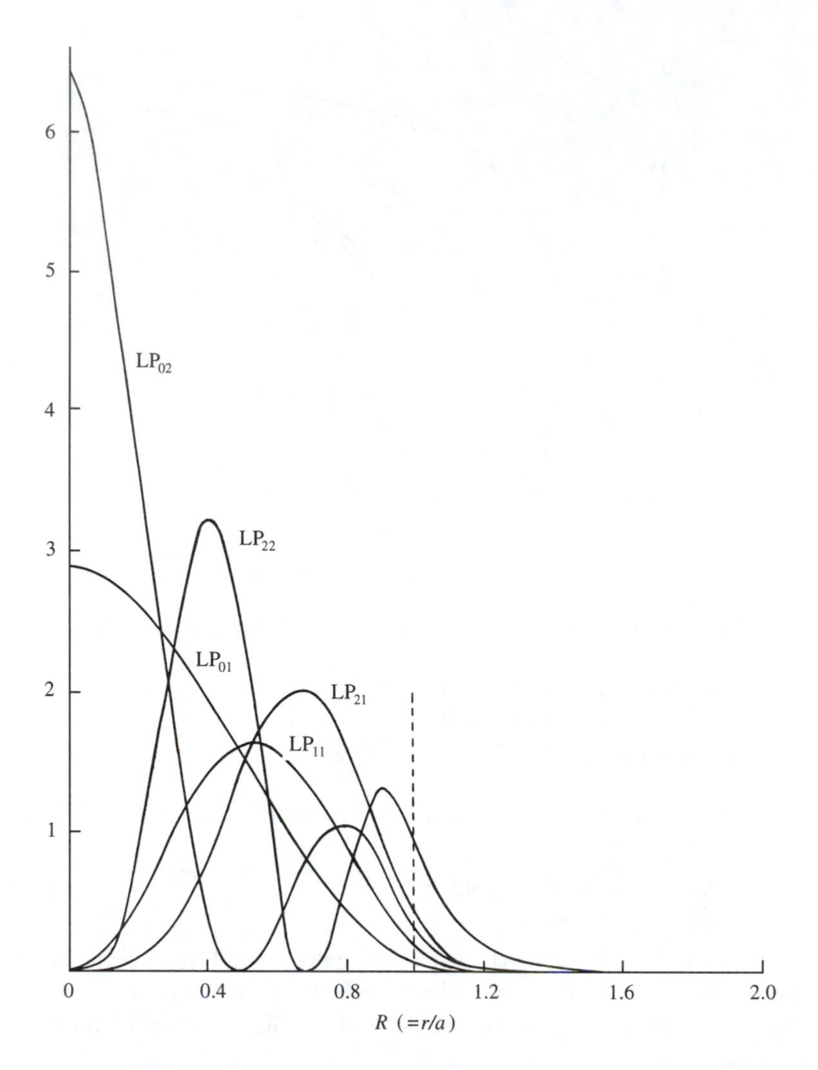

Fig. 8.6: Radial intensity distributions (normalized to the same power) of some low-order modes in a step index fiber for $V = 8$. Note that the higher order modes have a greater fraction of power in the cladding.

curves; the values of β can then be calculated by using the relation

$$\beta_{lm} = k_0\left[n_2^2 + b_{lm}\left(n_1^2 - n_2^2\right)\right]^{1/2}. \tag{8.51}$$

The radial field distributions and the schematic intensity patterns of some lower order modes are shown in Figures 8.6 and 8.7. The computer-generated field and intensity distributions for some of the modes of a $V = 8$ step index fiber are shown in Figures 8.8–8.10. Figures 8.8 and 8.9 correspond to $l = 0$ and $l = 1$, respectively; Figure 8.10 corresponds to $l = 2$ and $l = 4$. The white and black shadings in the field distribution represent the π phase reversal because of the r-dependence or ϕ-dependence of the modal field. Indeed, whenever the field passes through a zero there will be a reversal of phase. Obviously, the same will not be noticed in the intensity distribution. We may note the following points:

(i) The $l = 0$ modes are two-fold degenerate corresponding to two independent states of polarization.

(ii) The $l \geq 1$ modes are four-fold degenerate because, for each polarization, the ϕ dependence could be either $\cos l\phi$ or $\sin l\phi$.

Fig. 8.7: Schematic of the modal field patterns for some low-order modes in a step index fiber. The arrows represent the direction of the electric field.

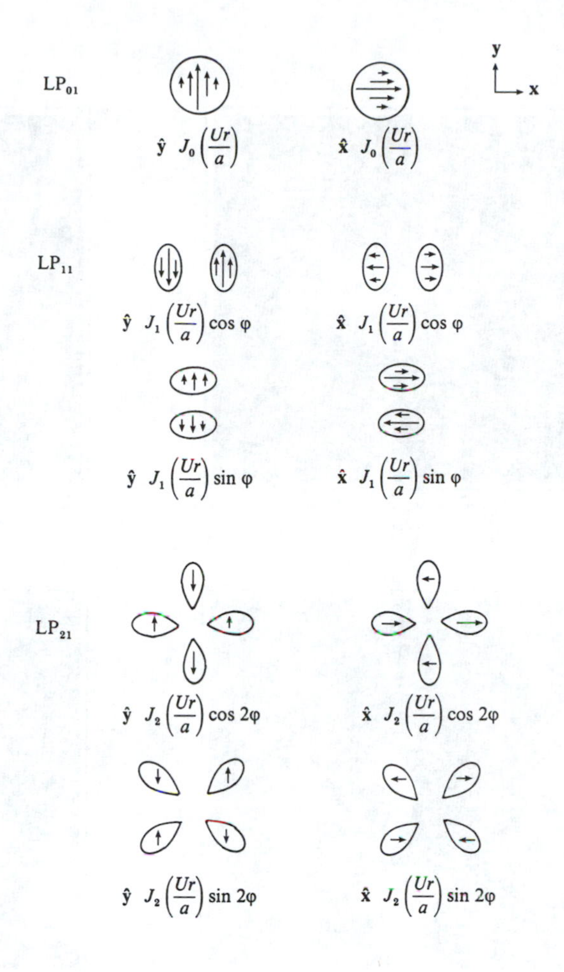

Further

$$\text{number of zeroes in the } \phi \text{ direction} = 2l \tag{8.52}$$

and

$$\text{number of zeroes in the radial direction (excluding } r = 0) = m - 1 \tag{8.53}$$

In Chapter 9 we show that when $V \gg 1$, the total number of modes is given by

$$N \approx V^2/2 \tag{8.54}$$

and such a fiber that supports a large number of guided modes is known as a multimode fiber. For a typical multimode step index fiber

$$n_1 = 1.47, \quad n_2 = 1.46, \quad a = 25\,\mu\text{m} \tag{8.55}$$

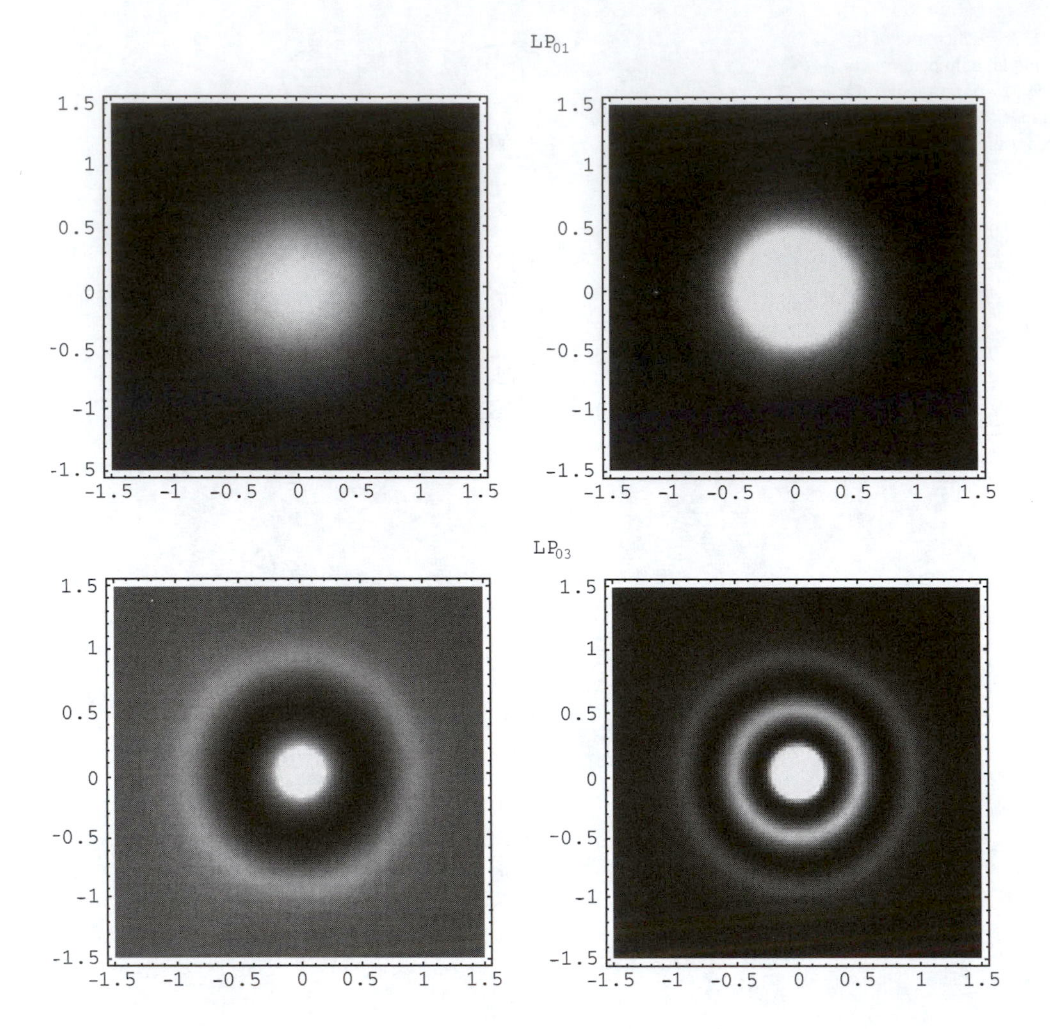

Fig. 8.8: Computer-generated field and intensity distributions of the LP_{01} and LP_{03} modes for a step index fiber with $V = 8$. The field distributions are on the left and the intensity distributions are on the right. The white and black shadings in the field distribution represent the π phase reversal because of the modal field passing through a zero.

and for $\lambda_0 = 0.8\,\mu$m, we obtain

$$V \approx 34$$

which would support approximately 580 modes.

8.4 Fractional modal power in the core

One of the important parameters associated with a fiber optic waveguide is the fractional power carried in the core. Now, the power in the core of the fiber is given by

$$\begin{aligned} P_{\text{core}} &= \text{const.} \int_0^a \int_0^{2\pi} |\psi|^2 r\, dr\, d\phi \\ &= \frac{2C}{J_l^2(U)} \int_0^a J_l^2\left(\frac{Ur}{a}\right) r\, dr \int_0^{2\pi} \cos^2 l\phi\, d\phi \end{aligned}$$

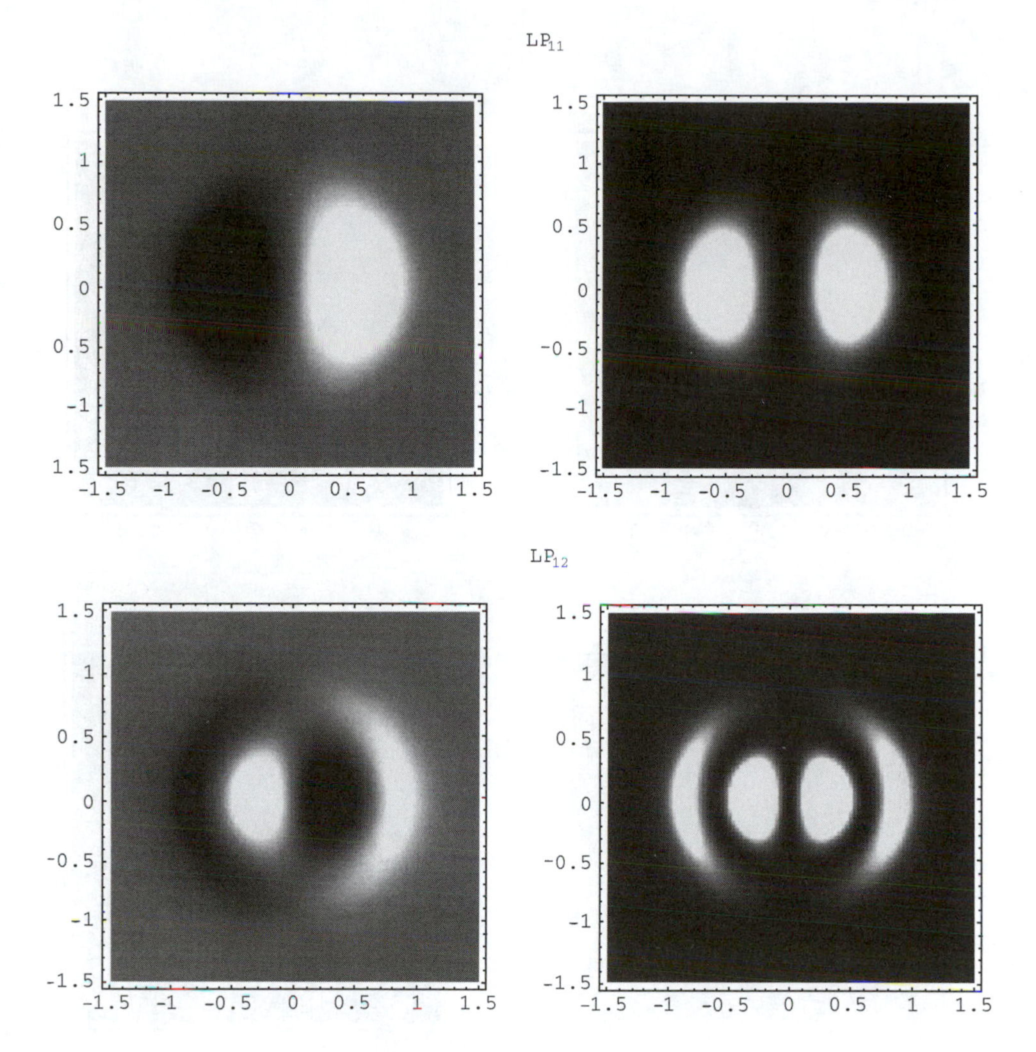

Fig. 8.9: Computer-generated field and intensity distributions of the LP_{11} and LP_{12} modes for a step index fiber with $V = 8$. The field distributions are on the left and the intensity distributions are on the right. The white and black shadings in the field distribution represent the π phase reversal because of the modal field passing through a zero.

or

$$P_{\text{core}} = C\frac{\pi a^2}{U^2}\frac{2}{J_l^2(U)}\int_0^U J_l^2(x)x\,dx$$

$$= C\pi a^2\left[1 - \frac{J_{l-1}(U)J_{l+1}(U)}{J_l^2(U)}\right] \tag{8.56}$$

where C is a constant and use has been made of standard integrals associated with Bessel functions. Similarly, the power in the cladding is given by

$$P_{\text{clad}} = \text{const.}\int_a^\infty\int_0^{2\pi} |\psi|^2 r\,dr\,d\phi$$

$$= C\pi a^2\left[\frac{K_{l-1}(W)K_{l+1}(W)}{K_l^2(W)} - 1\right] \tag{8.57}$$

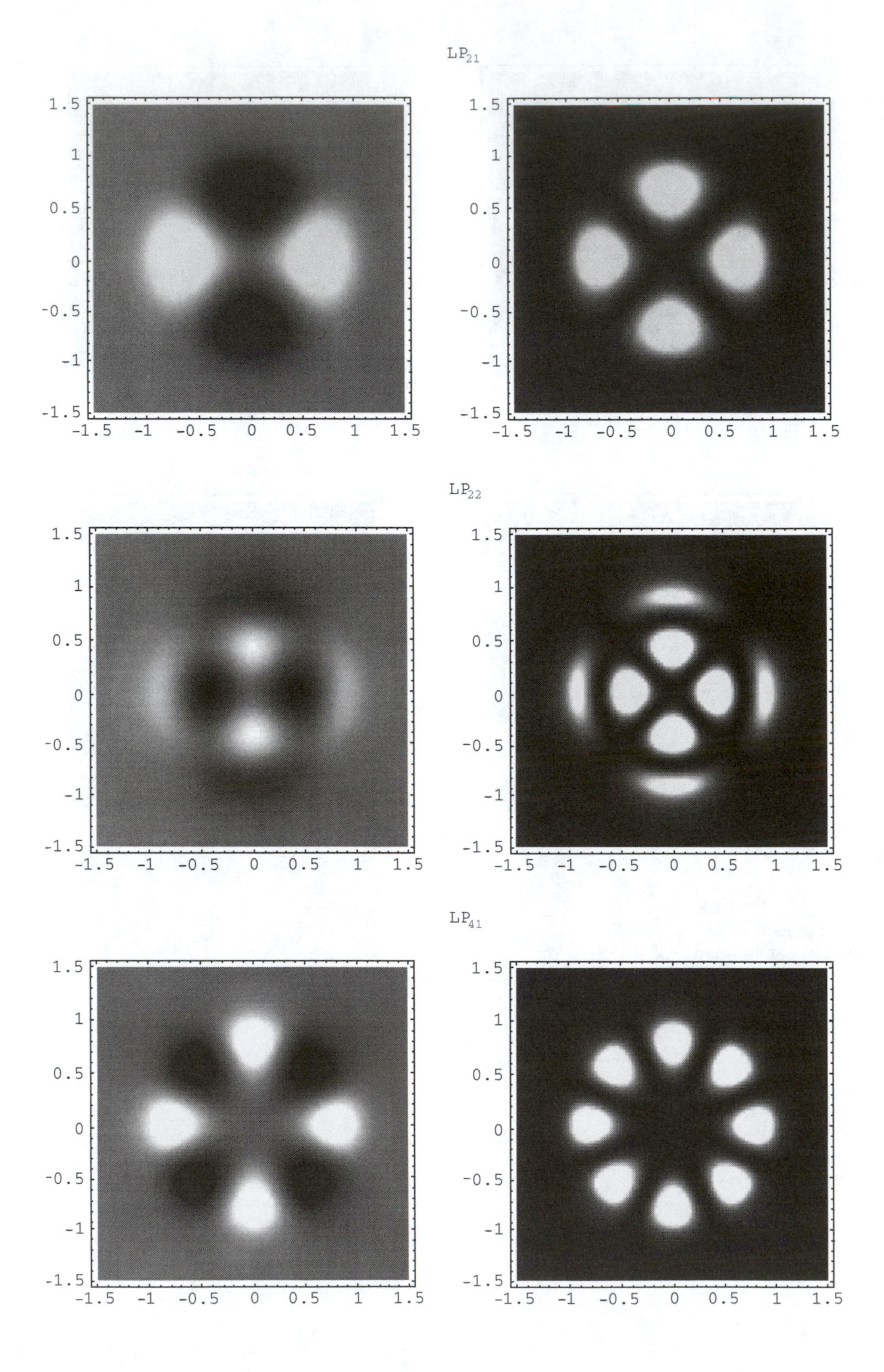

LP21
LP22
LP41
1.5
1
0.5
0
-0.5
-1
-1.5

Fig. 8.10: Computer-generated field and intensity distributions of the LP_{21}, LP_{22}, and LP_{41} modes for a step index fiber with $V = 8$. The field distributions are on the left and the intensity distributions are on the right. The white and black shadings in the field distribution represent the π phase reversal because of the modal field passing through a zero.

The total power is

$$\begin{aligned} P_{\text{tot}} &= P_{\text{core}} + P_{\text{clad}} \\ &= C\pi a^2 \frac{V^2}{U^2} \frac{K_{l+1}(W)K_{l-1}(W)}{K_l^2(W)} \end{aligned} \tag{8.58}$$

where use has been made of the equation

$$\frac{U^2 J_{l+1}(U) J_{l-1}(U)}{J_l^2(U)} = -W^2 \frac{K_{l+1}(W)K_{l-1}(W)}{K_l^2(W)} \tag{8.59}$$

which follows from the eigenvalue equations. The fractional power propagating in the core is thus given by

$$\eta = \frac{P_{\text{core}}}{P_{\text{tot}}} = \left[\frac{W^2}{V^2} + \frac{U^2}{V^2} \frac{K_l^2(W)}{K_{l+1}(W)K_{l-1}(W)} \right] \tag{8.60}$$

Thus, as the mode approaches cutoff

$$V \to V_c, \quad W \to 0, \quad U \to V_c \tag{8.61}$$

and if we now use the limiting forms of $K_l(W)$ given in footnote (5) on p. 137 of this chapter, we obtain

$$\eta \to \begin{cases} 0 & \text{for } l = 0 \text{ and } 1 \\ (l-1)/l & \text{for } l \geq 2 \end{cases} \tag{8.62}$$

In Figure 8.11 we have plotted the fractional power contained in the core and in the cladding as a function of V for various modes of a step index fiber. Note that the power associated with a particular mode is concentrated in the core for large values of V – that is, far from cutoff.

8.5 Single-mode fibers

Until now our discussion has been on the general modal analysis of a step index fiber. For highly multimoded fibers ($V \gtrsim 10$), one can use ray optics to describe their propagation characteristics. However, most fibers used today in fiber optic communication systems are single moded and for such fibers we have to use the modal analysis. In this section we study the various characteristics of single-mode fibers that are important in connection with their application in communication systems. The dispersion characteristics of single-mode fibers are discussed in Chapter 10.

8.5.1 The Gaussian approximation

The fundamental mode field distribution for a single-mode fiber is a very important characteristic that determines various important parameters such as splice loss at joints, launching efficiencies, bending loss, and so forth. In Figure 8.12 we have given some typical refractive index profiles of single-mode

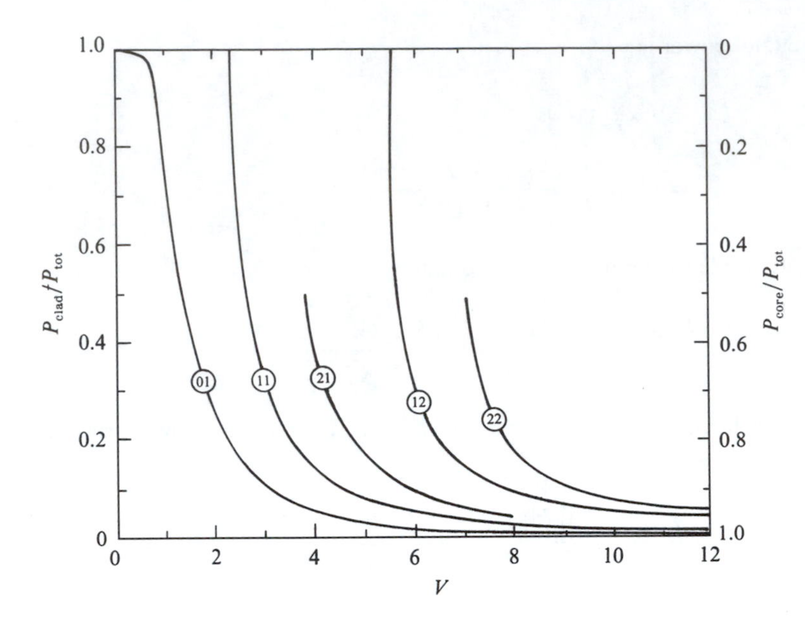

Fig. 8.11: Variation of the fractional power contained in the cladding with V for some low-order modes in a step index fiber. [Adapted from Gloge (1971).]

fibers and their corresponding fundamental mode field distributions. Also plotted in the figure are best-fit Gaussian functions and it can be seen that irrespective of the refractive index profile, the fundamental mode field distribution can be well approximated by a Gaussian function[6] that may be written in the form

$$\psi(r) = Ae^{-r^2/w^2} \tag{8.63}$$

where w is referred to as the spot size of the mode field pattern. The quantity $d = 2w$ is usually referred to as the mode field diameter (MFD). One of the criteria widely used to choose the value of w is that which leads to the maximum launching efficiency of the exact fundamental mode field by an incident Gaussian field – that is, one maximizes the quantity

$$\eta = \frac{\int_0^\infty e^{-r^2/w^2} R(r) r\, dr}{\left[\int_0^\infty e^{-2r^2/w^2} r\, dr \int_0^\infty R^2(r) r\, dr\right]^{\frac{1}{2}}} \tag{8.64}$$

where $R(r)$ represents the exact modal field. For example, for a step index fiber $R(r)$ can be expressed in terms of Bessel functions and one has the following empirical expression for w [Marcuse (1977)].

$$\frac{w}{a} \approx \left(0.65 + \frac{1.619}{V^{3/2}} + \frac{2.879}{V^6}\right); \quad 0.8 \lesssim V \lesssim 2.5 \tag{8.65}$$

where a is the core radius. The above empirical formula gives a value of w (as obtained by maximizing η) to within about 1%. For a typical single-mode

[6]Indeed, for an infinitely extended parabolic index medium, the fundamental modal field is exactly Gaussian (see Section 9.3).

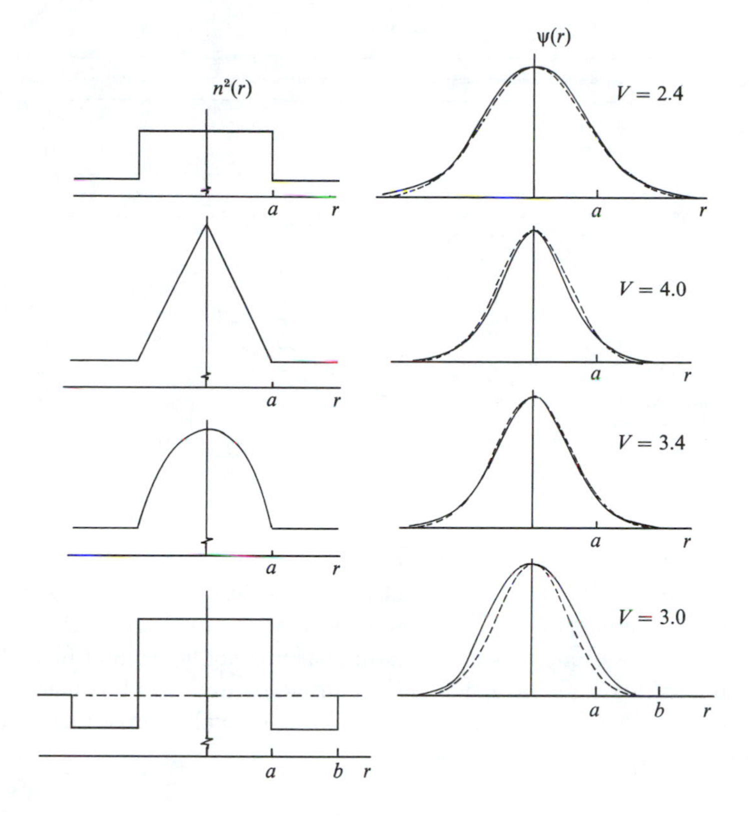

Fig. 8.12: Typical fundamental modal field shapes $\psi(r)$ for fibers with different profiles. The solid curves represent exact field variations and the dashed curves represent the best-fitted Gaussian function. [Adapted from Ghatak and Sharma (1986).]

fiber with $n_1 = 1.454$, $n_2 = 1.45$, $a = 4.46\,\mu$m, operating at 1300 nm, the mode field diameter is 10.0 μm. At 1500 nm the same fiber would have a mode field diameter of 11.2 μm (assuming the same values of n_1 and n_2 at 1500 nm).

Note also from equation (8.65) that two nonidentical single-mode fibers can have the same spot size at a given wavelength. Thus, a single-mode fiber with $V = 2.2$ and $a = 4.27\,\mu$m also has the same mode field diameter of 10.0 μm at 1300 nm as the previously considered fiber.

We should mention here that the different refractive index profiles shown in Figure 8.12 are of considerable technological importance; for example, as we discuss in Chapter 10, one can tailor the dispersion characteristics by a proper choice of refractive index variation inside the core.

8.5.2 Splice loss

One of the great advantages of the Gaussian approximation is that it gives us simple analytical expressions for losses at joints between two single-mode fibers. Figure 8.13 shows three common misalignments at a joint between two single-mode fibers. Even if none of the misalignments shown in Figure 8.13 exists, there will still be losses at the joint caused by nonidentical field distributions for dissimilar single-mode fibers. In this section we will use the Gaussian approximation to obtain the splice losses and estimate typical tolerances in the various misalignments for keeping the losses below practical limits.

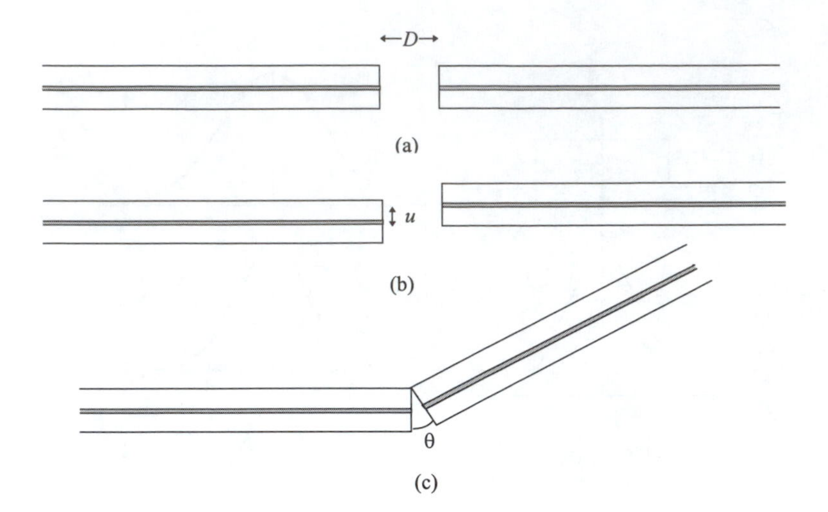

Fig. 8.13: (a), (b), and (c) correspond to longitudinal, transverse, and angular, misalignments at a joint between two single-mode fibers.

In the following we will derive formulae for the loss at a joint between two single-mode fibers. These formulae give the loss due to each separate misalignment in the absence of others. In general, the total loss in the simultaneous presence of more than one misalignment is not the sum of individual misalignment losses. If each individual contribution to loss is less than 1 dB, then the total loss is approximately the sum of individual losses calculated separately.

8.5.2.1 *Loss due to transverse misalignment*

As a specific example, we consider transverse misalignment of two single-mode fibers that are represented by Gaussian fundamental modes with spot sizes w_1 and w_2. Let us consider the direction of misalignment to be along the x direction. With respect to the coordinate axes fixed on the fiber, the normalized Gaussian modes can be represented by

$$\psi_1(x, y) = \left(\frac{2}{\pi}\right)^{1/2} \frac{1}{w_1} e^{-(x^2+y^2)/w_1^2} \tag{8.66}$$

$$\psi_2(x, y) = \left(\frac{2}{\pi}\right)^{1/2} \frac{1}{w_2} e^{-[(x-u)^2+y^2]/w_2^2} \tag{8.67}$$

where u is the transverse misalignment and the multiplying factors are such that

$$\int_{-\infty}^{+\infty} \int_{-\infty}^{+\infty} \psi_{1,2}^2 \, dx \, dy = 1$$

The fractional power that is coupled to the fundamental mode of the second fiber is given by

$$T = \left| \int_{-\infty}^{+\infty} \int_{-\infty}^{+\infty} \psi_1 \psi_2^* \, dx \, dy \right|^2 \tag{8.68}$$

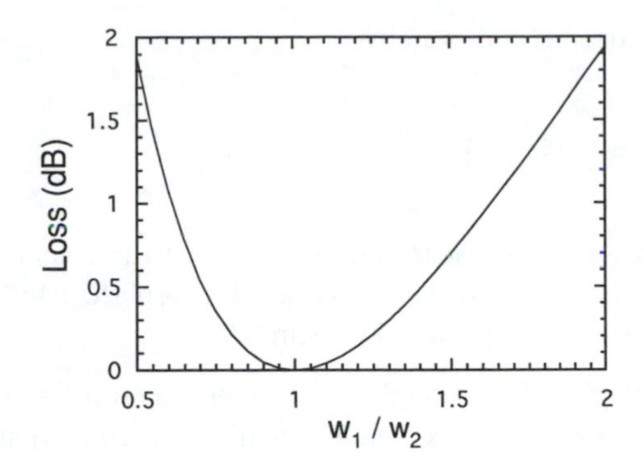

Fig. 8.14: Variation of loss at a perfectly aligned joint between two single-mode fibers with spot sizes w_1 and w_2 as a function of the ratio w_1/w_2.

Thus

$$T = \left| \frac{2}{\pi w_1 w_2} \int_{-\infty}^{+\infty} \int_{-\infty}^{+\infty} \exp\left[-x^2\left(\frac{1}{w_1^2} + \frac{1}{w_2^2}\right) + \frac{2xu}{w_2^2} - y^2\left(\frac{1}{w_1^2} + \frac{1}{w_2^2}\right) - \frac{u^2}{w_2^2} \right] dx\, dy \right|^2$$

$$= \left(\frac{2w_1 w_2}{w_1^2 + w_2^2}\right)^2 \exp\left[-\frac{2u^2}{(w_1^2 + w_2^2)}\right] \tag{8.69}$$

The maximum power coupling appears at $u = 0$ and the transmitted fractional power is

$$T_{\max} = \left(\frac{2w_1 w_2}{w_1^2 + w_2^2}\right)^2 \tag{8.70}$$

which is unity for two fibers having identical Gaussian fundamental modes. Thus, for a perfectly aligned joint between two single-mode fibers with spot sizes w_1 and w_2, the loss is given by

$$\alpha_n(\text{dB}) = -20 \log\left(\frac{2w_1\, w_2}{w_1^2 + w_2^2}\right) \tag{8.71}$$

Figure 8.14 shows a plot of α_n versus w_1/w_2. This loss will be less than 0.1 dB provided

$$0.86 < w_1/w_2 < 1.16 \tag{8.72}$$

Hence, even variations of 14% in the spot size result only in a loss of less than 0.1 dB.

For two identical fibers ($w_1 = w_2 = w$), the transverse offset loss varies as

$$T = e^{-u^2/w^2} \tag{8.73}$$

Thus, the loss in decibels is given by

$$\alpha_t(\text{dB}) = 4.34\left(\frac{u}{w}\right)^2 \tag{8.74}$$

Note that larger values of w lead to greater tolerance to transverse misalignment. (A more accurate expression for α_t is obtained if we replace w by the Petermann-2 spot size w_P; this is discussed in Problem 8.9.)

Example 8.3: For a single-mode fiber operating at 1300 nm, $w = 5\,\mu\text{m}$, and if α_t is to be below 0.1 dB, then from equation (8.74) we obtain

$$u < 0.76\,\mu\text{m}$$

Hence, for a low-loss joint, the transverse alignment is very critical and connections for single-mode fibers require precision matching and positioning for achieving low loss.

8.5.2.2 *Angular misalignment*

For an angular misalignment of θ (see Figure 8.13(c)) between the axes of two single-mode fibers with spot sizes w, the loss is given by (see Problem 8.10)

$$\alpha_a(\text{dB}) = 4.34\left(\frac{\pi n_l w\theta}{\lambda_0}\right)^2 \tag{8.75}$$

where n_l is the refractive index of the medium between the fiber ends, λ_0 is the free space wavelength, and θ is measured in radians. Note that smaller values of w lead to greater tolerance to angular misalignment.

Example 8.4: For a single-mode fiber with a spot size of 5 μm operating at 1300 nm, assuming $n_l = n_2 = 1.45$, if the splice loss due to angular misalignment is to be less than 0.1 dB, then

$$\theta \lesssim 0.5^\circ \tag{8.76}$$

In the presence of both transverse and angular misalignment, the total loss is approximately given by

$$\alpha(\text{dB}) = 4.34\left[\left(\frac{u}{w}\right)^2 + \left(\frac{\pi n_l w\theta}{\lambda_0}\right)^2\right] \tag{8.77}$$

Figure 8.15 shows the variation of normalized transverse misalignment (u/w) with respect to the normalized angular misalignment $n_l w\theta/\lambda_0$ with joint loss as a parameter.

If u_0 and θ_0 represent the transverse misalignment and angular misalignment required to introduce the same splice loss α_0, then from equations (8.74) and

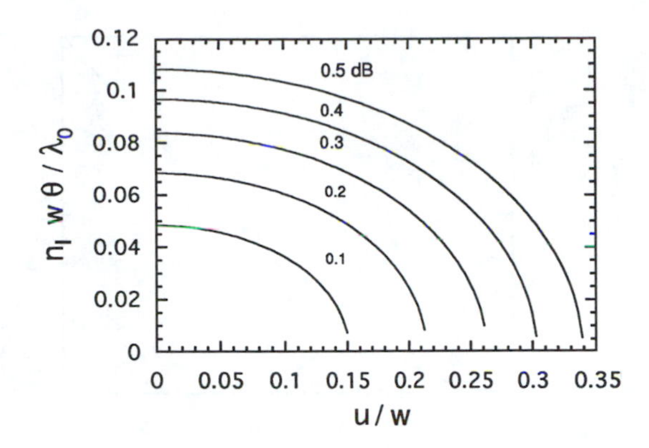

Fig. 8.15: Plot of variation of the normalized angular misalignment with respect to normalized transverse misalignment leading to different losses across a joint.

(8.75) we obtain

$$u_0 = \sqrt{\frac{\alpha_0}{4.34}}\, w \tag{8.78}$$

$$\theta_0 = \sqrt{\frac{\alpha_0}{4.34}}\, \frac{\lambda_0}{\pi n_l w} \tag{8.79}$$

Thus

$$u_0 \cdot \theta_0 = \frac{\lambda_0}{\pi n_l} \frac{\alpha_0}{4.34} \tag{8.80}$$

independent of the spot size w. Hence, any attempt to increase tolerance to transverse misalignment by increasing w will be accompanied by a tighter tolerance toward angular misalignment and vice versa. Since angular tolerances usually do not pose problems, it may be tempting to increase the spot size to large values. This is usually accompanied by an increased bending loss. Thus, there is an optimum value of spot size for a given operating wavelength. For 1300 nm, the optimum spot size is almost 5 μm.

It is interesting to note that the losses given by equations (8.71), (8.74), and (8.75) also give the coupling loss when a Gaussian beam from a laser excites a single-mode fiber. Hence, they can also be used to calculate excitation efficiency.

8.5.2.3 *Longitudinal misalignment*

For a longitudinal misalignment of D (see Figure 8.13(a)), the splice loss is given by (see Problem 8.11)

$$\alpha_l = 10\,\log(1 + \tilde{D}^2) \tag{8.81}$$

where

$$\tilde{D} = \frac{D\lambda_0}{2\pi n_l w^2} \tag{8.82}$$

Figure 8.16 shows the variation of α_l with $\tilde{D}$.

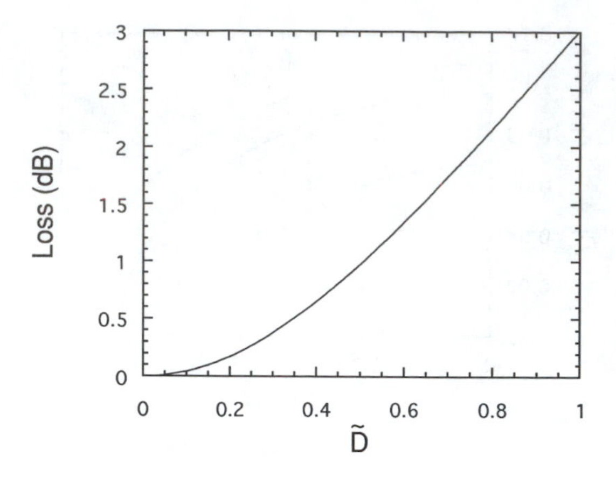

Fig. 8.16: Variation of loss at a joint with longitudinal misalignment with normalized separation $\tilde{D}$.

Example 8.5: Let us again consider a single-mode fiber with $w = 5\,\mu$m operating at 1300 nm with a longitudinal misalignment of 20 μm. Assuming $n_l = n_2 = 1.45$, we obtain $\tilde{D} = 0.114$ and

$$\alpha_l = 0.06\,\text{dB}$$

Thus, longitudinal misalignment is not a major source of loss.

The loss given by equation (8.81) considers only the loss caused due to the diffraction of light as it propagates from one fiber end to the other (see Figure 8.13(a)). In addition, there would also be Fresnel reflection loss at the ends of the launching and receiving fibers. This can be minimized by choosing a value of n_l close to that of core or cladding. If the fibers are spaced by an air gap, then the reflections at the output end of the launching fiber and the input end of the receiving fiber are larger and, in addition, if the fiber end faces are exactly parallel, multiple reflection effects (or Fabry–Perot effects) can dominate. In such a case the transmission through the splice oscillates as a function of the fiber end separation; the spatial period of oscillation will be approximately $\lambda/2$. Under some circumstances this could lead to feedback into a semiconductor laser source, resulting in laser instabilities.

8.5.2.4 Measuring spot size from splice loss measurements

Using equation (8.74) we can derive a method for the experimental determination of the fiber spot size w. By measuring the splice loss across a joint as a function of transverse misalignment and fitting the measured variation to the formula given by equation (8.74), the spot size w can be determined. One could in fact obtain w as a function of wavelength by using a white light source combined with wavelength filters or by using a tunable laser source. This measured dependence of w on λ_0 can be used to obtain an equivalent step index fiber model of the given single-mode fiber.

8.5.3 The Petermann-2 spot size

One of the very important characteristics of a single-mode fiber is the Petermann-2 spot size defined by the following equation [Petermann (1983),

Pask (1984)]

$$w_P = \left[\frac{2 \int_0^\infty \psi^2(r)\, r\, dr}{\int_0^\infty \left(\frac{d\psi}{dr}\right)^2 r\, dr} \right] \tag{8.83}$$

where $\psi(r)$ represents the transverse field pattern of the fiber. Loss across a splice for small transverse offsets is proportional to $1/w_P^2$ and is more accurately given by (see Problem 8.9 and equation (8.74))

$$\alpha\,(\text{dB}) = 4.34 \left(\frac{u}{w_P} \right)^2 \tag{8.84}$$

where u represents the transverse offset. Indeed, as we will discuss in Section 10.5, according to CCITT (Comité Consultatif Internationale Télégraphique et Téléphonique) recommendations, the value of w_P should lie between 4.5 and 5 μm. Furthermore, the waveguide dispersion is given by (see Problem 8.12)

$$D_w = \frac{\Delta\tau}{L\Delta\lambda_0} = \frac{\lambda_0}{2\pi^2 c n_2} \frac{d}{d\lambda_0} \left(\frac{\lambda_0}{w_P^2} \right) \tag{8.85}$$

(The waveguide dispersion is discussed in detail in Chapter 10.)

Now, for a step index fiber the integrals in equation (8.83) can be evaluated to obtain [Hussey and Martinez (1985)]

$$\frac{w_P}{a} = \sqrt{2} \frac{J_1(U)}{W J_0(U)} \tag{8.86}$$

which is a function of the parameter V; the quantity w_P/a has been tabulated in Table 8.1. Variations of w_P with core radius and Δ are discussed in Section 10.5. A convenient empirical formula for w_P is given by [Hussey and Martinez (1985), Neumann (1988, p. 227)]

$$\frac{w_P}{a} = \frac{w}{a} - \left(0.016 + \frac{1.567}{V^7} \right)$$

where w is the Gaussian spot size given by equation (8.65). The above formula is accurate to within about 1% for $1.5 \leq V \leq 2.5$.

8.5.4 The far-field pattern

In this section we will discuss the far-field technique, which is one of the commonly used techniques for characterization of a single-mode fiber in terms of its mode field diameter.

8.5.4.1 Far field of a Gaussian mode field

In Section 8.5.1 we discussed the Gaussian approximation for the fundamental mode of a single-mode fiber and showed that for conventional single-mode fibers the fundamental modal field can be accurately described by a Gaussian

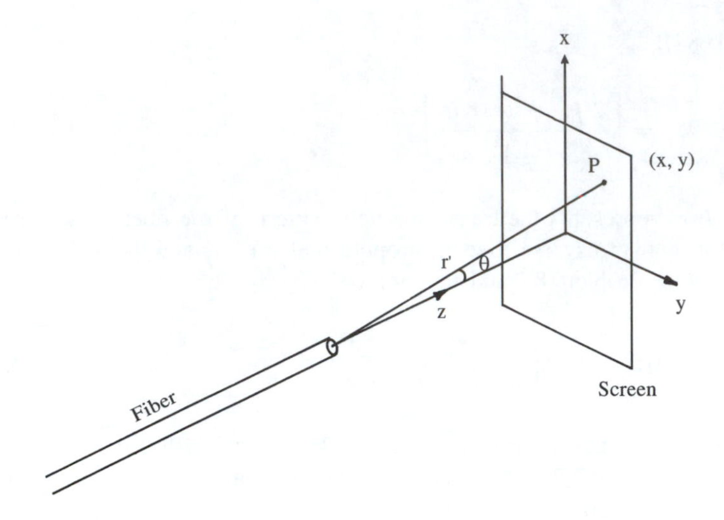

Fig. 8.17: The observation point P is at distance r' from the center of the exit end of the fiber and the direction of observation makes angle θ with the z-axis.

function of the form given by

$$\psi(r) = Ae^{-r^2/w^2} = Ae^{-(x^2+y^2)/w^2} \tag{8.87}$$

where A is a constant and w represents the (Gaussian) spot size. We will now calculate the far-field intensity distribution for a Gaussian modal field. For a near-field pattern $\psi(x, y)$, the far field pattern is given by (see, e.g., Ghatak and Thyagarajan (1989))

$$u = C \int_{-\infty}^{+\infty} \int_{-\infty}^{+\infty} \psi(\xi, \eta) e^{ik_0(l\xi + m\eta)}\, d\xi\, d\eta \tag{8.88}$$

where

$$l = \frac{x}{r'} \quad \text{and} \quad m = \frac{y}{r'} \tag{8.89}$$

represent the x and y direction cosines of the observation direction and $k_0(= 2\pi/\lambda_0)$ is the free space wave number. The quantities x and y represent the coordinates of the observation point P and r' is the distance of the point P from the axial point on the output end of the fiber (see Figure 8.17). Since the Gaussian near field has cylindrical symmetry, the far-field intensity distribution will also be cylindrically symmetric and will depend only on the angle θ (see Figure 8.17). Thus, we may calculate the field distribution along the x-axis for which

$$m = 0 \quad \text{and} \quad l = \frac{x}{r'} = \sin\theta \tag{8.90}$$

where θ is the angle made by the direction of observation with the z-axis. Substituting from equations (8.87) and (8.90) in equation (8.88) we obtain

$$u = AC \int_{-\infty}^{\infty} \int_{-\infty}^{\infty} \exp\left[-\frac{\xi^2 + \eta^2}{w^2}\right] e^{ik_0\xi \sin\theta}\, d\xi\, d\eta \tag{8.91}$$

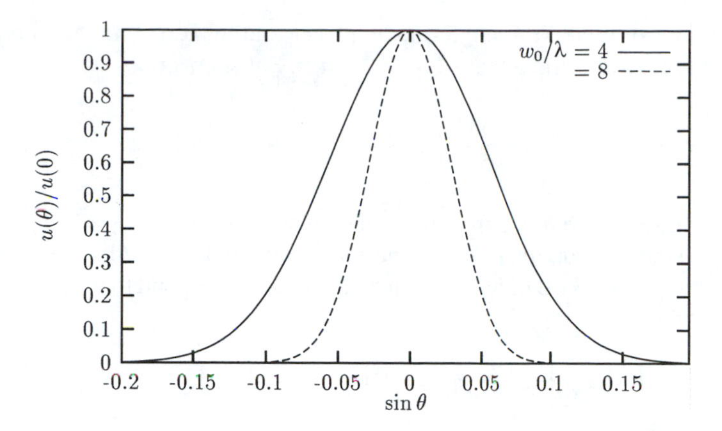

Fig. 8.18: The far-field amplitude distribution of a Gaussain mode having $w_0/\lambda_0 = 4$ and 8.

Using the standard integral

$$\int_{-\infty}^{+\infty} e^{-px^2+qx}\,dx = \sqrt{\frac{\pi}{p}}\, e^{q^2/4p} \tag{8.92}$$

equation (8.91) reduces to

$$u(\theta) = AC\pi w^2 \exp\left[-\frac{k_0^2 \sin^2\theta w^2}{4}\right] \tag{8.93}$$

The corresponding intensity distribution is

$$I(\theta) = I(0)\exp\left[-\frac{1}{2}k_0^2 w^2 \sin^2\theta\right] \tag{8.94}$$

Figure 8.18 shows a plot of the far-field distribution given by equation (8.94) corresponding to $w/\lambda_0 = 4$ and 8. Thus, the far-field pattern is peaked at $\theta = 0$ (i.e., along the axis), and the angle θ where the far-field amplitude falls to $1/e$ of the value along the axis is given by

$$\sin\theta_e = \frac{2}{k_0 w} = \frac{\lambda_0}{\pi w} \tag{8.95}$$

Thus, the measurement of θ_e gives w and hence the Gaussian mode field diameter $d(= 2w)$.

Note from equation (8.94) that the far-field intensity drops monotonically to zero as θ increases – that is, there are no side lobes. We later discuss the far field of a step index fiber and show the presence of side lobes.

Example 8.6: A conventional single-mode fiber operating at 1300 nm has a typical MFD of 10 μm – that is, $w = 5\ \mu$m. For such a fiber the angle θ_e is given by

$$\theta_e = \sin^{-1}\left(\frac{1.3}{5\pi}\right) \simeq 4.75^\circ$$

Example 8.7: A single-mode fiber operating at 633 nm has a typical MFD of 5 μm – that is, $w = 2.5\ \mu$m. For such a fiber

$$\theta_e = \sin^{-1}\left(\frac{0.633}{2.5\pi}\right) \simeq 4.62^\circ$$

Example 8.8: On measuring the far-field pattern at $\lambda_0 = 1.3\ \mu$m of a single-mode fiber, it is found that, at an angle of 2.74°, the intensity drops by 3 dB of its value on axis. Let θ_h correspond to the 3 dB point; then

$$\frac{1}{2} = \exp\left[-\frac{1}{2}k_0^2 w^2 \sin^2\theta_h\right]$$

or

$$k_0 w \sin\theta_h = \sqrt{2\ln 2}$$

Hence

$$w = \frac{\sqrt{2\ln 2}}{\left(\frac{2\pi}{1.3}\right)\sin(2.74^\circ)}$$

or

$$d = 2w \simeq 5.1\ \mu\text{m}$$

Thus, in the Gaussian approximation the mode field diameter is about 5.1 μm

8.5.4.2 *Far field of a step index fiber*

Earlier we obtained the far-field distribution of a Gaussian modal field that is an approximation to the actual modal field. Because the actual modal field of a step index fiber is known precisely, we will now discuss the far-field pattern of a step index single-mode fiber as well as the procedure to obtain the equivalent step index (ESI) parameters of a graded index fiber.

For a cylindrically symmetric structure (like the optical fiber), the fundamental mode distribution $\psi(r)$ depends only on the radial coordinate r. Consequently, the far-field pattern will also be cylindrically symmetric and is given by (see Appendix B)

$$u(\theta) = 2\pi C \int_0^\infty \psi(r) J_0(k_0 r \sin\theta) r\, dr \tag{8.96}$$

We can invert the above equation to obtain (see Problem 8.14)

$$\psi(r) \simeq \frac{k_0^2}{2\pi C} \int_0^{\pi/2} u(\theta) J_0(k_0 r \sin\theta) \sin\theta \cos\theta\, d\theta \tag{8.97}$$

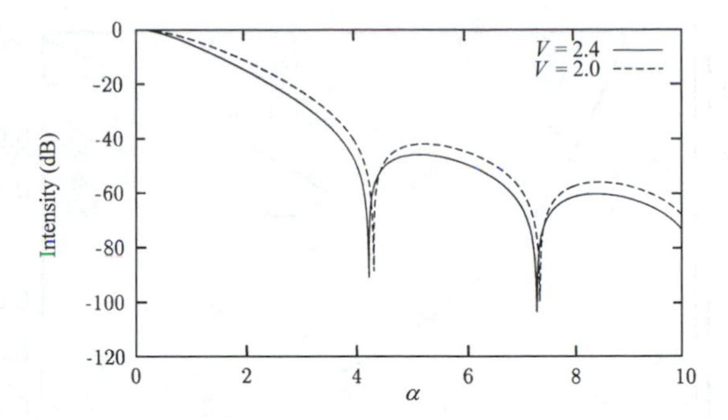

Fig. 8.19: The far-field intensity pattern of step index fibers having $V = 2$ and $V = 2.4$. These are universal curves valid for all step index fibers with $V = 2$ and $V = 2.4$; the actual angles will of course depend on the fiber parameters a and NA.

Thus, for a given $\psi(r)$ one can use equation (8.96) to determine $u(\theta)$ for all values of θ. On the other hand, if we measure $u(\theta)$ experimentally, one can determine $\psi(r)$ by using equation (8.97).

Now, in a step index fiber the fundamental modal field $\psi(r)$ is given by (see equation (8.25))

$$\psi(r) = \begin{cases} A\dfrac{J_0\left(\frac{Ur}{a}\right)}{J_0(U)} & 0 < r < a \\ A\dfrac{K_0\left(\frac{Wr}{a}\right)}{K_0(W)} & r > a \end{cases} \tag{8.98}$$

where U and W have been defined earlier (see equations (8.20) and (8.21)). If we substitute for $\psi(r)$ from equation (8.98) in equation (8.96) we obtain the following expression for the far-field intensity distribution (see Appendix B)

$$I(\theta) = \left|\frac{u(\theta)}{u(0)}\right|^2 = \begin{cases} \left\{\frac{U^2W^2}{(U^2-\alpha^2)(W^2+\alpha^2)}\left[J_0(\alpha) - \alpha J_1(\alpha)\frac{J_0(U)}{U J_1(U)}\right]\right\}^2 & \alpha \neq U \\ \left\{\frac{U^2W^2}{2V^2}\frac{1}{U J_1(U)}\left[J_0^2(\alpha) + J_1^2(\alpha)\right]\right\}^2 & \alpha = U \end{cases} \tag{8.99}$$

where

$$\alpha = k_0 a \sin\theta = \frac{2\pi}{\lambda_0} a \sin\theta \tag{8.100}$$

Equation (8.99) is normalized to unity at $\theta = 0$. Now, the quantities

$$U = V\sqrt{1 - b(V)} \quad \text{and} \quad W = V\sqrt{b(V)}$$

are completely determined from the value of V (see also Table 8.1); thus, for a given value of V, we can readily calculate U and W. Hence, for a given value of V of a step index fiber, if we plot the far-field intensity distribution $I(\theta)$ as a function of α, it will be an universal curve.

Figure 8.19 shows a typical plot of the far-field intensity pattern calculated from equation (8.99) for a step index fiber having $V = 2.0$ [$b = 0.41616$] and $V = 2.4$ [$b = 0.53003$]. Since the curves are universal they are valid for all

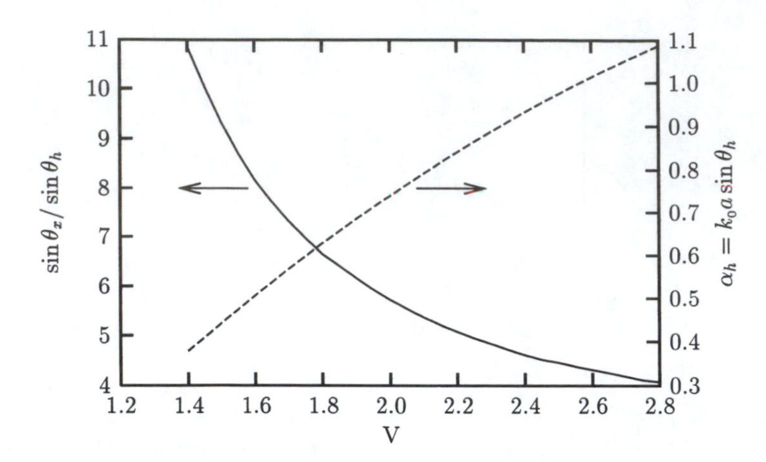

Fig. 8.20: Universal curves depicting the variation of $\sin\theta_x / \sin\theta_h$ and $k_0 a \sin\theta_h$ with V for step index fibers. The above curves can be used for characterizing single-mode fibers from a measurement of the far-field radiation pattern.

step index fibers having $V = 2$ and $V = 2.4$. Thus, if we have two different fibers operating at different wavelengths so that their V values are the same, then quantities like the angle at which the zeroes in far field appear will change, but the value of $k_0 \sin\theta$ at the zeroes remains the same. If θ_h represents the angle at which the far-field intensity falls by 3 dB and θ_x is the angle at which the intensity goes to its first zero (see Figure 8.19), then the quantities

$$\alpha_h = k_0 a \sin\theta_h \quad \text{and} \quad \alpha_x = k_0 a \sin\theta_x \tag{8.101}$$

depend only on the value of V. Thus, the ratio

$$\sigma_x = \frac{\alpha_x}{\alpha_h} = \frac{\sin\theta_x}{\sin\theta_h} \tag{8.102}$$

also depends only on the V number of the fiber. In Figure 8.20, we have plotted the universal curves corresponding to the variations of $\sin\theta_x / \sin\theta_h$ and $k_0 a \sin\theta_h$ with V. Indeed, one has the following approximate empirical formulae [Roy et al. (1991)]:

$$V = 8.039 - 2.347\sigma_x + 0.3329\sigma_x^2 - 0.0218\sigma_x^3 + 0.00054\sigma_x^4 \tag{8.103}$$

and

$$\alpha_h = k_0 a \sin\theta_h = -0.7858 + 0.994V - 0.1155V^2 \tag{8.104}$$

The error in the V value from equation (8.103) is less than 0.01 in the range $1.3 \leq V \leq 2.4$. Similarly, the error in α_h is less than 3.6×10^{-3} for the same range of V values.

The universal curves in Figure 8.20 and the above empirical formulae suggest the following experimental technique for obtaining the ESI parameters of a single-mode fiber:

(i) Using a known laser wavelength λ_0, measure the far-field diffraction pattern and determine the values of θ_h and θ_x.

(ii) Calculate $\sin\theta_x / \sin\theta_h$ and determine the V value either from the universal curve (Figure 8.20) or from the empirical relation (equation 8.103).

(iii) Knowing the value of V, calculate the value of $k_0 a \sin\theta_h$ either from the universal curve or from equation (8.104). Since the values of k_0 and $\sin\theta_h$ are known, the core radius a can be calculated.

(iv) Once the core radius and the V number are known, the NA can be determined.

The above procedure enables us to obtain an equivalent step index fiber that will have (for the given wavelength) the same values of θ_x and θ_h. For this "equivalent step index fiber" we can calculate the modal field (see equation (8.98)), which, for most practical fibers, should be close to the actual modal field of the given fiber.

Example 8.9: In an experiment using a single-mode fiber with a laser beam at 1.3 μm, the far-field angles θ_h and θ_x are 2.357° and 12.73°. We would like to determine the ESI parameters of the fiber. Now

$$\sigma_x = \frac{\sin\theta_x}{\sin\theta_h} = 5.358$$

implying $V = 2.111$. For $V = 2.111$, we have $\alpha_h = 0.798$, which implies $a = 4.01\ \mu$m. Since

$$V = \frac{2\pi}{\lambda_0} a\,\text{NA}$$

we have

$$\text{NA} = \frac{V\lambda_0}{2\pi a} \simeq 0.11$$

Hence, the ESI parameters of the given fiber are

$$a = 4.01\ \mu\text{m}, \quad \text{NA} = \sqrt{n_1^2 - n_2^2} = 0.11$$

8.5.4.3 Far-field root-mean-square (rms) mode field diameter

If $\Psi_f^2(q)$ represents the intensity distribution in the far-field of a single-mode fiber, then the far field rms mode field diameter is defined as

$$d_F = 2\left[\frac{2\int_0^\infty \Psi_f^2(q)\,q\,dq}{\int_0^\infty \Psi_f^2(q)\,q^3\,dq}\right]^{1/2} \tag{8.105}$$

where

$$q = k_0 \sin\theta = \frac{2\pi}{\lambda_0}\sin\theta \tag{8.106}$$

and θ is the far field angle (see Figure 8.17). In this section we will show that

$$d_F = 2w_P \tag{8.107}$$

Now, in a cylindrical system of coordinates, if $\psi(r)$ is the near-field pattern, the far-field pattern is given by (see Appendix B)

$$\Psi(q) = K\int_0^\infty \psi(r)J_0(qr)r\,dr \tag{8.108}$$

where K is a constant. To obtain the inverse transform of $\Psi(q)$, we multiply equation (8.108) by $J_0(qr')qdq$ and integrate to obtain

$$\int_0^\infty \Psi(q)J_0(qr')q\,dq = \int_0^\infty \psi(r)r\,dr\int_0^\infty J_0(qr')J_0(qr)q\,dq$$

Now, using the orthogonality relation

$$\int_0^\infty J_\nu(qr)J_\nu(qr')q\,dq = \frac{1}{r}\delta(r-r') \tag{8.109}$$

we obtain

$$\psi(r) = \int_0^\infty \Psi(q)J_0(qr)q\,dq \tag{8.110}$$

We now have

$$\begin{aligned}
\int_0^\infty \psi^2(r)r\,dr &= \int_0^\infty r\,dr\int_0^\infty \Psi(q)J_0(qr)q\,dq\int_0^\infty \psi(q')J_0(q'r)q'\,dq' \\
&= \int_0^\infty q\,dq\int_0^\infty q'\,dq'\,\Psi(q)\Psi(q')\int_0^\infty J_0(qr)J_0(q'r)r\,dr \\
&= \int_0^\infty q\,dq\int_0^\infty q'\,dq'\,\Psi(q)\Psi(q')\frac{1}{q}\delta(q-q') \\
&= \int_0^\infty \Psi^2(q)q\,dq
\end{aligned} \tag{8.111}$$

where we have used the orthogonality of Bessel functions (equation (8.109)). We also have from equation (8.110)

$$\begin{aligned}
\frac{d\psi}{dr} &= \int_0^\infty \Psi(q)J_0'(qr)q^2\,dq \\
&= -\int_0^\infty \Psi(q)J_1(qr)q^2\,dq
\end{aligned}$$

where prime denotes differentiation with respect to argument and we have used the relation $J_0'(\xi) = -J_1(\xi)$. Thus

$$\int_0^\infty \left(\frac{d\psi}{dr}\right)^2 r\,dr = \int_0^\infty r\,dr \int_0^\infty \Psi(q)J_1(qr)q^2\,dq \int_0^\infty \Psi(q')J_1(q'r)q'^2\,dq'$$

$$= \int_0^\infty q^2\,dq \int_0^\infty q'^2\,dq'\,\Psi(q)\Psi(q') \int_0^\infty J_1(qr)J_1(q'r)r\,dr$$

$$= \int_0^\infty \Psi^2(q)q^3 dq$$

where we have again used equation (8.109) with $\nu = 1$. Thus

$$d_F = 2\left[\frac{2\int_0^\infty \psi^2(r)r\,dr}{\int_0^\infty \left(\frac{d\psi}{dr}\right)^2 r\,dr}\right]^{1/2} = 2\left[\frac{2\int_0^\infty \Psi^2(q)q\,dq}{\int_0^\infty \Psi^2(q)\,q^3\,dq}\right]^{1/2} \tag{8.112}$$

proving that the far-field rms mode field diameter is exactly twice the Petermann-2 spot size.

Problems

8.1 Consider a step index fiber with

$$n_1 = 1.45, \quad \Delta = 0.003, \quad \text{and} \quad a = 3\,\mu\text{m}$$

For $\lambda_0 = 0.9\,\mu$m, 1.3 μm, and 1.55 μm, calculate the value of V and calculate (in each case) the values of b and β/k_0. Compare with the approximate values that you would have obtained using the empirical formula (equation (8.42))

[ANSWER: $b = 0.518,\ 0.282,\ 0.174$.]

8.2 In this and the following two problems, fiber 1 and fiber 2 are described by the following parameters (see equations (8.43) and (8.44))

Fiber 1: $n_2 = 1.45, \quad \Delta = 0.0064, \quad a = 3\,\mu\text{m}$

Fiber 2: $n_2 = 1.45, \quad \Delta = 0.010, \quad a = 2\,\mu\text{m}$

(a) Calculate the wavelengths at which the fibers will have $V = 1.8$.
(b) For $V = 2$, find the value of b and calculate the corresponding values of β/k_0.
(c) What would be the cutoff wavelengths for the two fibers.

8.3 Consider a step index fiber with $V = 8$.

(a) Using the values of the zeroes of $J_n(x)$, find the positions of zeroes and infinities of the LHS of the eigenvalue equations (equations (8.40) and (8.41)). Make qualitative plots of the LHS and RHS of the eigenvalue equations and determine the number of guided LP modes.
(b) Calculate the total number of modes and compare with the result of the formula $V^2/2$.
(c) Solve the eigenvalue equations to determine the normalized propagation constants of the guided modes.

(d) Consider two fibers described in Problem 8.2. In each case determine the value of λ_0 for which $V = 8$ and determine the corresponding values of β/k_0 and β.

(e) Using the values of the zeroes of $J_n(x)$, find the zeroes of the radial and angular intensity distributions and make qualitative plots.

[ANSWER:

(c) $b =$ 0.92881(LP$_{01}$), 0.63006(LP$_{02}$), 0.13211(LP$_{03}$);
0.81998(LP$_{11}$), 0.41046(LP$_{12}$);
0.67818(LP$_{21}$), 0.16870(LP$_{22}$), 0.50621(LP$_{31}$);
0.306618(LP$_{41}$), 0.082388(LP$_{51}$).
(Note that the last mode is very near to cutoff)
Total number of modes: 34

(d) $\lambda_0 =$ 0.3865 μm (fiber 1); 0.3221 μm (fiber 2)
$\beta/k_0 =$ 1.45156 (fiber 1); 1.45595 (fiber 2), and so forth.]

(for LP$_{12}$ mode)

8.4 Repeat the entire previous excercise for $V = 6.5$.

[ANSWER:

(c) see equation (8.45) for the values of b. Total number of modes : 24

(d) $\lambda_0 = 0.4757$ μm (fiber 1); 0.3963 μm (fiber 2)
$\beta/k_0 = 1.45166$ (fiber 1); 1.45260 (fiber 2), and so forth.]

(for LP$_{12}$ mode)

8.5 Consider 3 (step index fibers) with (i) $\Delta = 0.002$, (ii) $\Delta = 0.003$, and (iii) $\Delta = 0.004$. Assume $n_2 = 1.45$. In each case calculate the core radius so that the cutoff wavelength is 1150 nm.

8.6 For a typical step index fiber operating at 1330 nm we have

$$n_1 = 1.450840, \quad n_2 = 1.446918, \quad \text{and} \quad a = 4.1\ \mu\text{m}$$

Calculate w and w_P at $\lambda_0 = 1100$, 1300, and 560 nm.

[ANSWER: $w_P \approx 4.9\ \mu$m(1300), 5.7 μm(1560).]

8.7 Repeat the above calculations for a typical step index fiber operating at 1560 nm

$$n_1 = 1.457893, \quad n_2 = 1.446918, \quad \text{and} \quad a = 2.3\ \mu\text{m}$$

8.8 Measurement of the far-field pattern of a single-mode fiber at $\lambda_0 = 0.6328\ \mu$m gives $\theta_h = 1.942°$ and $\theta_x = 9.933°$. Obtain the characteristics of the fiber using equations (8.103) and (8.104).

[ANSWER: $a = 2.505\ \mu$m, NA = 0.089.]

8.9 For a transverse misalignment (say along the x-axis), between two identical single-mode fibers, the coupling efficiency is given by the overlap integral

$$T(u) = \left[\frac{\int\int \psi(x, y)\psi(x - u, y)dx\,dy}{\int\int \psi^2(x, y)dx\,dy} \right]^2 \tag{8.113}$$

where u represents the magnitude of the transverse misalignment and $\psi(x, y)$ represents the field associated with the fundamental mode. Show that, in general, the transverse misalignment splice loss is given by

$$\alpha\,(\text{dB}) \simeq 4.343 \left(\frac{u}{w_P}\right)^2 \tag{8.114}$$

where w_P is the Petermann-2 spot size defined through equation (8.83).

Solution: The transverse misalignment coupling efficiency (see equation (8.113)) is directly related to the Petermann-2 spot size. To show this, we first note that for small displacements u, we may write

$$\psi(x-u,y) = \psi(x,y) - u\frac{\partial\psi}{\partial x} + \frac{u^2}{2}\frac{\partial^2\psi}{\partial x^2} \tag{8.115}$$

Using

$$x = r\cos\theta, \quad y = r\sin\theta$$

and

$$r^2 = x^2 + y^2, \quad \tan\theta = \frac{y}{x}$$

We may write

$$\frac{\partial\psi}{\partial x} = \frac{d\psi}{dr}\frac{\partial r}{\partial x} = \cos\theta\,\frac{d\psi}{dr}$$

$$\frac{\partial^2\psi}{\partial x^2} = \cos^2\theta\,\frac{d^2\psi}{dr^2} + \frac{\sin^2\theta}{r}\,\frac{d\psi}{dr}$$

where we have used the fact that for a cylindrically symmetric structure the fundamental mode field pattern ψ depends only on the r coordinate. Substituting the above in equation (8.115) and using $dx\,dy = r\,dr\,d\theta$, equation (8.113) becomes

$$\sqrt{T(u)} = 1 - u\,\frac{\int_0^\infty \psi(r)\frac{d\psi}{dr}\,r\,dr \int_0^{2\pi}\cos\theta\,d\theta}{2\pi\int_0^\infty \psi^2(r)\,r\,dr}$$

$$+\frac{u^2}{2}\left[\frac{\int_0^\infty \psi\frac{d^2\psi}{dr^2}\,r\,dr\int_0^{2\pi}\cos^2\theta\,d\theta + \int_0^\infty \psi\frac{d\psi}{dr}\,dr\int_0^{2\pi}\sin^2\theta d\theta}{2\pi\int_0^\infty \psi^2(r)\,r\,dr}\right]$$

$$= 1 + \frac{u^2}{4}\left[\frac{\int_0^\infty \psi\frac{d^2\psi}{dr^2}\,r\,dr + \int_0^\infty \psi\frac{d\psi}{dr}dr}{\int_0^\infty \psi^2(r)\,r\,dr}\right] \tag{8.116}$$

where we have used

$$\int_0^{2\pi}\cos\theta d\theta = 0, \quad \int_0^{2\pi}\cos^2\theta\,d\theta = \pi = \int_0^{2\pi}\sin^2\theta\,d\theta$$

Now,

$$\int_0^\infty \psi\frac{d^2\psi}{dr^2}\,r\,dr = \psi\frac{d\psi}{dr}r\Big|_0^\infty - \int\frac{d\psi}{dr}\frac{d}{dr}(r\psi)dr$$

$$= -\int_0^\infty\left(\frac{d\psi}{dr}\right)^2 r\,dr - \int_0^\infty \psi\frac{d\psi}{dr}dr \tag{8.117}$$

where we have used the fact that in the limit $r \to \infty$, $r\psi \to 0$. Using equation (8.117), equation (8.116) becomes

$$\sqrt{T(u)} = \left[1 - \frac{u^2}{4} \frac{\int_0^\infty \left(\frac{d\psi}{dr}\right)^2 r\,dr}{\int_0^\infty \psi^2 r\,dr}\right]$$

or,

$$T(u) \simeq 1 - \frac{u^2}{w_P^2} \tag{8.118}$$

where w_P is the Petermann-2 spot size as defined by equation (8.83). From the above equation we readily obtain equation (8.114) for $u \ll w_P$. From equation (8.118) we find that if we measure $T(u)$ and fit it to a polynomial in u, then the coefficient of the u^2 term will give w_P. Note that we have only assumed the cylindrical symmetry of the single-mode fiber and, of course, the validity of the scalar wave equation.

8.10 Consider two single-mode fibers and assume that the fundamental modes of the two fibers can be represented by Gaussian distribution. Calculate the power transmission loss as a function of angular misalignment between the two fibers. Assume that the fiber ends are placed in a liquid of refractive index n_l.

Solution: We describe the fundamental modes of the two fibers by equations (8.66) and (8.67) with $u = 0$. Let us assume that there is a small angular misalignment of θ between the two fibers (see Figure 8.13(c)). To calculate the power transmission loss, we must transform the Gaussian beam of the first fiber into the coordinate system of the second fiber. If (x, y, z) and $(x', y,' z')$ represent the coordinate systems of the first and second fibers, respectively, then we have (see Figure 8.21)

$$\left.\begin{aligned} x &= x' \cos\theta + z' \sin\theta \\ y &= y' \\ z &= -x' \sin\theta + z' \cos\theta \end{aligned}\right\} \tag{8.119}$$

where we assume an angular misalignment in the x–z plane. Since the medium between the two fibers is of refractive index n_l, the Gaussian mode of the first fiber, as it emerges, will propagate approximately as

$$\psi_1(x, y, z) \approx \left(\frac{2}{\pi}\right)^{1/2} \frac{1}{w_1} e^{-(x^2+y^2)/w_1^2} e^{-ik_0 n_l z} \tag{8.120}$$

for small values of z so that we can neglect diffraction effects. We transform equation (8.120) into the (x', y', z') coordinate system and obtain for the incident beam at the input plane $z' = 0$ of the second fiber

$$\psi_1(x', y', z' = 0) = \left(\frac{2}{\pi}\right)^{1/2} \frac{1}{w_1} e^{-(x'^2+y'^2)/w_1^2} e^{ik_0 n_l x' \theta} \tag{8.121}$$

where we have assumed $\sin\theta \approx \theta$ and $\cos\theta \approx 1$. Calculation of the overlap of this field with the field of the second fiber and taking modulus squared gives us the power transmission coefficient as

$$T(\theta) = \left(\frac{2w_1 w_2}{w_1^2 + w_2^2}\right)^2 \exp\left[-\frac{k_0^2 n_l^2 \theta^2 w_1^2 w_2^2}{2\left(w_1^2 + w_2^2\right)}\right] \tag{8.122}$$

Fig. 8.21: Coordinate system for calculating loss at an angular misalignment of θ in the x–z plane.

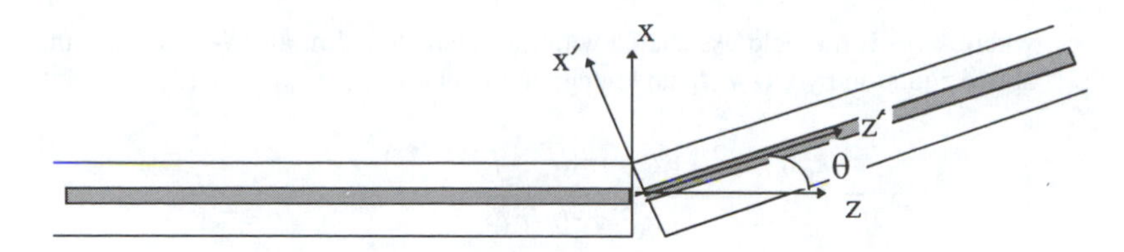

Thus, the coupled power decreases to e^{-1} of its maximum value (corresponding to $\theta = 0$) in an angle

$$\theta_e = \sqrt{2}\left(w_1^2 + w_2^2\right)^{1/2} / w_1 w_2 k_0 n_2 \tag{8.123}$$

For two identical fibers, $w_1 = w_2 = w$ and we readily get equation (8.75)

8.11 Obtain the splice loss between two identical single-mode fibers due to a longitudinal misalignment of D (see Figure 8.13(a)).

Solution: Let w represent the Gaussian spot size of the two fibers. The field at the exit of the first fiber is given by

$$\psi_0 = \sqrt{\frac{2}{\pi}} \cdot \frac{1}{w} \exp\left[-\frac{x^2 + y^2}{w^2}\right] \tag{8.124}$$

This output field will diffract over distance D before being incident on the receiving fiber. We may write for the field that is incident on the second fiber as (see Ghatak and Thyagarajan (1989), Section 5.4)

$$\psi_D = \sqrt{\frac{2}{\pi}} \cdot \frac{1}{w} \frac{ikw^2}{2D + ikw^2} \exp\left[-\frac{ik(x^2 + y^2)}{2\tilde{R}}\right] \exp\left[-\frac{x^2 + y^2}{\tilde{w}^2}\right] e^{-ikD} \tag{8.125}$$

where

$$\left.\begin{aligned} \tilde{R} &= D\left(1 + \frac{\pi^2 w^4}{\lambda^2 D^2}\right) \\ \tilde{w} &= w\left(1 + \frac{\lambda^2 D^2}{\pi^2 w^4}\right)^{1/2} \\ k &= k_0 n, \quad \lambda = \lambda_0 / n \end{aligned}\right\} \tag{8.126}$$

with n being the refractive index of the medium filling the region between the two fiber ends. The field of the second fiber is again given by equation (8.124). Thus, the transmission coefficient across the longitudinal misalignment is

$$T = \left| \int_{-\infty}^{\infty} \psi_0 \, \psi_D \, dx \, dy \right|^2 \tag{8.127}$$

Substituting for ψ_0 and ψ_D from equations (8.124) and (8.125) in equations (8.127) and carrying out straightforward integration we obtain

$$T = \left(1 + \frac{D^2 \lambda^2}{4\pi^2 w^4}\right)^{-1} \tag{8.128}$$

8.12 Obtain the relationship between waveguide dispersion and Petermann-2 spot size.

Solution: For $l = 0$, equation (8.16) can be written in the form

$$\frac{1}{r}\frac{d}{dr}\left(r\frac{d\psi}{dr}\right) + \left(k_0^2 n^2(r) - \beta^2\right)\psi(r) = 0$$

where $\psi(r)$ is the field associated with the fundamental mode. We multiply the above equation by $\psi(r) r\, dr$ and integrate to obtain

$$\beta^2 = \frac{k_0^2 \int_0^\infty n^2(r)\psi^2(r) r\, dr - \int_0^\infty \left(\frac{d\psi}{dr}\right)^2 r\, dr}{\int_0^\infty \psi^2(r) r\, dr} \tag{8.129}$$

In obtaining the above equation we have used the fact that $\psi(r)$ and $d\psi/dr \to 0$ as $r \to \infty$. Thus[7]

$$2\beta \frac{d\beta}{dk_0} = 2k_0 \frac{\int_0^\infty n^2(r)\, \psi^2(r) r\, dr}{\int_0^\infty \psi^2(r) r\, dr} \tag{8.130}$$

or

$$\begin{aligned}\frac{d\beta}{dk_0} &= \frac{k_0}{\beta} \frac{\beta^2 \int_0^\infty \psi^2(r) r\, dr + \int_0^\infty \left(\frac{d\psi}{dr}\right)^2 r\, dr}{k_0^2 \int_0^\infty \psi^2(r) r\, dr} \\ &= \frac{k_0}{\beta}\left[\frac{\beta^2}{k_0^2} + \frac{2}{k_0^2 w_P^2}\right]\end{aligned} \tag{8.131}$$

The group velocity v_g is therefore given by

$$\begin{aligned}\frac{1}{v_g} &= \frac{d\beta}{d\omega} = \frac{1}{c}\frac{d\beta}{dk_0} \\ &= \frac{\beta}{ck_0} + \frac{2}{c\beta k_0 w_P^2}\end{aligned} \tag{8.132}$$

Thus, the time taken for a pulse to traverse the length L of the fiber is given by

$$\tau = \frac{L}{v_g} = \frac{L}{2\pi c}\left[\beta\lambda_0 + \frac{2}{\beta}\left(\frac{\lambda_0}{w_P^2}\right)\right] \tag{8.133}$$

The waveguide dispersion is therefore given by

$$\begin{aligned}\Delta\tau &= \frac{d\tau}{d\lambda_0}\Delta\lambda_0 \\ &= \frac{L}{2\pi c}\left[\beta + \lambda_0\frac{d\beta}{d\lambda_0} - \frac{2}{\beta^2}\frac{d\beta}{d\lambda_0}\left(\frac{\lambda_0}{w_P^2}\right) + \frac{2}{\beta}\frac{d}{d\lambda_0}\left(\frac{\lambda_0}{w_P^2}\right)\right]\Delta\lambda_0\end{aligned} \tag{8.134}$$

[7]Equation (8.130) is rigorously correct – of course, we have neglected material dispersion. At first it may appear that $\psi(r)$ itself depends on the wavelength and we should take this into account. That equation (8.130) is rigorously correct can be seen from the fact that if ψ^2 contains N variational parameters α_i, then

$$\begin{aligned}\frac{d\beta^2}{dk_0} &= \frac{\partial\beta^2}{\partial k_0} + \sum_{i=1}^{N}\frac{\partial\beta^2}{\partial\alpha_i}\frac{d\alpha_i}{dk_0} \\ &= \frac{\partial\beta^2}{\partial k_0}\end{aligned}$$

since, by definition, the stationarity expression implies

$$\frac{\partial\beta^2}{\partial\alpha_i} = 0$$

Now using equation (8.132) we get

$$\frac{d\beta}{d\lambda_0} = \frac{d\beta}{d\omega} \cdot \frac{d\omega}{d\lambda_0} = \left[\frac{\beta\lambda_0}{2\pi c} + \frac{2}{2\pi c\beta}\frac{\lambda_0}{w_P^2}\right]\left(-\frac{2\pi c}{\lambda_0^2}\right)$$

or

$$\frac{d\beta}{d\lambda_0} = -\frac{\beta}{\lambda_0} - \frac{2}{\beta\lambda_0 w_P^2} \tag{8.135}$$

Substituting in equation (8.134) and simplifying we get

$$\Delta\tau = \frac{2L}{2\pi c\beta w_P^2}\left[\frac{2}{\beta^2 w_P^2} + w_P^2 \frac{d}{d\lambda_0}\left(\frac{\lambda_0}{w_P^2}\right)\right]\Delta\lambda_0 \tag{8.136}$$

For typical step index fibers, since $\beta \simeq k_0 n_2$ and $(w_P/\lambda_0) \gg 1$, the first term will always make a negligible contribution. Thus, we may write for the wavelength dispersion parameter

$$D_w = \frac{\Delta\tau}{L\,\Delta\lambda_0} = \frac{\lambda_0}{2\pi^2 c n_2}\frac{d}{d\lambda_0}\left(\frac{\lambda_0}{w_P^2}\right) \tag{8.137}$$

For power law profiles – such as step index, parabolic index, and triangular core index – β has been found to be accurately described by an empirical relation; this is discussed in Chapter 20.

8.13 Obtain the relationship between the near-field spot size and the angular misalignment splice loss.

Solution: Let $\psi(x, y)$ represent the modal field of both fibers and let us assume an angular misalignment θ in the x–z plane as shown in Figure 8.21. To find the coupling efficiency we must obtain the modal field of fiber 1 in the coordinate system fixed in fiber 2. Now from equation (8.119) we have for $z' = 0$

$$x = x'\cos\theta \simeq x' \tag{8.138}$$

$$z = -x'\sin\theta \simeq -x'\theta \tag{8.139}$$

for small θ. Thus, the field of fiber 1 at $z' = 0$ is

$$\begin{aligned}\psi_1(x', y', z' = 0) &= \psi(x', y')\,e^{-ik_0 z n_l} \\ &= \psi(x', y')\,e^{ik_0 n_l x'\theta} \\ &\simeq \psi(x', y')\left[1 + ik_0 n_l \theta x' - k_0^2 n_l^2 \theta^2 x'^2\right]\end{aligned} \tag{8.140}$$

where n_l is the refractive index of the medium between the two fibers and we have neglected any diffraction effects from the end of fiber 1 to the input of fiber 2. The coupling efficiency is given by

$$\begin{aligned}T(\theta) &= \left[\frac{\int\int \psi(x', y')\,\psi_1(x', y')dx'\,dy'}{\int\int \psi^2(x', y')dx'\,dy'}\right] \\ &= \left[1 - k_0^2 n_l^2 \theta^2 \frac{\int\int \psi^2(x', y')\,r^2\cos^2\phi\,dx\,dy}{\int\int \psi^2(x', y')\,r\,dr\,d\phi}\right]\end{aligned} \tag{8.141}$$

where we have used the relation $x' = r \cos\phi$. Changing from (x', y') to (r, ϕ) we have

$$T(\theta) = \left[1 - \frac{k_0^2 n_l^2 \theta^2}{2} \frac{\int \psi^2(r) r^3 dr}{\int \psi^2(r) r\, dr}\right]^2$$

or

$$T(\theta) \simeq 1 - \frac{k_0^2 n_l^2 \theta^2}{8} d_N^2 \tag{8.142}$$

where

$$d_N = 2\left[\frac{2\int_0^\infty \psi^2(r) r^3 dr}{\int_0^\infty \psi^2(r) r\, dr}\right]^{1/2} \tag{8.143}$$

Equation (8.142) describes the relationship between the near-field spot size d_N and the angular misalignment splice loss $T(\theta)$.

8.14 Obtain equation (8.97) from equation (8.96).

Solution: To invert equation (8.96) we make use of the following orthonormality relation satisfied by Bessel functions.

$$\int_0^\infty J_\nu(\alpha\rho)\, J_\nu(\alpha'\rho)\, \rho\, d\rho = \frac{1}{\alpha}\delta(\alpha - \alpha') \tag{8.144}$$

Multiplying both sides of equation (8.96) by $J_0\,(qr')$ and integrating over q, we get

$$\int_0^\infty u(q)\, J_0(qr')\, q\, dq = 2\pi C \int_0^\infty \psi(r)\, r\, dr \int_0^\infty J_0(qr)\, J_0(qr')\, q\, dq \tag{8.145}$$

which, using equation (8.144), becomes

$$\int_0^\infty u(q)\, J_0(qr')\, q\, dq = 2\pi C \psi(r') \tag{8.146}$$

from which equation (8.97) directly follows.

9

Propagation characteristics of graded index fibers

9.1 Introduction

Graded index multimode fibers find important applications in local area networks and short-haul communication systems. In Chapter 5 we used ray optics to study pulse dispersion in such fibers. In this chapter we carry out a modal analysis and study their propagation characteristics. We first discuss the propagation characteristics of an infinitely extended parabolic index fiber characterized by the following refractive index variation

$$n^2 = n_1^2\left[1 - 2\Delta\left(\frac{r}{a}\right)^2\right] \tag{9.1}$$

or, in Cartesian coordinates,

$$n^2 = n_1^2\left[1 - 2\Delta\frac{x^2 + y^2}{a^2}\right] \tag{9.2}$$

where Δ and a are constants and n_1 represents the axial refractive index. Although a profile given by equation (9.1) is unrealistic since $n^2 \to -\infty$ as $r \to \infty$, we will see that we can obtain analytic solutions for such a refractive index profile. Further, for well-guided modes in a realistic parabolic index core profile fiber, the analysis is quite accurate.

In Section 9.2 we solve the wave equation in Cartesian coordinates and obtain what are usually referred to as TEM_{mn} modes. In Section 9.3 we solve the same equation in cylindrical coordinates and obtain what are usually referred to as the LP_{lm} modes. Obviously, the LP_{lm} modes can be expressed as a linear combination of TEM_{mn} modes and conversely. In Section 9.4 we discuss (in the WKB approximation) the propagation characteristics of a fiber characterized by the power law profile. We show that the final result is almost identical to the one obtained by using ray analysis in Chapter 5.

9.2 Modal analysis of a parabolic index fiber

If we substitute equation (9.1) in the scalar wave equation we obtain

$$\nabla^2\Psi = \frac{n_1^2}{c^2}\left[1 - 2\Delta\left(\frac{x^2}{a^2} + \frac{y^2}{a^2}\right)\right]\frac{\partial^2\Psi}{\partial t^2} \tag{9.3}$$

We assume a modal solution of the form

$$\Psi(x, y, z, t) = \psi(x, y)\, e^{i(\omega t - \beta z)} \tag{9.4}$$

Then equation (9.3) becomes

$$\frac{\partial^2 \psi}{\partial x^2} + \frac{\partial^2 \psi}{\partial y^2} + \left\{k_0^2 n_1^2 \left[1 - 2\Delta\left(\frac{x^2}{a^2} + \frac{y^2}{a^2}\right)\right] - \beta^2\right\}\psi = 0 \tag{9.5}$$

We use the method of separation of variables and write

$$\psi(x, y) = X(x)\, Y(y) \tag{9.6}$$

If we substitute the above solution in equation (9.5) and divide by XY we obtain

$$\left(\frac{1}{X}\frac{d^2 X}{dx^2} - k_0^2 n_1^2 \frac{2\Delta}{a^2}\, x^2\right) + \left(\frac{1}{Y}\frac{d^2 Y}{dy^2} - k_0^2 n_1^2 \frac{2\Delta}{a^2}\, y^2\right) + \left(k_0^2 n_1^2 - \beta^2\right) = 0 \tag{9.7}$$

The variables have indeed separated out and we may write

$$\left(\frac{1}{X}\frac{d^2 X}{dx^2} - k_0^2 n_1^2 \frac{2\Delta}{a^2}\, x^2\right) = -K_1 \tag{9.8}$$

and

$$\left(\frac{1}{Y}\frac{d^2 Y}{dy^2} - k_0^2 n_1^2 \frac{2\Delta}{a^2}\, y^2\right) = -K_2 \tag{9.9}$$

where K_1 and K_2 are constants and

$$\beta^2 = k_0^2 n_1^2 - K_1 - K_2 \tag{9.10}$$

We now use the variables

$$\xi = \gamma x, \quad \eta = \gamma y \tag{9.11}$$

with

$$\gamma = \left[n_1 k_0 \frac{\sqrt{2\Delta}}{a}\right]^{1/2} = \frac{\sqrt{V}}{a} \tag{9.12}$$

where

$$V = k_0 n_1 a (2\Delta)^{1/2} \tag{9.13}$$

represents the waveguide parameter. Thus

$$d^2 X/d\xi^2 + (\Lambda_1 - \xi^2)\, X(\xi) = 0 \tag{9.14}$$

$$d^2 Y/d\eta^2 + (\Lambda_2 - \eta^2)\, Y(\eta) = 0 \tag{9.15}$$

where

$$\Lambda_1 = \frac{K_1}{\gamma^2} \tag{9.16}$$

and

$$\Lambda_2 = \frac{K_2}{\gamma^2} \tag{9.17}$$

For bounded solutions – that is, for $X(\xi)$ and $Y(\eta)$ to tend to zero as $\xi, \eta \to \pm\infty$ (i.e., $x, y \to \pm\infty$) – we must have

$$\Lambda_1 = 2m + 1; \quad m = 0, 1, 2, \ldots \tag{9.18}$$

and

$$\Lambda_2 = 2n + 1; \quad n = 0, 1, 2, \ldots \tag{9.19}$$

The corresponding modal distributions $X(x)$ and $Y(y)$ are the Hermite–Gauss functions (see Appendix A). Thus, we obtain the following expressions for mode profiles and corresponding propagation constants (cf. equations (7.118)–(7.122)).

$$\psi_{mn}(x, y) = \left[N_m H_m(\xi)\, e^{-\frac{1}{2}\xi^2}\right]\left[N_n H_n(\eta)\, e^{-\frac{1}{2}\eta^2}\right] \tag{9.20}$$

$$\beta_{mn} = k_0 n_1 \left[1 - \frac{2(m+n+1)}{k_0 n_1}\sqrt{\frac{2\Delta}{a^2}}\right]^{1/2} \tag{9.21}$$

where $m, n = 0, 1, 2, \ldots$ and

$$N_m = \left[\frac{\gamma}{2^m m! \sqrt{\pi}}\right]^{1/2} = \left[\frac{\sqrt{V}/a}{2^m m! \sqrt{\pi}}\right]^{1/2} \tag{9.22}$$

represents the normalization constant. Different values of m and n correspond to different modes of the fiber. Equation (9.21) is an analytic solution of the propagation constant of the (m, n)th mode in such a fiber. For a typical multimode parabolic index fiber,

$$n_1 = 1.46, \quad \Delta = 0.01, \quad a = 25\ \mu\text{m}$$

we have for $\lambda_0 \approx 0.8\ \mu$m

$$2(2\Delta)^{1/2}/k_0 n_1 a \approx 10^{-3}$$

Thus, for $m + n \ll 10^3$, one is justified in making a binomial expansion in equation (9.21) and if we retain only the first-order term we will obtain

$$\beta_{mn} \approx (\omega/c)n_1 - (m + n + 1)(2\Delta)^{1/2}/a \tag{9.23}$$

If we neglect the wavelength dependence of n_1 and Δ – that is, if we neglect material dispersion, we get the following expression for the group velocity

$$\frac{1}{v_g} = \frac{d\beta_{mn}}{d\omega} \approx \frac{n_1}{c} \tag{9.24}$$

implying that different modes in a parabolic index fiber travel with almost the same group velocity. In the language of ray optics, this corresponds to the fact that all rays take approximately the same amount of time to propagate through the fiber.

We should mention here that in an actual parabolic index core fiber the refractive index is given by equation (9.1) only for $r < a$ beyond which it has a constant value equal to

$$n = n_1(1 - 2\Delta)^{1/2}(= n_2) \quad \text{for } (x^2 + y^2) > a^2$$

and the solutions given by equations (9.20) and (9.21) are then not exact solutions for such a profile. However, even for the realistic profile, the solutions given by equations (9.20) and (9.21) are reasonably accurate for lower order modes in a multimoded waveguide. The guided modes will correspond to only those values of m and n for which

$$k_0 n_2 < \beta_{mn} < k_0 n_1 \tag{9.25}$$

The lowest order mode that corresponds to $m = 0, n = 0$ is given by

$$\psi_{00} = \frac{\gamma}{\sqrt{\pi}} e^{-\frac{1}{2}\gamma^2 r^2} = \frac{1}{a}\sqrt{\frac{V}{\pi}} e^{-\frac{1}{2}V(r/a)^2} \tag{9.26}$$

which has a Gaussian distribution. The spot size of the Gaussian is given by (see Section 8.5)

$$\frac{w}{a} = \sqrt{\frac{2}{V}} \tag{9.27}$$

The next mode is two-fold degenerate ($m = 0, n = 1$ and $m = 1, n = 0$) with the field distribution given by

$$\psi_{01} = \sqrt{\frac{2}{\pi}} \gamma^2 y e^{-\frac{1}{2}\gamma^2 r^2} \tag{9.28}$$

and

$$\psi_{10} = \sqrt{\frac{2}{\pi}} \gamma^2 x e^{-\frac{1}{2}\gamma^2 r^2} \tag{9.29}$$

Similarly, one can consider other higher order modes.

9.3 The LP_{lm} modes

In Section 8.2 we showed that the scalar modes of an optical fiber characterized by cylindrically symmetric refractive index distribution $[n = n(r)]$ are described by the transverse field profile (see equation (8.17))

$$\Psi(r, \phi, z, t) = e^{i(\omega t - \beta z)} R(r) \begin{Bmatrix} \cos l\phi \\ \sin l\phi \end{Bmatrix}; \quad l = 0, 1, 2, \ldots \tag{9.30}$$

where $R(r)$ is the solution of the radial part of the wave equation

$$\frac{d^2R}{dr^2} + \frac{1}{r}\frac{dR}{dr} + \left[k_0^2 n^2(r) - \beta^2 - \frac{l^2}{r^2}\right] R(r) = 0 \tag{9.31}$$

If we substitute equation (9.1) for $n^2(r)$, then the well-behaved solution of the above equation is given by (see, e.g., Ghatak et al. (1995), Example 7.2).

$$R_{lm}(r) = N_{lm} r^l \exp\left[-\frac{1}{2}\gamma^2 r^2\right] L_{m-1}^l(\gamma^2 r^2) \tag{9.32}$$

where

$$N_{lm} = \gamma^{l+1}\left[\frac{2(m-1)!}{\Gamma(l+m)}\right]^{1/2} \tag{9.33}$$

represents the normalization constant so that

$$\int_0^\infty |R_{lm}(r)|^2 \, r \, dr = 1 \tag{9.34}$$

The functions $L_n^k(x)$ are the associated Laguerre polynomials given by

$$L_n^k(x) = \sum_{p=0}^{n} (-1)^p \frac{\Gamma(n+k+1)}{(n-p)!\,\Gamma(p+k+1)\,p!} x^p \tag{9.35}$$

Thus

$$\left.\begin{aligned} L_0^k(x) &= 1, \quad L_1^k(x) = k + 1 - x \\ L_2^k(x) &= \tfrac{1}{2}(k+2)(k+1) - (k+2)x + \tfrac{1}{2}x^2 \end{aligned}\right\} \tag{9.36}$$

and so forth. The corresponding propagation constants are given by

$$\beta_{lm} = k_0 n_1 \left[1 - \frac{2(2m + l - 1)}{k_0 n_1}\sqrt{\frac{2\Delta}{a^2}}\right]^{1/2} \tag{9.37}$$

The LP_{01} mode is given by equation (9.26). Note that for LP_{lm} modes $l = 0, 1, 2, \ldots$ and $m = 1, 2, 3, \ldots$. Further,

$$\psi_{11}(r, \phi) = \sqrt{\frac{2}{\pi}}\gamma^2 r e^{-\frac{1}{2}\gamma^2 r^2} \begin{Bmatrix} \cos\phi \\ \sin\phi \end{Bmatrix} \tag{9.38}$$

where the subscripts now refer to the values of l and m. In writing the total field we assume normalization in both r and ϕ coordinates. Equation (9.38) describes the same field pattern given by equations (9.28) and (9.29). The next higher mode is given by

$$\psi_{21}(r,\phi) = \frac{\gamma}{\sqrt{\pi}}(\gamma^2 r^2)\, e^{-\frac{1}{2}\gamma^2 r^2} \begin{Bmatrix} \cos 2\phi \\ \sin 2\phi \end{Bmatrix} \tag{9.39}$$

$$\psi_{02}(r,\phi) = \sqrt{\frac{2}{\pi}}\gamma(\gamma^2 r^2 - 1)\, e^{-\frac{1}{2}\gamma^2 r^2} \tag{9.40}$$

Obviously, the above modal fields can be expressed as a linear combination of the $(m = 0, n = 2)$, $(m = 1, n = 1)$, and $(m = 2, n = 0)$ modes.

Example 9.1: *The elliptic profile*: One of the advantages of the solution in Cartesian coordinates is that it can readily be extended to elliptic fibers characterized by the following refractive index distribution.

$$n^2(x,y) = n_1^2\left[1 - 2\Delta\left\{\left(\frac{x}{a}\right)^2 + \left(\frac{y}{b}\right)^2\right\}\right] \tag{9.41}$$

Once again, we may use the method of separation of variables to obtain the modal profiles that would be given by equation (9.20) except that ξ and η would now be given by

$$\xi = \gamma_1 x; \quad \gamma_1 = \left[n_1 k_0 \frac{\sqrt{2\Delta}}{a}\right]^{1/2} \tag{9.42}$$

and

$$\eta = \gamma_2 y; \quad \gamma_2 = \left[n_1 k_0 \frac{\sqrt{2\Delta}}{b}\right]^{1/2} \tag{9.43}$$

Thus, the fundamental mode would be given by

$$\begin{aligned} \Psi_{0,0} &= \sqrt{\frac{\gamma_1\gamma_2}{\pi}}\, \exp\left[-\frac{1}{2}(\gamma_1^2 x^2 + \gamma_2^2 y^2)\right] \\ &= \sqrt{\frac{\gamma_1\gamma_2}{\pi}}\, \exp\left[-\frac{1}{2}n_1 k_0\sqrt{2\Delta}\left(\frac{x^2}{a} + \frac{y^2}{b}\right)\right] \end{aligned} \tag{9.44}$$

showing that the constant intensity profiles will have less ellipticity; the ratio of the major to minor axes will be $\sqrt{a/b}$.

9.4 Multimode fibers with optimum profiles

In Section 9.2 we have shown that for an infinitely extended square law profile the group velocity is almost independent of the mode number (see equation (9.24)). In this section we consider a more general class of refractive index profiles characterized by the following refractive index variation.

$$n^2(r) = \begin{Bmatrix} n_1^2[1 - 2\Delta(r/a)^q]; & 0 < r \le a \\ n_1^2(1 - 2\Delta) = n_2^2; & r \ge a \end{Bmatrix} \tag{9.45}$$

where a represents the radius of the core, Δ is the grading parameter, n_1 and n_2 represent the axial and cladding refractive indices, respectively, and q represents the exponent of the power law profile. In Figure 5.1 we plotted the refractive index variation as given by equation (9.45) for different values of q; $q = 2$ corresponds to a parabolic profile and $q = \infty$ corresponds to a step index profile. In Appendix C we have given a WKB analysis corresponding to the profile given by equation (9.45) and, if we label the propagation constants of the various modes as $\beta_1, \beta_2, \ldots$ (β_1 corresponds to the maximum value of β), we obtain (see Appendix F)

$$\beta_\nu^2 \approx k_0^2 n_1^2 [1 - 2\Delta(\nu/N)^{q/(q+2)}]; \quad \nu = 1, 2, \ldots, N \tag{9.46}$$

where

$$N = \frac{1}{2}\frac{q}{q+2}V^2 = \frac{q}{q+2}k_0^2 a^2 n_1^2 \Delta \tag{9.47}$$

represents the total number of guided modes and

$$V = k_0 a \left(n_1^2 - n_2^2\right)^{1/2} = k_0 a n_1 (2\Delta)^{1/2} \tag{9.48}$$

represents the normalized waveguide parameter. The WKB analysis is valid for highly multimode fibers with $V \gg 1$.

For a typical graded index fiber with $q = 2$

$$n_1 = 1.47, \quad \Delta = 0.01, \quad a = 25\,\mu\text{m}$$

so that at $\lambda_0 = 0.8\,\mu$m,

$$V \approx 41, \quad N \approx 420$$

The same fiber will support about 160 modes at 1.3 μm. In either case the fiber is highly multimoded and the WKB analysis is valid.

Example 9.2: For $q = 2$, equation (9.46) becomes

$$\beta_\nu^2 = k_0^2 n_1^2 - \frac{k_0 n_1 2(2\Delta)^{1/2}}{a}\nu^{1/2} \tag{9.49}$$

We write equation (9.21) as

$$\beta_{mn}^2 = k_0^2 n_1^2 - \frac{k_0 n_1 2(2\Delta)^{1/2}}{a}(m + n + 1) \tag{9.50}$$

which may be rewritten in the form

$$s = m + n + 1 = \frac{a}{k_0 n_1 2(2\Delta)^{1/2}}\left(k_0^2 n_1^2 - \beta_{mn}^2\right) \tag{9.51}$$

Thus

$$\begin{aligned}\nu &= \text{total number of modes above the propagation} \\ &\quad \text{constant } \beta_\nu = \beta_{mn} \\ &= 2 \times \text{total number of sets of integers } (m, n) \\ &\quad \text{so that } m + n + 1 < s \\ &= 2 \times [s + (s-1) + \cdots + 1] \\ &\approx (m + n + 1)^2 \end{aligned} \tag{9.52}$$

where the factor 2 is due to the two independent states of polarization and we have used the fact that when $m = 0, 1, 2, \ldots$ and so forth, n can take $s - 1, s - 2, \ldots$ values. Thus, for $q = 2$, equation (9.46) is almost identical to the result obtained by using equation (9.21).

In Appendix C the group velocity for different modes has been calculated and using this the time taken for the νth mode to propagate through distance L of the fiber can be expressed by

$$\begin{aligned} t_\nu &= \frac{L}{v_\nu} = L\frac{d\beta_\nu}{d\omega} \\ &\approx \frac{N_1 L}{c}\left[1 - \frac{q - 2 - \epsilon}{q + 2}\delta + \frac{3q - 2 - 2\epsilon}{2(q + 2)}\delta^2 + O(\delta^3)\right] \end{aligned} \tag{9.53}$$

where N_1 is the group index and

$$\epsilon = \frac{2k}{\Delta}\frac{d\Delta}{dk} = -\frac{2n_1}{N_1}\left[\frac{\lambda_0 \Delta'}{\Delta}\right] \tag{9.54}$$

leads to what is known as profile dispersion and

$$\delta = \Delta(\nu/N)^{q/(q+2)} \tag{9.55}$$

Obviously, $0 < \delta < \Delta$. In equation (9.54) primes denote differentiation with respect to λ_0.

We first neglect material dispersion – that is, we put $n_1' = \Delta' = 0$. We then obtain

$$t_\nu = \frac{n_1 L}{c}\left(1 + \frac{q - 2}{q + 2}\delta + \frac{3q - 2}{q + 2}\frac{\delta^2}{2} + \cdots\right) \tag{9.56}$$

which was first derived by Gloge and Marcatili (1973). Since $0 < \delta < \Delta$ and Δ is usually of the order of 0.01, unless q is very close to 2, we may neglect terms that are proportional to δ^2, δ^3, and so forth. Thus, for q not very close to 2 we retain only the term linear in δ and immediately find that the time taken increases monotonically as δ increases from 0 to Δ – that is, higher order modes

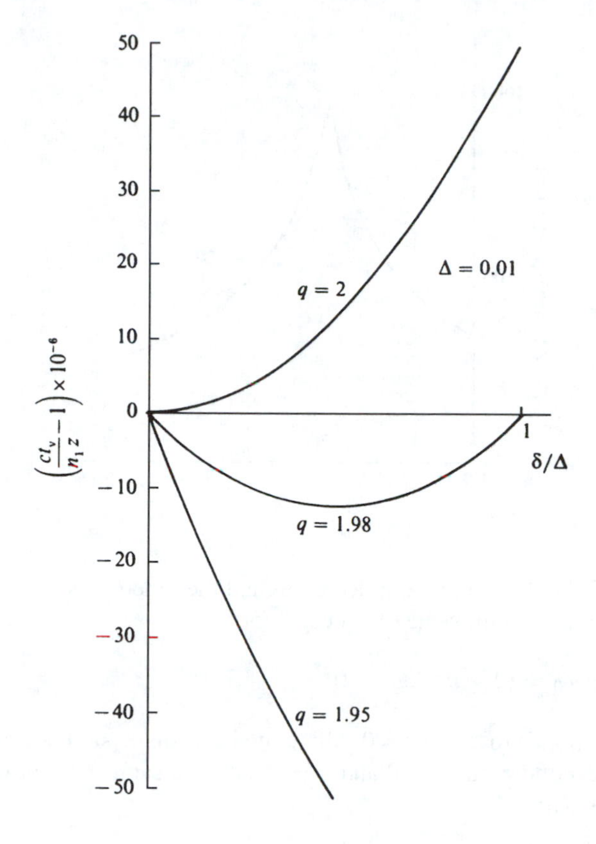

Fig. 9.1: Variation of $(c\,t_\nu/n_1 z - 1)$ versus δ for various values of q. The optimum profile corresponds to $q = 2 - 2\Delta$ when the modes corresponding to $\delta = 0$ and $\delta = \Delta$ take the same amount of time.

take a longer time to reach the output end of the fiber. Consequently, $t_{\max}$ and $t_{\min}$ correspond to $\delta = \Delta$ and $\delta = 0$, respectively, giving

$$t_{\max} - t_{\min} \approx \frac{n_1 L}{c}\frac{q-2}{q+2}\Delta; \quad \text{except for } q \approx 2 \tag{9.57}$$

For a step index fiber, $q = \infty$ and

$$\Delta\tau = t_{\max} - t_{\min} \approx (n_1 L/c)\Delta \tag{9.58}$$

which is the same as equation (3.13) when $n_1 \approx n_2$. For $q = 2$, equation (9.56) gives

$$t_\nu - \frac{n_1 L}{c} \approx (n_1 L/2c)\delta^2 \tag{9.59}$$

and t_ν again increases monotonically with δ (see Figure 9.1). Thus, since $0 < \delta < \Delta$, we obtain

$$\Delta\tau = t_{\max} - t_{\min} \approx (n_1 L/2c)\Delta^2 \tag{9.60}$$

Notice from equations (9.58) and (9.60) that for $\Delta = 0.01$, $\Delta\tau$ for a $q = 2$ fiber is 200 times smaller than the corresponding $\Delta\tau$ for a $q = \infty$ fiber.

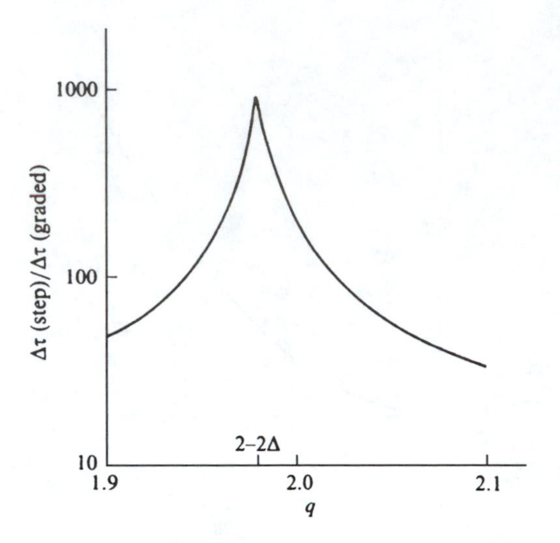

Fig. 9.2: Variation of the improvement of maximum time delay difference of a graded core fiber over a step index fiber as a function of q. Observe the very sharp variation of the improvement factor with q near the optimum value.

For $q \lesssim 2$ the delay time first decreases and then increases with δ (see Figure 9.1). The minimum value of t_ν occurs for

$$\delta = \delta_0 = (2 - q)/(3q - 2) \tag{9.61}$$

which corresponds to $dt_\nu/d\delta = 0$. Minimum pulse dispersion occurs when the modes corresponding to $\delta = 0$ and $\delta = \Delta$ take the same amount of time and this happens when

$$q = q_0 \approx 2 - 2\Delta \tag{9.62}$$

which represents the optimum profile for minimum intermodal dispersion.The corresponding pulse dispersion is given by

$$\Delta\tau = t_{\max} - t_{\min} \approx \frac{n_1 L}{c}\frac{\Delta^2}{8} \tag{9.63}$$

We may summarize that

$$\Delta\tau(q = \infty) = \frac{n_1}{c}\left(\frac{n_1}{n_2} - 1\right) L \approx \frac{n_1 \Delta}{c} L \quad \text{step profile} \tag{9.64}$$

$$\Delta\tau(q = 2) = \frac{n_1}{2c}\Delta^2 L \quad \text{parabolic profile} \tag{9.65}$$

$$\Delta\tau(q = 2 - 2\Delta) \approx \frac{n_1}{8c}\Delta^2 L \quad \text{optimum profile} \tag{9.66}$$

For $n_1 \approx 1.47$, $\Delta \approx 0.01$, we readily obtain

$$\text{Pulse dispersion} \approx \begin{cases} 50 \text{ ns/km}; & q = \infty, \text{ step profile} \\ 0.25 \text{ ns/km}; & q = 2, \text{ parabolic profile} \\ 0.06 \text{ ns/km}; & q = 1.98, \text{ optimum profile} \end{cases} \tag{9.67}$$

Thus, we obtain the same result as we obtained in Chapter 5 using ray analysis.

If we also take ϵ into account, one can show that minimum pulse dispersion occurs when

$$q = q_0 \approx 2 - 2\Delta + \epsilon\left(1 - \frac{1}{2}\Delta\right) \tag{9.68}$$

which shows the effect of material dispersion on the optimum profile. It should be noted that, in general, ϵ and Δ are functions of wavelength and therefore the optimum profile is also a function of wavelength. Indeed, q_0 is very sensitive to λ_0 for the most common germanium dioxide-doped silica glass [Olshansky and Keck (1976)].

Problems

9.1 Using the procedure given in Example 7.2, discuss the propagation of a Gaussian beam given by the following equation

$$\psi(x, y, z = 0) = \exp\left[-\frac{1}{2}\gamma^2\{(x - x_0)^2 + y^2\}\right]$$

[Assume β_{mn} to be given by equation (9.23).]

9.2 Show that field profiles given by equations (9.38)–(9.40) are normalized.

9.3 Express the modes given by equations (9.39) and (9.40) as a linear combination of the $(m = 0, n = 2)$, $(m = 1, n = 1)$, and $(m = 2, n = 0)$ modes.

9.4 Consider the $(m = 0, n = 3)$, $(m = 1, n = 2)$, $(m = 2, n = 1)$, and $(m = 3, n = 0)$ modes – all of them have the same propagation constants. Express them as a linear combination of the approximate LP_{lm} modes.

9.5 Consider an elliptic profile with $a/b = 9/4$. Plot the constant refractive index profiles and also the constant intensity profiles. Show that the intensity profiles have less ellipticity.

9.6 For any incident field distribution that is symmetric about the z-axis, the modes described by equation (9.32) with $l = 0$ can be used to study the propagation of a beam. Consider an incident Gaussian beam of the form

$$\psi(x, y, z = 0) = \frac{\alpha}{\sqrt{\pi}} \exp\left[-\frac{1}{2}\alpha^2 r^2\right]$$

Show that

$$\psi(x, y, z) = \sum_{m=1}^{\infty} A_m \left[\frac{\gamma}{\sqrt{\pi}} L_{m-1}(\gamma^2 r^2) \exp\left(-\frac{1}{2}\gamma^2 r^2\right)\right] e^{-i\beta_m z}$$

where

$$A_m = \frac{2w}{1 + w^2}\left(\frac{w^2 - 1}{w^2 + 1}\right)^m; \quad w = \frac{\alpha}{\gamma} \tag{9.69}$$

10

Waveguide dispersion and design considerations

10.1 Introduction

In Chapters 4 and 5 we discussed the dispersion mechanisms responsible for the broadening of a pulse when it propagates through a multimode optical fiber. The pulse broadening was mainly due to different group velocities of different modes in the waveguide; this is usually referred to as the *intermodal dispersion*. In addition, we have material dispersion (discussed in Chapter 6), which is due to the explicit dependence of the refractive index of the core and cladding on the wavelength λ_0.

Now, as shown in Section 8.3, for a step index fiber with

$$0 < V < 2.4048 \tag{10.1}$$

there is only one guided mode that can propagate through the fiber. Such single-mode fibers play a very important role in high-bandwidth optical fiber communication systems. Due to the presence of only a single mode, such fibers are free from intermodal dispersion. Pulse dispersion in such fibers is due to two mechanisms: (i) material dispersion, and (ii) waveguide dispersion. Even if the refractive indices of the core and cladding are independent of λ_0, we will still have waveguide dispersion, which is the subject matter of this chapter.

10.2 Expressions for group delay and waveguide dispersion

The normalized propagation constant b of a mode is defined by the following equation (see Section 8.3).

$$b = \frac{\frac{\beta^2}{k_0^2} - n_2^2}{n_1^2 - n_2^2} \tag{10.2}$$

Since, for guided modes β/k_0 lies between n_1 and n_2, b lies between 0 and 1. Further, as discussed in Section 8.3, for step index fibers, b depends only on the V value of the fiber. We rewrite equation (10.2) as

$$b = \frac{\frac{\beta}{k_0} - n_2}{n_1 - n_2} \frac{\frac{\beta}{k_0} + n_2}{n_1 + n_2}$$

$$\approx \frac{\frac{\beta}{k_0} - n_2}{n_1 - n_2} \tag{10.3}$$

where in the last step we have assumed n_1 to be very close to n_2, which is true for all practical fibers. Thus

$$\beta = \frac{\omega}{c}[n_2 + (n_1 - n_2)\, b(V)] \tag{10.4}$$

From the above equation we can see that even if n_1 and n_2 are independent of wavelength (i.e., if there is no material dispersion), $d\beta/d\omega$ will depend on ω due to the explicit dependence of b on V and hence on ω. Since $d\beta/d\omega$ represents the inverse of group velocity, this implies that the group velocity depends on ω even in the absence of material dispersion. This dispersion mechanism is referred to as *waveguide dispersion.*

We will now obtain explicit expressions for the group delay and waveguide dispersion. Using equation (10.4), the group velocity is given by

$$\frac{1}{v_g} = \frac{d\beta}{d\omega} = \frac{1}{c}[n_2 + (n_1 - n_2)\, b(V)] + \frac{\omega}{c}(n_1 - n_2)\frac{db}{dV} \cdot \frac{dV}{d\omega} \tag{10.5}$$

where we have assumed that n_1 and n_2 are independent of λ_0. Since

$$V = \frac{2\pi}{\lambda_0}\, a\sqrt{n_1^2 - n_2^2} = \frac{\omega}{c}\, a\sqrt{n_1^2 - n_2^2} \tag{10.6}$$

we have

$$\frac{dV}{d\omega} = \frac{V}{\omega} \tag{10.7}$$

implying

$$\frac{1}{v_g} = \frac{1}{c}[n_2 + (n_1 - n_2)\, b(V)] + \frac{1}{c}(n_1 - n_2)\, V\frac{db}{dV}$$

or,

$$\frac{1}{v_g} = \frac{n_2}{c} + \frac{n_1 - n_2}{c}\left[\frac{d}{dV}(bV)\right] \tag{10.8}$$

Thus, the time taken by a pulse to traverse length L of the fiber is given by

$$\tau = \frac{L}{v_g} = \frac{L}{c}\, n_2\left[1 + \Delta\,\frac{d}{dV}(bV)\right] \tag{10.9}$$

where

$$\Delta \equiv \frac{n_1^2 - n_2^2}{2n_1^2} \approx \frac{n_1 - n_2}{n_2} \tag{10.10}$$

Now for a source having a spectral width $\Delta\lambda_0$ the corresponding waveguide dispersion is given by

$$\Delta\tau_w = \frac{d\tau}{d\lambda_0}\,\Delta\lambda_0 \simeq \frac{L}{c}\, n_2\Delta\frac{d^2}{dV^2}(bV)\,\frac{dV}{d\lambda_0}\,\Delta\lambda_0$$

or

$$\Delta\tau_w \simeq -\frac{L}{c}n_2\Delta\left(\frac{\Delta\lambda_0}{\lambda_0}\right)\left(V\frac{d^2(bV)}{dV^2}\right) \tag{10.11}$$

where we have used the fact that [see equation (10.6)]

$$\frac{dV}{d\lambda_0} = -\frac{V}{\lambda_0}$$

We can rewrite equation (10.11) in the following form

$$\Delta\tau_w \simeq -\frac{n_2\,\Delta}{3\,\lambda_0} \times 10^7\left(V\,\frac{d^2(bV)}{dV^2}\right)\text{ ps/km·nm} \tag{10.12}$$

where λ_0 is measured in nanometers.

For the LP modes in a step index fiber we have [see, e.g., Adams (1981) Chapter 7]

$$\frac{d}{dV}[Vb(V)] = 1 - \frac{U^2}{V^2}[1 - 2\kappa_l] \tag{10.13}$$

and

$$V\frac{d^2}{dV^2}[Vb(V)] = \frac{2U^2\kappa_l}{V^2W^2}\left\{[3W^2 - 2\kappa_l(W^2 - U^2)] + W[W^2 + U^2\kappa_l][\kappa_l - 1]\left[\frac{K_{l-1}(W) + K_{l+1}(W)}{K_l(W)}\right]\right\} \tag{10.14}$$

where

$$\kappa_l = \kappa_l(W) = \frac{K_l^2(W)}{K_{l-1}(W)\,K_{l+1}(W)} \tag{10.15}$$

In Figure 10.1 we have plotted the variations of b, $(bV)'$, and $V(bV)''$ for the fundamental mode as a function of V for a step index fiber; these are universal curves. It can be seen that the waveguide dispersion is negative in the single-mode region. Since the material dispersion is positive for λ_0 greater than the zero material dispersion wavelength (see equation 6.8 and Figure 6.3), there is a wavelength at which the negative waveguide dispersion will compensate the positive material dispersion. At this wavelength the net dispersion of the single-mode fiber is zero and this wavelength is referred to as the zero dispersion wavelength. Single-mode fibers with zero (total) dispersion around 1300 nm are referred to as conventional single-mode fibers (CSFs) or nondispersion-shifted fibers (NDSFs). Most installed fiber optic systems today operate with such fibers. On the other hand, single-mode fibers with zero (total) dispersion around 1550 nm are referred to as dispersion-shifted fibers (usually abbreviated as DSFs) – see Section 10.4.

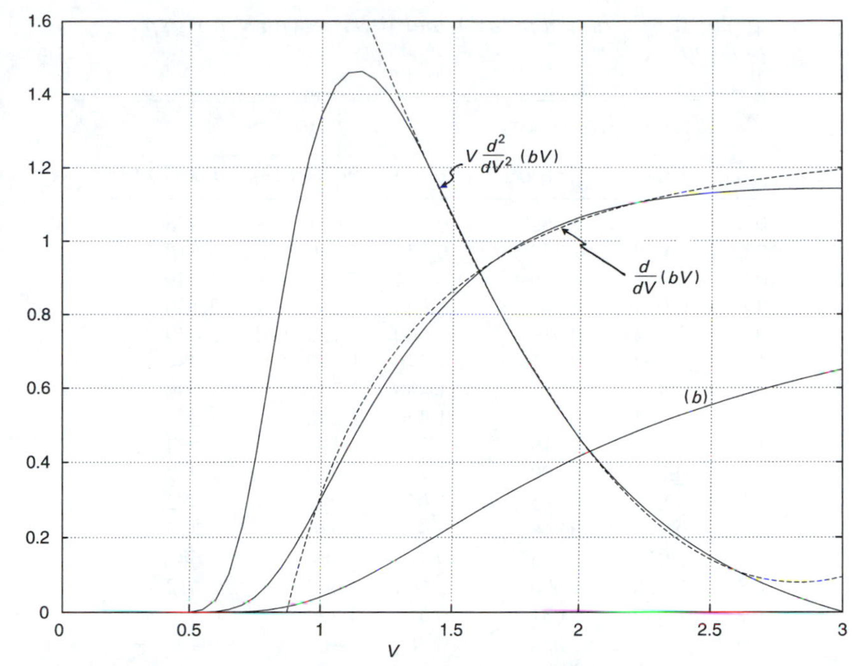

Fig. 10.1: The variations of $(Vb)'$ and $V(bV)''$ with V for the fundamental mode of a step index fiber. The dashed curve represents the corresponding variations as given by the empirical formulae equations (10.17) and (10.19). Also shown in the figure is the variation of b with V.

10.3 Empirical formula for step index fibers

For design purposes, it is often very convenient to use empirical formulae, which have, of course, restricted domains of applicability. A very simple empirical formula is given by (see Section 8.3)

$$b \simeq \left(A - \frac{B}{V}\right)^2 \tag{10.16}$$

with

$$A = 1.1428 \quad \text{and} \quad B = 0.996$$

Using the above equation one readily obtains

$$\frac{d}{dV}(bV) = A^2 - \frac{B^2}{V^2} \tag{10.17}$$

which has also been plotted (as a dashed curve) in Figure 10.1. Further,

$$V\frac{d^2}{dV^2}(bV) = \frac{2B^2}{V^2} \tag{10.18}$$

the last relation being accurate around $V \simeq 1.9$. A more accurate empirical formula is given by [Marcuse (1979)]

$$V\frac{d^2}{dV^2}(bV) \simeq 0.080 + 0.549(2.834 - V)^2 \tag{10.19}$$

Table 10.1. *Variation of $(bV)'$ and $V(bV)''$ with V for step index fibers*

	$\frac{d}{dV}(bV)$		$V(bV)''$		
V	Exact	Eq. (10.17)	Exact	Eq. (10.19)	Eq. (10.18)
0.5	0	−2.662	0.001	3.071	7.936
0.6	0.002	−1.450	0.034	2.820	5.510
0.7	0.019	−0.719	0.231	2.580	4.049
0.8	0.074	−0.244	0.634	2.351	3.100
0.9	0.174	0.081	1.060	2.133	2.449
1.0	0.302	0.314	1.346	1.927	1.984
1.1	0.437	0.486	1.458	1.731	1.640
1.2	0.564	0.617	1.440	1.546	1.378
1.3	0.676	0.719	1.343	1.372	1.174
1.4	0.771	0.800	1.209	1.209	1.012
1.5	0.849	0.865	1.063	1.057	0.882
1.6	0.913	0.918	0.919	0.916	0.775
1.7	0.965	0.963	0.785	0.786	0.687
1.8	1.006	1.000	0.664	0.667	0.612
1.9	1.039	1.031	0.556	0.559	0.550
2.0	1.065	1.058	0.462	0.462	0.496
2.1	1.086	1.081	0.380	0.376	0.450
2.2	1.102	1.101	0.309	0.301	0.410
2.3	1.114	1.118	0.248	0.237	0.375
2.4	1.124	1.134	0.195	0.183	0.344

which has also been plotted (as a dashed curve) in Figure 10.1. In Table 10.1 we have given the exact variation of b, $(bV)'$, and $V(bV)''$ [as obtained from equations (10.13) and (10.14)] along with the approximate values obtained by using the above empirical formulae.

Example 10.1: We consider a step index fiber with

$$n_1 = 1.450840, \quad n_2 = 1.446918, \quad a = 4.1\ \mu\text{m} \tag{10.20}$$

These values correspond to a typical step index fiber operating at 1300 nm. The corresponding value of Δ is 0.0027. Assuming that the refractive indices n_1 and n_2 are independent of wavelength, we can obtain the values of waveguide dispersion at different wavelengths by using Table 10.1 and equation (10.11).

Using the numerical values given in equation (10.20), we readily obtain

$$V = \frac{2\pi}{\lambda_0}\, a\sqrt{n_1^2 - n_2^2} \simeq \frac{2746.3}{\lambda_0}$$

where λ_0 is measured in nanometers. The cutoff wavelength of this fiber is 1142 nm. The results for waveguide dispersion are summarized[1]

[1]The values given in Tables 10.3 and 10.4 and in Figures 10.2–10.4 have been generated using a software developed by A.K. Ghatak, I.C. Goyal, and R.K. Varshney (unpublished work).

Table 10.2. *Dispersion characteristics of a step index fiber as characterized by equation (10.20)*

λ_0 (nm)	1100	1300	1560
V	2.497	2.113	1.761
$V(bV)''$	0.150	0.370	0.710
D_w (ps/km·nm)	−1.78	−3.72	−5.94
D_m (ps/km·nm)	−23.18	+1.58	+21.93
D_t (ps/km·nm)	−24.96	−2.14	+15.99

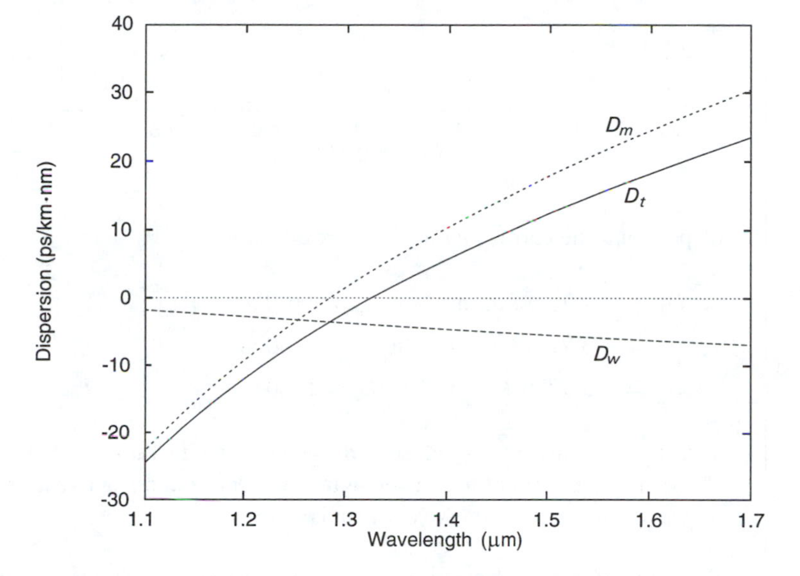

Fig. 10.2: Variation of material, waveguide, and total dispersion for a step index silica fiber characterized by equation (10.20). The zero dispersion wavelength is 1325 nm.

in Table 10.2. The values of $V(bV)''$ are exact; however, if we make linear interpolation of values given in Table 10.1, we would get quite accurate values of D_w.

The corresponding variation of material, waveguide, and total dispersion with wavelength is shown in Figure 10.2. The zero dispersion wavelength and the slope at the zero dispersion wavelength are 1325 nm and 0.083 ps/km·nm^2, respectively.

Example 10.2: In this example we will consider the effect on waveguide dispersion because of the refractive index variation with wavelength. We assume the core to be doped with GeO_2 and the cladding to be pure silica; then for a typical step index fiber (with $a = 4.1\ \mu$m)

$$n_1 = 1.45315; \quad n_2 = 1.44920 \quad \text{at } \lambda_0 = 1100 \text{ nm}$$

$$n_1 = 1.45084; \quad n_2 = 1.44692 \quad \text{at } \lambda_0 = 1300 \text{ nm}$$

$$n_1 = 1.44781; \quad n_2 = 1.44390 \quad \text{at } \lambda_0 = 1560 \text{ nm}$$

It may be seen that the corresponding values of Δ are 0.002714, 0.002698, and 0.002696, respectively; this is what is known as profile

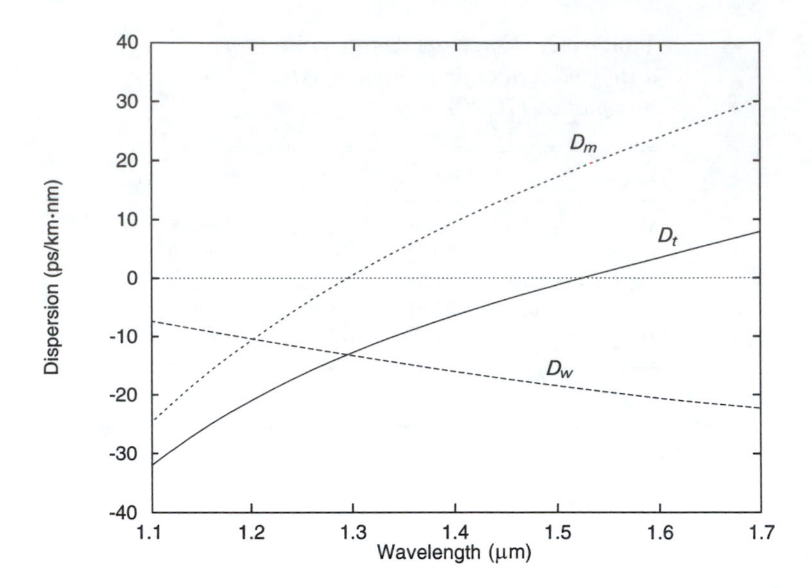

Fig. 10.3: Variation of material, waveguide, and total dispersion for a step index silica fiber characterized by equation (10.21). The zero dispersion wavelength is 1523 nm. Such a fiber is usually referred to as a DSF.

dispersion. The corresponding *exact* results are

$$\begin{aligned} D_w &= -1.75 \text{ ps/km·nm} \quad \text{at } \lambda_0 = 1100 \text{ nm} \\ &= -3.72 \text{ ps/km·nm} \quad \text{at } \lambda_0 = 1300 \text{ nm} \\ &= -5.97 \text{ ps/km·nm} \quad \text{at } \lambda_0 = 1560 \text{ nm} \end{aligned}$$

The above results are very close to those presented in Table 10.2. This shows that the effect of the wavelength dependence of refractive index on waveguide dispersion is very small.

Example 10.3: In this example, we consider a step index fiber characterized with a large value of Δ and a small core radius.

$$n_1 = 1.457893; \quad n_2 = 1.446918; \quad a = 2.3\,\mu\text{m} \tag{10.21}$$

Thus

$$\Delta = 0.0075 \quad \text{and} \quad V = \frac{2580.3}{\lambda_0}$$

with λ_0 in nanometers. The cutoff wavelength is 1073 nm. The dispersion characteristics of such a fiber are given in Figure 10.3 and Table 10.3. The germanium doping of the core is more than in Example 10.1 and, hence, the material dispersion is slightly different. The zero dispersion wavelength is now 1523 nm, because of which these are known as DSFs – see Section 10.4. The corresponding dispersion slope is 0.048 ps/km·nm^2.

Example 10.4: In this example, we consider a step index fiber with an extremely large value of Δ so that the fiber has a very large negative dispersion at $\lambda_0 \approx 1550$ nm. Such fibers are usually referred to as

Table 10.3. *Dispersion characteristics of a step index fiber as characterized by equation (10.21)*

λ_0 (nm)	1100	1300	1560
V	2.346	1.985	1.654
$V(bV)''$	0.223	0.476	0.845
D_w (ps/km·nm)	−7.33	−13.24	−19.59
D_m (ps/km·nm)	−25.04	+0.59	+21.5
D_t (ps/km·nm)	−32.37	−12.65	+1.91

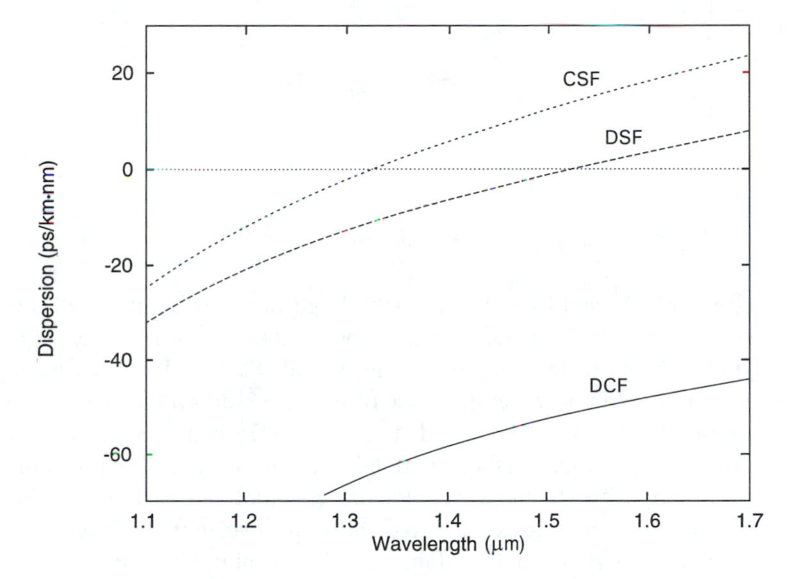

Fig. 10.4: Variation of total dispersion with wavelength for step index fibers discussed in Examples 10.1, 10.3, and 10.4, which correspond to CSFs, DSFs, and DCFs, respectively.

dispersion-compensating fibers (DCFs) (see Chapter 15). We assume

$$n_1 = 1.476754; \quad n_2 = 1.446918; \quad a = 1.5\ \mu\text{m} \tag{10.22}$$

Thus

$$\Delta = 0.02 \quad \text{and} \quad V = \frac{2783.6}{\lambda_0}$$

where λ_0 is in nanometers. The cutoff wavelength is 1158 nm. For such a large value of Δ, the germanium doping of the core would be much larger than in previous examples. Assuming the validity of the Sellemeier formula given in Chapter 6, we have carried out numerical calculations to obtain

$$\text{Total dispersion} \simeq -97, -67, \text{and} -50 \text{ ps/km·nm}$$

for $\lambda_0 = 1100$, 1300, and 1560 nm, respectively. Thus, the fiber is now characterized by a large negative dispersion at $\lambda_0 \approx 1560$ nm (see Figure 10.4).

10.4 Dispersion-shifted fibers

In Figure 3.12 we showed the loss spectrum corresponding to an extremely low-loss fiber, and as can be seen the loss attains a minimum value of about 0.2 dB/km at 1550 nm. Thus, operating at $\lambda_0 = 1300$ nm, the system would be limited by the loss in the fiber. Since the lowest loss lies at $\lambda_0 \approx 1550$ nm, if the zero dispersion wavelength could be shifted to the $\lambda_0 \approx 1550$-nm region, one could have both minimum loss and very low dispersion. This would lead to very high bandwidth systems with very long (~ 100 km) repeaterless transmission. The shift of the zero total dispersion wavelength to a region around 1550 nm can indeed be accomplished by changing the fiber parameters.

To illustrate this we first calculate the material dispersion at $\lambda_0 \simeq 1550$ nm. Now (for pure silica)

$$\frac{d^2 n_1}{d\lambda_0^2} = -4.2 \times 10^{-3}\ \mu\text{m}^{-2} \quad \text{at } \lambda_0 \approx 1550\,\text{nm}$$

so that

$$D_m = \Delta\tau_m / L\Delta\lambda_0 \approx 22 \text{ ps/km·nm}$$

(see also Tables 10.2 and 10.3). To cancel this large material dispersion, we must increase the magnitude of waveguide dispersion that (as can be seen from equation (10.11)) can be achieved by increasing Δ while keeping V approximately constant; this would imply that the value of the core radius a has to decrease. For example, Figure 10.3 corresponds to $\Delta \simeq 0.0075$ and $a = 2.3\,\mu$m. For such a fiber, at $\lambda_0 \simeq 1523$ nm the material dispersion is ~19 ps/km·nm, which is almost exactly offset by the waveguide dispersion of -19 ps/km·nm. Thus, $\lambda_0 \simeq 1523$ nm is referred to as the *zero dispersion wavelength.*

For a typical single-mode fiber fabricated by Corning Glass Works in the United States [quoted by Blank, Bickers, and Walker (1985)]

Attenuation: 0.215 dB/km at $\lambda_0 = 1550$ nm
0.230 dB/km at $\lambda_0 = 1520$ nm
Dispersion: Zero dispersion wavelength: 1550 nm
Dispersion slope = 0.075 ps/nm^2·km (maximum)
$\Rightarrow$ dispersion < 0.75 ps/nm·km for $1540 \text{ nm} < \lambda_0 < 1560$ nm

Using multimode lasers, Blank et al. (1985) could achieve a 140-Mbit/s system for a *repeaterless link* of 222.8 km; the 222.8-km span consists of 37 spliced fiber lengths with a total span loss of 50.6 dB. We should mention here that repeater spacings $\gtrsim 250$ km are of tremendous interest because about 40% of undersea systems are less than 250 km in length and, hence, use of such fiber optic communication systems would not require any repeaters.

Figure 10.4 shows the variation of total dispersion of the single-mode step index fibers discussed in Examples 10.1, 10.3, and 10.4. Example 10.1 corresponds to the CSF having zero dispersion around $\lambda_0 \approx 1300$ nm; Example 10.3 corresponds to DSFs having zero dispersion around $\lambda_0 \approx 1550$ nm, and Example 10.4 corresponds to DCFs having large negative dispersion around $\lambda_0 \approx 1550$ nm.

We should mention here that for step index DSFs, a high value of Δ would imply a higher dopant (GeO_2) concentration in the core, because of which the attenuation would also increase. One of the methods to avoid this additional loss is to reduce the concentration of GeO_2 in the core by making a triangular grading of the refractive index. This results in a higher waveguide dispersion with lower attenuation. Such triangular core fibers were among the first dispersion-shifted designs that were demonstrated [Saifi et al. (1982)]. The attenuation in such fibers can be reduced to as low as 0.21 dB/km at $\lambda_0 \sim 1550$ nm [Pearson et al. (1984)]. The main disadvantage of these fibers is their short cutoff wavelength, which was around 0.85 μm in the initial designs. This leads to higher sensitivity to bending. Attempts have been made to increase the cutoff wavelength by making a trapezoidal profile or by making a depressed index cladding around the core. With the latter design, it has been possible to increase the cutoff wavelength to 1.1 μm while the losses at $\lambda_0 \approx 1.55$ μm were 0.24 dB/km [Shang et al. (1985)]. Another approach to design bend optimized DSFs is based on dual-shape core (DSC) index fibers [Ohasi, Kuwaki, and Tanaka (1986); Tewari, Pal, and Das (1992)], in which the core consists of two sections of different index variations.

Another type of fiber design is based on fluorine doping, in which the core is of pure silica and the cladding is doped with fluorine to reduce the index. Obviously, the main advantage of these fibers is their lowest loss since core is made of pure silica. A loss of 0.15 dB/km at 1550 nm has been attained [Csencsits et al. (1984)]. However, the zero dispersion wavelength in such fibers cannot be shifted to 1.55 μm. For more details on single-mode fiber characteristics, readers may look up Sharma (1995), Pal (1995), and references therein.

10.5 Design considerations

We next consider several issues in connection with design considerations of a single-mode fiber. We restrict our analysis to step index fibers.

10.5.1 Single-mode operation

For a (step index) fiber to be single moded we must have

$$V = \frac{2\pi}{\lambda_0} a \sqrt{n_1^2 - n_2^2} < 2.4048 \tag{10.23}$$

Thus

$$a\sqrt{2\Delta} < 0.26\,\lambda_0 \tag{10.24}$$

where we have assumed $n_1 \simeq 1.45$.

The corresponding domains of single-mode operation for $\lambda_0 = 1.3$ μm and 1.55 μm are shown in Figure 10.5. Obviously, for the fiber to be single moded at 1.3 μm as well as at 1.55 μm, we must consider the curve corresponding to

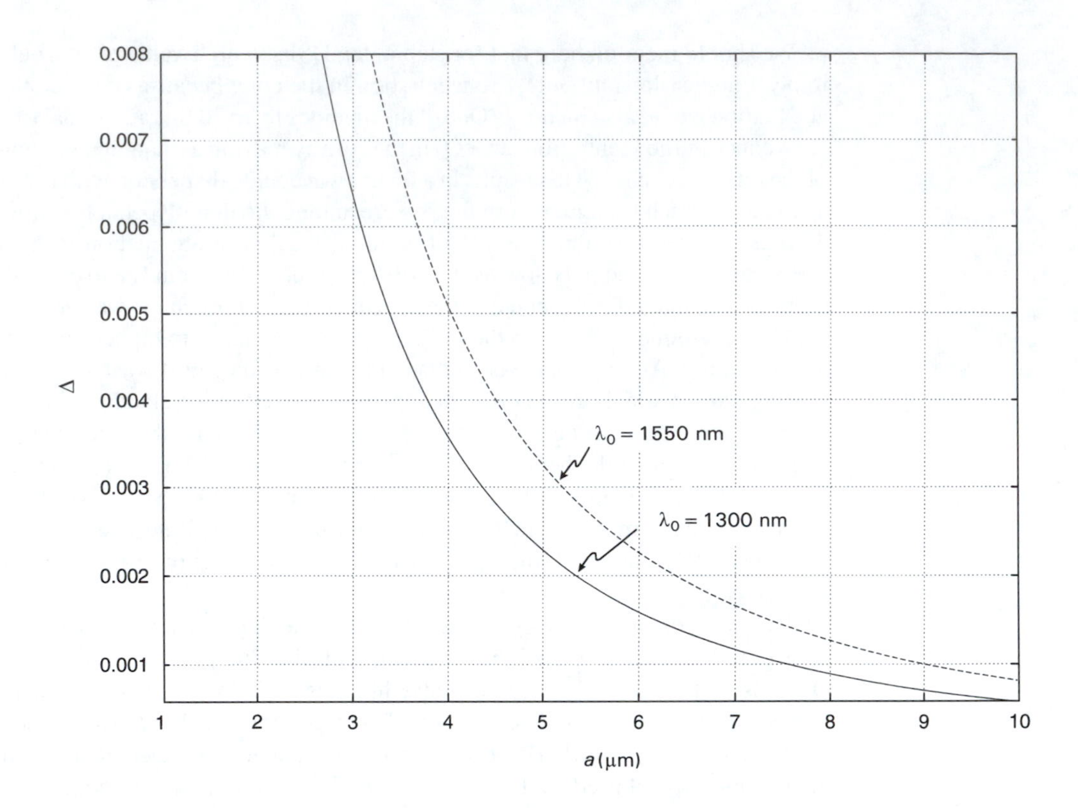

Fig. 10.5: The region to the left of the curves represents the domain of single-mode operation for a step index fiber.

$\lambda_0 = 1.3\,\mu$m. Thus, assuming $\lambda_0 = 1.3\,\mu$m we must, for example, have

$$\begin{aligned} a &< 5.34\,\mu\text{m} \quad \text{for } \Delta = 0.002 \\ a &< 3.8\,\mu\text{m} \quad \text{for } \Delta = 0.004 \\ a &< 3.1\,\mu\text{m} \quad \text{for } \Delta = 0.006 \end{aligned} \tag{10.25}$$

and

$$a < 2.7\,\mu\text{m} \quad \text{for } \Delta = 0.008$$

10.5.2 Splice loss

In any communication system there are usually a large number of splice joints and often there are small transverse offsets at the splice. Obviously, from the design point of view, the fiber parameters should be such that the transverse offset splice loss should be small. In Problem 8.9 we showed that the splice loss at a transverse offset of u is given by

$$\alpha\,(\text{dB}) \approx 4.34 \left(\frac{u}{w_P}\right)^2 \tag{10.26}$$

where w_P is the Petermann-2 spot size, which was tabulated in Table 8.1. For example, for the splice loss to be less than 0.1 dB for a 0.5-μm offset at the splice, we must have

$$w_P > 3.30\,\mu\text{m} \tag{10.27}$$

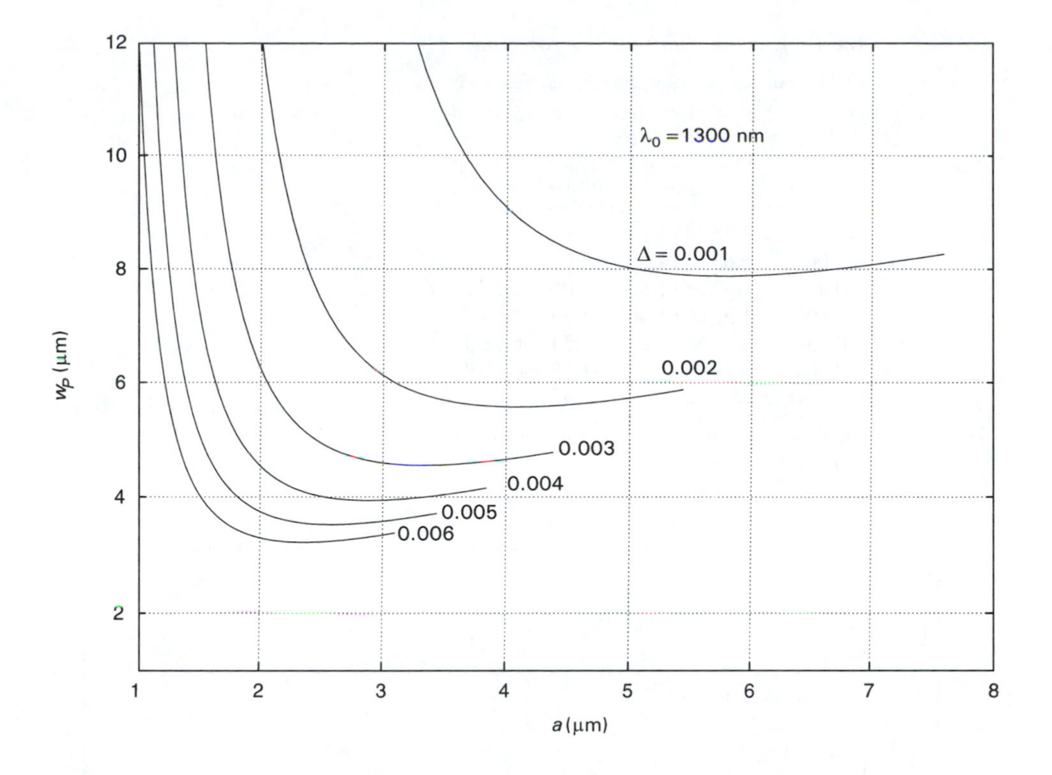

Fig. 10.6: Variation of w_P with the core radius for step index fibers with $\Delta = 0.001, 0.002, 0.003, 0.004, 0.005$, and 0.006; $\lambda_0 = 1300$ nm.

Now, according to the CCITT recommendations, a single-mode fiber (to be used in a communication system) must have[2]

$$4.5\,\mu\text{m} \lesssim w_P \lesssim 5\,\mu\text{m} \tag{10.28}$$

In Figures 10.6 and 10.7 we have plotted w_P as a function of the core radius a for different values of Δ at $\lambda_0 = 1300$ nm and $\lambda_0 = 1550$ nm, respectively. Obviously, for the fiber to operate both at 1300 nm and at 1550 nm, we need to consider only the curves corresponding to $\lambda_0 = 1300$ nm.

Now, for each value of Δ, the curves are plotted for $a < a_0$ beyond which the fiber is not single moded (see, e.g., equation (10.25)). Now, if we use the condition given by equation (10.28), we find (for $\lambda_0 = 1300$ nm) that for $\Delta \lesssim 0.002$, w_P is always greater than $5\,\mu$m and, hence, such low values of Δ are not permissible. On the other hand, for higher values of Δ, there is an upper limit for the core radius, which is tabulated in Table 10.4.

10.5.3 Bend loss

Bend loss represents an important characteristic in the design of a single-mode fiber. The bend loss coefficient (in a step index fiber) is given by [Snyder and

[2] The upper limit on w_P is imposed so that there is no significant splice loss due to nonmatching of the spot sizes between fibers from different vendors.

Table 10.4. *Upper limit for the core radius*

Δ	$\lambda_0 = 1300$ nm	$\lambda_0 = 1550$ nm
	For $5.0\,\mu$m $> w_P >$ $4.5\,\mu$m, the core radius should lie between	
0.003	2.45 and 4.36[a]	—
0.004	1.83 and 2.02	2.66 and 4.51[a]
0.005	1.52 and 1.63	2.06 and 2.35
0.006	1.32 and 1.40	1.75 and 1.90
0.007	—	1.54 and 1.65
0.008	—	1.38 and 1.47

[a]This value is determined from the cutoff condition (see Figure 10.5).

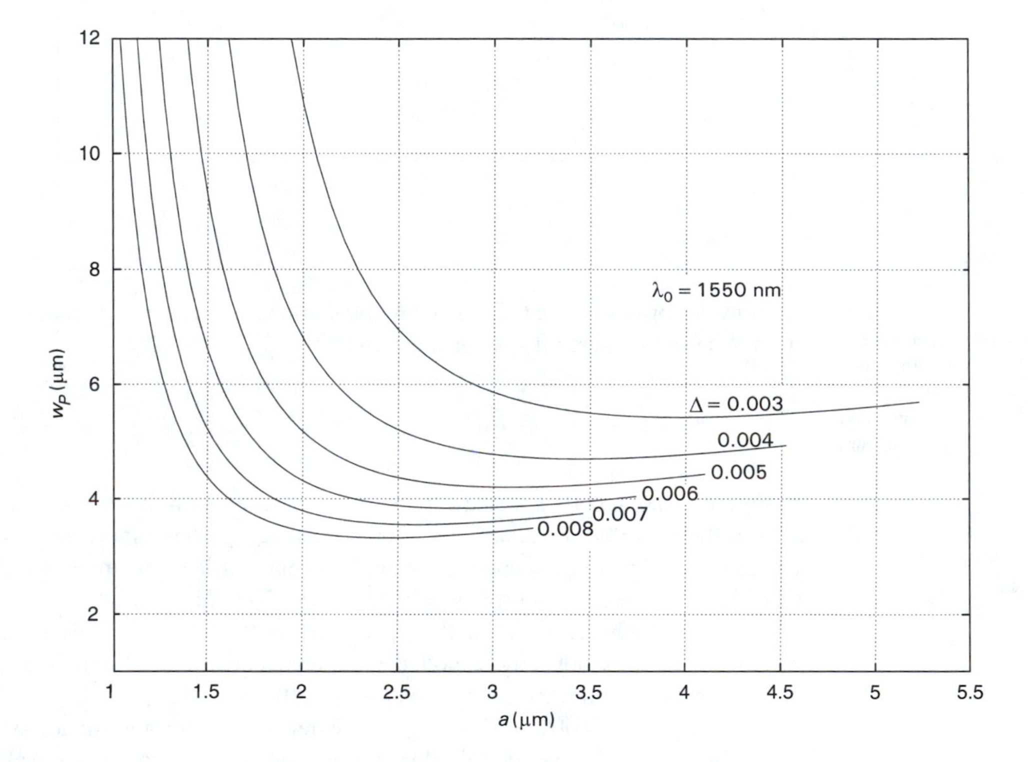

Fig. 10.7: Variation of w_P with the core radius for step index fibers with $\Delta = 0.003, 0.004, 0.005, 0.006, 0.007$, and 0.008; $\lambda_0 = 1550$ nm.

Love (1983), p. 481]

$$\alpha\,(\text{dB/m}) = 4.343\left(\frac{\pi}{4aR_c}\right)^{1/2}\left[\frac{U}{VK_1(W)}\right]^2\frac{1}{W^{3/2}}\exp\left[-\frac{2W^3}{3k_0^2a^3n_1^2}\,R_c\right] \tag{10.29}$$

where R_c is the radius of curvature of the bend and other symbols have their usual meaning.

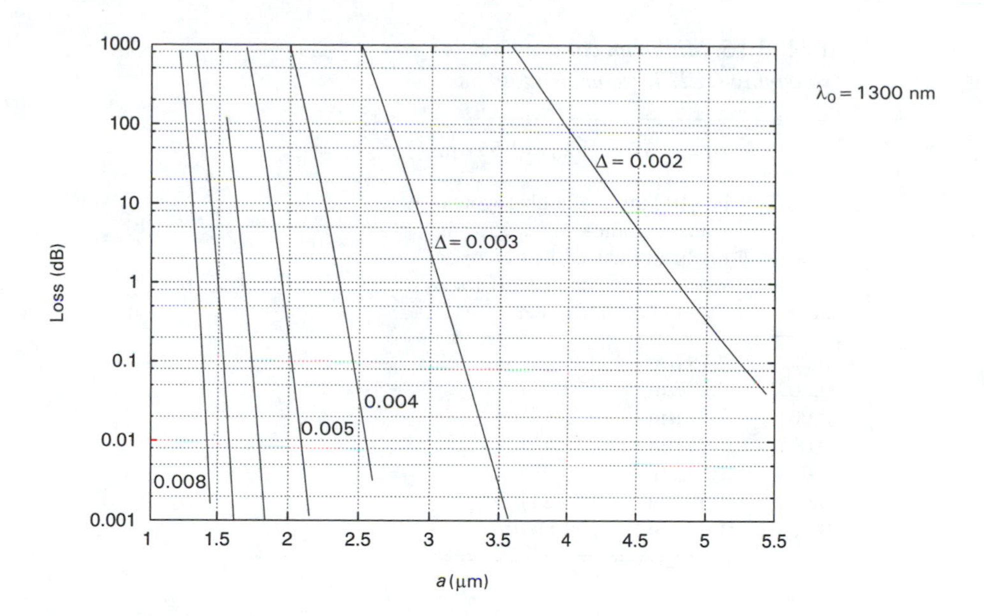

Fig. 10.8: Bend loss as a function of the core radius a (for $\lambda_0 = 1300$ nm and different values of Δ) corresponding to 100 turns of the fiber wound with a radius of 3.75 cm.

For a given step index fiber if V lies between 1.5 and 2.5, then we can use Table 8.1 to calculate the bend loss coefficient. In an actual optical transmission system, the 1300-nm fibers are designed so that at a later date, they may be used for transmitting signals at 1550-nm wavelength. In view of this, CCITT has recommended that

> the bend induced loss increase at 1550 nm of 100 turns of fiber wound with a radius of 3.75 cm should be less than 1 dB.

This requirement is determined by typical dimensions of joint boxes wherein an extra length of fiber is usually provided to cater to later demands.

In Figures 10.8 and 10.9 we have plotted the bend loss as a function of the core radius a for $R = 3.75$ cm with 100 turns corresponding to $\lambda_0 = 1300$ nm and 1550 nm, respectively. For given values of a and Δ, the bending loss is higher at 1550 nm. The minimum core radius satisfying the CCITT condition has been tabulated in Table 10.5.

Problems

10.1 Consider three step index fibers with

(i) $\Delta = 0.002$
(ii) $\Delta = 0.003$
(iii) $\Delta = 0.004$

Assume $n_2 = 1.45$.

(a) In each case, calculate the core radius so that the cutoff wavelength is 1150 nm.
(b) In the domain $1200 < \lambda_0 < 1600$ nm, calculate material dispersion for pure silica (see Table 6.1) and waveguide dispersion using Table 10.1 and obtain the zero dispersion wavelength.
(c) In each case, calculate the Gaussian spot size w and the Petermann-2 spot size w_P for the corresponding zero dispersion wavelength.

Table 10.5. *Minimum core radius satisfying CCITT recommendations*

Δ	$\lambda_0 = 1.3\ \mu\text{m}$	$\lambda_0 = 1.55\ \mu\text{m}$
	For 100 turns of a fiber wound with a radius of 3.75 cm, the bend-induced loss increase would be less than 1 dB for a core radius greater than	
0.002	4.79 μm	6.30 μm
0.003	3.07 μm	3.87 μm
0.004	2.35 μm	2.93 μm
0.005	1.95 μm	2.41 μm
0.006	1.69 μm	2.08 μm
0.007	1.50 μm	1.84 μm
0.008	1.35 μm	1.66 μm

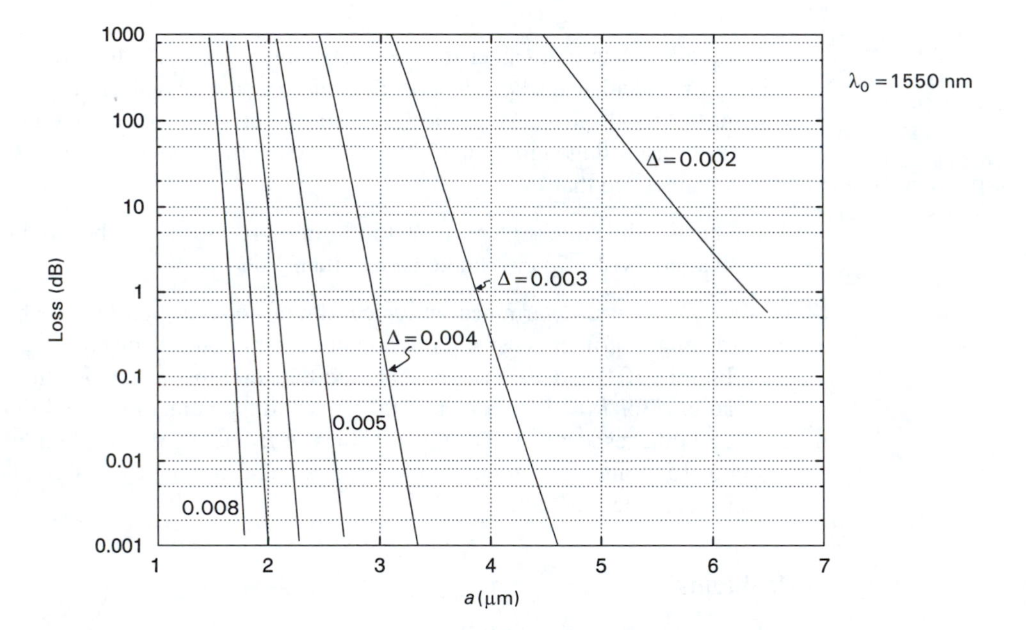

Fig. 10.9: Bend loss as a function of the core radius a (for $\lambda_0 = 1550$ nm and different values of Δ) corresponding to 100 turns of the fiber wound with a radius of 3.75 cm.

10.2 (a) For the step index fibers discussed in Examples 10.1, 10.3, and 10.4, calculate using Table 8.1 (and linear interpolation) the value of the Petermann-2 spot size w_P for $\lambda_0 = 1150$ nm, 1300 nm, 1400 nm, and 1560 nm. In each case, calculate the splice loss for a 0.5-μm offset.

[ANSWER: w_P (in μm) $\approx$ (4.5, 4.9, 5.2, 5.7); (2.6, 2.9, 3.1, 3.4); (1.6, 1.8, 1.9, 2.1).]

10.3 Consider the step index fiber discussed in Examples 10.1 and 10.3. For $\lambda_0 =$ 1100 nm, 1300 nm, and 1560 nm, use the empirical formula given by equation (10.16) to calculate waveguide dispersion and use Table 6.1 to calculate material dispersion. Calculate total dispersion at the above wavelengths and compare with the results given in Tables 10.2 and 10.3.

10.4 Consider a step index silica fiber operating at 1330 nm.

(a) Assuming the cutoff wavelength to be 1150 nm, calculate the value of the product $a\sqrt{\Delta}$ (assume $n_2 \approx 1.45$).

(b) Calculate the value of V at the operating wavelength and use Table 10.1 (with linear interpolation) to calculate $V(bV)''$.

(c) Use Table 6.1 (with linear interpolation) to calculate material dispersion and calculate the value of Δ (and therefore the value of a) to obtain zero dispersion at 1330 nm.

(d) Use Table 8.1 and linear interpolation to calculate the Petermann-2 spot size w_P.

(e) Use equation (10.29) (along with Table 8.1) to calculate bend loss for 100 turns of the fiber wound with a radius of 3.75 cm.

[ANSWER: (a) 0.215 μm (b) 2.079, 0.397 (c) 5.39 ps/km·nm $\Delta \simeq 0.0037$; $a = 3.5\,\mu$m (d) 4.2 μm.]

10.5 Consider a step index silica fiber operating at 1550 nm.

(a) Assuming the cutoff wavelength to be 1150 nm, calculate the value of the product $a\sqrt{\Delta}$ (assume $n_2 \approx 1.45$).

(b) Calculate the value of V at the operating wavelength and use Table 10.1 (with linear interpolation) to calculate $V(bV)''$.

(c) Use Table 6.1 (with linear interpolation) to calculate material dispersion and calculate the value of Δ (and therefore the value of a) to obtain zero dispersion at 1550 nm.

(d) Use Table 8.1 and linear interpolation to calculate the Petermann-2 spot size w_P.

(e) Use equation (10.29) (along with Table 8.1) to calculate bend loss for 100 turns of the fiber wound with a radius of 3.75 cm.

[ANSWER: (a) 0.2053 μm (b) 1.707, 0.776 (c) 21.9 ps/km·nm $\Delta \simeq 0.00905$; $a = 2.16\,\mu$m (d) $w_P \simeq 3.1\,\mu$m.]

11

*Sources for optical fiber communication**

11.1 Introduction

The most important and widely exploited application of optical fiber is its use as the transmission medium in an optical communication link. The basic optical fiber communication system consists of a transmitter, an optical fiber, and a receiver. The transmitter has a light source, such as a laser diode, which is modulated by a suitable drive circuit in accordance with the signal to be transmitted. Similarly, the receiver consists of a photodetector, which generates electrical signals in accordance with the incident optical energy. The photodetector is followed by an electronic amplifier and a signal recovery unit.

Among the variety of optical sources, optical fiber communication systems almost always use semiconductor-based light sources such as light-emitting diodes (LEDs) and laser diodes because of the several advantages such sources have over the others. These advantages include compact size, high efficiency, required wavelength of emission, and, above all, the possibility of direct modulation at high speeds.

In this chapter, we discuss the mechanism of light generation, basic device configurations, and relevant output characteristics of the light source. In Section 11.2 we discuss the basic requirements that the source should meet to be suitable for use in an optical fiber communication system. In Section 11.3 we briefly present an elementary account of the principle of operation of a laser. In Section 11.4 we discuss basic semiconductor physics relevant to the operation of a semiconductor laser followed by the device structure and characteristics in Section 11.5. Finally, in Section 11.6 we briefly discuss the characteristics of LEDs that are relevant to a fiber optic communication link.

11.2 Communication requirements

The choice of an optical source for a particular application is determined by the requirements that it should meet. High-speed communication links employing optical fibers normally deal with high-speed digital signals at bit rates in excess of 1 Gb/s and repeater spacings of several tens of kilometers. The source should thus meet the following three most important requirements:

(i) The source wavelength should correspond to the low-loss windows of silica – namely, around 1.30-μm and 1.55-μm wavelengths. For a

*A major portion of this chapter has been very kindly written by Dr. Raj Shenoy.

given power level at the transmitter, lower fiber losses would lead to larger repeater spacings (i.e., the propagation distance after which the signal level needs to be boosted to facilitate error-free detection at the receiver).

(ii) The spectral linewidth of the source should be as small as possible, typically $\lesssim$1 nm. This is very important because the magnitude of temporal dispersion is directly proportional to the length of the fiber and the linewidth of the source (see Chapter 10).

(iii) It should be possible to modulate the source at speeds in excess of several Gb/s. To meet this requirement, one can either choose a suitable source that can be directly modulated at the desired rate or use an external modulator in tandem with a source that gives steady power output.

In addition to the above, it is very desirable that the source be efficient, compact, reliable, durable, and inexpensive to meet the other important requirements of economic viability and acceptability.

Almost all these requirements are ideally met by semiconductor laser diodes based on the quaternary material InGaAsP. In fact, the intense research and development activities that led to the development of efficient laser diodes were greatly motivated by the need for developing suitable sources for optical communication.

In the following two sections we first outline the principle of operation of a general laser and apply the founding concepts to a semiconductor p–n junction laser. The objective is to provide a basic understanding of the device structure and the output characteristics; essential results are quoted, wherever necessary, without actually deriving them. Interested readers can find the details in standard textbooks on lasers and semiconductor lasers [e.g., Agrawal and Dutta (1993), Yariv (1991), Ghatak and Thyagarajan (1989)].

11.3 Laser fundamentals

LASER is an acronym for *light amplification by stimulated emission of radiation*. Therefore, our first task is to understand what is meant by *stimulated emission* and under what conditions one can achieve amplification of light by stimulated emission. *Laser* – the device – may be defined as a highly monochromatic, coherent source of optical radiation. In this sense, it is analogous to an electronic oscillator, which is a source of electromagnetic waves in the lower frequency range of the electromagnetic spectrum. A laser consists of an *active medium* that is capable of providing optical amplification and an *optical resonator* that provides the necessary optical feedback (see Figure 11.1). The active medium may be a collection of atoms, molecules, or ions in the solid, liquid, or gaseous form, and we may address it as an *atomic system*. The optical resonator in its simplest form consists of two plane or spherical mirrors aligned suitably to confine the optical energy as light propagates back and forth between the mirrors. We first review the basic emission and absorption processes in an atomic system and then discuss the conditions for light amplification and laser oscillation.

11.3.1 Absorption and emission of radiation

In general, an atomic system is characterized by discrete energy levels, and the constituent atoms/molecules can exist in one of the allowed levels or *states*. In

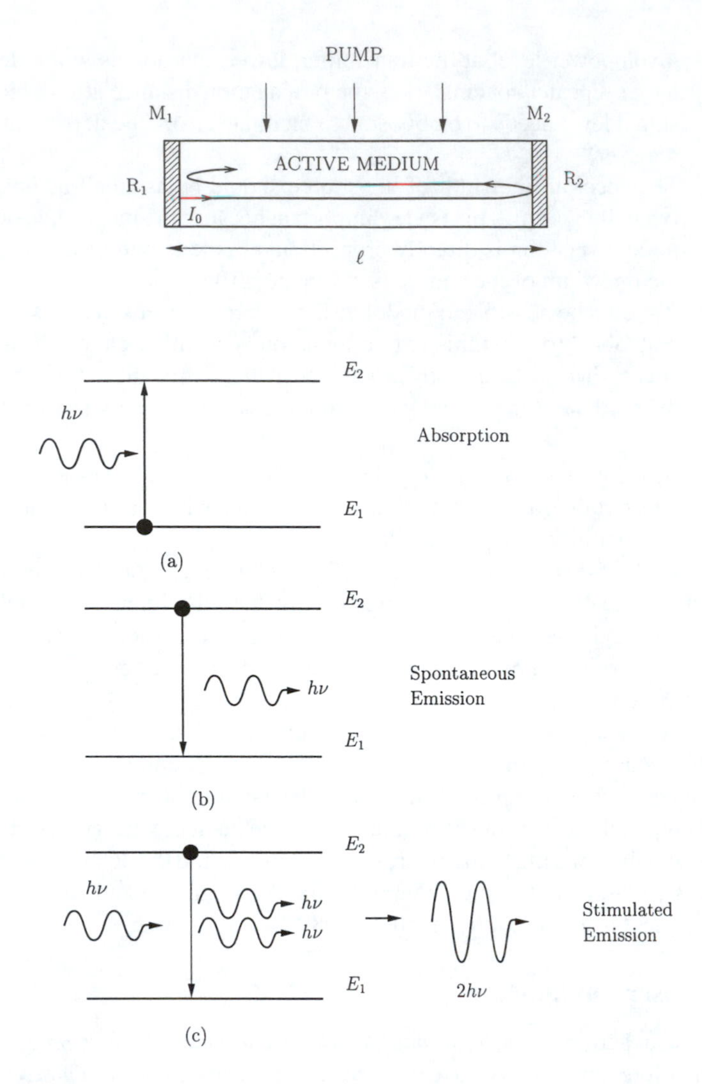

Fig. 11.1: A laser consists of an active medium capable of providing optical amplification placed inside an optical resonator that provides optical feedback. The pump is an external source of energy that converts the medium from an attenuating to an amplifying state.

Fig. 11.2: Radiation at frequency $\nu = (E_2 - E_1)/h$ can interact with an atomic system consisting of levels E_1 and E_2 through (a) absorption, (b) spontaneous emission, and (c) stimulated emission.

the energy domain, atoms/molecules can make upward or downward transitions between any two allowed states by absorbing or releasing, respectively, an amount of energy equal to the difference between the two energy levels. Thus, if we consider two levels of an atomic system that participate in an interaction with optical radiation of photon energy $h\nu = E_2 - E_1$, then there could be three different types of interactions (see Figure 11.2):

(a) Absorption: An atom in level 1 can absorb photons of frequency $(E_2 - E_1)/h$ and make an upward transition to the higher energy level. The rate of absorption depends on the number of atoms present in the lower level 1 and the energy density of radiation present in the system.

(b) Spontaneous emission: An atom in an excited level can make a downward transition spontaneously (i.e., on its own) by emitting a photon corresponding to the energy difference between the two levels. The rate of such transitions depends only on the number of atoms present in the excited level.

(c) Stimulated emission: An atom in an excited level can also make a downward transition in the presence of an external radiation of frequency $(E_2 - E_1)/h$. As in the case of absorption, the rate of such stimulated emissions or induced emissions depends on the energy density of the external radiation and the number of atoms in the excited level. The most important aspect of this type of transition, however, is the fact that the emitted radiation is coherent with the stimulating radiation. Thus, stimulated emission leads to a coherent amplification of the incident stimulating radiation and is the process responsible for amplification of optical radiation in a laser.

Transitions involving emission or absorption of photons are known as *radiative transitions*. Atoms can also make an upward or downward transition without involving a photon. For example, an excited atom in the gaseous state can come down to a lower level through an inelastic collision with the walls of the container. Similarly, an excited atom in a solid can make a downward transition by giving the excess energy to the lattice. These are known as *nonradiative transitions*.

For an atomic system in thermal equilibrium, the occupation probabilities of various atomic energy levels at any temperature T are given by the Maxwell–Boltzmann statistics.

$$P(E_i) \propto g_i e^{-E_i/k_B T} \tag{11.1}$$

where i refers to the level under consideration, k_B is the Boltzmann constant, and g_i is the *degeneracy* of the level. For nondegenerate energy levels, $g_i = 1$, whereas it is equal to the number of sublevels (that have the same energy value) for a degenerate level. Thus, for an atomic system with nondegenerate energy levels, the population density in the various levels under thermal equilibrium is given by

$$N_1 = Ce^{-E_1/k_B T} \tag{11.2}$$

$$N_2 = Ce^{-E_2/k_B T} \tag{11.3}$$

and so on, where the constant C depends on the total number of atoms in the system. Using the above equations, we get

$$\frac{N_2}{N_1} = e^{-(E_2 - E_1)/k_B T} \tag{11.4}$$

For E_2 greater than E_1, N_2 is less than N_1. In other words, an atomic system in thermal equilibrium has its lower energy level more densely populated than a state with higher energy, and the population has an exponential dependence on energy.

Example 11.1: Let us consider an atomic system with the ground level E_1 and the first excited level E_2 separated by an energy gap corresponding to a wavelength of 694 nm. For such a pair of levels,

$$E_2 - E_1 = h\nu = h\frac{c}{\lambda} = \frac{6.626 \times 10^{-34} \times 3 \times 10^8}{694 \times 10^{-9}} \simeq 2.86 \times 10^{-19}\,\text{J}$$

For a temperature of 300 K,

$$k_B T = 1.38 \times 10^{-23} \times 300 \simeq 4.14 \times 10^{-21}\ \text{J}$$

Thus

$$N_2/N_1 = e^{-69} \simeq 10^{-30}$$

Note that at 300 K there is hardly any atom in the excited state of this atomic system when compared with the number of atoms in the ground state!

11.3.2 Condition for amplification by stimulated emission

In a nondegenerate atomic system, in the presence of an external radiation, the probability of absorption per atom is the same as the probability of stimulated emission per atom. Therefore, if the number of atoms in the upper level is more than the number of atoms in the lower level, the rate of stimulated emission will exceed the rate of absorption. In other words, there will be a net emission of radiation, leading to light amplification.

If $u(\nu)\,d\nu$ represents the radiation energy per unit volume between ν and $\nu + d\nu$, then

$$\text{Rate of absorption} = Bu(\nu)N_1 \tag{11.5}$$

and

$$\text{Rate of emission} = AN_2 + Bu(\nu)N_2 \tag{11.6}$$

where A and B are the Einstein coefficients and represent proportionality constants associated with the spontaneous and stimulated emissions, respectively. At steady state, the rate of absorption must be equal to the rate of emission – that is,

$$Bu(\nu)N_1 = AN_2 + Bu(\nu)N_2 \tag{11.7}$$

If we ignore spontaneous emission for the moment, since we are interested in amplification process due to stimulated emission, then it is clear that for the rate of emission to exceed the rate of absorption of photons (i.e., for net emission or light amplification), the number of atoms per unit volume (N_2) in the upper level must be greater than the number of atoms per unit volume (N_1) in the lower level. Since this does not correspond to the normal population distribution in the atomic system in thermal equilibrium, the condition in which a higher level is more populated than a lower level is known as *population inversion. Indeed, population inversion is the condition for light amplification.*

The next step would be to understand how to achieve population inversion in an atomic system. We have seen above that at thermal equilibrium the population of various levels is given by the Boltzmann distribution. However, if we irradiate an atomic system by an external radiation of appropriate frequency,

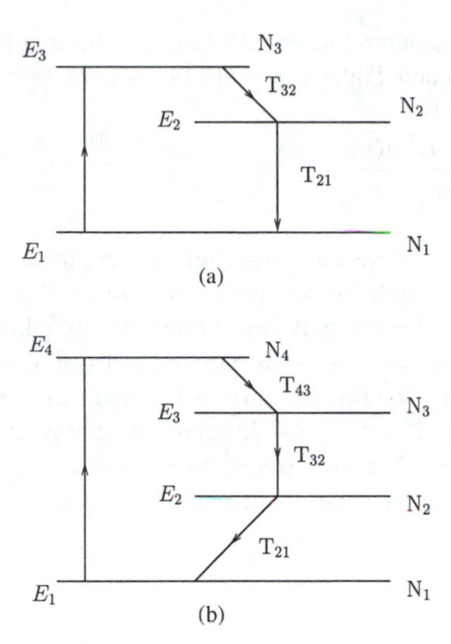

Fig. 11.3: (a) A three-level laser system and (b) a four-level laser system.

this population distribution changes, and in some atomic systems, under certain conditions, it is possible to achieve population inversion between a pair of levels.

Consider the three levels – the ground state and two excited states – of an atomic system shown in Figure 11.3(a). If we irradiate this system with a radiation of frequency $\nu = (E_3 - E_1)/h$, atoms can get excited from level 1 to level 3. Excitation of an atomic system by an external means is often referred to as *pumping*. The excited atoms in level 3 can make a downward transition to level 1 by spontaneous and stimulated emissions. Also, the transitions either may be one-step direct transition from level 3 to level 1 or may involve two steps: an initial transition from level 3 to level 2, followed by a second transition from level 2 to level 1. In the latter case, at a given pumping rate, if the rate of transition T_{32} from level 3 to level 2 exceeds that between level 2 and level 1 (i.e., T_{21}), then atoms would start accumulating in level 2, and the population of level 2 would increase with the pumping rate. When the pumping rate exceeds a certain threshold, the steady-state population in level 2 becomes higher than that in level 1. In this state of population inversion, radiation at frequency $\nu_l = (E_2 - E_1)/h$ will get amplified by the phenomenon of stimulated emission.

The above scheme of transitions to realize laser action is known as a three-level pumping scheme, and the atomic system is identified as a *three-level system*. Because population inversion is to be achieved between an excited state (level 2) and the ground state (level 1), considering the fact that the population of the ground level at thermal equilibrium is very much larger than that in an excited level, the required threshold pumping power for population inversion is very large (~kW). A more efficient scheme employed in most lasers is the four-level scheme illustrated in Figure 11.3(b). Here, the population inversion and the lasing transitions occur between levels 3 and 2. If the transition rates T_{43} and T_{21} are large compared with the transition rate T_{32}, then population inversion can be achieved between levels 3 and 2 even with low to moderate pumping powers.

The small signal gain coefficient of a laser amplifier at frequency ν is given by [see, e.g., Ghatak and Thyagarajan (1989), Section 8.3]

$$\gamma(\nu) = \frac{(c/n)^2}{8\pi\tau_{sp}}\frac{g(\nu)}{\nu^2}\Delta N \tag{11.8}$$

where τ_{sp} is the spontaneous emission lifetime (i.e., the average time spent by an atom in the excited state before making a downward transition); ΔN is the population inversion density between the two levels taking part in the laser transition – that is, $\Delta N = N_2 - N_1$ in a three-level system and $\Delta N = N_3 - N_2$ in a four-level system. The function $g(\nu)$ is known as the *normalized lineshape function* and is defined so that $g(\nu)\,d\nu$ represents the probability that an atom interacts with radiation between frequencies ν and $\nu + d\nu$. Since the atomic system will interact with radiation at one frequency or the other, we must have

$$\int_0^\infty g(\nu)\,d\nu = 1 \tag{11.9}$$

The function $g(\nu)$ also gives the absorption or emission spectrum of the atomic system. In general, this function is sharply peaked in the form of a Lorentzian or Gaussian around a central resonant frequency. The gain coefficient at any frequency and the bandwidth of the laser amplifier are essentially determined by $g(\nu)$.

Example 11.2: If $N_1 > N_2$, ΔN is negative, and $\gamma(\nu)$ given by equation (11.8) gives the absorption coefficient of the medium. As an example, let us consider ruby (which is Al_2O_3 doped with Cr^{3+} ions), in which the Cr^{3+} ions take part in absorption and emission around 694.3 nm. At thermal equilibrium at 300 K, $N_1 \simeq 1.6 \times 10^{19}$ cm^{-3}, $N_2 \simeq 0$, $n = 1.76$, $\tau_{sp} = 3 \times 10^{-3}$ s, $g(\nu_0) \simeq 6.9 \times 10^{-12}$ s, and the absorption coefficient at 694.3 nm can be evaluated from equation (11.8) as $\simeq$2.3 cm^{-1}.

For achieving a gain coefficient of 10^{-2} cm^{-1}, the required population inversion density is 7×10^{16} cm^{-3}.

11.3.3 Laser oscillation

In the above section we saw how optical amplification can be achieved by having population inversion. If such an optical amplifier is provided with an optical feedback, one would realize an optical oscillator or a source of optical radiation. This is nothing but the laser. The necessary optical feedback is provided by placing the active medium in an optical resonator composed of two high-reflectivity mirrors, separated by a suitable distance l (see Figure 11.1). The mirrors may be planar or spherical and may either be in the form of discrete components located outside the gain medium or be attached to the ends of the gain medium. Usually, one of the mirrors (say M_1) is almost 100% reflecting (i.e, $R_1 \simeq 1.0$) and the other mirror (M_2) is partially reflecting (i.e., $R_2 < 1.0$), so that a small fraction of optical energy comes out of the resonator; obviously, from the resonator's point of view, this output is a loss but forms the useful output of the laser for the users.

Consider the optical resonator as shown in Figure 11.1. In the absence of the gain medium, there is no gain in the resonator, and the resonator is referred to as a *passive resonator*; it is a lossy device because of the finite reflectivities (R_1, $R_2 < 1$) of the mirrors at the ends of the resonator and the propagation loss associated with the light bouncing back and forth between the mirrors. The propagation loss consists of loss mechanisms such as scattering loss in the medium and diffraction loss due to finite size of the mirrors. In the presence of pumping, there would be both amplification and attenuation of radiation at the lasing transition within the resonator. Let I_0 be the irradiance of the light beam leaving the mirror M_1 and propagating toward M_2. If α represents the average loss coefficient per unit length of the resonator due to all mechanisms other than the finite reflectivities of the mirrors, and if γ represents the small signal gain coefficient per unit length of the resonator, then the irradiance of the beam after one complete round trip is given by

$$I_1 = I_0 e^{-\alpha l} e^{\gamma l} R_2 e^{-\alpha l} e^{\gamma l} R_1$$

The factors multiplying I_0 on the RHS of the above equation indicate the various processes occurring in one complete round trip of the light beam. We may rewrite the above equation as

$$I_1 = I_0 R_1 R_2 e^{2(\gamma - \alpha) l} \tag{11.10}$$

Therefore, if the radiation has to build up and sustain in the resonator, we must have

$$R_1 R_2 e^{2(\gamma - \alpha) l} \geq 1 \tag{11.11}$$

The equality sign in the above equation corresponds to the situation wherein the round trip loss is exactly compensated by the amplification provided by the gain medium, so that $I_1 = I_0$. This is the *threshold condition* for laser oscillation and the corresponding value of the gain coefficient is known as the *threshold gain coefficient*, γ_{th}. Thus, at threshold we have

$$R_1 R_2 e^{2(\gamma_{\text{th}} - \alpha) l} = 1$$

or

$$\gamma_{\text{th}} = \alpha - \frac{1}{2l} \ln R_1 R_2 \tag{11.12}$$

When the gain coefficient becomes larger than the threshold value, the "greater than" sign in equation (11.11) applies, and the radiation builds up after every round trip. However, as one would expect, it cannot go on building up, and soon *saturation effects* take over and the gain drops down to its threshold value. The saturation effect can be explained qualitatively as follows. At a given pumping rate, if $\gamma > \gamma_{\text{th}}$, the increasing power density at the laser frequency induces more and more stimulated emissions, which results in a faster depletion of the atoms from the excited state, thereby reducing the population inversion ΔN and, hence, the gain. Thus, laser oscillations start when the gain coefficient exceeds its threshold value, and when the laser oscillates in the steady state, the gain coefficient gets clamped at the threshold value. A knowledge of the passive resonator losses, therefore, enables the estimation of the threshold gain coefficient through equation (11.12), which in turn leads to the estimation of the threshold population inversion through equation (11.8).

$$\Delta N_{\text{th}} = \gamma_{\text{th}} \frac{8\pi\tau_{\text{sp}}}{(c/n)^2} \frac{\nu^2}{g(\nu)} \tag{11.13}$$

A knowledge of the threshold population inversion is essential to estimate the necessary pumping power because the magnitude of inversion, ΔN, depends on the pumping rate of atoms to the excited state.

Example 11.3: Consider a resonator with $\alpha = 0$, $R_1 = 0.99$, $R_2 = 0.9$, and $l = 10$ cm. The corresponding threshold gain coefficient is 5.8×10^{-3} cm^{-1}. The length of the active medium required for light amplification by a factor of 2 is $\simeq$120 cm.

Example 11.4: For a semiconductor laser, $l = 300\,\mu$m, $R_1 = R_2 \simeq 0.32$, and $\alpha = 10$ cm^{-1}. We thus have $\gamma_{\text{th}} \simeq 48$ cm^{-1}. Note the very large threshold gain compared with that in Example 11.3. In this case the length of the active medium required for amplification by a factor of 2 is only 0.15 mm!

11.3.4 Resonator modes

In the above section we obtained the threshold gain coefficient necessary to overcome the losses in the cavity, and we were primarily concerned with the equation for optical energy inside the resonator. The pair of mirrors that provides for optical feedback in a laser forms an optical resonator.

An optical resonator is characterized by *transverse* and *longitudinal* modes. The transverse modes refer to specific transverse field distributions that reproduce themselves after a round trip through the cavity. The fundamental transverse mode of a laser resonator can be approximately described by a Gaussian field pattern of the form [see e.g., Ghatak and Thyagarajan (1989), Chapter 9].

$$E(x, y) = E_0 \exp\left(-\frac{x^2 + y^2}{w_0^2}\right) \tag{11.14}$$

where x and y represent the coordinates transverse to the axis of the resonator and w_0 is the mode spot size. The fundamental (Gaussian) mode is characterized by a uniform phase front and has the least diffraction divergence (see Section 2.8). The higher order transverse modes have phase reversals and zeroes across the beam. Thus, most lasers are designed to operate in the fundamental transverse mode.

Since the optical waves in the cavity are bouncing back and forth between the two mirrors, it is necessary that when a wave returns after one round trip it finds itself in phase with the wave existing on the initial plane. Only under such a condition can standing optical waves exist in the cavity. (Recall standing waves on a string fixed at two ends.) Thus, the cavity will support only those frequencies for which the round trip phase shift is an integral multiple of 2π.

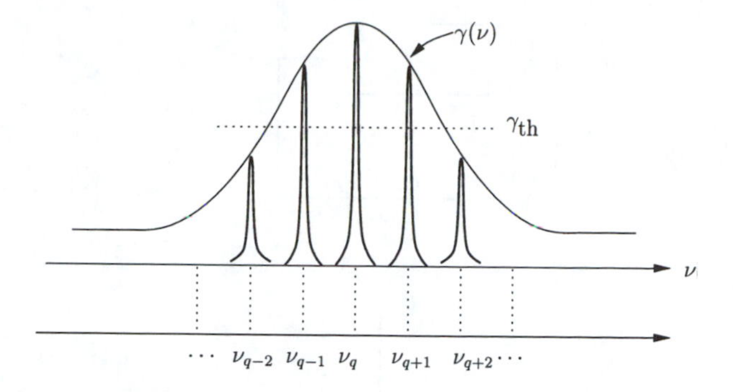

Fig. 11.4: The amplifying medium provides gain at different frequencies, the gain coefficient $\gamma(\nu)$ being described by equation (11.8). The cavity has a loss that normally is constant over the frequency range shown. Oscillation can take place only in a set of discrete frequencies as determined by equation (11.15) and for which gain exceeds loss.

If n represents the refractive index of the medium (assumed to fill the entire cavity) and l is the cavity length, then the oscillation can take place only at frequencies ν_q satisfying

$$\frac{2\pi\nu_q}{c}n2l = q2\pi; \quad q, \text{ an integer}$$

or

$$\nu_q = q \cdot \frac{c}{2nl} \tag{11.15}$$

These oscillation frequencies represent the various longitudinal modes of the cavity.

We have seen earlier that the amplifying medium is characterized by a gain spectrum determined by $\gamma(\nu)$. The presence of the cavity implies that within this spectrum only those frequencies that satisfy equation (11.15) can oscillate. Thus, as the pumping is increased and as one reaches threshold, the longitudinal mode closest to the gain peak will be the first to start oscillating. If the pumping is further increased, other adjacent modes may also begin to oscillate, leading to multilongitudinal mode oscillation (see Figure 11.4). The spectral width of the laser in such a case would of course be larger than in a single longitudinal mode oscillation.

11.4 Semiconductor laser: basics

In its simplest form, a semiconductor laser consists of a forward biased p–n junction, formed in a direct bandgap semiconductor (see Figure 11.5). The recombination of injected carriers – namely, electrons and holes – in the junction region results in the emission of photons. The cleaved ends of the laser structure act like mirrors, forming a Fabry–Perot resonator along the p–n junction. The other two ends (in the perpendicular direction) are saw-cut to reduce reflections from these ends and prevent lasing along the perpendicular direction. Typical dimensions of a discrete laser-diode chip are shown in Figure 11.5(b). When the forward current through the diode exceeds a critical value known as the *threshold current*, optical gain in the resonator due to stimulated emissions overcomes the losses in the resonator, leading to net amplification and eventually to steady-state laser oscillations.

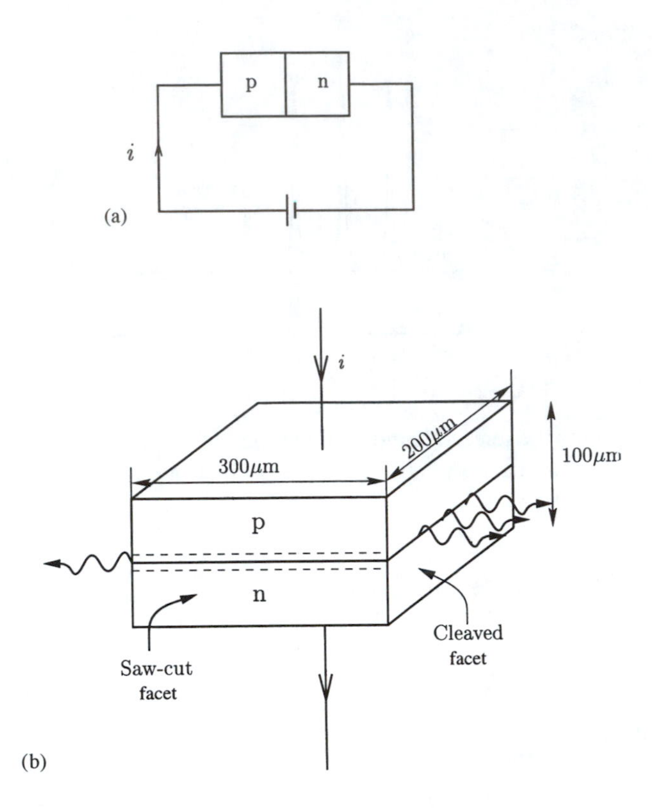

Fig. 11.5: (a) In its simplest form a semiconductor laser consists of a forward biased p–n junction formed in a direct bandgap semiconductor. (b) Two ends are cleaved and the laser output emerges from these ends. The other two surfaces are saw-cut to arrest any oscillation along that direction. Typical dimensions are also shown.

11.4.1 Energy bands and carrier distribution in semiconductors

Matter consists of atoms, and in the solid state of matter atoms are packed very closely with interatomic spacings of the order of a few angstroms. Since the electron cloud surrounding the positively charged nucleus of an atom also has a spatial spread of the same or comparable order, the electron distribution and the associated energy levels get perturbed in the solid state compared with those in an isolated atom. This leads to formation of *energy bands* in solids unlike in the case of an isolated atom or an atomic gas, which is characterized by well-separated discrete energy levels. The highest energy band, in a solid, that is completely filled or occupied by electrons at 0 K is known as the *valence band,* and the next higher band that is partially occupied or vacant is known as the *conduction band.* The forbidden gap or the energy gap between these two energy bands is indicative of several important electrical and optical properties of the solid.

Semiconductors, as the name indicates, refer to materials that have electrical conductivity values between those of good conductors and insulators. Most semiconductors are crystalline solids formed by some of the elements from group II to group VI of the periodic table of the elements and typically have energy gaps in the range 0.25–2.5 eV. In an intrinsic semiconductor at 0 K, the valence band is completely full and the conduction band is completely empty of electrons. At any other temperature the occupational probabilities of the allowed states by electrons in a semiconductor are given by the Fermi distribution:

$$f(E) = \frac{1}{1 + e^{-(E-E_f)/k_B T}} \tag{11.16}$$

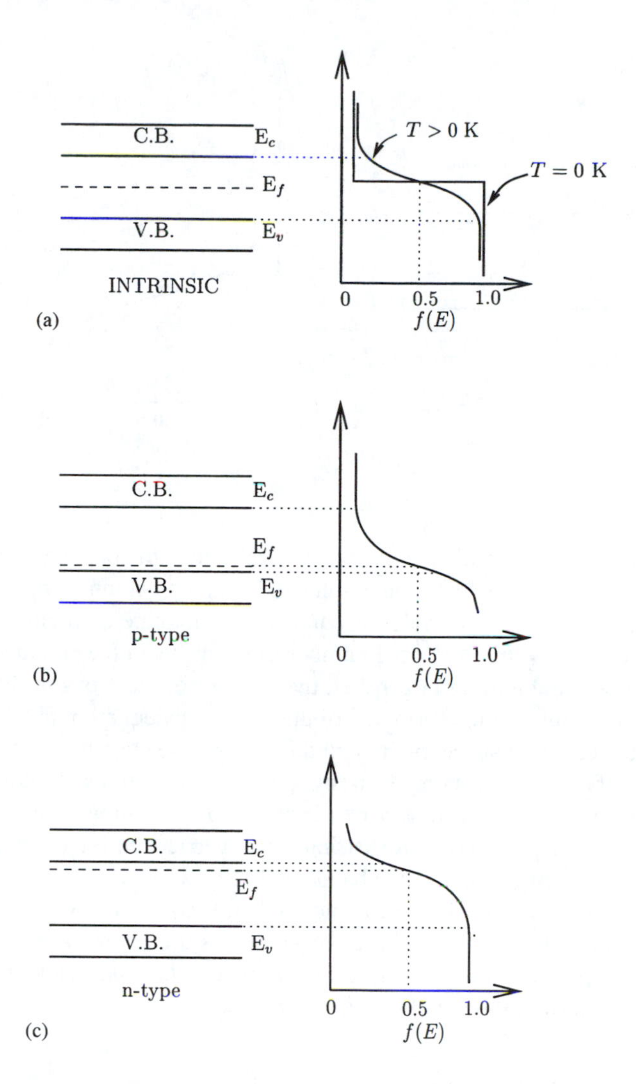

Fig. 11.6: Fermi distribution in (a) an undoped, (b) a p-type, and (c) an n-type semiconductor.

where E_f is a constant known as the *Fermi energy* or the *Fermi level*; it represents the energy value at which the probability of occupation of electrons is 0.5 (if there existed an allowed state at this energy value). In a pure (undoped) semiconductor, the number of holes in the valence band is equal to the number of electrons in the conduction band, and the Fermi level lies approximately midway between the top of the valence band and the bottom of the conduction band (see Figure 11.6(a)). In a p-type (doped) semiconductor, there are a greater number of holes in the valence band than electrons in the conduction band, and therefore the electron distribution function shifts toward the valence band; in other words, the Fermi level, in this case, is situated nearer to the valence band edge (see Figure 11.6(b)). Similarly, in an n-type semiconductor, there are a greater number of electrons in the conduction band than holes in the valence band, and the Fermi level lies nearer to the conduction band edge (see Figure 11.6(c)).

The above discussion of carrier distribution pertains to thermal equilibrium in the absence of any external excitation of the semiconductor. However, if we excite the semiconductor by external means – say by irradiating the

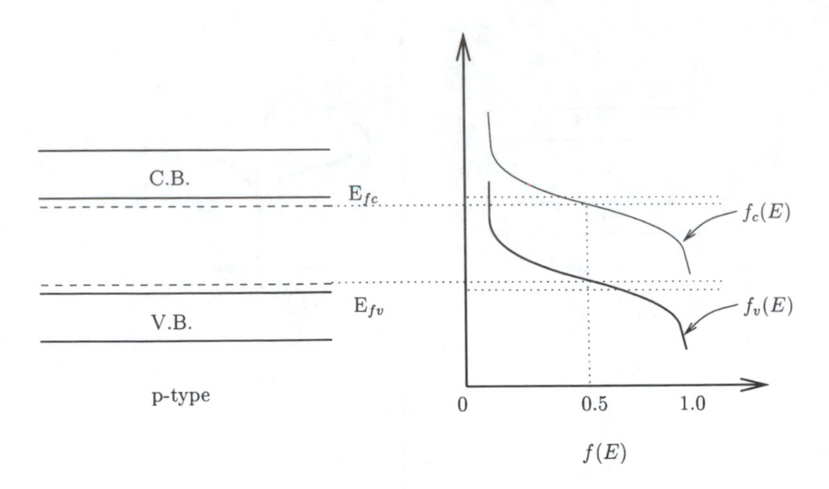

Fig. 11.7: Diagram showing quasi-Fermi levels E_{fc} and E_{fv} in a semiconductor.

semiconductor with a radiation of energy greater than the bandgap – more and more electrons can go to the conduction band, leaving behind a greater number of vacancies (holes) in the valence band. If we imagine a situation wherein the excitation is so strong that the number of electrons in the conduction band becomes comparable to or larger than the number of electrons in the valence band, it is obvious that the Fermi distribution given by equation (11.16) cannot describe the carrier distribution in both the bands. However, because the relaxation time of carriers for intraband transitions (i.e., transitions within a band) is much smaller than the band-to-band electron–hole recombination time, there is a quasi-stationary steady-state distribution of carriers within each band. This carrier distribution within the two bands can be described by defining two different Fermi levels – one for the conduction band, E_{fc}, and one for the valence band, E_{fv} (see Figure 11.7). These are known as the *quasi-Fermi levels*, and the semiconductor is then said to be in *quasi-equilibrium*. Thus, the carrier probability distribution in the two bands is given by

$$f_c(E) = \frac{1}{1 + e^{-(E-E_{fc})/k_B T}} \tag{11.17}$$

for the conduction band, and

$$f_v(E) = \frac{1}{1 + e^{-(E-E_{fv})/k_B T}} \tag{11.18}$$

for the valence band. Quasi-Fermi levels play an important role in determining the gain coefficient of a semiconductor laser amplifier.

11.4.2 Absorption and emission in a semiconductor

Consider a crystalline semiconductor in which the constituent atoms are arranged in a regular periodic lattice. The motion of an electron in such a crystalline solid in any particular direction can be viewed as the motion of a negatively charged particle in a periodically varying electrostatic potential due to the positively charged atomic nuclei. If we assign a wave function ψ and a wave vector $\mathbf{k}$ to the electron and solve the equation of motion, the solution

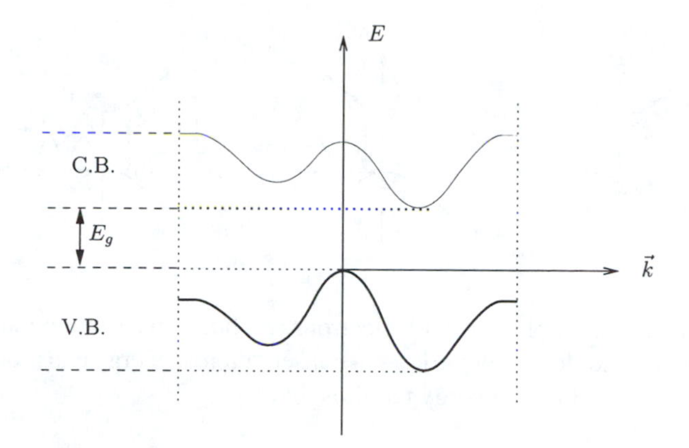

Fig. 11.8: E-**k** diagram of a semiconductor. E_g represents the bandgap. The figure corresponds to an indirect bandgap semiconductor.

yields a periodic dependence of the electron energy on the wave vector. If we restrict ourselves to one period, the corresponding variation of energy as a function of **k** (see Figure 11.8) is known as the E-**k** diagram of a semiconductor. Semiconductors in which the minimum of the conduction band and maximum of the valence band occur at the same **k** value are known as *direct bandgap* semiconductors, whereas semiconductors in which the minimum of the conduction band and the maximum of valence band occur at two different values of **k** (as shown in Figure 11.8) are known as *indirect bandgap* semiconductors. The reason for the above nomenclature is obvious if we recall the definition of bandgap – the difference in energy between the conduction band minimum and the valence band maximum irrespective of their positions in the E-**k** diagram.

We next consider the phenomenon of absorption and emission of photons in a semiconductor. We know that there are electrons and holes in the valence band of a semiconductor, wheareas the conduction band has electrons and unoccupied vacant states. The holes accumulate near the top of the valence band (which are low-energy states for the holes) and the electrons in the conduction band accumulate near the bottom of the band. As in the case of an atomic system, there can be three different processes involving a photon, an electron, and a hole in a semiconductor:

(a) An electron in the valence band can make an upward transition (in energy) to the conduction band by absorbing a photon of energy $h\nu$ so that $E_2 - E_1 = h\nu$, where E_1 and E_2 are the energies associated with the initial and final states of the electron in the valence and conduction bands, respectively (see Figure 11.9(a)).

(b) An electron in the conduction band can recombine with a hole in the valence band by spontaneously emitting a photon corresponding to the energy difference between its initial and final states. This is the spontaneous emission process (see Figure 11.9(b)).

(c) An electron in the conduction band can also undergo a stimulated transition to the valence band in the presence of a photon of appropriate energy (see Figure 11.9(c)). The emitted photon in such a transition is fully coherent with the inducing photons. As described earlier, this is the process that can lead to coherent amplification of an incident radiation.

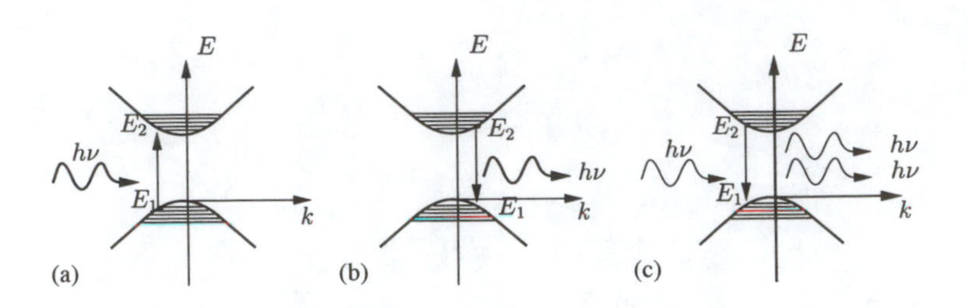

Fig. 11.9: The processes of (a) absorption, (b) spontaneous emission, and (c) stimulated emission in a direct bandgap semiconductor.

The interaction of photons with electrons and holes in a semiconductor, described above, should satisfy the laws of conservation of energy and momentum. Thus, the conservation of energy requires that

$$E_2 - E_1 = h\nu \tag{11.19}$$

If $\mathbf{k_1}$ and $\mathbf{k_2}$ represent the wave vectors associated with the electrons in the valence band and the conduction band, respectively, then the conservation of momentum requires that

$$\hbar\mathbf{k_1} + \hbar\mathbf{k}_\nu = \hbar\mathbf{k_2} \tag{11.20}$$

for both the absorption and the emission process. Here $k_\nu = 2\pi\nu/c = 2\pi/\lambda$ is the wave vector associated with the photon, and $\hbar = h/2\pi$, h being Planck's constant. Equation (11.20) can be written as

$$\frac{2\pi}{\Lambda_1} + \frac{2\pi}{\lambda} = \frac{2\pi}{\Lambda_2} \tag{11.21}$$

where Λ_1 and Λ_2 are the de Broglie wavelengths associated with the electron states. Since the average de Broglie wavelength of electrons in a semiconductor is typically ~10 Å, and the wavelength of light is 2–3 orders of magnitude larger, it follows that

$$\Delta k = k_2 - k_1 \approx 0 \tag{11.22}$$

This is known as the *k-selection rule*, and it implies that the allowed transitions between the conduction band and the valence band are "vertical transitions" in the E-$\mathbf{k}$ diagram. This, however, does not completely rule out the possibility of occurrence of transitions that are not vertical in the E-$\mathbf{k}$ diagram. Indeed, transitions that do not conserve the momentum of an electron before and after the transition (i.e., $k_1 \neq k_2$), but satisfy the law of conservation of momentum, also occur with the participation of *phonons*. Phonons are quanta of lattice vibrations, with typical energy per phonon in the range 0.01–0.1 eV; however, the momentum associated with phonons can be quite large and comparable to that of electrons. The probability of occurrence of phonon-assisted interactions of photons with electrons and holes in a semiconductor is much smaller compared with the occurrence of allowed (vertical) transitions. This is primarily because of the involvement of an additional "entity" – phonons in this case – that should participate in an appropriate quantity to offset simultaneously the energy mismatch and the momentum mismatch associated with a $\Delta k \neq 0$ transition.

If we now recall the fact that the electrons accumulate near the bottom of the conduction band and the holes near the top of the valence band, it is clear that the emission process is highly probable in a direct bandgap semiconductor. For this reason, almost all efficient semiconductor photon sources are fabricated by using direct bandgap semiconductors such as GaAs and InP. In the rest of this chapter, therefore, we focus only on direct bandgap semiconductors.

11.4.3 Optical gain in a semiconductor

The gain coefficient for amplification of radiation of frequency ν by stimulated emission in a semiconductor is given by [see, e.g., Saleh and Teich (1991)]

$$\gamma(\nu) = \frac{(c/n)^2}{8\pi\tau_r}\frac{\rho(\nu)}{\nu^2}\Delta P \tag{11.23}$$

where τ_r is the radiative (electron–hole) recombination time, and $\rho(\nu)$ is the reduced density of states, or the *optical joint density* of states. $\Delta P = P_e(\nu) - P_a(\nu)$ is the difference of the probabilities for emission and absorption at the optical frequency ν and is known as the Fermi inversion factor. For amplification, $\gamma(\nu) > 0$, which requires $\Delta P > 0$ – that is, the probability of emission has to be greater than the probability of absorption. This is the necessary condition for light amplification in a semiconductor.

The *density of states* in a semiconductor refers to the number of allowed electron states per unit volume per unit energy interval. Thus, for example, if $\rho_c(E)$ represents the density of states in the conduction band, then $\rho_c(E)\,dE$ gives the number of states per unit volume between energy values E and $E + dE$. A similar definition holds good for the density of states $\rho_v(E)$ in the valence band. When a photon interacts with an electron and a hole in a semiconductor, the number of states available for the interaction is limited by the requirement of energy conservation. For example, if the photon energy is just about equal to the bandgap, then the electron states well above the conduction band edge and well below the valence band edge cannot participate in the interaction. The optical joint density of states takes into account the number of states available in both the conduction and the valence bands with which a photon of energy $h\nu$ can interact and is given by

$$\rho(\nu) = \frac{(2m_r)^{3/2}}{\pi\hbar^2}(h\nu - E_g)^{1/2}; \quad h\nu \geq E_g \tag{11.24}$$

where E_g is the bandgap and m_r is the reduced mass of the carriers. Note that only photons with energy $h\nu > E_g$ can participate in the emission and absorption process.

Equation (11.23) is very similar to equation (11.8) for the gain coefficient of general lasers discussed in Section 11.3, with the Fermi inversion factor ΔP taking up the place of population inversion ΔN. We therefore look for the conditions leading to a positive value of ΔP.

Let E_1 and E_2 be the energies of two allowed electron states in the valence band and the conduction band, respectively. If $h\nu = E_2 - E_1$ is the energy of an incident photon then for absorption of the photon an electron should be present in the state with energy E_1 in the valence band, and there should be an unoccupied allowed state at energy E_2 in the conduction band. Thus, the probability of absorption is given by

$$P_a(\nu) = f(E_1)[1 - f(E_2)] \tag{11.25}$$

Note that since $f(E)$ represents the probability of occupation of a state with energy E, $[1 - f(E)]$ gives the probability that a state with energy E remains unoccupied. Similarly, for emission of a photon due to an electron–hole recombination, it requires that an electron exists at energy level E_2 in the conduction band and there exists a hole at energy level E_1 in the valence band. Therefore, the probability of emission is given by

$$P_e(\nu) = f(E_2)[1 - f(E_1)] \tag{11.26}$$

Note that for a semiconductor in thermal equilibrium at practical temperatures, the carrier distribution is given by the Fermi function, and therefore both $f(E_1)$ and $[1 - f(E_2)]$ are much larger than $f(E_2)$ and $[1 - f(E_1)]$. Thus, $P_e(\nu) < P_a(\nu)$, which implies that optical amplification is not possible in a semiconductor at thermal equilibrium.

We next consider a semiconductor in quasi-equilibrium. The probabilities of absorption and emission are now given by

$$P_e(\nu) = f_c(E_2)[1 - f_v(E_1)] \tag{11.27}$$

and

$$P_a(\nu) = f_v(E_1)[1 - f_c(E_2)] \tag{11.28}$$

where the functions f_c and f_v represent the quasi-Fermi distributions in the conduction band and the valence band, respectively (see Figure 11.7). For net emission, we must have

$$f_c(E_2)[1 - f_v(E_1)] > f_v(E_1)[1 - f_c(E_2)] \tag{11.29}$$

which, using equations (11.17) and (11.18), simplifies to the condition

$$E_{fc} - E_{fv} > h\nu \tag{11.30}$$

Since $h\nu \geq E_g$ for radiative interactions in a semiconductor, it follows that

$$E_{fc} - E_{fv} > E_g \tag{11.31}$$

that is, when the separation between the quasi-Fermi levels in a semiconductor exceeds the bandgap, then for all frequencies ν that satisfy equation (11.30) it is possible to have light amplification.

The above condition is equivalent to the requirement of population inversion in an atomic system. Note that to satisfy equation (11.31), the quasi-Fermi levels should lie close to or into the respective bands. This implies that there is a large concentration of electrons in the conduction band and holes in the valence band, which is unlike the normal carrier distribution in a semiconductor at thermal

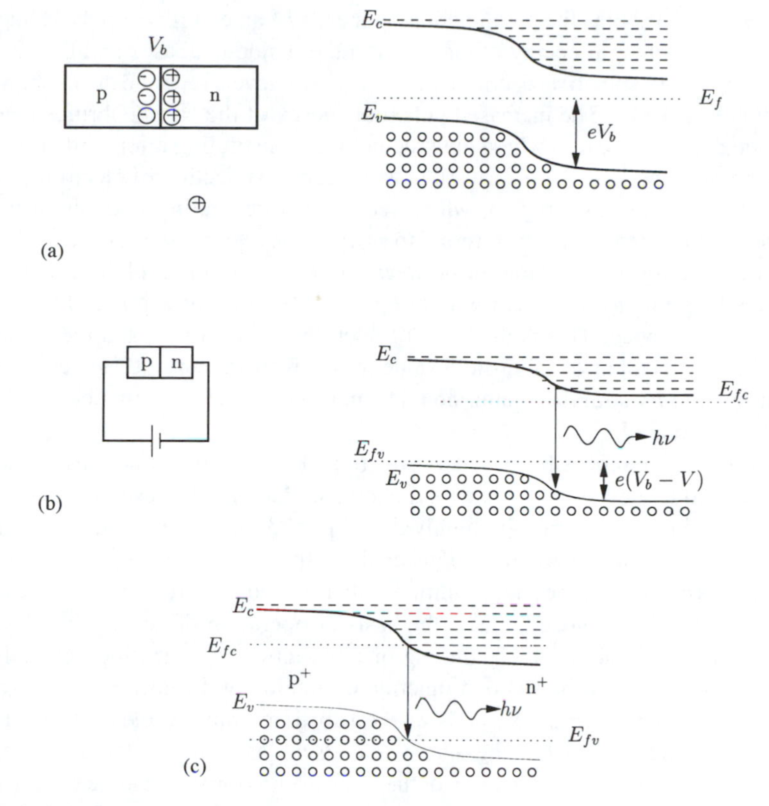

Fig. 11.10: (a) Unbiased p–n junction and (b) forward biased p–n junction. When the p–n junction is forward biased, one can create a situation satisfying equation (11.31) in the depletion region and thus achieve optical amplification. (c) A forward biased highly doped p–n junction.

equilibrium. In this sense, it is indeed equivalent to population inversion, though it does not necessarily have more electrons in the conduction band.

Our next task, as in the case of general lasers discussed in Section 11.3, is to look for implementation schemes that satisfy the gain condition. One of the most convenient methods for achieving this is to employ a forward biased p–n junction.

11.4.4 Gain in a forward biased p–n junction

Consider a p–n junction formed between a p-doped and an n-doped semiconductor, as shown in Figure 11.10(a). Because of different carrier concentrations of electrons and holes in the p and n regions, electrons from the n region diffuse into the p region, and holes from the p region diffuse into the n region. The diffusion of these carriers across the junction leads to a built-in potential difference between the positively charged immobile ions in the n side and the negatively charged immobile ions in the p side of the junction. This built-in potential V_b lowers the potential energy of electrons in the n side with respect to the potential energy of electrons in the p side, which is represented by "bending of the energy bands" near the p–n junction, as shown in Figure 11.10(b). Note that the Fermi levels on both sides of the p–n junction are aligned at the same energy value. This is necessary because, in the absence of any applied external energy source, the charge neutrality in the material requires that the probability of finding an electron should be the same everywhere, and therefore only one Fermi function should be able to describe the carrier distribution. In this case, there will be no net current in the medium.

If we forward bias the p–n junction by means of an external supply voltage V, then the potential energy of electrons in the n side increases and the band moves up. The band offset decreases, and the Fermi levels separate out as shown in Figure 11.10(b). The increased potential energy of the carriers brings them into the depletion region where they recombine, constituting a forward current through the junction. Thus, forward biasing leads to injection of electrons and holes into the junction region, which recombine, generating photons in this process. This phenomenon is referred to as *injection electroluminescence*. Note that even though the separation between the quasi-Fermi levels is less than the bandgap energy E_g, there is light emission because of a forward current through the device. This is the basis of operation of an LED, and the device therefore does not have a threshold value for the forward current. However, for amplification by stimulated emission, leading to laser action, equation (11.31) has to be satisfied.

It is usually not possible to satisfy equation (11.31) in p–n junctions formed between moderately doped p- and n-type semiconductors. However, if one starts with a p–n junction formed by highly doped p- and n-type semiconductors, in which the Fermi levels are located inside the respective bands, application of a strong bias can lead to the gain condition (satisfying equation (11.31)), as shown in Figure 11.10(c). Indeed, this is the basis of operation of an injection laser diode. As mentioned at the beginning of this section, a laser diode basically consists of a forward biased p–n junction of a suitably doped direct bandgap semiconductor material. Two ends of the substrate chip are cleaved to form mirror-like end faces, while the other two ends are saw-cut, so that the optical resonator is formed in the direction of the cleaved ends only. The large refractive index difference at the semiconductor ($n \sim 3.5$) and air ($n \simeq 1.0$) interface provides a reflectance of about 30% ($R = 0.3$), which is good enough to sustain laser oscillations in most semiconductor laser diodes. This is primarily due to the large gain coefficients per unit length achievable in semiconductor p–n junctions.

Figure 11.11 shows a typical variation of the gain coefficient γ with photon energy for GaAs corresponding to different injected carrier densities. In the figure 1.424 eV corresponds to 0.87 μm and 1.5 eV corresponds to 0.827 μm. Note that for a given injected carrier density, gain is available over a certain frequency band. For example, for an injected carrier density of 2.5×10^{18} cm^{-3}, the peak gain is about 132 cm^{-1} and extends up to a photon energy of $\sim$1.483 eV corresponding to 0.836 μm. The corresponding variation of peak gain with carrier density is shown in Figure 11.12.

> **Example 11.5:** Let us consider a GaAs laser diode with $L = 500\,\mu$m, $R_1 = R_2 = 0.3$, and $\alpha = 5$ mm^{-1}. The threshold gain required is (see equation (11.12))
>
> $$\gamma \simeq 74\,\text{cm}^{-1}$$
>
> This value of gain requires (see Figure 11.12) an injected carrier density of $\sim 2.02 \times 10^{18}$ cm^{-3}.

Now at steady state, the rate at which excess carriers (Δn) are injected must equal the rate of recombination. At threshold this rate is just the spontaneous

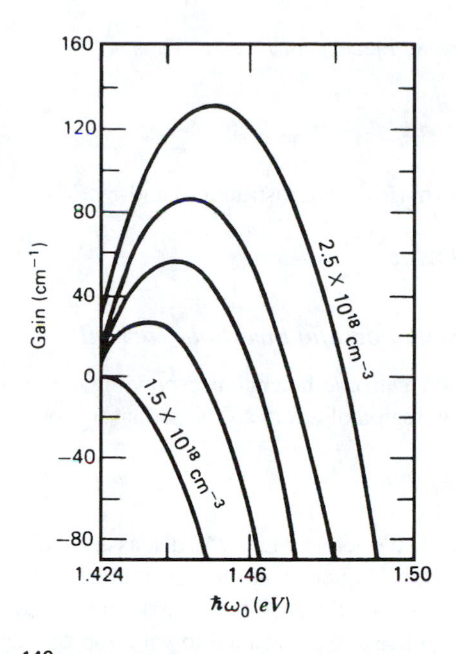

Fig. 11.11: A typical variation of gain versus photon energy in a GaAs semiconductor device with injected carrier density as a parameter. [After Yariv (1989).]

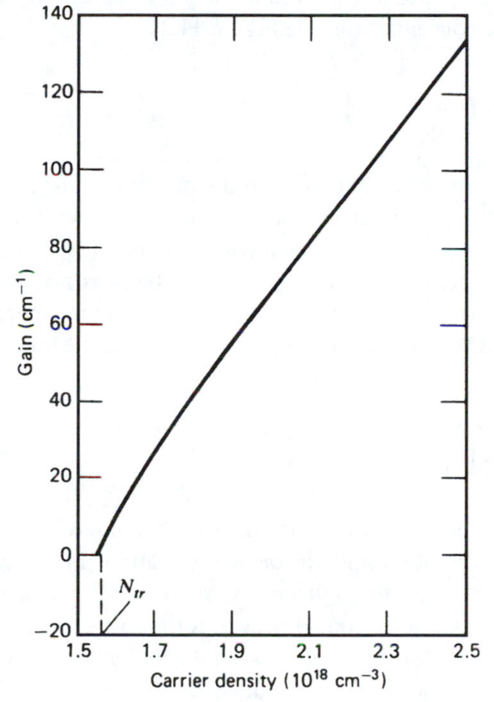

Fig. 11.12: Variation of peak gain with the injected carrier density corresponding to Figure 11.11. The temperature is 300 K. [After Yariv (1989).]

recombination rate. If τ is the spontaneous lifetime then

$$R_{sp} = \frac{\Delta n A d}{\tau} \tag{11.32}$$

where R_{sp} is the spontaneous recombination rate, A is the area of cross section, and d is the thickness of the gain region. If J represents the current density, then the rate of current injection is $J \cdot A/e$. Thus

$$J = \frac{\Delta ned}{\tau} \tag{11.33}$$

Typically, $d \simeq 0.1\ \mu$m (for a heterostructure) and $\tau \simeq 4$ ns. Thus

$$J_{th} = 808\ \text{KA/cm}^2$$

11.4.5 Laser oscillation and threshold current

Let $\gamma(\nu)$ represent the gain coefficient corresponding to the amplifier response in the recombination region of a forward biased laser diode – that is

$$I_{\text{out}}(l) = I_{\text{in}}(0)\, e^{\gamma(\nu)l} \tag{11.34}$$

where l is the length of the active region. As discussed above, the gain varies in a complex manner with frequency and the magnitude of the injection current through the diode. However, the peak gain coefficient γ_p (corresponding to the peak of the gain response curve) is found to vary approximately linearly with the excess carrier concentration (see Figure 11.12).

$$\gamma_p = \alpha_a \left(\frac{\Delta n}{\Delta n_T} - 1 \right) \tag{11.35}$$

where α_a is the absorption coefficient of the material in the absence of current injection, Δn is the excess carrier concentration in the active region due to the injection current, and Δn_T is its value corresponding to "transparency." (Note that when $\Delta n = \Delta n_T, \gamma_p = 0$, implying that there is no loss or gain in the medium; in other words, the medium is transparent for the input frequency.) Using equation (11.33) we can rewrite equation (11.35) in the following alternative forms

$$\gamma_p = \alpha_a \left(\frac{J}{J_T} - 1 \right) = \alpha_a \left(\frac{i}{i_T} - 1 \right) \tag{11.36}$$

where, as before, subscript T refers to the transparency values and i is the forward (injection) current through the device. Equation (11.36) clearly indicates that only when the current through the device exceeds the transparency value i_T is there gain in the active region. However, for laser oscillations to take place, the gain must be at least equal to the loss in the optical resonator, including the loss due to the useful output from the device.

If α represents the intrinsic loss coefficient in the active medium, and if R_1 and R_2 are the reflectivities at the cleaved ends of the device, then the total loss coefficient is given by (cf. equation (11.12))

$$\alpha_{tot} = \alpha - \frac{1}{2l} \ln(R_1 R_2) \tag{11.37}$$

When the current through the diode reaches a value i_{th} (above i_T) so that $\gamma_p = |\alpha_{tot}|$, then the losses in the resonator are exactly compensated by the gain

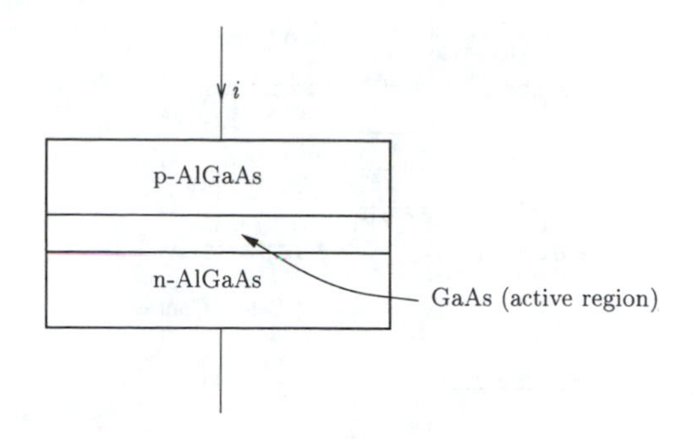

Fig. 11.13: A double heterostructure laser in which a lower bandgap semiconductor is sandwiched between two larger bandgap semiconductors. This leads to both current and optical confinement, the latter due to a refractive index difference between the lower bandgap (larger refractive index) and the larger bandgap (lower refractive index).

in the active region. Thus, at threshold, we have

$$\gamma_p = |\alpha_{tot}| = \alpha_a \left(\frac{i_{th}}{i_T} - 1 \right) \tag{11.38}$$

or

$$i_{th} = i_T \left(1 + \frac{|\alpha_{tot}|}{\alpha_a} \right) \tag{11.39}$$

Any increase in the forward current will result in a net gain in the active region leading to light amplification, and the feedback from the cleaved ends will eventually lead to gain saturation and steady-state oscillations of the laser. The fractional energy transmitted at the cleaved ends forms the useful output from the device. Equation (11.39) clearly indicates that the threshold current is larger than the transparency current; however, for most lasers $|\alpha_{tot}|$ is much smaller than α_a, and typically the ratio is ~0.1. This implies that the threshold current is only slightly more than the transparency current.

From equation (11.35) we see that for a given material the transparency value of the carrier concentration is inversely proportional to the thickness (d) of the junction region. Typical values of d for a p–n junction are a micron or more depending on the dopant concentrations of the p side and the n side. The corresponding values for the transparency current densities (J_T) are of the order of kiloamperes per square centimeter. If we somehow reduce d to a much smaller value, it would lead to a smaller value for the transparency current density. Indeed, this can be achieved by employing heterostructures.

11.4.6 Heterostructure lasers

The basic laser structure shown in Figure 11.5 is referred to as a *homojunction laser*, invented in 1962. In this device the p–n junction is formed by using the same semiconductor on both sides of the junction. Lasing in these devices can be achieved only in a pulsed operation since the threshold current values are in the range of a few amperes to tens of amperes, which could lead to catastrophic damage of the device, if operated continuously. The basic configuration in all present-day laser diodes is a *double heterostructure* wherein a thin layer of a suitable semiconductor is sandwiched between two layers of a higher bandgap material, forming two heterojunctions (see Figure 11.13). This provides for what are known as "carrier confinement" and "optical confinement" in the junction region and results in moderate to low values for the threshold current.

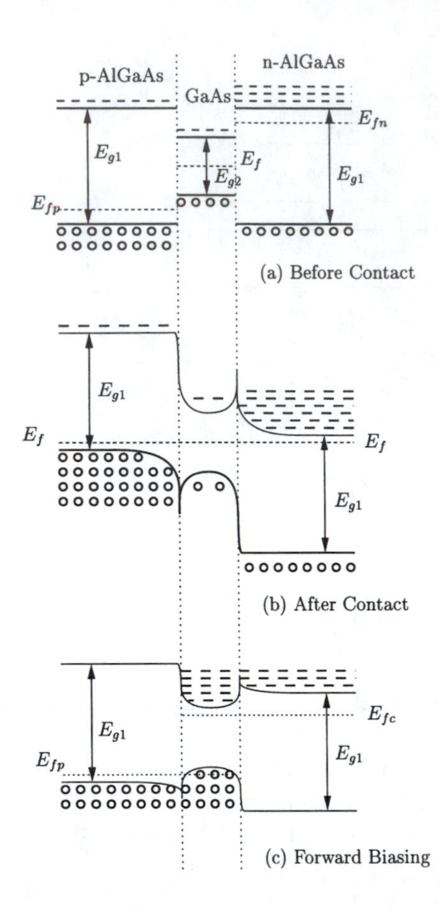

Fig. 11.14: Energy band diagram corresponding to (a) the three regions of a double heterostructure laser when they are not in contact, (b) when they are in contact and under no bias, and (c) under forward bias.

The basic heterostructure employed in the fabrication of laser diodes consists of a thin layer (approximately 0.1 μm thick) of a direct bandgap material. The material system and the compositions of the layers are suitably chosen to permit lattice-matched epitaxial growth of the layers. *Epitaxy* refers to growth-on-top – that is, layer by layer deposition on the surface and lattice-matched growth refers to the process wherein the spacing between constituent atoms in the lattice of each layer is the same. This type of growth results in structures with very little built-in strain and defects in the lattice [see, e.g., Agrawal and Dutta (1993)].

Consider a heterostructure formed by a thin layer of GaAs sandwiched between two layers of p- and n-doped $Al_xGa_{1-x}As$ as shown in Figure 11.13. The band gap of GaAs is about 1.42 eV at room temperature, whereas that of $Al_xGa_{1-x}As$ increases (from 1.42 eV when $x = 0$) with an increasing fraction of aluminum. Figure 11.14(a) shows the energy band diagram corresponding to the three regions when they are not in contact. Figures 11.14(b) and (c) show the energy band diagram of the composite before and after forward biasing, respectively. As can be seen from the figure, the potential barriers at the two heterojunctions restrict the flow of electrons from the n-AlGaAs to the p-AlGaAs and holes from the p-AlGaAs to the n-AlGaAs layers, respectively. This results in a large concentration of accumulated carriers in the thin GaAs layer and leads to a large number of carrier recombinations and photon emission. Note that the energy of the emitted photons will be around the bandgap energy of GaAs, and therefore these photons will not be absorbed by the AlGaAs layers, which have a higher bandgap.

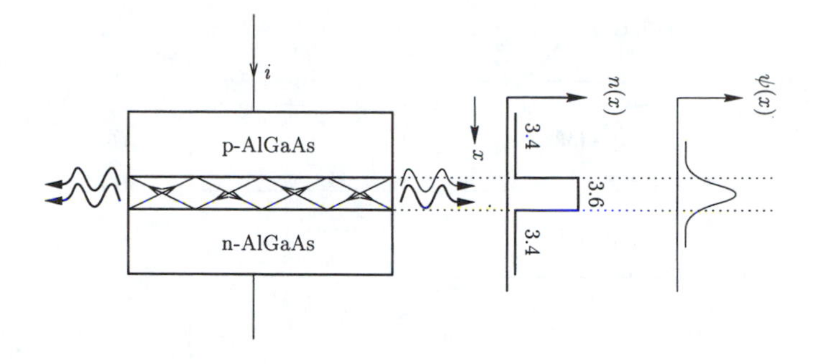

Fig. 11.15: The double heterostructure results in optical confinement due to the formation of a waveguide. The lower bandgap GaAs has a higher refractive index compared with the higher bandgap AlGaAs regions. $n(x)$ and $\psi(x)$ represent the refractive index profile and the modal field profile, respectively.

A fortunate situation leads to yet another advantage of using double heterostructures in the realization of laser diodes. For a given material system (GaAs/AlGaAs in this case), the composition having a larger bandgap is characterized by a lower refractive index. This makes the active region into an optical waveguide, leading to what is known as optical confinement. Figure 11.15 illustrates the optical waveguiding effect: GaAs has a refractive index of about 3.6 (around 0.8-μm wavelength), whereas AlGaAs is characterized by a lower refractive index of $\sim$3.4. Thus, the active region forms the guiding film and the AlGaAs layers form the cladding of the waveguide. Guided light propagates back and forth within the laser resonator formed by the active layer and the cleaved ends. $\psi(x)$ and $n(x)$ in Figure 11.15 represent, respectively, the mode profile and the refractive index profile in the transverse direction.

The combined effect of these three major advantages of a heterojunction laser – namely, carrier confinement, optical confinement, and lower absorption losses – leads to low threshold current ($\sim$ tens of milliamperes) and high overall efficiency of the device.

As discussed earlier, by using a double heterostructure (DH) design, the optical radiation gets confined in the direction perpendicular to the junction. Thus, such a DH structure behaves as a planar optical waveguide and the waveguide will support only a discrete set of modes (see Chapter 7). Typically, the refractive index difference between the guiding region and the surrounding area is 0.2 and for a typical thickness of 0.2 μm, the corresponding V value is

$$
\begin{aligned}
V &= \frac{2\pi}{\lambda_0} \cdot d\sqrt{n_1^2 - n_2^2} \\
&\simeq 1.75
\end{aligned}
$$

where we have used $\lambda_0 = 0.85\ \mu$m, $n_1 = 3.6$, and $n_2 = 3.4$. Thus, since $V < \pi$, the structure will support only a single transverse mode (see Chapter 7).

Due to the planar geometry, the light beam has no confinement parallel to the junction and, hence, it will diffract and spread over the entire width of the laser. Because of the spreading, the threshold currents are high and also the emission pattern is not stable with variation in current. To overcome these problems, laterally confined semiconductor lasers were developed. In these lasers, in addition to guidance in the direction perpendicular to the junction, the optical beam is also guided parallel to the junction plane.

There are two main types of laterally confined lasers depending on the guiding mechanism employed: *gain-guided lasers* and *index-guided lasers* (see Figure 11.16).

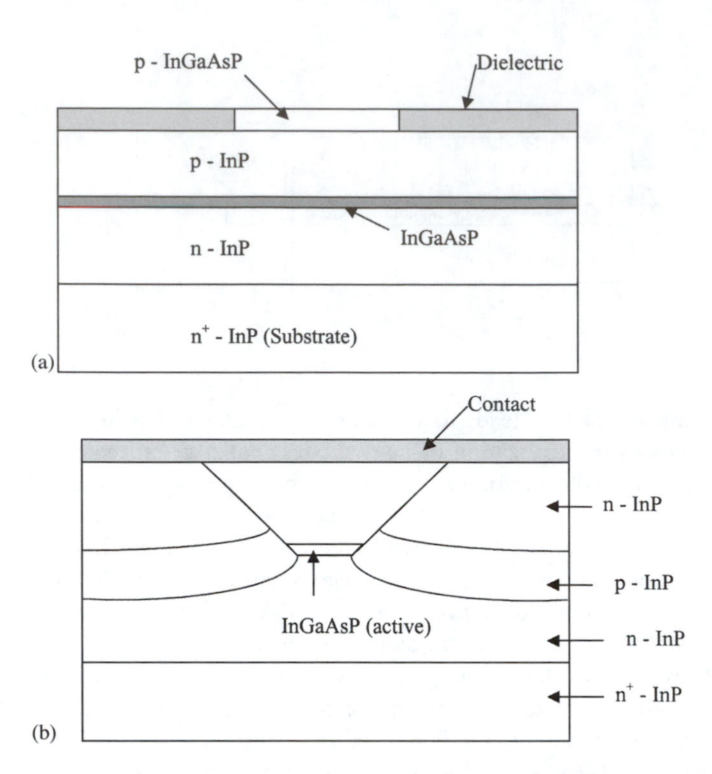

Fig. 11.16: (a) Gain-guided and (b) buried heterostructure index-guided laser structures.

In the case of gain-guided diodes, one limits the current injection over a narrow stripe. This can be achieved, for example, by coating an insulating layer such as SiO_2 on the uppermost semiconductor layer, leaving an opening for current injection. Due to this kind of injection, the carrier density is largest just under the opening and decreases away from it in the direction parallel to the junction. Because of this, the gain is also a function of lateral position and this gain variation leads to a confinement of optical energy in the lateral direction. Such lasers are hence referred to as gain-guided lasers. Since the gain distribution changes with injection current, the transverse mode profile of the laser is not very stable with changing current.

In contrast to gain-guided lasers, in the case of the index-guided lasers, a real index step is provided even in the lateral direction. Figure 11.12(b) shows a typical buried heterostructure (BH) laser in which a strong lateral guidance is provided by having a lower index surrounding the gain region. In such BH lasers, the active region has typical dimensions of $0.1\ \mu m \times 1\ \mu m$ with typical refractive index steps of 0.2–0.3. Because of the strong guidance provided by the large index step, the output is a single transverse mode and is very stable with respect to current variations. In comparison to gain-guided lasers, BH lasers are more expensive because of additional processing in their fabrication. Most fiber optical communication systems today employ BH lasers as transmitter sources.

11.4.7 Choice of materials

Recall that silica-based optical fibers have low-loss windows around 1300-nm and 1550-nm wavelengths. Therefore, to operate at these wavelengths, one has to choose a direct bandgap semiconductor whose bandgap energy corresponds to a wavelength in the low-loss window. Accordingly, the material system that is widely used in fabrication of the laser diodes for optical communication in the 1300-nm and 1550-nm window is InGaAsP/InP (InP, indium phosphide).

InP is a III–V binary compound semiconductor with a direct bandgap of 1.35 eV at 300 K (III–V indicates that the constituent elements belong to group III and group V of the periodic table of the elements). The quaternary compound $In_{1-x}Ga_xAs_yP_{1-y}$ is formed by replacing fraction x of In atoms by Ga, which is another member of group III, and fraction y of P by As, which is another member of group V. Thus, the semiconductor $In_{1-x}Ga_xAs_yP_{1-y}$ is also a III–V compound. However, the bandgap can now be tailored by selecting the appropriate composition of the constituent elements. Further, $In_{1-x}Ga_x\ As_yP_{1-y}$ is lattice matched to InP when the ratio $x/y \simeq 0.45$. Therefore, one can choose the composition x and y so that the bandgap corresponds to the desired wavelength of operation. Indeed, InGaAsP lasers can be realized to emit light at any particular wavelength in the range 1.0–1.65 μm (see Example 11.6).

We may mention here that another material system that has been studied extensively and used in the fabrication of laser diodes is AlGaAs/GaAs. By choosing an appropriate composition of the ternary compound $Al_xGa_{1-x}As$, it is possible to fabricate laser diodes emitting at any desired wavelength in the range 0.78–0.88 μm. These lasers are widely employed in consumer applications such as compact disc (CD) players and CD drives in computers and laser printers and for communication systems in the 850-nm wavelength region.

Example 11.6: The quaternary compound semiconductor $In_{1-x}Ga_x\ As_yP_{1-y}$ is lattice matched to the binary compound semiconductor InP whenever $x = 0.45y$. The corresponding bandgap of InGaAsP is given by

$$E_g(y) = 1.35 - 0.72y + 0.12y^2 \text{ eV} \tag{11.40}$$

Let us first determine the composition of the active layer for semiconductor lasers designed to operate at 1.30-μm and 1.55-μm wavelengths.

The InGaAsP active layer of the double heterostructure should have a composition so that the bandgap corresponds to the operating wavelength. The bandgap energy and the emission wavelength are related by

$$\lambda_g(\mu\text{m}) = \frac{1.24}{E_g(\text{eV})} \tag{11.41}$$

Thus, for $\lambda_g = 1.30\ \mu$m, $E_g = 0.954$ eV and for $\lambda_g = 1.55\ \mu$m, $E_g = 0.800$ eV. Using these values of E_g in equation (11.40), and keeping in mind that the value of y cannot be greater than 1, we get

$$y = 0.611 \quad \text{for } \lambda_0 = 1.30\ \mu\text{m}$$

$$y = 0.898 \quad \text{for } \lambda_0 = 1.55\ \mu\text{m}$$

The corresponding compositions of InGaAsP are

$$In_{0.726}Ga_{0.274}As_{0.611}P_{0.389} \quad \text{for } \lambda_0 = 1.30\ \mu\text{m}$$

and

$$In_{0.596}Ga_{0.404}As_{0.898}P_{0.102} \quad \text{for } \lambda_0 = 1.55\ \mu\text{m}$$

To determine the shortest and longest wavelengths at which lattice-matched double-heterostructure InGaAsP lasers can operate, we note that the longest wavelength corresponds to the smallest bandgap, $E_g = 0.75$ eV, which occurs for $y = 1$, and $x = 0.45$; thus

$$\lambda_{\max} = \frac{1.24}{0.75} \simeq 1.653\ \mu\text{m}$$

Similarly, the shortest wavelength corresponds to the largest bandgap, $E_g = 1.35$ eV, which occurs for $y = 0$ and $x = 0$; thus

$$\lambda_{\min} = \frac{1.24}{1.35} \simeq 0.918\ \mu\text{m}$$

11.5 Laser diode characteristics

There are many operating characteristics of laser diodes that are of primary importance in its application as a source in a fiber optic communication system. Some of the major performance characteristics of laser diodes and LEDs are given in Table 11.1.

11.5.1 Laser threshold

As discussed in Section 11.4.5, a laser is characterized by the presence of a threshold. Figure 11.17 shows a typical variation of output power from a laser diode as a function of the current passing through the diode. For comparison, we have also shown the output power dependence of an LED. We note that, below the threshold current, the output power is low and, as the current passing through the diode crosses the threshold value, the output power increases significantly. Indeed, the slope of the curve above threshold is much larger than that below threshold. The emission appearing below threshold is mainly due to the spontaneous transitions, whereas above threshold it is primarily due to stimulated emission.

An important parameter specifying a laser diode is the *slope efficiency*, which is the slope of the light–current curve above the threshold current (see Figure 11.17). If dI represents the change in the forward current through the diode, then the increase in the number of electrons injected per unit time into the laser is dI/e, where e is the electronic charge. If dP represents the corresponding increase in light power output, then the additional number of photons exiting the laser is $dP/h\nu$, where ν is the frequency of the radiation. We define the differential *external quantum efficiency* of the laser as

$$\eta_D = \frac{dP/h\nu}{dI/e} = \frac{e}{h\nu} \cdot \frac{dP}{dI} \tag{11.42}$$

Table 11.1. *Typical performance characteristics of laser diodes and LEDs*

	LEDs			Lasers			
Wavelength	Class	800–850	1300	Class	800–850	1300	1500
Material		GaAlAs	InGaAsP		GaAlAs	InGaAsP	InGaAsP
Spectrum width (nm)		30–60	50–150	MM	1–2	2–5	2–10
Line width SM (MHz)				SM FP		150	150
				DFB		10–30	10–30
Output power (mW)		0.5–4.0	0.4–0.6	BH	2–8	1.5–8	1.5–8
Drive current (mA)		50–150	100–150	BH	10–40	25–130	
Rise time (ns)	Surface	4–14					
	Edge	2–10	2.5–10	BH	0.3–1	0.3–0.7	0.3–0.7
Modulation frequency (GHz)		0.08–0.15	0.1–0.3	BH	2–3	2–3	2–3
Beam width (half)							
Parallel	Surface	120–180°					
Perpendicular	Surface	120–180°					
Parallel	Edge	180°		BH	10–25°	10–30°	10–30°
Perpendicular	Edge	30–70°		BH	20–35°	30–40°	30–40°
Lifetime (million hours)		1–10	50–1000		1–10	0.5–50	0.5–50

Note: MM = multimode, SM = single mode, FP = Fabry–Perot, DFB = distributed feedback, BH = buried heterostructure. [Table adapted from Hoss and Lacy (1993).]

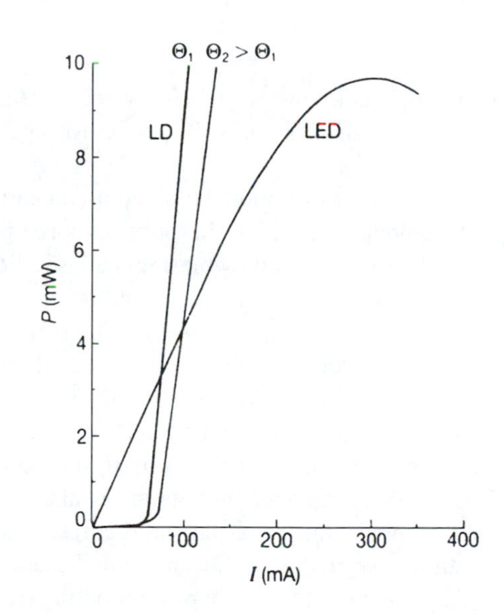

Fig. 11.17: A typical variation of output power with current for a laser diode (LD) and an LED.

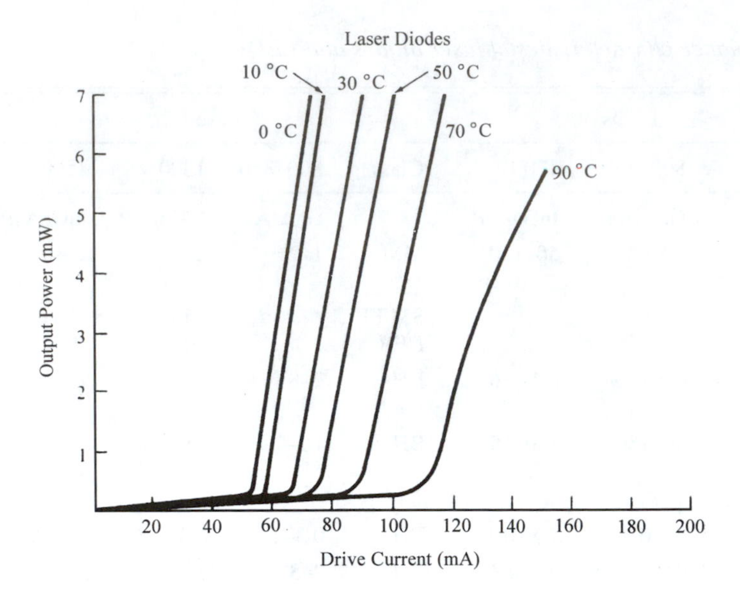

Fig. 11.18: Effect of temperature on the output power versus drive current. Note that the threshold current increases with an increase in temperature.

Typical cw (i.e., continuous wave) laser diodes have η_D values between 0.25 and 0.6. The quantity dP/dI is the slope efficiency and is specified in mW/mA. Threshold currents lie typically in the range 25–250 mA and output powers are typically in the range 1–10 mW.

One of the major aspects of laser diodes is a strong dependence of the threshold current and the output power on temperature. Figure 11.18 shows a typical output power variation with the diode current for different temperatures. We note that the threshold current depends critically on temperature. This dependence is approximately described by the relation

$$I_{th}(T) = I_0 \, e^{T/T_0} \tag{11.43}$$

where I_0 is a constant and T_0 is known as the characteristic temperature of the diode. Typically, the increase in I_{th} is 0.6–1% per °C for GaAlAs lasers and 1.2–2% per °C for GaInAsP lasers.

As we discuss in Chapter 13, most optical fiber communication systems use digital transmission techniques, in which the optical source is modulated to generate optical pulses. This is achieved by biasing the laser diode to a current slightly above I_{th}. Thus, in the absence of any signal, the optical power output is very small. This corresponds to a digital "zero." The signal in the form of current pulses adds to the bias current, thus generating a high-output power corresponding to a digital "1" (see Figure 11.19). The amplitude of the signal current pulses is adjusted not to exceed the maximum rated current value.

Biasing the diode near threshold helps in turning on the diode faster. In addition, the required signal current for modulation is also lower if the laser diode is biased near threshold. Since the threshold current itself depends on the operating temperature, either the diodes are cooled thermoelectrically to maintain a constant temperature or the output power (exiting from the other end of the diode) is monitored by a photodetector and the bias current is adjusted to give the same output optical power.

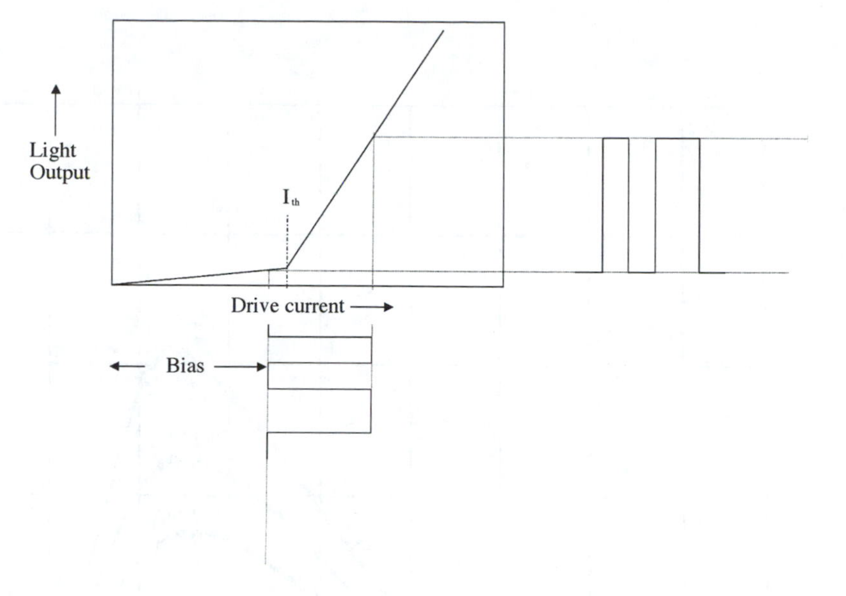

Fig. 11.19: For digital modulation of laser diodes, they are biased close to the threshold value. The current pulses lead to optical pulsing of the diode.

11.5.2 *Output spectrum*

As discussed earlier, when the input current is below threshold, the laser diode behaves like an LED and the output is mainly due to spontaneous emissions and, hence, the spectrum is broad. As the current increases beyond threshold, the frequencies having a larger gain and smaller cavity loss begin to oscillate and the spectrum changes significantly. The output power also increases. As the current is further increased, the output spectrum becomes sharper and the total output power also increases. Figure 11.20 shows a typical output spectrum from a laser diode oscillating below threshold and above threshold.

In a multilongitudinal mode laser, the output spectrum consists of a series of wavelengths (see Figure 11.20(b)). The oscillating wavelengths are determined by the cavity resonance condition (see equation (11.15))

$$\nu = \frac{c}{2n(\nu)l} \cdot q; \quad q = 1, 2, 3, \ldots$$

where $n(\nu)$ is the refractive index of the semiconductor material at frequency ν. Two adjacent oscillating frequencies corresponding to $q = q_0$ and $q_0 + 1$ are given by

$$\nu = \frac{c}{2n(\nu)l} \cdot q_0 \tag{11.44}$$

$$\nu + \Delta\nu = \frac{c}{2n(\nu + \Delta\nu)l} \cdot (q_0 + 1) \tag{11.45}$$

For $\Delta\nu \ll \nu$, we may write

$$n(\nu + \Delta\nu) = n(\nu) + \Delta\nu \cdot \frac{dn}{d\nu} + \cdots \tag{11.46}$$

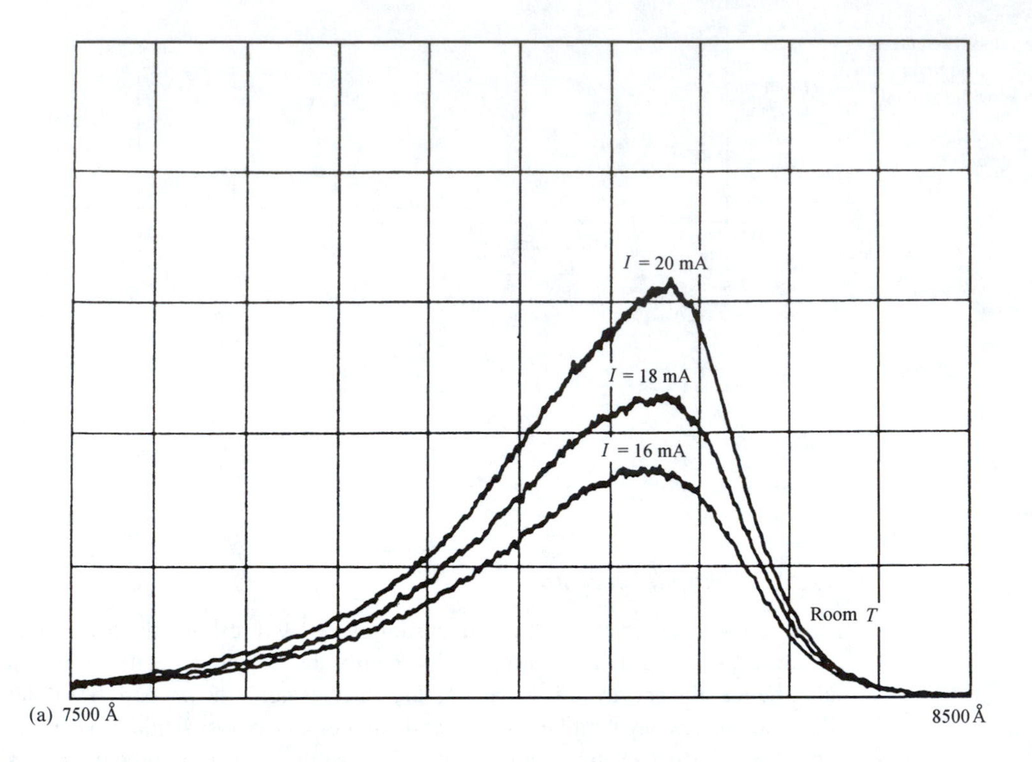

I = 20 mA
I = 18 mA
I = 16 mA
Room T
(a) 7500 Å
8500 Å

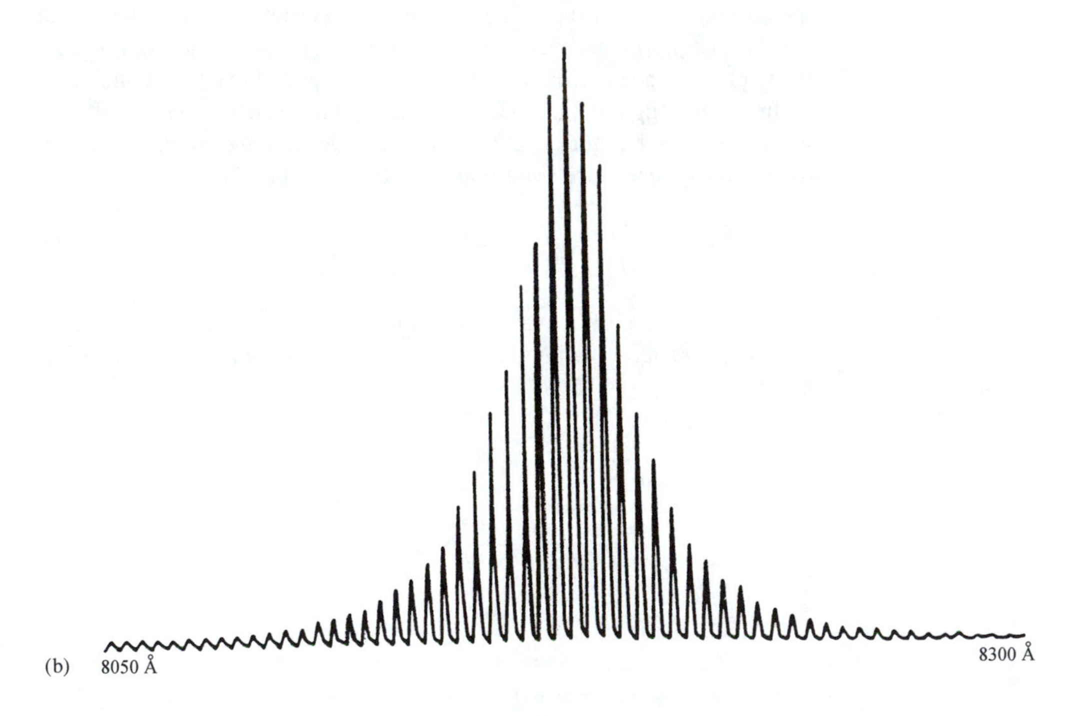

(b) 8050 Å
8300 Å

Fig. 11.20: (a) The output spectrum of a GaAs/AlGaAs laser below threshold. The emission is mainly due to spontaneous transitions. (b) The output spectrum when the laser oscillates above threshold. Note the multilongitudinal modes of oscillation. [After Yariv (1991).]

Substituting in equation (11.45) and neglecting higher order terms, we have for the intermodal spacing

$$\Delta\nu = \frac{c}{2nl}\left(1 + \frac{\nu}{n}\frac{dn}{d\nu}\right)^{-1} \tag{11.47}$$

Typically, $n = 3.6$, $l = 250\,\mu$m, $\frac{\nu}{n}\frac{dn}{d\nu} \simeq 0.38$ [Kressel and Butler (1977)] and we have

$$\Delta\nu \simeq 125 \text{ GHz}$$

For wavelength $\lambda_0 = 0.85\,\mu$m, the corresponding wavelength spacing is

$$\Delta\lambda = \frac{\lambda^2}{c}\Delta\nu = 0.3\,\text{nm}$$

In the above discussion we saw how a laser diode generally oscillates simultaneously in a number of frequencies giving a wider spectrum of the output. A larger frequency spread leads to a greater pulse dispersion in a fiber optic communication system (see Chapter 13). To reduce dispersion, we must use lasers that oscillate in only a single longitudinal mode. There are many ways to achieve single longitudinal mode oscillation.

Since the gain spectrum has a finite spectral width, one of the methods could be to increase the longitudinal mode spacing to have only one longitudinal mode within the gain spectrum. For example, for a gain spectral width of 5 nm, the intermode spacing has to be larger than 5 nm. Thus, at a wavelength of 1300 nm, this corresponds to an intermode frequency spacing of about 890 GHz. Thus, using equation (11.47), we must have

$$L < \frac{c}{2n\Delta\nu}\left(1 + \frac{\nu}{n}\frac{dn}{d\nu}\right)^{-1} \simeq 34\,\mu\text{m}$$

Such small devices create problems in handling and also, because the volume of the gain medium is restricted, the corresponding power outputs are limited.

A more reliable and efficient method to achieve single longitudinal mode emission is to introduce components or mechanisms into the laser cavity that result in a loss for all longitudinal modes except one. In this way, only one of the longitudinal modes for which the gain exceeds the losses would be able to oscillate. Some of these techniques include the use of an external cavity, using gratings at the end sections of the optical cavity itself (distributed Bragg reflector (DBR)), or even integrating the grating over the entire cavity region; these techniques are referred to as distributed feedback (DFB) structures.

In Chapter 21 we discuss periodic waveguides and we show that, by providing a periodic perturbation (in thickness or refractive index) along the propagation direction, one can achieve a highly selective wavelength reflection. Such a wavelength selective reflection can be used to provide low-loss feedback at only one oscillation frequency and higher losses at other frequencies.

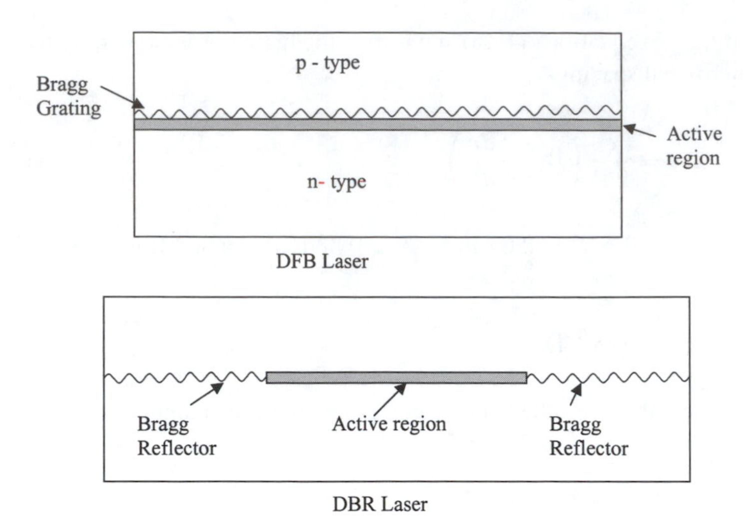

Fig. 11.21: (a) DFB and (b) DBR laser structures.

If n_{eff} represents the effective index of the propagating mode in the waveguide forming the laser, then for efficient Bragg reflection we must have (see Chapter 21)

$$2 \cdot \frac{2\pi}{\lambda_B} \cdot n_{eff} = \frac{2\pi}{\Lambda}$$

or

$$\Lambda = \frac{\lambda_B}{2n_{eff}} \tag{11.48}$$

where Λ is the period of the perturbation and λ_B satisfying the above equation is known as the Bragg wavelength. If a periodic modulation with a spatial period given by equation (11.48) is provided, then the reflectivity is strongest for the wavelength λ_B and the periodic structure acts like a mirror (see also Section 17.9). For a wavelength of 1550 nm, assuming $n_{eff} \simeq 3.5$, we obtain the required spatial period as 221 nm. Gratings with such short periods are usually fabricated by using holographic techniques.

For wavelengths away from λ_B, the reflectivity drops very sharply and thus such gratings act as highly selective wavelength mirrors.

The very strong wavelength dependence of reflectivity of a periodic waveguide can be used to fabricate single-frequency lasers. The Bragg gratings could be placed outside the gain region, in which case they replace the cleaved ends as wavelength selective mirrors (see Figure 11.21(b)). Such a structure is referred to as a DBR laser. The Bragg grating can also be integrated along with the gain region, in which case at every point in the laser cavity one has reflection and transmission and thus the feedback is distributed throughout the entire length of the cavity. Such a laser is referred to as a DFB laser (see Figure 11.21(a)). In DFB lasers, normally the grating is formed in the waveguide layer just above the active layer. The end facets are usually antireflection coated to avoid any reflections. Since the wavelength selective element is integrated along with the laser, such lasers are reproducible and are also more stable with time and external conditions.

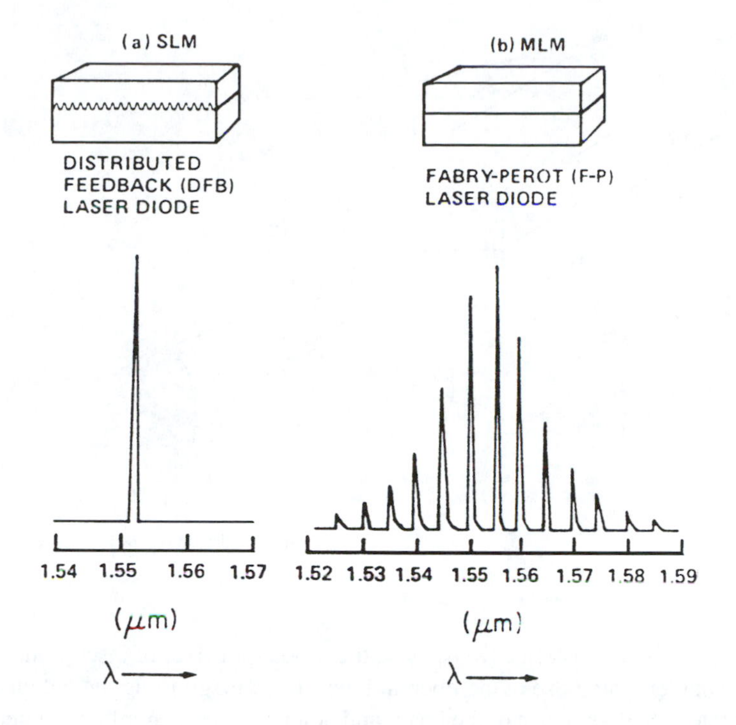

Fig. 11.22: Output spectra of (a) a DFB single longitudinal mode (SLM) laser diode and (b) a Febry–Perot (F-P) multilongitudinal mode (MLM) laser diode. [After Lin (1989).]

Figure 11.22 shows a comparison between typical output spectra of a single longitudinal mode DFB laser and a multilongitudinal mode FP laser.

11.5.3 Radiation pattern

As discussed in Section 11.4.6, in a laser structure, the optical radiation is confined in both the lateral and the transverse directions by an index step. Thus, the guiding region acts like an optical waveguide. As discussed in Chapters 7 and 8, an optical waveguide is characterized by various transverse modes of propagation. Under certain conditions the waveguide can be made to support only a single transverse mode. Most laser structures for use in optical fiber communication operate under a single transverse mode condition. The waveguide in a laser is similar to a rectangular cross-section waveguide and one can indeed obtain the field distribution of the fundamental mode as well as its effective index. One may approximate the mode field distribution by a Gaussian function with two different widths along the transverse (w_T) and lateral (w_L) directions as

$$\psi(x, y) = A \exp\left(-\frac{x^2}{w_L^2} - \frac{y^2}{w_T^2}\right) \tag{11.49}$$

where x and y represent axes parallel and perpendicular to the junction plane. Typically, for BH lasers $w_T \simeq 0.5$–$1\,\mu$m and $w_L \simeq 1$–$2\,\mu$m.

The corresponding far-field pattern will be elliptical with a larger divergence in a plane perpendicular to the junction (see Figure 11.23). The divergences parallel and perpendicular to the junction plane are typically 5–10° and 30–50°, respectively. Such large divergences pose problems in coupling light into single-mode fibers. For efficient coupling, one can use lenses to transform the output

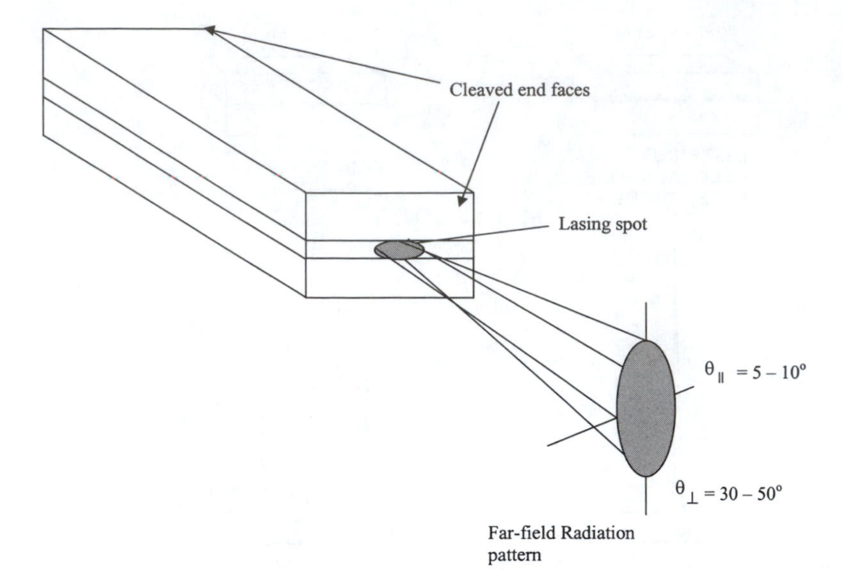

Fig. 11.23: The far-field pattern from a laterally confined laser diode has an elliptical cross section.

from the laser. This is effective because the mode spot sizes of laser diodes are much smaller than those of the fiber and, hence, by magnifying the output one can reduce the divergence of the beam and achieve a good coupling efficiency. For maximum efficiency, the Gaussian field profile of the laser diode should be made to match that of the fiber mode. The coupling lenses can be external to the fiber or can be formed at the tip of the fiber itself by etching. Using such techniques one can achieve about 50% coupling efficiency to single-mode fibers.

11.5.4 Modulation of semiconductor lasers

To encode information into the laser beam, the optical output of the laser must be modulated. One of the unique attractions of a semiconductor laser is the possibility of directly modulating the output of the laser by modulating the external current.

When the external current through the laser diode is changed, this results in a changed electron–hole (e–h) population inside the laser cavity. This changed *e–h* population changes the gain, which in turn changes the output power from the laser. The dynamics of such a modulation is determined by many factors – important among them, the carrier recombination times and the photon lifetime of the cavity. The carrier recombination lifetime due to spontaneous recombination is typically 1 ns in GaAs-based materials, whereas the stimulated carrier lifetime depends on the density of photons (i.e., the optical energy) within the cavity and is of the order of 10 ps. The photon lifetime is the average time that a photon spends inside the cavity before either escaping from the cavity or being absorbed or scattered. The cavity photon lifetime is [see, e.g., Ghatak and Thyagarajan (1989), Chapter 9]

$$\tau_{ph} = \frac{n_0}{c\left(\alpha - \frac{1}{2l}\ln R_1 R_2\right)} = \frac{n_0}{c\gamma_{th}} \qquad (11.50)$$

where γ_{th} (see equation (11.12)) represents the threshold gain coefficient. Typically, $\gamma_{th} \simeq 50\ \text{cm}^{-1}$ (see Example 11.4) and using $n_0 \simeq 3.5$, we obtain $\tau_{ph} \simeq 2$ ps. The upper limit to the modulation capability is thus set by this photon lifetime.

For digital modulation, we need to pulse the laser diode. When a current pulse is applied to the laser, electrons and holes get injected into the laser cavity. This leads to the spontaneous generation of radiation that then takes part in stimulating further emissions. The increased radiation density in the cavity then stimulates a larger number of *e–h* recombination, which tends to reduce the *e–h* population. The net output power is a result of a dynamic interaction between *e–h* recombination and photon lifetimes. If the laser diode is completely turned off after each pulse, then the spontaneous recombination time will limit the modulation rate. There is a delay between the onset of laser emission and the current pulse. In view of this, for high-speed modulation lasers, diodes are biased close to threshold (see Figure 11.19). This, of course, results in a smaller extinction ratio between bits 1 and 0 in a pulse code modulation system.

Because of the short stimulated recombination lifetimes, semiconductor lasers can be modulated to very high speeds of 20 GHz or more.

11.5.5 Frequency chirping

When a laser diode is modulated, the current modulation results in a changing carrier concentration in the laser cavity. Since the refractive index of the semiconductor material of the cavity depends on the carrier concentration, this current modulation leads to a change in the laser mode frequency – that is, within the pulse the frequency changes with time. This phenomenon is called *chirping*. The mode frequency is blue-shifted (increase in frequency) near the leading edge and red-shifted (decrease in frequency) near the trailing edge of the optical pulse. Such a chirped pulse has a much broader frequency spectrum than a corresponding unchirped pulse (see Chapter 15). Chirping can increase the spectral width from less than about 0.01 nm under cw operation to about 0.2 nm under modulation. This increased spectral bandwidth will lead to an increased pulse broadening when it propagates through an optical fiber and, hence, will limit the bit rate.

For very high speed communications (~10 Gb/s) the chirping of optical pulses can be avoided by employing continuous wave (cw) laser diodes and using an external electrooptic modulator to modulate the output.

11.6 LED characteristics

An LED is a forward biased p–n junction in which *e–h* recombination leads to the generation of optical radiation through the process of spontaneous emission. The structure of an LED is similar to that of a laser diode except that there is no cavity for feedback. Unlike the emission from a laser diode, which is primarily due to stimulated emissions, the emission from an LED is due to spontaneous recombinations and the output from an LED differs significantly from that of a laser diode.

Figure 11.17 also shows the variation of output power from an LED as a function of the drive current. Unlike the laser diode, there is no threshold and the output power increases smoothly as a function of current. At large currents

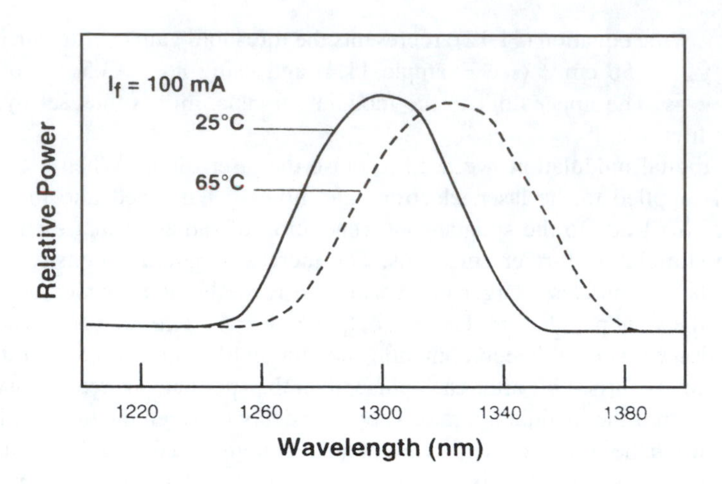

Fig. 11.24: Output optical spectrum of a typical high-power 1300-nm LED.

the output power saturates. The total power outputs from LEDs can be a few milliwatts.

Figure 11.24 shows the output spectrum of a typical LED. Again, unlike a laser, the spectrum is quite broad (typically 30–80 nm wide). Such a large spectral width leads to a large dispersion when LEDs are used as sources in optical fiber communication systems.

Because spontaneous emission is random and appears along all directions, the output from an LED is not directional. Output beam angles may be typically in the range 30° perpendicular to the junction, to about 120° parallel to the junction. Such a broad emission pattern implies that coupling into single-mode fibers will be very inefficient. On the other hand, the coupling efficiency into multimode fibers can be significant. For example, with an LED emitting 1 mW of optical power, one can approximately launch 40 μW ($= -14$ dBm) of power in multimode fibers and about 10 μW ($= -20$ dBm) in single-mode fibers.

The modulation capabilities of LEDs are limited by the carrier recombination time and the capacitance of the device. Unlike laser diodes, in the case of LEDs, the rise times are in the range of 2–10 ns, giving a 3-dB modulation bandwidth of 30–180 MHz.

Problems

11.1 The bandgap energy E_g of $Ga_{1-x}Al_xAs$ depends on x through the following approximate equation.

$$E_g(x < 0.37) = (1.424 + 1.247x)\,\text{eV}$$

Calculate the bandgap energy and the corresponding cutoff wavelength for $x =$ 0.2 and 0.3.

11.2 When a laser diode is modulated, the refractive index of the cavity also changes due to carrier injection. If the fractional change in refractive index is 10^{-7}, what is the corresponding fractional change in wavelength?

Solution: From equation (11.44), we have

$$\lambda_0 = \frac{2nL}{q}; \quad q = \text{integer}$$

where n is the refractive index. Thus, the change $\Delta\lambda_0$ in wavelength of oscillation for a change Δn of refractive index is

$$\frac{\Delta\lambda_0}{\lambda_0} = \frac{\Delta n}{n}$$

For $\lambda_0 = 1300$ nm and $\Delta n/n = 10^{-7}$, we have

$$\Delta\lambda_0 = 1.3 \times 10^{-4}\,\text{nm}$$

The corresponding frequency shift is

$$\Delta\nu = \frac{c}{\lambda_0^2}\Delta\lambda_0 = 23\,\text{MHz}$$

11.3 Consider a semiconductor laser operating at 900 nm. If the spectral width over which gain is available is 15 nm, what is the maximum cavity length for single longitudinal mode operation? Assume the refractive index $n = 3.6$.

11.4 The forward current through a GaAsP red LED emitting at 670-nm wavelength is 30 mA. If the internal quantum efficiency of GaAsP is 0.1, what is the optical power generated by the LED?

Solution:

$$P_{opt} = \eta_i \left(\frac{i}{e}\right)\frac{hc}{\lambda} = \eta_i \cdot i \cdot \frac{1.24}{\lambda\,(\mu\text{m})}$$

$$\approx 5.5\,\text{mW}$$

Note that this is not the output power of the LED, since only a fraction of this will exit from the LED.

11.5 The threshold current of a particular laser diode doubles when its temperature is increased by 50°C. Determine the characteristic temperature of the laser.

Solution: If I_{th1} and I_{th2} represent the threshold current at temperatures T_1 and $T_1 + 50$ then using equation (11.43) we have

$$\frac{I_{th1}}{I_{th2}} = \frac{1}{2} = \frac{e^{T_1/T_0}}{e^{(T_1+50)T_0}}$$

Thus

$$T_0 = \frac{50}{\ln 2} \simeq 72\,\text{K}$$

11.6 Calculate the threshold gain for a semiconductor laser with a length of 250 μm, $\alpha = 5\,\text{mm}^{-1}$, and end faces of reflectives 90% and 30%. What is the corresponding photon lifetime?

11.7 A He–Ne laser has a gain bandwidth of 1700 MHz. What should be the cavity length so that the laser oscillates in a single longitudinal mode?

12

Detectors for optical fiber communication

12.1 Introduction

An optical detector is a device that converts light signals into electrical signals, which can then be amplified and processed. Such detectors are one of the most important components of an optical fiber communcation system and dictate the performance of a fiber optic communication link.

There are many different types of photodetectors such as photomultiplier tubes, vacuum photodiodes, pyroelectric detectors, and semiconductor photodiodes. Semiconductor photodiodes are the most commonly used detectors in optical fiber systems since they provide good performance, are compatible with optical fibers (being small in size), and are of relatively low cost. These photodiodes are made generally from semiconductors such as silicon or germanium or from compound semiconductors such as GaAs, InGaAs, etc.

In this chapter we briefly discuss the basic principle of operation of two commonly used photodiodes – namely, PIN (p-doped, intrinsic, and n-doped layers) diode and avalanche photodiodes (APD) – and study their important characteristics that are of particular relevance to optical fiber communication systems.

12.2 Principle of optical detection

The basic principle behind photodetection using semiconductors is optical absorption. When light is incident on a semiconductor, the light may or may not get absorbed depending on its wavelength. If the energy $h\nu$ of a photon of the incident light beam is greater than the bandgap of the semiconductor, then it can be absorbed, leading to generation of e–h pairs (see Section 11.4.2). When an electric field is applied across the semiconducting material, the photogenerated e–h pairs are swept away, leading to a photo current in the external circuit.

If E_g is the bandgap of the semiconductor, then the maximum wavelength of absorption (also referred to as the cutoff wavelength) is given by

$$\lambda_c = \frac{hc}{E_g} \tag{12.1}$$

Substituting for h (=6.63×10^{-34} J·s) and c, the cutoff wavelength in micrometers is given by

$$\lambda_c \simeq \frac{1.24}{E_g(\text{eV})} \tag{12.2}$$

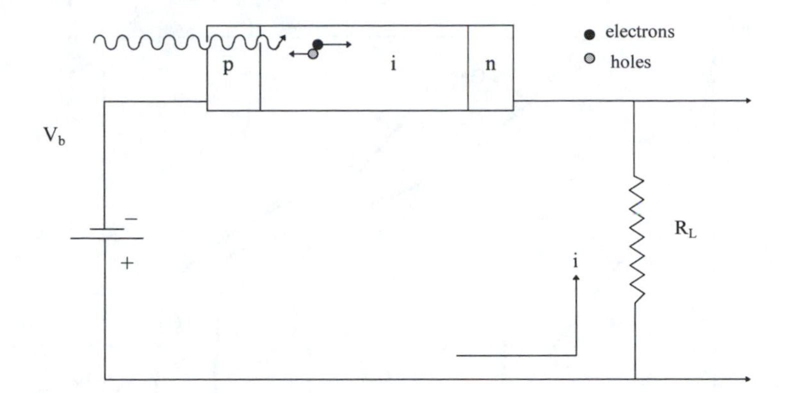

Fig. 12.1: A PIN photodiode consisting of an intrinsic semiconducting layer sandwiched between p-doped and n-doped layers. In the photoconductive mode the diode is reverse biased.

The bandgap energies are 1.11 eV for silicon, 0.67 eV for germanium, 1.43 eV for GaAs, and 0.75 eV for InGaAs. Thus, the corresponding cutoff wavelengths are 1.13 μm, 1.85 μm, 0.87 μm, and 1.65 μm, respectively, implying that these semiconductors can be used for detection of light below the above cutoff wavelengths.

Photodetectors are made out of p–n junctions. There are two different detection modes available – namely, photovoltaic and photoconductive. In the photovoltaic mode, electrons and holes that are generated by absorption of the incident light are collected at either end of the junction, leading to a potential difference. A current flows if the device is loaded. In this mode the photodetector is an unbiased diode. In contrast, in the photoconductive mode, the diode is under reverse bias and the *e*–*h* pairs generated by light absorption are separated by the high electric field in the depletion layer. Such a drift of carriers induces a current in the outer circuit. Photodiodes operating in the photoconductive mode have a faster response because of the carriers being swept away by the electric field and are the ones used in fiber optic communication. We shall hence restrict our attention to this mode of operation.

Photodetectors used in fiber optic communication systems fall under two categories: PIN and APD. In both the devices, the *e*–*h* pairs are generated in the depletion region and these are swept away by the applied electric field. In contrast to PIN, APDs have an inbuilt current gain.

In the following sections we discuss the most important characteristics of PIN and APDs. Noise in photodetection is also an important aspect and is discussed in Section 13.3.

12.3 PIN photodetector

The most common semiconductor photodetector is the PIN photodiode, which consists of an intrinsic (very lightly doped) semiconductor sandwiched between a p-doped and an n-doped region (see Figure 12.1). The PIN photodiode is normally subjected to a reverse bias, as shown in Figure 12.1. Since the intrinsic (i) region has no free charges, its resistance is high and hence most of the voltage across the diode appears across the i region. The i region is usually wide so that incoming photons have a greater probability of absorption in the i region rather than in the p or n regions. Since the electric field is high in the i region, any *e*–*h* pairs generated in this region are immediately swept away by the

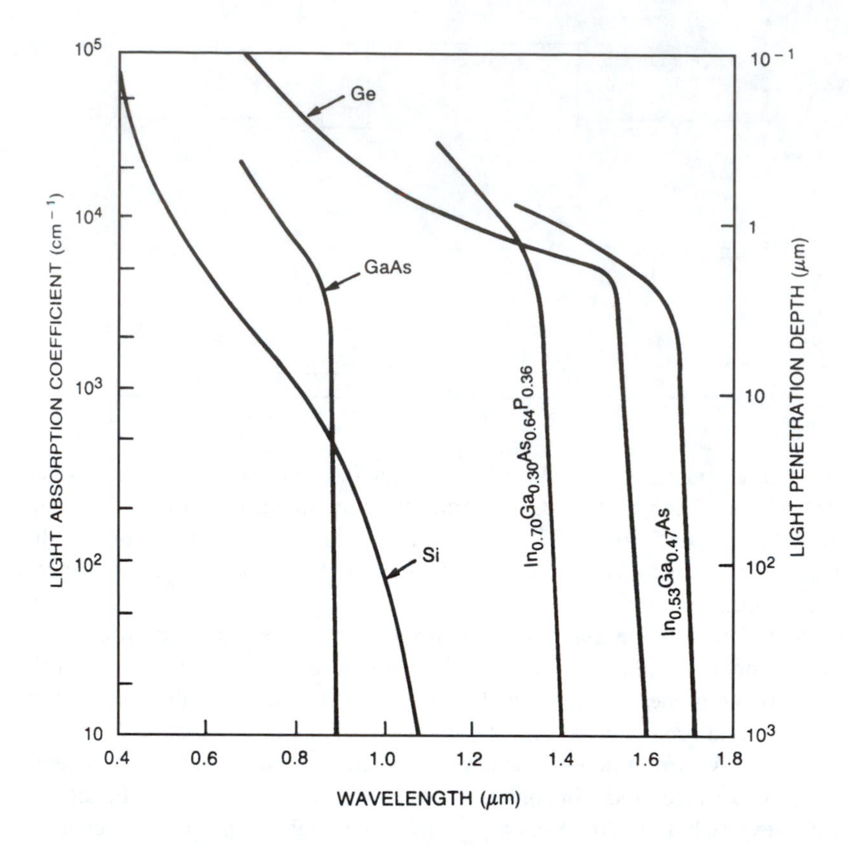

Fig. 12.2: Wavelength dependence of the absorption coefficient for different important semiconductor materials used in fiber optic communications. [After Campbell (1989).]

field. e–h pairs generated in the p and n regions have to first diffuse into the depletion region before being swept away. Also, these e–h pairs may suffer recombination, resulting in a reduced current.

12.3.1 Responsivity and quantum efficiency

The absorption of optical radiation in the semiconductor material is described by

$$P(z) = P_0[1 - e^{-\alpha(\lambda)z}] \tag{12.3}$$

where P_0 is the optical power at $z = 0$ (the incident optical power), $P(z)$ is the optical power absorbed over distance z, and $\alpha(\lambda)$ is the wavelength-dependent absorption coefficient. Figure 12.2 shows the λ dependence of $\alpha(\lambda)$ for some typical photodiode materials.

We note from Figure 12.2, that near cutoff α rises much more rapidly for GaAs, InGaAs, and InGaAsP than for silicon and germanium. This is because Si and Ge have an indirect bandgap, whereas the others have a direct bandgap. We also note that typical absorption coefficients are in the range of 10^3–10^5 cm^{-1}.

Let w represent the width of the depletion region. An optical power P_0 incident on the photodetector first suffers a partial reflection at the air–semiconductor surface before entering the detector. If R represents the reflection coefficient, then the optical power entering the detector is $P_0(1 - R)$. The optical power absorbed in a distance w will then be

$$P_0(1 - R)[1 - e^{-\alpha(\lambda)w}]$$

If ν is the frequency of the incident light, then the number of photons absorbed per unit time will be

$$\frac{P_0}{h\nu}(1 - R)(1 - e^{-\alpha w})$$

Since each absorbed photon leads to generation of an *e–h* pair, the above expression also gives the number of *e–h* pairs generated per unit time. Assuming that only a fraction ζ of the *e–h* pairs contributes to the photo current (the remaining having been lost due to recombination), the photo current is

$$I = \frac{e}{h\nu}(1 - R)\zeta(1 - e^{-\alpha w})\, P_0 \tag{12.4}$$

where e is the magnitude of the electronic charge.

From equation (12.4) we can define two important quantities – namely, quantum efficiency η and responsivity ρ.

The quantum efficiency η is the ratio of the number of *e–h* pairs generated to the number of incident photons. Thus

$$\eta = \frac{I/e}{P_0/h\nu} = (1 - R)\zeta(1 - e^{-\alpha w}) \tag{12.5}$$

The responsivity ρ is the photo current generated per unit optical power and is usually specified in A/W.

$$\rho = \frac{I}{P_0} = \frac{\eta e}{h\nu} = \frac{e\zeta}{h\nu}(1 - R)(1 - e^{-\alpha w}) \tag{12.6}$$

Substituting for e (=1.6×10^{-19} C) and h (=6.63×10^{-34} J·s), equation (12.6) can be rewritten as

$$\rho = \frac{\lambda_0}{1.24}\eta \tag{12.7}$$

with λ_0 measured in micrometers.

Example 12.1: For light of wavelength 0.8 μm, the absorption coefficient of silicon $\alpha \simeq 10^5$ m^{-1}. Since the refractive index of Si is approximately 3.5, for an uncoated Si photodiode the reflection coefficient

$$R = \left(\frac{3.5 - 1}{3.5 + 1}\right)^2 \simeq 0.31$$

For a depletion layer width of 20 μm, we obtain (assuming $\zeta = 1$)

$$\eta \simeq 0.6$$

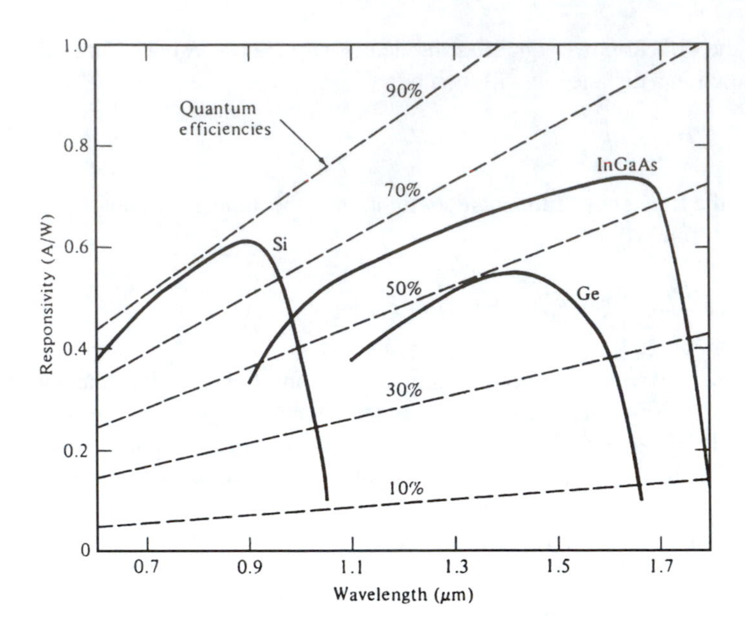

Fig. 12.3: Spectral dependence of responsivity ρ and quantum efficiency η for the three important semiconductor photodiode materials, Si, Ge, and InGaAs. [After Keiser (1991).]

The corresponding responsivity is

$$\rho = \frac{\eta e}{h\nu} = 0.39 \text{ A/W}$$

If the detector surface is antireflection coated, then $R = 0$ and we have $\eta \simeq 0.87$ and $\rho \simeq 0.57$ A/W.

As mentioned earlier, the long wavelength cutoff of a photodetector is caused by the fact that the energy of the incident photons is less than the bandgap. At the lower wavelength side, the response of the photodetector cuts off as a result of a very large value of α. This large absorption coefficient results in their absorption very close to the photodetector surface where the *e–h* recombination time is very short. Thus, the photogenerated *e–h* pairs recombine within the detector itself before they can contribute to the current in the circuit.

Figure 12.3 shows the responsivity of Si, Ge, and InGaAs photodetectors as a function of wavelength. As evident, Si is ideal for detection in the region of 850 nm (the I window of fiber optic systems) and InGaAs is the preferred detector in the 1.3-μm and 1.55-μm wavelength regions (II and III windows).

The presence of the intrinsic region between the p and n regions of the PIN photodetector improves the performance of the photodetector compared with a simple p–n junction detector. Thus, because of the large depletion region, most photons can get absorbed in this region. Due to the presence of a strong electric field, the carriers drift rapidly (without suffering many recombinations), resulting in a good quantum efficiency and, hence, responsivity. The width of the intrinsic region cannot be made too large since the carriers then would take longer to drift to the terminals and thus lower the speed of response of the photodetector (see Section 12.3.2). For silicon and germanium, indirect bandgap semiconductors, the widths are typically 20–50 μm, whereas for InGaAs, a direct bandgap semiconductor, the width is typically 3–5 μm.

Figure 12.1 shows a simple reversed biased operation of a PIN photodiode with R_L being the load resistance. In the absence of any light falling on the photodetector, the entire bias voltage drops across the photodiode and the voltage across R_L is zero. When light with power P falls on the photodiode, it leads to a current ρP, where ρ is the responsivity. This current leads to a drop in voltage across R_L of

$$V_R = \rho P R_L \tag{12.8}$$

and the bias voltage across the photodiode reduces. We see from equation (12.8) that the voltage V_R across the load is proportional to the optical power P falling on the photodetector. Thus, a measurement of V_R gives us the optical power P. The maximum value of P that is measurable in such a fashion is when $V_R = V_b$ – that is

$$P_{\max} = \frac{V_b}{\rho R_L} \tag{12.9}$$

Beyond this value the photodiode gets saturated. Thus, by choosing an appropriate value of R_L one can operate the photodiode over an optical power range 0 to $P_{\max}$. Decreasing R_L to increase the detection range of course reduces the sensitivity given by

$$V_R/P = \rho R_L \tag{12.10}$$

Example 12.2: Consider a Si PIN detector with $\rho = 0.5$ A/W. We assume a reverse bias of 20 V and a load resistor of 100 Ω. In such a case

$$P_{\max} = 400 \text{ mW}$$

and

$$\frac{V_R}{P} = 50 \text{ mV/mW}$$

On the other hand, increasing the load resistor to 10 kΩ gives

$$P_{\max} = 4 \text{ mW}$$

and

$$\frac{V_R}{P} = 5 \text{ V/mW}$$

Thus, increasing R_L increases the sensitivity while reducing the range of detection.

12.3.2 *Speed of response*

The speed of response and, hence, the bandwidth of a photodetector is dependent on three primary factors: the transit time of the photo-generated carriers through the depletion region; the electrical frequency response as determined by the RC time constant, which depends on the diode capacitance; and the slow diffusion of carriers generated outside the depletion region.

The transit time of carriers across the depletion region of width w is given by

$$\tau_t = \frac{w}{v_d} \tag{12.11}$$

where v_d is the carrier drift velocity. The smaller is w, the smaller is τ_t and the lesser will be the limitation due to transit time. This requirement of smaller w is contrary to that required to achieve larger quantum efficiency.

As an example we consider an intrinsic region of width 20 μm in a Si PIN detector. The drift velocity of electrons is typically 10^5 m/s. Thus, the time taken to cross the intrinsic region is 200 ps. In InGaAs PIN detectors, the typical widths are about 5 μm, leading to transit times of 50 ps. These numbers correspond approximately to the photodiode rise times.

Apart from transit time limitations, photodiode capacitance may also play a significant role. Thus, if the diode area is A and the depletion layer width is w, then the junction capacitance is

$$C_d = \frac{\epsilon A}{w} \tag{12.12}$$

where ϵ is the permittivity of the semiconductor. In a circuit as shown in Figure 12.1, the speed of response is given by the RC time constant. Indeed, the rise time (10–90%) is given by (see Example 13.1)

$$\begin{aligned} t_r &= 2.19 R_L C_d \\ &= 2.19 R_L \frac{\epsilon A}{w} \end{aligned} \tag{12.13}$$

Decreasing w to reduce transit time would increase the capacitive rise time, which would have to be balanced by a decreased R_L. The bandwidth of the photodiode as determined by R_L and C_d is

$$\Delta f = \frac{1}{2\pi R_L C_d} \tag{12.14}$$

Clearly, to achieve small rise times, photodetectors must have small area, large w, and small R_L.

Example 12.3: As an example, let us consider a Si PIN detector with a diameter of 500 μm and $w = 20\ \mu$m. Using $\epsilon = 10.5 \times 10^{-13}$ F/cm, we have

$$C_d = \frac{\epsilon A}{w} \simeq 4\,\text{pF}$$

Table 12.1. *Typical performance characteristics of detectors*[a]

	Silicon		Germanium		InGaAs	
Parameter	PIN	APD	PIN	APD	PIN	APD
Wavelength range (nm)	400–1100		800–1800		900–1700	
Peak (nm)	900	830	1550	1300	1300 (1550)	1300 (1550)
Responsivity ρ (A/W)	0.6	77–130	0.65–0.7	3–28	0.63–0.8 (0.75–0.97)	
Quantum efficiency(%)	65–90	77	50–55	55–75	60–70	60–70
Gain (M)	1	150–250	1	5–40	1	10–30
Excess noise factor[b] (x)	–	0.3–0.5	–	0.95–1	–	0.7
Bias voltage (−V)	45–100	220	6–10	20–35	5	<30
Dark current[b] (nA)	1–10	0.1–1.0	50–500	10–500	1–20	1–5
Capacitance (pF)	1.2–3	1.3–2	2–5	2–5	0.5–2	0.5
Rise time (ns)	0.5–1	0.1–2	0.1–0.5	0.5–0.8	0.06–0.5	0.1–0.5

[a] Adapted from Hoss and Lacy (1993).
[b] Discussed in Section 13.3.

Thus, if $R_L = 1000\ \Omega$, then $t_r \simeq 8.8$ ns. The corresponding bandwidth is $\Delta f = 40$ MHz. Decreasing R_L to 100 Ω reduces t_r to 0.88 ns and the bandwidth goes up to 400 MHz.

We see from above that for achieving a small rise time and, hence, a large bandwidth, R_L should be small. Recalling our discussion in Section 12.3.1, we note that to increase the quantum efficiency we need to increase the thickness of the depletion region. This, on the other hand, increases the response time and lowers its bandwidth. A compromise is usually made by choosing depletion widths between $1/\alpha$ and $2/\alpha$.

In Section 11.4.6 we discussed how heterostructures are formed between semiconductors with different bandgaps. Use of heterostructures in detectors can also enhance their performance. Thus, for detection in the 1.3- to 1.6-μm region one uses InGaAs with a cutoff wavelength of about 1.65 μm as the intrinsic sandwiched absorption layer between two InP layers (p-type and n-type) on either side. Since InP has a cutoff wavelength of 0.92 μm, light in the wavelength region 1.3–1.6 μm is absorbed only in the InGaAs region. Because of this, one can eliminate any component to the photocurrent due to the diffusion of carriers (which is much slower than drift) and thus enhance the speed of response. Also by choosing the InGaAs layer to be several micrometers thick, the quantum efficiency can be significantly increased.

Table 12.1 lists some of the characteristics of Si, Ge, and InGaAs PIN and APDs.

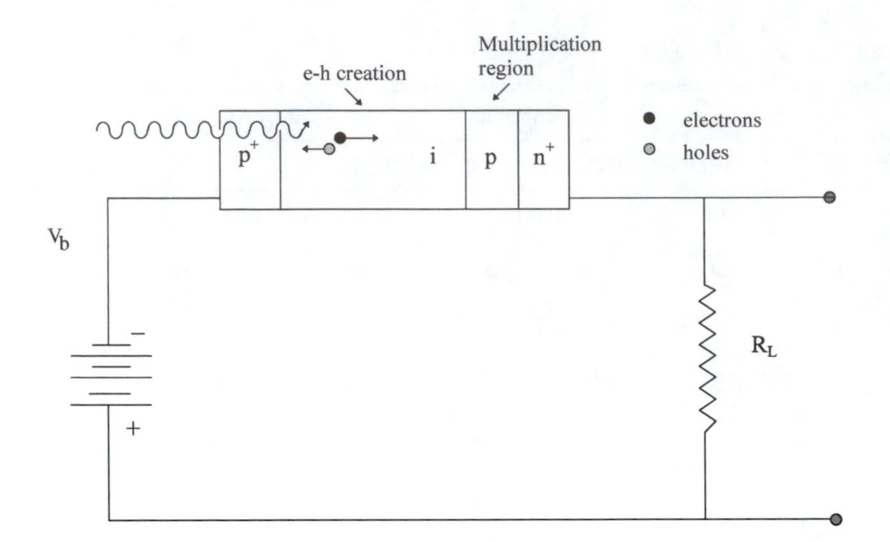

Fig. 12.4: An APD that is reversed biased to a high voltage. Photons absorbed in the i region create *e–h* pairs that on acceleration and collision produce an avalanche of *e–h* pairs, leading to an internal multiplication primarily in the p region.

12.4 Avalanche photodiodes (APDs)

The APD is a photodiode with an internal current gain that is achieved by having a large reverse bias. In an APD the absorption of an incident photon first produces an *e–h* pair just like in a PIN. The large electric field in the depletion region causes the charges to accelerate rapidly. Such charges propagating at high velocities can give a part of their energy to an electron in the valence band and excite it to the conduction band. This results in an additional *e–h* pair. These in turn can further accelerate and create more *e–h* pairs. This process leads to an avalanche multiplication of the carriers.

For avalanche multiplication to take place, the diode must be subjected to large electric fields. Thus, in APDs one uses several tens of volts to several hundred volts of reverse bias.

APDs differ from PIN diode designs in having an additional *p*-type layer between the intrinsic and a highly doped *n* region (see Figure 12.4). The *e–h* pairs are still generated in the i region but the avalanche multiplication takes place in the *p*-type region. APDs that have avalanche multiplication of just one type of charge carrier have superior noise characteristics.

If M represents the multiplication factor, then for an APD, we have

$$R_{\mathrm{APD}} = M\,R = M\frac{\eta e}{h\nu} \tag{12.15}$$

We should mention here that M in equation (12.15) is an average value and the gain factor M itself fluctuates around the mean. This leads to an additional noise (also referred to as excess noise) in APDs (see Section 13.3.3). Thus, there is an optimum multiplication factor to achieve best operation. Figure 12.5 shows a typical variation of M with the applied reverse bias voltage.

In the 850-nm range, silicon APDs typically require bias voltages of around 250 V to achieve an optimum gain of about 100, whereas PIN devices require bias of only 10–50 V. At longer wavelengths (1300–1500 nm) the APDs require typical bias of 20–30 V for typical gains of 10–30, whereas PIN devices require only 5–15 V.

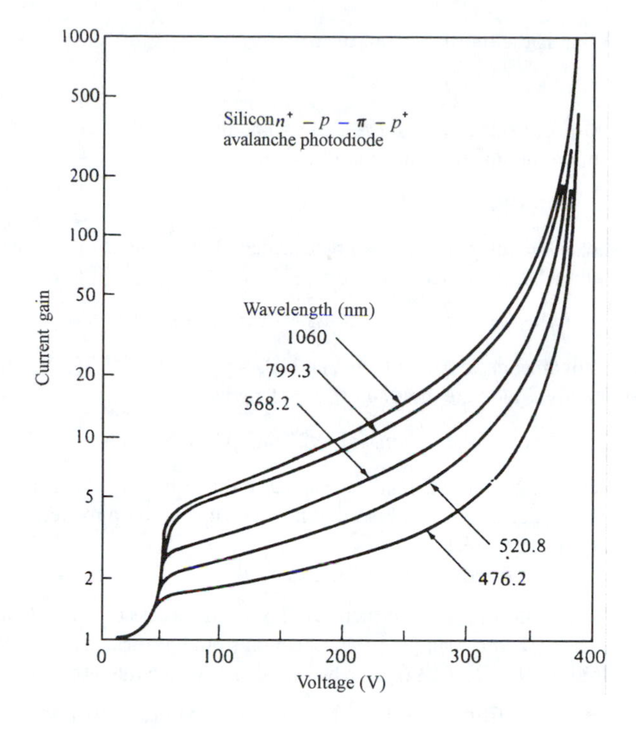

Fig. 12.5: Typical variation of current gain with the applied bias voltage of a Si APD for different wavelengths. [After Melchoir et al. (1978).]

Problems

12.1 The bandgap energy of $In_{0.53}Ga_{0.47}As$ is 0.73 eV. Obtain the corresponding cutoff wavelength

[ANSWER: $\lambda_c = 1.65\,\mu$m.]

12.2 Consider a Si PIN photodiode with a depletion layer width of 20 μm and a diameter of 200 μm. The absorption coefficient at 800 nm is 10^3 cm^{-1}.

(a) Assuming antireflection-coated surface, obtain the quantum efficiency.
(b) If the load resistance is 10 kΩ, obtain the RC time constant.
(c) Compare this with the drift time for a typical carrier velocity of 10^5 m/s.

12.3 Consider an InGaAs PIN photodiode with a quantum efficiency of 0.6. Calculate the responsivity at 1300 and 1550 nm. Why is the responsivity larger at 1550 nm?

12.4 An optical power of -40 dBm is incident on a photodetector with $\rho = 0.65$ A/W. Calculate the current that is generated.

12.5 Consider a Si PIN photodiode operating in a circuit as shown in Figure 12.1. If the absorption coefficient at 800 nm is 10^5 m^{-1}, then assuming the photodiode to be antireflection coated, what should be the depletion layer thickness for a quantum efficiency of 0.5 and 0.7? (Assume $\zeta = 1$.)

Solution: From equation (12.5), we have

$$w = \frac{1}{\alpha}\ln\left(\frac{1}{1-\eta}\right)$$

Thus, for $\eta = 0.5$ and 0.7, we must have $w = 6.9\,\mu$m and 12 μm.

12.6 Consider a PIN photodiode with $w = 20\,\mu$m and a carrier velocity of 5×10^4 m/s.

(a) What is the transit time of the carriers?

[ANSWER: 0.4 ns.]

(b) If the dielectric constant $K = 11.7$ and the photodiode diameter is 1 mm, obtain the junction capacitance.

[ANSWER: 4 pF.]

(c) For what values of the load resistance will the detector rise time be determined by the transit time?

[ANSWER: $R_L < 0.4 \times 10^{-9}/2.19\,C \simeq 46\,\Omega$.]

12.7 The bandgap energies of GaP, InP, and AlAs are 2.25 eV, 1.35 eV, and 2.16 eV, respectively. Calculate their cutoff wavelengths.

[ANSWER: GaP: 550 nm, InP: 920 nm, AlAs: 570 nm.]

12.8 Calculate the maximum load resistance that can be used with a photodiode having a capacitance of 2 pF so that it can be operated up to frequencies of (i) 10 MHz and (ii) 1 GHz.

12.9 Silicon has a refractive index of 3.5.

(a) What is the optimum refractive index and thickness of a thin dielectric film to be coated for suppressing reflection from the surface?

(b) If SiO_2 ($n = 1.46$) is used, what should be the thickness for least reflection?

12.10 A power of -3 dBm at 1300 nm is launched in an optical fiber with an attenuation of 0.4 dB/km. After transmission through 50 km of the fiber, the output is coupled to a PIN photodetector with a quantum efficiency of 0.7. What is the optical power falling on the photodetector and the corresponding photocurrent?

12.11 What is the quantum efficiency of a Si PIN photodiode at 800 nm under normal illumination with $\alpha = 10^5\ \text{m}^{-1}$ and $d = 10\ \mu\text{m}$? Assume the refractive index of Si to be 3.5. Obtain the corresponding responsivity.

12.12 A uniform intensity beam of radius 1 mm and power 10 mW falls on the photodetector of Problem 12.11 with a sensitive detection area of 100 μm. Calculate the current generated in the photodiode.

12.13 A photodetector has a responsivity of 0.5 A/W at 1300 nm and at 1500 nm. Calculate the number of *e–h* pairs generated if the photodiode is illuminated by 1 mW of 1300-nm radiation and 1500-nm radiation.

12.14 The bandgap of $In_{0.14}Ga_{0.86}As$ is 1.15 eV. What is the corresponding cutoff wavelength?

12.15 A Si PIN photodiode with a diameter of 0.5 mm is to be used with a load resistor of 100 Ω. Estimate the thickness of the intrinsic region so that the diode response is fast. Obtain the corresponding responsivity if the diode is provided with antireflection coating. Assume $\epsilon = 10.4 \times 10^{-10}$ F/m and $v_d = 10^7$ m/s.

13

Design considerations of a fiber optic communication system

13.1 Introduction

In the preceding chapters we discussed the characteristics of optical fibers, optical sources, and optical detectors. These form the three basic units of any optical fiber communication system. In this chapter we discuss how these basic elements can be put together to build a simple point-to-point optical fiber communication link.

Let us assume that we need to transmit information between two points (see Figure 13.1). The separation between these points could range anywhere from less than a kilometer in the case of computer data links to several thousands of kilometers, for instance, in transoceanic links. In all such links, there would be a transmitter that could be either an LED or a laser diode, the transmission path consisting of optical fibers that could be either multimode or single-mode fibers, and the optical receiver that could be a PIN or APD followed by the detection electronics. The choice of the components would depend on the distance as well as on the bit rate. When the separation between the points is greater than about 50–100 km, then because of attenuation in the link or pulse dispersion, it becomes necessary to use regenerators that consist of a receiver–transmitter combination. These regenerators detect the pulse stream before the power becomes too low or the pulses become unresolvable and retime, reshape, and regenerate (and, hence, the name 3R repeater) a new set of optical pulses to be transmitted over the next part of the link. For links that are limited by loss rather than by dispersion, one can use optical amplifiers in place of regenerators. These optical amplifiers amplify the optical signals in the optical domain itself without converting them into an electrical signal and, hence, have many attractive features (see Chapter 14). Of course, we cannot cascade such optical amplifiers indefinitely since these do not compensate for dispersion and also add noise. Thus, at such points in a long-haul link we would have to use electronic regenerators.

In the design of a fiber optic link, we usually have to carry out two analyses to ensure that the system performs to our requirements: power budgeting and rise time budgeting. Power budgeting ensures that enough power is being received at the receiver so that the error in detector is below a specified limit. Rise

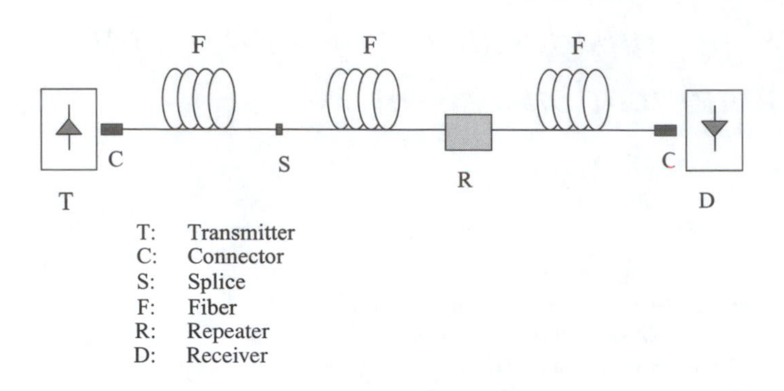

Fig. 13.1: A simple point-to-point fiber optic communication link consisting of a transmitter, connectors, splice, optical fiber, receiver, and regenerator.

time budgeting ensures that the overall bandwidth of the system is capable of handling the bit rates at which the system is supposed to operate.

In Section 13.2 we briefly discuss analog and digital modulation, and in Section 13.3 we discuss some of the important noise mechanisms in the detection process. Section 13.4 discusses the concept of bit error rate and how it fixes the minimum optical power required to fall on a detector. In Section 13.5 we discuss power budgeting and rise time budgeting, and finally in Section 13.6 we look at the limitations to a fiber optic system due to attenuation and dispersion.

13.2 Analog and digital modulation

We begin by considering the example of transmission of audio signals over a large distance. When we speak, we produce sound waves that mostly vary in frequency from about 20 Hz to about 4000 Hz. To transmit the sound signals over a large distance, the audio signal is made to modulate a radio wave – that is, some parameter of the radio wave like amplitude, frequency, or phase is made to change in a manner proportional to the audio signal. Thus, the modulated radio wave has coded into it the audio waves (which is the information to be sent). The radio wave here carries the information and, hence, is called a carrier wave. The audio wave modulates the radio wave and, hence, is called the modulating wave.

There are various ways of modulating the radio waves. For example, one may modulate the amplitude of the radio wave in accordance with the signal. Such a scheme results in what is known as amplitude modulation and is depicted in Figure 13.2. Similarly, instead of modulating the amplitude of the radio wave, one may modulate the frequency of the radio wave in accordance with the audio wave. This results in frequency modulation and is shown in Figure 13.3.

Let us first consider the amplitude-modulating scheme. It can be shown that when a carrier wave of frequency ν_c is modulated with a modulating wave of frequency ν_m, then the modulated wave consists of waves of frequencies $\nu_c - \nu_m$ and $\nu_c + \nu_m$ in addition to waves of frequency ν_c. The waves at frequencies $\nu_c - \nu_m$ and $\nu_c + \nu_m$ are said to lie in the lower side band and upper side band, respectively. In general, the modulating wave may consist of waves of frequencies from 0 to ν_m. Thus, on modulating, one obtains two bands of frequencies lying from $\nu_c - \nu_m$ to ν_c (the lower side band) and another lying between ν_c and $\nu_c + \nu_m$ (the upper side band). Since both the side bands contain the information, one may transmit only one of the side bands. Such a system is called single side-band transmission. One can extract the information out of the signal by again "mixing" the received wave with the frequency ν_c. Thus, when the

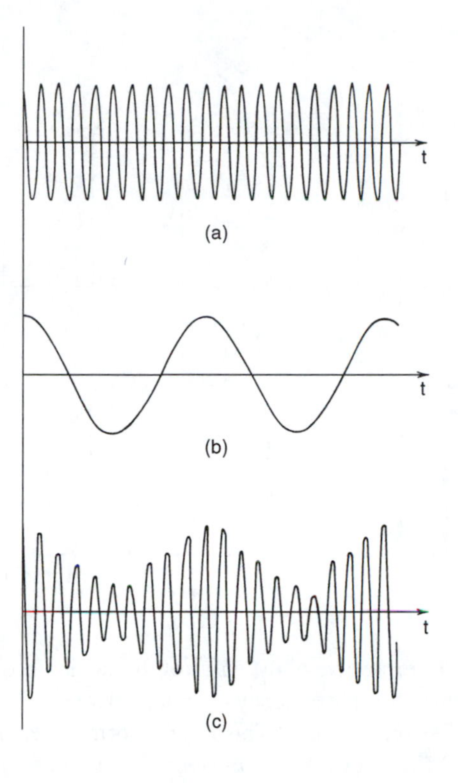

Fig. 13.2: When the carrier wave shown in (a) is amplitude modulated by the modulating wave shown in (b), one obtains an amplitude-modulated wave as shown in (c).

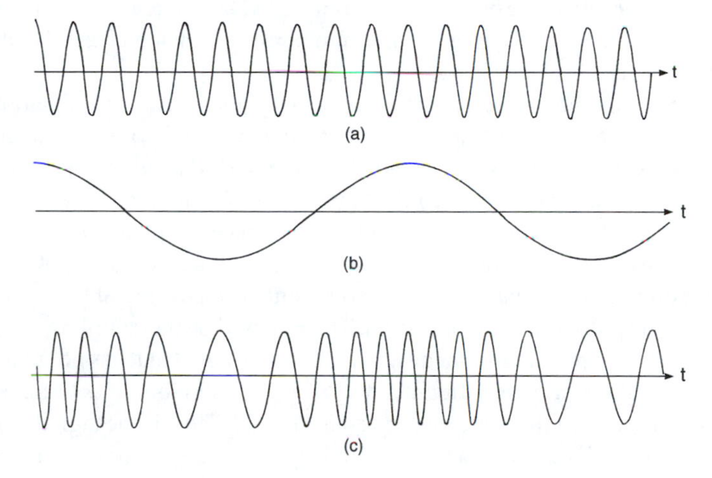

Fig. 13.3: When the carrier wave shown in (a) is frequency modulated by the modulating wave shown in (b), one obtains a frequency-modulated wave as shown in (c).

wave with frequency $\nu_c - \nu_m$ is mixed with a wave of frequency ν_c, one will obtain waves of frequencies ν_m, ν_c, and $2\nu_c - \nu_m$. One of these corresponds to ν_m, the modulating signal.

Since the frequency of the speech signal may lie anywhere between 0 and 4000 Hz, it becomes clear that, if the carrier wave has a frequency of, say, 100,000 Hz, then, for example, in the upper side band transmission, we must transmit frequencies between 100,000 and 104,000 Hz. Thus, a band of at least 4000 Hz must be reserved for one speech signal. Hence, between carrier frequencies of 100,000 and 500,000 Hz, we can at most send (500,000–100,000)/4000 = 100 independent speech signals simultaneously. Since the same bandwidth of

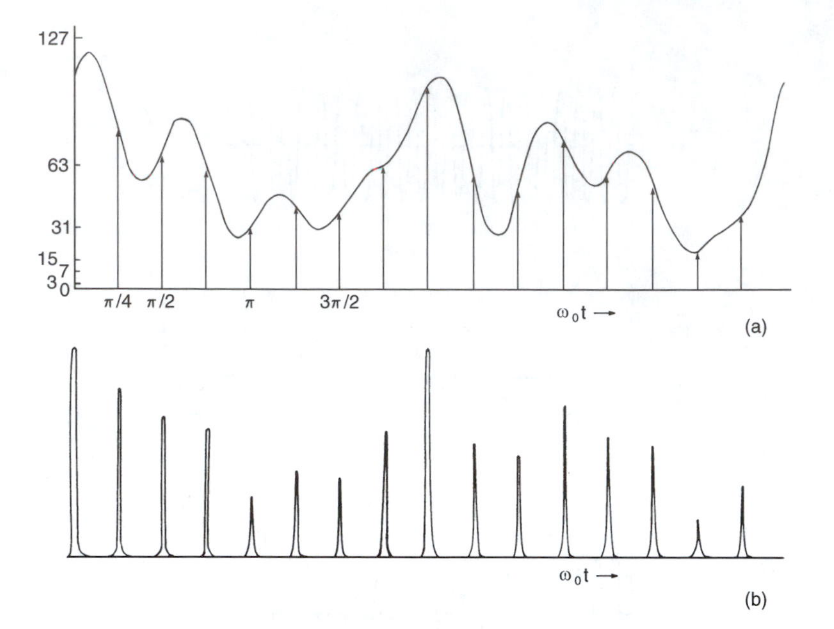

Fig. 13.4: (a) The figure shows a time-varying signal given by equation (13.1), whose maximum frequency component ν_m is $4\omega_0/2\pi$. Hence, by the sampling theorem, the signal can be completely determined by sampling the signal at time intervals of $1/2\nu_m = \pi/4\omega_0$, which is shown as vertical lines. (b) In the pulse amplitude modulation system, the amplitude of the signal at the sampled times is represented by the amplitude of the pulses.

4000 Hz is required irrespective of the value of the carrier frequency, it becomes clear that, in a higher carrier frequency channel between 10^{15} and 5×10^{15} Hz, one can send 10^{12} speech signals. This is an enormous capacity indeed. The bandwidth of 4000 Hz that we have considered is enough for intelligibility of speech. But for music the bandwidth is about 20 kHz. For television, the bandwidth is about 6 MHz. Thus, a smaller number of television channels exist in the same carrier frequency band.

The above method of sending simultaneously more than one signal along the same channel by assigning different frequency bands for each channel is referred to as frequency division multiplexing. Notice that all of the signals are overlapping in the time domain while they are nonoverlapping in the frequency domain. The different signals are separated at the receiver by making use of filters that transmit only the frequency band corresponding to the signal of interest.

The modulation system that is most commonly used in optical fiber communication is the pulse modulation system. This is based on the sampling theorem, according to which a band-limited signal – that is, a signal that has no frequency components above a certain frequency (say ν_m) – is uniquely determined by its values at uniform time intervals less than $1/2\nu_m$. Thus, if the signal value is specified every $1/2\nu_m$ s, the sampling rate is given by $2\nu_m$ samples per second. Consequently, instead of sending the complete signal continuously, one may send the sampled values at a rate of only $2\nu_m$ samples per second. Hence, for example, since speech signals mostly contain frequencies up to 4 kHz, if the speech signal is specified at regular intervals of 1/8000 s – that is, if the signal is sampled at the rate of 8000 samples per second – then the signal can be completely retrieved from these sampled values. Thus, a simple system involving pulse modulation could be one that produces pulses of light at a rate of 8000 times per second with the amplitude of the pulse being proportional to the value of the signal at that time. This is referred to as pulse amplitude modulation and is depicted in Figure 13.4.

In addition to the pulse amplitude modulation technique, there exist other techniques of coding the signal value into the samples. Thus, one could modulate the width of the pulses in accordance with the signal value while keeping the pulse amplitude constant; this leads to pulse width modulation. Similarly, one could alter the position of the pulse within the time interval, leading to pulse position modulation.

Another very important mode of transmission of the sampled values of the signal is called the pulse code modulation (PCM). In such a scheme, instead of varying the amplitude of a pulse in proportion to the value of the signal, the signal is approximated by a new signal that is obtained by the process of quantization. The quantized value is represented by a code formed by a pattern of identical pulses. In binary coding, the signal value is first converted to the binary code and then the ones and zeroes in the code are represented by the presence or absence of a pulse. Thus, in a coding scheme having seven-bit[1] coding length, the signal value at a particular time is coded in the form of seven pulses, all the pulses being of the same height and width.

As an example, we consider a signal that is of the form

$$f(t) = 15[4 + \sin \omega_0 t + \sin(1.5\omega_0 t) + \cos(2\omega_0 t) + \cos(3\omega_0 t) + \sin(4\omega_0 t)] \tag{13.1}$$

where ω_0 is a constant. The maximum frequency in the above signal is

$$\nu_m = \frac{4\omega_0}{2\pi} \tag{13.2}$$

Thus, according to sampling theorem, the sampling rate must be (at least)

$$2\nu_m = \frac{4\omega_0}{\pi} \text{ samples/s} \tag{13.3}$$

Hence, the samples must be taken every $1/2\nu_m$ s – that is, the time interval between two samples is

$$t = \frac{1}{2\nu_m} = \frac{\pi}{4\omega_0} \text{ s} \tag{13.4}$$

In Figure 13.4(a) we have drawn the signal given by equation (13.1) as a function of $\omega_0 t$, and, as can be seen, the sample values are being taken every $\pi/4\omega_0$ s.

To consider the PCM system involving seven bits one must first code each of the sampled signal values into a binary code of seven bits. Figure 13.5(a) gives the signal value at the sampling times and the corresponding binary code. To understand this, we note that the signal value of 105 corresponds to the binary

[1]The word bit is derived from binary digit.

Time, t	Signal Value	Binary Equivalent in Seven Levels	Pulse Code Waveform
0	105	1101001	
$\frac{\pi}{4\omega_0}$	84	1010100	
$\frac{\pi}{2\omega_0}$	70	1000110	
$\frac{3\pi}{4\omega_0}$	64	1000000	
$\frac{\pi}{\omega_0}$	30	0011110	
$\frac{5\pi}{\omega_0}$	44	0101100	

(a)

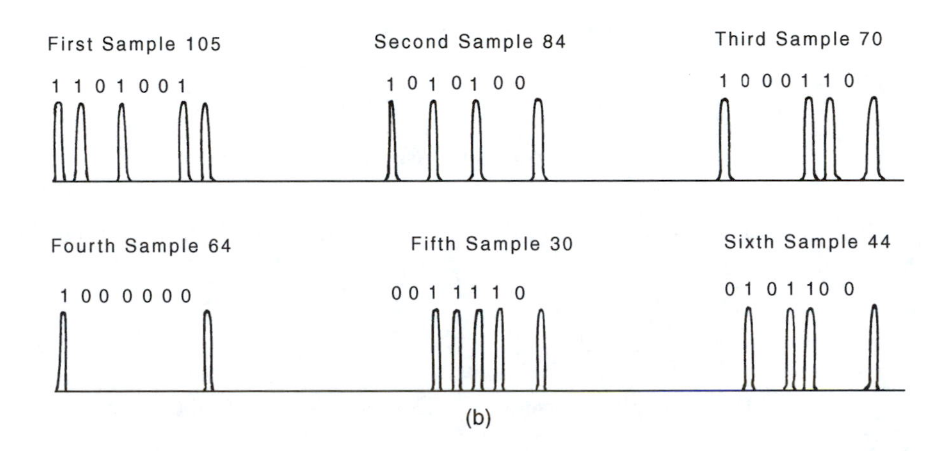

(b)

Fig. 13.5: (a) Pulse code wave form of sample signal depicted in Figure 13.4(a). (b) In the PCM using binary coding involving seven levels, the sample values are first converted into the binary equivalent as shown in (a). Then the ones and zeroes in the binary coding may be represented by the presence or absence of a pulse.

equivalent 1101001 from the following logic:

	Remainder
2 | 105	1
2 | 52	0
2 | 26	0
2 | 13	1
2 | 6	0
2 | 3	1
1	

The binary equivalent of 105 is formed by successively dividing by 2, then listing the final quotient and the remainders from bottom up. Hence, the binary equivalent of 105 is 1101001.

The maximum signal value that can be coded into a binary equivalent (of seven bits) is $2^7 - 1 = 127$. Thus, the signal values can be coded into intervals of unity in a total amplitude of 127. This therefore corresponds to coding into equally spaced levels and becomes less accurate at low-amplitude values. To overcome this, one often uses the logarithm scale coding scheme (companding) [see, e.g., Cattermole, 1969].

In Figure 13.5(b), the binary-coded sampled values have been depicted in the form of pulses. It may be noted here that in Figure 13.5(b), at every sample,

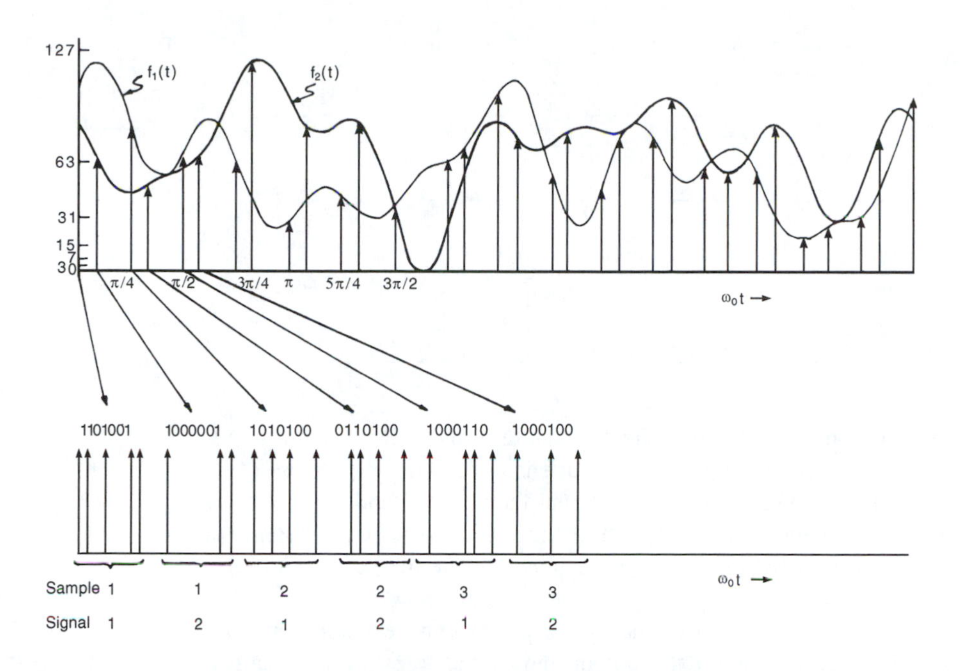

Fig. 13.6: Curves 1 and 2 correspond to two different signals to be transmitted along the same channel simultaneously. Each signal is first sampled, then coded into a binary code, and finally intermingled so that the two signals appear at different times. The interspaced series of pulses is then sent along the channel. Such a scheme in which the different signals occupy different time intervals is called time division multiplexing (TDM).

there are eight pulses instead of seven; the eighth pulse is required for signaling and supervisory information at the receiver. Thus, for the above signal, there would be $8 \times 4\omega_0/\pi$ pulses per second; the first factor of 8 corresponds to the fact that each sample is being replaced by eight pulses and the second factor corresponds to the sampling rate.

Now, coming back to the case of speech signal, since the maximum frequency is 4000 Hz, the sampling rate should be 8000 samples per second. If each sample is replaced by eight pulses, one would have $8 \times 8000 = 64{,}000$ pulses per second, which is also written as 64,000 bits/s or 64 Kb/s.

Hence, corresponding to each signal, one would have a series of pulses appearing 64,000 times per second. Simultaneous transmission of many signals is done by intermingling the sample values of the various signals so that the pulses of different signals propagate at different times. Figure 13.6 shows how two signals can be sent through the same channel on a time-sharing basis. Such a scheme of multiplexing is known as TDM. In such a scheme all the signals occupy the same frequency band but are separated in time. This is in contrast to frequency division multiplexing, in which the various signals occupy the same region in the time domain while they occupy different regions in frequency domain.

The number of independent messages that can be sent simultaneously in a system employing TDM would be determined by the condition that the pulses at the output be well resolved in time so that information can be retrieved back. Thus, pulse broadening would ultimately limit the information capacity of the system.

If one compares the pulse amplitude modulation (PAM) and PCM systems, it is apparent that the PCM system requires more bandwidth for the transmission of the same information since, corresponding to each pulse in the PAM system, one has eight pulses in the PCM system. This disadvantage of the PCM system is offset by the fact that a system employing PCM is more

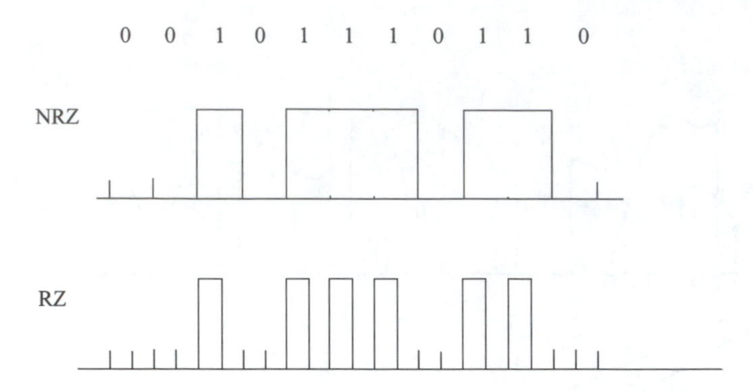

Fig. 13.7: The digital pulse sequence of 00101110110 in NRZ and RZ formats.

immune to noise and interference effects because in the PCM system the receiver has only to detect the presence or absence of a pulse regardless of its amplitude value and width. Thus, all external factors that tend to distort the amplitude or the shape of the pulses have no effect on a PCM system; of course, the distortion should not be so much as to make the pulses unresolvable.

It can be seen from the above that if the pulses can be made very narrow and still are resolvable at the output end of the transmitting channel, then the information capacity of the system will be much higher. If the pulses are of infinitesimal width (i.e., the sampling is by impulses), then the bandwidth of the sampled signal can be shown to be infinite. On the other hand, if the pulses are of finite width, the resulting sampled signal has negligible energy content at higher frequencies. In fact, the bandwidth required for transmission becomes smaller with an increase in pulse width. It therefore appears that sampling with finite width pulses is superior to impulse sampling since it requires a smaller bandwidth. But if the pulses have a finite width, then the pulse requires a longer time interval for transmission and only a smaller number of signals can be transmitted simultaneously on a time-sharing basis.

In optical communication systems, the pulse sequence is formed by turning on and off an optical source such as a laser diode or LED. The presence of the light pulse would correspond to a binary 1 and the absence to a binary 0. The two commonly used techniques for representing the digital pulse train are the nonreturn to zero (NRZ) and the return to zero (RZ) formats. In the case of NRZ, the duration of each digital pulse is equal to the period, whereas in the RZ case, the pulse duration is shorter than the period (typically the duration is half the period). Figure 13.7 shows the same digital pattern in the NRZ and RZ formats. The choice of the scheme depends on several factors such as synchronization, drift, and so forth. For example, a long sequence of 1 (or 0) bits would generate a constant signal in the NRZ scheme and would pose problems in terms of extraction of timing information for electronic processing. Such problems are usually overcome by using line-coding techniques. As an example, we could represent 1s and 0s by pulses of duration $T/2$ (T being the interpulse separation) as shown in Figure 13.8. This coding scheme is referred to as Manchester coding and solves the problem of d.c. wander of baseline for threshold detection. For other line-coding schemes and their relative merits and demerits,

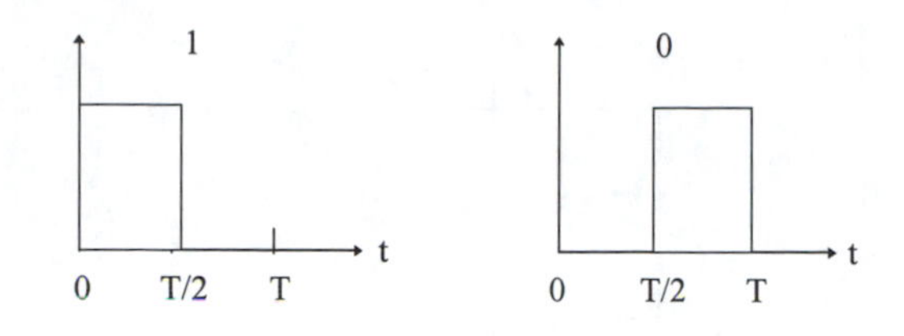

Fig. 13.8: Pulses corresponding to bit 1 and bit 0 in Manchester coding.

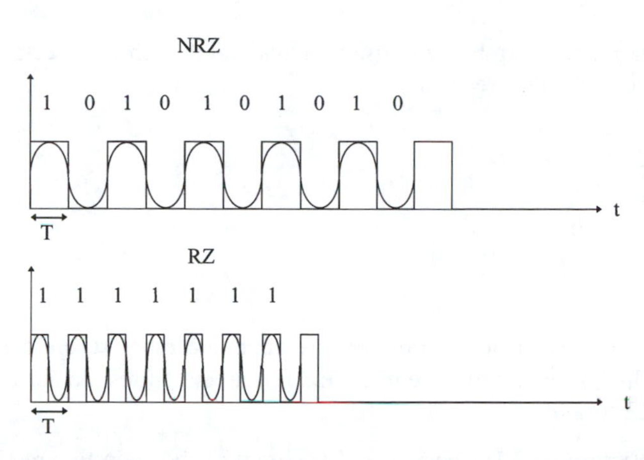

Fig. 13.9: An alternating sequence of 1s and 0s corresponds to the maximum rate of change in NRZ, whereas a series of 1s gives the fastest change in RZ. The sinusoidal curves correspond to the fundamental frequency component in the pulse sequence.

readers are referred to any standard text on digital communication [e.g., Lathi (1989)].

One of the major differences between NRZ and RZ pulse sequence is in the bandwidth requirement. To appreciate this, we first note that a sequence of 1s in RZ would correspond to the fastest changes, whereas an alternating sequence of 1s and 0s in NRZ would correspond to the fastest change.

The fundamental frequency component[2] in the RZ pulse sequence of 1s is $1/T$ as can be seen from the sinusoidal distribution plotted in Figure 13.9. Similarly, in the case of NRZ, the fundamental frequency component in the alternating sequence of 1s and 0s is $1/2T$ (see Figure 13.9). Hence, a system having a bandwidth of at least $1/T$ and $1/2T$ should pass the RZ and NRZ sequence, respectively, without too much deterioration. Thus, if we represent the bit rate by B, then $B = 1/T$ and the bandwidth Δf required by RZ and NRZ schemes are

$$\begin{aligned} \Delta f &\simeq B;\ RZ \\ &\simeq \frac{B}{2};\ NRZ \end{aligned} \tag{13.5}$$

Another familiar method of representing the bandwidth requirement is through the parameter called rise time T_r, which is the time taken by a system to rise from 10% to 90% of the maximum value for a step function input. The rise time and bandwidth are related through (see Example 13.1)

$$T_r = \frac{0.35}{\Delta f} \tag{13.6}$$

[2]This can be seen easily by expanding the pulse sequence in a Fourier series.

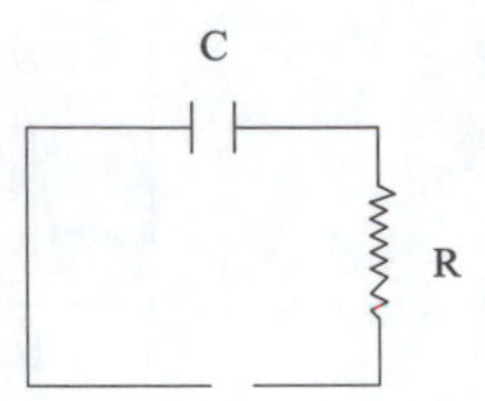

Fig. 13.10: A simple RC circuit.

An inverse relationship between rise time and bandwidth is expected for any linear system. We thus see that

$$T_r \simeq \frac{0.35}{B}; \; RZ$$
$$\simeq \frac{0.70}{B}; \; NRZ \tag{13.7}$$

Thus, the bandwidth or rise time requirements are more severe for the RZ format, which is expected since in the RZ format the pulses are narrower than in the NRZ format.

Example 13.1: Consider an RC circuit as shown in Figure 13.10 and to which a step voltage V_0 is applied at $t = 0$. It is well known that the corresponding voltage across the capacitance increases according to the following equation

$$V_c(t) = V_0(1 - e^{-t/RC}) \tag{13.8}$$

Thus, the rise time is

$$\begin{aligned} T_r &= RC(\ln 10 - \ln 1/0.9) = RC \ln 9 \\ &\simeq 2.2RC \end{aligned} \tag{13.9}$$

The corresponding 3-dB bandwidth of the RC circuit is given by

$$\Delta f = \frac{1}{2\pi RC} \tag{13.10}$$

Hence, from equations (13.9) and (13.10) we have

$$T_r = \frac{2.2}{2\pi \Delta f} \simeq \frac{0.35}{\Delta f} \tag{13.11}$$

Example 13.2: For sending a 2.5-Gb/s pulse rate, we would require a bandwidth of 2.5 GHz in the RZ format and a bandwidth of 1.25 GHz in the NRZ format, clearly showing reduced bandwidth requirements of the NRZ format.

Digital communication systems have many advantages over analog systems. One of the most important is the fact that, using regenerative repeaters, one can send digital signals over very long distances without much addition of noise. This is brought about primarily because in digital systems, in contrast to analog systems, at the repeaters, one needs to detect only the presence or absence of a pulse rather than measure the pulse shape. This decision can be made with reasonable accuracy even if the pulses are distorted and noisy. Thus, at the receiver, new clean pulses are generated and transmitted to the next repeater station. Hence, this process prevents any accumulation of distortion and noise along the path.

13.3 Noise in detection process

We have seen in the previous chapter that when light falls on a photodetector, $e-h$ pairs are generated, which give rise to an electrical current. This conversion process from light to electrical current is accompanied by the addition of noise. The two most important noise mechanisms in a photodetector circuit are the shot noise and thermal noise.

13.3.1 Shot noise

Shot noise arises from the fact that an electric current is made up of a stream of discrete charges – namely, electrons – which are randomly generated. Thus, even when a photodetector is illuminated by constant optical power P, due to the random generation of e–h pairs, the current will fluctuate randomly around an average value determined by the average optical power P.

Since shot noise current is random, its average value is zero. We can hence define a mean square shot noise current that can be shown to be given by [see, e.g., Yariv (1991)]

$$\overline{i_{NS}^2} = 2eI\Delta f \tag{13.12}$$

where e is the electron charge, I is the average current generated by the detector, and Δf is the bandwidth over which the noise is being considered. Since the photo current I itself depends on the incident optical power, the shot noise increases with an increase in incident optical power.

We need to mention here that even in the absence of any optical power, all photodetectors generate some current I_d, which arises from thermally generated carriers. This is known as dark current and increases with increase in temperature. Taking this current into account, we may write the total shot noise generated by the photodetector as

$$\overline{i_{NS}^2} = 2e(I + I_d)\Delta f \tag{13.13}$$

Typical dark current values for different detectors are

$$\begin{aligned} I_d &\simeq 1-10 \text{ nA} && \text{silicon} \\ &\simeq 50-500 \text{ nA} && \text{germanium} \\ &\simeq 1-20 \text{ nA} && \text{InGaAs} \end{aligned} \tag{13.14}$$

Example 13.3: As an example, let us consider a silicon PIN photodiode operating at 850 nm with a typical dark current of 1 nA. For an input optical power of 1 μW, with a responsivity of 0.65 A/W,

$$I = R \cdot P \simeq 0.65\, \mu\text{A}$$

The typical dark current of such a detector is 1 nA, which is very small compared with the signal current of 650 nA and, hence, may be neglected in the shot noise calculation. Hence, for a detector bandwidth of 100 MHz

$$\overline{i_{NS}^2} = 2 \times 1.6 \times 10^{-19} \times 0.65 \times 10^{-6} \times 10^8$$
$$\simeq 2.08 \times 10^{-17}\ \text{A}^2$$

The rms shot noise current is $\sqrt{\overline{i_{NS}^2}} \simeq 4.6$ nA.

13.3.2 Thermal noise

Thermal noise (also referred to as Johnson noise or Nyquist noise) arises in the load resistor of the photodiode circuit due to random thermal motion of electrons. In fact, electrons in any resistor are never stationary but have random motions within the resistor. Since motion of electrons constitutes a current, this random thermal motion leads to the presence of a random current in the resistor. Since the electron motion is random, the average of this current is zero. This thermal noise adds to the signal current generated by the photodetector. The mean square thermal noise current in a load resistor R_L is given by [see, e.g., Yariv (1991)]

$$\overline{i_{NT}^2} = \frac{4k_B T \Delta f}{R_L} \tag{13.15}$$

where $k_B = 1.38 \times 10^{-23}$ J/K is the Boltzmann constant, T is the absolute temperature, and Δf is the bandwidth of detection.

Note that the thermal noise is independent of the incident optical power, unlike shot noise. Also, increasing R_L reduces the thermal noise. R_L cannot be increased indefinitely since the detector bandwidth is determined by R_L (see Section 12.3.2).

Example 13.4: Typical parameter values are $R_L = 500\,\Omega$, $\Delta f =$ 100 MHz, $T = 300$ K, giving

$$\overline{i_{NT}^2} = 3.3 \times 10^{-15}\ \text{A}^2$$

Thus, the rms thermal noise current is 5.75×10^{-8} A. Comparing with Example 13.3, we find that, in such a situation, thermal noise dominates over shot noise.

13.3.3 Signal-to-noise ratio (SNR)

One of the most important parameters in detection is the SNR. It is defined by

$$\text{SNR} = \frac{\text{average signal power}}{\text{total noise power}} \tag{13.16}$$

If P represents the optical power incident on a photodetector with a responsivity R, then the signal current is RP and the electrical signal power is proportional to R^2P^2.

The total noise power is proportional to the total mean square noise current, which is the sum of shot noise and thermal noise terms. Thus

$$\mathrm{SNR} = \frac{R^2P^2}{2e(I + I_d)\Delta f + 4k_BT/R_L \cdot \Delta f} \tag{13.17}$$

In the above equation defining SNR, usually one of the noise terms in the denominator (shot noise or thermal noise) dominates depending upon the operating conditions.

Thus, under shot noise-limited operation, the SNR is given by

$$\mathrm{SNR} = \frac{R^2P^2}{2e(RP + I_d)\Delta f} \quad \text{(shot noise limited)} \tag{13.18}$$

where we have used $I = RP$.

On the other hand, the thermal noise-limited SNR is given by

$$\mathrm{SNR} = \frac{R^2P^2R_L}{4k_BT\Delta f} \quad \text{(thermal noise limited)} \tag{13.19}$$

We assume that the minimum detectable optical power corresponds to the situation when the signal power and noise power are equal. This optical signal power is referred to as noise equivalent power or NEP and is usually quoted in units of $\mathrm{W}/\sqrt{\mathrm{Hz}}$. Assuming that the minimum power will be so low that $I_d \gg I$, we obtain an expression for NEP by putting SNR = 1 in equation (13.17).

$$\mathrm{NEP} = \frac{1}{R}\left(2eI_d + \frac{4k_BT}{R_L}\right)^{1/2} \frac{\mathrm{W}}{\sqrt{\mathrm{Hz}}} \tag{13.20}$$

The above equation shows that for small R_L, the thermal noise dominates. Small values of R_L are needed for large detector bandwidths.

Example 13.5: For a typical silicon detector operating at 850 nm

$$R = 0.65\ \mathrm{A/W}, \quad I_d \simeq 1\ \mathrm{nA}$$

Assuming an R_L of 1000 Ω, $T = 300$ K, we find

$$\mathrm{NEP} \simeq 6.3 \times 10^{-12}\ \mathrm{W}/\sqrt{\mathrm{Hz}}$$

If in this detector, the dark current is the major noise term, then

$$\mathrm{NEP} \simeq \frac{\sqrt{2eI_d}}{R}$$

$$\simeq 2.75 \times 10^{-14}\ \mathrm{W}/\sqrt{\mathrm{Hz}}$$

Example 13.6: We consider a silicon PIN photodiode with $R = 0.65$ A/W, $I_d \simeq 1$ nA, $R_L = 1000\ \Omega$ operating at 850 nm. If the incident optical power is 500 nW and the bandwidth is 100 MHz, then

$$\text{Signal current } I = RP$$
$$= 0.65 \times 5 \times 10^{-7}$$
$$= 0.325\,\mu\text{A}$$

rms shot noise current due to signal

$$= (2eRP\Delta f)^{1/2} \simeq 3.2\,\text{nA}$$

rms shot noise current due to dark current

$$= (2eI_d\Delta f)^{1/2} \simeq 0.18\,\text{nA}$$

rms thermal noise current is

$$= \left(4\frac{k_B T}{R_L}\Delta f\right)^{1/2} \simeq 40.7\,\text{nA}$$

Hence, for this detector the thermal noise current is about 12 times greater than the signal shot noise current and about 225 times greater than dark current.

The SNR corresponding to this incident power is

$$\text{SNR} = 63 \simeq 18\,\text{dB}$$

The above discussion is valid for PIN detectors that do not have any internal gain. As discussed in Section 12.4, APDs have an internal gain due to the avalanche process. Hence, if M represents the internal gain then for an input optical power P the signal current is

$$I = MRP \tag{13.21}$$

and the electrical signal power is proportional to $M^2R^2P^2$.

The thermal noise for APDs is the same as for a PIN and is again given by equation (13.15).

With regard to shot noise, we first note that since the avalanche multiplication process is random, the multiplication factor M is itself random and the M appearing in equation (13.21) is only an average value. Thus, for an APD, the shot noise can be written as

$$\overline{i_{NS}^2} = M^{2+x}\,2e(RP + I_d)\,\Delta f \tag{13.22}$$

where the excess noise factor by which the shot noise term has increased with respect to the signal power is $M^{2+x}/M^2 = M^x$; $x = 0$ corresponds to no excess noise. Typically, for silicon APDs $x \simeq 0.3$, for InGaAs APDs $x \simeq 0.7$, and for germanium APDs $x \simeq 1.0$.

In the case of an APD, we have

$$\mathrm{SNR} = \frac{M^2 R^2 P^2}{2eM^{2+x}(RP + I_d)\,\Delta f + \dfrac{4k_B T}{R_L}\cdot \Delta f} \tag{13.23}$$

Example 13.7: Let us consider an APD with $M = 50$ and with $x = 0$ (i.e., no excess noise). If we again use the same parameters as in Example 13.6, then in this case

$$\begin{aligned}\text{Signal current} &= 50 \times 0.65 \times 5 \times 10^{-7}\\ &\simeq 16.25\,\mu\mathrm{A}\end{aligned}$$

rms shot noise current due to signal

$$= (2eM^2 RP\Delta f)^{1/2} \simeq 161\,\mathrm{nA}$$

Similarly, rms dark current noise is

$$= (2eM^2 I_d \Delta f)^{1/2} \simeq 8.9\,\mathrm{nA}$$

The thermal noise remains the same as before and is equal to 40.7 nA.

The SNR in this case is given by

$$\begin{aligned}\mathrm{SNR} &= \frac{(16.25 \times 10^3)^2}{[(161)^2 + (8.9)^2 + (40.7)^2]}\\ &\simeq 9548 \simeq 39.8\,\mathrm{dB}\end{aligned}$$

a very significant improvement over the 18-dB SNR of PIN (see Example 13.6).

If thermal noise is dominant over shot noise, then

$$\mathrm{SNR} = \frac{M^2 R^2 P^2 R_L}{4k_B T\Delta f} \tag{13.24}$$

and we see that the SNR has indeed improved by a factor M^2 with respect to a PIN detector (see equation 13.19). On the other hand, if M becomes large, then the detector gets shot noise limited and if we neglect thermal noise, we obtain

$$\mathrm{SNR} = \frac{R^2 P^2}{2eM^x (RP + I_d)\,\Delta f} \tag{13.25}$$

which is worse than a PIN detector due to the excess noise M^x (see equation 13.18). Thus, under thermal noise dominant operation, an APD has better SNR and, hence, is more attractive than a PIN.

We can rewrite equation (13.23) in the form

$$\mathrm{SNR} = \frac{R^2 P^2}{2eM^x (RP + I_d)\,\Delta f + \dfrac{4k_B T}{R_L}\Delta f M^{-2}} \tag{13.26}$$

from which we note that the second term in the denominator (corresponding to thermal noise) reduces with increasing M and the first term (corresponding to shot noise) increases with increasing M. Thus, there is an optimum value of the multiplication factor M to achieve maximum SNR. Maximum SNR will correspond to a minimum in denominator of equation (13.26) with respect to M. Differentiating the denominator and equating to zero, we obtain the optimum M value as

$$M_{op}^{x+2} = \frac{4k_BT}{xeR_L(RP + I_d)} \tag{13.27}$$

Note that the optimum M value depends on the input power if the dark current noise is small.

Example 13.8: Let us consider a silicon APD operating at 300 K, with $R_L = 1000\,\Omega$ and an input optical power of 100 nW. Typically, $R = 0.65$ A/W and $x = 0.3$. Thus, neglecting dark current, we have from equation (13.27)

$$M_{op} \simeq 42$$

The corresponding SNR for $\Delta f = 100$ MHz is

$$\begin{aligned} \text{SNR} &= 577 \\ &\simeq 27.6\,\text{dB} \end{aligned}$$

The corresponding SNR of a PIN (with $M = 1$) would be

$$\text{SNR} \simeq 2.5 \simeq 4 \text{ dB}$$

Thus, in the present case, the SNR of the APD is improved by 23.6 dB in comparison to PIN.

Example 13.9: As another example, let us consider a germanium APD with a responsivity $R = 0.45$ A/W at 1300 nm operating at 300 K. For germanium $x = 1$ and thus

$$M_{opt} = \left[\frac{4k_BT}{eR_L(RP + I_d)}\right]^{1/3}$$

If $P = 500$ nW, neglecting I_d, we have for $R_L = 1000\,\Omega$,

$$M_{opt} \simeq 7.7$$

which is lower than for silicon.

Example 13.10: For a typical InGaAs APD operating at 1550 nm

$$R \simeq 0.6\,\text{A/W}, \quad x = 0.7, \quad T = 300\,\text{K}, \quad R_L = 1000\,\Omega$$

Neglecting dark current, we have for $P = 100$ nW

$$M_{opt} \simeq 18$$

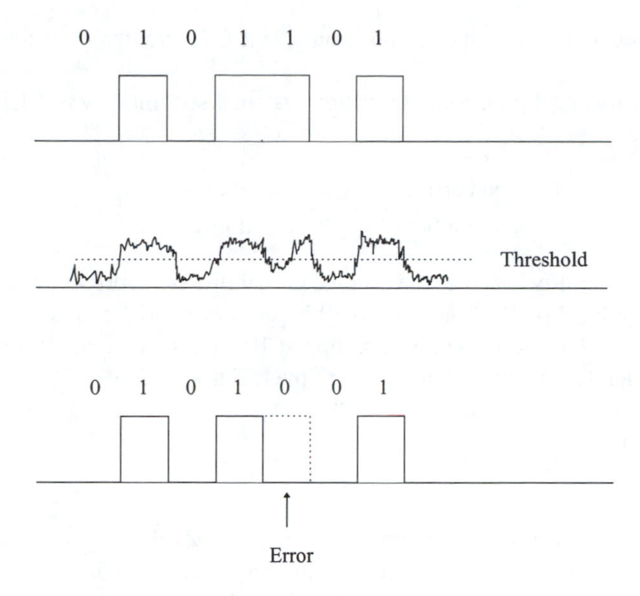

Fig. 13.11: The transmitted pulse sequence 0101101, the received signal corrupted by noise and dispersion, and the reformed pulse sequence. Note that due to the presence of noise, one bit has been read incorrectly.

13.4 Bit error rate (BER)

The purpose of a regenerator or a receiver is to sample the incoming optical pulses at a rate equal to the bit rate of transmission and at each of the samples to decide whether it corresponds to a one or a zero. This decision is usually performed by setting a threshold level, and any signal above the set threshold is taken as one and any signal below the set threshold is taken as zero. Once this is done, the signal stream consisting of ones and zeroes is retrieved in a receiver or is again used to drive an optical source to recreate the optical pulse stream as in a regenerator.

It is now obvious that if there is insufficient optical power in the received optical pulses (due to fiber attenuation), if there has been a large dispersion, or if too much noise is added by the detector, then there could be errors in the retrieved information or in the reformed optical pulse stream. Thus, although the optical pulse stream starts as a fresh sequence of clean pulses, at every regenerator, one would, in principle, add errors to the signal. It is desirable to keep the error rates below 10^{-9} or 10^{-12} at every regenerator or at the receiver in any practical communication system.

In retrieving the information at the receiver or in the regenerator, it is also necessary that the pulses are sampled at the correct rate. Thus, the decision points must remain in correct phase with respect to the incoming pulse train; otherwise, some information may be lost. The clock pulses necessary for the decision operation are usually derived from the incoming pulse train itself. Thus, it is necesssary to ensure that the incoming pulse train has sufficient energy content at the frequency corresponding to the bit rate; otherwise, other techniques are used to obtain this information from other frequency components in the input signal.

Figure 13.11 shows the transmitted pulse sequence consisting of 0101101, the received signal corrupted by noise and dispersion, and the reformed pulse sequence. As shown, due to the presence of noise, the transmitted bits may be read incorrectly if the received signal at the position of bit 1 is below threshold or if the received signal at the position of bit 0 is greater than the threshold.

The quality of a digital communication system is specified by its BER, which is defined as

$$\mathrm{BER} = \frac{\text{bits read erroneously in a duration } \tau}{\text{Total number of bits received in } \tau} \tag{13.28}$$

The BER essentially specifies the average probability of incorrect bit identification. Thus, a BER of 10^{-9} means that 1 bit out of every 10^9 bits is, on average, read incorrectly. If the system is operating at 100 Mb/s – that is, 10^8 pulses per second – then to receive 10^9 pulses, the time taken would be

$$\frac{10^9}{10^8} \simeq 10\,\mathrm{s}$$

which is the average time for an error to occur. On the other hand, if the BER is 10^{-6}, then, on average, an error would occur every 0.01 s, which is unacceptable.

It is now obvious that the higher the SNR, the lower would be the corresponding BER. For most PIN receivers, the noise is dominated by thermal noise, which, as we have seen earlier, is independent of the signal current. Thus, the noise in bit 1 and bit 0 is the same and in such a case the optimum setting of threshold value is at the midpoint of the one and zero levels and the BER is related to SNR through the equation [see, e.g., Palais (1988)]

$$\mathrm{BER} = \frac{1}{2}\left[1 - \mathrm{erf}\left(\frac{\sqrt{\mathrm{SNR}}}{2\sqrt{2}}\right)\right] \tag{13.29}$$

where erf represents the error function. For $x > 3$, a very good approximation to $\mathrm{erf}(x)$ is

$$\mathrm{erf}(x) = 1 - \frac{1}{\sqrt{\pi}x} e^{-x^2} \tag{13.30}$$

Thus, for $\mathrm{SNR} \geq 72$, equation (13.29) can be approximated by

$$\mathrm{BER} \simeq \left(\frac{2}{\pi \cdot \mathrm{SNR}}\right)^{1/2} e^{-\mathrm{SNR}/8} \tag{13.31}$$

For achieving a BER of 10^{-9}, equation (13.31) predicts an SNR $\simeq$ 144 or 21.6 dB.

Figure 13.12 shows the dependence of BER on SNR as given by equation (13.29). Note that since the curve decreases sharply beyond an SNR of 15 dB, one can achieve a large improvement in detection even with small increments in SNR beyond 15 dB.

Under thermal noise-limited operation, one needs a certain minimum SNR for achieving a specific BER (see equation (13.29)). Using equation (13.19) for the SNR for a PIN detector under thermal noise-limited operation, we obtain the following equation giving us the minimum optical power for achieving a

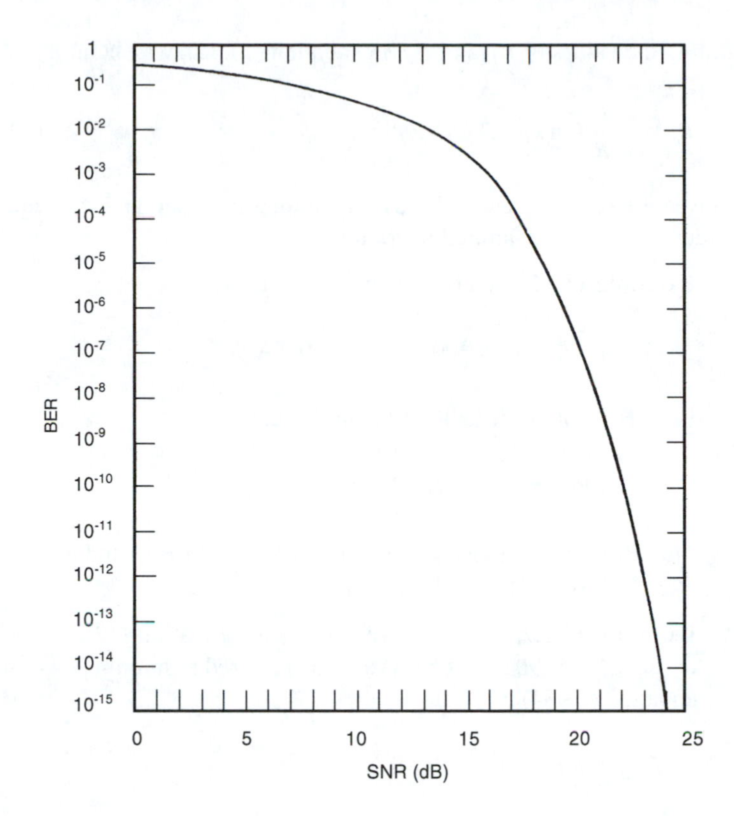

Fig. 13.12: Dependence of BER on SNR as given by equation (13.29).

certain SNR

$$\frac{R^2 P_{min}^2 R_L}{4k_B T \Delta f} = \text{SNR}$$

or

$$P_{min} = \frac{1}{R}\left(\frac{4k_B T \Delta f}{R_L}\text{SNR}\right)^{1/2} \tag{13.32}$$

Assuming $R_L = 160\,\Omega$ so that, with a detector capacitance of 1 pF, the corresponding bandwidth $\Delta f = 1/(2\pi R_L C) \simeq 1$ GHz, we have for $R = 0.5$ A/W, SNR = 144 (for a BER of 10^{-9}),

$$P_{min} \simeq 8.0\,\mu\text{W} \simeq -21\,\text{dBm}$$

Equation (13.32) can be put in terms of bit rate B by noting that, for NRZ systems, the bandwidth $\Delta f = B/2$ (see Section 13.2). Also, if C represents the photodiode capacitance, then

$$\Delta f = \frac{1}{2\pi R_L C} = \frac{B}{2}$$

or

$$R_L = \frac{1}{\pi c B} \tag{13.33}$$

Substituting the values of R_L and Δf in equation (13.32), we obtain

$$P_{min} = \frac{B}{R}(2\pi k_B T C \cdot \text{SNR})^{1/2} \tag{13.34}$$

which gives the receiver sensitivity as a function of bit rate and the required SNR under thermal noise-limited operation.

Example 13.11: As a typical example we have

$$C = 1\,\text{pF}, \quad T = 300\,\text{K}, \quad R = 0.5\,\text{A/W}$$

For a BER of 10^{-9}, SNR=144 and we obtain

$$P_{min}\,(\text{nW}) \simeq 3.87 \times B\,\text{Mb/s}$$

Thus, for 100 Mb/s, $P_{\text{min}} \simeq 0.39\,\mu\text{W}$ ($\simeq -34.1$ dBm) and for 1 Gb/s, $P_{\text{min}} \simeq 3.87\,\mu\text{W}$ ($\simeq -24.1$ dBm).

Example 13.12: For a BER of 10^{-6}, from equation (13.31), the required SNR is 90 (or 19.5 dB) and the required minimum power from equation (13.34) is

$$P_{min}\,(\text{nW}) \simeq 3.06 \times B\,\text{Mb/s}$$

Thus, for 100 Mb/s, $P_{\text{min}} \simeq 0.306\,\mu\text{W}$ ($\simeq -35.1$ dBm). Comparing with the previous example, we note that the BER can be reduced from 10^{-6} to 10^{-9} by increasing the power level by just 1 dB. Thus, even small reductions of loss in a system can result in significant improvements in detection.

In the case of APDs, if thermal noise remains dominant, then we see from equation (13.24) that the SNR is improved by a factor M^2. Thus, from equation (13.34) we find that P_{min} for the case of APDs reduces by a factor M. Since M is typically in the range 20–50, the improvement could be about 15 dB. In reality, since the shot noise also increases with increasing M, there is an optimum value of M for achieving maximum improvement.

In the preceding discussion we assumed a thermal noise-limited detection. The ultimate in detection sensitivity is provided by a shot noise-limited operation. For an ideal detector (with no thermal noise, no dark current, and unity quantum efficiency) one can show that [see, e.g., Agrawal (1993)]

$$\text{BER} = \frac{e^{-N_P}}{2} \tag{13.35}$$

where N_P is the average number of photons in bit 1. For a BER of 10^{-9}, equation (13.35) gives us $N_P = 21$. Thus, the corresponding minimum optical power in the bit 1 (of duration T) is given by

$$P = \frac{N_P h\nu}{T}$$

For an NRZ system, since the bit rate B and the pulse duration T are related by $B = 1/T$, we have

$$P = h\nu N_P B \tag{13.36}$$

For a bit rate of 1 Gb/s, at 1550 nm, we have

$$P \simeq 2.69\,\text{nW} \simeq -55.7\,\text{dBm}$$

The limit $N_P = 21$ is referred to as the quantum limit. Most receivers operate at 20 dB higher levels with $N_P \simeq 2000$.

In the above discussions P represents the peak power. The relationship is also sometimes written in terms of average power. Since 1s and 0s are equally likely to occur $P_{av} = P/2$. Thus

$$P_{av} = \frac{N_P h\nu B}{2} \tag{13.37}$$

We may note that for an RZ system, the pulse duration is $T/2 = 1/2B$ and, since the pulse occupies only one-half of the bit period, and 1s and 0s are equally likely $P_{av} = P/4$. Thus, even for an RZ system, P_{av} required under quantum-limited operation is still given by equation (13.37).

13.5 System design

The simplest fiber optic communication link is the point-to-point link wherein a transmitter at one end sends information along an optical fiber link to a receiver at the other end. The design of such a system involves many aspects such as the type of source to be used (LED or LD), the kind of fiber to be employed (multimode or single mode), and the photodetector (PIN or APD). The choice of various components depends on the distance between the transmitter and the receiver station as well as the information rate. Apart from these, issues of cost, reliability of components, possibility of upgrading, and so forth are also of importance.

In earlier sections we have seen how the power level falling on a detector determines the performance of the link in terms of the BER. Also, the rise times of the source, fiber, and the detector will determine the bandwidth available for transmission. The design of a fiber optic system is usually carried out using power budgeting and rise time budgeting.

13.5.1 Power budgeting

A typical point-to-point fiber optic link is shown in Figure 13.1. The source emitting a power P_i couples light into the optical fiber. At this point there is a coupling loss. The light propagates through the fiber, suffering transmission loss. In the link, one has splices and connectors where there are power losses. Finally, the light reaches the detector (in the regenerator or receiver), where again the beam suffers loss. If P_i and P_0 are the powers emitted by the source and that falling on the detector, then the total loss in decibels is given by

$$\text{Loss} = 10\log\left(\frac{P_0}{P_i}\right) \tag{13.38}$$

Apart from the actual losses suffered, while designing a link one usually incorporates a margin of 6–8 dB to account for losses from splices or components that may have to be added at a future date and also to allow for any deterioration of components due to aging.

If all losses are expressed in decibels, then the power received by the receiver for a source power P_i (in dBm) is

$$P_0 = P_i - N_c l_c - N_s l_s - L\alpha_t \tag{13.39}$$

where l_c is the connector loss and N_c represents the number of connectors, l_s is the loss at every splice and N_s represents the number of splices, α_t is the fiber transmission loss (in dB/km), and L represents the total fiber length in kilometers.

We have seen earlier that for achieving a certain BER, there is a minimum value of power (P_{min}) falling on the detector. Thus, if P_m represents the power margin (typically, 6–8 dB), then we must have

$$P_0 - P_m > P_{min} \tag{13.40}$$

Example 13.13: Let us assume that we need to install a link 40 km long with a fiber having a loss of 0.5 dB/km. Let us assume that the receiver sensitivity is −39 dBm. There are four splices with the loss at each splice being 0.5 dB and two connectors with a loss of 1 dB each. Assuming a margin of 6 dB, the source power must exceed

$$P_{min} + P_m + 2l_c + 4l_s + L\alpha_t = (-39 + 6 + 2 \times 1 + 4 \times 0.5 + 40 \times 0.5)\,\text{dBm}$$

$$-9\,\text{dBm} = 0.13\,\text{mW}$$

Example 13.14: Let us consider a pigtailed LED source at 850 nm emitting 50 μW from the pigtail and a silicon PIN detector with $R = 0.65$ A/W, $C = 5$ pF. Let us assume that we wish to transmit information at 20 Mb/s. For this bit rate, assuming thermal noise-limited operation with an SNR of 144 (giving BER $= 10^{-9}$), we obtain the receiver sensitivity as (see equation (13.34))

$$P_{min} \simeq 1.32 \times 10^{-7}\,\text{W} \simeq -38.8\,\text{dBm}$$

The power emitted by the source is 50 μW $= -13$ dBm. Thus, the permissible loss between the transmitter and receiver (including margin) is $38.8 - 13 = 25.8$ dB. If the system margin is 8 dB and connector loss is assumed to be 2 dB, the available transmission loss is 15.8 dB. If a fiber with a loss of 2.5 dB/km (at 850 nm) is used, then the maximum permissible link length is ≃6.3 km.

If in this system a laser emitting a power of 0 dBm (=1 mW) is used, then the maximum permissible fiber length would be ≃11.5 km.

13.5.2 Rise time budgeting

Rise time budgeting is a convenient analysis to determine whether the proposed system will be able to operate properly at the required bit rate on account of dispersion of the link and limited speed of response of the transmitter and the receiver.

As discussed in Section 13.2, the rise time of a device is the time taken for the response to increase from 10% to 90% of the final output value when the input is changed abruptly in a step-like fashion. The total rise time τ_s of a combination of the various elements of the link is given approximately in terms of individual rise times τ_i by the equation

$$\tau_s \simeq \left(\sum_{i=1}^{N} \tau_i^2\right)^{1/2} \tag{13.41}$$

In a fiber optic communication system, the total link rise time is determined by the rise times of the transmitter (τ_t), the fiber link (τ_f), and the receiver (τ_r). Thus, the system rise time is

$$\tau_s \simeq \left(\tau_t^2 + \tau_f^2 + \tau_r^2\right)^{1/2} \tag{13.42}$$

In Chapters 11 and 12 we discussed the rise time of the optical source and of the optical detector.

In the case of the rise time of the fiber, we approximate the fiber rise time by the pulse dispersion. Thus, the rise times for different types of fibers can be written as follows

multimode step index fibers

$$\tau_{im} = \frac{n_1 \Delta}{c} L \tag{13.43}$$

multimode graded fibers

parabolic index

$$\tau_{im} = \frac{n_1}{2c} \cdot \Delta^2 L \tag{13.44}$$

optimum profile

$$\tau_{im} = \frac{n_1}{8c} \cdot \Delta^2 L \tag{13.45}$$

(where the subscript *im* stands for intermodal)

material dispersion

$$\begin{aligned} \tau_m &\simeq 85 L \Delta\lambda \text{ ps } (\lambda_0 \sim 850 \text{ nm}) \\ &\lesssim 0.5 L \Delta\lambda \text{ ps } (\lambda_0 \sim 1300 \text{ nm}) \\ &\simeq 20 L \Delta\lambda \text{ ps } (\lambda_0 \sim 1500 \text{ nm}) \end{aligned} \tag{13.46}$$

with L in kilometers and $\Delta\lambda$ in nanometers.

The total fiber rise time for multimode fibers is (neglecting waveguide dispersion)

$$\tau_f = \left(\tau_{im}^2 + \tau_m^2\right)^{1/2} \tag{13.47}$$

For single-mode fibers, we have (since $\tau = D \cdot L \cdot \Delta\lambda$),

$$\begin{aligned} \tau_f &\simeq D \cdot L\Delta\lambda \\ &\simeq 2L\Delta\lambda(\lambda_0 \sim 1300\ \text{nm},\ \lambda_z \sim 1300\ \text{nm}) \\ &\simeq 16L\Delta\lambda\,(\lambda_0 \sim 1550\ \text{nm},\ \lambda_z \sim 1300\ \text{nm}) \\ &\simeq 2L\Delta\lambda(\lambda_0 \sim 1550\ \text{nm},\ \lambda_z \sim 1550\ \text{nm}) \end{aligned} \tag{13.48}$$

with $D \simeq 2$ ps/km · nm at 1300 nm and $\sim$ 16 ps/km·nm at 1550 nm with fibers with $\lambda_z \sim$ 1300 nm. Also $D \sim 2$ ps/km·nm at 1550 nm for fibers with $\lambda_z \sim$ 1550 nm. In the above equations $\Delta\lambda$ represents the source spectral width, L the fiber length, λ_0 the operating wavelength, and λ_z the zero dispersion wavelength.

Once the total rise time of the link is calculated using equation (13.42), one can then obtain the maximum permissible bit rate through the link as (see equation (13.7))

$$\begin{aligned} B &\lesssim \frac{0.35}{\tau_s};\ \text{RZ} \\ &\lesssim \frac{0.7}{\tau_s};\ \text{NRZ} \end{aligned} \tag{13.49}$$

(1) We first note that for multimode step index fibers, the pulse spread is very nearly independent of the spectral width of the source, since the intermodal dispersion is very much larger than material dispersion. Taking typical values of $n_1 = 1.46,\ \Delta = 0.01$, we find that the rise time due to the fiber is $\tau_f \simeq 50$ ns/km. Using equation (13.49) we can see that the maximum bit rate that such a fiber can handle is $0.7/50 \times 10^{-9} \simeq 14$ Mb·km/s.

(2) We next consider a parabolic index multimode fiber for which the intermodal dispersion is given by (see equation (13.44))

$$\tau_{im} = \frac{n_1}{2c}\Delta^2 L$$

For $n_1 \simeq 1.46,\ \Delta = 0.01$, we have $\tau_{im} \simeq 0.24$ ns/km. For an operating wavelength of 850 nm, the material dispersion is given by equation (13.46). If the source is a laser diode with a spectral width of $\simeq$1 nm, then the contribution of material dispersion is 85 ps/km and the total rise time of the fiber $\simeq (0.24^2 + 0.085^2)^{1/2} \simeq 0.25$ ns/km, giving a maximum permissible bit rate of $0.7/0.25 \times 10^{-9} \simeq 2.8$ Gb·km/s. On the other hand, if the source is an LED with $\Delta\lambda \simeq 25$ nm, then $\tau_m \simeq$ 2.125 ns/km and then the total fiber rise time is $(0.25^2 + 2.125^2)^{1/2} \simeq$ 2.14 ns/km. Thus, in this case the rise time is limited due to material dispersion and the maximum bit rate is $0.7/2.14 \times 10^{-9} \simeq 300$ Mb·km/s.

The above estimation neglects contribution due to source and detector rise times. Indeed, if we assume a typical source (LED) rise time of 5 ns and a detector rise time of 1 ns, the total system rise time for a fiber length of 10 km is

$$[(2.14 \times 10)^2 + 5^2 + 1^2]^{1/2} \simeq 22\,\text{ns}$$

giving a maximum bit rate (for 10 km) of 32 Mb/s.

The I generation fiber optical communication systems used 850-nm LEDs (with $\Delta\lambda \sim 25$ nm), with fiber losses of $\simeq$3 dB/km, repeater spacing of $\sim$10 km, and bit rates of 45 Mb/s.

(3) If the operating wavelength is around 1300 nm, the material dispersion becomes very small and, thus, even with $\Delta\lambda \sim 100$ nm, the rise time due to material dispersion $\simeq 0.05$ ns/km, which is negligible compared with the intermodal dispersion of 0.15 ns/km.

Again, assuming a source (LED) rise time of 5 ns and a detector rise time of 1 ns, the total rise time for 30 km is $[(0.15 \times 30)^2 + 5^2 + 1^2]^{1/2} \simeq 6.8$ ns, giving a maximum bit rate of about 100 Mb/s.

The II generation fiber optic systems used 1300-nm LEDs (with $\Delta\lambda \sim 25$ nm), with fibers possessing losses $\sim$1 dB/km, repeater spacings of 30 km, and bit rates of $\sim$45 Mb/s.

(4) The next shift was to use single-mode fibers in place of multimode fibers and operating around 1300 nm where the dispersion passes through zero. Assuming a typical dispersion of 2 ps/km·nm at the operating wavelength, using a laser diode with $\Delta\lambda = 2$ nm, the fiber rise time is only 4 ps/km. If the source and detector rise times are assumed to be given by 0.5 ns each, then assuming a fiber length of 50 km, the total system rise time is $(0.2^2 + 0.5^2 + 0.5^2)^{1/2} \simeq 0.73$ ns, corresponding to a maximum bit rate of 0.96 Gb/s.

The III generation fiber optic systems used 1300-nm laser diodes (with $\Delta\lambda \sim 2$ nm), with single-mode fibers having losses $\sim$ 0.8 dB/km. The repeater spacing was about 40 km and the bit rate was about 500 Mb/s.

(5) When the operating wavelength is shifted to 1550 nm, the total dispersion can again be made very small ($\lesssim$2 ps/km·nm) with dispersion-shifted fibers (DSFs). The fiber loss at this wavelength is $\sim$0.25 dB/km. Thus, the IV generation fiber optic systems use 1550-nm laser diodes with DSFs and repeater spacings of 100 km. The systems operate at 2.5 Gb/s and higher by using high-speed laser diodes and photodetectors.

Table 13.1 gives the evolution of the various generations of fiber optic systems.

Example 13.15: Let us consider the rise time budgeting of a 400-Mb/s NRZ transmission over 100 km with a BER of 10^{-9}. Now, since $B = 400$ Mb/s, the total system rise time is (under NRZ transmission)

$$\tau_s = \frac{0.7}{B} \simeq 1.75\,\text{ns} \tag{13.50}$$

Table 13.1. *Typical characteristics of fiber optic communication systems at different stages*

Generation	Date	Bit rate	Type of fiber	Loss (dB/km)	Repeater spacing (km)
I (0.8–0.9 μm)	1977	$\sim$45 Mbit/s	Multimode (graded index)	$\sim$3	$\sim$10
II (1.3 μm)	1981	$\sim$45 Mbit/s	Multimode (graded index)	$\sim$1	$\sim$30
III (1.3 μm)	At present	$\sim$2.5 Gb/s	Single mode	$\leq$0.5	$\sim$40
IV (1.55 μm)	At present	$\geq$10 Gb/s	Single mode	<0.3	$\geq$100

Note: Futuristic system: WDM, solitons. Infrared fibers ($\lambda_0 > 2\ \mu$m); extremely low loss ($<10^{-2}$ dB/km); $\Rightarrow$ repeater spacing >1000 km.

Even if we allocate all the rise time to 100 km of fiber, this implies a pulse dispersion of less than $1.75 \times 10^{-9}/100 = 17.5$ ps/km. Surely, multimode fibers cannot be used and the link has to be based on single-mode fibers. Because, at 1300 nm, the typical fiber loss is 0.4 dB/km, 100 km of fibers would result in 40 dB of loss (apart from connector and splice losses), which is too large. Hence, it would be necessary to use 1550-nm transmission.

A 1300-nm zero dispersion fiber has about 16 ps/km·nm at 1550 nm. In this case one cannot employ the multifrequency semiconductor lasers since their spectral width is typically 4 nm and, clearly, 100 km of the fiber would lead to a pulse dispersion of $16 \times 100 \times 4 = 6.4$ ns, which is much larger than the permissible value of 1.75 ns for a 400-Mb/s system. Hence, one has to use single-frequency laser diodes, which have typical spectral widths of 0.15 nm. Using such lasers, the pulse dispersion due to the fiber would be $16 \times 100 \times 0.15 = 0.24$ ns, which is much smaller than the total permissible rise time of 1.75 ns.

If we assume a typical rise time of 1 ns for the laser transmitter, then the permissible rise time of the photodiode is

$$\begin{aligned}\tau_r &= \left(\tau_s^2 - \tau_f^2 - \tau_t^2\right)^{1/2} \\ &= (1.75^2 - 0.24^2 - 1^2)^{1/2} \\ &\simeq 1.42\,\text{ns}\end{aligned}$$

If the capacitance of the photodiode is C and the load resistance is R_L, then [see equation (13.9)]

$$\tau_r \simeq 2.2 R_L C$$

giving us

$$R_L \simeq 645\,\Omega$$

assuming $C = 1$ pF.

For power budgeting let us assume that the pigtail laser diode has an output power of 1 mW (= 0 dBm). If the system has two connectors (with 1 dB loss at each connector) and 10 splices of 0.1 dB each, assuming a fiber attenuation of 0.25 dB/km at 1550 nm, the power reaching the detector would be

$$P_{rec} = (0 - 2 - 10 \times 0.1 - 100 \times 2.5)\,\text{dBm}$$

$$= -28\,\text{dBm}$$

At the bit rate of 400 Mb/s with $R = 0.65$ A/W, $C = 1$ pF, and an SNR = 144 (for BER of 10^{-9}), the sensitivity of PIN [from equation (13.34)] is -29.2 dBm. Thus, there is a margin of only 1.2 dB. An APD can give an improvement of 10 dB (under thermal noise-limited conditions) and, hence, can be used in the present link.

13.6 Maximum transmission distance due to attenuation and dispersion

In the following we obtain the maximum unrepeatered link length as imposed by fiber attenuation and dispersion. These are not fundamental limits as they can be surpassed by using components such as optical amplifiers, dispersion compensators, etc.

13.6.1 Attenuation limit

As discussed in Section 13.5, for the transmitted data represented by a digital pulse sequence to be detected with a BER of less than a certain value (typically 10^{-9}), there must be a minimum number of photons per bit of information. If this number is N_P, then for a bit rate B, the minimum average optical power received is [see equation (13.37)]

$$P_r = \frac{N_P B h\nu}{2} \tag{13.51}$$

where $h\nu$ is the energy of each received photon.

If α represents the loss coefficient of the fiber in dB/km, then for an incident power P_i, the optical power at any length L of the fiber is given by

$$P_0 = P_i\, 10^{-\alpha L/10} \tag{13.52}$$

Thus, if the received power is to be at least P_r, then the maximum permissible length would be

$$L_{max} = \frac{10}{\alpha} \log\left(\frac{P_i}{P_r}\right)$$

$$= \frac{10}{\alpha} \log\left(\frac{2P_i}{N_P B h\nu}\right) \tag{13.53}$$

Note that the maximum permissible loss-limited length decreases with increased bit rate.

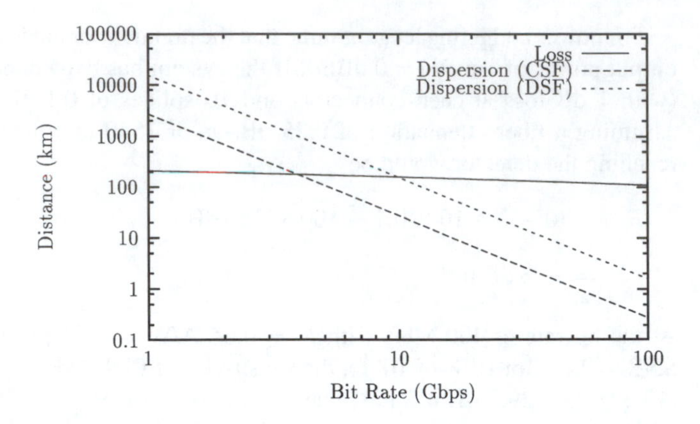

Fig. 13.13: Maximum transmission distance as determined by attenuation and dispersion.

Example 13.16: Let us first consider a system operating at 1300 nm with a fiber loss of $\alpha \simeq 0.4$ dB/km. Assuming an input power of 0 dBm ($P_i = 1$ mW) and $N_P = 1000$, the maximum length for a bit rate of 2.5 Gb/s is

$$\begin{aligned} L_{max} &= \frac{10}{0.4} \log\left(\frac{2 \times 10^{-3} \times 1.3 \times 10^{-6}}{1000 \times 2.5 \times 10^{9} \times 6.63 \times 10^{-34} \times 3 \times 10^{8}} \right) \\ &\simeq 93\,\text{km} \end{aligned}$$

Example 13.17: For a system operating at a 1550 nm, $\alpha \simeq 0.2$ dB/km and assuming $P_i = 1$ mW, $N_P = 1000$, the maximum repeaterless link length at 2.5 Gb/s is

$$L_{max} \simeq 190\,\text{km}$$

The above two examples clearly show the advantage of using a 1550-nm wavelength where the fiber attenuation is minimum for silica fibers.

Substituting $P_i = 1$ mW, $N_P = 1000$, $\alpha = 0.2$ dB/km, for an operating wavelength of 1550 nm, equation (13.53) becomes

$$L_{max} \simeq (660 - 50 \log B)\,\text{km} \tag{13.54}$$

The limit as determined by the above equation is plotted in Figure 13.13.

Example 13.18: At 850 nm, if we assume a typical fiber loss of 2.5 dB/km, the loss limited length at 100 Mb/s is

$$L_{max} \simeq 20\,\text{km}$$

13.6.2 Dispersion limit

We have seen earlier that, apart from attenuation, broadening also limits the repeaterless span length of a fiber optic communication system. Pulse broadening causes adjacent pulses to overlap, resulting in increased errors. A commonly used criterion for the maximum allowed pulse dispersion is

$$\Delta\tau \leq \frac{T}{4} \tag{13.55}$$

where T is the bit duration. In terms of bit rate $B(= 1/T)$, equation (13.55) can be rewritten as

$$4\,\Delta\tau B \leq 1 \tag{13.56}$$

The above condition can be shown to imply that, at least for Gaussian pulse shapes, more than 95% of the pulse energy lies within the bit slot T. The condition imposed by equation (13.56) is more conservative than equation (13.49).

We consider single-mode fibers for which the dispersion is given by

$$\Delta\tau = DL\Delta\lambda \tag{13.57}$$

where D is the dispersion coefficient, L is the fiber length, and $\Delta\lambda$ represents the spectral width of the source.

We first assume that the spectral width of the source is much larger than the spectrum due to modulation. This is true in the case of modulation of a multilongitudinal mode laser.

Substituting for $\Delta\tau$ in equation (13.56), we obtain the bit rate length product as

$$B \cdot L \leq \frac{1}{4D\Delta\lambda} = \frac{250}{D(\text{ps/km·nm})\Delta\lambda(\text{nm})}\ \text{Gb/s·km} \tag{13.58}$$

For a conventional single-mode fiber with zero dispersion at 1300 nm, we take as typical parameters

$$\Delta\lambda = 1\ \text{nm}, \quad D \sim 1\ \text{ps/km·nm}$$

Thus

$$B \cdot L < 250\ \text{Gb/s·km}$$

which implies that at 2.5 Gb/s, the maximum repeater spacing is 100 km.

Operating this fiber at 1550 nm, where $D \sim 16$ ps/km·nm, we have (assuming $\Delta\lambda \simeq 1$ nm),

$$B \cdot L \lesssim 15\ \text{Gb/s·km}$$

Thus, at 2.5 Gb/s, the maximum unrepeated length is only 6 km. This shows the enormous reduction in repeaterless length due to increased dispersion.

We next assume that the laser is a single-frequency laser (such as a DFB laser) and assume that the spectral width due to modulation is much larger than the inherent source spectral width. Now, if τ_0 represents the temporal full width of the input pulse (assumed to be unchirped and Fourier transform limited), then the spectral width is

$$\Delta\nu \sim \frac{1}{\tau_0} \tag{13.59}$$

which in terms of wavelength becomes

$$\Delta\lambda \simeq \frac{\lambda_0^2}{c\tau_0} \tag{13.60}$$

Assuming a NRZ pulse sequence (see Section 13.2),

$$\tau_0 = \frac{1}{B} \tag{13.61}$$

Using equations (13.60), (13.61) and (13.57) in equation (13.56), we obtain

$$B^2 \cdot L \lesssim \frac{c}{4D\lambda_0^2} \tag{13.62}$$

Note that, in this case, doubling the bit rate would reduce the unrepeated length by a factor of 4 in contrast to the case in which the source spectral width is large (equation (13.58)), wherein doubling the length reduces B by only a factor of 2.

Again, consider a conventional single-mode fiber operating at 1300 nm with $D \sim 1$ ps/km·nm,

$$B^2 L \lesssim 4.4 \times 10^4\,(\text{Gb/s})^2\text{·km}$$

Thus, at 2.5 Gb/s, $L \lesssim 7040$ km and at 10 Gb/s, $L \lesssim 440$ km.

For the 1300-nm fiber operating at 1550 nm, $D \sim 16$ ps/km·nm and,

$$B^2 L \lesssim 2750\,(\text{Gb/s})^2\text{·km}$$

Thus, at $B = 2.5$ Gb/s, $L \lesssim 440$ km and for $B = 10$ Gb/s, $L \lesssim 27.5$ km.

If we use DSFs operating at 1550 nm, we then have $D \sim 1$ ps/km·nm and

$$B^2 L \lesssim 3.12 \times 10^4\,(\text{Gb/s})^2\text{·km}$$

Thus, at $B = 2.5$ Gb/s, $L \lesssim 4992$ km and for $B = 10$ Gb/s, $L \lesssim 312$ km.

The limits on unrepeated length as imposed by pulse dispersion in single-mode fibers is plotted in Figure 13.13. One can see that using DSF, even at 10 Gb/s, optical fiber systems are loss limited rather than dispersion limited.

Problems

13.1 Consider an InGaAs PIN photodiode with the following specifications as 1300 nm:

$$R = 0.8\ \text{A/W}, \quad C = 1\ \text{pF}, \quad I_d = 5\ \text{nA}$$

If the load resistance is 1000 Ω and the incident power is 500 nW, obtain the signal current and the various noise currents for bandwidth $\Delta f = 20$ MHz. What is the corresponding SNR?

13.2 For the same photodiode of Problem 13.1, consider an incident signal of 500 nW at 1550 nm at which $R = 1$ A/W and all other parameters remain the same. Obtain the signal and various noise currents and the corresponding SNR.

13.3 Consider a silicon APD with responsivity $R = 0.6$ A/W, gain $M = 100$, excess noise factor $x = 0.5$, dark current $I_d = 1$ nA, load resistor $R_L = 1$ kΩ, capacitance $C = 2.0$ pF, and an operating bandwidth of $\Delta f = 1$ MHz. Obtain the signal current and the various noise current terms.

13.4 Consider a silicon APD with $R = 0.65$ A/W, $I_d = 1$ nA, $R_L = 1000\,\Omega$, $\Delta f = 100$ MHz, $x = 0.3$, and gain $M = 100$. Calculate the SNR for input powers of (i) $P = 100$ nW and (ii) $P = 10\,\mu$W.

Solution: Using equation (13.26) we obtain

(1) for $P = 100$ nW

$$\text{SNR} \simeq 493 \simeq 27\,\text{dB}$$

(2) for $P = 10\,\mu$W

$$\text{SNR} \simeq 51000 \simeq 47\,\text{dB}$$

13.5 Consider a germanium APD with $R = 0.65$ A/W, $I_d = 10$ nA, $R_L = 1000\,\Omega$, $\Delta f = 100$ MHz, $T = 300$ K, and $x = 1.0$ with an input power of 100 nW. Calculate the SNR with magnifications of 5, 10, 20, and 30. Note that as M increases beyond an optimum value, SNR decreases.

13.6 A fiber optic communication system is to operate with a graded index multimode fiber at 850 nm. The source is an LED with a spectral width of 20 nm and a coupled optical power of −15 dBm. Calculate the loss-limited length and dispersion-limited length at 10 Mb/s if the fiber has an attenuation coefficient of 4 dB/km (including splice losses) and the receiver sensitivity is −50 dBm.

Solution: Since operation is at 850 nm with graded index multimode fibers, the dispersion limitations would be mainly due to material dispersion. At 850 nm, the material dispersion coefficient is ~852 ps/km·nm, which with $\Delta\lambda = 20$ nm gives a dispersion of 1.7 ns/km.

For the system to operate at 10 Mb/s with NRZ, the rise time should be less than $0.7/10 \times 10^6 = 70$ ns. Thus, neglecting the rise times of source and detector, the dispersion-limited length would be $70/1.7 \simeq 41$ km.

With an input power of −15 dBm and a receiver sensitivity of −50 dBm, the maximum permissible loss = 35 dB, which, with a loss figure of 4 dB/km, gives a maximum repeater spacing (without any margin) of 8.5 km. Thus, the system would be loss limited.

13.7 Consider a fiber optic system based on multimode graded index fibers with a parabolic index core operating at 850 nm. Assume $n_1 = 1.46$, $\Delta = 0.01$, and the fiber loss is ≃3 dB/km. The source is an LED with $\Delta\lambda = 25$ nm, a rise time of 5 ns, and a power output of 50 μW from a 50-μm core pigtail. If the receiver is a PIN photodetector with a sensitivity of −35 dBm and a rise time of 1 ns, calculate the maximum repeaterless link length as determined by loss as well as by dispersion.

13.8 Calculate the maximum loss-limited and dispersion-limited distances of a link operating at 850 nm at 100 Mb/s using the following components:

(1) Source: GaAlAs laser diode with 0-dBm fiber-coupled power, $\Delta\lambda \simeq 2$ nm, rise time = 1 ns.

(2) Detector: Si APD with a sensitivity of −50 dBm, rise time ≃2 ns, capacitance of 1 pF.

(3) Fiber: Parabolic index multimode fiber with $\Delta = 0.01$, $n_1 = 1.46$, and a loss of 3.5 dB/km at 850 nm.
(4) Two connectors with a loss of 1 dB each.
(5) Splice every 2 km with a splice loss of 0.1 dB per splice.

13.9 Consider a parabolic index multimode fiber ($\Delta = 0.01$, $n_1 = 1.46$) operating at the zero material dispersion wavelength. If we wish to transmit at 100 Mb/s, what is the maximum unrepeated distance using

(1) InGaAs LED with -13-dBm fiber-coupled power with $\Delta\lambda = 25$ nm, rise time $\simeq$3 ns.
(2) InGaAs PIN diode with -35-dBm sensitivity, rise time of 0.5 ns, capacitance of 0.5 pF.
(3) Fiber attenuation of 0.5 dB/km at 1300 nm.
(4) Two connectors with a loss of 1 dB each and splice every 5 km with a splice loss of 0.1 dB per splice.

13.10 Consider the design of a single-mode fiber link at 1300 nm operating at 565 Mb/s. The source is a GaInAsP laser with a fiber-coupled power of -3 dBm, spectral width of 2 nm, and a rise time of 0.3 ns. The detector is an InGaAs PIN with a responsivity of 0.7 A/W, a dark current of 1 nA, a capacitance of 0.5 pF, and a rise time of 0.3 ns. Assuming an average fiber attenuation (including splice and connector loss) of 0.5 dB/km, obtain the maximum permissible repeaterless transmission distance.

14

Optical fiber amplifiers

14.1 Introduction

As discussed earlier, optical fiber communication systems are ultimately limited by either loss or dispersion. Loss leads to power levels of received signal that cannot be detected within tolerable errors, whereas dispersion leads to overlapping of adjacent pulses of light with resultant loss in resolution and, hence, information. Figure 14.1 shows a typical wavelength dependence of loss and dispersion of a single-mode fiber for both a CSF with zero dispersion around 1300 nm and a DSF with zero dispersion around 1550 nm. As evident from the figure, the effect of loss can be minimized by operating at the minimum loss wavelength around 1550 nm, whereas dispersion can be minimized by operating at the zero dispersion wavelength. Using DSFs has advantages of both minimum loss and zero dispersion. Figure 14.2 shows the maximum permissible unrepeatered length as a function of bit rate as determined by loss or by dispersion for both a CSF and a DSF (see Chapter 13 for a detailed discussion of this figure). The pulses are assumed to be Fourier transform limited. As can be seen from the figure, even at bit rates of 2.5 Gb/s, a system operating with CSFs is limited by loss rather than by dispersion. For DSF, loss-limited operation extends to almost 10 Gb/s.

In long-haul fiber optic communication systems, the effects of loss and pulse dispersion are normally overcome by using periodically spaced electronic repeaters. In these repeaters the input optical signal is first detected and converted to electrical signals. These electrical signals are processed (reshaped and retimed) to remove the effects of pulse dispersion and then amplified to drive an optical source, thus regenerating the pulse train. Such repeaters are referred to as 3R repeaters (retiming, reshaping, and regeneration) (Figure 14.3). Hence, the signal emerging from the optical source at the repeater is almost as good as at the start of the link and can be sent again through the next segment of the

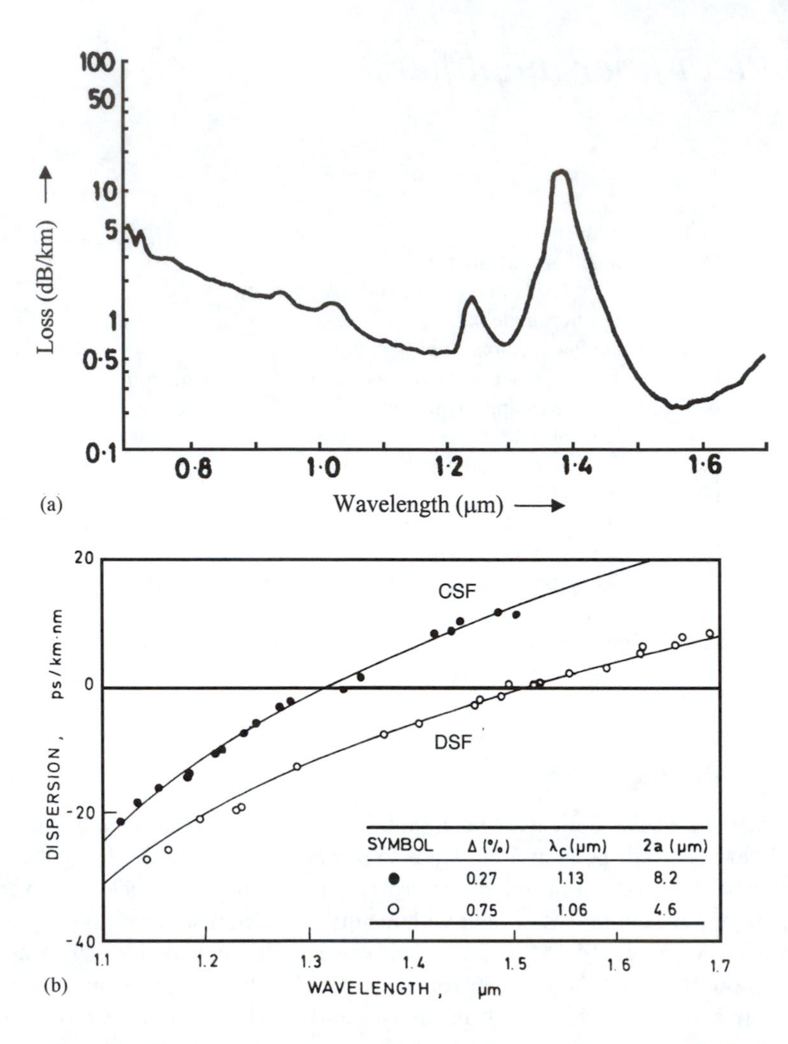

Fig. 14.1: Typical wavelength dependence of (a) loss and (b) dispersion of a single-mode fiber for both a CSF with zero dispersion around 1300 nm and a DSF with zero dispersion around 1550 nm. [(a) Adapted from Miya et al. (1979). (b) Adapted from Kimura (1988).]

link. In this way almost error-free transmission over thousands of kilometers can be achieved.

In dispersion-limited systems, the spacing between repeaters is dictated by overlapping of neighboring light pulses (and, hence, a loss of information) rather than by loss of signal power. In such a situation regeneration of the optical signal is necessary.[1] On the other hand, in loss-limited systems the repeater spacing is dictated by signal power loss rather than by loss of resolution. For such loss-limited systems it would indeed be advantageous to have optical amplifiers that directly amplify the optical signals without going through the complicated process of conversion to electrical signal and back again to optical signal. Indeed, as discussed in Chapter 10, dispersion can be minimized by operating at the zero dispersion wavelength or by using dispersion-compensating schemes (see Chapter 15). Further, if soliton pulses are used in the data stream, pulse dispersion can be completely eliminated (see Chapter 16) and the system will be loss limited.

[1]In Chapter 15 we show that such dispersion effects can also be optically compensated.

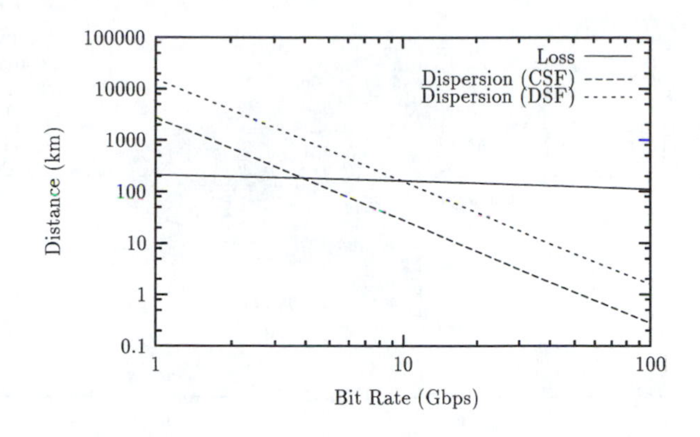

Fig. 14.2: Maximum possible unrepeatered length as a function of bit rate as determined by fiber attenuation and by dispersion for operation at 1550 nm. —–, Loss limit; -----, dispersion limit for CSF; and · · ·, for DSF corresponding to Fourier transform limited pulses.

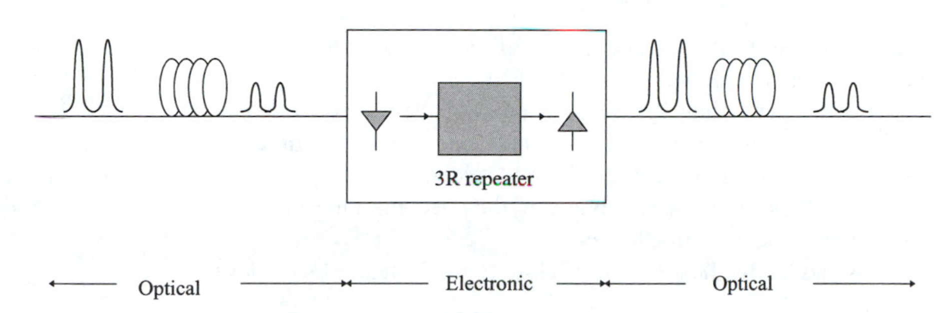

Fig. 14.3: An electronic repeater performing retiming, reshaping, and regeneration of the optical pulse train.

One of the most important optical amplifiers is the erbium-doped fiber amplifier (EDFA) (Figure 14.4). The EDFA was first reported in 1987 and the progress in this technology has been so rapid that American Telephone and Telegraph has already put into service a terrestial system (since July 1993), and many more systems employing EDFAs are being installed. Indeed, the transatlantic and transpacific submarine systems planned for installation will all be using EDFAs.

Figure 14.4 shows a typical configuration of an EDFA. A WDM multiplexes the light from a high-power pump laser diode (wavelength of 980 nm or 1480 nm) and the signal to be amplified (in the wavelength region 1530–1570 nm) into an erbium-doped silica fiber. Because of absorption of the pump laser light by the erbium ions in the doped fiber, the erbium-doped fiber becomes an optical amplifier for light waves in the wavelength region 1530–1570 nm (see Section 14.2). Because of this, the incoming optical signal (lying in the region 1530–1570 nm) gets amplified as it propagates through the doped fiber. The amplified signal passes through an optical isolator (which is inserted to avoid any back reflections) and a wavelength filter to filter out wavelengths other than the signal wavelength. Using such an EDFA, one can achieve gains as large as 40–45 dB.

Using optical fiber amplifiers in a fiber optic communication link has many advantages.

(1) Since the information-carrying signals are directly amplified in the optical domain without conversion to the electrical domain, a long-haul system employing fiber amplifiers is bit rate transparent – that is, the amplifier will work efficiently even at higher bit rates, which

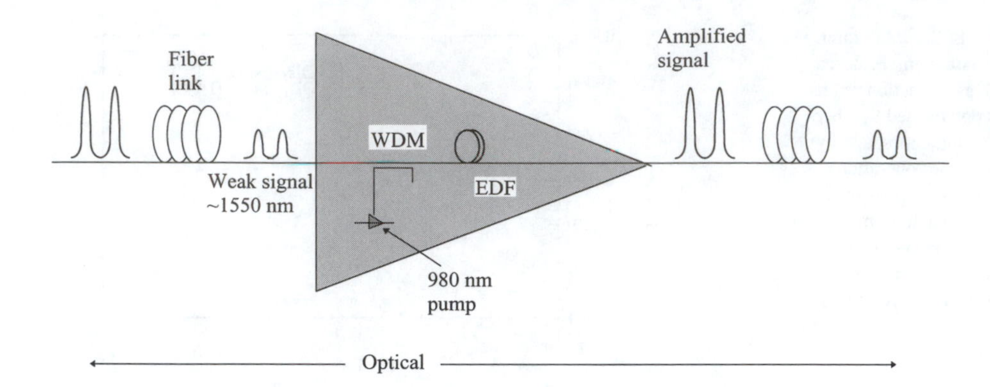

Fig. 14.4: An optical amplifier in which the input optical pulses are amplified in the optical domain itself without any conversion to the electrical domain. EDF, erbium-doped fiber; WDM, wavelength division multiplexer.

may become necessary at a later date. This is in contrast to electronic repeaters, which work only at the designed bit rate.

(2) Since EDFAs have a large-gain bandwidth (i.e., they can provide gain over a large spectral bandwidth ~ 40 nm), one can use a single amplifier in wavelength division multiplexed systems.
(3) Fiber amplifiers can be easily spliced to the telecommunication fiber link with minimal insertion losses.
(4) The noise added by the amplifier is close to the lowest possible level (3–4 dB).
(5) The gain provided by EDFA is polarization insensitive.

Following are some of the drawbacks of EDFAs.

(1) Currently, they are limited to 1550-nm systems only. For 1300-nm systems, efforts are under way to develop amplifiers based on other rare earth elements such as praseodymium.
(2) They require high pump powers (50–100 mW).
(3) Very short lengths are not possible.

Today EDFAs are finding many applications in lightwave systems. These include (Figure 14.5)

(a) Power amplifiers or booster amplifiers to boost the signal power exiting from a semiconductor laser before launching into the transmission fiber. Such power amplifiers can increase span length in transmission systems and also can compensate for splitting losses in networks.
(b) Preamplifiers for enhancing the receiver sensitivity.
(c) In-line amplifiers to boost the signal level periodically along the transmission path.

There are some differences in the characteristics of the three types of fiber amplifiers mentioned above.

In the case of power amplifiers, the input signal levels are quite high (-3 dBm to 0 dBm) as they are placed immediately after the transmitter and, thus, the amplifier operates in the signal saturation region (see Section 14.8), wherein the gain is reduced in comparison with the small signal regime due to gain saturation. For such amplifiers, the most important factor is the obtainable signal

Fig. 14.5: Various configurations of an EDFA. (a) Power amplifier to boost the source power, (b) preamplifier to improve receiver sensitivity, (c) in-line amplifier to amplify signals en route, and (d) a long-haul link using a power, in-line amplifier, and a preamplifier.

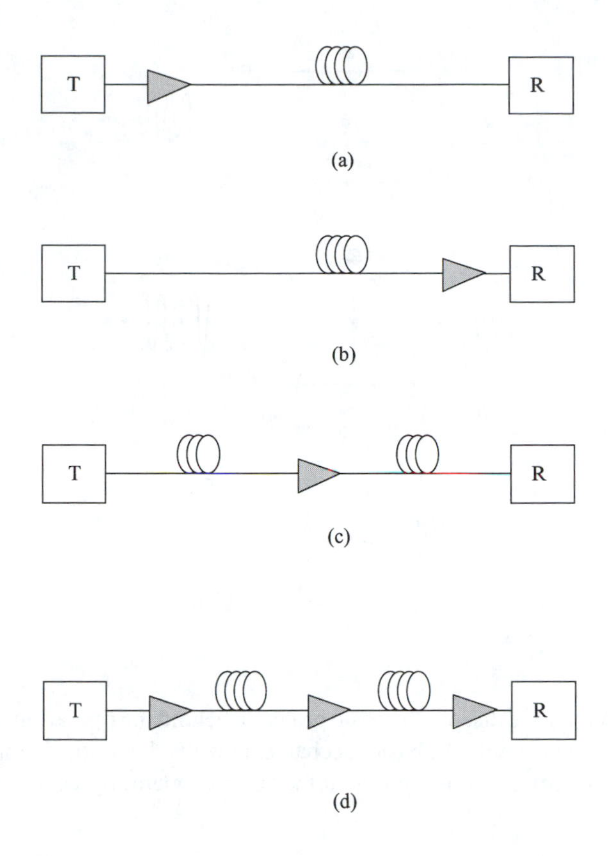

output power, which should be high. In such amplifiers, the high-output power is achieved at the expense of a reduced gain and an increased noise figure (see Section 14.10). For better operation characteristics, one could use backward pumping or even bidirectional pumping – that is, either the pump beam could propagate in a direction opposite the signal beam or there could be two pump beams propagating along both directions in the fiber. Power amplifiers do not need stringent requirements of noise and optical filtering.

The preamplifier, which is placed just before the receiver, is used to increase the receiver sensitivity and is usually operated in the unsaturated regime since the input signal powers are very small (≈ -40 dBm to -50 dBm). Since this operates with low-input signal powers, the amplifier should have a very good noise figure (see Section 14.10) as well as low insertion loss components. Narrowband optical bandpass filters are usually used to filter the broad amplified spontaneous emission spectrum falling outside the signal wavelength band. Also, 980-nm pumping provides better noise figures than 1480-nm pumping. In addition, a codirectional pumping configuration provides better noise performance than a contradirectional pumping scheme. Use of preamplifiers in conjunction with power amplifiers can result in significant increases in available power budget in a fiber optic communication system.

In-line amplifiers that are located along the optical fiber link are characterized by moderate gain and a moderate noise figure. Optical bandpass filters are used with each of the in-line amplifiers to reduce accumulation of amplified spontaneous emission noise.

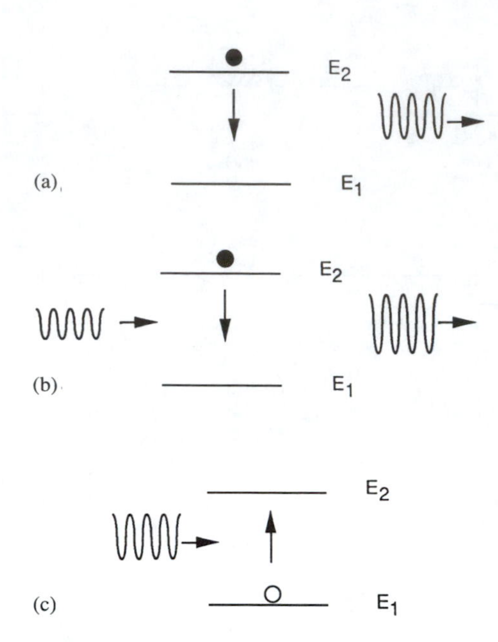

Fig. 14.6: Three different ways in which an atom interacts with radiation. (a) Spontaneous emission, (b) stimulated emission, and (c) absorption. The frequency of the radiation $\nu_c \simeq (E_2 - E_1)/h$.

In this chapter we discuss the basic principle behind the operation of EDFAs. In Section 14.2 we study the basic mechanism involved in optical amplification and in subsequent sections we obtain the various operating characteristics of EDFAs.

14.2 Optical amplification

It is well known that an atom or an ion is characterized by a discrete set of energy levels in which it can exist. Let E_1 represent the ground state (the lowest possible energy level) and E_2 one of the excited states of the atom (see Figure 14.6). Let N_1 and N_2 be the number of atoms per unit volume in levels E_1 and E_2, respectively.

Generally, atoms lying in energy level E_2 can make a transition to level E_1 with a certain probability and in this process emit radiation at a frequency

$$\nu_c = \frac{E_2 - E_1}{h} \tag{14.1}$$

with the corresponding free space wavelength given by

$$\lambda_c = \frac{hc}{E_2 - E_1} \tag{14.2}$$

Since this process can occur even in the absence of any radiation, this is called *spontaneous emission* (see Figure 14.6(a)). The rate of (spontaneous) transitions from E_2 to E_1 is proportional to N_2 and thus

$$\frac{dN_2}{dt} = -A_{21}N_2 = -\frac{N_2}{t_{sp}} \tag{14.3}$$

The constant A_{21} is called the Einstein's A coefficient and depends on the energy level pair; $t_{sp}(=\frac{1}{A_{21}})$ is called the *spontaneous emission* lifetime due to the transition $E_2 \rightarrow E_1$.

In contrast to the spontaneous emission process, atoms lying in E_2 can also be stimulated to emit radiation by incident radiation at frequency ν_c. This is called stimulated emission (see Figure 14.6(b)).

The important difference between spontaneous and stimulated emission is that, although the former is random and incoherent, the latter process is phase coherent with the incident radiation. This characteristic is used in the amplification of optical signals and in lasers for the generation of coherent radiation.

An atom lying in level E_1 can also absorb radiation at frequency ν_c and get excited to level E_2 (see Figure 14.6(c)). This is the phenomenon of stimulated absorption or simply absorption.

The processes of absorption and stimulated emission are induced by the incident radiation. To calculate the rate of such emissions and absorptions, it is necessary to introduce the concept of cross section.

Let N_1 and N_2 represent the number of atoms per unit volume that are in energy states E_1 and E_2, respectively. We assume a monochromatic light beam (of frequency ν and intensity I_ν) to be propagating through the medium. Thus

$$\phi_\nu = \frac{I_\nu}{h\nu} \tag{14.4}$$

represents the number of photons crossing a unit area per unit time. Let $(dN_1/dt)_{\text{abs}}$ represent the rate of change of N_1 due to absorption (to level E_2). Now, $(dN_1/dt)_{\text{abs}}$ is proportional to N_1 and to ϕ_ν. Thus, we may write

$$\begin{aligned}\left(\frac{dN_1}{dt}\right)_{\text{abs}} &= -\sigma_a(\nu)\phi_\nu \cdot N_1 \\ &= -\sigma_a(\nu)\frac{I_\nu}{h\nu} \cdot N_1 \end{aligned} \tag{14.5}$$

where the constant of proportionality $\sigma_a(\nu)$ is known as the absorption cross section (having dimensions of area).

Similarly, the rate of decrease of N_2 due to stimulated emissions is

$$\left(\frac{dN_2}{dt}\right)_{\text{st. em}} = -\frac{\sigma_e(\nu)\, I_\nu}{h\nu} \cdot N_2 \tag{14.6}$$

where $\sigma_e(\nu)$ is the emission cross section.

For isolated atoms with nondegenerate energy levels $\sigma_a(\nu) = \sigma_e(\nu) = \sigma(\nu)$. Erbium ions in the silica matrix can be approximately described as a three-level system provided the transitions are characterized by different absorption and emission cross sections.[2]

[2]This is because in the case of erbium ions in the silica matrix, each of the energy levels is a band consisting of a multiplicity of closely lying energy levels (brought about by Stark splitting due to crystal electric field). One can show that the Stark split three-level system of erbium can be described as a nondegenerate three-level system provided the transitions are characterized by different absorption and emission cross sections [Desurvire (1994)]. This is due to the rapid thermalization of the ions within each band resulting in an almost constant population *distribution* within each band. Thus, they can be treated as levels rather than as bands.

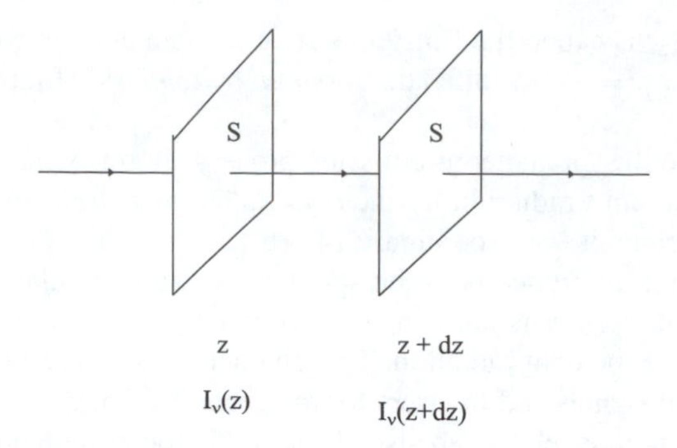

Fig. 14.7: A plane wave propagating through a medium consisting of a collection of atoms. The intensity of the beam changes from $I_\nu(z)$ at plane z to $I_\nu(z+dz)$ at plane $(z+dz)$ due to the interaction with the atomic population.

To obtain an equation for the rate of change of intensity of a monochromatic beam as it propagates through a collection of atoms, we let $I_\nu(z)$ and $I_\nu(z+dz)$ represent the intensities of the beam at z and $z+dz$ (see Figure 14.7). Considering an area of cross section S, the number of atoms in levels E_1 and E_2 in the volume between z and $z+dz$ is $N_1 S\, dz$ and $N_2 S\, dz$, respectively. The number of upward transitions (absorptions) per unit time is

$$\frac{\sigma_a(\nu) I_\nu(z) N_1 S\, dz}{h\nu}$$

and the corresponding number of stimulated emissions per unit time is

$$\frac{\sigma_e(\nu) I_\nu(z) N_2 S\, dz}{h\nu}$$

Hence, the energy absorbed per unit time in the volume $S\, dz$ is

$$\frac{I_\nu(z)}{h\nu}(\sigma_a N_1 - \sigma_e N_2) S\, dz\, h = \sigma_a(\nu) I_\nu (N_1 - \eta N_2) S\, dz \tag{14.7}$$

where

$$\eta(\nu) = \frac{\sigma_e(\nu)}{\sigma_a(\nu)} \tag{14.8}$$

In the wavelength range 1.41–1.57 μm, η for erbium ions in silica matrix lies in the range 0.1–2. Figure 14.8 shows the measured wavelength dependence of absorption and emission cross sections of a typical erbium doped silica fiber and Table 14.1 gives the corresponding numerical values.

In writing equation (14.7), we have neglected the contribution due to spontaneous emission that is responsible for the noise in the amplifier (see Section 14.10).

The above energy absorbed per unit time must be equal to the net energy entering the volume $S\, dz$, which is given by

$$I(z)S - I(z+dz)S = \left(I(z) - I(z) - \frac{dI}{dz}\, dz \right) S$$

$$= -\frac{dI}{dz} S\, dz$$

Table 14.1. *Absorption (σ_{sa}) and emission (σ_{se}) cross sections at different wavelengths corresponding to a typical erbium-doped silica fiber. Numerical values correspond to Figure 14.8*

Wavelength (nm)	σ_a ($\times 10^{-25}$ m^2)	σ_e ($\times 10^{-25}$ m^2)
1500.3	2.257	1.133
1505.5	2.403	1.340
1509.7	2.553	1.514
1514.9	2.744	1.884
1520.2	3.365	2.489
1525.4	4.421	3.495
1530.6	5.379	4.709
1535.9	4.644	4.644
1540.1	3.154	3.503
1545.3	2.850	3.386
1550.5	2.545	3.410
1555.8	2.229	3.057
1560.0	1.859	2.801
1565.2	1.303	2.180
1570.4	0.934	1.717
1575.7	0.759	1.303
1579.8	0.654	1.133
1585.1	0.576	0.978
1590.3	0.503	0.889
1595.6	0.459	0.804
1600.8	0.442	0.727
1605.0	0.402	0.670
1610.2	0.378	0.609
1615.5	0.345	0.544
1619.6	0.325	0.487
1624.9	0.292	0.426
1630.1	0.276	0.369
1635.3	0.252	0.309
1640.6	0.252	0.268

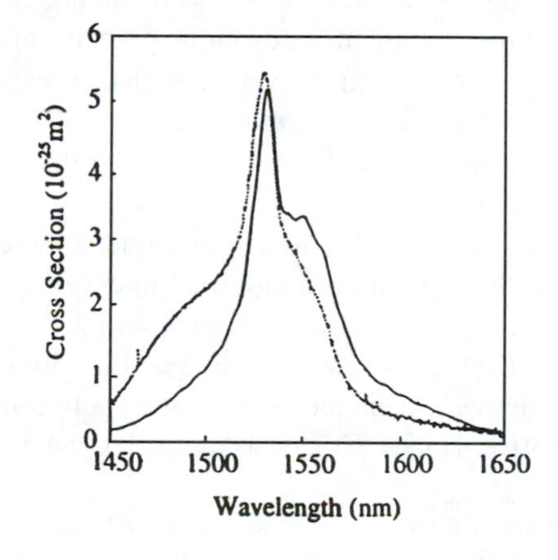

Fig. 14.8: Experimentally obtained emission (solid curve) and absorption (dashed curve) cross sections of an erbium doped Ge/A1/P silica fiber. [After Pedersen (1994).]

Hence, equating the energy absorbed and the net energy entering the volume $S\,dz$ we have

$$\begin{aligned}\frac{dI_\nu(z)}{dz} &= -\sigma_a(\nu)(N_1 - \eta N_2)I_\nu(z) \\ &= -\alpha(\nu)I_\nu(z)\end{aligned} \tag{14.9}$$

where

$$\alpha(\nu) = \sigma_a(\nu)(N_1 - \eta N_2) \tag{14.10}$$

is the absorption coefficient.

Equation (14.9) tells us that for light amplification (i.e., for $dI/dz > 0$) we must have $\eta N_2 > N_1$. For nondegenerate levels $\eta = \sigma_e/\sigma_a = 1$ and the above condition reduces to $N_2 > N_1$. This is called population inversion. It may be noted that if $\eta > 1$, light amplification can take place even if $N_2 < N_1$.

Under normal conditions $\alpha(\nu) > 0$ and the light beam gets attenuated. Thus, when light at a frequency $(E_2 - E_1)/h$ passes through such a medium, it will trigger more absorptions than emissions and therefore the light wave will get attenuated. On the other hand, if one can create a condition in which $\alpha(\nu) < 0$, then an incident light wave at frequency $\nu = (E_2 - E_1)/h$ will trigger more stimulated emissions than absorptions. This results in an amplification of the beam rather than attenuation (see Chapter 11).

14.3 Energy levels of erbium ions in silica matrix

At the heart of the EDFA is a short length (5–50 m) of silica optical fiber in which the core of the fiber is doped with about 200 mole ppm of erbium (an optically active rare earth element) corresponding typically to an erbium concentration of about 10^{25} ions/m^3 (see Section 14.13). Figure 14.9 shows the energy level diagram of the Er^{3+} ion in silica host glass. Each energy level is split into a multiplicity of levels due to the electric field of adjacent ions in the glass matrix and due to the amorphous nature of the silica glass matrix. The energy difference between the ground level and successive excited energy levels corresponds to wavelengths around 1530 nm, 980 nm, 800 nm, 670 nm, and so forth. Figure 14.10 shows a typical absorption spectrum of an erbium-doped fiber showing strong absorptions at the wavelengths corresponding to various energy levels. Note that peak absorption coefficients are around a few dB/m compared with typical attenuation figures of 0.2–0.5 dB/km in a silica fiber used in communication.

When a laser beam corresponding to, for example, a wavelength of 980 nm or 1480 nm is passed through an erbium-doped fiber, then the erbium ions in the ground level $E_1(^4I_{15/2})$ absorb this radiation and get excited to the upper levels $E_2(^4I_{13/2})$ and $E_3(^4I_{11/2})$, respectively. For Er^{3+} ions in silica host, all transitions are nonradiative[3] except the last transition between E_2 and E_1, which is almost 100% radiative. Hence, Er^{3+} ions excited to the higher energy levels

[3] A nonradiative transition is one in which the energy released by the atomic system is in a form other than electromagnetic radiation – for example, in the form of heat given to the host lattice.

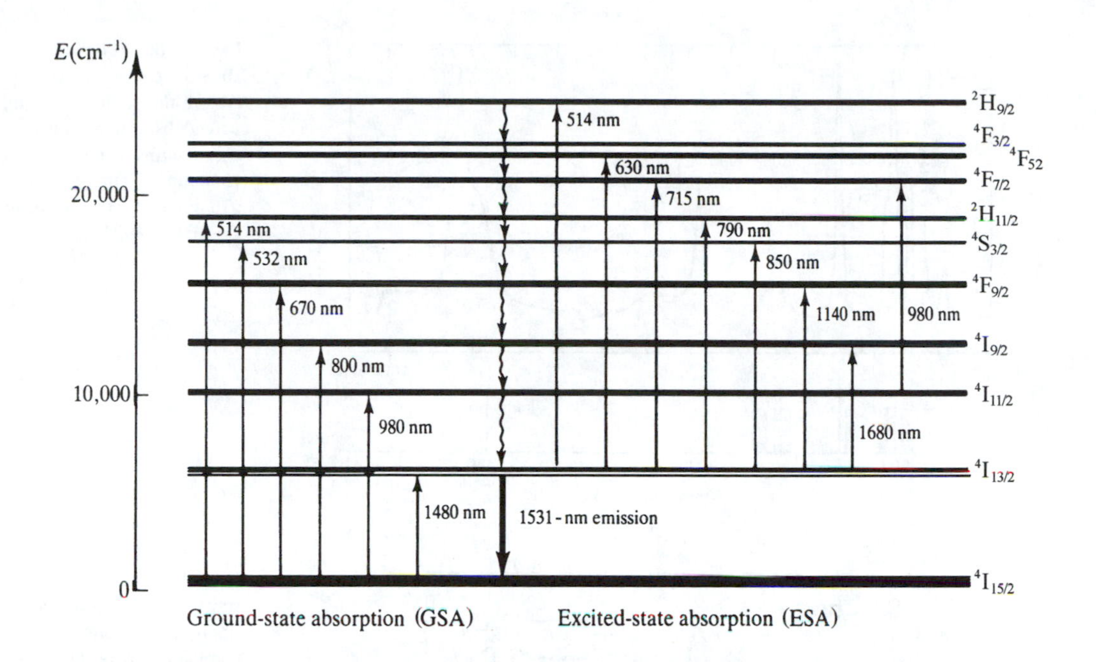

Fig. 14.9: Energy levels of Er^{3+} ion in a silica matrix. Each energy level is broadened into a band due to the electric field of adjacent ions and due to the amorphous nature of silica matrix. [After Desurvire (1994).]

quickly relax down to the level marked E_2 from which they undergo either spontaneous or stimulated emission to the level E_1.

Figure 14.8 shows a typical variation of absorption and emission with wavelength corresponding to the energy levels $^4I_{13/2}$ and $^4I_{15/2}$. Note that compared with the absorption spectrum, the emission spectrum is slightly shifted to longer wavelengths. This is due to the rapid thermalization of Er^{3+} ions in the $^4I_{13/2}$ band and consequent emission from the bottom of the $^4I_{13/2}$ level to $^4I_{15/2}$. This difference in the absorption and emission spectra is taken into account by assuming $\sigma_a \neq \sigma_e$ (see Section 14.2). Figure 14.11 shows the emission spectra of erbium doped fibers with different codopants. It can be seen that the spectrum depends on the codopants used and in particular, the flattening of the spectrum with aluminum codoping. This can be important from the point of view of WDM applications (see Section 14.11).

For low pump powers, although the Er^{3+} ions are getting excited to the E_2 level, population inversion may not exist because of spontaneous emission. Thus, in such a case a signal beam at 1550 nm will get attenuated (due to absorption from E_1 to E_2) rather than being amplified. As the pump power increases, the rate of excitation increases and at some power level one can achieve population inversion between E_2 and E_1 and in such a situation a signal around 1550 nm will get amplified rather than absorbed. This is the basic principle behind optical amplification by the erbium-doped fiber.

Figure 14.12 shows a schematic of the variation of pump and signal powers along the length of a doped fiber. Near the input end, the pump power is high, leading to a population inversion and, hence, signal amplification. As the propagation distance increases, $P_p(z)$ reduces and, beyond a certain length, the pump power is insufficient to create an inversion and, hence, the signal starts to get attenuated.

The pump bands corresponding to 800 nm, 980 nm, and 1480 nm are interesting since high-power semiconductor lasers are available at these wavelengths.

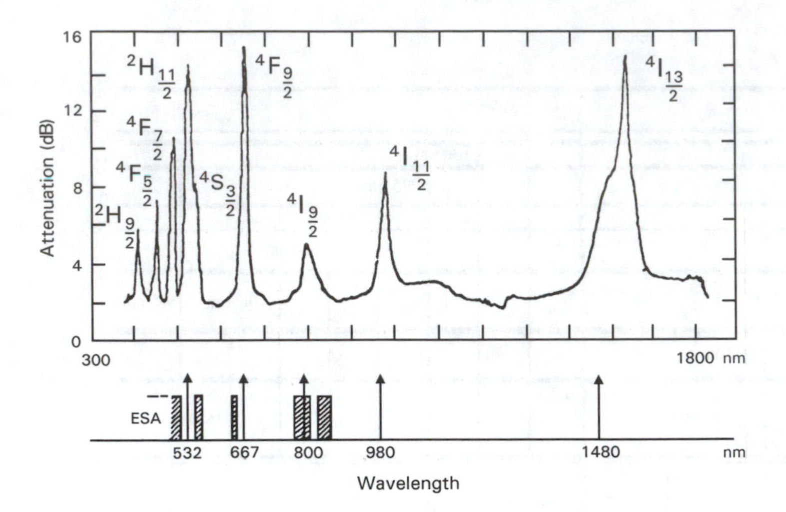

Fig. 14.10: A typical absorption spectrum of an erbium-doped fiber showing strong absorptions at many wavelengths including 1530 nm, 980 nm, 800 nm, 670 nm, and so forth. [After Desurvire (1992).]

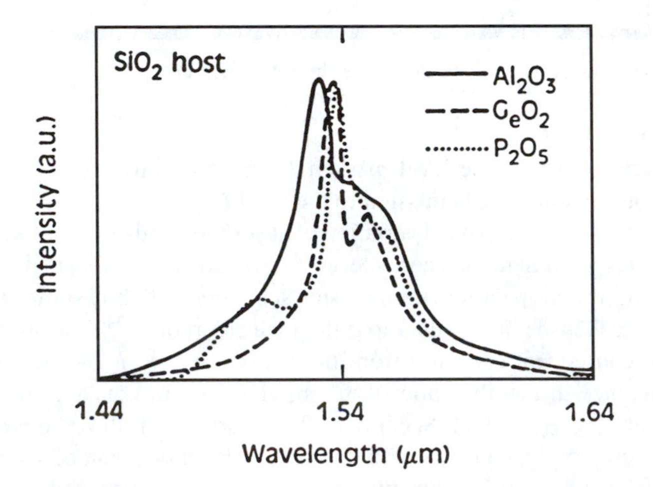

Fig. 14.11: Emission spectra of erbium-doped fibers with different codopants. [After Desurvire (1994).]

One drawback of the pump wavelengths of 800 nm and other lower wavelengths is what is termed excited state absorption (ESA). Due to the multiplicity of energy levels of Er^{3+} ions, ions sitting in the level E_2 can also absorb the pump radiation and get excited to higher levels. This reduces the pump efficiency significantly since it affects the population inversion directly and, hence, the gain. Hence, the pump bands at 980 nm and 1480 nm for which there is no ESA are the ones more commonly used.

In the next section we will write down the equations for optical amplification in an erbium-doped fiber by representing it as a three-level system and study the various characteristics such as gain, pump power, and signal power variations with fiber length, and so forth.

14.4 Amplifier modeling

As seen earlier, although pumping of erbium-doped fiber is possible at many wavelengths, the main pump wavelengths that are used are the 980-nm and the 1480-nm bands as there is no excited state absorption corresponding to

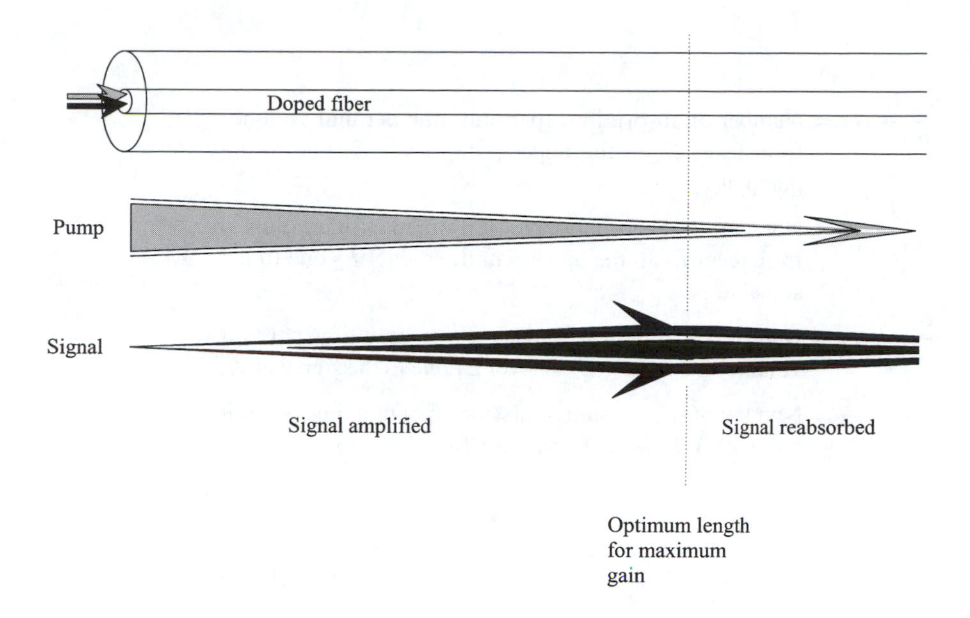

Fig. 14.12: Schematic diagram showing variation of pump and signal powers with distance along the doped fiber.

both these wavelengths. The 980-nm pump band offers higher pump efficiency compared with the 1480-nm pump and also has better noise characteristics. As evident from Figure 14.9, pumping at 980 nm corresponds to a three-level system. In the following we will write down a simplified form of the rate equations corresponding to such a system and obtain some of the important features of EDFAs.

We consider an erbium-doped silica fiber in which the core (or a fraction of it) is doped with erbium with an ion density of $N_t(r)$. We will also assume that the doped fiber is single moded at both the pump (980 nm) and signal wavelengths ($\sim$1550 nm) so that all parameters such as intensities, population densities of each level and so forth are independent of the azimuthal coordinate ϕ.

Let $N_1(r, z)$ and $N_2(r, z)$ be the population densities (number of ions/unit volume) of Er^{3+} ion in the ground state $E_1(^4I_{15/2})$ and the upper amplifier level $E_2(^4I_{13/2})$, respectively. Here r represents the cylindrical radial coordinate and z is along the fiber axis. Pumping by 980 nm takes the Er^{3+} ions from the ground state E_1 to the upper pump level E_3 from which the ions relax very rapidly to the upper amplifier level E_2. Since the relaxation rate of level E_3 is very rapid, we may assume that level E_3 is unpopulated and, hence, write

$$N_1(r, z) + N_2(r, z) \simeq N_t(r) \tag{14.11}$$

Let $I_p(r, z)$ and $I_s(r, z)$ represent the intensity distributions of the pump and signal beams. Let σ_{pa}, σ_{sa}, and σ_{se} denote the absorption cross section at pump, absorption cross section at the signal, and emission cross section at the signal, respectively. We may then write for the rate of change of population of the ground level E_1 as

$$\frac{dN_1}{dt} = -\frac{\sigma_{pa} I_p}{h\nu_p} N_1 - \frac{\sigma_{sa} I_s}{h\nu_s} N_1 + \frac{\sigma_{se} I_s}{h\nu_s} N_2 + \frac{N_2}{t_{sp}} \tag{14.12}$$

where

$$\frac{\sigma_{pa} I_p}{h\nu_p} N_1 = \text{Number of absorptions (per unit time per unit volume) from level } E_1 \text{ to the upper pump level } E_3 \text{ due to the pump at } \nu_p.$$

$$\frac{\sigma_{sa} I_s}{h\nu_s} N_1 = \text{Number of absorptions (per unit time per unit volume) from level } E_1 \text{ to the upper amplifier level } E_2 \text{ due to the signal at } \nu_s.$$

$$\frac{\sigma_{se} I_s}{h\nu_s} N_2 = \text{Number of stimulated emissions (per unit time per unit volume) from level } E_2 \text{ to level } E_1 \text{ due to the signal at } \nu_s.$$

$$\frac{N_2}{t_{sp}} = \text{Number of spontaneous emissions (per unit time per unit volume) from level } E_2 \text{ to level } E_1.$$

Since

$$\eta_s = \frac{\sigma_{se}}{\sigma_{sa}} \tag{14.13}$$

Equation (14.12) can be written in the form

$$\frac{dN_1}{dt} = -\frac{\sigma_{pa} I_p}{h\nu_p} N_1 + \frac{\sigma_{sa} I_s}{h\nu_s} (\eta_s N_2 - N_1) + \frac{N_2}{t_{sp}} \tag{14.14}$$

At steady state,

$$\frac{dN_1}{dt} = 0$$

and elementary manipulations give us

$$\frac{N_2(r,z)}{N_1(r,z)} = \frac{\tilde{I}_p + \dfrac{\tilde{I}_s}{(1+\eta_s)}}{1 + \dfrac{\eta_s}{(1+\eta_s)} \tilde{I}_s} \tag{14.15}$$

where

$$\tilde{I}_s(r,z) = \frac{I_s(r,z)}{I_{s0}}; \; I_{s0} = \frac{h\nu_s}{\sigma_{sa} t_{sp}(1+\eta_s)} = \frac{h\nu_s}{(\sigma_{sa} + \sigma_{se}) t_{sp}} \tag{14.16}$$

$$\tilde{I}_p(r,z) = \frac{I_p(r,z)}{I_{p0}}; I_{p0} = \frac{h\nu_p}{\sigma_{pa} t_{sp}} \tag{14.17}$$

The physical significance of I_{p0} is discussed in Problem 14.17.

Note that since the pump and signal waves are propagating as modes of the fiber, they have an intensity distribution decreasing radially away from the fiber axis. Hence, for any given z value the ratio N_1/N_2 is also r dependent.

Using equations (14.11) and (14.15), we obtain

$$N_2(r,z) = \frac{\tilde{I}_p + \dfrac{\tilde{I}_s}{1+\eta_s}}{1+\tilde{I}_p+\tilde{I}_s} N_t \tag{14.18}$$

$$N_1(r,z) = \frac{1 + \dfrac{\eta_s}{1+\eta_s}\tilde{I}_s}{1+\tilde{I}_p+\tilde{I}_s} N_t \tag{14.19}$$

which represent the steady-state populations of the energy states E_1 and E_2, respectively. Note that $N_1(r,z) + N_2(r,z) = N_t(r)$ independent of z.

Equations (14.9) and (14.10) tell us that for amplification we must have $\eta_s N_2 - N_1 > 0$. Using equations (14.18) and (14.19) we have

$$\eta_s N_2(r,z) - N_1(r,z) = \frac{[\eta_s \tilde{I}_p(r,z) - 1]}{1+\tilde{I}_p(r,z)+\tilde{I}_s(r,z)} N_t \tag{14.20}$$

Hence, for amplification (at a particular value of r, z) we must have $\tilde{I}_p > 1/\eta_s$ or

$$I_p(r,z) > I_{pt} = \frac{1}{\eta_s} I_{p0} \tag{14.21}$$

where I_{pt} is known as the threshold pump intensity. Thus, a minimum pump intensity is needed at any value of (r, z) to achieve amplification.

Example 14.1: For a typical erbium-doped fiber pumped at 980 nm

$$\sigma_{pa} = 3.1 \times 10^{-25}\ \text{m}^2$$

$$t_{sp} = 12 \times 10^{-3}\ \text{s}$$

giving

$$I_{p0} = \frac{h\nu_p}{\sigma_{pa} t_{sp}} \simeq 5.46 \times 10^7\ \text{W/m}^2 \tag{14.22}$$

(a) For a signal wavelength of 1536 nm,

$$\sigma_{sa} \simeq 4.644 \times 10^{-25}\ \text{m}^2$$

$$\sigma_{se} \simeq 4.644 \times 10^{-25}\ \text{m}^2$$

giving $\eta_s = 1$ and

$$\text{Threshold pump intensity} = \tfrac{1}{\eta_s} I_{p0}$$

$$(\text{for } \lambda_s = 1536\ \text{nm}) = 5.46 \times 10^7\ \text{W/m}^2 \tag{14.23}$$

Further,

$$I_{s0} = \frac{h\nu_s}{(\sigma_{sa}+\sigma_{se})t_{sp}} \simeq 1.16 \times 10^7\ \text{W/m}^2$$

(b) For a signal wavelength of 1550 nm,

$$\sigma_{sa} \simeq 2.545 \times 10^{-25} \text{ m}^2$$

$$\sigma_{se} \simeq 3.410 \times 10^{-25} \text{ m}^2$$

giving $\eta_s = 1.3398$ and

$$\text{Threshold pump intensity } \simeq 4.07 \times 10^7 \text{ W/m}^2 \quad (\text{for } \lambda_s = 1550 \text{ nm}) \tag{14.24}$$

(c) For a signal wavelength of 1580 nm,

$$\eta_s = \frac{\sigma_{se}}{\sigma_{sa}} \simeq \frac{1.133 \times 10^{-25}\text{m}^2}{0.654 \times 10^{-25}\text{m}^2} \simeq 1.732$$

Thus,

$$\text{Threshold pump intensity } \simeq 3.15 \times 10^7 \text{ W/m}^2 \quad (\text{for } \lambda_s = 1580 \text{ nm}) \tag{14.25}$$

This shows that as a given erbium-doped fiber is pumped harder and harder, population inversion and, hence, gain is first achieved at longer wavelengths and then at shorter wavelengths (see Section 14.9).

14.4.1 *Variation of pump and signal powers with length*

In Section 14.2 we obtained an equation describing the variation of the intensity of a wave as it propagates through a medium (see equation (14.9)). Since the pump wave at frequency ν_p corresponds to transitions between E_1 and E_3 and since the population of the E_3 level is negligible, we can write for the rate of change of pump intensity (cf. equation (14.9))

$$\frac{dI_p}{dz} = -\sigma_{pa} N_1(r, z) I_p(r, z) \tag{14.26}$$

Similarly, the change of signal intensity with z is described by the equation

$$\frac{dI_s}{dz} = \sigma_{sa}(\eta_s N_2 - N_1) I_s(r, z) \tag{14.27}$$

In the case of an optical fiber, we should describe the amplification in terms of signal and pump *powers* rather than in terms of *intensities* since the propagating modes at the pump and signal wavelengths are characterized by transverse intensity profiles – that is, I_p and I_s – which are also functions of the transverse coordinate r. If we assume that both the pump and signal radiations exist in the fundamental mode of the doped fiber, then we can write

$$I_p(r, z) = P_p(z) f_p(r) \tag{14.28}$$

$$I_s(r, z) = P_s(z) f_s(r) \tag{14.29}$$

where $P_p(z)$ and $P_s(z)$ are the z-dependent powers at the pump and signal wavelengths and the quantities $f_p(r)$ and $f_s(r)$ represent the transverse dependence of the modal intensity patterns of the pump and signal waves, respectively, and are normalized so that

$$2\pi \int_0^\infty f_p(r) r\, dr = 1 \tag{14.30}$$

$$2\pi \int_0^\infty f_s(r) r\, dr = 1 \tag{14.31}$$

Thus, the pump power at any value of z is given by

$$\int_0^\infty \int_0^{2\pi} I_p(r, z) r\, dr\, d\phi = P_p(z) \tag{14.32}$$

Similarly, the signal power at any z value is

$$\int_0^\infty \int_0^{2\pi} I_s(r, z) r\, dr\, d\phi = P_s(z) \tag{14.33}$$

The pump power propagating at any value of z is given by

$$P_p(z) = 2\pi \int_0^\infty I_p(r, z) r\, dr \tag{14.34}$$

Thus,

$$\begin{aligned} \frac{dP_p}{dz} &= 2\pi \int_0^\infty \frac{dI_p}{dz} r\, dr \\ &= -2\pi \sigma_{pa} \int N_1(r, z)\, I_p(r, z) r\, dr \end{aligned} \tag{14.35}$$

We now assume that the fiber is doped with erbium with a uniform concentration N_0 (ions/m^3) up to a radius b (which in general could be different from core radius a). Thus

$$\begin{aligned} N_t(r) &= N_0; \quad 0 \le r \le b \\ &= 0; \quad r > b \end{aligned} \tag{14.36}$$

Using equations (14.36) and (14.19) in equation (14.35), we obtain

$$\begin{aligned} \frac{dP_p}{dz} &= -2\pi \sigma_{pa} N_0 \int_0^b \frac{1 + \frac{\eta_s}{1+\eta_s} \cdot \tilde{I}_s}{1 + \tilde{I}_p + \tilde{I}_s} I_p r\, dr \\ &= -2\pi \sigma_{pa} N_0 \int_0^b \frac{1 + \frac{\eta_s}{1+\eta_s} \cdot \frac{P_s(z) f_s(r)}{I_{s0}}}{1 + \frac{P_p(z) f_p(r)}{I_{p0}} + \frac{P_s(z) f_s(r)}{I_{s0}}} P_p(z) f_p(r) r\, dr \end{aligned} \tag{14.37}$$

Similarly, for the signal power we have (using equation (14.27))

$$\frac{dP_s}{dz} = \frac{d}{dz}\left[2\pi \int_0^\infty I_s(r, z) r\, dr\right]$$

$$= 2\pi \int_0^\infty \frac{dI_s}{dz} r\, dr$$

$$= 2\pi\sigma_{sa} \int_0^\infty (\eta_s N_2 - N_1) I_s r\, dr \tag{14.38}$$

or (using equation (14.20))

$$\frac{dP_s}{dz} = 2\pi\sigma_{sa} N_0 \int_0^b \frac{\eta_s \dfrac{P_p(z)}{I_{p0}} f_p(r) - 1}{1 + \dfrac{P_p(z) f_p(r)}{I_{p0}} + \dfrac{P_s(z) f_s(r)}{I_{s0}}} P_s(z) f_s(r) r\, dr \tag{14.39}$$

Equations (14.37) and (14.39) together describe the evolution of the pump and signal powers along the doped fiber. To solve these equations, we have to know the pump and signal intensity distributions in the transverse plane. For a step index fiber, $f_p(r)$ and $f_s(r)$ can be written in terms of Bessel functions. Since the transverse intensity patterns closely resemble a Gaussian pattern, the Gaussian envelope approximation for $f_p(r)$ and $f_s(r)$ is frequently employed. Using this approximation one can analytically integrate the RHS of equations (14.37) and (14.39) under a certain approximation and thus simplify the solution.

14.5 Gaussian envelope approximation

One of the simplest models describing the transverse intensity pattern of the modes is the Gaussian envelope approximation in which we assume

$$f(r) = \frac{1}{\pi\Omega^2} e^{-r^2/\Omega^2} \tag{14.40}$$

where Ω is determined by the characteristics of the fiber. The multiplying factor in equation (14.40) is chosen to normalize $f(r)$.

For a step index fiber, Ω is approximately given by (see Appendix D)

$$\Omega = a J_0(U) \frac{V}{U} \frac{K_1(W)}{K_0(W)} \tag{14.41}$$

where U, W, and V are parameters characterizing a single-mode fiber and are defined by

$$U = a\left(k_0^2 n_1^2 - \beta^2\right)^{1/2}; \quad W = a\left(\beta^2 - k_0^2 n_2^2\right)^{1/2} \tag{14.42}$$

$$V = k_0 a \sqrt{n_1^2 - n_2^2} \tag{14.43}$$

β is the propagation constant of the mode, n_1 and n_2 are the core and cladding refractive indices, and a is the core radius.

For a given V, W can be obtained (for a step index fiber) using the following approximate empirical relationship (see Chapter 8)

$$W = 1.1428V - 0.996 \tag{14.44}$$

which is accurate for $1.5 < V < 2.5$.

Corresponding to the pump and signal wavelengths, one can obtain the respective values of U, V, and W and subsequently the parameter Ω.

Example 14.2: As an example, we consider a fiber with $a = 1.5\ \mu$m, NA $= 0.24$. Thus, corresponding to a pump wavelength of 980 nm and signal wavelength of 1550 nm, we have

$$\begin{aligned} V_p &= 2.308 & V_s &= 1.459 \\ W_p &= 1.6416 & W_s &= 0.6713 \\ U_p &= 1.6224 & U_s &= 1.2954 \end{aligned}$$

giving

$$\begin{aligned} \Omega_p &\simeq 1.20\,\mu\text{m} \\ \Omega_s &\simeq 1.70\,\mu\text{m} \end{aligned}$$

The above approximation is referred to as the Gaussian envelope approximation [Desurvire (1994)] in contrast to the Gaussian mode approximation discussed in Chapter 8. The Gaussian mode approximation provides the best fit to the power launched into a mode, whereas the Gaussian envelope approximation provides a better match with the radial intensity distributions of the pump and signal.

Figure 14.13 show the actual Bessel field pattern and the Gaussian envelope approximation at the pump (980 nm) and signal (1550 nm) wavelengths for a typical erbium-doped fiber with

$$a = 1.5\ \mu\text{m}, \quad \text{NA} = 0.24$$

It is obvious that the Gaussian envelope approximation gives a good fit to the intensity distribution at both the pump and signal wavelengths.

The Gaussian envelope approximation can be used for the pump and signal wavelengths to simplify equations (14.37) and (14.39). We will consider this in the next section.

14.6 Solutions under the Gaussian envelope approximation

Under the Gaussian envelope approximation we may write

$$f_p(r) = \frac{1}{\pi \Omega_p^2} e^{-r^2/\Omega_p^2} \tag{14.45}$$

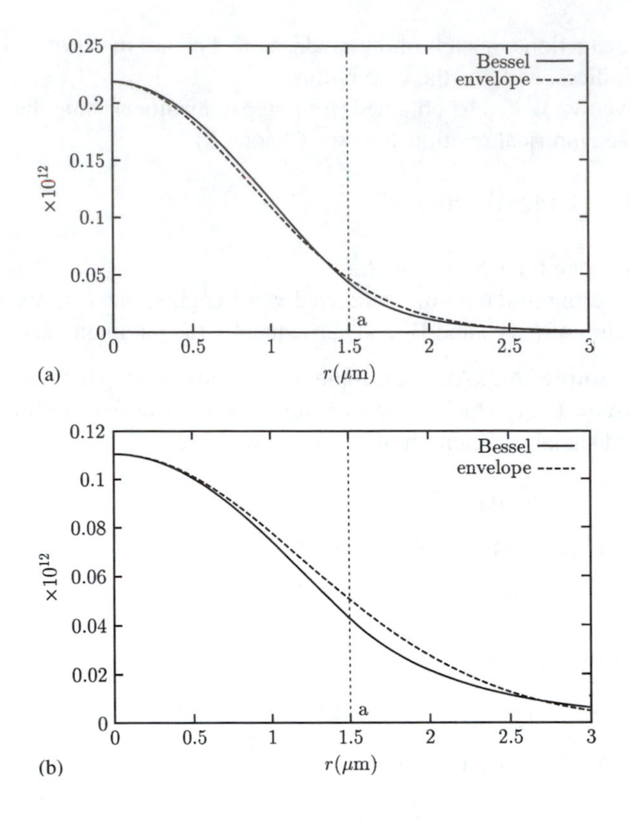

Fig. 14.13: Comparison of the modal intensity patterns corresponding to (a) pump (980 nm) and (b) signal (1550 nm) wavelengths of a fiber with $a = 1.5\,\mu\text{m}$, NA $= 0.24$. The solid curve corresponds to the actual Bessel function, and the dashed curve is the Gaussian envelope approximation.

and

$$f_s(r) = \frac{1}{\pi\Omega_s^2}\, e^{-r^2/\Omega_s^2} \tag{14.46}$$

We will evaluate the integrals in equations (14.37) and (14.39) assuming that $\Omega_s \simeq \Omega_p$. Under this approximation we may write

$$\frac{dP_p}{dz} = -\frac{2}{\Omega_p^2}\sigma_{pa}N_0\, P_p(z) \int_0^b \left[\frac{1 + ue^{-r^2/\Omega_p^2}}{1 + we^{-r^2/\Omega_p^2}}\right] e^{-r^2/\Omega_p^2} r\, dr \tag{14.47}$$

where

$$u = \frac{\eta_s}{1+\eta_s}\,\frac{P_s(z)}{P_{s0}} \tag{14.48}$$

$$w = \frac{P_p(z)}{P_{p0}} + \frac{P_s(z)}{P_{s0}} \tag{14.49}$$

with

$$P_{s0} = \pi\Omega_s^2\, I_{s0} \tag{14.50}$$

and

$$P_{p0} = \pi\Omega_p^2\, I_{p0} \tag{14.51}$$

Note that we have assumed $\Omega_s \simeq \Omega_p$ in writing the expression for $f_s(r)$; this approximation should not lead to much error as the quantity inside the square brackets in equation (14.47) is slowly varying in comparison to the Gaussian function outside the square brackets. Now

$$\int_0^b \frac{1+ue^{-r^2/\Omega^2}}{1+we^{-r^2/\Omega^2}} e^{-r^2/\Omega^2} r\,dr$$

$$= \frac{u}{w}\left[\left(\frac{w}{u}-1\right)\int_0^b \frac{e^{-r^2/\Omega^2} r\,dr}{1+we^{-r^2/\Omega^2}} + \int_0^b e^{-r^2/\Omega^2} r\,dr\right]$$

$$= \left(1-\frac{u}{w}\right)\frac{\Omega^2}{2w}\ln\left(\frac{1+w}{1+we^{-b^2/\Omega^2}}\right) + \frac{u}{w}\frac{\Omega^2}{2}(1-e^{-b^2/\Omega^2})$$

Thus,

$$\frac{dP_p}{dz} = -\frac{\sigma_{pa} N_0 P_p(z)}{w}\left[\left(1-\frac{u}{w}\right)\ln\left\{\frac{1+w}{1+we^{-b^2/\Omega_p^2}}\right\} + u\left(1-e^{-b^2/\Omega_p^2}\right)\right] \tag{14.52}$$

Similarly, we can carry out the integration in equation (14.39) to obtain

$$\frac{dP_s}{dz} = \frac{\sigma_{sa} N_0 P_s(z)}{w}\left[-\left(1+\frac{v}{w}\right)\ln\left\{\frac{1+w}{1+we^{-b^2/\Omega_s^2}}\right\} + v\left(1-e^{-b^2/\Omega_s^2}\right)\right] \tag{14.53}$$

where

$$v = \eta_s \frac{P_p(z)}{P_{p0}} \tag{14.54}$$

and in the evaluation of the integral we have assumed $\Omega_p \simeq \Omega_s$. Note that whereas dP_p/dz is always negative, dP_s/dz can be either positive or negative, leading to either amplification or attenuation of the signal.

We will first solve equations (14.52) and (14.53) under the small signal approximation, which will be followed by the solution in the signal saturation regime.

14.7 Small signal approximation

In the small signal approximation

$$P_s(z) \ll \frac{I_{s0}}{f_s(r=0)} = \pi\Omega_s^2 I_{s0} = P_{s0} \tag{14.55}$$

and equations (14.52) and (14.53) reduce to

$$\frac{dP_p}{dz} = \sigma_{pa} P_{p0} N_0 \ln\left[\frac{1+\frac{P_p(z)}{P_{p0}}e^{-b^2/\Omega_p^2}}{1+\frac{P_p(z)}{P_{p0}}}\right] \tag{14.56}$$

$$\frac{dP_s}{dz} = \sigma_{sa} N_0 P_s \frac{P_{p0}}{P_p} \left[\frac{\eta_s P_p}{P_{p0}} (1 - e^{-b^2/\Omega_s^2}) + (1 + \eta_s) \ln \left\{ \frac{1 + \frac{P_p}{P_{p0}} e^{-b^2/\Omega_s^2}}{1 + \frac{P_p}{P_{p0}}} \right\} \right] \tag{14.57}$$

Example 14.3: We consider a typical erbium-doped fiber with the following characteristics:

core radius	a	$= 1.64\ \mu\text{m}$
NA		$= 0.21$
Doping radius	b	$= 1.64\ \mu\text{m}$
t_{sp}		$= 12 \times 10^{-3}$ s
Doping concentration		$= 6.8 \times 10^{24}\ \text{m}^{-3}$ ($\simeq$ 120 mole ppm)
Pump absorption cross section	σ_{pa}	$= 2.17 \times 10^{-25}\ \text{m}^2$
Signal absorption cross section	σ_{sa}	$= 2.57 \times 10^{-25}\ \text{m}^2$
Signal emission cross section	σ_{se}	$= 3.41 \times 10^{-25}\ \text{m}^2$
Pump wavelength	λ_p	$= 980$ nm
Signal wavelength	λ_s	$= 1550$ nm (14.58)

For these values of parameters, the Gaussian envelope approximation gives us

$$\Omega_p = 1.35\ \mu\text{m}$$

$$\Omega_s = 1.96\ \mu\text{m} \tag{14.59}$$

Also

$$I_{p0} = 7.81 \times 10^7\ \text{W/m}^2$$

Using these values we obtain

$$P_{p0} = 0.41\ \text{mW}$$

Now, under the small signal approximation,[4] the variation of pump power, signal power, and gain are described by equations (14.56) and (14.57); the corresponding numerical results (for an amplifier characterized by equation (14.58)) are plotted[5] in Figures 14.14–14.18.

Figure 14.14 shows the variation of pump power with the length of the fiber for different input pump power levels of 3, 5, and 7 mW. As expected, the pump power monotonically reduces as it propagates along the fiber. It is worthwhile to note that the decrease is initially almost linear when the pump power is high and becomes exponential for lower pump powers (see Problem 14.17).

Figure 14.15 shows the variation of signal power with length for an input signal power of $P_s(0) = 1\ \mu$W. The signal power increases initially and reaches a maximum value before beginning to reduce. The corresponding variation of

[4] The validity of the small signal approximation is discussed in Problem 14.6.

[5] Calculations were carried out by Dr. Parthasarathy Palai.

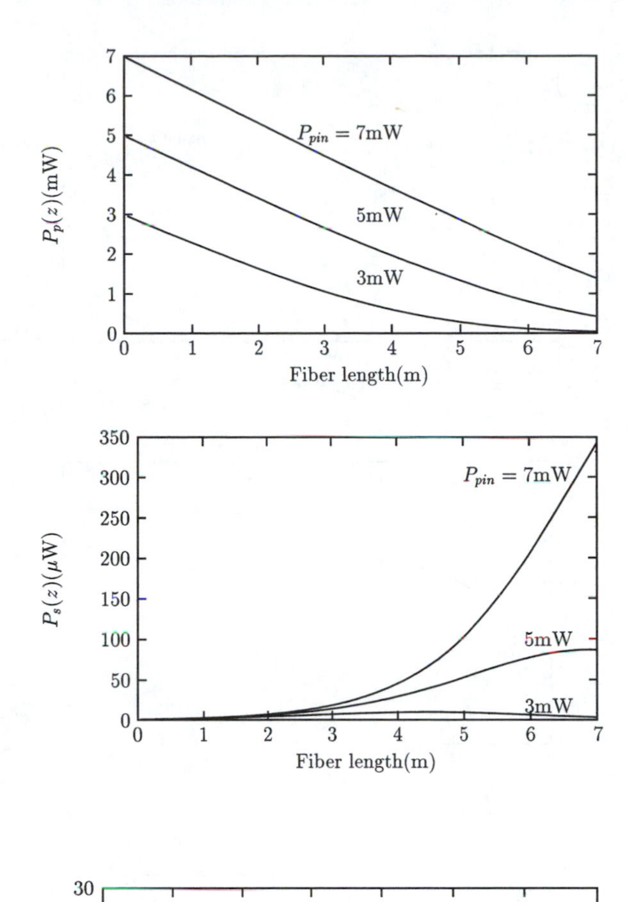

Fig. 14.14: Variation of pump power $P_p(z)$ with z corresponding to input pump powers of 3, 5, and 7 mW for an erbium-doped fiber described by the parameters given in equation (14.58). Note that around the input, the pump power is high and the decrease is almost linear, whereas for large z it becomes exponential.

Fig. 14.15: Variation of signal power $P_s(z)$ with z for an input signal power of 1 μW. Initially the signal power increases with z due to the presence of inversion. Beyond an optimum length (which is $\sim$ 7 m for $P_p(0) = 5$ mW), the signal gets attenuated. As is evident from the figure, the optimum length depends on the input pump power. The fiber parameters are those given in equation (14.58).

Fig. 14.16: The variation of gain with z corresponding to Figure 14.15. For maximum gain, the length of the fiber must be chosen equal to the optimum value. The fiber parameters are those given by equation (14.58).

gain is shown in Figure 14.16. It can be seen that for every input pump power level, there is an optimum length of the fiber for achieving maximum gain. For example, for $P_p(0) = 5$ mW, the optimum length is about 7 m. This behavior can be easily understood from the fact that, as the pump light propagates through the fiber, it gets absorbed by the fiber and the pump power level keeps falling monotonically with z (Figure 14.14). Thus, the inversion keeps reducing with z, which in turn reduces the gain at any value of z. At some z value the pump power drops below the critical value, wherein the fiber becomes attenuating rather than amplifying. Any fiber length beyond this point reabsorbs the amplified signal, thus reducing the gain. Indeed, Figure 14.17 shows the variation of the optimum length with input pump power.

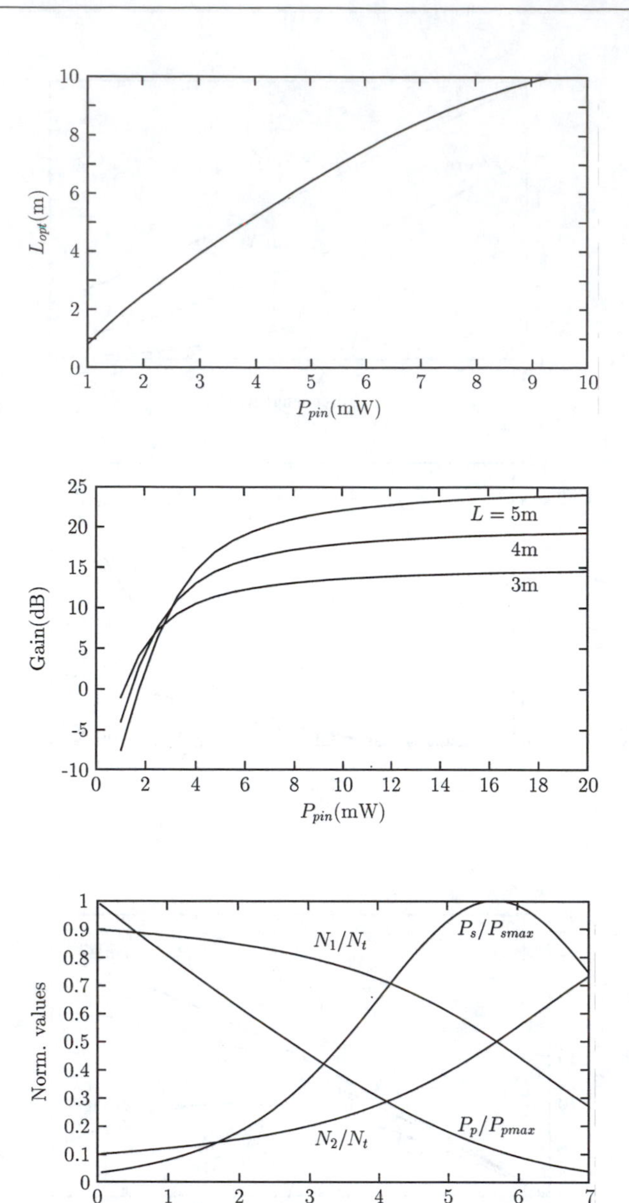

Fig. 14.17: Variation of optimum length with the input pump power for the fiber parameters given in equation (14.58).

Fig. 14.18: Variation of gain with input pump power for different lengths of the erbium-doped fiber. Note that the gain saturates to different values depending on the length of the fiber.

Fig. 14.19: Variation of N_1/N_t, N_2/N_t, P_p/P_{pmax}, and P_s/P_{smax} as a function of z for the fiber specifiied by equation (14.58). P_{pmax} and P_{smax} correspond to maximum values of P_p and P_s, respectively.

Figure 14.18 shows the variation of gain with the input pump power for different fiber lengths. Note that for any given fiber length there is a threshold pump power for transparency of the fiber. At this pump power, the signal gets neither attenuated nor amplified. Beyond this threshold pump power, the gain increases with increasing pump power, finally saturating at large pump powers. The saturation behavior is essentially due to the fact that as the pump power is increased, more and more erbium ions get inverted and, for large pump powers, almost the entire fiber is inverted. Hence, there would be no more increase in inversion and, hence, gain.

Figure 14.19 shows the variation of normalized values of N_1, N_2, P_p, and P_s as a function of z. Note that the signal power attains its maximum value close to $N_2 \simeq N_1$ and then starts to decrease.

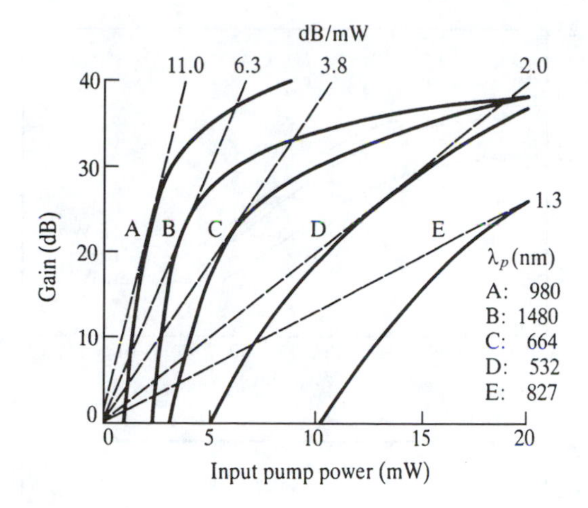

Fig. 14.20: Dependence of gain on input pump power corresponding to different pump wavelengths. The largest gain coefficient of 11 dB/mW corresponds to a 980-nm pump. [After Desurvire (1994).]

The largest value of the ratio of gain to pump power is referred to as the *gain coefficient* and is measured in dB/mW. This is one of the figures of merit of an amplifier and depends on the pump laser wavelength. Figure 14.20 shows the variation of gain with pump power corresponding to different pump laser wavelengths and shows that the maximum gain efficiency of 11 dB/mW has been achieved with a 980-nm pump laser. The maximization of gain coefficient can be achieved through proper optimization of the fiber parameters (core radius and NA) and confinement of Er ions within the fiber core.

14.8 Signal saturation

We first note that the quantity $(\eta_s N_2 - N_1)$ depends on I_s (see equation (14.20)), decreasing as I_s increases. This reduction leads to a reduction in the gain (see equation (14.27)). Thus, if the input signal is very high or if the gain is high, then the amplified signal can lead to saturation of the gain. This is referred to as signal-induced gain saturation. Thus, for large signal powers the gain is expected to decrease below the small signal value.

In the previous section we discussed the characteristics of the amplifier under the small signal approximation – that is, $I_s \ll I_{s0}$ at all z. This is true for low signal input powers and low gains. If the signal input becomes large or if the gain becomes high so that $I_s(z)$ becomes comparable to or larger than I_{s0}, then we must solve equations (14.52) and (14.53) to obtain the evolution of pump and signal powers with z in the presence of signal saturation.

Figure 14.21(a) shows a typical plot of gain versus input signal power, Figure 14.21(b) shows the corresponding gain variation with output signal power, and Figure 14.21(c) shows the variation of output amplified signal power with input signal power for the same fiber parameters as given by equation (14.58).

Note that as the input (or the output) signal power increases, the gain decreases from the small signal value. The output signal power where the gain is reduced by 3 dB compared with the small signal value is called *saturation output power*. Typically, the saturation output power is around 10 mW. The

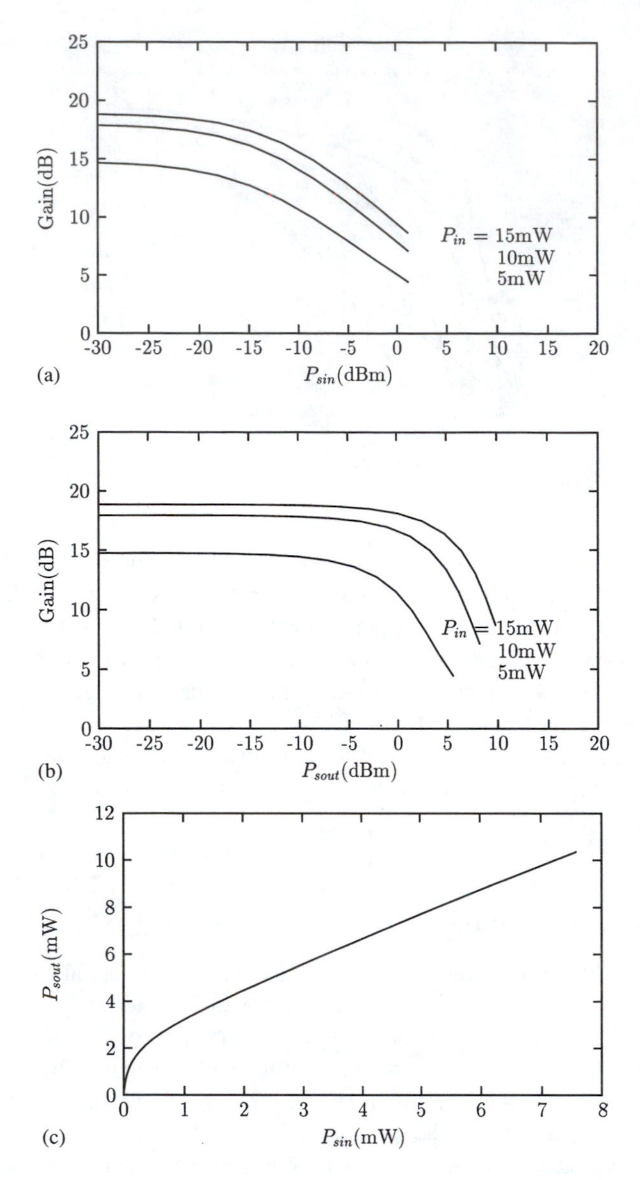

Fig. 14.21: Variation of gain with (a) input signal power, (b) output signal power, and (c) output amplified signal power with input signal power for an erbium-doped fiber described by equation (14.58) and for different input pump powers. Note that, for small input signal powers, the gain is almost independent of the signal power and for large signal powers the gain decreases.

saturation output power is an important parameter specifying a power amplifier.

14.9 Gain spectrum and gain bandwidth

Since the absorption and emission cross sections depend on the signal wavelength (see Figure 14.8 and Table 14.1), the gain of an erbium-doped fiber will depend on the signal wavelength. Figure 14.22 shows a typical spectral dependence of gain of an erbium-doped fiber amplifier pumped at 980 nm with different pump powers. The parameter γ is defined as

$$\gamma = \frac{P_p(0)}{P_{p0}} = \frac{P_p(0)}{A_p}\left(\frac{\sigma_p t_{sp}}{h\nu_p}\right) \tag{14.60}$$

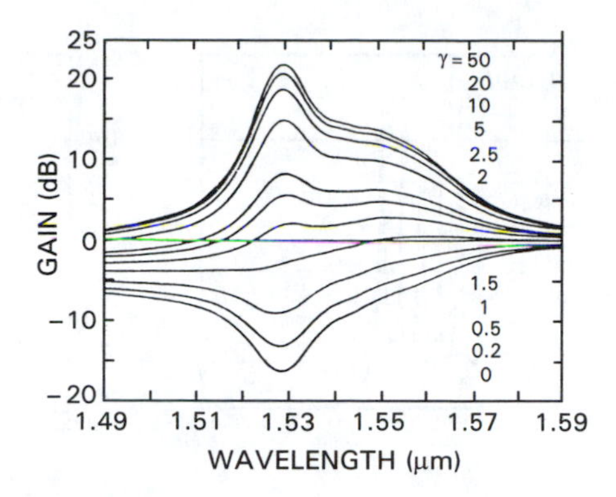

Fig. 14.22: Typical variation of gain with signal wavelength of a 980-nm pumped EDFA for different input pump powers. The parameter γ is defined in equation (14.60); $\gamma = 0$ corresponds to zero input pump power and increasing value of γ to increasing pump powers. [After Desurvire (1990).]

where $P_p(0)$ is the input pump power. Thus, $\gamma = 0$ corresponds to an unpumped fiber and the large value of γ represents the high pump power regime.

The figure clearly shows that for a given input pump power, the fiber exhibits gain for wavelengths greater than a specific wavelength and exhibits attenuation at wavelengths shorter than this specific wavelength. As the pump power increases, the specific wavelength for which the net gain is zero (i.e., there is neither amplification nor attenuation) moves toward shorter wavelengths.

The fact that for a given pump power, there is net amplification on the long-wavelength side and a net attenuation on the short-wavelength side of a specific wavelength can be understood because absorption from the ground level is smaller at longer wavelengths than at shorter wavelengths and fluorescence is larger at longer wavelengths than at shorter wavelengths (see Figure 14.8). For a signal wavelength of 1550 nm, the fiber becomes transparent for $\gamma \simeq 1$ – that is, for $P_p(0) \simeq P_{p0}$. For large pump powers, the gain does not change appreciably due to complete inversion in the medium. Figure 14.22 also shows that apart from a peak around 1535 nm, the gain is almost uniform over a bandwidth of approximately 25 nm or so. Thus, such an amplifier is capable of amplifying simultaneously optical signals over a band of wavelengths in a region of about 25 nm. This is very interesting from the point of view of application in WDM systems (see Section 14.11).

Example 14.4: Let us calculate the gain bandwidth in the frequency domain corresponding to a gain bandwidth of 30 nm in the wavelength domain centered around 1550 nm. Since $\nu = c/\lambda_0$, we have

$$\frac{\Delta\nu}{\nu} = \frac{\Delta\lambda_0}{\lambda_0}$$

Hence,

$$\Delta\nu = \frac{\nu}{\lambda_0}\Delta\lambda_0 = \frac{c}{\lambda_0^2}\Delta\lambda_0 \simeq 3.7 \text{ THz}$$

which is an enormous bandwidth. This huge bandwidth can be fully exploited using WDM schemes.

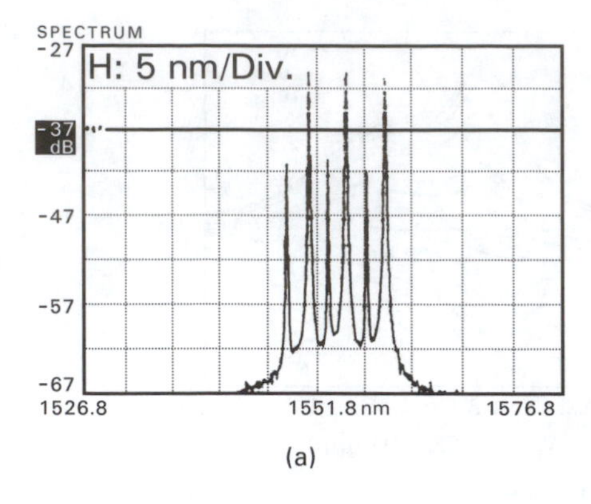

(a)

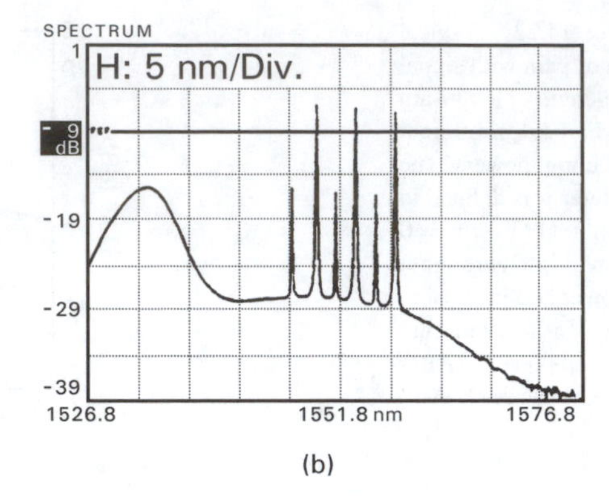

(b)

Fig. 14.23: (a) Spectrum of the signal input into an EDFA and (b) the corresponding amplifier output. Note the generation of noise over the entire fluorescence band due to ASE. [After Chen et al. (1995).]

14.10 Noise in EDFA

We have seen that amplification of the optical signal takes place by stimulated emission from an upper energy level to the ground level in Er^{3+} ions. The ions sitting in the higher energy state can also emit radiation by the process of spontaneous emission. This spontaneous emission appears over the entire fluorescent band and is completely incoherent with respect to the signal beam. Some of this spontaneously emitted radiation is coupled into the LP_{01} mode of the erbium-doped fiber and as it propagates through the amplifying fiber (in both forward and backward directions) gets amplified just like the signal. The resultant optical radiation is referred to as amplified spontaneous emission (ASE). This incoherent radiation also propagates with the signal and interferes with the signal when it is detected by a photodetector. This results in the generation of noise that ultimately limits the receiver sensitivity of an optical fiber transmission system.

Figure 14.23 shows the spectrum of an input signal and the output amplified spectrum from an EDFA. It is obvious that along with the amplification process, the EDFA also generates a background spontaneous emission noise over the entire gain band.

The noise in the wavelength regions outside of the signal spectrum can be filtered by using optical wavelength filters with a pass band coinciding with the signal spectrum. On the other hand, the noise added in the region of signal spectrum interferes with the detection process and results in the noise of the amplifier.

The noise figure F of an amplifier is defined as the ratio of the input signal to noise ratio ($(\mathrm{SNR})_i$) to that of the output signal to noise ratio ($(\mathrm{SNR})_o$).

$$F = \frac{(\mathrm{SNR})_i}{(\mathrm{SNR})_o} \tag{14.61}$$

It can be shown that for high gain amplifiers [see, e.g., Yariv (1991), Desurvire (1994)]

$$F \simeq 2\frac{N_2}{(N_2 - N_1)} \tag{14.62}$$

where N_1 and N_2 are the population densities of the lower and upper amplifier levels.

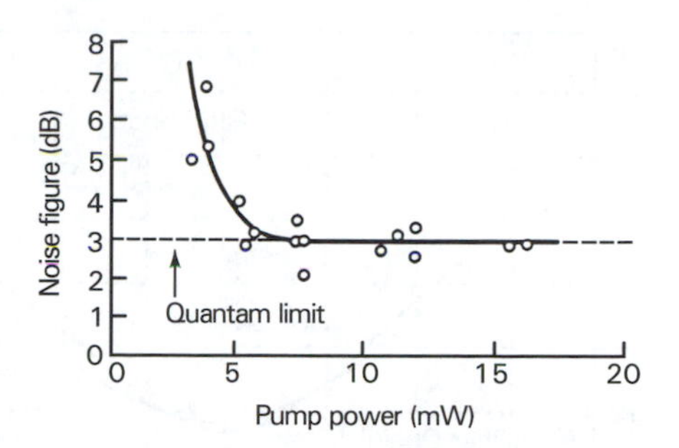

Fig. 14.24: Dependence of noise figure of an amplifier pumped at 980 nm on the input pump power. [After Laming and Payne (1990).]

Minimum noise figure is obtained when $N_2 \gg N_1$ (i.e., the population density of the lower level is negligible compared with that of the upper level) and F is equal to 2. This corresponds to the theoretical limit of noise figure and in the logarithmic scale is 3 dB. An F value of 2 implies that the SNR at the output of the amplifier is degraded by a factor of 2 with respect to the SNR at the input.

For the case of pumping with a 980-nm pump, one can achieve the condition $N_2 \gg N_1$ and, hence, a noise figure of 3 dB; on the other hand, since the 1480-nm pump wavelength corresponds to the same levels used for amplification, the pump itself induces downward transitions (in addition to upward transitions) and in this case it is not possible to have $N_2 \gg N_1$. Thus, the noise figure with the 980-nm pumping scheme is lower than with the 1480-nm pumping scheme. This is one of the advantages of using a 980-nm pump.

Figure 14.24 shows the dependence of noise figure on the input pump power at 980 nm for an input signal power of 1 μW at 1535 nm and shows that as the pump power increases beyond a certain value (for a given length of doped fiber, which was 11 m in this case), the noise figure reaches the theoretical limit of 3 dB. As the input pump power increases, there is a greater amount of population inversion and, hence, greater gain. As the inversion increases, the factor $N_2/(N_2 - N_1)$ in equation (14.62) decreases and beyond a certain gain (typically, 15 dB), the noise figure reaches the lowest possible value.

14.11 EDFAs for WDM transmission

There are two routes to increasing the capacity of a fiber optic communication system. One involves increasing the bit rate, which would require sophisticated time division multiplexing and demultiplexing components. The other route involves using multiple signal wavelengths to carry different channels, also referred to as WDM. The second route is not attractive using conventional electronic repeaters since each WDM channel would need an individual opto-electronic repeater. On the other hand, the appearance of EDFAs has made the latter choice very attractive and today WDM looks to be the most promising route to enhancing the capacity.

Since one of the major advantages of an EDFA over the conventional electronic amplifier is the ability of the EDFA to simultaneously amplify many signals at different wavelengths, EDFAs can be advantageously used for expanding the capacity of the fiber optic system using WDM. Such multiplexing schemes are limited by a variety of phenomena, including fiber nonlinearities (in which four-wave mixing is the most significant), wavelength-dependent gain of the EDFAs, and accumulation of noise in the channels.

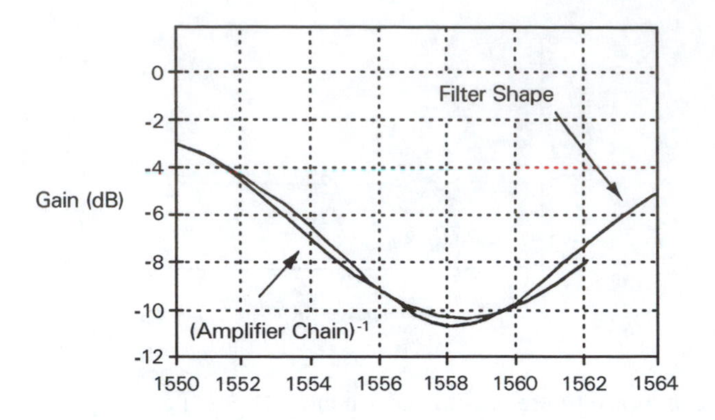

(a)

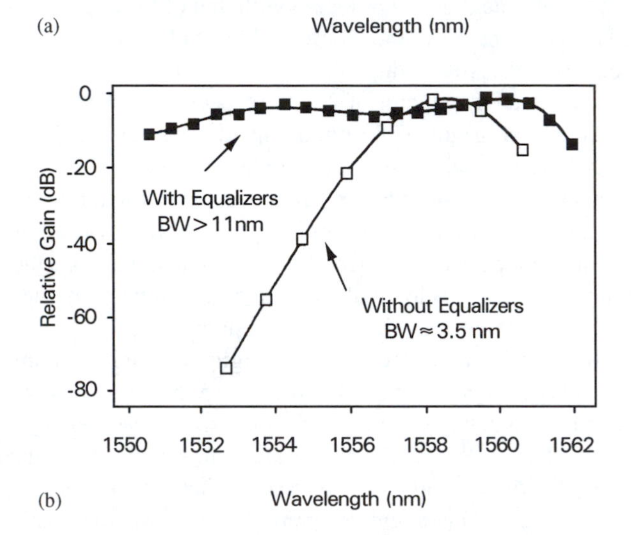

(b)

Fig. 14.25: (a) The variation of the inverse spectral response with wavelength and the corresponding filter spectral variation of a long-period grating. (b) The relative spectral gain variation for a 6300-km amplifier chain with and without passive gain equalization. [After Bergano and Davidson (1996).]

Unfortunately, the inherent wavelength dependence of the gain and noise characteristics of EDFAs implies that different wavelengths in the signal will have different gain and noise characteristics and, hence, will suffer from an imbalance in the SNRs. Thus, to employ erbium-doped fibers in WDM systems, it is very important that the gain and noise characteristics of the various wavelengths should be almost the same. For a conventional EDFA, the gain spectrum is, in general, not flat (see Figure 14.22) and there is a significant spectral variation of gain. There are many techniques that currently are being studied for gain flattening in EDFAs, including aluminum codoping [Yoshida, Kurvano, and Iwashita (1995)], preemphasis of the signal levels to compensate for the differential gain [Chraplyvy, Nagel, and Tkach (1992)], using optical bandpass and notch filters [Shimojoh et al. (1996)], using erbium-doped fluoride fibers [Miyajima et al. (1994)], using blazed Bragg gratings [Kashyap, Wyatt, and Mckee (1993)], using a twin-core doped fiber [Zervas and Laming (1995)], and using long-period gratings [Vengsarkar et al. (1996a,b)].

As an example, Figure 14.25(a) shows the filter shape of a long-period fiber Bragg grating (see also Section 17.11) compared with the inverse gain spectrum of an amplifier chain made up of ten EDFA spans; Figure 14.25(b) shows the corresponding relative gain variation of the chain with and without the filter.

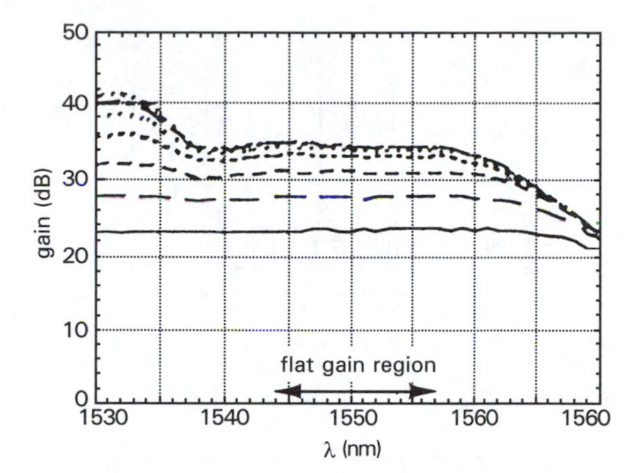

Fig. 14.26: Variation of gain with signal wavelength of a high Al concentration erbium-doped fiber. The gain flattened region extends from 1544 to 1557 nm. [After Yoshida et al. (1995).]

It can be seen that with the filter, the gain bandwidth of the chain has been increased from 3.5 nm to over 11 nm, showing a three-fold improvement.

Another solution for gain flattening involves doping the active fiber with a high concentration of aluminum [Yoshida et al. (1995)]. With 2.9 weight % of Al concentration, an extremely flat gain characteristic with gain excursion of less than 0.02 dB/nm/stage (in a multiamplifier system) from 1544 to 1557 nm has been demonstrated (see Figure 14.26). The flat gain characteristic was also verified by conducting a 16-channel, 100-GHz spaced WDM experiment.

Figure 14.27 shows the input and output spectra of 10 WDM channels with each channel operating at 10 Gb/s after propagation through 1200 km of fibers with 11 EDFAs in the link (repeater spacing of 100 km). The result shows the signal excursion is less than 2 dB and also the absence of any four-wave mixing (FWM) components.

One of the major problems that a WDM system has to counteract is the effect of fiber nonlinearities on the propagating signals. Among all nonlinear effects, FWM is the most deleterious. FWM is caused by the mixing of three different signal frequencies to generate a fourth signal frequency. Thus, if ω_1, ω_2, and ω_3 are three signal frequencies propagating through the fiber, FWM results in the generation of frequencies given by

$$\omega_1 \pm \omega_2 \pm \omega_3$$

If any of these frequencies coincides with an existing signal frequency, then it will result in cross talk between the different channels. Different techniques have evolved to reduce the effects of FWM, including operation at a wavelength away from zero dispersion wavelength, which would reduce the efficiency of FWM because of nonphase matching of different frequency components. Indeed, such nonzero dispersion fibers are becoming very important in system applications. Another route is to use unequal channel spacing, which would not reduce FWM but would not lead to any cross talk. Yet another technique referred to as dispersion management is to use fibers possessing positive and negative dispersion coefficients in the link so that, although the total dispersion in the link may be very small, the different wavelengths are not phase matched in each fiber link, thus reducing the effects of FWM.

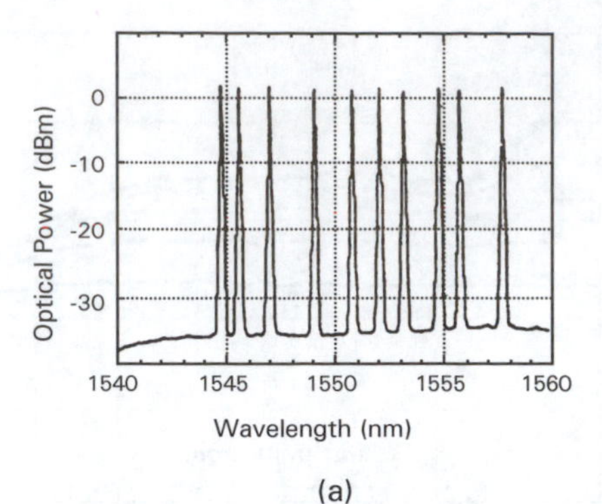

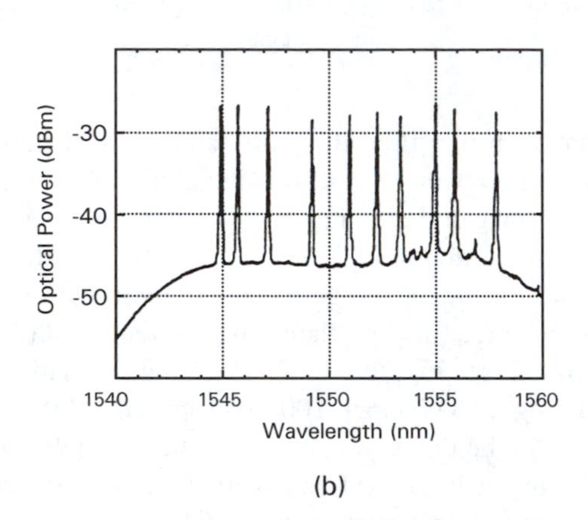

Fig. 14.27: Optical spectrum of 10 WDM channels each operating at 10 Gb/s (a) before and (b) after transmission through 1200 km of fiber with 11 EDFAs in the link. [After Yoshida et al. (1996).]

14.12 Comparison with plane wave case

For an infinitely extended medium, doped uniformly with erbium with a concentration of N_0 ions/m^3, with plane waves of intensity $I_p(z)$ and $I_s(z)$ propagating through the medium, the populations N_1 and N_2 would be independent of r and equations (14.26) and (14.27) then would be independent of r. In such a case, considering area S of the beams, we can transform equations (14.26) and (14.27) in terms of powers $P_p(=I_pS)$ and $P_s(=I_sS)$ to

$$\frac{dP_p}{dz} = -\sigma_{pa} N_1 P_p \tag{14.63}$$

$$\frac{dP_s}{dz} = \sigma_{sa}(\eta_s N_2 - N_1) P_s \tag{14.64}$$

which under small signal power approximation and using equations (14.19) and (14.20) can be written as

$$\frac{dP_p}{dz} = \frac{-\sigma_{pa} N_0 P_p}{1 + \tilde{P}_p} \tag{14.65}$$

$$\frac{dP_s}{dz} = \sigma_{sa} N_0 P_s \frac{\eta_s \tilde{P}_p - 1}{\tilde{P}_p + 1} \tag{14.66}$$

where $\tilde{P}_p = P_p/SI_{p0} = \tilde{I}_p$. When the pump power is also small, we may assume $\tilde{P}_p \ll 1$ and equations (14.65) and (14.66) become

$$\frac{dP_p}{dz} = -\sigma_{pa}N_0P_p; \qquad \frac{dP_s}{dz} = -\sigma_{sa}N_0P_s \tag{14.67}$$

which can easily be integrated to give

$$P_p(z) = P_p(0)\,e^{-\sigma_{pa}N_0\,z}; \quad P_s(z) = P_s(0)\,e^{-\sigma_{sa}N_0\,z} \tag{14.68}$$

thus giving the pump and signal absorption coefficients of $\alpha_p' = \sigma_{pa}\,N_0$ and $\alpha_s' = \sigma_{sa}\,N_0$.

On the other hand, for a doped fiber, as discussed in Section 14.7, the z variation of pump and signal powers (under small pump and signal power approximation) are given by (from equations (14.56) and (14.57))

$$\frac{dP_p}{dz} = -\sigma_{pa}N_0\zeta_pP_p \tag{14.69}$$

$$\frac{dP_s}{dz} = -\sigma_{sa}N_0\zeta_sP_s \tag{14.70}$$

where $\zeta_p = (1 - e^{-b^2/\Omega_p^2})$ and $\zeta_s = (1 - e^{-b^2/\Omega_s^2})$. From equations (14.69) and (14.70) we get the effective absorption coefficients of $\alpha_p = \sigma_{pa}N_0\zeta_p$ and $\alpha_s = \sigma_{sa}\,N_0\,\zeta_s$ for the case of a fiber. Hence, comparing with the plane wave case, we note that the factors ζ_p and ζ_s take account of the fact that the doping is not uniform ($N_t = 0$ for $r > b$) and that the pump and signal waves have a transverse intensity distribution. Indeed, if we let $b \to \infty$, then $\zeta_p = \zeta_s = 1$ and the fiber case (for small pump and signal powers) reduces to the plane wave case.

14.13 Relation between mole ppm, weight ppm, and ions/cm^3 of Er concentration

The erbium doping concentration in a doped fiber is specified usually in terms of mole ppm (parts per million) or weight ppm or in terms of number of Er^{3+} ions per cm^3. In this section we will show a relationship between these. Let M represent the mole ppm of Er_2O_3 in a doped pure silica fiber. This implies that in 1 mole of the composite material there is (1–10^{-6} M) mole of SiO_2 and 10^{-6} M mole of Er_2O_3. Since the molecular weight of SiO_2 is 60.1 and that of Er_2O_3 is 382.6, the weight of 1 mole of doped SiO_2 is

$$(1 - 10^{-6}\text{ M})\,60.1 + \text{M} \times 10^{-6}\,382.6 \simeq 60.1\text{ g}$$

Since 1 mole of any substance contains 6.02×10^{23} molecules, 60.1 g of the composite material will contain 6.02×10^{23} molecules. If D (g/cc) is the mass density of the doped material, then this implies that $60.1/D$ cm^3 of the material contains 6.02×10^{23} molecules. The weight of Er_2O_3 in this volume is 382.6×10^{-6} M g. Now, 382.6 g of Er_2O_3 contains 6.02×10^{23} molecules of Er_2O_3. Hence, 382.6×10^{-6} M g of Er_2O_3 will contain 6.02×10^{17} M molecules of

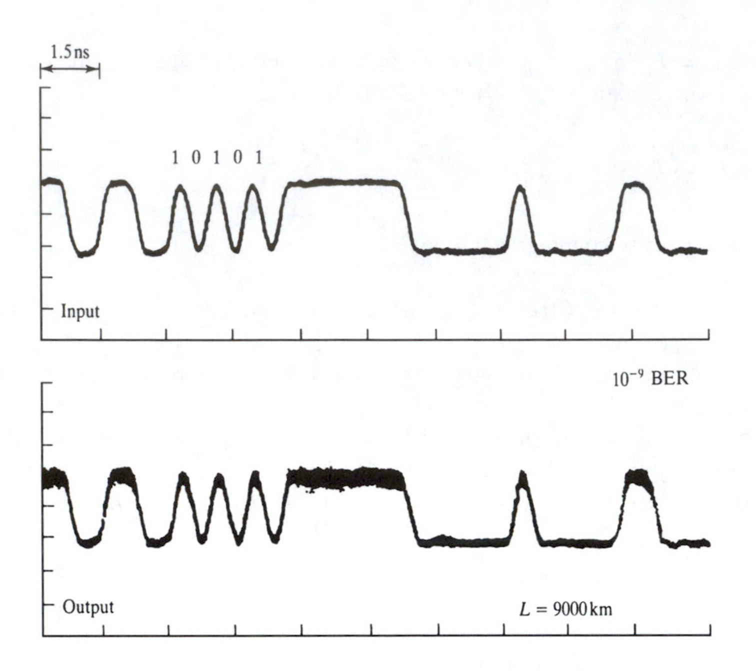

Fig. 14.28: The upper trace shows a pseudorandom bit sequence at 5 Gb/s in NRZ format at the input, and the lower trace shows the corresponding output after the pulses have propagated through 9000 km of fiber with 274 EDFAs. Note that the output pulses are almost free from distortion and noise even after 9000 km of propagation. [After Desurvire (1994).]

Er_2O_3. Since every molecule of Er_2O_3 contains two ions of Er^{3+}, the number of Er^{3+} ions in this will be 12.04×10^{17} M. Thus, $60.1/D$ cm^{-3} of the composite material contains 12.04×10^7 M ions of Er^{3+}. Hence, the Er^{3+} ion density is

$$\frac{1.204 \times 10^{18} \times D \times M}{60.1} \text{ ions/cm}^3$$

Hence,

$$C[\text{concentration in ions/m}^3] \simeq 5.73 \times 10^{22} \text{ M (mole ppm)} \tag{14.71}$$

where we have used the density D of SiO_2, which is about 2.86 g cm^{-3}.

Thus, 100 mole ppm corresponds to about 5.73×10^{24} ions/m^3.

14.14 Demonstration of some EDFA systems

There has been tremendous progress in the demonstration of various high-capacity long-distance optical fiber communication systems, and new record figures are being achieved at a tremendous pace. In this section we will briefly mention some representative system demonstrations.

Using 274 EDFAs over a 9000-km fiber optic link, groups at AT & T in the United States and KDD in Japan have demonstrated error-free performance at 5 Gb/s using NRZ signals. Figure 14.28 shows the pulse sequence at the input and the output (after propagating through 9000 km of fiber) and 274 EDFAs. It is obvious that the pulses are almost undistorted and are almost free from noise.

Figure 14.29 shows an experiment conducted by British Telecom demonstrating the broadcast of signals at 39.81 Gb/s (using 16 DFB lasers operating

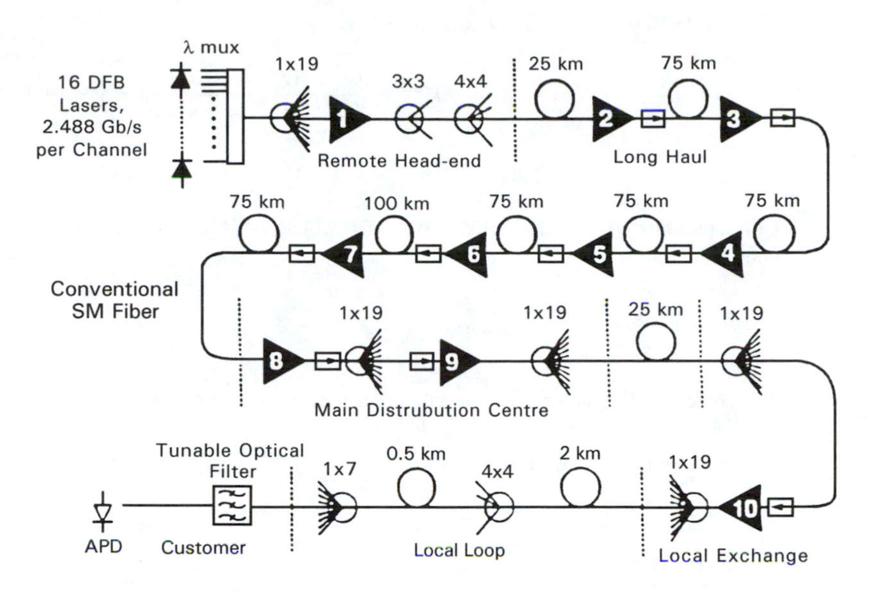

Fig. 14.29: The broadcast distribution network demonstrated by British Telecom for transmission to 43.8 million potential users within a 527-km range at 39.81 Gb/s using 16 DFB lasers, each sending signals at 2.448 Gb/s. [After Forrestier et al. (1991).]

at slightly different wavelengths, each one carrying a data rate of 2.488 Gb/s) to 43.8 million potential users within a 527-km range. This experiment shows the immense potential offered by EDFAs in future broadcast networks.

Transoceanic systems installed across the Atlantic ocean in 1996 and the Pacific ocean in 1997 will offer 5 Gb/sec capacity per fiber. The latter system is expected to transmit 200-ps pulses over the 9000-km distance between North America and Japan without any regeneration.

Experimental demonstration of speeds in excess of 1 Tb/s are becoming common [Chraplyvy et al. (1996), Onaka et al. (1996)]. Recently, a group from NEC in Japan has reported a record data transmission at a rate of 2.6 Tb/s over 120 km (European Conference on Optical Communication, 1996). The system used WDM transmission with 132 channels lying between 1529.03 nm and 1563.86 nm spaced at 33.3 GHz. Gain-flattened EDFAs were used in this record transmission experiment.

With loss taken care of by optical amplification, one has to design systems to overcome polarization mode dispersion and various nonlinear effects such as self-phase modulation, FWM in WDM systems, and so forth.

Problems

14.1 Consider an erbium-doped medium with 5×10^{24} m^{-3} Er^{3+} ions per unit volume. Using the absorption cross sections listed in Table 14.1, obtain the number of absorptions per unit time per unit volume if light of power 1 mW with a cross-sectional area of 1 mm^2 and wavelengths of 1535.9 nm and 1570.4 nm is incident on the medium. Obtain the corresponding absorption coefficients (assuming $N_2 \simeq 0$).

Solution: A 1-mW beam with a cross-sectional area of 1 mm^2 corresponds to

$$I = \frac{10^{-3}}{10^{-6}} = 10^3 \text{ W/m}^2$$

For $\lambda = 1535.9\,\text{nm}$, $\sigma_{sa} = 4.644 \times 10^{-25}\,\text{m}^2$ and

$$\left(\frac{dN_1}{dt}\right)_{\text{abs}} = 4.644 \times 10^{-25} \times \frac{10^3 \times 1535.9 \times 10^{-9}}{6.62 \times 10^{-34} \times 3 \times 10^8} \times 5 \times 10^{24}$$

$$\simeq 1.8 \times 10^{22}\ \text{m}^{-3}\text{s}^{-1}$$

Corresponding to $\lambda = 1570.4$ nm, $\sigma_{sa} = 0.934 \times 10^{-25}\,\text{m}^2$ and we obtain similarly

$$\left(\frac{dN_1}{dt}\right)_{\text{abs}} = 3.69 \times 10^{21}\ \text{m}^{-3}\ \text{s}^{-1}$$

The corresponding absorption coefficients are given by (see equation (14.10))

$$\alpha(1535.9\,\text{nm}) = 4.644 \times 10^{-25} \times 5 \times 10^{24}$$

$$\simeq 2.322\,\text{m}^{-1}$$

which is equivalent to ≈ 1 dB/m.

$$\alpha(1570.4\,\text{nm}) = 0.934 \times 10^{-25} \times 5 \times 10^{24}$$

$$= 0.4670\,\text{m}^{-1}$$

which is equivalent to ≈ 0.2 dB/m.

14.2 For the fiber parameters given by equation (14.58), assuming low signal power, obtain the threshold pump power required for amplification of the signal at any value of z.

Solution: To have amplification at any z value, dP_s/dz must be greater than zero. Thus, from equation (14.57) we obtain for the threshold pump power to have amplification

$$\frac{P_p}{P_{p0}}\eta_s(1 - e^{-b^2/\Omega_s^2}) + (1 + \eta_s)\ln\left[\frac{1 + \frac{P_p}{P_{p0}}e^{-b^2/\Omega_s^2}}{1 + \frac{P_p}{P_{p0}}}\right] = 0 \tag{14.72}$$

For a given fiber, we have to solve the above transcendental equation to obtain the value of P_p/P_{p0}. For the fiber parameters given by equation (14.58), the solution of equation (14.72) gives us

$$\frac{P_p}{P_{p0}} \simeq 0.95$$

which shows that for the signal wavelength of 1550 nm, amplification at any value of z can be achieved at $P_p = 0.95\,P_{p0}$. Note that this threshold pump power depends on signal wavelength through η_s and Ω_s.

14.3 For the same fiber considered in Problem 14.2, calculate the threshold pump powers corresponding to a signal wavelength of 1530 nm (for which $\sigma_{sa} = 5.25 \times 10^{-25}\ \text{m}^2$, $\sigma_{se} = 4.36 \times 10^{-25}\ \text{m}^2$). Assume Ω_s to be the same as in Problem 14.2.

Solution: Using equation (14.72), we obtain the threshold pump power required for amplification at any value of z as

$$P_p \simeq 1.51\, P_{p0}$$

Note that this is much larger than that obtained in Problem 14.2 for a signal wavelength of 1550 nm. The difference arises primarily because of the difference in η_s value.

14.4 Consider an erbium-doped fiber with $\Omega_p = 1.591\ \mu$m, $\Omega_s = 2.288\ \mu$m, and a core radius of 2.5 μm. Estimate the threshold pump power required for achieving amplification at any value of z for doping radii $b = a$, $0.5a$, $0.25a$, and $0.1a$ for a signal wavelength of 1550 nm for which $\eta_s = 1.327$. Assume $\sigma_{pa} = 2.17 \times 10^{-25}$ m^2, $t_{sp} = 12 \times 10^{-3}$ s, $\lambda_p = 980$ nm.

Solution: For the given values of parameter

$$P_{p0} = \pi \Omega_p^2 I_{p0} \simeq 0.62\,\text{mW}$$

Further, solving equation (14.72) for P_p for different values of b, we obtain the required threshold pump powers

(i) $b = a$ $\quad P_p \simeq 1.206\, P_{p0} \simeq 0.75$ mW
(ii) $b = 0.5a$ $\quad P_p \simeq 0.862\, P_{p0} \simeq 0.53$ mW
(iii) $b = 0.2a$ $\quad P_p \simeq 0.772\, P_{p0} \simeq 0.48$ mW
(iv) $b = 0.1a$ $\quad P_p \simeq 0.76\, P_{p0} \simeq 0.47$ mW

Note that the threshold pump power required for amplification decreases with decreasing b.

Indeed, for $b \to 0$, one can show that the solution of equation (14.72) is given by

$$P_p \simeq P_{p0}/\eta_s$$

For a given fiber, this limit corresponds to $P_p \simeq 0.754\, P_{p0} \simeq 0.467$ mW. Note that the threshold pump power required for signal amplification is independent of doping concentration; however, the gain will depend on the concentration of the erbium ions.

14.5 Assuming the pump intensity distribution to be Gaussian as given by equation (14.45) with $\Omega_p = 1.591\ \mu$m in an erbium-doped fiber of core radius 2.5 μm with the entire core doped with 2.44×10^{24} erbium ions/m^3, at what pump power will the entire core cross section have $N_2 \geq N_1$? Assume $\eta_s = 1$.

Solution: From equation (14.21) we note that for obtaining $N_2 \geq N_1$, the local pump intensity should satisfy

$$I_p(r) > I_{p0}$$

Now, due to the Gaussian nature of the pump intensity distribution, the local intensity is maximum on axis and decreases with increasing r. For achieving inversion in the entire core, we must have

$$I_p(r = a) = P_p f_p(r = a) \geq I_{p0}$$

Using $I_{p0} = 5.46 \times 10^7$ W/m^2 (see Example 14.1) and equation (14.51), we get

$$P_p \geq \frac{I_{p0}}{f_p(r=a)} = \pi \Omega_p^2 \, e^{a^2/\Omega_p^2} \cdot I_{p0}$$

$$\simeq 5.13\,\text{mW}$$

Thus, at $P_p < 5.13$ mW, only a fraction of the doped core will be inverted while the remaining portion would have $N_2 < N_1$ and, hence, would be absorbing.

14.6 In the small signal approximation, we assume that over the entire amplifier length $\tilde{I}_s = I_s/I_{s0} \ll 1$. For a typical value of I_{s0} of 4.07 $\times 10^7$ W/m^2 (see Example 14.1) and an Ω_s of 1.96 μm, determine what this implies in terms of signal power.

Solution: Since the maximum intensity is at $r = 0$, we shall consider the implications of this at $r = 0$. Now from equations (14.29) and (14.46), we obtain at $r = 0$,

$$I_s\,(0, z) = P_s/\pi \Omega_s^2$$

For

$$I_s\,(0, z) \ll I_{s0}$$

we have

$$P_s(z) = \pi \Omega_s^2 \, I_s\,(0, z) \ll \pi \Omega_s^2 \, I_{s0}$$

Substituting the values of I_{s0} and Ω_s, this implies

$$P_s(z) \ll 0.49\,\text{mW}$$

Hence, for input power levels of less than $\sim 1\,\mu$W and gains of less than about 20 dB, the small signal approximation will work well.

14.7 Maximum efficiency of energy conversion from pump to signal is reached when every pump photon results in one signal photon due to stimulated emission. Assuming such a condition, estimate the maximum efficiency of conversion for pump wavelengths of 980 nm and 1480 nm and a signal wavelength of 1550 nm.

Solution: If ν_p and ν_s are the pump and signal frequencies, then the maximum efficiency is given by

$$\eta_m = \frac{h\nu_s}{h\nu_p} = \frac{\lambda_p}{\lambda_s}$$

$$\simeq 63\% \text{ for } \lambda_p = 980 \text{ nm}$$

$$\simeq 95\% \text{ for } \lambda_p = 1480 \text{ nm}$$

14.8 An erbium-doped fiber has a doped core with a dopant concentration of 10^{24} m^{-3} with $\Omega_p = 1.59$ μm, $\Omega_s = 2.29$ μm. Given that the absorption cross sections at 980 nm and 1535 nm are 1.57×10^{-25} m^2 and 1.75×10^{-25} m^2, obtain the absorption coefficient of the fiber at these two wavelengths for low light levels.

Solution: For low light levels (which are conventionally used for measuring the absorption coefficient of a fiber) we may assume that only a very small fraction of atoms get excited from the ground state. Thus, we may assume $N_1 \simeq$ constant $= 10^{24}$ m^{-3}. The power in the light beam varies according to the following equation (see equations (14.69) and (14.70))

$$\frac{dP}{dz} \simeq -\sigma N_0 \zeta P$$

where $P = P_p$ or P_s, $\sigma = \sigma_{pa}$ or σ_{sa}, and $\zeta = \zeta_p$ or ζ_s for the pump and signal, respectively. Thus

$$P(z) = P(0)\, e^{-\sigma N_0 \zeta z}$$

The absorption coefficient in dB/m is given by

$$\alpha = -10 \log\left[\frac{P\,(1\text{ m})}{P(0)}\right] = 10\sigma\zeta N_0 \log_{10} e$$

$$\simeq 0.62 \text{ dB/m at 980 nm}$$

$$\simeq 0.53 \text{ dB/m at 1550 nm}$$

Note that if the entire fiber cross section is doped, then $b \to \infty$ and $\zeta_p \simeq 1$, $\zeta_s \simeq 1$. For this case the absorption coefficients would be 0.68 dB/m and 0.76 dB/m at the pump and signal wavelengths, respectively.

14.9 Consider an erbium-doped fiber containing 20 ppm of erbium, which corresponds to a concentration of 4.41×10^{23} ions/m^3. Assuming well-confined mode, obtain the absorption coefficient of the fiber at 1535 nm, given that the absorption cross section at 1535 nm is 5.57×10^{-25} m^2. ($\alpha = -10\sigma N_t \log_{10} e \simeq 1$ dB/m [Ohashi and Tsubshawa (1991)].

14.10 An optical amplifier can become a source of radiation – that is, a laser if optical feedback is provided. Such a feedback is provided by placing the amplifier between a pair of mirrors. The reflections from the bare ends of a fiber amplifier could act as feedback, forcing the fiber amplifier into oscillation. Estimate the threshold small signal gain at which the reflections from the fiber ends make the fiber amplifier into an oscillator.

Solution: The refractive index of the fiber core is about 1.46. Assuming a perfectly perpendicular cut, the reflection coefficient from each fiber end would approximately be

$$R \simeq \left(\frac{1.46 - 1}{1.46 + 1}\right)^2 \simeq 0.035$$

If an input signal power P_s^{in} becomes P_s^{out} at the output of the amplifying fiber, then the single pass gain G is

$$G = 10 \log \frac{P_s^{\text{out}}}{P_s^{\text{in}}}$$

or

$$P_s^{\text{out}} = P_s^{\text{in}} 10^{G/10} = P_s^{\text{in}} e^{0.2303\, G}$$

Laser oscillation threshold corresponds to an exact compensation of loss by gain. Assuming the only loss to be the mirror reflections, the threshold corresponds to

$$P_s^{\text{in}} e^{0.2303\, G} \cdot R \cdot e^{0.2303\, G} R = P_s^{\text{in}}$$

or

$$G = -\frac{1}{0.2303} \ln R$$

For $R = 0.035$, this gives $G \simeq 14.6$ dB. Such gains are easily obtainable in an erbium-doped fiber amplifier. Thus, at $G \gtrsim 14.6$ dB, the fiber amplifier may begin to oscillate to form a fiber laser just from the feedback provided by Fresnel reflections from the cleared fiber end.

14.11 Consider an erbium-doped fiber with a core radius of 1.9 μm, NA = 0.20 doped in the entire core with erbium with a concentration of 1000 ppm by weight.

(a) Calculate the cutoff wavelength of the fiber.
(b) Obtain the Gaussian envelope mode sizes at 980 nm and 1530 nm.
(c) Assuming $\sigma_{pa} = 3.1 \times 10^{-25}$ m^2, $\sigma_{sa} = 7 \times 10^{-25}$ m^2, obtain the absorption coefficient of the fiber at 980 nm and 1530 nm.
(d) Obtain the transparency pump power for a fiber length of 20 m.
(e) For this length calculate the pump power required for a small signal gain of 20 dB.

14.12 Find the characteristics of an erbium-doped fiber with the same characteristics as in Problem 14.11 except that the same doping is up to a radius of only 0.75 μm and 0.5 μm. Compare the performance with the case in which the entire core is doped.

14.13 Consider an erbium-doped fiber with $a = 1.8$ μm, NA = 0.19, and whose entire core cross section is doped with 4370 ppm by weight of Er.

(a) Obtain the cutoff wavelength.
(b) Obtain the absorption coefficient at 980 nm and at 1535 nm.

14.14 Consider a step index erbium-doped fiber with a core diameter of 2.2 μm and NA of 0.3 doped within the entire core with Er_2O_3 concentration of 40 ppm.

(a) Calculate the cutoff wavelength of the fiber.

[ANSWER: $\lambda_c = 0.86$ μm]

(b) Calculate the Er concentration in ions/m^3.

[ANSWER: 2.29×10^{24}m^{-3}]

(c) Obtain the corresponding small signal absorption coefficients of the fiber at 980 nm and 1532 nm. Assume $\sigma_{pa} = 3.1 \times 10^{-25}$ m^2 and $\sigma_{sa} = 7 \times 10^{-25}$ m^2.
(d) Obtain P_{p0} and P_{s0}.

Solution:

$$\begin{aligned} V\,(0.98\ \mu\text{m}) &= 2.116, & U\,(1.532\ \mu\text{m}) &= 1.353 \\ U\,(0.98\ \mu\text{m}) &= 1.5668, & V\,(1.532\ \mu\text{m}) &= 1.2361 \\ W\,(0.98\ \mu\text{m}) &= 1.4222, & W\,(1.532\ \mu\text{m}) &= 0.5502 \\ \Omega_p &= 0.925\ \mu\text{m} & \Omega_s &= 1.359\ \mu\text{m} \end{aligned}$$

$$\alpha_p = \sigma_p\, N_t\, \eta_p = 3.1 \times 10^{-25} \times 2.29 \times 10^{24} \times 0.7569,$$
$$R_p = 1.75 \times 10^{11}\ \text{m} = 0.5373\ \text{m}^{-1} = 2.3\ \text{dB/m}$$

$$\alpha_s = \sigma_s\, N_t\, \eta_s = 7 \times 10^{-25} \times 2.29 \times 10^{24} \times 0.4806$$
$$\simeq 0.7704\ \text{m}^{-1} \simeq 3.3\ \text{dB/m}$$

14.15 Calculate the minimum splice loss at 1550 nm achievable between an erbium-doped fiber with a core radius of 1.55 μm, NA = 0.22, and (a) a conventional SMF with an MFD of 10 μm and (b) a dispersion-shifted SMF with an MFD of 7 μm.

Solution: Using Gaussian approximation, the splice loss between two fibers with Gaussian spot sizes w_1 and w_2 is given by (see Chapter 8)

$$\text{Loss} = 20\ \log\left(\frac{w_1^2 + w_2^2}{2w_1 w_2}\right)$$

The given EDF has a V value of 1.38 at 1.55 μm. Using Marcuse's formula for spot size of a fiber (see Chapter 8), we obtain $w_2 \simeq 3.2\ \mu$m. Thus, the splice loss for the conventional SMF would be $\simeq$ 0.84 dB and for the DSF would be 0.03 dB.

Splice loss reduction in the optimization of an erbium-doped fiber amplifier is a subject of great importance [see, e.g., Zheng, Hulten, and Rylander (1994).]

14.16 Show that if the concentration of Er_2O_3 is specified as P ppm by weight, then this corresponds to approximately a $9.0 \times 10^{21}\ P\ \text{m}^{-3}$ concentration of Er^{3+}.

14.17 Consider propagation of only the pump beam through an erbium-doped medium. In such a case, using equations (14.19) and (14.26), we obtain

$$\frac{dI_p}{dz} = -\frac{\sigma_{pa}}{1 + I_p/I_{p0}} N_t I_p \tag{14.73}$$

Obtain the solution to the above equation when $I_p \ll I_{p0}$ and when I_p is comparable to I_{p0}.

Solution: For $I_p \ll I_{p0}$, we obtain

$$\frac{dI_p}{dz} = -\sigma_{pa} N_t I_p$$

whose solution is

$$I_p(z) = I_p\,(0)\, e^{-\sigma_{pa}\, N_t\, z}$$

showing an exponential decrease in intensity.

For I_p comparable to I_{p0}, we rewrite equation (14.73) as

$$\left(\frac{1}{\tilde{I}_p} + \frac{1}{I_{p0}}\right) dI_p = -\sigma_{pa} N_t \, dz$$

On integrating the above equation, we obtain

$$\ln\left[\frac{I_p(z)}{I_p(0)}\right] + \frac{1}{I_{p0}}\left[I_p(z) - I_p(0)\right] = -\sigma_{pa} N_t z \qquad (14.74)$$

For $I_p \gg I_{p0}$, equation (14.74) becomes

$$\frac{dI_p}{dz} = -\sigma_{pa} N_t$$

whose solution is

$$I_p(z) = I_p(0) - \sigma_{pa} N_t z$$

showing a linear decrease in pump intensity.

Figure 14.14 shows the linear decrease at large pump powers and an exponential decrease at small pump powers.

15

Dispersion compensation and chirping phenomenon

15.1 Introduction

In recent years there has been considerable work on dispersion-compensating fibers (DCFs), which are being used extensively for upgrading the installed 1310-nm optimized optical fiber links for operation at 1550 nm. In Section 15.2 we discuss the basic principle behind dispersion compensation, and in Section 15.3 we discuss the characteristics of DCFs. In Section 15.4 we discuss dispersion management to overcome the effects of nonlinear interactions in WDM systems. In Section 15.5 we show explicitly that broadening of an unchirped (Fourier transform limited) pulse is accompanied by a corresponding frequency chirp within the pulse – that is, the instantaneous frequency within the pulse does not remain constant but changes continuously from the leading edge to the trailing edge of the pulse. We also show that a chirped pulse can be made to undergo compression if passed through a proper dispersive medium. In Chapter 16 we show how such a chirping and pulse broadening can be compensated by using the nonlinear properties of the optical fiber leading to solitons.

15.2 Dispersion compensation

Let us consider a pulse (of spectral width $\Delta\lambda_0$) propagating through a fiber characterized by the propagation constant β. The spectral width $\Delta\lambda_0$ could be due to either the finite spectral width of the source itself or the finite duration of a Fourier transform-limited pulse. In Chapter 6 we considered the propagation of such a pulse and showed that the group velocity of the pulse is given by

$$\frac{1}{v_g} = \frac{d\beta}{d\omega} \tag{15.1}$$

For a conventional single-mode fiber with zero dispersion around 1300 nm, a typical variation of v_g with wavelength is shown by the solid curve in Figure 15.1. As can be seen from the figure, v_g attains a maximum value at the zero dispersion wavelength and on either side it monotonically decreases with

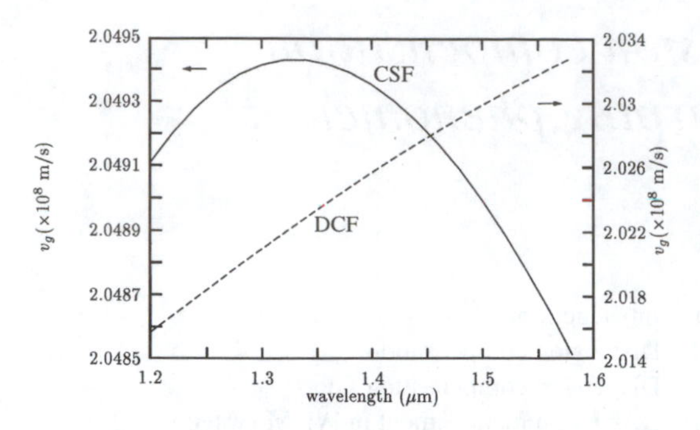

Fig. 15.1: Typical variation of v_g with λ_0 for a CSF (conventional single-mode fiber) and a DCF. At the operating wavelength, the CSF has (small) positive dispersion and the DCF has (large) negative dispersion. Notice that the fractional variation of v_g is much more for the DCF.

wavelength. Thus, if the central wavelength of the pulse is around 1.55 μm (see Figure 15.1), then the red components of the pulse (i.e., longer wavelengths) will travel slower than the blue components (i.e., smaller wavelengths) of the pulse. Because of this the pulse will get broadened. The leading edge (which appears earlier) of the output pulse is blue shifted and the trailing edge is red shifted.

Now, after propagating through such a fiber for a certain length L_1, we allow the pulse to propagate through another fiber where the group velocity varies, as shown by the dashed curve in Figure 15.1. The red components will now travel faster than the blue components and the pulse will tend to reshape itself into its original form. This is the basic principle behind dispersion compensation. Now, the total dispersion of a single-mode fiber is given by (see Chapter 8)

$$D_t = D_m + D_w = -\frac{2\pi c}{\lambda_0^2}\frac{d^2\beta}{d\omega^2} \tag{15.2}$$

Thus, $d^2\beta/d\omega^2 < 0$ implies operation at $\lambda_0 > \lambda_z$ and conversely. Let $(D_t)_1$ and $(D_t)_2$ be the dispersion coefficient of the first and second fiber, respectively. Thus, if the lengths of the two fibers (L_1 and L_2) are such that

$$(D_t)_1 L_1 + (D_t)_2 L_2 = 0 \tag{15.3}$$

then the pulse emanating from the second fiber will be identical to the pulse entering the first fiber.

To explicitly understand this, in Figure 15.2(a) we show the broadening of an unchirped pulse as it propagates through a fiber characterized by

$$(D_t)_1 > 0 \quad (\lambda_0 > \lambda_z)$$

Thus, because of the physics discussed above, the pulse gets broadened and chirps, the front end of the pulse gets blue shifted, and the trailing edge of the pulse gets red shifted (the details are given in Section 15.5). The pulse is said to be negatively chirped. If such a negatively chirped pulse is now propagated through another fiber of length L_2 characterized by

$$(D_t)_2 < 0$$

then the chirped pulse will get compressed (see Figure 15.2(b)), and, if the length satisfies equation (15.3), then the pulse dispersion will be exactly compensated.

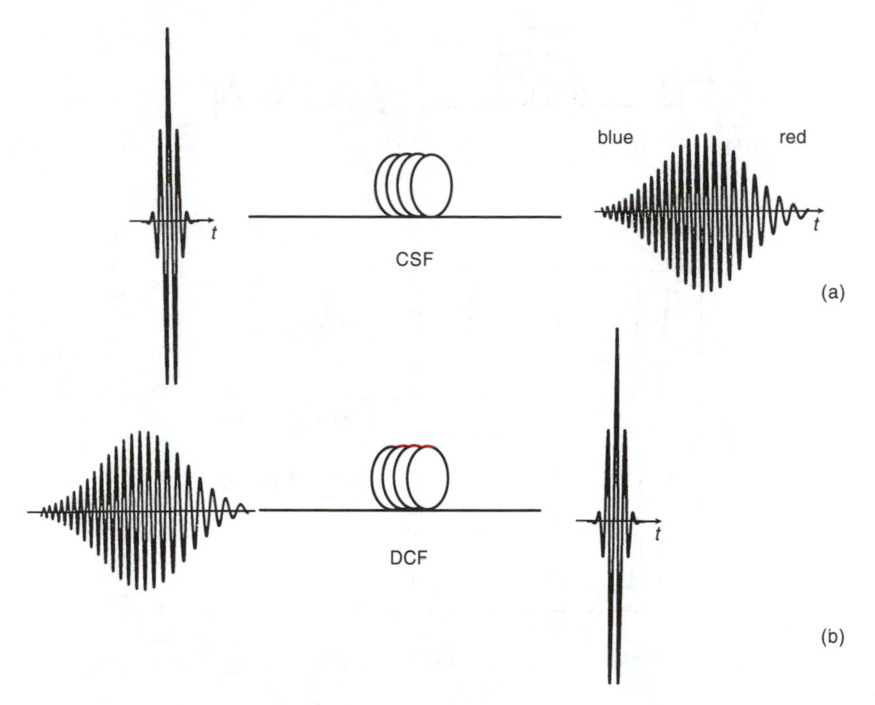

Fig. 15.2: The basic principle of dispersion compensation.

15.3 Dispersion compensating fibers

As discussed in Section 10.2, CSFs are characterized by large ($\sim$5–6 μm) core radii and zero dispersion occurs around 1300 nm (see Figure 10.2). Operation around $\lambda_0 \simeq 1300$ nm thus leads to very low pulse broadening, but the attenuation is higher than at 1550 nm (see Figure 14.1(a)). Thus, to exploit the low-loss window around 1550 nm, new fiber designs were developed that had zero dispersion in the 1550-nm wavelength region. These fibers are referred to as DSFs and have typically a triangular refractive index profiled core. Using DSFs operating at 1550 nm, one can achieve zero dispersion as well as minimum loss in silica-based fibers (see Figure 10.3).

Now, in many countries, tens of millions of kilometers of CSFs already exist in the underground ducts operating at $\lambda_0 \simeq 1300$ nm. One could increase the transmission capacity by operating these fibers at 1550 nm and using WDM techniques and optical amplifiers. But, then there will be significant residual (positive) dispersion. On the other hand, replacing these fibers by DSFs would involve huge costs. As such, in recent years, there has been considerable work in upgrading the installed 1310-nm optimized optical fiber links for operation at 1550 nm. This is achieved by developing fibers with very large negative dispersion coefficients, a few hundred meters to a kilometer, which can be used to compensate for dispersion over tens of kilometers of the fiber in the link.

Compensation of dispersion at a wavelength around 1550 nm in a 1310-nm optimized single-mode fiber can be achieved by specially designed fibers whose dispersion coefficient (D) is negative and large at 1550 nm. These types of fibers are known as DCFs.

Since the DCF has to be added on to an existing fiber optic limit, it would increase the total loss of the system and, hence, would pose problems in detection at the end. The length of the DCF required for compensation can be reduced by

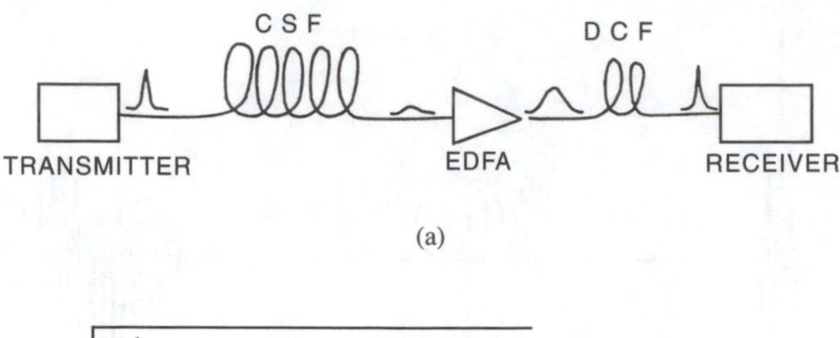

(a)

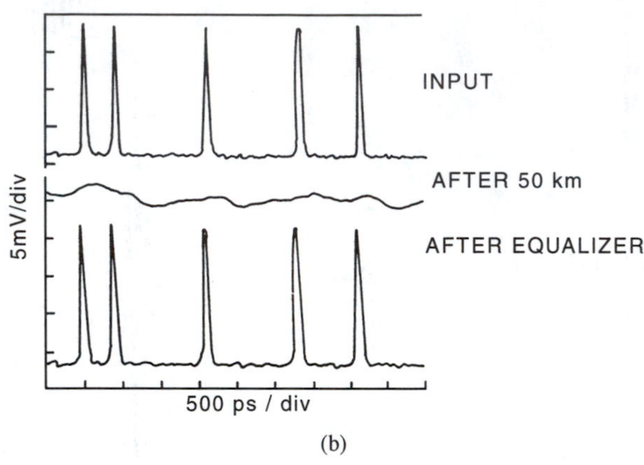

(b)

Fig. 15.3: (a) A schematic of dispersion compensation scheme in a CSF system operating at 1550 nm using a DCF. (b) A typical result showing the performance of a dispersion compensator for a 2.5-Gb/s bit pattern. [After Poole et al. (1994).]

having fibers with very large negative dispersion coefficients. Thus, there has been considerable research effort to achieve DCFs with very large (negative) dispersion coefficients.

As an example, if we consider propagation in a 50-km length fiber (i.e., $L = 50$ km) with $D = +16$ ps/km·nm, then to compensate the dispersion by a 2-km-long fiber we must have

$$D' = -400 \text{ ps/km·nm}$$

The higher the dispersion coefficient of the compensating fiber, the smaller will be the required length of the compensating fiber. Figure 15.3 shows the waveforms at the input to a 50-km conventional single-mode fiber, the output without the dispersion compensator, and the output with a DCF with $D = -548$ ps/km·nm and of length 1.44 km. Note that without the compensating fiber, no information can be retrieved while the DCF fully restores the pulses.

To achieve a very high negative value of D, the core of the compensating fiber has to be doped with relatively high GeO_2 compared with the conventional fibers. Unfortuantely, the total fiber loss (α) increases because of this doping. Hence, for DCFs a measure of the dispersion compensation efficiency is given by the figure of merit (FOM), which is defined as the ratio of the dispersion coefficient to the total loss and has a unit of ps/(dB-nm)

$$\text{FOM(ps/(dB·nm))} = |D|/\alpha \tag{15.4}$$

A typical refractive index profile of DCF is shown in Figure 15.4, which has $D \sim -300$ ps/(km-nm) and FOM $\sim$ 400 ps/(dB·nm) [Hawtof, Berkey, and Antos (1996)].

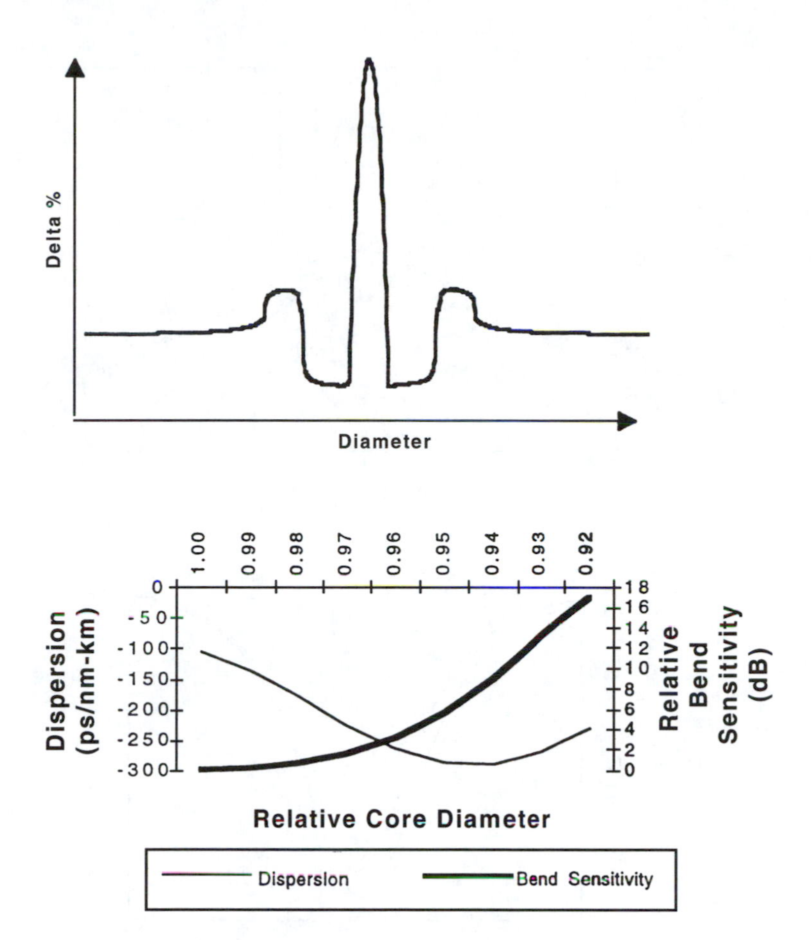

Fig. 15.4: Typical refractive index profile and the corresponding dispersion of a DCF. [After Hawtof et al. (1996).]

In a recent paper [Thyagarajan et al. (1996)], a novel DCF design capable of providing very high dispersion values has been proposed. It consists of two highly asymmetric concentric cores (see the inset in Figure 15.5(a)). It has been shown that this design can provide a very high value of $D(\sim -5000$ ps/(km-nm)) with proper choice of parameters. Figure 15.5(a) shows the variation of effective index ($= \beta / k_0$) with wavelength. Far away on either side of 1550 nm, the fundamental mode index is very close to those of the individual fiber modes of the inner and outer cores. On the other hand, close to the phase matching wavelength of 1550 nm, the mode index of the composite fiber changes rapidly because of a strong coupling between the two individual modes of the inner and outer core. Due to a strong refractive index asymmetry between the two cores, there is a rapid change in the slope of the wavelength variation of the fundamental index. This leads to a large value of D around 1550 nm (see Figure 15.5(b)).

In a recent paper, Onaka et al. (1996) have demonstrated 1.1-Tb/s (55 wavelengths $\times$ 20 Gb/s) WDM transmission over 150 km of a 1.3-μm zero-dispersion single-mode fiber; wideband EDFAs and DCFs with a negative dispersion slope were used (see Figure 15.6).[1]

[1] Optical amplification of the signal in the EDFA is accompanied by noise due to ASE generated in the amplifier. Optical filters are used to filter the ASE spectrum, but the noise generated in the frequency band of the signal does result in a decreased SNR. EDFAs can provide typical gains of 30–40 dB and saturated output powers of about 100 mW around the 1550-nm wavelength region (see Chapter 14).

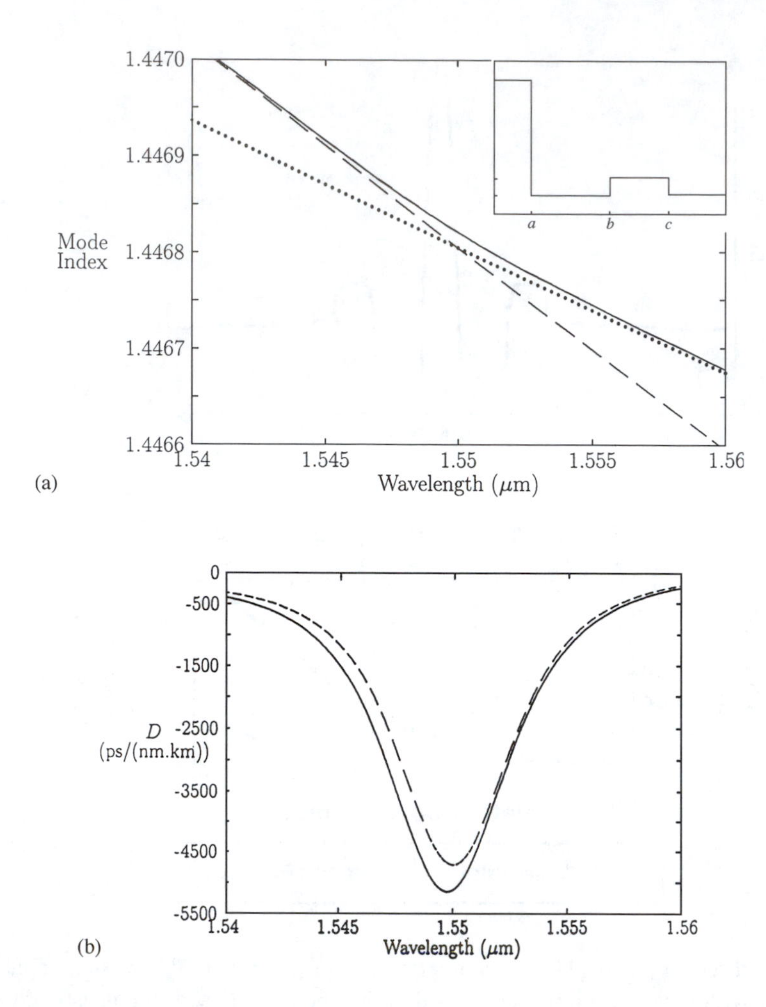

Fig. 15.5: (a) Variation of the mode index with wavelength for the refractive index profile (RIP) of the dual-core DCF shown in the inset. The solid curve corresponds to composite step RIP design; dashed and dotted curves correspond to two separate fibers, one corresponding to a fiber with inner core with step RIP and the other with outer core, respectively. (b) Variation of D with wavelength. Solid and dashed curves correspond to step and parabolic RIPs, respectively. [After Thyagarajan et al. (1996).]

15.4 Dispersion management in WDM systems

Transmission performances of long-haul optical transmission systems using different signals can be limited by the presence of dispersion, nonlinearity, and noise. For long-haul systems the nonlinear refractive index of the fiber can couple different signal channels at different wavelengths and can also couple the signal with noise. Until recently, the idea of using the system around the zero dispersion wavelength to achieve maximum bandwidth was prevalent. However, when the system is operated at the zero dispersion wavelength, the signal and noise from the amplifiers will have similar velocities, leading to a large interaction length, which will enhance the nonlinear interactions between the channels and noise components. This deleterious interaction can be alleviated by using the system not at zero dispersion wavelength but slightly away from it so that the chromatic dispersion present will reduce the phase matching or, equivalently, the interaction length. This is the basic concept of dispersion management in long-haul systems where the nonlinear interaction is controlled by tailoring the accumulated dispersion to keep the interaction length small and also the end-to-end dispersion small.

In an example discussed by Bergano and Davidson (1996), an eight-channel 5-Gb/s WDM transmission experiment was performed. Here eight signals with

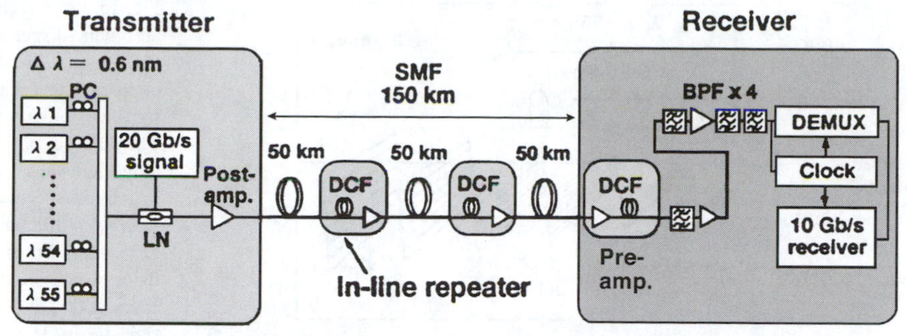

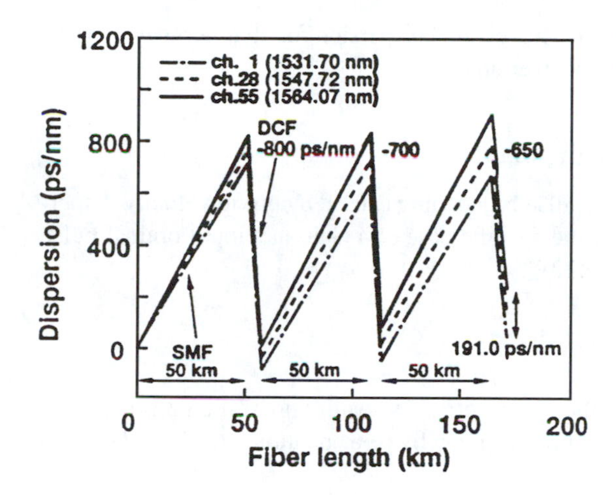

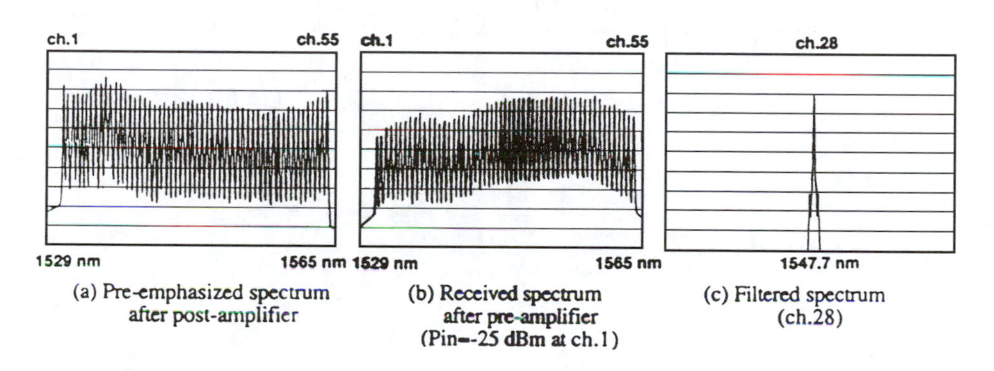

Fig. 15.6: (a) Experimental setup for 55-wavelength WDM transmission. (b) The corresponding dispersion map. (c) 55-wavelength spectra (H: 3.6 nm/div, V: 5 dB/div, Res: 0.1 nm). [Adapted from Onaka et al. (1996).]

wavelengths from 1556 nm to 1560 nm were transmitted over 900 km of a single-mode fiber with $D = -2$ ps/(km-nm); the zero dispersion wavelength was 1585 nm. This ensures that the signals are not traveling at similar velocities. The total accumulated dispersion over 900 km of fiber is compensated by 100 km of CSF with $\lambda_0 = 1310$ nm.

One can notice (see Figure 15.7) that the accumulated dispersion of all channels at the end of the link (1000 km) does not become zero as the conventional fiber with $\lambda_0 = 1310$ nm has a dispersion slope opposite, but not exactly equal, to that of the 900-km fiber used so that the compensation of dispersion will be

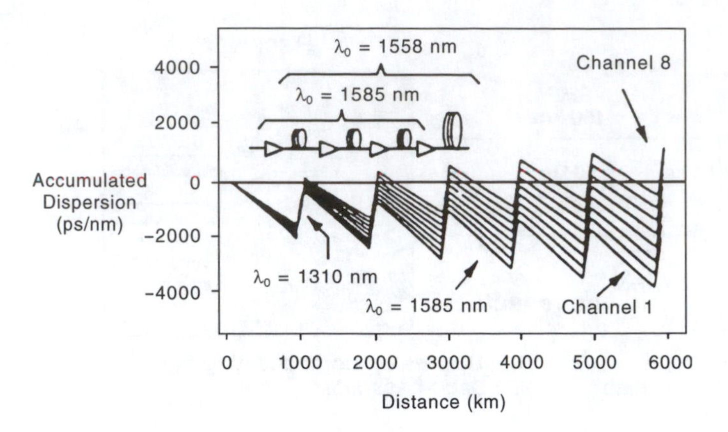

Fig. 15.7: Acccumulated chromatic dispersion versus transmission distance for eight channels of a WDM transmission experiment. The majority of the amplifier spans use a negative dispersion fiber with $\lambda_0 \approx 1585$ nm and $D \approx -2$ ps/km-nm. The dispersion is compensated every 1000 km using a CSF (i.e., $\lambda_0 = 1310$ nm). [After Bergano and Davidson (1996).]

exact at one only wavelength. The residual dispersion for all other wavelengths can be compensated at the receiver end.

15.5 Dispersion and chirping

In Section 6.3 we discussed pulse broadening caused by group velocity dispersion in an optical fiber. For the specific case of a Gaussian input optical pulse of the form (see equation (6.22))

$$E(z=0,t) = E_0 e^{-t^2/\tau_0^2} e^{i\omega_0 t} \tag{15.5}$$

we showed in Section 6.4 that the electric field variation of the output optical pulse at distance z along the fiber is given by (see equations (6.32) and (6.36))

$$E(z,t) = \frac{E_0}{(1+\sigma^2)^{1/4}} \exp\left[-\frac{\left(t-\frac{z}{v_g}\right)^2}{\tau^2(z)}\right] \exp[i(\Phi(z,t) - \beta(\omega_0)z)] \tag{15.6}$$

where

$$\begin{aligned}
\Phi(z,t) &= \omega_0 t + \kappa\left(t - \frac{z}{v_g}\right)^2 - \frac{1}{2}\tan^{-1}(\sigma) \\
\sigma &= \frac{2\alpha z}{\tau_0^2} \\
\kappa &= \frac{\sigma}{(1+\sigma^2)\tau_0^2} \\
\tau^2(z) &= \tau_0^2(1+\sigma^2) \\
\alpha &= \left.\frac{d^2\beta}{d\omega^2}\right|_{\omega=\omega_0}
\end{aligned} \tag{15.7}$$

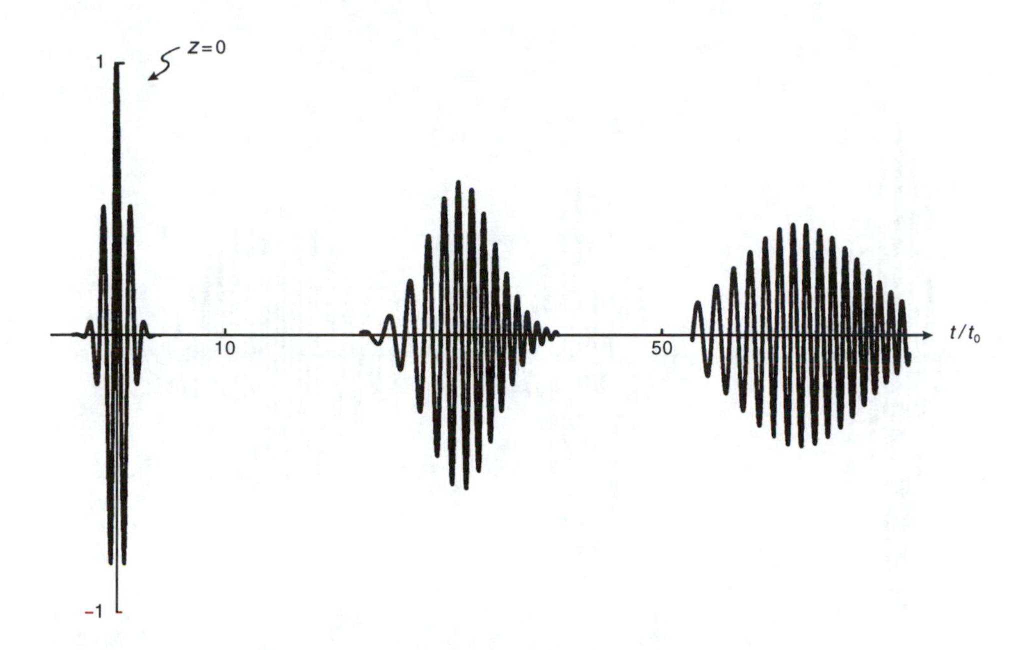

Fig. 15.8: Calculated temporal variation of the electric field of an optical pulse with $\omega_0\tau_0 = 7$ corresponding to different values of Z. Note that as the pulse propagates, it broadens in time domain and also gets chirped.

In Section 6.4 we discussed the corresponding intensity distribution of the output pulse and saw that the envelope of the pulse broadens due to a finite value of α. This phenomenon is termed group velocity dispersion (GVD).

From the phase term in equation (15.6) it follows that the oscillations within the output pulse are not periodic. Now for a time varying function of the form

$$g(t) = Ae^{i\zeta(t)} \tag{15.8}$$

we can define an instantaneous angular frequency as

$$\omega(t) = \frac{d\zeta}{dt} \tag{15.9}$$

Thus, if $\zeta(t) = \omega_0 t$, then $\omega = \omega_0$ – that is, the wave is a pure sinusoid. From equation (15.6) we obtain for the instantaneous frequency within the pulse envelope

$$\omega(t) = \frac{\partial\Phi}{\partial t} = \omega_0 + 2\kappa\left(t - \frac{z}{v_g}\right) \tag{15.10}$$

Thus, the instantaneous frequency within the pulse envelope changes with time. Such a pulse is termed a *chirped pulse*.

If α is positive (i.e., $d^2n/d\lambda_0^2 > 0$) and the instantaneous frequency within the optical pulse increases with time, the leading edge of the pulse corresponds to $t < z/v_g$ and the trailing edge corresponds to $t > z/v_g$. Thus, the leading edge of the pulse has a frequency lower than ω_0 and the trailing edge has a frequency higher than ω_0. Similarly, for $d^2n/d\lambda_0^2 < 0$, the leading edge will be upshifted and the trailing edge will be downshifted. Figure 15.8 shows the calculated temporal variation of the electric field of the optical pulse for the

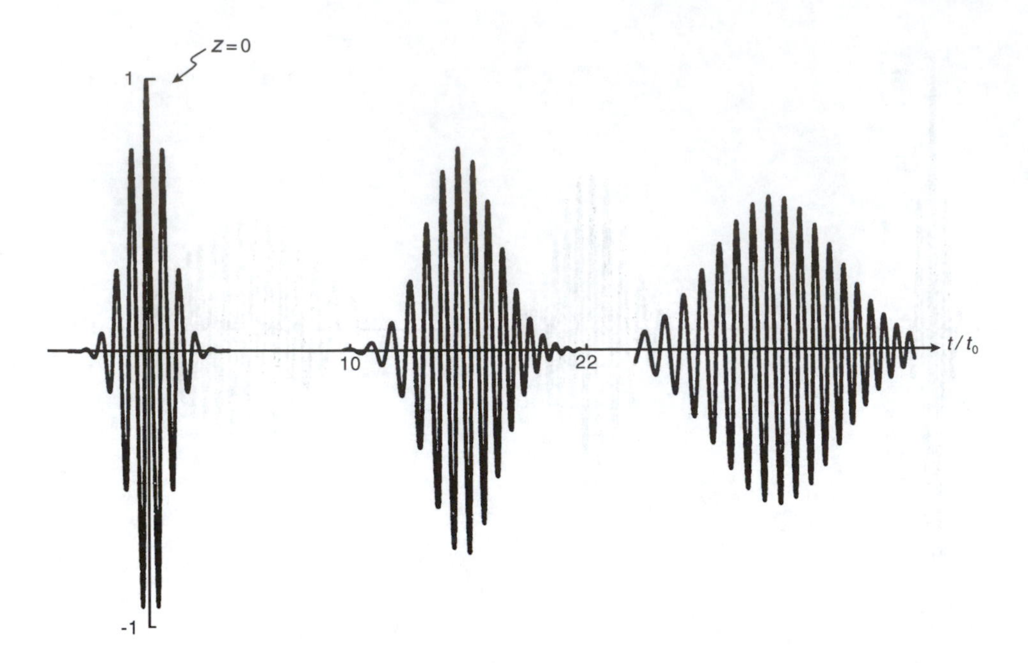

Fig. 15.9: Calculated temporal variation of the electric field of an optical pulse with $\omega_0\tau_0 = 11.2$ corresponding to different values of Z. Note that as the pulse propagates, it broadens in time domain and also gets chirped.

following values of various parameters.

$$\omega_0\tau_0 \simeq 7$$

$$\frac{v_g}{v_p} = \frac{v_g}{[\omega_0/k(\omega_0)]} = 0.99; \quad \frac{\alpha v_g}{\tau_0} = 6$$

$$\sigma = 2\left(\frac{\alpha v_g}{\tau_0}\right) Z; \quad Z = \frac{z}{v_g\tau_0}$$

Figure 15.9 shows the corresponding variation for

$$\omega_0\tau_0 \simeq 11.2$$

and other parameters having the same values. Both figures clearly show the phenomenon of chirping associated with broadening caused due to group velocity dispersion. In Figures 15.8 and 15.9, the parameters have been chosen to distinctly show chirping within the pulse; for actual values of parameters, the chirping leads to frequency variations of only $10^{-4}\ \nu_0$ within the pulse.

The fact that the output broadened pulses should be chirped can be understood from the following argument. Since we are not considering any frequency-dependent attenuation or gain and the system is linear, the optical spectrum of the input and the output pulses should be the same. Thus, analyzing the input and the output pulse by a Fabry–Perot interferometer (or an optical spectrum analyzer) would show no difference in the spectra. Since the output pulse envelope is broader and it has the same frequency spectrum as the output, *it should be chirped*. It is worthwhile to note that the inverse relationship between the pulse temporal width τ_0 and the corresponding spectra width $\Delta\nu$,

$$\Delta\nu \sim \frac{1}{\tau_0} \tag{15.11}$$

is valid only for unchirped pulses, which are also referred to as Fourier transform-limited pulses. For a chirped pulse, the spectrum can be much greater than the inverse of its temporal width. At the same time, for a given spectral width $\Delta\nu$, the shortest pulse width that one can achieve is that satisfying equation (15.11). This fact is used in pulse compression (see Section 15.6).

As discussed in Section 6.4, the output pulse is broadened to a new temporal width of

$$\tau(z) = \tau_0(1+\sigma^2)^{1/2} \tag{15.12}$$

To get a numerical estimate of broadening and chirping in an actual optical fiber, we consider a 10-ps pulse at $\lambda_0 = 1.55\,\mu$m propagating through pure silica. For pure silica at $\lambda_0 = 1.55\,\mu$m,

$$\frac{d^2n}{d\lambda_0^2} \simeq -4.2 \times 10^{-3}\,\mu\text{m}^{-2} = -4.2 \times 10^9\,\text{m}^{-2}$$

We neglect waveguide dispersion and consider only material dispersion (see Problem 15.1) to obtain

$$\alpha = \frac{\lambda_0}{2\pi c^2}\left(\lambda_0^2 \frac{d^2n}{d\lambda_0^2}\right) \simeq -2.77 \times 10^{-26}\,\text{s}^2/\text{m}$$

which corresponds to a dispersion coefficient of

$$D = -\frac{2\pi c\alpha}{\lambda_0^2} \simeq 21.7\ \text{ps/km}\cdot\text{nm}$$

For $z = 50$ km, we have

$$\sigma = \frac{2\alpha z}{\tau_0^2} \simeq 27.7$$

Thus

$$\tau(z) \simeq 277\ \text{ps}$$

If ω_0 and $\omega_0 + \Delta\omega$, respectively, represent the instantaneous frequencies at the center of the pulse ($t = z/v_g$) and at $t = z/v_g + 277$ ps, then

$$\Delta\omega = 2\kappa\tau(z)$$

or

$$\Delta\nu = \frac{\kappa\tau(z)}{\pi} \simeq 3.18 \times 10^{10}\ \text{Hz}$$

which of course is very small compared with the pulse center frequency of 1.94×10^{14} Hz.

15.6 Compression of a chirped pulse

Let us consider an unchirped Gaussian pulse given by

$$E(t) = E_0\, e^{-t^2/\tau_0^2}\, e^{i\omega_0 t} \tag{15.13}$$

The spectral distribution of such a pulse is obtained by taking a Fourier transform

$$\begin{aligned} A(\omega) &= \frac{1}{2\pi}\int_{-\infty}^{\infty} E(t)\, e^{-i\omega t}\, dt \\ &= \frac{E_0}{2\sqrt{\pi}}\tau_0 \exp\left[-\frac{(\omega-\omega_0)^2\tau_0^2}{4}\right] \end{aligned} \tag{15.14}$$

The corresponding full width at half maximum (FWHM) of the power spectral density $|A(\omega)|^2$ is

$$\text{FWHM} = \Delta\omega_f = \frac{2}{\tau_0}\sqrt{2\ln 2} \tag{15.15}$$

For this spectral width, the shortest pulse width is the FWHM of the pulse represented by equation (15.13) and is given by

$$\text{FWHM} = \Delta\tau_f = \tau_0\sqrt{2\ln 2} \tag{15.16}$$

From equations (15.15) and (15.16) we have

$$\Delta\omega_f \Delta\tau_f = 4\ln 2 \simeq 2.8$$

As discussed earlier, when such a pulse propagates through a linear dispersive medium (such as an optical fiber), it gets broadened in the time domain while remaining unchanged in its spectral width. Such a broadened pulse is chirped and its temporal width $\Delta\tau_f'$ satisfies the relationship

$$\Delta\tau_f' > \frac{4\ln 2}{\Delta\omega_f}$$

One can compress this chirped pulse by propagating it through another linear dispersive medium but with a dispersion having a sign opposite that of the first medium (see Problem 6.9). Thus, if the chirping is due to propagation through a fiber operating above its zero dispersion wavelength, one can achieve compression by propagating through another fiber operating below the zero dispersion wavelength. This concept is indeed being exploited as an optical equalizer or dispersion compensator in fiber optic communication systems. Note that having started from an unchirped pulse, if the spectrum of the pulse is not changed during propagation, one cannot compress it below the starting pulse width.

To achieve pulse compression beyond the initial pulse width, one has to first broaden the spectral content of the pulse (without any broadening in the time domain) and then by using dispersion (with a proper sign) the temporal width of the pulse can be reduced below the starting pulse width. This can be done by using nonlinear effects in fibers as discussed in Chapter 16.

As an example, let us assume that we start with an unchirped pulse whose FWHM is 100 fs. The corresponding spectral width is

$$\Delta\nu_f = \frac{2\ln 2}{\pi\,\Delta\tau_f} \simeq 4.41 \times 10^{12}\,\text{Hz}$$

If the corresponding wavelength is 1 μm, the carrier frequency is

$$\nu_0 = \frac{c}{\lambda_0} = 3 \times 10^{14}\,\text{Hz}$$

Thus, $\Delta\nu_f/\nu_0 \simeq 0.015$. If we now wish to compress the pulse to say $\Delta\tau_f' = 10$ fs, then we must increase the spectrum of the pulse to

$$\Delta\nu_f' = \frac{2\ln 2}{\pi\,\Delta\tau_f'} \simeq 4.4 \times 10^{13}\,\text{Hz}$$

that is, a ten-fold increase in spectrum. Indeed, in Chapter 16 we show that such increases in optical spectrum can be achieved by using nonlinear optical effects.

15.7 Propagation of a chirped pulse through a linear dispersive medium

We now demonstrate explicitly how a chirped pulse can get compressed as it propagates through a dispersive medium with a specific sign of dispersion coefficient. We consider a chirped Gaussian pulse at $z = 0$

$$\begin{aligned} E(z=0,t) &= E_0 \exp\left[-\frac{t^2}{\tau^2}(1+ig)\right] e^{i\omega_0 t} \\ &= E_0 e^{-t^2/\tau^2} \exp\left[i\left(\omega_0 t - \frac{t^2}{\tau^2} g\right)\right] \end{aligned} \tag{15.17}$$

where g is known as the chirp parameter; $g < 0$ corresponds to up chirp in which the instantaneous frequency increases from the leading to the trailing edge, whereas $g > 0$ corresponds to down chirp when the opposite occurs. The spectrum $A(\omega)$ of the pulse is given by

$$\begin{aligned} A(\omega) &= \frac{E_0}{2\pi} \int \exp\left[-\frac{t^2}{\tau^2}(1+ig)\right] e^{i(\omega_0-\omega)t}\,dt \\ &= \frac{E_0\tau}{2\sqrt{\pi}(1+ig)^{1/2}} \exp\left[-\frac{(\omega-\omega_0)^2\tau^2}{4(1+ig)}\right] \end{aligned} \tag{15.18}$$

The corresponding power spectrum is given by

$$S(\omega) = |A(\omega)|^2 = \frac{E_0^2\tau^2}{4\pi(1+g^2)^{1/2}} \exp\left[-\frac{(\omega-\omega_0)^2\tau^2}{2(1+g^2)}\right] \tag{15.19}$$

with a FWHM width of

$$\Delta\omega_f = \frac{2\sqrt{2\ln 2}\sqrt{1+g^2}}{\tau} \tag{15.20}$$

Note that the FWHM of the pulse given by equation (15.17) is

$$\Delta\tau_f = \tau\sqrt{2\ln 2}$$

Thus, for the pulse

$$\Delta\omega_f\ \Delta\tau_f = 4\ln 2\sqrt{1+g^2} \tag{15.21}$$

Hence, for the chirped pulse the product $\Delta\omega_f\ \Delta\tau_f$ is always greater than the minimum value $4\ln 2$.

As seen earlier, a pulse with the spectral width given by equation (15.20) can be compressed to a Gaussian pulse with a minimum pulse width of

$$\begin{aligned}\Delta\tau_{\min} &= \frac{4\ln 2}{\Delta\omega_f} = 4\ln 2\frac{\tau}{2\sqrt{2\ln 2}\sqrt{1+g^2}} \\ &= \sqrt{2\ln 2}\frac{\tau}{(1+g^2)^{1/2}}\end{aligned} \tag{15.22}$$

For a pulse with an up chirp ($g < 0$) the leading edge has a lower frequency than the trailing edge; thus for compression, the dispersion should be such that higher frequency components travel faster than lower frequency components and, hence, can "catch up" with the low-frequency components. A medium in which higher frequency components travel faster than lower frequency components corresponds to negative GVD. Similarly, a down chirped pulse needs to be passed through a positive GVD medium for pulse compression (see Problem 6.9).

Figure 15.10 shows the compression of a down chirped pulse as it propagates through a linear dispersive medium (the figure corresponds to positive linear dispersion).

15.8 Dispersion compensation of a Gaussian pulse

In Section 6.4 we showed that an input pulse represented by

$$E_0\, e^{i\omega_0 t}\, e^{-t^2/\tau_0^2} \tag{15.23}$$

after propagating through a fiber of length L becomes

$$E_0\sqrt{\frac{\tau_0^2}{\tau_0^2 + 2i\alpha L}}\ e^{i\omega_0 t}\, e^{-t^2/(\tau_0^2+2i\alpha L)} \tag{15.24}$$

where α represents the dispersion property of the fiber and is given by equation

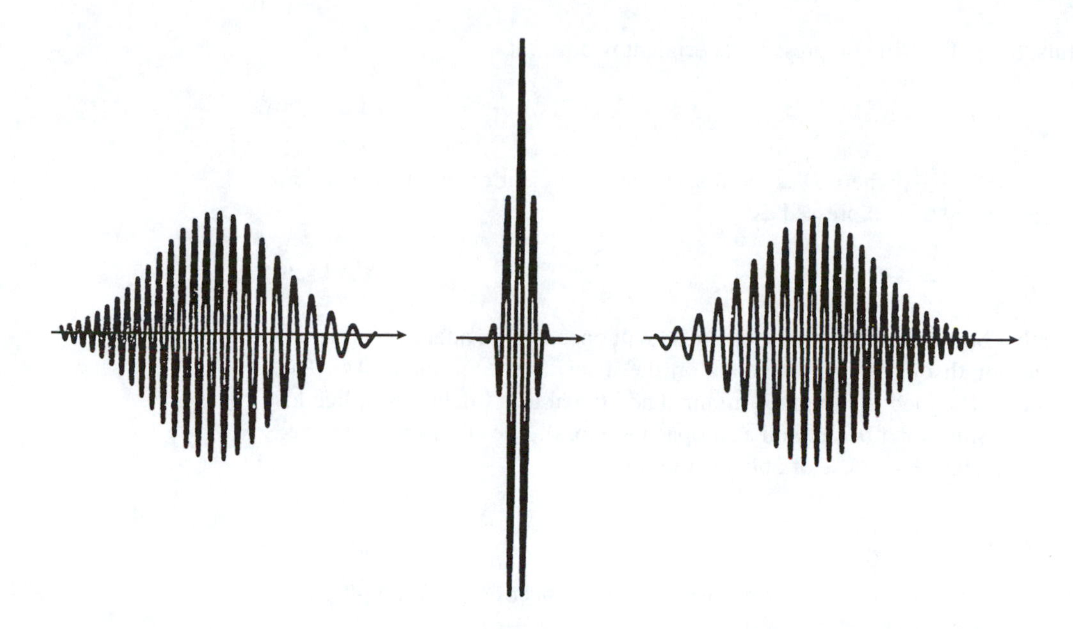

Fig. 15.10: Compression of a chirped pulse as it propagates through a linear dispersive medium. After compressing to a minimum width, the pulse again undergoes broadening.

(15.7). Thus, propagation through the fiber changes

$$\tau_0^2 \to \tau_0^2 + 2i\alpha L \tag{15.25}$$

and

$$E_0 \to E_0\sqrt{\frac{\tau_0^2}{\tau_0^2 + 2i\alpha L}} \tag{15.26}$$

Hence, if we consider a chirped input pulse given by

$$\left(E_0\sqrt{\frac{\tau_0^2}{\tau_0^2 + 2i\alpha L}}\right) e^{i\omega_0 t}\, e^{-t^2/(\tau_0^2+2i\alpha L)} \tag{15.27}$$

to propagate through a fiber of length L' and dispersion parameter α', then using the recipe given by equations (15.25)–(15.26), the output pulse would be

$$\begin{aligned}
&\left(E_0\sqrt{\frac{\tau_0^2}{\tau_0^2 + 2i\alpha L}}\right)\left(\sqrt{\frac{\tau_0^2 + 2i\alpha L}{\tau_0^2 + 2i\,(\alpha L + \alpha' L')}}\right) e^{i\omega_0 t} \\
&\qquad \times \exp\left[-\frac{t^2}{\tau_0^2 + 2i(\alpha L + \alpha' L')}\right] \\
&\qquad = E_0\sqrt{\frac{\tau_0^2}{\tau_0^2 + 2i(\alpha L + \alpha' L')}} e^{i\omega_0 t}\, \exp\left[-\frac{t^2}{\tau_0^2 + 2i\,(\alpha L + \alpha' L')}\right]
\end{aligned} \tag{15.28}$$

Thus, the pulse will compress to its original width τ_0 if

$$\alpha' L' = -\alpha L \tag{15.29}$$

Since $\alpha = -\frac{\lambda_0^2 D}{2\pi c}$, where D is the dispersion coefficient of the fiber, equation (15.29) can also be expressed as

$$D' L' = -DL \tag{15.30}$$

Note that beyond the length L', the pulse will once again undergo broadening.

The length of the DCF depends on the dispersion coefficient. The DCF should be designed to introduce minimal additional loss (including splice loss to the transmission fiber as well as propagation loss). The additional loss needs to be compensated by use of optical amplifiers.

Problems

15.1 (a) If we neglect waveguide dispersion and consider only material dispersion then, we may write

$$\beta \approx \frac{\omega}{c}\, n(\omega)$$

Show that

$$D = D_m = -\frac{2\pi c}{\lambda_0^2}\frac{d^2\beta}{d\omega^2} = -\frac{1}{c\lambda_0}\left(\lambda_0^2 \frac{d^2 n}{d\lambda_0^2}\right)$$

(b) If we neglect material dispersion and consider only waveguide dispersion then (for step index fibers)

$$D = D_w = -\frac{2\pi c}{\lambda_0^2}\frac{d^2\beta}{d\omega^2} \simeq -\frac{n_2\Delta}{c\lambda_0}\left(V\frac{d^2(bV)}{dV^2}\right)$$

where the parameters are defined in Chapter 10.

15.2 When one uses WDM systems, one often operates at a wavelength slightly lower than the zero dispersion wavelength (see Section 15.4). For the fiber parameters discussed in Example 10.3 (i.e., for $a = 2.3\,\mu$m, $n_2 = 1.447$, $\Delta = 0.0075$), use Tables 6.1 and 10.1 to calculate the material, waveguide, and total dispersion at $\lambda_0 = 1540$ nm. What would be the total dispersion for a 100-km length of the fiber?

15.3 Consider the silica fiber discussed in Problem 10.1. It had zero dispersion at $\lambda_0 = 1330$ nm. If the fiber is operated at 1540 nm, using Tables 6.1 and 10.1 calculate the material dispersion, waveguide dispersion, and total dispersion. How much length of such a fiber would be required to compensate for the negative dispersion calculated in the previous problem?

15.4 Using Table 6.1, plot the group velocity v_g [see equation (6.4)] as a function of wavelength and show that it attains a maximum value at $\lambda_0 \simeq 1.3\,\mu$m.

16

Optical solitons

16.1 Introduction

In Chapters 6, 10, and 15, while considering the broadening of an optical pulse propagating through an optical fiber, we treated the optical fiber as a linear medium – that is, the intensities associated with the propagating optical pulse were assumed to be so small that there was no significant effect on the propagation characteristics of the waveguide. In actual practice, all media exhibit nonlinear effects. In the case of silica optical fibers, one of the manifestations of the nonlinearity is the intensity-dependent refractive index according to the following equation

$$n = n_0 + n_2 I \tag{16.1}$$

where n_0 is the linear refractive index of silica (for low intensity levels), n_2 is the nonlinear refractive index coefficient, and $I = P/A_{eff}$ is the effective intensity within the medium with P being the power carried by the mode and A_{eff} the effective area of the fiber mode. For single-mode silica fibers operating at 1550 nm

$$n_0 \simeq 1.46, \quad n_2 \simeq 3.2 \times 10^{-20}\,\mathrm{m^2/W}, \quad A_{eff} \simeq 50\,\mu\mathrm{m}^2 \tag{16.2}$$

Thus, when an optical pulse travels through the fiber, the higher intensity portions of the pulse encounter a higher refractive index of the fiber compared with the lower intensity regions. This intensity-dependent refractive index leads to the phenomenon known as self-phase modulation (SPM). The primary effect of SPM is to broaden the spectrum of the pulse while keeping the temporal shape unaltered. The phenomenon of SPM is discussed in Section 16.2, and in Section 16.3 we discuss the spectral broadening (i.e., generation of additional frequencies) due to SPM.

This spectral broadening of the pulse without a corresponding increase in temporal width leads to a frequency chirping of the pulse. Indeed, for silica optical fibers for which n_2 is positive, the frequencies in the trailing edge of the pulse are upshifted and those in the leading edge are downshifted with respect to the center frequency of the pulse. This broadening of the pulse spectrum generates new frequencies in the pulse and will ultimately lead to an increased broadening through the phenomenon of dispersion.

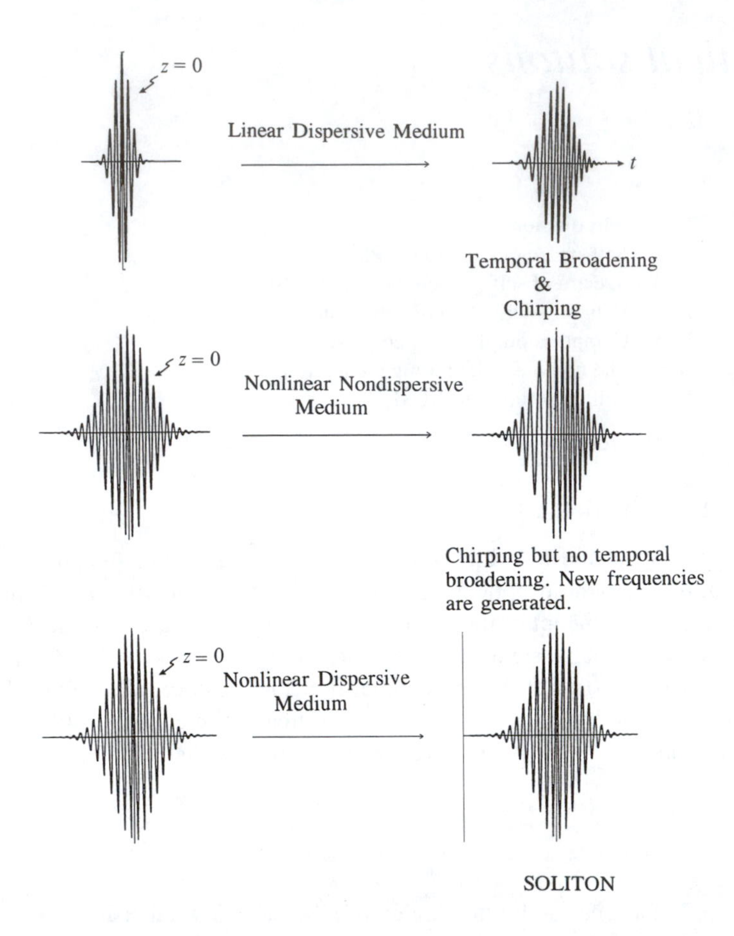

Fig. 16.1: (a) An optical pulse propagating through a linear dispersive medium undergoes temporal broadening as well as chirping. For operation at a wavelength greater than zero dispersion wavelength, the instantaneous frequency decreases with increasing time. (b) A pulse propagating through a nonlinear nondispersive medium undergoes no temporal broadening but undergoes only chirping. Note that the chirpings in (a) and (b) are of opposite sign. (c) A soliton is a pulse propagating through a nonlinear dispersive medium that broadens in neither the temporal domain nor the spectral domain.

In silica optical fibers, if the operating wavelength is above the zero dispersion wavelength, then higher frequencies travel faster than lower frequencies and pulse broadening in the absence of any nonlinear effect is accompanied by a chirp within the pulse – that is, within the pulse, the instantaneous frequency decreases with increasing time.

SPM leads to a chirping with lower frequencies in the leading edge and higher frequencies in the trailing edge, which is just opposite the chirping caused by linear dispersion in the wavelength region above the zero dispersion wavelength; thus, by a proper choice of pulse shape (a hyperbolic secant shape) and the power carried by the pulse, we can indeed compensate one effect with the other (see Section 16.4). In such a case the pulse would propagate undistorted by a mutual compensation of dispersion and SPM. This is schematically shown in Figure 16.1. Such a pulse would broaden neither in the time domain (as in linear dispersion) nor in the frequency domain (as in SPM) and is called a soliton. Since a soliton pulse does not broaden during its propagation, it has tremendous potential for applications in super high bandwidth optical communication systems.

For a soliton propagating around a 1550-nm wavelength, the peak power in the pulse (in mW) and pulse duration are related through the equation (see Section 16.4)

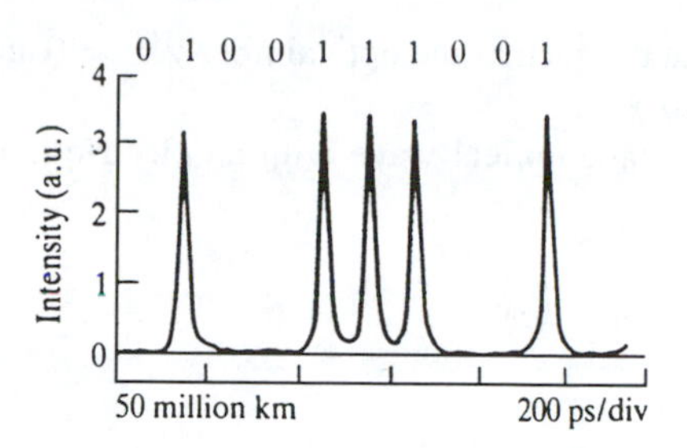

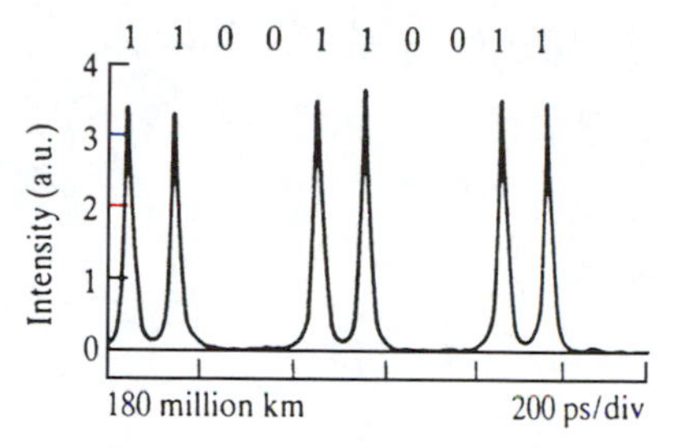

Fig. 16.2: Transmitted data patterns ⟨1100110011⟩ consisting of soliton pulses as received after 50 million km and 180 million km (15-min delay time) of single-mode fiber in a recirculating fiber loop experiment. [After Nakazawa (1994).]

$$P \simeq \frac{1.5 \times 10^3 D}{\tau_f^2} \tag{16.3}$$

where D is the dispersion coefficient in ps/km·nm and τ_f is the pulse FWHM[1] in picoseconds. Thus, for a 10-ps soliton operating in a dispersion-shifted fiber with $D = 1$ ps/km·nm, the required peak power will be approximately 15 mW.

Figure 16.2 shows the results of experiments on dispersionless soliton propagation over 180 million kilometers at a data rate of 10 Gb/s. The results show the limitless possibilities that soliton systems offer us.

The emergence of EDFAs that optically compensate any fiber attenuation and optical solitons that use dispersion and nonlinearity against each other simultaneously compensating both effects are truly revolutionizing the field of optical fiber telecommunications. These developments are expected to make terabit communication systems over hundreds of thousands of kilometers a reality.

16.2 Self-phase modulation (SPM)

In a linear medium the electric polarization is assumed to be a linear function of the electric field

$$P = \epsilon_0 \chi E \tag{16.4}$$

where, for simplicity, a scalar relation has been written. The quantity χ is termed the linear dielectric susceptibility. At high optical intensities (or equivalently at high optical fields) all media behave nonlinearly – that is, the relation expressed in equation (16.4) is approximate and one has

$$P = \epsilon_0 \chi E + \epsilon_0 \chi^{(2)} E^2 + \epsilon_0 \chi^{(3)} E^3 + \cdots \tag{16.5}$$

[1] For a Gaussian pulse given by equation (16.26), $\tau_f = \tau_0 \sqrt{2 \ln 2} \simeq 1.18 \tau_0$.

For noncrystalline media such as the optical fiber $\chi^{(2)} = 0$ and the lowest order nonlinearity is due to $\chi^{(3)}$.

If we consider a plane optical wave with an electric field variation of the form

$$E = E_0 \cos(\omega t - kz) \tag{16.6}$$

then

$$P = \epsilon_0 \chi E_0 \cos(\omega t - kz) + \epsilon_0 \chi^{(3)} E_0^3 \cos^3(\omega t - kz) \tag{16.7}$$

Now

$$\cos^3\theta = \frac{1}{4}(\cos 3\theta + 3\cos\theta) \tag{16.8}$$

Hence

$$P = \epsilon_0 \left(\chi + \frac{3}{4}\chi^{(3)} E_0^2\right) E_0 \cos(\omega t - kz) + \epsilon_0 \frac{\chi^{(3)}}{4} E_0^3 \cos 3(\omega t - kz) \tag{16.9}$$

The second term on the RHS corresponds to third harmonic generation, which is negligible in optical fibers due to phase mismatch between frequencies ω and 3ω. The polarization at frequency ω is

$$P = \epsilon_0 \left(\chi + \frac{3}{4}\chi^{(3)} E_0^2\right) E_0 \cos(\omega t - kz) \tag{16.10}$$

For a plane wave given by equation (16.6), the intensity is given by

$$I = \frac{1}{2} c\, \epsilon_0\, n_0 E_0^2 \tag{16.11}$$

where n_0 is the refractive index of the medium at low fields. Hence

$$P = \epsilon_0 \left(\chi + \frac{3}{2}\frac{\chi^{(3)}}{c\epsilon_0 n_0} I\right) E_0 \cos(\omega t - kz) \tag{16.12}$$

The general relationship between polarization and refractive index is given by

$$P = \epsilon_0 (n^2 - 1) E_0 \cos(\omega t - kz) \tag{16.13}$$

Comparing equations (16.12) and (16.13), we see that the nonlinear term containing $\chi^{(3)}$ leads to an intensity-dependent refractive index

$$n^2 = 1 + \chi + \frac{3}{2}\frac{\chi^{(3)}}{c\epsilon_0 n_0} I \tag{16.14}$$

Since the last term in the above equation is usually very small even for very intense light beams, we may approximate by a Taylor series expansion

$$\begin{aligned} n &\simeq n_0 + \frac{3}{4}\frac{\chi^{(3)}}{c\epsilon_0 n_0^2} I \\ &= n_0 + n_2 I \end{aligned} \tag{16.15}$$

where

$$n_0^2 = 1 + \chi \tag{16.16}$$

and

$$n_2 = \frac{3}{4}\frac{\chi^{(3)}}{c\epsilon_0 n_0^2} \tag{16.17}$$

is the nonlinear coefficient.

For fused silica fibers

$$n_0 \simeq 1.46, \quad n_2 \simeq 3.2 \times 10^{-20}\ \mathrm{m^2/W}$$

If we consider the propagation of a mode carrying 100 mW of power in a single-mode fiber with an effective mode area $\simeq 50\ \mu\mathrm{m}^2$, then the resultant intensity is $2 \times 10^9\ \mathrm{W/m^2}$ and the change in refractive index due to nonlinear effects is

$$\Delta n = n_2 I \simeq 6.4 \times 10^{-11} \tag{16.18}$$

Although this change in refractive index seems too small, due to very long interaction lengths (10–10,000 km) in an optical fiber, the accumulated effects become significant. In fact, it is this small nonlinear term that is responsible for the formation of solitons.

If P is the power carried by a mode in an optical fiber, since the propagating mode has a transverse intensity distribution, I in equation (16.15) represents the effective intensity within the fiber that can be approximately written as

$$I \simeq \frac{P}{A_{eff}} \tag{16.19}$$

where A_{eff} represents the effective area of the fiber mode. If one uses the Gaussian approximation for the fundamental mode (see Section 8.5.1), then

$$A_{eff} = \pi w_0^2 \tag{16.20}$$

where w_0 is the Gaussian spot size of the mode. Thus, for a fiber we may write

$$n = n_0 + n_2 \frac{P}{A_{eff}} \tag{16.21}$$

This change in refractive index leads to a corresponding change in the effective index of the mode. Thus, if β_0 is the propagation constant in the linear case, then the new propagation constant can approximately be written as

$$\beta \simeq \beta_0 + \frac{k_0 n_2 P}{A_{eff}} \tag{16.22}$$

Hence, an incident wave of the form $Ae^{i\omega_0 t}$ would emerge as

$$Ae^{i(\omega_0 t - \beta z)} = A \exp\left[i\left(\omega_0 t - \beta_0 z - \frac{k_0 n_2 P}{A_{eff}} z\right)\right] \tag{16.23}$$

If the input wave is a pulse with a power variation given by $P(t)$, then the output phase dependence would be

$$\exp\left[i\left(\omega_0 t - \frac{k_0 n_2 P(t)}{A_{eff}} z - \beta_0 z\right)\right] \tag{16.24}$$

Since $P(t)$ is a function of time, the output pulse is chirped. This is termed SPM wherein the power variation within the pulse leads to its own phase modulation.

As in the previous chapter, we can define an instantaneous frequency within the pulse

$$\omega(t) = \omega_0 - \frac{k_0 n_2 z}{A_{eff}} \frac{dP}{dt} \tag{16.25}$$

As an example, if we consider an input Gaussian pulse given by

$$E(z = 0, t) = E_0 \, e^{-t^2/\tau_0^2} \, e^{i\omega_0 t} \tag{16.26}$$

then after propagating through length L of an optical fiber, the pulse becomes

$$E(z = L, t) = E_0 \, e^{-\left(t - \frac{L}{v_g}\right)^2 / \tau_0^2} \times \exp\left[i\left(\omega_0 t - \beta_0 L - \frac{k_0 n_2 P(t)}{A_{eff}} L\right)\right] \tag{16.27}$$

where β_0, v_g, and A_{eff} represent, respectively, the propagation constant, group velocity, and effective area of the fundamental mode of the fiber; we have neglected the effects of dispersion. The quantity $P(t)$ represents the temporal variation of the power in the pulse, which is given by

$$P(z, t) = P_0 \exp\left[-\frac{2\left(t - \frac{L}{v_g}\right)^2}{\tau_0^2}\right] \tag{16.28}$$

Thus, the instantaneous frequency is given by (see equation (16.25))

$$\begin{aligned} \omega(t) &= \omega_0 + \frac{k_0 n_2 z}{A_{eff}} \left[4 \frac{(t - z/v_g)}{\tau_0^2} P_0 \exp\left\{-\frac{2\left(t - \frac{z}{v_g}\right)^2}{\tau_0^2}\right\}\right] \\ &= \omega_0 + \frac{k_0 n_2 z}{A_{eff}} \frac{4T}{\tau_0^2} P_0 \, e^{-2T^2/\tau_0^2} \end{aligned} \tag{16.29}$$

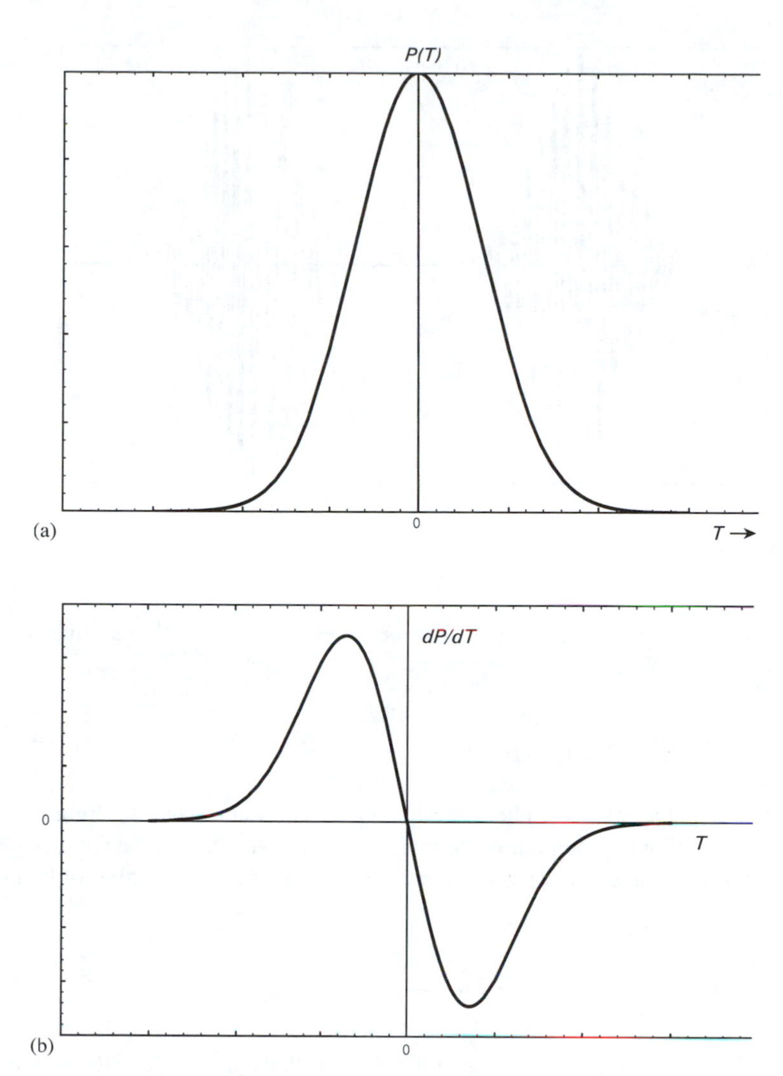

Fig. 16.3: For a pulse with power varying with time as shown in (a), dP/dt varies as shown in (b). Since $dP/dt > 0$ for $t < z/v_g$, the frequency in the leading edge is less than ω_0 [see equation (16.25)]. Similarly, the frequency in the trailing edge is greater than ω_0.

where

$$T = t - \frac{z}{v_g} \tag{16.30}$$

represents the time in the moving frame. The above equations clearly show that nonlinearity leads to the spectral broadening of the pulse leaving the pulse envelope unchanged. Figures 16.3(a) and (b) show the variation of $P(t)$ and dP/dt for a Gaussian pulse. The leading edge of the pulse ($t < z/v_g$) corresponds to $dP/dt > 0$ and the instantaneous frequency is downshifted from ω_0, whereas the trailing edge ($t > z/v_g$) corresponds to $dP/dt < 0$ and the instantaneous frequency is upshifted from ω_0. Figure 16.4 shows the real part of equation (16.27) corresponding to an unchirped pulse ($L = 0$) and a chirped pulse for $L > 0$. One can clearly see the chirping phenomenon generated as a result of SPM.

Note that GVD leads to broadening of the pulse in the time domain keeping the spectral content the same, whereas SPM leads to broadening of the spectrum keeping the temporal distribution unaltered.

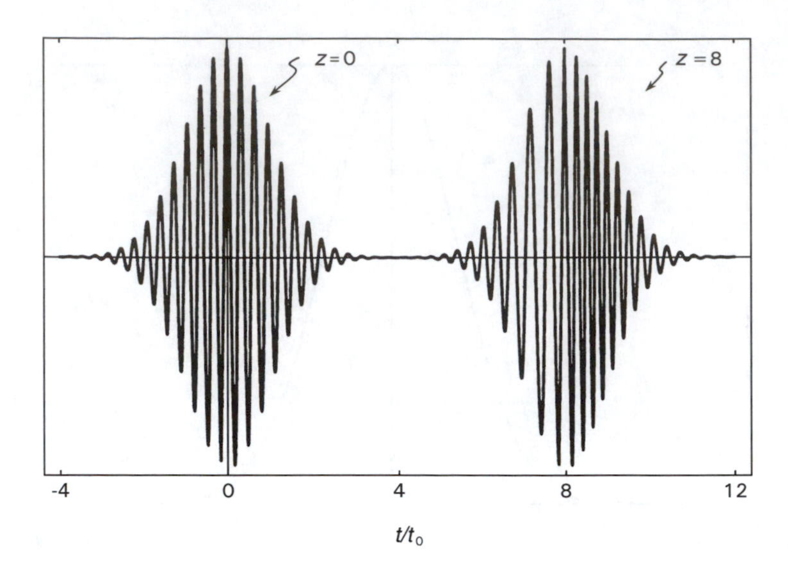

Fig. 16.4: Time variation of the real part of equation (16.27) representing the actual electric field variation of the optical pulse corresponding to the input and output.

From equation (16.25) we have for the excess spectral width resulting from SPM

$$\Delta\nu_{SPM} \simeq \frac{n_2 L_{eff}}{\lambda_0 A_{eff}} \frac{dP}{dt} \tag{16.31}$$

where L_{eff} is the effective fiber length over which the propagating light pulse (which is getting exponentially attenuated) can be assumed to have an approximately constant average intensity. If α is the loss coefficient of a fiber of length L, then

$$L_{eff} \simeq \int_0^L e^{-\alpha z}\, dz = \frac{1 - e^{-\alpha L}}{\alpha} \tag{16.32}$$

Thus, if $\alpha L \ll 1$, then $L_{eff} \simeq L$ and if $\alpha L \gg 1$, then $L_{eff} \simeq 1/\alpha$. At 1550 nm, $\alpha \simeq 0.2$ dB/km and for $\alpha L \gg 1$, $L_{eff} \simeq 20$ km.

If we assume

$$\frac{dP}{dt} \simeq \frac{P}{\tau_0} \sim P \Delta\nu_P \tag{16.33}$$

where τ_0 is the input pulse width that is the inverse of the pulse spectral width $\Delta\nu_P$ and P is the pulse peak power, then for $\Delta\nu_{SPM} = \Delta\nu_P$ we have

$$P L_{eff} \simeq \frac{\lambda_0 A_{eff}}{n_2} \sim 2.5 \times 10^3 \text{ mW·km}$$

where for silica fibers operating at 1550 nm we have used $n_2 \simeq 3.2 \times 10^{-20}$ m²/W, $A_{eff} \simeq 50\ \mu\text{m}^2$. Hence, if optical amplifiers are used to compensate any loss, then for $L_{eff} = 1000$ km, a pulse peak power of 3 mW will lead to a spectral broadening by a factor of two.

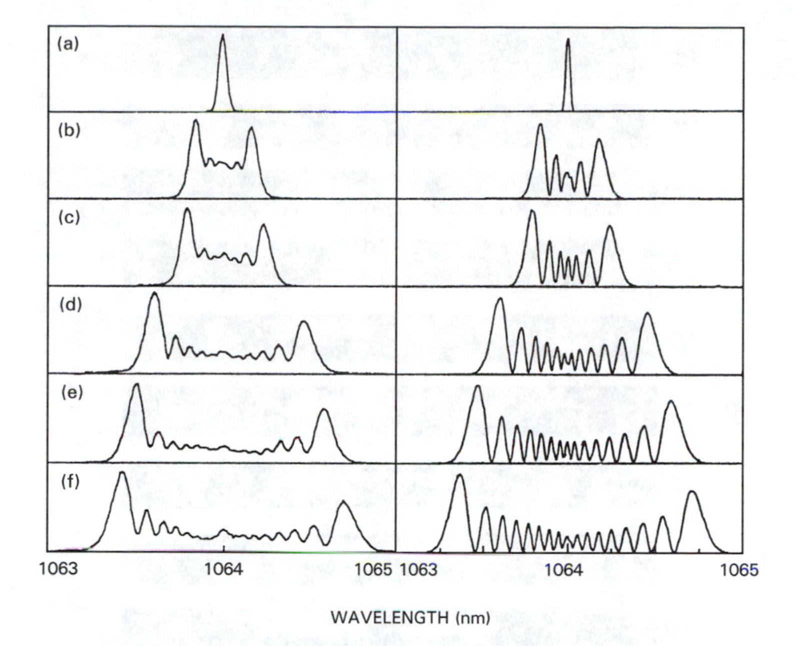

Fig. 16.5: Digital-intensity spectral curves of 1064 nm laser pulses propagating in a 1-m, 2.5 μm core optical fiber with different peak powers. The left hand column shows the experimental results, and the right hand column displays the numerical simulations. (a) Input laser, (b) $P_0 = 1800$ W, (c) $P_0 = 2300$ W, (d) $P_0 = 3900$ W, (e) $P_0 = 4900$ W (f) $P_0 = 5700$ W. [Adapted from Wang et al. (1994)].

16.3 Spectra of self-phase modulated pulses

In the previous section we showed that SPM leads to chirping of the pulse and that, because of the intensity-dependent refractive index, new frequencies are generated in the pulse. The spectral power distribution can be obtained by evaluating the Fourier transform of the temporal pulse distribution:

$$\tilde{E}\,(z = L, \omega) = \int E\,(z = L, t)\; e^{-i\omega t}\, dt \tag{16.34}$$

where $E\,(z = L, t)$ is given by equation (16.27). Figure 16.5 shows the measured and calculated power spectra ($= |\tilde{E}(z = L, \omega)|^2$) corresponding to an input laser at 1.064 μm propagating through a fiber of core diameter 2.5 μm and of length 1 m. The various figures correspond to increasing input power with (a) corresponding to the input laser pulse. It can be seen that as the intensity of the input laser pulse increases, the spectrum broadens. Obviously, for a given input power the spectrum will broaden with an increase in length of fiber (neglecting attenuation). The large intensity oscillations within the spectra are due to interference effects. Figure 16.6 shows a corresponding experimentally measured video display of the output spectra. One can see a close match between the ones calculated and those measured experimentally.

16.4 Heuristic derivation of soliton power

In the previous chapter we have seen that dispersion produces chirping in a pulse. For a Gaussian input pulse, the instantaneous frequency within the pulse

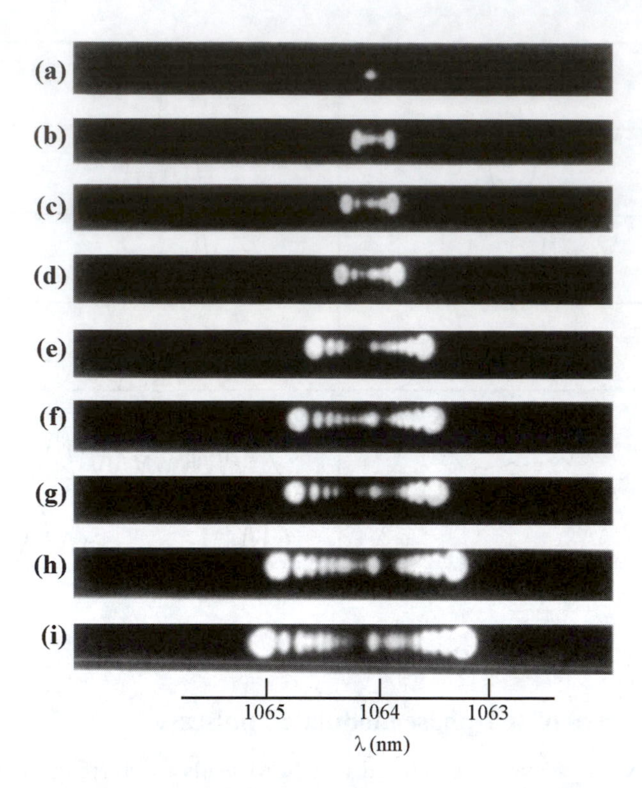

Fig. 16.6: Measured video display of output spectra of a pulse at 1.064 μm after propagating through 1 m of a fiber for varying input powers. The measured spectra closely match the theoretically estimated one on the basis of SPM. [After Wang et al. (1994) Photograph Courtesy Professor R. R. Alfano.]

envelope is given by (see equation (15.10)).

$$\omega_d(t) = \omega_0 + \frac{2\sigma}{(1+\sigma^2)\tau_0^2}\left(t - \frac{z}{v_g}\right) \tag{16.35}$$

where

$$\sigma = \frac{2\alpha\, z}{\tau_0^2} \tag{16.36}$$

and

$$\alpha = \frac{\lambda_0^3}{2\pi\, c^2}\frac{d^2 n}{d\lambda_0^2} \tag{16.37}$$

Similarly, we saw in Section 16.2 that, because of intensity-dependent refractive index, the pulse gets chirped without any broadening. The corresponding instantaneous frequency (close to the center of the pulse) is given by [see equation (16.29)]

$$\omega_{nl}(t) = \omega_0 + \frac{4\,k_0\, n_2\, z}{\tau_0^2}\left(t - \frac{z}{v_g}\right)\frac{P_0}{A_{eff}} \tag{16.38}$$

where P_0 is the peak power carried by the pulse and A_{eff} is the effective area of the fiber mode.

If the chirping effects due to the dispersion and nonlinearity cancel each other, then we would have a pulse that remains unaltered both in time and in frequency domains. This would be the soliton.

We first note that to cancel the two chirps, σ in equation (16.35) has to be negative. Thus, such a soliton can be formed only in the negative group velocity dispersion regime.[2]

For a soliton to be formed, we require that over infinitesimal propagation distances the chirpings produced by dispersion and nonlinearity cancel each other. Thus, in equation (16.35) we assume $\sigma \ll 1$ and equate the chirping due to dispersion and nonlinearity to obtain

$$\frac{4k_0 n_2 z}{\tau_0^2} \frac{P_0}{A_{eff}} = -\frac{2\sigma}{\tau_0^2} = -\frac{4\alpha z}{\tau_0^4}$$

or

$$P_0 = \frac{|\alpha| A_{eff}}{k_0 n_2 \tau_0^2} \tag{16.39}$$

The dispersion coefficient D of the optical fiber is given by (see equation (6.8))

$$D = -\frac{2\pi c}{\lambda_0^2} \alpha = -\frac{\lambda_0}{c} \frac{d^2 n}{d\lambda_0^2} \tag{16.40}$$

Thus, equation (16.39) gives the following approximate equation for the peak power in the pulse.

$$P_0 \simeq \frac{\lambda_0^3 |D| A_{eff}}{4\pi^2 c n_2 \tau_0^2} \tag{16.41}$$

A more rigorous analysis (see Section 16.7) gives

$$P_0 \simeq 0.776 \frac{\lambda_0^3 |D|}{\pi^2 c n_2 \tau_f^2} A_{eff} \tag{16.42}$$

where τ_f is the FWHM of the pulse, which is given by

$$\tau_f = \tau_0 \sqrt{2 \ln 2} \simeq 1.18\, \tau_0$$

In terms of FWHM, equation (16.41) becomes

$$P_0 \tau_f^2 \simeq \frac{0.35\, \lambda_0^3 |D|\, A_{eff}}{\pi^2 c n_2} \tag{16.43}$$

which is different from the more accurate calculation (equation (16.42)) by a

[2] In the positive group velocity regime one can form what are termed as dark solitons.

factor of about 2. For silica fibers operating at 1550 nm,

$$\lambda_0 = 1550\,\text{nm}, \quad n_2 \simeq 3.2 \times 10^{-20}\,\text{m}^2/\text{W}, \quad A_{eff} \simeq 50\,\mu\text{m}^2$$

we get (using equation (16.42))

$$P_0\,\tau_f^2 \simeq 1.5 \times 10^3 D \tag{16.44}$$

where P_0, τ_f, and D are measured in mW, ps, and ps/km·nm, respectively. For a CSF (with zero dispersion wavelength at 1300 nm), $D \simeq 18$ ps/km·nm (at 1550 nm) and we obtain

$$P_0\,\tau_f^2 \simeq 2.7 \times 10^4\,\text{mW}\cdot\text{ps}^2$$

Thus, the power required for the formation of a soliton with an FWHM of 10 ps is

$$P_0 \simeq 270\,\text{mW}$$

On the other hand, for a DSF (with zero dispersion wavelength around 1550 nm), we may assume $D \simeq 1$ ps/km·nm and we obtain

$$P_0\tau_f^2 = 1500 \quad \text{mW}\cdot\text{ps}^2$$

Hence, the power required for a 10-ps soliton is

$$P_0 \simeq 15\,\text{mW}$$

For transmission bit rates below 20 Gb/s, the soliton pulse width is typically 20 ps and the required peak power is only a few mW, which is easily achievable with laser diodes.

Figure 16.7 shows experimental results on transmission of 55-ps pulses at 4 Gb/s measured after propagation through 125 and 310 km of CSF with zero dispersion at 1300 nm. It can be seen that at low power levels the pulses, after propagating through the fiber, overlap considerably and, indeed, after 310 km of propagation are completely unresolvable. By increasing the power level to that required for soliton (19-mW peak power), the pulses are restored to their original width even after 310 km of propagation.

As an example, we consider the experiment of Taga et al. (1994) using DSFs operating at 1559 nm. The values of various parameters were

$$\tau_f = 30\,\text{ps}$$

$$\lambda_{\text{op}} = 1559\,\text{nm}$$

$$D = 0.6\,\text{ps/km}\cdot\text{nm}$$

$$P_0 = -7\,\text{dBm} \simeq 0.2\,\text{mW}$$

where P_0 represents the peak power of the pulse. From equation (16.44) we get $P_0 \simeq 1$ mW, which gives the correct order. The discrepancy may be due to a smaller effective mode area.

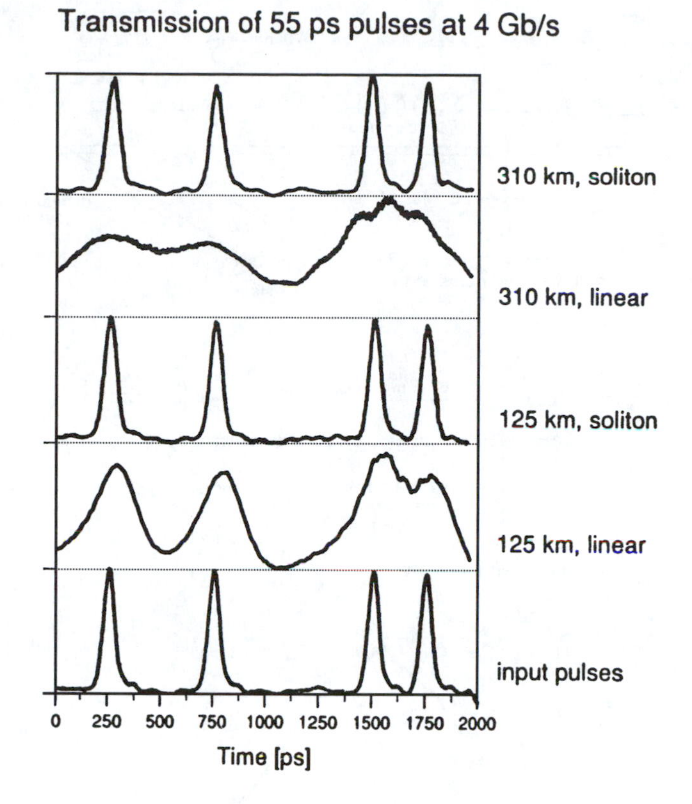

Fig. 16.7: Experimental results of transmission of 55-ps pulses at 4 Gb/s measured after propagation through 125 km and 310 km of CSF with zero dispersion at 1300 nm. At low power levels the pulses are completely unresolvable after 310 km of propagation, whereas by increasing the power level to that required for soliton, the pulses are fully restored. [After Christiansen et al. (1994).]

Since solitons use nonlinear effects to compensate for pulse dispersion, the pulse must have sufficient peak power, as discussed earlier. Actual optical fibers possess loss and, thus, even though the launched optical pulse may have sufficient power to form a soliton, as it propagates through the fiber it will suffer attenuation, leading to reduced power, finally resulting in nonsoliton-like behavior – that is, suffering dispersion. For retaining the soliton nature of propagation in the entire link, it is very important to compensate this loss. The development of the EDFAs (see Chapter 14) has finally resolved this problem and has made soliton communication a reality. EDFAs capable of amplification of signals around 1550 nm (the lowest loss wavelength of silica-based fibers) are placed every 30–40 km and have to compensate for losses of 7.5–10 dB (assuming a typical loss figure of 0.25 dB/km at 1550 nm). With this, propagation of 80 Gb/s over 80 km [Iwatsuki et al. (1993)], 10 Gb/s over 1 million km [Nakazawa (1994)], have been demonstrated.

16.5 Compression of a chirped pulse

In the previous chapter we showed that a (positively) chirped pulse, while propagating through a dispersive medium (characterized by negative dispersion), can be made to undergo compression (see Figure 15.3). SPM can indeed be used to achieve chirping of the pulse, which can then be passed through a dispersive system to achieve pulse compression. Figure 16.8 shows the experimental arrangement of Nikolaus and Grischowsky (1983) showing pulse compression. A 5.9-ps, 2-kW optical pulse is first propagated through an optical

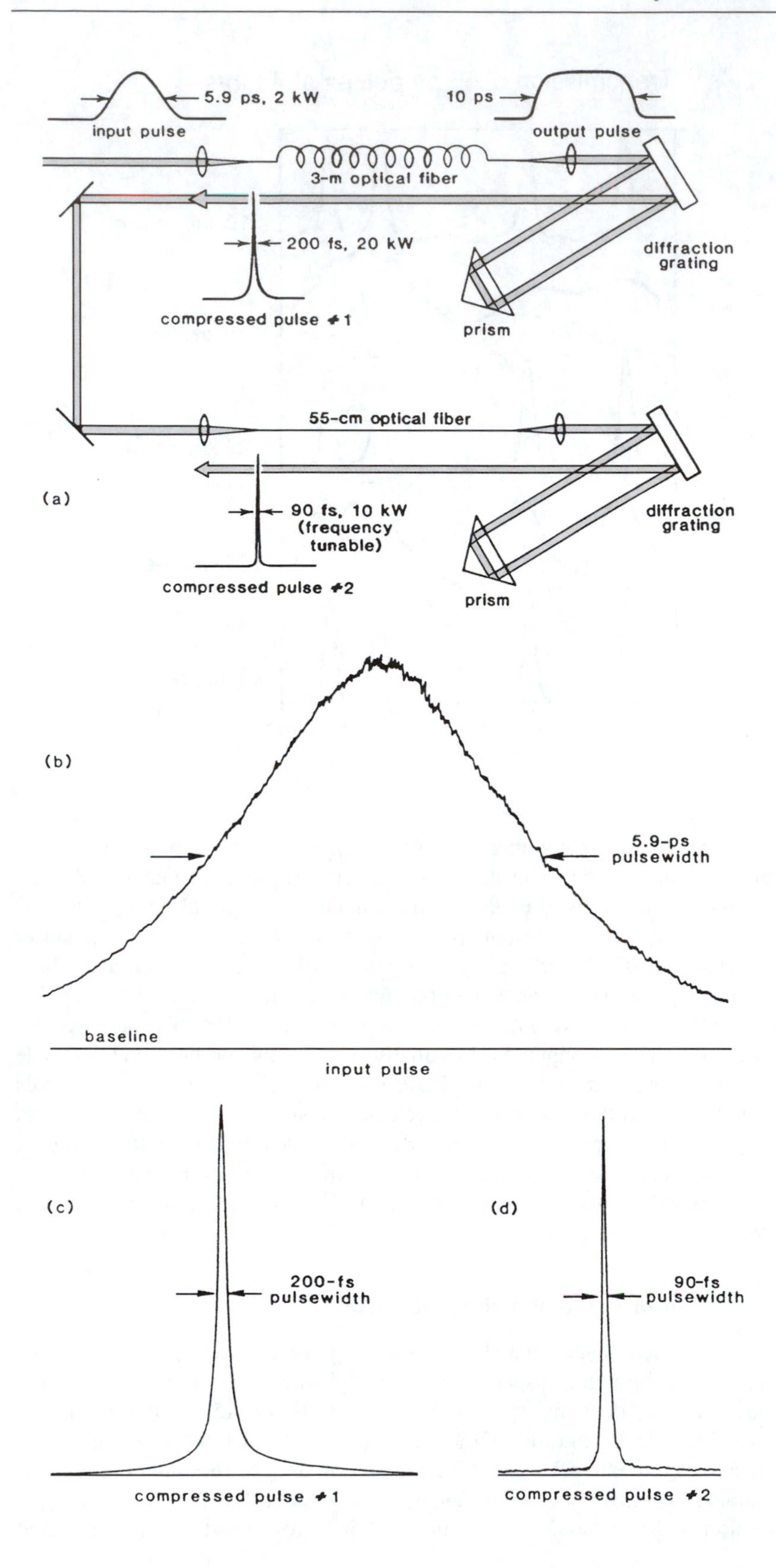

Fig. 16.8: Experimental arrangement for achieving pulse compression. With this arrangement an input pulse of 5.9 ps was compressed to 90 fs using spectral broadening through SPM in the optical fiber and pulse compression using a grating-prism arrangement. [After Nikolaus and Grischowsky (1983).]

fiber; the SPM leads to frequency chirping (due to dispersion, there is a slight broadening also – the 5.9-ps pulse broadens to 10 ps). The chirped pulse is allowed to fall on a grating prism pair as shown in Figure 16.8; the anomalous GVD is caused by the angular dispersion of the grating-prism pair [see, e.g., Agrawal (1989)]. The compressed pulse of width 1 ps is again passed through an optical fiber and a grating-prism pair. The final output is a 90-fs (10 kW) pulse.

16.6 The nonlinear Schrödinger equation (NLSE)

The evolution of an optical pulse propagating through a nonlinear dispersive medium is approximately governed by the following equation.

$$-i\left(\frac{\partial f}{\partial z}+\frac{1}{v_g}\frac{\partial f}{\partial t}\right)-\frac{1}{2}\alpha\frac{\partial^2 f}{\partial t^2}+\Gamma\,|f|^2 f=0 \tag{16.45}$$

where

$$\frac{1}{v_g}=k'=\left.\frac{dk}{d\omega}\right|_{\omega=\omega_0} \tag{16.46}$$

$$\alpha=k''=\left.\frac{d^2k}{d\omega^2}\right|_{\omega=\omega_0} \tag{16.47}$$

$$\Gamma=\frac{1}{2}\,\omega_0\epsilon_0\,n_0n_2 \tag{16.48}$$

and $f\,(z,t)$ represents the envelope term of the pulse (cf. equation (6.32)).

$$E(z,t)=\underbrace{e^{i[\omega_0 t-k(\omega_0)z]}}_{\text{Phase term}}\ \underbrace{f(z,t)}_{\text{Envelope term}} \tag{16.49}$$

with

$$k\,(\omega_0)=\frac{\omega_0}{c}\,n_0 \tag{16.50}$$

The second term on the LHS of equation (16.45) is proportional to α and represents the dispersion term; the last term on the LHS corresponds to the nonlinear term. In the following we justify equation (16.45).

16.6.1 Propagation in absence of dispersion and nonlinearity

To understand the physics of pulse evolution described by equation (16.45), we first neglect the terms representing second-order dispersion and nonlinearity to obtain

$$\frac{\partial f}{\partial z}+\frac{1}{v_g}\frac{\partial f}{\partial t}=0 \tag{16.51}$$

We go over to a moving frame and change the variable set (z,t) to (z,T)

where

$$T = t - \frac{z}{v_g} \tag{16.52}$$

$$z = z$$

Thus, equation (16.51) takes the form[3]

$$\frac{\partial f(z, T)}{\partial z} = 0 \tag{16.53}$$

The above equation has the general solution

$$f = f_0(T) = f_0\left(t - \frac{z}{v_g}\right) \tag{16.54}$$

Equation (16.54) indicates that the pulse propagates without any distortion with the group velocity v_g. If we multiply equation (16.51) by f^* and the complex conjugate of equation (16.51) by f and then add, we get

$$\frac{\partial |f|^2}{\partial z} + \frac{1}{v_g}\frac{\partial |f|^2}{\partial t} = 0 \tag{16.55}$$

implying that the pulse energy also propagates with the same velocity. Thus, in the absence of dispersion and nonlinearity, the pulse propagates without any change.

16.6.2 Propagation in presence of dispersion only

We next include the second-order term in equation (16.45) but neglect the nonlinear term to obtain (in the moving frame)

$$-i\frac{\partial f(z, T)}{\partial z} - \frac{1}{2}\alpha\frac{\partial^2 f(z, T)}{\partial T^2} = 0 \tag{16.56}$$

Using the method of separation of variables, one can readily obtain the general

---

[3]

$$\begin{aligned}\frac{\partial f(z,t)}{\partial z} &= \frac{\partial f(z,T)}{\partial z} + \frac{\partial f(z,T)}{\partial T}\frac{\partial T}{\partial z}\\ &= \frac{\partial f(z,T)}{\partial z} - \frac{1}{v_g}\frac{\partial f(z,T)}{\partial T}\end{aligned}$$

$$\begin{aligned}\frac{\partial f(z,t)}{\partial t} &= \frac{\partial f(z,T)}{\partial z}\frac{\partial z}{\partial t} + \frac{\partial f(z,T)}{\partial T}\frac{\partial T}{\partial t}\\ &= \frac{\partial f(z,T)}{\partial T}\end{aligned}$$

solution of the above equation; the result is

$$f(z, T) = \int A(\Omega)\, e^{i\left(\Omega T - \frac{1}{2}\alpha\Omega^2 z\right)}\, d\Omega \tag{16.57}$$

Obviously $A(\Omega)$ represents the frequency spectrum of the input pulse.

We obtained the above equation in Section 6.4 and evaluated it for a Gaussian temporal pulse. The pulse underwent broadening and chirping due to dispersion.

16.6.3 Propagation in presence of nonlinearity only

We next neglect the second-order dispersion term and analyze the effect of nonlinearity alone on the zeroeth-order solution (equation (16.54)). Neglecting the dispersion term and keeping only the nonlinear term, equation (16.45) gives

$$-i\left(\frac{\partial f}{\partial z} + \frac{1}{v_g}\frac{\partial f}{\partial t}\right) + \Gamma\,|f|^2 f(z, t) = 0 \tag{16.58}$$

In the moving frame, the above equation becomes

$$-i\frac{\partial f(z, T)}{\partial z} + \Gamma\,|f|^2 f(z, T) = 0 \tag{16.59}$$

If we multiply the above equation by f^* and its complex conjugate by f and subtract we obtain

$$\frac{\partial\,|f|^2}{\partial z} = 0 \tag{16.60}$$

which has the general solution

$$|f|^2 = F(T) = F\left(t - \frac{z}{v_g}\right) \tag{16.61}$$

Thus, if the nonlinearity is weak enough (so that v_g remains intensity independent), the absolute square of the wave envelope $|f|^2$ retains its shape as it propagates through the fiber. We therefore look for a solution of equation (16.59) in the form

$$f(z, T) = f_0(T)\, e^{-i\phi(z,T)} \tag{16.62}$$

where $f_0(T)$ and $\phi(z, T)$ are assumed to be real functions. From equations (16.59) and (16.62) we obtain

$$\frac{\partial\phi}{\partial z} = \Gamma\,|f_0(T)|^2 \tag{16.63}$$

Hence

$$\phi(z, T) = \phi_0 + \Gamma\,|f_0(T)|^2\, z \tag{16.64}$$

where

$$\phi_0 \equiv \phi(T, 0) \tag{16.65}$$

We can always put $\phi_0 = 0$, which yields

$$f(z, T) = f_0(T) \exp[-i\,\Gamma |f_0(T)|^2 z] \tag{16.66}$$

implying that nonlinearity leads to phase modulation that is directly proportional to the pulse intensity and the distance of propagation (see Figure 16.4). This is the phenomenon of SPM. Thus, the electric field is given by

$$E(z, t) = f_0\left(t - \frac{z}{v_g}\right) \times \exp\left[i\left\{\omega_0 t - \Gamma\left|f_0\left(t - \frac{z}{v_g}\right)\right|^2 z - k(\omega_0)\,z\right\}\right] \tag{16.67}$$

Corresponding to this SPM there is a frequency modulation given by

$$\Delta\omega = -\Gamma\, z \frac{\partial}{\partial t}\left[\left|f_0\left(t - \frac{z}{v_g}\right)\right|^2\right] \tag{16.68}$$

which is what we had obtained in Section 16.2. It is clear from the above analysis that nonlinearity leads to the broadening of the spectrum, leaving the pulse envelope unchanged.

16.7 Soliton solution to NLSE

The analysis carried out in the previous section suggests that it may be possible to completely balance the frequency modulations induced by dispersion and nonlinearity and obtain a soliton solution of the NLSE. To obtain the soliton solution we rewrite equation (16.45) as

$$-i\left(\frac{\partial f}{\partial z} + \frac{1}{v_g}\frac{\partial f}{\partial t}\right) - \frac{1}{2}\alpha\frac{\partial^2 f}{\partial t^2} + \Gamma\,|f|^2 f = 0 \tag{16.69}$$

Once again, going over to the moving frame we obtain

$$-i\frac{\partial f}{\partial z} - \frac{1}{2}\alpha\frac{\partial^2 f}{\partial T^2} + \Gamma\,|f|^2 f = 0 \tag{16.70}$$

We look for a soliton-shaped solution of the form

$$f(z, T) = E_0 \psi(T)\, e^{-i\phi(z)} \tag{16.71}$$

where E_0 represents the peak electric field and the envelope function $\psi(T)$ is assumed to be a real function of T. The phase term is assumed to be independent of time so that there is no *chirping*. Substituting equation (16.71) in equation

(16.70) we obtain

$$-\frac{d\phi}{dz} E_0 \psi(T) - \frac{1}{2} \alpha E_0 \frac{d^2\psi}{dT^2} + \Gamma E_0^3 \psi^3(T) = 0$$

or

$$\begin{aligned} -\frac{d\phi}{dz} &= \frac{1}{2} \alpha \frac{1}{\psi(T)} \frac{d^2\psi}{dT^2} - \Gamma E_0^2 \psi^2(T) \\ &= \frac{1}{2} K \end{aligned} \tag{16.72}$$

Since the left-hand side depends only on the z coordinate and the right-hand side depends on T, in the last step we have set each side equal to a constant $\frac{1}{2}K$. Thus

$$\phi(z) = -\frac{1}{2} K z \tag{16.73}$$

where we have neglected an unimportant constant of integration. Rewriting equation (16.72) we get

$$\frac{d^2\psi}{dT^2} - \frac{2\Gamma}{\alpha} E_0^2 \psi^3(T) - \frac{K}{\alpha} \psi(T) = 0 \tag{16.74}$$

We multiply the above equation by $2\, d\psi/dT$ and rearrange to obtain

$$\frac{d}{dT}\left[\left(\frac{d\psi}{dT}\right)^2 - \frac{\Gamma}{\alpha} E_0^2 \psi^4(T) - \frac{K}{\alpha} \psi^2(T)\right] = 0 \tag{16.75}$$

Thus

$$\left(\frac{d\psi}{dT}\right)^2 = \frac{\Gamma}{\alpha} E_0^2 \psi^4(T) + \frac{K}{\alpha} \psi^2(T) + C \tag{16.76}$$

where C is the constant of integration. Now, for a localized soliton we should expect

$$\lim_{T \to \pm\infty} \psi(T) = 0 \tag{16.77}$$

and

$$\lim_{T \to \pm\infty} \frac{d\psi}{dT} = 0 \tag{16.78}$$

Therefore

$$C = 0 \tag{16.79}$$

implying

$$\left(\frac{d\psi}{dT}\right)^2 = \frac{\Gamma E_0^2}{\alpha}\,\psi^4(T) + \frac{K}{\alpha}\,\psi^2(T) \tag{16.80}$$

For a localized soliton, $\psi(T)$ must have a maximum and without any loss of generality we may choose the value of E_0 so that the maximum value of $\psi(T)$ is unity. Since at the maximum value $(d\psi/dT) = 0$, we must have

$$\psi = 1 \qquad \text{when } (d\psi/dT) = 0 \tag{16.81}$$

implying

$$\frac{\Gamma E_0^2}{\alpha} + \frac{K}{\alpha} = 0$$

or

$$K = -\Gamma E_0^2 \tag{16.82}$$

Since n_2 (and therefore Γ) is a positive quantity, K must be negative. Using the above value of K, we get

$$\frac{d\psi}{dT} = \gamma\,\psi\sqrt{1-\psi^2} \tag{16.83}$$

where

$$\gamma = \left(\frac{K}{\alpha}\right)^{1/2} = \left(-\frac{\Gamma}{\alpha}\right)^{1/2} E_0 \tag{16.84}$$

Thus

$$\int \frac{d\psi}{\psi\sqrt{1-\psi^2}} = \gamma \int dT \tag{16.85}$$

We can readily integrate the above equation by making the substitution

$$\psi = \operatorname{sech}\theta \tag{16.86}$$

Thus

$$\int d\theta = \gamma\,T$$

or

$$\psi(T) = \operatorname{sech}\theta = \operatorname{sech}(\gamma\,T) \tag{16.87}$$

Thus, the soliton solution of equation (16.69) is given by

$$f(z,t) = E_0 \operatorname{sech}\left[\gamma\left(t - \frac{z}{v_g}\right)\right] e^{-igz} \tag{16.88}$$

where

$$g = -\frac{1}{2}\alpha\,\gamma^2 = \frac{1}{2}\Gamma\,E_0^2 \tag{16.89}$$

The solution given by equation (16.88) can also be verified by direct substitution in equation (16.69); the solution represents a soliton of amplitude E_0.

At $z = 0$ – that is, at the input – the soliton is given by

$$f(z, T) = E_0 \operatorname{sech}(\gamma t) \tag{16.90}$$

The FWHM of the above pulse is obtained by solving

$$\operatorname{sech}^2 \gamma t_0 = \frac{1}{2}$$

and is given by

$$\tau_f = 2 t_0 = \frac{2}{\gamma} \ln\left(1 + \sqrt{2}\right) \simeq \frac{1.7627}{\gamma}$$

Now, using equations (16.84), (16.48), and (16.40) and the following expression of peak soliton power

$$P_0 = \frac{1}{2} \epsilon_0 n_0 c |E_0|^2 A_{eff}$$

we obtain

$$P_0 = 0.776 \frac{\lambda_0^3 |D| A_{eff}}{\pi^2 c n_2 \tau_f^2} \tag{16.91}$$

consistent with equation (16.42).

Problems

16.1 Show by direct substitution that

$$f(z, t) = E_0 \operatorname{sech}[\gamma(t - z/v_g)] e^{-igz} \tag{16.92}$$

is a solution of the NLS equation given by equation (16.69). Show that the values of γ and g are given by equation (16.89).

16.2 Show that the general solution of equation (16.56) is given by equation (16.57).

16.3 (a) Equation (16.56) takes into account terms up to the second order in dispersion. Show that the corresponding equation taking into account the third-order dispersion will be given by

$$-i \frac{\partial f}{\partial z} - \frac{1}{2} \alpha \frac{\partial^2 f}{\partial T^2} + \frac{i}{6} \kappa \frac{\partial^2 f}{\partial T^3} = 0 \tag{16.93}$$

where

$$\kappa = k''' = \left. \frac{d^3 k}{d\omega^3} \right|_{\omega=\omega_0} \tag{16.94}$$

(b) Show that the general solution of equation (16.93) is given by

$$f(z, T) = \int A(\Omega)\, e^{i\left(\Omega T - \frac{1}{2}\alpha\Omega^2 z - \frac{1}{6}\kappa\Omega^3 z\right)}\, d\Omega \tag{16.95}$$

17

Single-mode fiber optic components

17.1 Introduction

With the increased penetration of optical fibers into the subscriber loop and their applications in various different kinds of sensors and other optical processing applications, there is a growing demand for in-line fiber optic components capable of performing various functions such as modulation, splitting, filtering, etc. Normally these functions are performed by taking the light out of the fiber, processing it using bulk optical components, and then coupling the light back into the fiber. This would involve interruption of the light beam as it propagates in the fiber and thus would lead to high optical insertion loss, problems of stability of the components, and larger size and perhaps would be costlier. These problems can be overcome by using in-line fiber optic components in which the processing is performed without taking the light out of the fiber and which are completely compatible with the transmission medium – namely, the optical fiber.

There are many different kinds of fiber optic components that are used in many applications in fiber optic communication systems, fiber optic sensors, fiber optic local area networks, etc. These can be broadly classified as follows:

- Amplitude/intensity components
  - Couplers
  - Splitters
  - Amplifiers
  - Attenuators
  - Reflectors
- Phase components
  - Phase shifters
  - Phase modulators
- Polarization components
  - Polarizers
  - Polarization splitters
  - Polarization controllers

Wavelength components
- Wavelength filters
- Wavelength division multiplexers/demultiplexers

Frequency components
- Frequency shifters
- Filters

Active components
- Fiber amplifiers

In this chapter we discuss some of the most important in-line fiber optic components, including the fiber optic directional coupler, the fiber optic polarizer, fiber optic polarization controller, and Bragg fiber gratings. In Chapter 14 we have discussed fiber optic amplifiers in detail.

17.2 The optical fiber directional coupler

The optical fiber directional coupler is the guided wave equivalent of a bulk optic beam splitter and is one of the most important in-line fiber components. It is based on the fact that the modal field of the guided mode extends far beyond the core–cladding interface. Thus, when two fiber cores are brought sufficiently close to each other laterally so that their modal fields overlap, then the modes of the two fibers become coupled and power can transfer periodically between the two fibers. If the propagation constants of the modes of the individual fibers are equal, then this power exchange is complete. On the other hand, if their propagation constants are different, then there is still a periodic, but incomplete, exchange of power between the fibers.

Directional couplers have many interesting applications in power splitting, wavelength division multiplexing/demultiplexing, polarization splitting, fiber optic sensing, and so forth. In this section we briefly outline the coupling phenomenon and discuss some of its applications.

17.2.1 Principle

To understand the basic working mechanism of a directional coupler, let us first consider the simpler case of a directional coupler formed by a pair of identical symmetric single-mode planar waveguides (see Figure17.1). The coupled waveguide system can be viewed as a single waveguide with two cores. This system will have two modes, the fundamental being the symmetric mode and the first excited being the antisymmetric mode. These two modes will have different propagation constants. When power is incident on one waveguide, then it excites a linear combination of the symmetric and the antisymmetric modes. The excitation would be such as to add the lobes in one waveguide and cancel in the other (see Figure 17.2). Since the propagation constants of the two modes are unequal, as the fields propagate through the system, they develop a phase difference. When the accumulated phase difference is π, then the superposition of these two modal fields will result in a cancellation in the first waveguide and an addition in the second. Further propagation over an equal length will result in phase difference 2π, leading to a power transfer back to the first waveguide. Thus, the power exchanges periodically between the two waveguides.

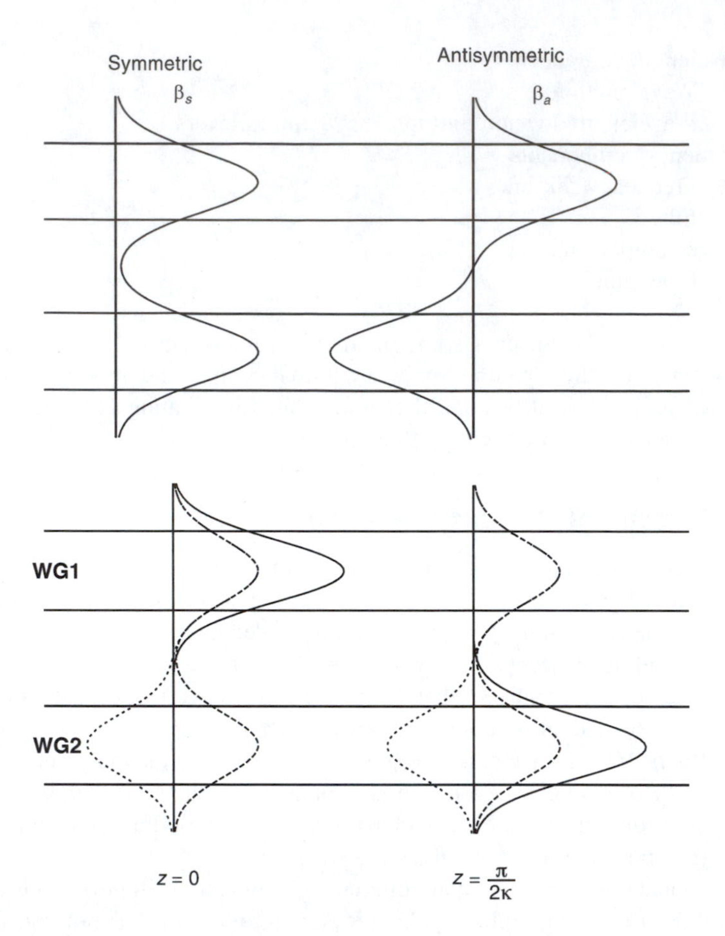

Fig. 17.1: A directional coupler formed by a pair of identical symmetric single-mode planar waveguides. The symmetric and antisymmetric mode fields of the composite structure are also shown.

Fig. 17.2: At $z = 0$ power is launched into waveguide 1, which excites the symmetric and antisymmetric modes. Since $\beta_s \neq \beta_a$, the two modes develop a phase difference as they propagate. When the phase difference is π, then the superposition will cancel in waveguide 1 and add in waveguide 2.

The above picture is in terms of a beating between the normal modes of the composite directional coupler structure. An equivalent picture is to treat the system as a coupled waveguide system. It is shown in Appendix E that when the two interacting waveguides – be it fibers, planar or channel waveguides – have the same propagation constant, then the power transfer is complete. On the other hand, for waveguides with unequal propagation constants, the power transfer is incomplete.

17.2.2 Power exchange

Consider a directional coupler formed of two, in general, nonidentical single-mode fibers supporting the LP_{01} modes with propagation constants β_1 and β_2. In Appendix E we have shown that if $P_1(0)$ is the power launched into fiber 1 at $z = 0$, then at any value of z the powers propagating in the two fibers are given by

$$\frac{P_1(z)}{P_1(0)} = 1 - \frac{\kappa^2}{\gamma^2} \sin^2 \gamma z \tag{17.1}$$

$$\frac{P_2(z)}{P_1(0)} = \frac{\kappa^2}{\gamma^2} \sin^2 \gamma z \tag{17.2}$$

where

$$\gamma^2 = \kappa^2 + \frac{1}{4}(\Delta\beta)^2 \tag{17.3}$$

and

$$\Delta\beta = \beta_1 - \beta_2 \tag{17.4}$$

In the above equations κ is called the coupling coefficient and is a measure of the strength of interaction between the two fibers, which depends on the fiber parameters, the separation between the cores, and the wavelength of operation (see Section 17.2.3). The parameter $\Delta\beta$ is referred to as the *phase mismatch*.

Note from equations (17.1) and (17.2) that

$$P_1(z) + P_2(z) = P_1(0) \tag{17.5}$$

independent of z. This is nothing but a statement of conservation of power.

If the two fibers are separated by a very large distance (large compared with the mode size), then there would be no interaction between the two fibers. In such a case, $\kappa = 0$ and equations (17.1) and (17.2) give us

$$P_1(z) = P_1(0) \quad P_2(z) = 0 \tag{17.6}$$

that is, there is no exchange of power. We now consider some special cases:

(a) Phase-matched case: Let us first consider a directional coupler made up of two fibers with identical propagation constants. For such a case $\Delta\beta = 0$ and equations (17.1) and (17.2) reduce to

$$P_1(z) = P_1(0)\cos^2 \kappa z \tag{17.7}$$

$$P_2(z) = P_1(0)\sin^2 \kappa z \tag{17.8}$$

Figure 17.3 shows the variation of the powers in the two fibers as a function of z. From the above equations and Figure 17.3 we note that there is a periodic exchange of power between the two fibers. At

$$z = 0, \frac{\pi}{\kappa}, \frac{2\pi}{\kappa}, \ldots = \frac{m\pi}{\kappa}; \quad m = 0, 1, 2, \ldots. \tag{17.9}$$

$P_1(z) = P_1(0)$ and $P_2(z) = 0$ – that is, the entire power is in the input fiber. At

$$z = \frac{\pi}{2\kappa}, \frac{3\pi}{2\kappa}, \frac{5\pi}{2\kappa} \ldots$$

$$= \left(m + \frac{1}{2}\right)\frac{\pi}{\kappa}; \quad m = 0, 1, 2, \ldots. \tag{17.10}$$

$P_1(z) = 0,\ P_2(z) = P_1(0)$ and the entire power is in the other fiber.

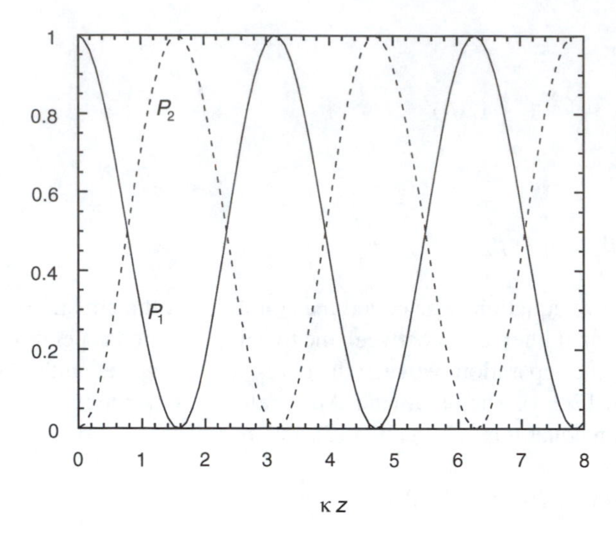

Fig. 17.3: Variation of powers in the two fibers in a directional coupler as a function of z when the two fibers have the same propagation constant.

The minimum distance at which the power completely transfers from the input fiber to the other fiber is given by

$$z = L_c = \frac{\pi}{2\kappa} \tag{17.11}$$

and is referred to as the coupling length. Strong interaction implies a large value of κ and, hence, a small coupling length.

For typical single-mode fibers operating at a wavelength of 1.3 μm, $\kappa \simeq 0.8\ \text{mm}^{-1}$ to $0.3\ \text{mm}^{-1}$, leading to a coupling length of $\simeq$2–5 mm (see Section 17.2.3).

One of the obvious applications of such a fiber directional coupler is as a power divider. Thus, for example, if we form a directional coupler with two identical fibers and choose a length of interaction L so that $L = \pi/4\kappa$, then for power P_0 launched at the input end in fiber 1, the powers in each waveguide at $z = L$ will be

$$P_1 = \frac{1}{2}P_0, \quad P_2 = \frac{1}{2}P_0$$

that is, it acts as a 3-dB power splitter. Obviously, since κ itself is wavelength dependent, the device would work as a perfect 3-dB splitter only at a certain chosen operating wavelength.

By appropriately choosing the value of κL, we can fabricate couplers with an arbitrary splitting ratio.

(b) Non-phase-matched case: Let us now consider the case when $\beta_1 \neq \beta_2$. For this case, equations (17.1) and (17.2) describe the evolution of power in both waveguides as a function of z. In Figure 17.4 we have plotted the z variation of $P_2(z)$ for $\Delta\beta/2\kappa = 0.1$, 1, and 5. For clarity, we have plotted only $P_2(z)$; $P_1(z)$ would, of course, be $P_1(0) - P_2(z)$. From Figure 17.4 we note the following:

(i) If $\Delta\beta \neq 0$, there is an incomplete transfer of power. In fact, the maximum fractional power that is transferred from the input

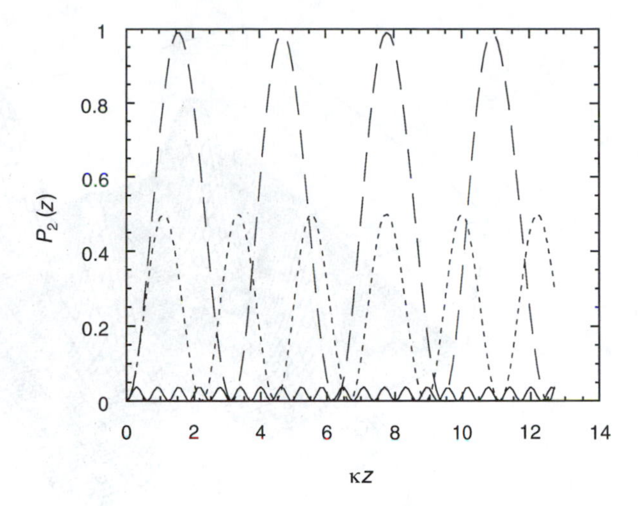

Fig. 17.4: Variation of power in the coupled fiber plotted as a function of κz for different values of relative phase mismatch $\Delta\beta/2\kappa$. For $\Delta\beta \neq 0$, the power transfer is always incomplete. The larger the value of $\Delta\beta/2\kappa$, the smaller is the fractional power transfer. The three curves correspond to $\Delta\beta/2\kappa = 0.1$ (long dash), 1.0 (small dash) and 5.0 (solid).

fiber to the coupled fiber is given by

$$\eta_{\max} = \frac{P_{2,max}}{P_1(0)} = \left(\frac{\kappa^2}{\gamma^2}\sin^2\gamma z\right)_{\max}$$

$$= \frac{\kappa^2}{\gamma^2} = \frac{1}{1 + (\Delta\beta/2\kappa)^2} \tag{17.12}$$

Thus, for complete power exchange, we must have $\Delta\beta = 0$; the larger the ratio $\Delta\beta/2\kappa$, the smaller is the fractional power transfer. For example, for $\Delta\beta/2\kappa = 0.1$, 1, and 5, the maximum fractional power transferred is 0.99, 0.5, and 0.04. Hence, very little exchange of power will take place between two highly non-phase-matched (large $\Delta\beta$) fibers even if their cores lie close to each other (i.e., large κ) as long as $\Delta\beta/2\kappa \gg 1$.

As an example, let us consider a coupler with $\kappa \simeq 0.2$ mm^{-1}. If we require that $\eta_{\max}$ should be less than 1%, then

$$\Delta\beta > 4 \text{ mm}^{-1}$$

If λ_0 is the wavelength of operation and n_e is the effective index of the mode of the fiber, then

$$\Delta\beta = \Delta\left(\frac{2\pi}{\lambda_0}n_e\right) = \frac{2\pi}{\lambda_0}\Delta n_e \tag{17.13}$$

Hence, we obtain for $\lambda_0 = 1.3\ \mu$m

$$\Delta n_e \geq 8 \times 10^{-4}$$

Compare this with the typical core–cladding refractive index difference value of 3×10^{-3} in single-mode fibers.

(ii) Note also from Figure 17.4 that for larger $\Delta\beta/2\kappa$ values, the oscillations in power become more and more rapid with z. This

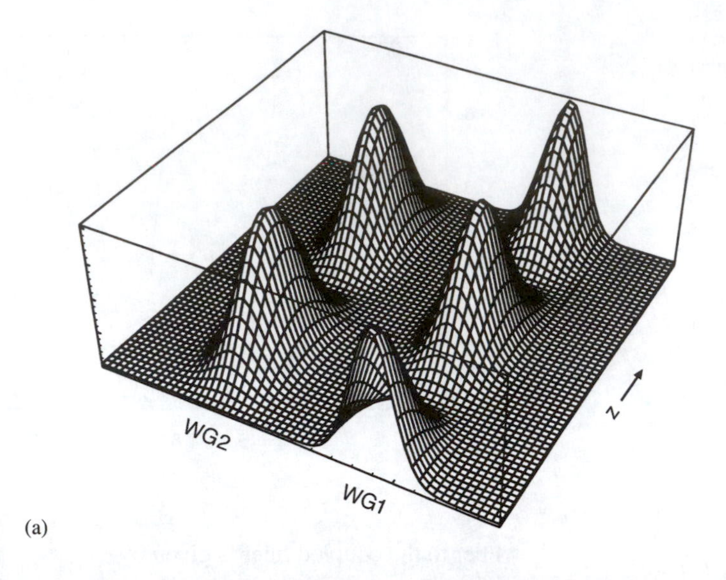

(a)

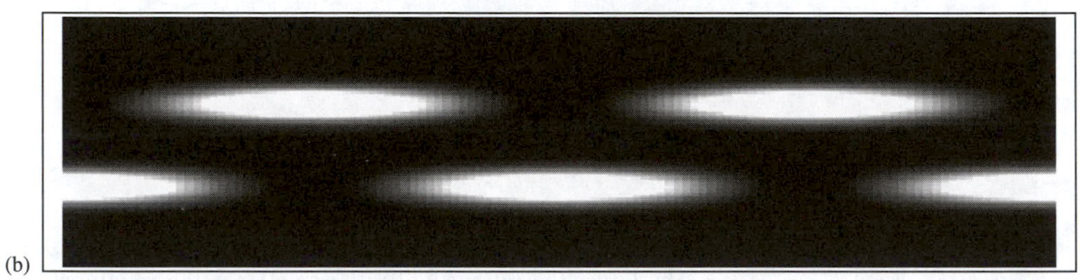

(b)

Fig. 17.5: (a) Three-dimensional plot of variation of power with propagation length in a directional coupler with $\Delta\beta = 0$. Note that the power exchange between the two waveguides (WG1 and WG2) is complete. (b) The corresponding density plot.

effect is used in integrated optics for realizing optical switches [see, e.g., Ghatak and Thyagarajan, 1989], in wavelength multiplexers/demultiplexers (see Section 17.7.2), in polarization splitting using birefringent fibers, and so forth.

Figure 17.5(a) shows a three-dimensional plot of variation of transverse intensity pattern along the propagation direction for a pair of identical planar waveguides. The corresponding density plot clearly showing complete power transfer is given in Figure 17.5(b). Figures 17.6(a) and (b) show the corresponding plots for a pair of nonidentical planar waveguides.

17.2.3 Coupling coefficient of identical fiber directional couplers

By using the fields of the LP_{01} modes of a step index fiber, one can obtain the following approximate expression for κ for identical fibers [Snyder (1972)]

$$\kappa(d) = \frac{\lambda_0}{2\pi n_1} \frac{U^2}{a^2 V^2} \frac{K_0(Wd/a)}{K_1^2(W)} \tag{17.14}$$

where λ_0 is the free space wavelength, n_1 and n_2 are the core and cladding refractive indices, respectively, of the fiber, a is the fiber core radius, d is the separation between the fiber axes, and $K_\nu(x)$ is the modified Bessel function of

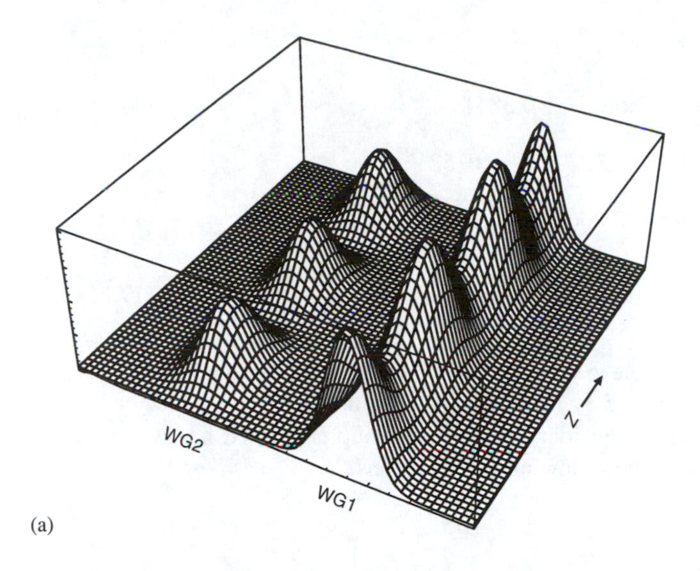

(a)

(b)

Fig. 17.6: (a) Three-dimensional plot of variation of power with propagation length in a directional coupler with $\Delta\beta \neq 0$. Note the incomplete power transfer between the two waveguides (WG1 and WG2). (b) The corresponding density plot.

order ν (see Chapter 8) and

$$U = k_0 a \left(n_1^2 - n_e^2\right)^{1/2} \tag{17.15}$$

$$W = k_0 a \left(n_e^2 - n_2^2\right)^{1/2} \tag{17.16}$$

$$V = \sqrt{U^2 + W^2} = k_0 a \sqrt{n_1^2 - n_2^2} \tag{17.17}$$

$$k_0 = \frac{2\pi}{\lambda_0}$$

and

$$n_e = \frac{\beta}{k_0}$$

is the mode effective index.

A simple and accurate empirical relation for κ for a directional coupler made up of identical step index fibers is given by [Tewari and Thyagarajan (1986)]

$$\kappa(d, V) = \frac{\pi}{2} \frac{\sqrt{\delta}}{a} e^{-(A + B\bar{d} + C\bar{d}^2)} \tag{17.18}$$

where

$$A = 5.2789 - 3.663V + 0.3841V^2 \quad (17.19)$$

$$B = -0.7769 + 1.2252V - 0.0152V^2 \quad (17.20)$$

$$C = -0.0175 - 0.0064V - 0.0009V^2 \quad (17.21)$$

$$\delta = \frac{n_1^2 - n_2^2}{n_1^2}, \quad \tilde{d} = \frac{d}{a} \quad (17.22)$$

Equation (17.18) is accurate to within 1% of the value given by equation (17.14) for the practical range of $1.5 \leq V \leq 2.5$ and $2.0 \leq \tilde{d} \leq 4.5$.

As an example, let us consider a directional coupler formed between two single-mode fibers with the following specifications.

$$\begin{aligned} n_1 &= 1.4532 \\ n_2 &= 1.45 \\ a &= 5.0\,\mu\text{m} \\ \lambda_0 &= 1.3\,\mu\text{m} \end{aligned} \quad (17.23)$$

Using the above numbers we get

$$V = 2.329 \quad \text{and} \quad \delta = 0.0044$$

Thus, using equation (17.18) we obtain

$$\kappa = 20.8 \exp[-(-1.1693 + 1.9945\tilde{d} - 0.0373\tilde{d}^2)]\ \text{mm}^{-1}$$

For a fiber core center to center spacing of 12 μm, we obtain

$$\kappa = 0.694\ \text{mm}^{-1}$$

Thus, the corresponding coupling length is

$$L_c = \frac{\pi}{2\kappa} \simeq 2.26\ \text{mm}$$

Figure 17.7 shows the variation of κ with d/a for a fiber specified by equation (17.23).

Example 17.1: For the directional coupler considered above, for a length of interaction of $L_c/2 = 1.132$ mm, the coupled power at 1.3 μm will be

$$P_2(\lambda_0 = 1.3\,\mu\text{m}) = \sin^2(\kappa \cdot L_c/2) = 0.5$$

Thus, it will behave as a 3-dB coupler for 1.3 μm. If the same coupler is used at 1.35 μm, then at 1.35 μm (from equation 17.18)

$$\kappa \simeq 0.7422\ \text{mm}^{-1}$$

where we have neglected the wavelength dependence of n_1 and n_2. Thus, for the coupler of length 1.132 mm, which behaves as a 3-dB

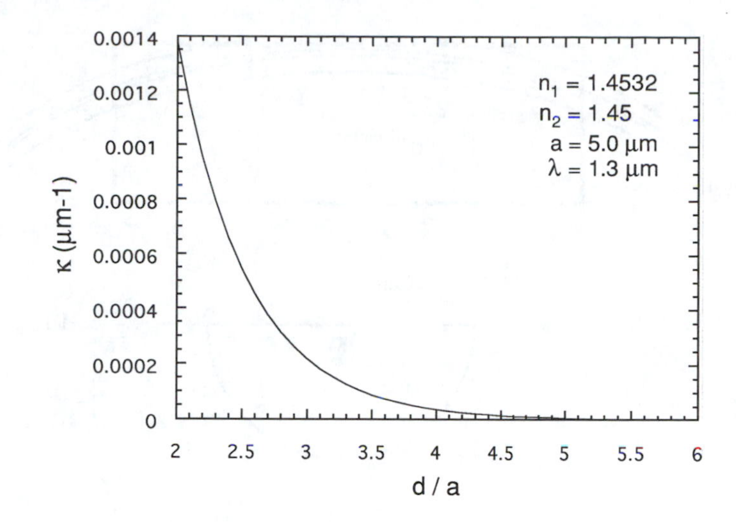

Fig. 17.7: Variation of κ with d/a for a fiber directional coupler consisting of fibers specified by equation (17.23).

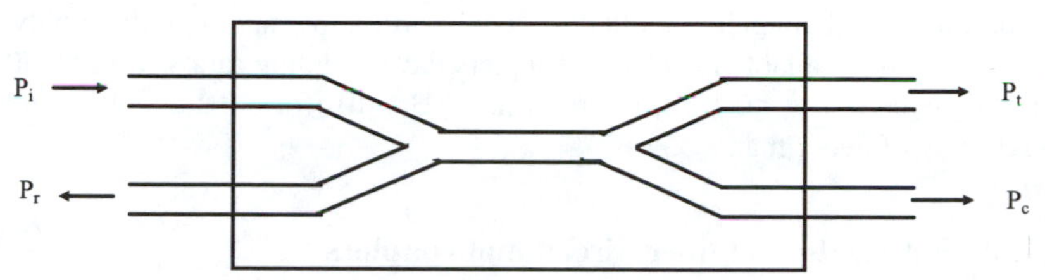

Fig. 17.8: For an input power P_i, the transmitted power, coupled power, and back-coupled power are P_t, P_c, and P_r. An ideal directional coupler would have $P_r = 0$ and $P_t + P_c = P_i$.

coupler at 1.3 μ m, the coupled power at 1.35 μm will be

$$P_2(\lambda_0 = 1.35\ \mu\text{m}) = 55.5\%$$

thus showing the strong wavelength dependence of the coupler.

17.2.4 Practical parameters of a coupler

If power P_i is launched in the input port of a directional coupler as shown in Figure 17.8 and if the transmitted power, coupled power, and back-coupled power are P_t, P_c, and P_r, respectively, then the various characteristics of the coupler are

$$\text{Coupling ratio } R(\%) = \frac{P_c}{P_c + P_t} \times 100 \tag{17.24}$$

$$R(\text{dB}) = 10 \log\left[\frac{P_t + P_c}{P_c}\right]$$

$$\text{Excess loss } L_i(\text{dB}) = 10 \log\left[\frac{P_i}{P_t + P_c}\right] \tag{17.25}$$

$$\text{Insertion loss} = 10 \log\left(\frac{P_i}{P_c}\right)$$

$$= \text{Coupling ratio} + \text{excess loss} \tag{17.26}$$

$$\text{Directivity } D(\text{dB}) = 10 \log\left[\frac{P_r}{P_i}\right] \tag{17.27}$$

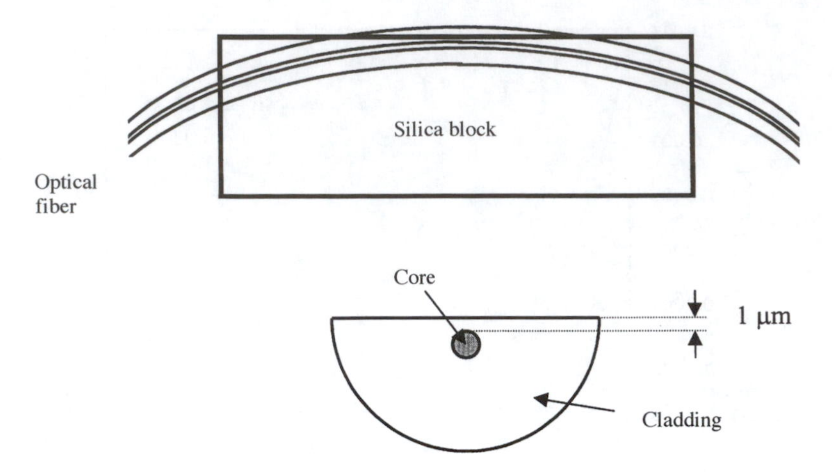

Fig. 17.9: A polished fiber half block fabricated by side polishing the cladding.

Good directional couplers should have low insertion loss and high directivity. Commerically available directional couplers have coupling ratios from 50/50 to 1/99, excess loss ≤ 0.1 dB, insertion loss ≤ 3.4 dB (for 3-dB coupler), and directivity of better than -55 dB.

17.3 Fabrication of fiber directional couplers

Practical single-mode fibers have a thick cladding to isolate the light propagating in the core. Hence, to fabricate a directional coupler, it is necessary to remove a major portion of the cladding so that the cores can be brought sufficiently close for the coupling process. Two main techniques have been developed to accomplish this. In the following, we briefly describe the methods.

17.3.1 Polished fiber couplers

Polished fiber couplers rely on exposing the core of the fiber by mechanically polishing off the cladding along one side of the fiber. To achieve this, the fiber is first bonded (using, e.g., ultraviolet cure epoxy) on a curved groove fabricated on a fused silica block by use of, for example, a diamond impregnated wire saw. The groove depth at the center of the block is slightly greater than the cladding diameter of the fiber; the substrate (along with the fiber) is then ground and polished by standard mechanical polishing techniques so as to almost expose the core (see Figure 17.9).

One of the standard methods to verify the proximity of the core to the polished surface is to observe the change in the transmission through the polished surface when a drop of index matching liquid with a refractive index slightly greater than the core is placed on the polished region. If the core is far from the polished surface, then there is almost no change in transmission. On the other hand, if the core is very close to the polished surface, the transmission drops to a very small value because of leakage of light from the core to the higher refractive index liquid. From the drop in power, one can estimate the proximity of the core–cladding interface to the polished surface.

A directional coupler is formed by mating two such polished fiber blocks (see Figure 17.10). Usually the space between the substrates is filled with an index matching liquid.

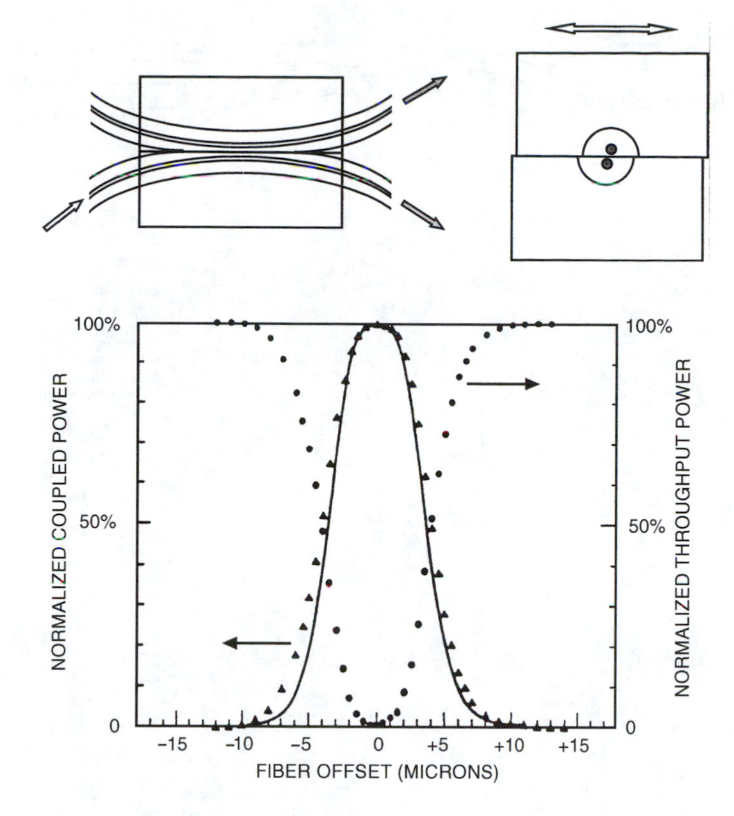

Fig. 17.10: A fiber directional coupler made up of two side-polished fiber half blocks. Tuning is achieved by one block against the other.

Fig. 17.11: Experimental points and theoretical curve (solid line) showing the tunability of a side-polished fiber directional coupler. The figure corresponds to $\lambda_0 = 633$ nm. [After Digonnet and Shaw (1982).]

One of the principal advantages of a polished fiber directional coupler is tunability (Figure 17.11). By laterally moving one block with respect to the other block (Figure 17.10), one can change the fiber core separation. This in turn changes κ (see equation 17.14) and, hence, the power coupled (see equation 17.8). Hence, such couplers are also referred to as tunable directional couplers.

Polished fiber directional couplers exhibit excellent characteristics. Their coupling ratio can be tuned continuously from $\simeq$0 to 100. Their directivity is usually very high; values as small as −70 dB have been realized. The insertion loss of such couplers can also be as low as 0.005 dB. Also, such couplers are almost polarization insensitive – that is, the variation in coupling ratio can be less than 0.5% as a function of the state of polarization of the input light. However, their performance could be highly temperature sensitive because of temperature dependence of the refractive index of the liquid filling the space between the two polished fiber blocks.

One of the interesting features of the polished fiber half block is that the evanescent field of the propagating mode is accessible through the polished surface. Thus, by choosing appropriate materials as cover media on the polished fiber surface, one can design several in-line fiber optic devices (see Section 17.5).

If, instead of using ordinary circular core optical fibers, one uses polarization maintaining fibers, one can achieve polarization maintaining couplers or polarization splitting couplers. Commercial polarization maintaining couplers with a polarization extinction ratio better than −25 dB and excess loss less than 0.05 dB are available. In such couplers, in addition to the usual aspects of fabrication of couplers, it is very important to achieve correct alignment

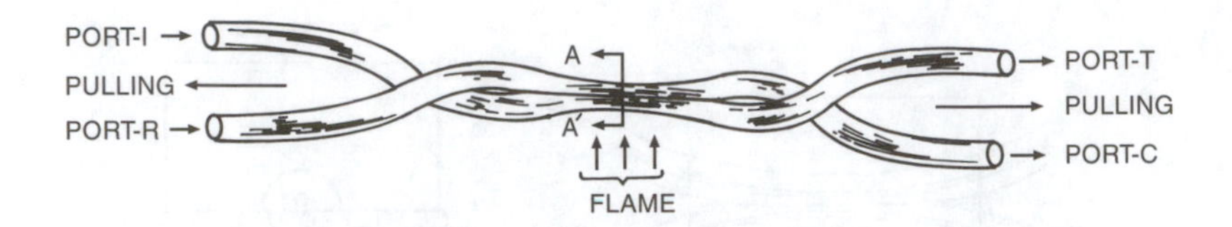

Fig. 17.12: Schematic experimental setup for fabrication of fused fiber directional couplers. The outputs at Port T and C are used for on-line control of the fabrication process.

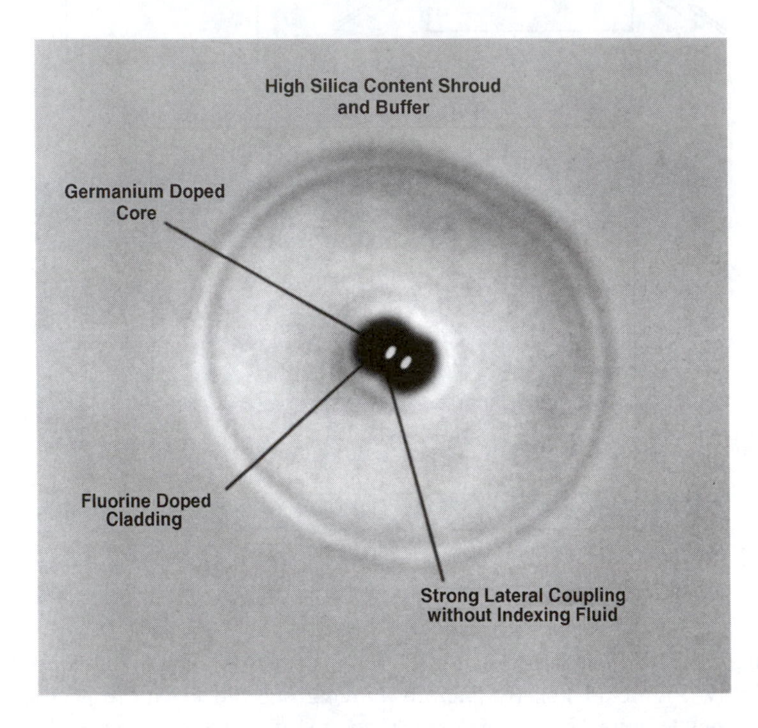

Fig. 17.13: The cross section in the fused region of a fused fiber coupler composed of two polarization maintaining fibers. Note the proximity of the two cores. [Adapted from the data sheet provided by Andrew Corporation, Orland Park, IL, USA.]

of the birefringence axes of the polarization maintaining fibers. Such couplers are useful components of fiber optic sensors based on polarization effects, in coherent optical communication systems, and so forth.

17.3.2 Fused couplers

Although polished fiber couplers have excellent characteristics, fabricating them is a time-consuming operation. In contrast, fused directional couplers are easier to fabricate and their fabrication can be automated without much difficulty. Fused couplers are fabricated by first slightly twisting two single-mode fibers (after removing their protective coating) and then heating and pulling them so that the fibers fuse laterally with one another and are also tapered (see Figure 17.12). Heating can be accomplished by using an oxybutane flame or miniature electrical heating elements.

Figure 17.13 shows the cross section of a fused fiber coupler composed of polarization maintaining fibers. One can see the two cores that lie close to each other. The coupling ratio is monitored on line as the fibers are fused and drawn. Figure 17.14(a) shows a typical variation of the power exiting from the fiber (in which input power is coupled) as a function of drawing length. The variation of output power from the other port is complementary to the curve shown. As can be seen, the exiting light power oscillates as a function of the drawn length

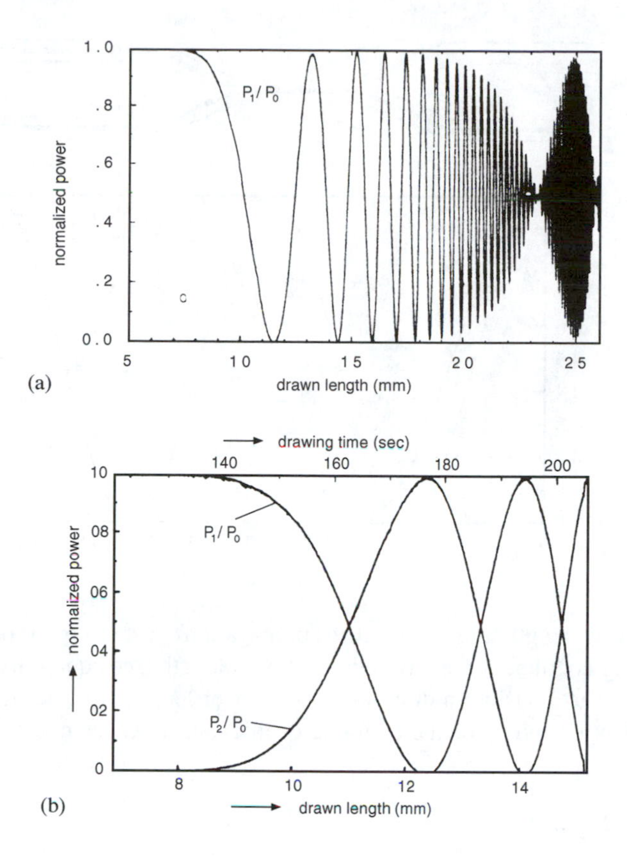

Fig. 17.14: (a) Typical variation of normalized power exiting from the input fiber as a function of drawing length. Note the periodic oscillation (with decreasing period) of the power output. [After Eisenmann and Weidel (1988).] (b) Measured normalized power exiting the two fibers versus the drawn length. [After Eisenmann and Weidel (1988).]

with complete power transfer. At larger drawn lengths, the small difference in the power transfer characteristics of the two polarization states results in a decrease in oscillation amplitude. Figure 17.14(b) shows the normalized power exiting from both the output ports versus drawn length showing a sinusoidal dependence with decreasing period.

Fused fiber couplers find wide applications in local area networks, WDM applications, etc. Excess loss of such a coupler is typically less than 0.1 dB with directivities in excess of −55 dB.

17.4 Applications

17.4.1 Power dividers

As discussed earlier, one of the most important applications of a fiber directional coupler is as a power divider. In many applications, such as in local area networks or in fiber optic sensing, it is necessary to split or combine optical beams. Such a fiber optic directional coupler forms an ideal component since it is compact and possesses low loss. Application of such couplers to sensors is discussed in Chapter 18.

17.4.2 Wavelength division multiplexers/demultiplexers

Another very important application of such couplers is in wavelength division multiplexing/demultiplexing. As discussed earlier, fiber directional couplers are

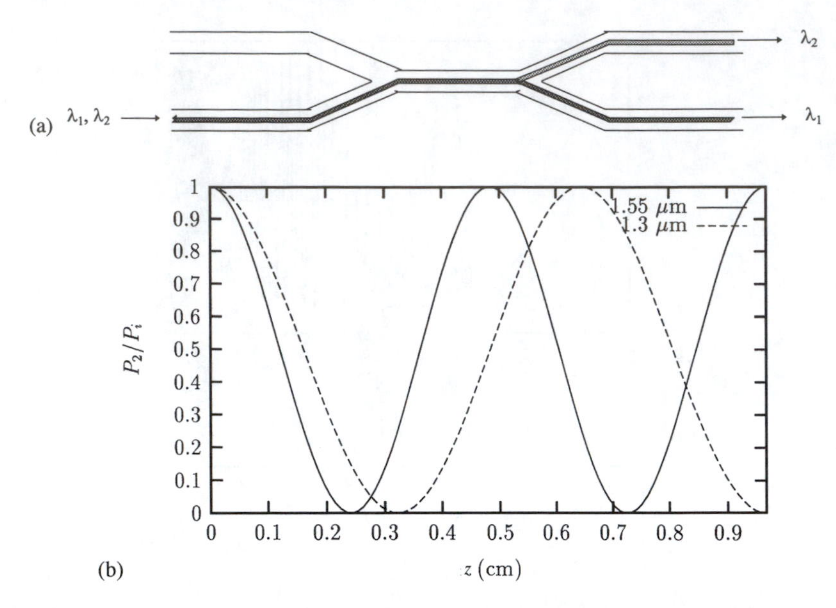

Fig. 17.15: (a) A directional coupler as a wavelength division demultiplexer. The length L is chosen so that for wavelengths λ_1 and λ_2, equations (17.28) and (17.29) are satisfied. In such a case, light of wavelength λ_1 exits from the input fiber, whereas that of wavelength λ_2 exits from the coupled fiber. (b) The corresponding calculated variation of the normalized power exiting from the second fiber with z for a fiber directional coupler made with fibers characterized by equation (17.32).

in general wavelength sensitive since the propagation constants of the modes and the coupling coefficient κ are functions of wavelength (see equations (17.18)–(17.22)). Let us consider a directional coupler of length L made of identical fibers and let κ_1 and κ_2 be the coupling coefficients at wavelengths λ_1 and λ_2 so that

$$\kappa_1 L = m\pi \tag{17.28}$$

and

$$\kappa_2 L = \left(m - \frac{1}{2}\right)\pi \tag{17.29}$$

In such a case, if light beams at wavelengths λ_1 and λ_2 are launched simultaneously in the input fiber, then for light of wavelength λ_1

$$P_2(\lambda_1, L) = P_1 \sin^2(\kappa_1 L) = 0 \tag{17.30}$$

and for light of wavelength λ_2

$$P_2(\lambda_2, L) = P_1 \sin^2(\kappa_2 L) = P_i \tag{17.31}$$

Thus, light of wavelength λ_1 will exit from the input fiber and that of wavelength λ_2 will exit from the other fiber (see Figure 17.15). Such a device forms a wavelength demultiplexer. The same device can also operate as a wavelength multiplexer.

As an example of a wavelength demultiplexing coupler, let us consider a directional coupler made of identical fibers with the following specifications.

$$\begin{aligned} n_1 &= 1.4525 \\ n_2 &= 1.45 \\ a &= 5.6\,\mu\text{m} \end{aligned} \tag{17.32}$$

Assuming no material dispersion (i.e., n_1 and n_2 to be independent of wavelength), the V number of the fiber at $\lambda_0 = 1.55\,\mu$m and 1.3 μm are 1.934 and 2.306, respectively. For $\tilde{d} = 2.473$ – that is, core center to center spacing of 13.85 μm – using κ given by equation (17.18) we obtain

$$\kappa_1 = \kappa(1.55\,\mu\text{m}) = 6.496\,\text{cm}^{-1} \tag{17.33}$$

and

$$\kappa_2 = \kappa(1.30\,\mu\text{m}) = 4.872\,\text{cm}^{-1} \tag{17.34}$$

Note that κ_1 is larger than κ_2 due to greater field penetration in the cladding of the field at 1.55 μm compared with 1.3 μm.

If we consider an interaction length of 9.67 mm, we obtain

$$\kappa_1 L \simeq 2\pi \quad \text{and} \quad \kappa_2 L \simeq \frac{3}{2}\pi \tag{17.35}$$

Since the coupling length is given by $\pi/2\kappa$ (see equation 17.11), the coupler has four coupling lengths at 1.55 μm and three coupling lengths at 1.3 μm (see Figure 17.15). Thus, if light of wavelengths 1.55 μm and 1.3 μm is incident on the same fiber at the input of the coupler, light corresponding to 1.55 μm will exit from the same fiber, whereas that at 1.3 μm will exit from the other fiber.

The above example of wavelength demultiplexer was based on two identical fibers and used the variation of the coupling coefficient with wavelength. Such couplers usually have a large channel wavelength separation for moderate interaction lengths. The wavelength selectivity can be significantly enhanced by making couplers with highly dissimilar fibers [Marcuse (1985)]. If the two fibers are chosen so that their modal propagation constants are equal only at a particular wavelength, then strong power exchange occurs only at this wavelength. At a nearby wavelength, the modal propagation constants will be unequal (i.e., $\Delta\beta \neq 0$ – see equation (17.4)) and there would be very little exchange of power. Thus, by choosing a proper interaction length, one can achieve efficient wavelength multiplexing/demultiplexing. Such couplers can exhibit much narrower channel spacings compared with identical fiber directional coupler multiplexers.

Wavelength multiplexing/demultiplexing couplers form very important components in the rapidly emerging area of fiber optic amplifiers (see Chapter 14) as well as in communication systems using wavelength division multiplexing schemes for increased bandwidth in transmission. Figure 17.16 shows the variation of insertion loss with wavelength of a typical commercially available 980/1550-nm single-mode WDM. The device exhibits an excess loss of 0.3 dB, an insertion loss of 0.5 dB, and an isolation of about 20 dB.

17.4.3 Wavelength flattened couplers

The coupling ratio of fused tapered directional coupler power splitters is quite wavelength dependent. Thus, a 50/50% coupler at 1300 nm can have a coupling ratio anywhere from 80/20% to 99/1% depending on the fusion. For couplers made from identical single-mode fibers, the coupling ratio has the strongest variation with wavelength around the 3 dB point due to the sinusoidal dependence.

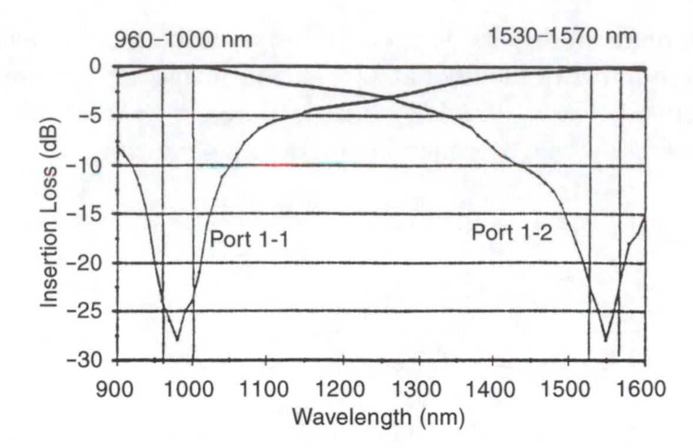

Fig. 17.16: A typical variation of insertion loss versus wavelength for the two output ports of a 980/1550-nm WDM. [Adapted from the data sheet of MP Fiber Optics, USA.]

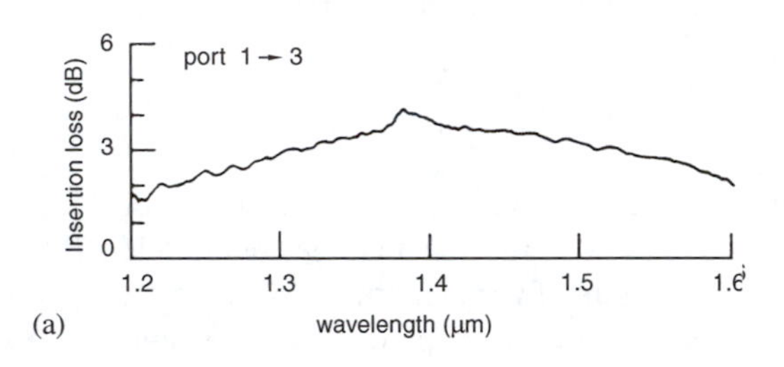

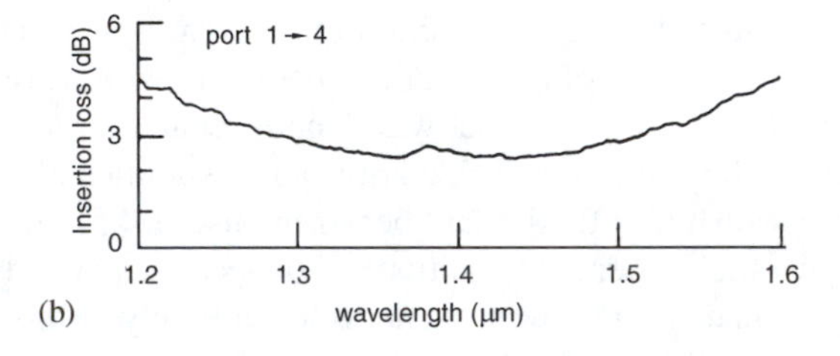

Fig. 17.17: Variation of insertion loss with wavelength of a wavelength flattened directional coupler formed of two nonidentical fibers. [After Mortimore (1985).]

Devices fabricated to operate near the maximum or minimum values of the sinusoidal dependence will have the least sensitivity. Thus, if a directional coupler is made with two nonidentical fibers, then the maximum fractional power coupled from the input port to the coupled port will be less than unity. The maximum power coupled is given by equation (17.12). Thus, if $\Delta\beta = 2\kappa$, then $P_{\max} = 1/2$. Couplers fabricated in this fashion are expected to show much lower wavelength dependence. Figure 17.17 shows a typical wavelength dependence of a fused fiber coupler fabricated from a pair of identical fibers in which one of the fibers is slightly pretapered before fusion. The pretapering of one of the fibers makes the fibers nonidentical and, thus, the coupler formed exhibits less wavelength dependence.

Commercial wideband couplers operating in the wavelength region from 1260 to 1580 nm with a uniformity of better than 0.85 dB and excess loss less than 0.1 dB are available.

17.5 Other polished fiber half-block devices

In Section 17.3.1 we discussed a technique to fabricate fiber directional couplers by a side-polishing technique. A polished fiber half block prepared in this fashion has a significant fraction of the evanescent field of the propagating mode close to the polished surface. Such a direct access to the propagating fiber mode leads to many interesting and important applications of such polished half blocks, including fiber polarizers, polarization splitters, wavelength filters, modulators, evanescent field sensors, and so forth. In the following, we discuss some fiber devices based on polished fiber half blocks.

17.5.1 Fiber polarizers

In many applications, expecially in fiber optic sensors or in integrated optics, it is necessary to choose the state of polarization of the light entering the device. Since nominally circular core fibers do not maintain the polarization state of the propagating beam, it is necessary to be able to polarize the light before it enters the fiber device. Various in-line fiber optic polarizers have been developed, some of which are discussed below.

(a) Metal-clad polarizers: It is well known that a metal–dielectric interface can support a TM guided wave (which is polarized perpendicular to the metal dielectric interface), which is also referred to as the surface plasmon mode. Because a metal has an imaginary component of dielectric constant, such a surface plasmon mode is lossy. At the same time, the metal–dielectric interface cannot support any guided TE wave since one cannot satisfy the required boundary conditions at the dielectric–metal interface. Now, if a polished fiber half block that is polished almost up to the core of the fiber is coated with a metal, then the fiber mode with polarization perpendicular to the metal interface (TM) will suffer losses due to the metal, whereas the parallel polarization (TE) will suffer much less loss. Using direct coating of gold or aluminum, one can achieve reasonable losses in the TM polarization.

Because the metal film supports a lossy plasmon mode, the attenuation of the TM polarization can be considerably increased by using an intermediate buffer layer, since in such a case we can consider the metal-coated fiber device as a directional coupler between the fiber mode and the metal-clad plasmon mode. Changing the buffer refractive index and the thickness leads to changes in the plasmon mode and fiber mode effective indices. Thus, by varying the buffer refractive index and thickness, phase matching between the plasmon mode and the fiber mode may be attained, leading to increased attenuation at a specific buffer thickness/refractive index. The existence of a critical buffer thickness was experimentally demonstrated by Thyagarajan et al. (1985), and Figure 17.18 shows the corresponding typical dependence of the attenuation on buffer thickness, as demonstrated on an integrated optical waveguide. One can see that attenuations as high as 200 dB/cm are possible.

Compared with the TM polarization, because there is no corresponding plasmon mode with polarization parallel to the metal interface (TE polarization), the fiber mode with its polarization parallel to the metal interface suffers much less attenuation. This leads to a polarizer action of the device.

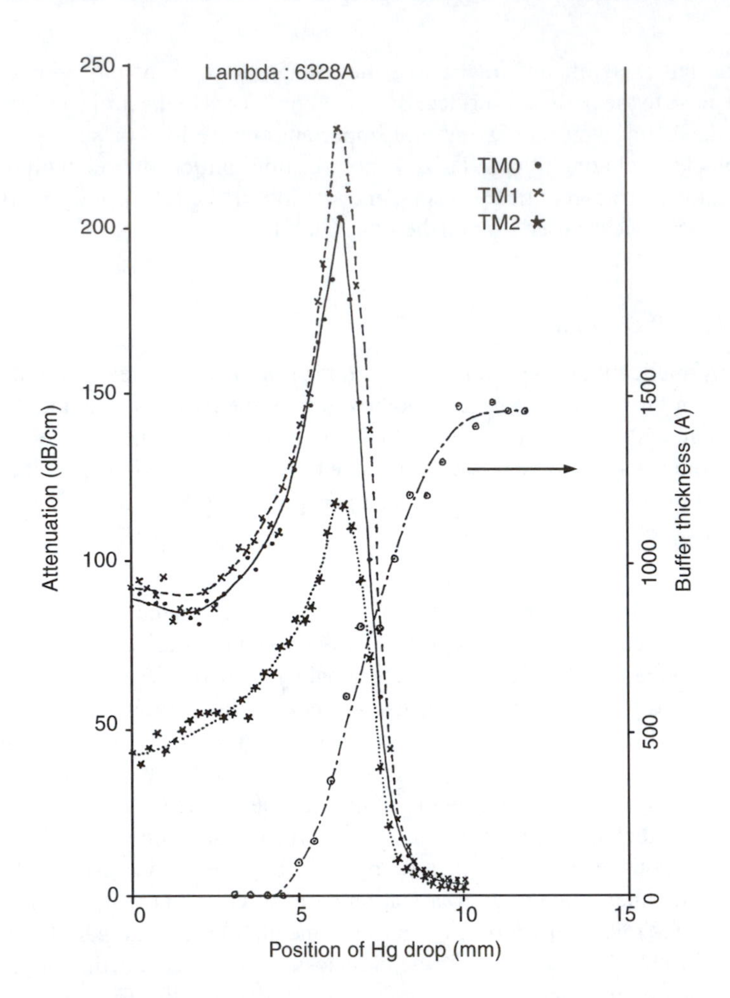

Fig. 17.18: Measured attenuation of TM polarization with the buffer thickness as induced by a metal (in this case, mercury). [After Thyagarajan et al. (1985).]

Fiber-based polarizers based on plasmon mode interactions using metal coating on fibers have typical extinction ratios of 30–40 dB and insertion losses of less than 1 dB [Dyott, Bello, and Handreak (1987)].

(b) Thin metal-clad polarizers: In the conventional metal-clad polarizer discussed above, the thickness of the metal layer is usually very large compared with the field penetration and, hence, the outer metal surface not in contact with the waveguide does not play any role. In another class of metal-clad polarizers, called thin metal-clad polarizers, the thickness of the metal layer is only about 10–15 nm [Johnstone et al. (1988)]. In such a case, the depth of penetration of the plasmon mode is larger than the metal thickness and the behavior of the polarizer is also affected by the refractive index of the overlay medium placed above the polarizer. If the refractive index of the overlay is varied, then one observes that at a specific overlay refractive index the loss of the perpendicular polarization component can become very large and, at the same time, have a low loss for the parallel component. This behavior can be explained on the basis of resonant coupling of power from the fiber mode to the plasmon mode and subsequent leakage of power to the overlay [Thyagarajan et al. (1990)]. Such polarizers have

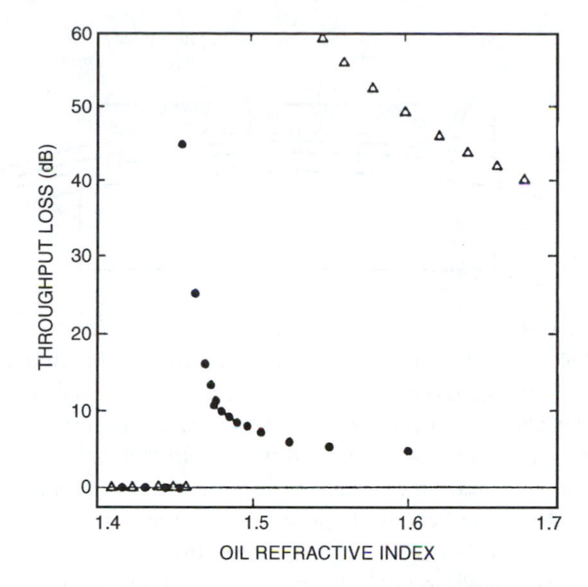

Fig. 17.19: Variation of loss of two side-polished fibers as a function of the refractive index of the overlay medium. [After Burns (1994).]

been demonstrated to have extinction ratios greater than 50 dB and insertion loss less than 0.5 dB at 1300 nm. [Johnstone et al. (1988), (1990)].

Using another polished half block on a thin metal-coated polished half block, one can indeed design a fiber polarization splitter [Thyagarajan, Diggavi, and Ghatak (1988a)]. In this case, the TM mode resonantly couples from one fiber to the other via the plasmon mode and the other polarization remains in the first fiber.

(c) Birefringent crystal-clad polarizer: Another very interesting technique to realize an in-line fiber polarizer involves the use of a birefringent crystal on a side-polished fiber. When a side-polished fiber is covered with a medium, transmission of the polished fiber depends on the refractive indices of the cover medium. Thus, if the refractive index of the cover is less than the fiber mode effective index, then total internal reflection continues to take place and there is no additional attenuation. On the other hand, if the refractive index of the cover is higher than the core index, then power from the fiber mode leaks away due to transmission to the high index layer (see also Chapter 24). If the polishing exposes the core of the fiber, this leakage is simply due to partial reflection at the core–overlay interface. On the other hand, if a finite thickness of cladding is still left or if a thin low refractive index dielectric buffer layer is deposited on the polished fiber surface, then this phenomenon can be explained by the phenomenon of frustrated total internal reflection. The modes formed in this fashion are called leaky modes (see Chapter 24), and by a proper analysis one can indeed calculate the leakage loss as a function of various parameters for optimization purposes [Ghatak, (1985), Thyagarajan, Diggavi, and Ghatak (1987a)].

For maximum rate of attenuation, the refractive index of the cover medium should be higher than, but close to, the effective index of the fiber mode. Figure 17.19 shows the loss for two side-polished fibers as

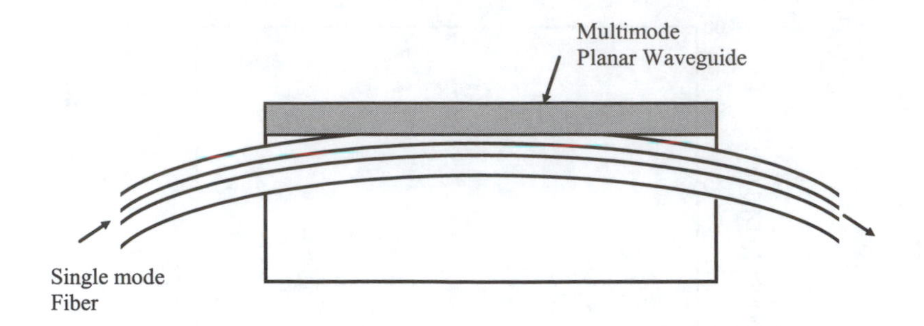

Fig. 17.20: A typical multimode planar waveguide overlay device in which a side-polished fiber is covered by a multimode planar waveguide.

a function of the refractive index of the overlay medium, in this case oil. The peak loss close to the fiber core refractive index is apparent.

The above principle can indeed be used to fabricate a polarizer [Bergh, Lefevre, and Shaw (1980)]. If a birefringent crystal is used as the overlay and if the effective index of the fiber mode lies between the ordinary and extraordinary refractive indices of the crystal, then by properly orienting the crystal, one of the polarizations would "see" an overlay with an index lower than the effective index and, hence, would be guided. The other polarization would "see" an overlay index greater than the effective index and, hence, would leak away.

Using the above principle, very high extinction ratio (60 dB) fiber polarizers have been demonstrated.

17.6 Multimode overlay fiber devices

Figure 17.20 shows a typical multimode overlay fiber device in which a multimode planar dielectric waveguide is coated on a side-polished fiber half block. This device functions as a directional coupler formed between a single-mode fiber and a multimode planar waveguide. We discussed in Section 17.2.1 that in a directional coupler, when the effective indices of the modes of the two interacting waveguides are equal, then a strong exchange of power from one waveguide to the other takes place. In the present configuration, the fiber is single moded, whereas the planar waveguide is multimoded. Thus, depending on the parameters of the fiber and the multimode overlay and the wavelength, the effective index of the fiber may or may not coincide with the effective index of a particular mode of the overlay waveguide. The effective index of the fiber mode as well as the indices of the multimode overlay will depend on wavelength and they will also be characterized by different modal dispersions. Thus, for a given multimode overlay device, as we vary the wavelength, there would be specific wavelengths at which the effective indices of the fiber mode and one of the modes of the multimode planar waveguide will become identical. At such wavelengths, the power propagating in the optical fiber can efficiently couple to the planar overlay waveguide, and if the length is close to the coupling length, then there will be a drop in transmission through the fiber. At other wavelengths, when the modes are not phase matched, there will be little coupling of power and the single-mode fiber will transmit almost the entire power. Hence, such a multimode overlay covered polished fiber half block can act as a wavelength filter or what is referred to as a channel dropping filter.

Figure 17.21 shows a typical experimentally measured transmission spectrum through a side-polished single-mode fiber covered with a 7.6-μm-thick

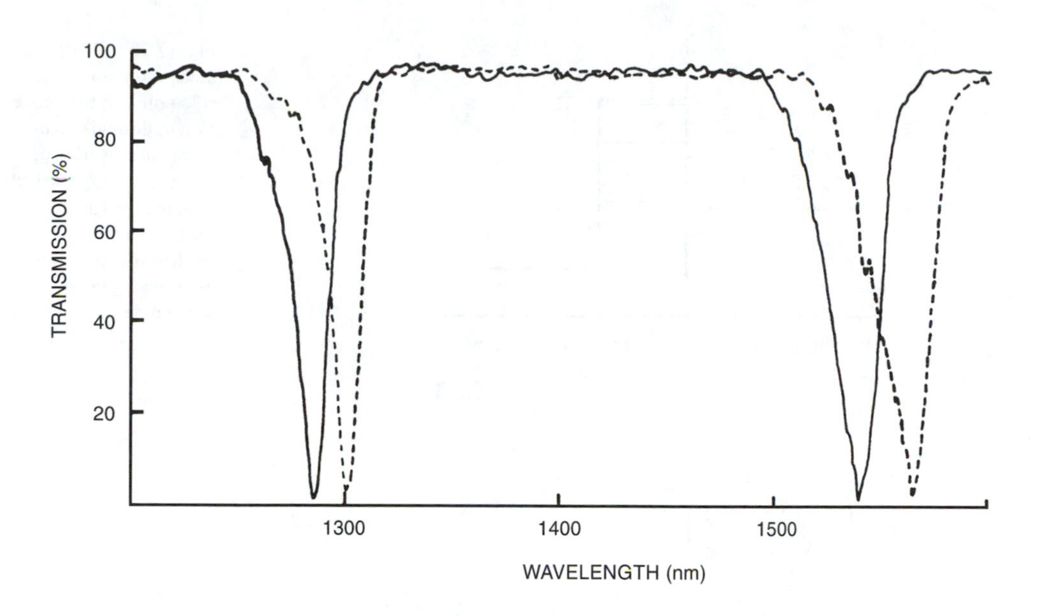

Fig. 17.21: Variation of transmission through the single-mode fiber of a side-polished fiber covered by a 7.6-μm-thick BK7 glass planar waveguide for both TE (solid) and TM (dashed) polarizations. (Johnstone, W., Moodie D. G., McCallion K., and Thursby G., personal communication.)

overlay of BK7 glass ($n_0 = 1.504$). The minimum transmission wavelengths of TE and TM polarizations are different due to the geometric birefringence of the multimode planar waveguide. As is evident, the transmission through the single-mode fiber drops at certain specific wavelengths determined by phase matching between the fiber mode and one of the modes of the multimode planar waveguide overlay. Thus, such a device acts as a channel dropping filter wherein a chosen wavelength is filtered (dropped off) from the transmitting fiber. This wavelength can be tuned by varying the overlay parameters such as the thickness or refractive index or even the refractive index of the medium on top of the overlay.

One can obtain the approximate positions of the resonant coupling between the fiber mode and the multimode planar overlay by noting the following.

> The multimode overlay medium usually has a refractive index n_0 greater than the fiber core index n_1. The modes of the multimode overlay planar waveguide have effective indices between n_0 and n_2 (the cladding index), with the fundamental mode being closest to n_0 and the highest order mode closest to n_2 (see Figure 17.22). Since the fiber mode index n_{ef} lies between n_1 and n_2, as we change the overlay parameters or the wavelength, phase matching will take place between the fiber mode and one of the highest order modes of the planar waveguide, which is close to cutoff.

Now, the equation determining the effective indices n_{eo} of the modes of the planar overlay waveguide is given by [see, e.g., Ghatak and Thyagarajan (1989)]

$$\frac{2\pi}{\lambda_0} d\left(n_0^2 - n_{eo}^2\right)^{1/2} = \phi_1 + \phi_2 + m\pi; \quad m = 0, 1, 2, \ldots \tag{17.36}$$

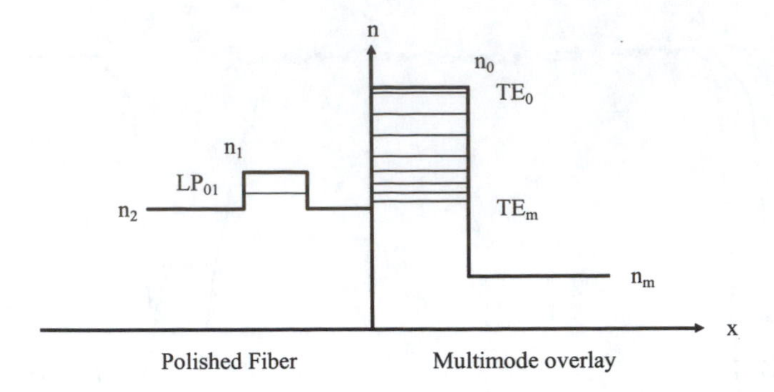

Fig. 17.22: The refractive index profile of a side-polished fiber covered by a multimode planar waveguide overlay device. Phase matching takes place between the LP_{01} mode of the fiber and one of the modes close to cutoff of the multimode planar waveguide.

where

$$\phi_1 = \tan^{-1}\left(\gamma_1\sqrt{\frac{n_{eo}^2 - n_2^2}{n_o^2 - n_{e0}^2}}\right) \tag{17.37}$$

$$\phi_2 = \tan^{-1}\left(\gamma_2\sqrt{\frac{n_{eo}^2 - n_s^2}{n_o^2 - n_{e0}^2}}\right) \tag{17.38}$$

$$\gamma_1 = \gamma_2 = 1 \text{ for TE polarization} \tag{17.39}$$

$$\gamma_1 = \frac{n_o^2}{n_2^2} \quad \text{and} \quad \gamma_2 = \frac{n_o^2}{n_s^2} \text{ for TM polarization} \tag{17.40}$$

d is the thickness of the multimode waveguide and n_s is the refractive index of the medium covering the overlay.

For efficient coupling between the fiber mode and the planar waveguide mode, we must have $n_{ef} = n_{eo}$. The precise positions of resonances can thus be obtained from the following equation, which is obtained by replacing n_{eo} with n_{ef} in equation (17.36)

$$\frac{2\pi}{\lambda_0} d\sqrt{n_o^2 - n_{ef}^2} = \tan^{-1}\left(\gamma_1\sqrt{\frac{n_{ef}^2 - n_2^2}{n_o^2 - n_{ef}^2}}\right) + \tan^{-1}\left(\gamma_2\sqrt{\frac{n_{ef}^2 - n_s^2}{n_o^2 - n_{ef}^2}}\right) + m\pi \tag{17.41}$$

Thus, the precise positions of the resonances depend on the following.

(i) The wavelength – both a direct dependence as well as an indirect one through the dependence of the refractive indices of the media and effective index on wavelength. This feature is used in its application as channel dropping filter.

(ii) The refractive index n_o of the multimode overlay. This dependence is used in the realization of in-line fiber modulators, in which the overlay is an electrooptic medium such as lithium niobate and whose refractive index can be changed by applying an external electric field [Johnstone et al. (1991)].

(iii) The refractive index n_s of the medium covering the overlay (see Figure 17.22). This dependence can be used for sensing refractive index changes in the cover medium by measuring the changes in the resonance positions [Johnstone et al. (1992), Raizada and Pal (1996)]. Sensitivities of better than 10^{-5} in refractive index changes have been demonstrated [Johnstone, Fawcett, and Yuiv (1994)].

17.7 Bandpass filters

In the above discussion, the multimode overlay covered polished half block acts as a channel dropping filter, wherein specific wavelengths are removed from the fiber. Another class of filters that can be realized by using multimoded overlays on polished fiber half blocks are the bandpass filters [McCallion, Johnstone, and Fawcett (1994), McCallion et al. (1995)]. Unlike the channel dropping filters, in this case the fiber is side polished to such an extent into the core of the fiber that the polished fiber region is below cutoff. Thus, in the absence of the multimode overlay, there would be high attenuation of the input light. When a high index multimoded planar overlay waveguide is deposited on the polished block, then the wavelength response of the device shows a periodic bandpass characteristic rather than as discussed in Section 17.6. The response of one such device with an overlay index of 1.65 and of thickness 23 μm is shown in Figure 17.23. It can be seen that the device transmits strongly at certain wavelengths, and in the intermediate region the transmission drops to very small values.

The resonant wavelengths are approximately given by [McCallion et al. (1995)].

$$\lambda_r = \frac{2\pi d\left(n_0^2 - n_2^2\right)^{1/2}}{m\pi + \phi_s} \tag{17.42}$$

where

$$\phi_s = \tan^{-1} \gamma \sqrt{\frac{n_2^2 - n_s^2}{n_0^2 - n_2^2}} \tag{17.43}$$

n_2, n_0, and n_s are the refractive indices of the fiber cladding, the overlay medium, and the medium above the overlay (usually air); $\gamma = 1$ for TE and $(n_0/n_s)^2$ for TM polarizations. Equation (17.42) can be obtained from equation (17.41) under the assumption that $n_{ef} \simeq n_2$ (which is a reasonably good approximation for weakly guiding fibers). Equation (17.42) tells us that by appropriately choosing n_s and d one can tailor the peak transmission wavelengths. Indeed, by choosing an overlay that is electrooptic (i.e., whose refractive index can be varied by applying an external electric field), one can tune the resonant wavelengths. McCallion et al. (1994) and Creaney, Johnstone, and MaCallion (1996) have reported such a tunable in-line fiber optic bandpass filter by using thin $LiNbO_3$ substrate as an overlay. Resonant wavelength shifts of 110 V/nm have been demonstrated [Creaney et al. (1996)].

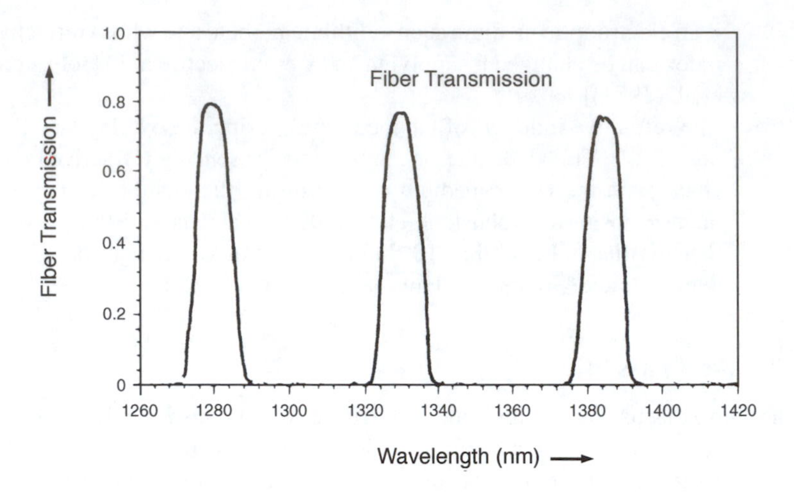

Fig. 17.23: A bandpass characteristic is obtained if a multimode planar waveguide is coated on an overpolished side-polished fiber. [After McCallion et al. (1994).]

17.8 Fiber polarization controllers

Circular core fibers whose axes are straight are not birefringent – that is, the two orthogonally polarized LP_{01} modes have the same effective indices. Bending such a fiber introduces stresses in the fiber and makes the fiber linearly birefringent with the fast and slow axes in the plane and perpendicular to the plane of the loop, respectively. The bending-induced birefringence of a single-mode silica fiber is given by [Ulrich, Rashleigh, and Eikhoff (1980)]

$$\Delta n_{eff} = n_{ex} - n_{ey} = -C\left(\frac{b}{R}\right)^2 \tag{17.44}$$

where n_{ex} and n_{ey} represent the effective indices of the LP_{01} modes polarized in the plane and perpendicular to the plane of the bend, respectively, b is the outer radius of the fiber, R is the radius of the loop, and C is a constant that depends on the fiber material and the elastooptic properties of the fiber. For silica fibers, $C \simeq 0.133$ at 633 nm.

Equation (17.44), tells us that the smaller the loop radius, the larger is the birefringence. Note that any bending will also introduce attenuation and, hence, very small bend radii are not very practical.

Example 17.2: Let us consider a silica fiber of outer radius $b = 62.5\ \mu$m bent into a circular loop of radius 30 mm. The birefringence of the fiber at 633 nm is then

$$\Delta n_{eff} = -0.133\left(\frac{0.0625}{30}\right)^2 \simeq 5.77 \times 10^{-7}$$

which is indeed very small compared with the core–cladding indices difference.

Although the induced birefringence is very small, by having the two polarizations propagate over a long fiber length, one can obtain large phase shifts. Thus, if the fiber is coiled around N loops of radius R, then the bend-induced phase difference between the two polarizations is

$$\Delta\phi = \frac{2\pi}{\lambda_0}\Delta n_{eff} 2\pi RN \tag{17.45}$$

Substituting for Δn_{eff} from equation (17.44), we obtain

$$\Delta\phi = \frac{4\pi^2}{\lambda_0} C \frac{b^2}{R} N \tag{17.46}$$

where we have disregarded an unimportant negative sign. For achieving phase differences of π (corresponding to a half-wave plate) or $\pi/2$ (corresponding to a quarter-wave plate), we must have

$$R(\Delta\phi = \pi) \simeq \frac{4\pi C b^2 N}{\lambda_0}$$

$$R\left(\Delta\phi = \frac{\pi}{2}\right) \simeq \frac{8\pi C b^2 N}{\lambda_0}$$

Example 17.3: For simulating a quarter-wave plate at $\lambda = 633$ nm, using bend-induced birefringence, if we have a single loop ($N = 1$), then

$$R = \frac{8\pi \times 0.133 \times (62.5 \times 10^{-6})^2}{(0.633 \times 10^{-6})} \simeq 2.1\,\text{cm}$$

Using the same loop radius of 2.1 cm, we can simulate a half-wave retardation plate using two loops ($N = 2$). Bend-induced linear birefringence can be used to build in-line polarization controllers [Lefevre (1980)]. Figure 17.21 shows an in-line fiber polarization controller that utilizes bend-induced birefringence. It consists of three fiber birefringence components; the first and the last are quarter-wave retarders and the central one is a half-wave retarder. The bent fibers are fixed at points marked A, B, C, and D. The three fiber loops are free to rotate as shown in Figure 17.24. A rotation of each of the loops will rotate the principal axes of the birefringent fiber sections with respect to the input polarization state. This is analogous to rotation of a classical bulk half-wave or a quarter-wave plate with respect to the incident light. Thus, rotation of the three loops is equivalent to the rotations of a combination of a $\lambda/4$, $\lambda/2$, and $\lambda/4$ plate. One can show that with this combination, any input polarization state can be transformed to any other output polarization state.

The polarization controller described above is used in many applications such as in fiber optic sensors where control of the state of polarization of the light propagating through the fiber is required.

Polarization controllers operating over a wavelength range of 1250–1600 nm with optical insertion loss variations of less than 0.004 dB

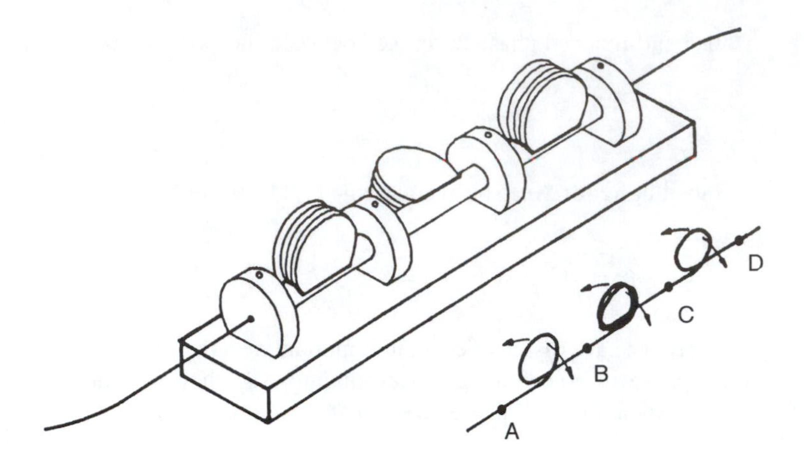

Fig. 17.24: Fiber polarization controller consisting of a sequence of $\lambda/4$, $\lambda/2$, and $\lambda/4$ in-line fiber retarders.

over the band are commercially available. Such polarization controllers are extremely important components in the measurement of polarization dependence of optical devices such as optical isolators, EDFAs, and so forth.

17.9 Fiber Bragg gratings

When a germanium-doped silica core fiber is exposed to ultraviolet (UV) radiation (with wavelength around 240 nm), this leads to a permanent change in refractive index of the germanium-doped region. This is termed photosensitivity and, using such an exposure, it is possible to obtain refractive index changes as large as 10^{-3} in germanium-doped silica fibers.

Now, if the fiber is exposed to a pair of interfering uv beams as shown in Figure 17.25, then in regions of constructive interference and, hence, high UV intensity the local refractive index will increase. At the same time, in regions of destructive interference, where the intensity of UV light is negligible, there is no index change. Therefore, an exposure to an interference pattern will result in a periodic refractive index modulation along the length of the fiber, the period of modulation being exactly equal to the spacing between the interference fringes (see Figure 17.25). When light is made to propagate through such a fiber with a periodically modulated refractive index, under certain conditions, the propagating light beam can get strongly coupled to the mode propagating in the backward direction (see Chapter 21). This happens when we satisfy the Bragg condition – namely, that the difference in the propagation constants of the two coupled modes must be equal to the spatial frequency of the grating (see Chapter 21).

$$\beta_g - (-\beta_g) = K = \frac{2\pi}{\Lambda} \tag{17.47}$$

where β_g is the propagation constant of the forward propagating guided mode, $-\beta_g$ is that of the backward propagating guided mode, and Λ is the spatial period of the grating. Figure 17.26 shows the corresponding vector diagram wherein the propagation vectors and the grating vector satisfy the condition given by

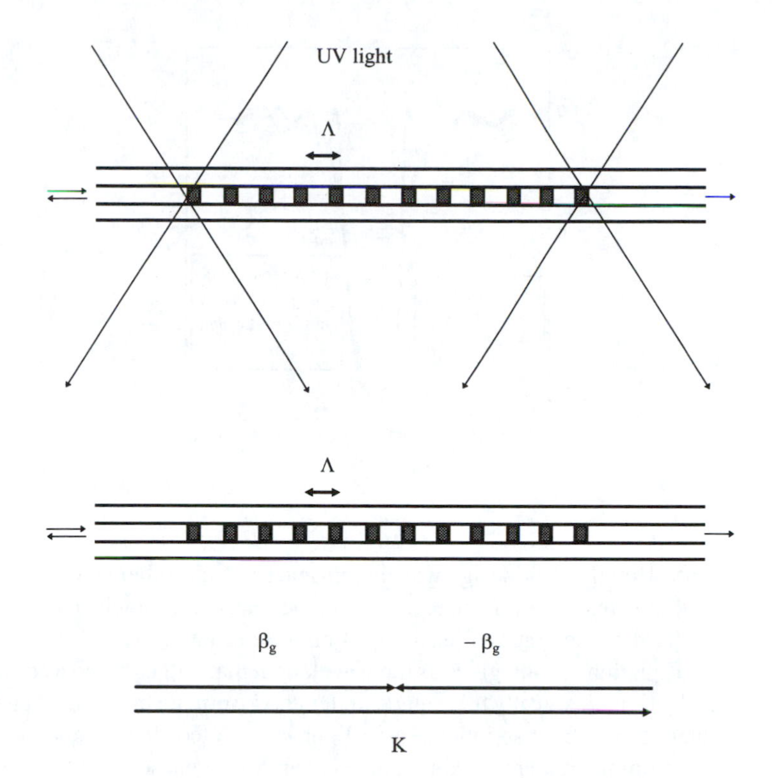

Fig. 17.25: Exposing a germanosilicate fiber to the interference pattern formed between two UV beams leads to the formation of a permanent refractive index grating in the core of the fiber. The period of the grating can be controlled by choosing the angle between the interfering beams.

Fig. 17.26: Vector diagram showing the Bragg condition to be satisfied for strong reflection; β_g is the propagation constant of the LP_{01} mode and $K = 2\pi/\Lambda$ is the grating vector.

equation (17.47). If n_{eff} is the effective index of the mode, then equation (17.47) can be written as

$$2\frac{2\pi}{\lambda_B}n_{eff} = \frac{2\pi}{\Lambda}$$

or

$$\lambda_B = 2\Lambda\, n_{eff} \tag{17.48}$$

where λ_B is called the Bragg wavelength – that is, the wavelength that satisfies the Bragg condition. The Bragg wavelength depends on the effective index as well as the grating period.

Example 17.4: Let us consider a step index fiber with $n_2 = 1.45$, $a = 3\,\mu$m, and NA $= 0.1$. The corresponding cutoff wavelength of the LP_{11} mode is $\lambda_c = 0.784\,\mu$m. At 850 nm, the effective index of the LP_{01} mode can be calculated by the approximate formula given in Chapter 8, and this gives us $n_{eff} \simeq 1.4517$. The spatial grating period required for inducing a strong reflection using this fiber at 850 nm is

$$\Lambda = \frac{\lambda_B}{2\,n_{eff}} \simeq 0.293\,\mu\text{m}$$

Note that the required grating period is less than the optical wavelength.

Figure 17.27 shows a typical transmittance of a 4.8-mm-long fiber grating having an index modulation of 2.6×10^{-4}. The grating reflects

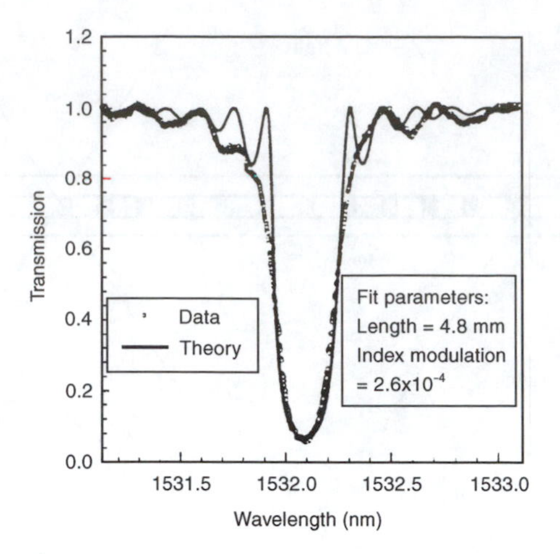

Fig. 17.27: Transmission spectrum of a fiber Bragg grating and the corresponding theoretical fit. The length of the grating is 4.8 mm and the index modulation corresponds to 2.6×10^{-4}. [After Dong et al. (1996).]

very strongly at the design wavelength and transmits other wavelengths without much loss. Hence, such a device acts as a notch filter. The FWHM of the grating filter is ~0.4 nm (see Example 17.7).

Equation (17.48) gives us the wavelength that will undergo a strong reflection. Physically this can be understood from the fact that, at each change in refractive index, some light is reflected. If the reflections from points that are a spatial period apart are in phase, then the various multiple reflections add in phase leading to a strong reflection. This happens when we satisfy equation (17.48). Using a coupled mode theory for contradirectional coupling due to a periodic perturbation, one can show that the reflection coefficient is given by (see Section 21.7)

$$R = \tanh^2 \kappa L \qquad (17.49)$$

where L is the length of the fiber grating and κ is the coupling coefficient, which is approximately given by (see Appendix F)

$$\kappa \simeq \frac{\pi \Delta n\, I}{\lambda_B} \qquad (17.50)$$

where I is a transverse overlap integral of the modal distribution with the region where the grating is formed – for example, over the core of the fiber. The overlap integral $I(< 1)$ accounts for the fact that only a part of the light propagating within the core interacts with the periodic refractive index variation; the field in the cladding is traveling in a region of uniform refractive index. [For plane wave interaction in an infinitely extended medium with a periodic index modulation, $I = 1$, see, e.g., Ghatak and Thyagarajan (1989).]

The bandwidth of the reflection spectrum, which in this case is defined as the wavelength spacing between the two reflection minima on either side of the central peak, is approximately given by [see e.g., Ghatak and Thyagarajan (1989)]

$$\Delta\lambda = \frac{\lambda_B^2}{\pi\, n_{eff} L} (\kappa^2 L^2 + \pi^2)^{1/2} \qquad (17.51)$$

Example 17.5: Let us assume that we need to have a fiber Bragg grating at 800 nm that should have a reflectivity of 90% with a length of 25 mm. Then, from equation (17.49), we have

$$\tanh \kappa L = \sqrt{0.9}$$

giving us

$$\kappa = \frac{1}{2L} \ln \left(\frac{1 + \sqrt{0.9}}{1 - \sqrt{0.9}} \right) \simeq 0.073\,\text{mm}^{-1}$$

Assuming $I = 0.5$, equation (17.50) gives us

$$\Delta n \simeq 3.72 \times 10^{-5}$$

which is very reasonable. The corresponding bandwidth is

$$\Delta\lambda = 0.02\,\text{nm}$$

Example 17.6: To achieve the same reflectivity of 90% with a 10-mm-long grating, the required coupling coefficient would be $\kappa \simeq 0.183$ mm^{-1}, which (assuming $I \simeq 0.5$) gives us $\Delta n \simeq 9.3 \times 10^{-5}$. The corresponding bandwidth would be 0.05 nm.

Example 17.7: From Figure 17.27, showing the measured transmission spectrum, we note that the peak reflectivity is about 0.93, corresponding to a Bragg wavelength of 1532.1 nm. Since the peak reflectivity is given by $\tanh^2 \kappa L$ (see equation 17.49), we obtain

$$\tanh \kappa L = \sqrt{0.93}$$

which gives

$$\kappa L \simeq 2$$

Since $L = 4.8$ mm, we obtain from equation (17.50), the effective index modulation of

$$\Delta n_{eff} = \Delta n I = \frac{\kappa \lambda_B}{\pi} \simeq 2 \times 10^{-4}$$

Since $I < 1$, the actual index modulation is larger and has been reported to be 2.6×10^{-4}. We can also estimate the bandwidth of the reflection spectrum from equation (17.51), which, using the above parameters, gives

$$\Delta\lambda \simeq 0.4\,\text{nm}$$

which matches closely with the measured spectrum shown in Figure 17.27.

Example 17.8: Let us assume that we wish to fabricate a Bragg grating filter at 1550 nm with a peak reflection of 99% and a bandwidth of 1 nm. Since $R = 0.99$, we have from equation (17.49)

$$\kappa L = 2.993$$

Since $\lambda_B = 1550$ nm, $\Delta\lambda = 1$ nm, and $n_{eff} \simeq 1.45$, we have from equation (17.51)

$$L = \frac{\lambda_B^2}{\pi\, n_{eff} \Delta\lambda} (\kappa^2 L^2 + \pi^2)^{1/2}$$

$$= 2.29\,\text{mm}$$

Using equation (17.50) for κ we have

$$\Delta n I \simeq 6.45 \times 10^{-4}$$

Taking a typical value of 0.75 for I, we obtain the required Δn as

$$\Delta n \simeq 8.6 \times 10^{-4}$$

Example 17.9: Assuming a sinusoidal refractive index modulation that is uniform within the core of the fiber, I will correspond to the fractional power in the core of the fiber. In such a case, one can approximately write [Appendix F]

$$I \simeq 1 - \exp\left[-2\left(\frac{a}{w_0}\right)^2\right]$$

where a is the fiber core radius and w_0 is the mode spot size, which under the Gaussian approximation, is given by equation (8.63).

If we consider a fiber with $a = 5\,\mu$m and NA $= 0.09$ with a Bragg grating to reflect 1.3-μm radiation, we obtain

$$V = \frac{2\pi}{\lambda_0} a \cdot NA \simeq 2.1749$$

Thus, $w_0/a \simeq 1.182$ and $I \simeq 0.76$.

Fiber gratings are finding wide applications in fiber optic communications and sensing. One of the most interesting applications is as narrow-band notch filters to reflect a chosen wavelength. Such reflectors can be used to stabilize distributed Bragg reflector lasers [Bird et al. (1991)], as reflectors for fiber lasers [Kashyap et al. (1990)], and as WDM components [Juma (1996)]. Multiple gratings at different positions on the fiber can be designed to reflect different wavelengths, and different gratings with overlapping reflection spectra may be designed to generate tailored transmission spectra. A pair of identical gratings separated by a certain length of the fiber forms a Fabry–Perot filter and can be designed to form a narrow-band transmission filter [Morey, Ball, and Meltz

(1994)]. Chirped fiber Bragg gratings in which the grating period varies along the fiber length can be used for dispersion compensation [Ouellette (1987)]. Fiber gratings can be used in gain flattening of EDFAs (see Section 17.11). Sensing is another area that is finding wide application. At 1300 nm, the temperature coefficient of wavelength shift is $\sim$0.009 nm/°C, and the strain coefficient is $\sim$0.001 nm/μ strain. Thus, from the changes in the Bragg wavelength, one can measure temperature and/or strain [Patrick et al. (1996)].

17.10 Fabrication of fiber gratings

Many different techniques have been developed for the fabrication of fiber gratings. UV light from a laser (excimer or frequency-doubled argon) is split into two beams with a beam splitter and the two beams are then made to interfere (see Figure 17.25) by using two beam steering mirrors [Meltz, Morey, and Glenn (1989)]. The angle between the two interfering beams would decide the period of the interference fringes and, hence, the grating (see Problem 17.7). The fiber is placed in the region of interference and exposed to the beam. Cylindrical lenses can be used to produce a line focus on the fiber. Such an arrangement has also been used for writing gratings with a single UV pulse while the fiber is being drawn [Askins et al. (1992), Archambault et al. (1993)]. Since the grating periods are less than a micrometer, the interference pattern must be very stable over the exposure time.

In the interference pattern method described above, it is quite difficult to write gratings of a precise wavelength. A simpler technique involves the use of phase masks. A phase mask is produced by exposing a silica mask plate to an electron beam followed by etching to produce a surface relief grating with the required period. When the UV laser beam is incident on the mask, it diffracts into the $+1$, 0, and -1 orders. The $+1$ and -1 order diffracted beams are made to interfere on the fiber by placing the fiber just behind the phase mask, as shown in Figure 17.28. This arrangement is highly stable and is also very compact.

Another technique involves creating refractive index increases by tightly focusing a UV laser beam at periodic positions along the fiber. This is referred to as the point-by-point writing technique [Hill et al. (1991), Bilodeau et al. (1991)]. Since very tight focusing is difficult, this technique is usually applicable to form gratings with periodicities of greater than several tens of micrometers. An advantage of this method is that the writing beam need not have a very high degree of coherence.

One can increase the photosensitively of doped silica fibers by treating the fiber with hot hydrogen [Lemaire et al. (1993)]. This is done by exposing the fiber to high-pressure hydrogen gas (20–750 atmospheres) at temperatures ranging from 21°C to 75°C. This has been shown to lead to index changes as large as 5.9×10^{-3}.

The formation of gratings in the fiber require UV light around a wavelength of 240 nm. The lasers that have been used for the fabrication are excimer lasers (KrF emitting 248 nm), 244 nm of frequency-doubled Ar–ion laser operating at 488 nm, or the frequency-tripled output of the 1060-nm Nd:YAG laser. Typical average power levels of 4–20 mW and exposure times typically of 5 min are required for fabrication. Single-shot fabrication typically requires pulse energies of a few hundred millijoules in about 20 ns.

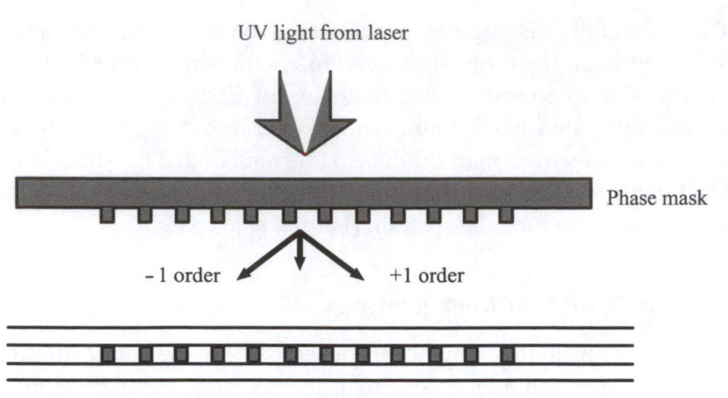

Fig. 17.28: Fabrication of fiber gratings by exposing the fiber to the fringe pattern formed by passing the UV laser radiation through a phase mask. The phase mask is a binary grating in which the groove profile and depth are specially tailored to diffract most of the incident light into +1 and −1 order waves.

Recent experiments have demonstrated that even phosphorous-doped silica fibers that are sensitized by deuterium fiber gratings can be formed by UV exposure [Strasser (1996)]. These fibers, indeed, show better characteristics in terms of index modulation and are easily integrated in fibers like Er/Yb-codoped fiber lasers that do not contain germanium.

17.11 Long-period fiber Bragg gratings

In Section 17.9 we described Bragg gratings that couple from a forward propagating mode to a backward propagating mode (contradirectional coupling). As discussed, such gratings have periods of a micrometer or less. Fiber gratings can also be used to couple light from one mode to another propagating along the same direction (codirectional coupling). If n_{e1} and n_{e2} are the effective indices of the two modes that need to be coupled, then the required grating period is obtained from the Bragg condition (see Chapter 21).

$$\frac{2\pi}{\lambda_0} n_{e1} - \frac{2\pi}{\lambda_0} n_{e2} = \frac{2\pi}{\Lambda}$$

or

$$\Lambda = \frac{\lambda_0}{(n_{e1} - n_{e2})} \tag{17.52}$$

The corresponding vector diagram is shown in Figure 17.29. If n_{e1} and n_{e2} correspond to two forward propagating guided modes of the fiber (for example, LP_{01} and LP_{11} modes or two orthogonally polarized modes of a birefringent fiber), then $(n_{e1} - n_{e2})_{max} = \Delta n$, the index difference between core and cladding. Since Δn lies in the range 0.005 – 0.01, the required period lies between 100 and 200 times the optical wavelength – that is, periodicities greater than 100 μm. Such gratings are referred to as long-period gratings and have been a subject of study in recent years [Vengsarkar et al. (1996a, 1996b), Patrick et al. (1996)].

One of the interesting classes of long-period gratings involves coupling from the guided LP_{01} mode to the forward propagating cladding modes – that is, those modes that are guided by total internal reflection at the cladding–air interface.

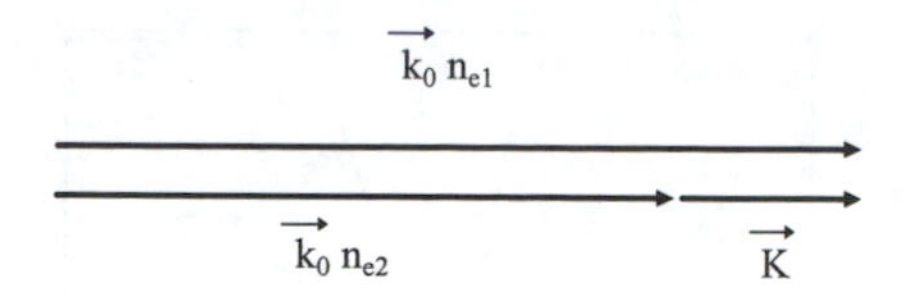

Fig. 17.29: Vector diagram showing Bragg coupling from one mode into another mode propagating in the same direction. In this case, length of $\vec{K}$ is much smaller than in the use of reflection (see Figure 17.26).

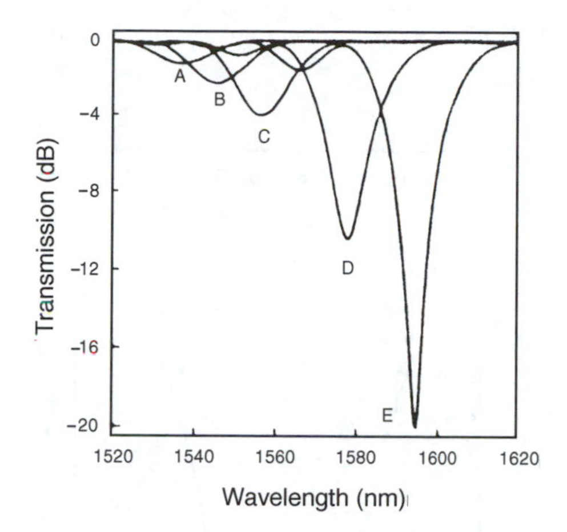

Fig. 17.30: Transmission spectrum of a long-period grating at 1-min intervals, starting from 1 min (curve A). Note that as exposure increases, there is a decrease in the transmission at resonance and also a drift of the center wavelength to longer wavelengths. [After Vengsarkar et al. (1996b).]

Such cladding modes are lossy due to the large scattering losses at the cladding–air interface as well bends and other perturbations. Since the periodic coupling process is wavelength selective, this coupling acts as a wavelength selective loss element that is finding applications in many areas such as gain flattening of EDFAs (see Chapter 14), sensors, etc.

For the fabrication of long-period gratings, a hydrogen-loaded germanosilicate single-mode fiber is exposed by a UV laser beam (from a KrF laser emitting at 248 nm) through an amplitude mask made of chrome-plated silica; the mask has periodic openings corresponding to the pattern required. Typical exposure energies are 250 mJ/pulse at a repetition frequency of 20 Hz for about 5–10 min. The transmission spectrum is usually monitored in real time as the grating gets written in the fiber. Figure 17.30 shows the change in the transmission spectrum as a function of exposure time. The peak loss increases with time as the index modulation increases, and the peak wavelength also increases because of an increase in the average refractive index in the grating region. These gratings are usually annealed at 150°C for a few hours to outgas the unreacted hydrogen as well as to anneal any unstable UV-created site.

A typical transmission spectrum of a band rejection filter is shown in Figure 17.31. The corresponding grating period was $\Lambda = 402\,\mu$m and the length was 25.4 mm. The maximum loss is 32 dB at $\lambda_p = 1517$ nm with $\Delta\lambda$ (3 dB) $\simeq 22$ nm. The insertion loss at an off-resonance wavelength is less than 0.2 dB. These gratings correspond to $\Delta n \sim 5 \times 10^{-4}$.

By concatenating two or more gratings with different transmission spectra, one can generate almost any required transmission spectrum. This ability is finding applications in gain flattening of EDFAs. As discussed in Chapter 14, EDFAs are characterized by a particular gain spectrum that is not flat – that is, the

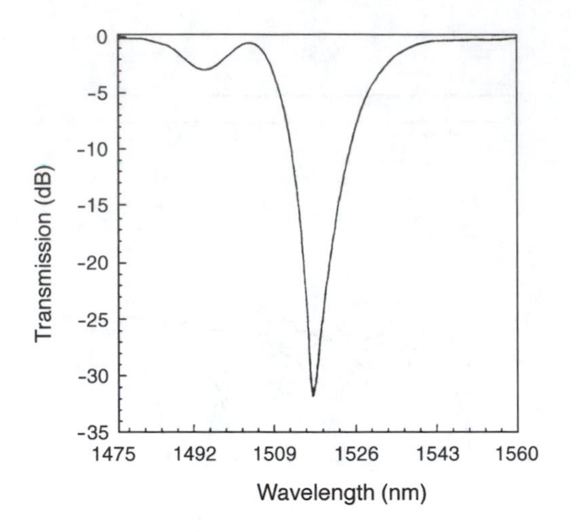

Fig. 17.31: Transmission spectrum of a band rejection filter made from a long-period grating of length 2.54 cm and a period of 402 μm. The peak loss is 32 dB at 1517 nm with an insertion loss of 0.2 dB [After Vengsarkar et al. (1996a).]

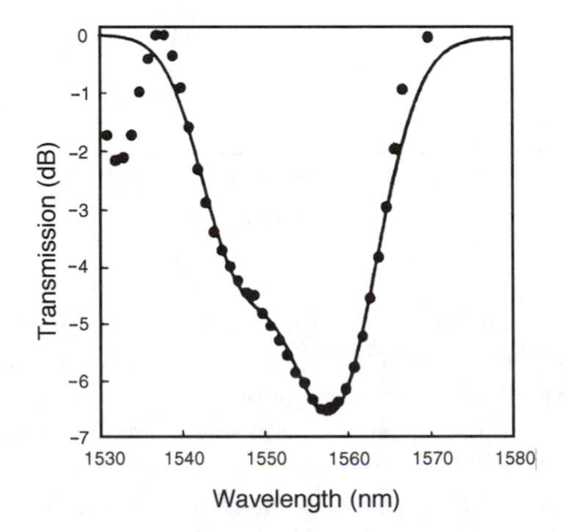

Fig. 17.32: Comparison of inverted erbium amplifier gain spectrum and the corresponding transmission spectrum of a Bragg fiber filter obtained by concatenating two long-period gratings. [After Vengsarkar et al. (1996b).]

gain is a strong function of signal wavelength. This can pose serious problems when EDFAs are used in wavelength division multiplexed systems as different signal wavelengths will have different gains, and an optimum design for all signal wavelengths is not possible. Hence, there is a great deal of interest in flattening the gain of EDFAs for WDM applications (see, e.g., Bergano (1996)].

Figure 17.32 shows a comparison of inverted erbium amplifier gain spectrum and the corresponding transmission spectrum of a Bragg fiber filter obtained by concatenating two long-period grating filters. This filter used with an EDFA can provide a flat gain (with gain variations of < 0.2 dB) over a 25- to 30-nm band.

Problems

17.1 Consider a fiber with $a = 3.5\ \mu$m and $\Delta n = 0.007$. Assuming $n_2 = 1.45$, obtain the grating period required for a Bragg wavelength of 800 nm.

17.2 Consider a directional coupler formed by a pair of identical step index single-mode fibers with $a = 4\ \mu$m and NA $= 0.11$. Assuming $n_1 \simeq 1.45$ and a core-to-core separation of 12 μm, obtain the coupling length at 1.3 μm and at 1.5 μm. For a coupler with a length equal to the coupling length at 1.3 μm, what fraction of power would be exiting from the two fibers at 1.5 μm?

17.3 Consider a fiber optic directional coupler with an interaction length (equal to the coupling length) of 5 mm.

(a) Obtain the corresponding coupling coefficient.
(b) What should be the value of κ so that a 5-mm-long coupler behaves as a 3-dB coupler?

17.4 Consider a single-mode fiber with $a = 3.5\mu$m and $\Delta n = 0.007$.

(i) Obtain the cutoff wavelength
(ii) Calculate the mode effective index at 1550 nm assuming pure silica cladding.
(iii) Calculate the required grating period for a Bragg wavelength of 1550 nm.
(iv) If you had assumed the effective index to be that for pure silica (instead of the estimated effective index), what would have been the period? For this period, what would be the actual Bragg wavelength?

17.5 Assuming the effective index of the fiber mode to be approximately equal to the refractive index of pure silica, obtain the grating period required for a Bragg wavelength of 800 nm and at 1550 nm. Use the Sellemier formula given in Chapter 6 for estimating the refractive index of pure silica at these wavelengths.

17.6 Consider a single-mode fiber with a core diameter of 3 μm and NA of 0.3. What should be the Bragg grating period for a resonance wavelength of 1532.1 nm (see Figure 17.27)?

17.7 Two plane waves at free space wavelength λ_{uv} propagating at an angle θ with each other are made to interfere in a medium of refractive index n_{uv}. What will be the fringe spacing?

Solution: Let us consider the two waves to propagate at angles $\pm\theta/2$ with respect to the x-axis in the x–z plane. If the interference pattern is viewed on the plane $z = 0$, then the intensity pattern is given by

$$I = I_0 \left| e^{ikz\sin\theta/2} + e^{-ikz\sin\theta/2} \right|^2$$

where $k = \frac{2\pi}{\lambda_{uv}} n_{uv}$ and I_0 is the intensity of each of the interfering waves. Thus

$$\begin{aligned} I &= 4I_0 \sin^2(kz\sin\theta/2) \\ &= 4I_0 \sin^2\left(\frac{2\pi}{\lambda_{uv}} n_{uv} z \sin\frac{\theta}{2}\right) \end{aligned} \tag{17.53}$$

Thus, the periodicity of the interference pattern is

$$\Lambda = \frac{\lambda_{uv}}{n_{uv}\sin(\theta/2)} \tag{17.54}$$

A fiber exposed to the above interference pattern would have a refractive index modulation with the period Λ.

As an example, let us calculate the angle θ required for forming a reflection grating at a reading wavelength λ_R. For this we need a grating period of (see equation 17.48)

$$\Lambda = \frac{\lambda_R}{2n_R} \tag{17.55}$$

where n_R is the effective index of the fiber mode at the reading wavelength λ_R. Thus, using equation (17.55) in equation (17.54) we obtain

$$\sin\frac{\theta}{2} = 2\frac{\lambda_{uv}}{\lambda_R}\frac{n_R}{n_{uv}} \tag{17.56}$$

For $\lambda_R = 1550$ nm, $n_R \simeq 1.45$, and for $\lambda_{uv} \simeq 244$ nm, $n_{uv} \simeq 1.508$, and we obtain

$$\theta \simeq 35.2^\circ$$

which is easily achieved.

17.8 Consider a multimoded fiber overlay device. When the fiber mode is phase matched to one of the modes of the multimode planar waveguide and the interaction length equals the coupling length, then when neglecting interaction with other modes of the multimode guide no power will exit from the fiber. Calculate the fractional error in the interaction length that is permissible if the loss is to be greater than 99%.

Solution: If L_c denotes the coupling length then, at any z, the power remaining in the fiber is

$$P_f = \cos^2 \kappa z = \cos^2\left(\frac{\pi}{2}\frac{z}{L_c}\right)$$

where $\kappa = \pi/2L_c$. If $z = L_c$, $P_f = 0$. For $P_f < 0.01$, we must have

$$\cos^2\left(\frac{\pi}{2}\frac{z}{L_c}\right) < 0.01$$

or

$$\cos\left(\frac{\pi}{2}\frac{z}{L_c}\right) < 0.1$$

If we write

$$z = L_c \pm \Delta L = L_c\left(1 \pm \frac{\Delta L}{L_c}\right)$$

then we have

$$\sin\frac{\pi}{2}\frac{\Delta L}{L_c} < 0.1$$

or

$$\frac{\Delta L}{L_c} < \frac{0.2}{\pi} \simeq 0.064$$

where we have assumed $\Delta L \ll L_c$.

Thus, if the coupling length is 5 mm, then for the peak loss in the device to be greater than 99%, the length of the device should lie between 4.68 mm and 5.32 mm.

18

Single-mode optical fiber sensors

18.1 Introduction

Although the major application of optical fibers has been in telecommunications, there is a growing application of optical fibers in sensing applications for measurement of various physical and chemical variables, including pressure, temperature, magnetic field, current, rotation, acceleration, displacement, chemical concentration, pH, and so forth. Such fiber optic sensors are finding applications in industrial process control, the electrical power industry, automobiles, and the defense sector.

One of the main advantages of a fiber optic sensor stems from the fact that optical fibers are purely dielectric and thus can be easily used in hazardous areas where conventional electrically powered sensors would not be safe. In addition, fiber optic sensors are immune to electromagnetic interference, have greater geometric versatility (i.e., they can be configured into a variety of arrangements to suit the application), and should have a very short response time. They can be multiplexed into various configurations and the information from various sensors can be transmitted over long distances by optical fibers. They can also be configured to provide spatially distributed measurements of external parameters.

In a typical optical fiber sensor, light from a source such as a laser diode or LED is guided by an optical fiber to the sensing region. Some property of the propagating light beam gets modulated by the external measurand such as pressure, temperature, magnetic field, and so forth. The modulated light beam is then sent via another (or the same) optical fiber for detection and processing. The modulation could be in terms of the intensity, phase, state of polarization, or frequency. Using multimode fibers for sensing applications leads to less sensitive but simpler, low-cost solutions. On the other hand, with single-mode fibers and laser diodes, orders of magnitude improvements in sensitivity are possible but require more sophisticated optical components as well as processing.

There are many different types of sensors using multimode as well as single-mode fibers. Sensors based on single-mode fibers are much more sensitive than multimode fiber sensors. In this chapter we briefly outline three basic single-mode fiber optic sensors – namely, the Mach–Zehnder interferometric sensor, the fiber optic current sensor, and the optical fiber gyroscope. For more details, readers may consult Dakin and Culshaw (1988) and Grattan and Meggitt (1995).

18.2 Mach–Zehnder interferometric sensor

One of the most sensitive arrangements of a fiber optic sensor is the Mach–Zehnder interferometric sensor arrangement shown in Figure 18.1 (see also

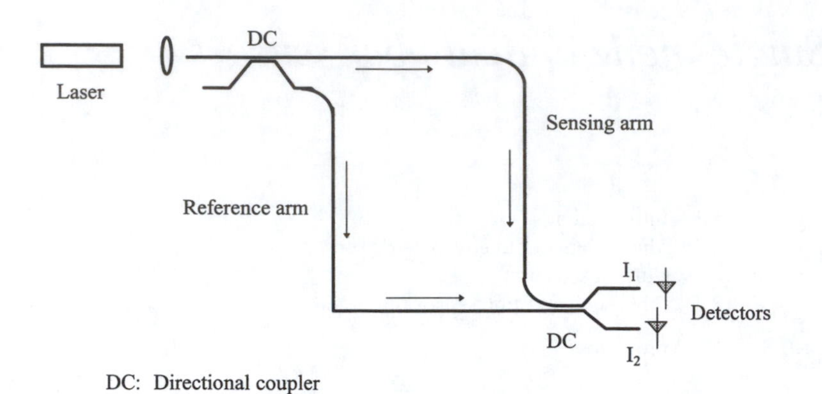

Fig. 18.1: The Mach–Zehnder interferometric arrangement for sensing. DC represents a fiber optic directional coupler.

Section 2.6.1). Light from a laser is passed through a 3-dB fiber optic coupler (see Section 17.2), which splits the beam equally into the two single-mode fiber arms. After traversing the fiber lengths, the two fibers form inputs to another 3-dB coupler, which helps in superposing the two beams. In a bulk Mach–Zehnder interferometer, the two 3-dB couplers are beam splitters. The two outputs of the output coupler arms are detected and processed. One of the arms of the interferometer is the sensing arm and is usually coated with a material that is sensitive to the parameter of interest, and the other arm, called the reference arm, is shielded from the external perturbation. When an external parameter acts on the sensor it alters the phase of the light propagating through the sensing arm by changing the refractive index and/or the length of the sensing arm. At the same time, since the light propagating in the reference arm is shielded, the external perturbation has no influence on the phase. Thus, as the two beams enter the second 3-dB coupler, the powers exiting from the two output arms will be determined by the phase difference between the two beams. A measurement of the output intensities gives us the parameter to be measured.

Indeed, if ϕ_1 and ϕ_2 are the phases of the two beams as they enter the output 3-dB coupler, then one can show that

$$I_1 = I_0 \cos^2 \frac{\Delta\phi}{2} \tag{18.1}$$

and

$$I_2 = I_0 \sin^2 \frac{\Delta\phi}{2} \tag{18.2}$$

where I_0 is the input intensity, I_1 and I_2 are the output intensities from arms 1 and 2 (see Figure 18.1), and

$$\Delta\phi = \phi_1 - \phi_2 \tag{18.3}$$

Thus, if the signal and reference arms introduce identical phase shifts, then $\Delta\phi = 0$ and all power exits from output 1. Similarly, for a phase difference of π, all power exits from output 2. For other values the power gets divided into both arms. Note from equations (18.1) and (18.2) that

$$I_1 + I_2 = I_0 \tag{18.4}$$

which is a constant.

Figure 18.2 shows the variation of I_1 and I_2 with $\Delta\phi$. Since the external perturbation usually induces very small changes $\Delta\phi$ (milli- to microradians)

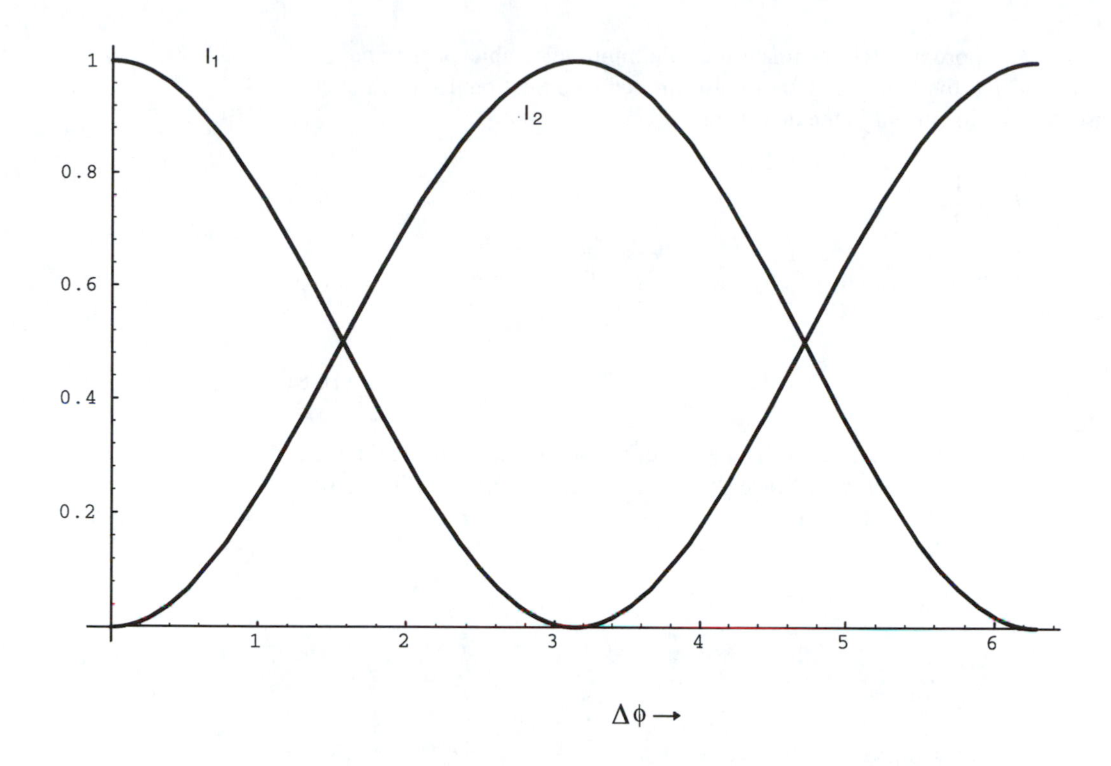

Fig. 18.2: Variation of outputs I_1 and I_2 with $\Delta\phi$.

between the two arms, we note from Figure 18.2 that if the sensor is operated around the minima (or maxima), then the intensity modulation due to change in $\Delta\phi$ will be very small. The most sensitive point of operation is around a phase difference corresponding to $(2m+1)\frac{\pi}{2}$. This is referred to as quadrature point. Due to temperature and pressure fluctuations, the phase difference $\Delta\phi$ may drift with time. This would lead to the point of operation shifting away from the quadrature point, leading to signal fading. Usually an active controller is placed in the reference arm to maintain the operation at a desired point. This can be accomplished by wrapping a length of the reference arm on a piezoelectric cylinder. When a voltage is applied across the cylinder, the cylinder expands and stretches the fiber, leading to a controlled $\Delta\phi$.

If the interferometric sensor is operated at the quadrature point then

$$\Delta\phi = \frac{\pi}{2} + \delta \tag{18.5}$$

where $\pi/2$ is the fixed bias and δ is the phase change induced by the measurand; the phase change δ is usually very small compared with π. Using equation (18.5) in equation (18.2) and assuming $\delta \ll \pi$, we can approximately write

$$\begin{aligned} I_2 &= I_0 \sin^2\left(\frac{\pi}{4} + \frac{\delta}{2}\right) \\ &\simeq \frac{I_0}{2}(1+\delta) \end{aligned} \tag{18.6}$$

Thus, under such a case, the intensity variations in I_2 (and similarly in I_1) are linearly related to the phase change δ.

We can approximately estimate the minimum detectable phase change δ assuming that the detector is shot noise limited (see Section 13.3). In such a case, the signal current in the detector is

$$i_s = \frac{1}{2} I_0 \rho \delta \tag{18.7}$$

where ρ is the responsivity of the detector. Now the shot noise current is (see Section 13.3)

$$i_{sN} = (2ei\Delta f)^{1/2} \tag{18.8}$$

where e is electronic charge, Δf is the detection bandwidth, and i is the mean photocurrent in the detector. Since the mean optical intensity falling on the photodetector is $I_0/2$, we have

$$i = \frac{1}{2} \rho I_0 \tag{18.9}$$

Thus

$$i_{sN} = (e\rho I_0 \Delta f)^{1/2} \tag{18.10}$$

Hence, the SNR is given by

$$\text{SNR} = \frac{i_s}{i_{sN}} = \frac{1}{2}\left(\frac{\rho I_0}{e\Delta f}\right)^{1/2} \delta \tag{18.11}$$

Defining the minimum detectable phase change to correspond to an SNR of unity we have

$$\delta_{min} = 2\left(\frac{e\Delta f}{\rho I_0}\right)^{1/2} \tag{18.12}$$

As an example, we consider a detector with $\rho = 0.5$ A/W, $I_0 = 1$ mW, and for a detector bandwidth of 1 Hz, we have

$$\begin{aligned} \delta_{min} &= 2\left(\frac{1.6 \times 10^{-19} \times 1}{0.5 \times 10^{-3}}\right)^{1/2} \\ &\simeq 3.6 \times 10^{-8} \text{ rad} \end{aligned}$$

This shows that extremely small phase changes are detectable. In actual systems, the detector is not shot noise limited, and the minimum detectable phase shift is larger than predicted by equation (18.12). The other noise factors include laser phase and intensity noise. Even in the presence of these noise sources, measurement of phase changes of 10^{-6} rad is possible.

Example 18.1: A phase change of 10^{-6} rad corresponds to a change in fiber length of

$$\Delta l = \frac{\Delta\phi}{\left(\frac{2\pi}{\lambda_0}\cdot n\right)} \simeq \frac{10^{-6}\times 0.633\times 10^{-6}}{2\pi\times 1.45}$$
$$\simeq 7\times 10^{-14}\,\text{m}$$

assuming $\lambda_0 \simeq 0.633\,\mu$m. This is indeed remarkable.

When the measurand is applied on the signal arm, then it changes the phase of the light propagating through the arm. Thus, if the phase is given by

$$\phi = \frac{2\pi}{\lambda_0}\cdot n_{eff}L \tag{18.13}$$

where λ_0 is the free space wavelength, n_{eff} and L are the LP_{01} mode effective index and the fiber length, then

$$\Delta\phi = \frac{2\pi}{\lambda_0}[n_{eff}\Delta L + L\Delta n_{eff}] \tag{18.14}$$

where ΔL and Δn_{eff} are the changes in the length of the fiber and the mode effective index, respectively. Thus, the fractional change is

$$\frac{\Delta\phi}{\phi} = \frac{\Delta L}{L} + \frac{\Delta n_{eff}}{n_{eff}} \tag{18.15}$$

In equation (18.15), the first term – the length change – is the dominant term. The second term (caused because of the strain in the fiber) is the change in effective index due to changes in the core and cladding refractive indices and changes in the core diameter (and, hence, V). The variation brought about by core diameter variation is negligible and, hence, we may replace n_{eff} by n, the refractive index of silica.

The phase change per unit length of sensing fiber due to changes in pressure is approximately given by

$$\frac{\Delta\phi}{L\Delta P} = \frac{2\pi}{\lambda_0}\left[n\frac{\Delta L}{L\Delta P} + \frac{\Delta n}{\Delta P}\right] \tag{18.16}$$

For silica fibers[1]

$$n\frac{\Delta L}{L\Delta P} \simeq -1.53\times 10^{-11}\,\text{rad/Pa} \tag{18.17}$$

and

$$\frac{\Delta n}{\Delta P} \simeq 1.08\times 10^{-11}\,\text{rad/Pa} \tag{18.18}$$

[1] 1 Pa = 1 N/m^2.

Thus, at 633 nm,

$$\frac{\Delta\phi}{L\Delta P} \simeq -4.5 \times 10^{-5}\ \text{rad/Pa}\cdot\text{m}$$

Sound pressure levels (SPLs) are usually measured in decibels with respect to a reference pressure level P_r.

$$\text{SPL} = 20\log\left(\frac{P}{P_r}\right)\ \text{dB re}\ P_r \tag{18.19}$$

Note the factor 20 instead of 10 – since sound intensity is proportional to P^2 and thus an increase in pressure by $\sqrt{2}$ would double the intensity. The reference pressure is usually chosen either as 1 μPa or to correspond to the threshold of hearing, which is 2×10^{-5} Pa.

Example 18.2: Sound levels with respect to the threshold of hearing are specified as decibels. Thus, 0 dB would correspond to threshold of hearing [see, e.g., Halliday, Resnick, and Walker (1993)]. A whisper at 1 m is 20 dB; the corresponding pressure P_w is given by

$$20 = 20\log\left(\frac{P_w}{2 \times 10^{-5}}\right)$$

or

$$P_w = 2 \times 10^{-4}\ \text{Pa}$$

Similarly, normal conversation corresponds to 60 dB or a pressure of 2×10^{-2} Pa, and the threshold of pain is 120 dB or a pressure of 20 Pa.

Example 18.3: Let us consider an MZ interferometer with a 100-m-long sensing arm consisting of a coated optical fiber with a sensitivity of 3.4×10^{-4} rad/Pa/m. Sound at the threshold of hearing falling on the sensing arm would induce a phase change of

$$\begin{aligned}\Delta\phi &= 3.4 \times 10^{-4} \times 2 \times 10^{-5} \times 100 \\ &\simeq 6.8 \times 10^{-7}\ \text{rad}\end{aligned}$$

which is measurable.

Example 18.4: If $\delta_{min} = 3.6 \times 10^{-8}$ rad, then the minimum detectable pressure for 1-m length of sensing arm using a coated fiber with a sensitivity of 3.4×10^{-4} rad/Pa/m is

$$\begin{aligned}P_{min} &= \frac{3.6 \times 10^{-8}}{3.4 \times 10^{-4}}\ \text{Pa} \\ &\simeq 10^{-4}\ \text{Pa}\end{aligned}$$

If this fiber is wound on a hollow polyethylene cylinder, one can increase the sensitivity to 10^{-6} Pa, which is below the threshold of hearing.

As we saw earlier, the pressure-induced phase changes in a bare silica fiber are about 4.5×10^{-5} rad/Pa/m. One can indeed increase the sensitivity to pressure by using special coatings with materials such as rubber, silicone, and plastics. For example, a thick coating of PTFE (a plastic) can increase the pressure sensitivity to 3.4×10^{-4} rad/Pa/m (a ten-fold increase). The sensitivities can be further increased by two orders of magnitude by winding the jacketed fiber on hollow cylinders made of polyethylene [McMohan and Cielo (1979)]. One can also reduce the sensitivity to pressure by using metal coatings. Thus, a 10-μm-thick nickel coating or a 96-μm-thick aluminum coating has been shown to reduce the pressure sensitivity to zero. Using such coatings one can thus completely desensitize the reference arm from the measurement [Giallorenzi et al. (1982)].

Fiber optic pressure sensors are finding applications as hydrophones for detection of underwater sound waves. One of the great advantages in such an application is the possibility to configure them as omnidirectional or highly directional receivers.

Any measurand that changes the phase of one of the arms with respect to the other can be measured by using the Mach–Zehnder arrangement. The sensitivity to various measurands is

$$\text{temperature}\left(\frac{\delta\phi}{L\Delta T}\right) \simeq 100\,\text{rad/K/m} \tag{18.20}$$

$$\text{Axial load}\left(\frac{\delta\phi}{L\Delta F}\right) \simeq 2 \times 10^{4}\,\text{rad/N/m} \tag{18.21}$$

$$\text{Linear strain} \simeq 10^{-7}\,\text{rad/m/unit strain} \tag{18.22}$$

18.3 Current sensors

Measurement of currents with optical fibers is very attractive since optical fibers are dielectric and, hence, electrical isolation is not a problem. Such sensors are also immune to electromagnetic interference, have a high bandwidth, and potentially have a lower cost alternative. One of the most important fiber optic current sensors uses the Faraday effect in glass. The Faraday effect refers to the rotation of the plane of polarization of a linearly polarized wave traveling through a medium in which a strong magnetic field is applied along the direction of propagation. This rotation can be thought of as being brought about by the fact that, in the presence of a longitudinal magnetic field, the medium becomes circularly birefringent – that is, the right circularly and left circularly polarized waves travel with different velocities.

The amount of rotation is proportional to the applied magnetic field and the length of interaction. Mathematically we may write for the rotation

$$\theta = V \int \vec{H} \cdot d\vec{l} \tag{18.23}$$

where θ is the rotation of the plane of polarization, $\vec{H}$ is the applied magnetic field, and V is called the Verdet constant. For silica, $V \simeq 2.64 \times 10^{-4}$ deg/A $\simeq 4.6 \times 10^{-6}$ rad/A.

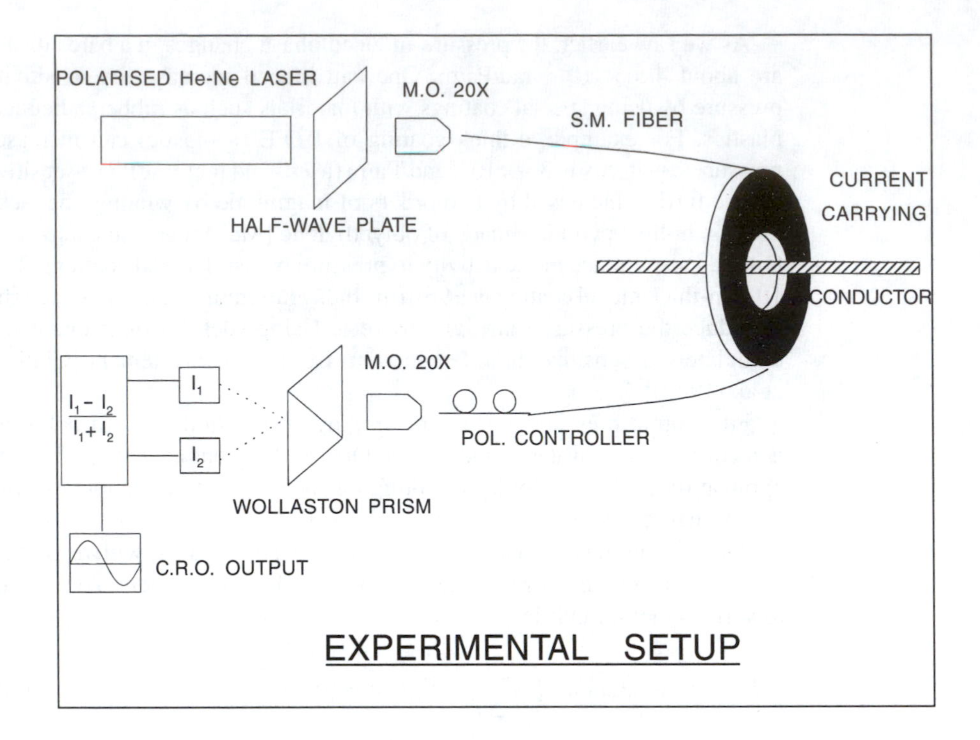

Fig. 18.3: A typical arrangement to measure current by using Faraday rotation in an optical fiber.

Figure 18.3 shows a typical arrangement of a fiber optic current sensor based on Faraday effect. It consists of a single-mode fiber wound helically around the current-carrying conductor. The magnetic field associated with the current leads to a rotation of the plane of polarization of the light propagating through the fiber. If there are N turns of fiber around the conductor, then by Ampere's law

$$\int \vec{H} \cdot d\vec{l} = NI \tag{18.24}$$

where I is the current through the conductor. Thus

$$\theta = VNI \tag{18.25}$$

The rotation of the plane of polarization is detected by first passing the output through a Wollaston prism, which breaks up the incident light into two orthogonal components whose intensities change when the plane of polarization rotates.

Note that the rotation θ is dependent only on the current enclosed by the fiber loop and is independent of the shape of the loop as well as any current sources lying outside the loop.

Example 18.5: Let us consider 30 turns of fiber wound around the current conductor. For a current of 1 A, the rotation is

$$\theta \simeq 7.92 \times 10^{-3} \text{ deg}$$

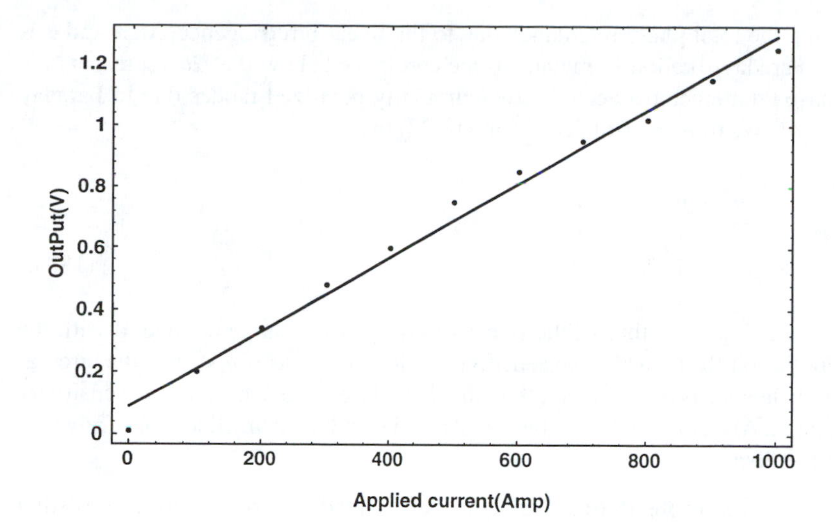

Fig. 18.4: A typical variation of output signal with current. Note the linear response and possibility of measuring large currents.

Each of the outputs of the Wollaston prism is detected separately, and then one obtains the following quantity by electronically processing the outputs of the photodetector.

$$R = \frac{I_1 - I_2}{I_1 + I_2} = K\theta \tag{18.26}$$

where I_1 and I_2 are the intensities of the two orthogonally polarized components and K is a constant. The ratio R is proportional to the rotation θ. The above processing makes the output independent of received light power and, hence, of laser drift.

Figure 18.4 shows a typical measured output signal with applied current. The response is very linear and shows the capability of such a sensor to measure large currents.

In the discussion above we have assumed that the fiber perfectly maintains the state of polarization (SOP) and the only change (rotation) is due to the applied magnetic field. In practice, this is not true since real fibers do possess some random birefringence that leads to a change in SOP even in the absence of the magnetic field. In addition, bending the fiber introduces linear birefringence; the smaller the loop radius the larger is the birefringence (see Section 17.8). In fact, in the presence of linear as well as circular birefringence, the signal R is given by

$$R = 2\theta \frac{\sin \Delta}{\Delta} \tag{18.27}$$

where

$$\Delta^2 = 4\theta^2 + \delta^2 \tag{18.28}$$

with

$$\delta = \frac{2\pi}{\lambda_0} \Delta n_{eff} 2\pi RN \tag{18.29}$$

being the total phase retardance due to the linear birefringence Δn_{eff} and θ is the Faraday rotation in radians. (One can indeed show that 2θ represents the phase retardance between the two circularly polarized modes due to Faraday effect.) We thus see from equation (18.27) that

$$\begin{aligned} R &\simeq 2\theta \frac{\sin\delta}{\delta} \qquad \delta \gg \theta \\ &\simeq \sin 2\theta \qquad \delta \ll \theta \end{aligned} \tag{18.30}$$

Thus, if $\delta \gg \theta$ – that is, the linear birefringence is large compared with the circular birefringence – the sensitivity is low, whereas if $\delta \ll \theta$ – the circular birefringence is much greater than the linear birefringence – then the sensitivity is large. Also, in such a case the signal is independent of any linear birefringence in the fiber.

Example 18.6: Let us consider a Faraday current sensor consisting of 30 loops of a single-mode fiber of radius 62.5μm bent around a circular former of radius 20 cm. In such a case (see equation (17.44))

$$\Delta n_{eff} = -0.133 \frac{b^2}{R^2}$$

where b is the fiber radius and R is the loop radius. Thus

$$\delta = \frac{5.25}{\lambda_0} \frac{b^2}{R} N$$

For the present case

$$\delta \simeq 4.86 \, \text{rad}$$

The corresponding Faraday rotation for a current of 1 A is

$$\begin{aligned} \theta &= VNI \\ &= 2.64 \times 10^{-4} \times \frac{\pi}{180} \times 30 \times 1 \\ &\simeq 1.4 \times 10^{-5} \, \text{rad} \end{aligned}$$

which is indeed very small compared with the linear birefringence!

One of the methods to reduce the effect of linear birefringence is to introduce an additional circular birefringence, which can be brought about by twisting the fiber that introduces a circular birefringence in the fiber [Graindorge et al. (1982)]. If the twist rate is such that the total circular birefringence is comparable to linear birefringence, then the effect of linear birefringence is very much reduced.

Now, for silica optical fibers, the rotation Ω of plane of polarization produced by a twist rate ζ over a length z is given by [Ulrich and Simon (1979)].

$$\Omega = 0.073 \, \zeta z \tag{18.31}$$

Thus, to reduce the effect due to linear birefringence, the required twist rate is such that

$$0.073\,\zeta z \gg 5.25\frac{b^2}{\lambda_0 R}N$$

Using $z = 2\pi RN$, we have

$$\zeta \gg 11.44\frac{b^2}{\lambda_0 R^2} \tag{18.32}$$

Using $b = 62.5\ \mu$m, $R = 30$ cm, $\lambda_0 = 0.633\ \mu$m, we have

$$\zeta \gg 0.78\ \text{rotations/m}$$

Twisted fibers can also be annealed to eliminate any residual linear birefringence [Rose, Ren, and Day (1996)].

18.4 Fiber optic rotation sensor (gyroscope)

The fiber optic rotation sensor based on the Sagnac effect is shown in Figure 18.5. It consists of a loop of polarization maintaining single-mode optical fiber connected to a pair of directional couplers, a polarized source, and a detector. Light from the source is split into two equal parts at the coupler close to the fiber loop (DC2), one part traveling clockwise and the other traveling anticlockwise through the loop. After traversing the loop the beams are recombined at the same coupler, and the portion returning to the other coupler DC1 then has 50% of it detected and processed. Assuming that the directional couplers are lossless, we first note that if the entire arrangement is not rotating, then the times taken by the clockwise and anticlockwise propagating beams through the loop would be same. Since a directional coupler introduces a $\pi/2$ phase difference between the two output arms, if the loop does not introduce any phase difference then the coupler DC2 gives an additional phase difference of $\pi/2$, resulting in all the power returning toward DC1.

Now, if the entire arrangement rotates with angular velocity Ω, then a phase difference given by [Burns (1994)]

$$\Delta\phi = \frac{8\pi NA\Omega}{c\lambda_0} \tag{18.33}$$

is introduced between the two beams. Here N is the number of fiber turns in the loop, A is the area enclosed by one turn (which need not be circular), and λ_0 is the free space wavelength of light.

One way to picture the Sagnac effect is by noting that when the entire setup is undergoing rotation (say clockwise) about an axis perpendicular to the loop, by the time the clockwise beam propagates through the loop the output coupler will have rotated and, hence, it will have to travel an extra path length before coupling out. On the other hand, the anticlockwise propagating beam will have to traverse a smaller path length to reach the output coupler. This difference leads to a phase difference given by equation (18.33).

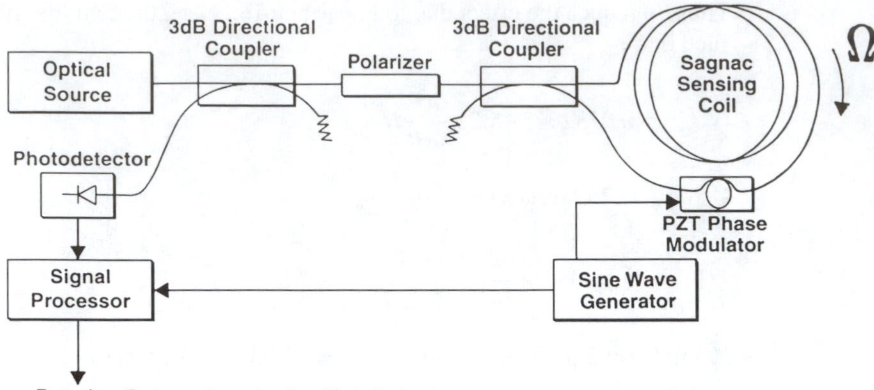

Fig. 18.5: An all-fiber optic gyroscope. [After Allen et al. (1994).]

To get a feeling for the phase shifts involved, let us consider a fiber optic gyroscope with a coil diameter of 10 cm and having 1500 turns (corresponding to a total fiber length of $\pi DN \simeq 470$ m) operating at 850 nm. The corresponding phase difference is

$$\Delta\phi = 1.16\,\Omega\,\text{rad}$$

If Ω corresponds to earth's rotation speed (15°/hr), then the corresponding phase change is

$$\Delta\phi = 8.4 \times 10^{-5}\ \text{rad}$$

a small shift indeed.

The output intensity from the Sagnac loop is given by

$$I = \frac{I_0}{2}(1 + \cos\Delta\phi) \tag{18.34}$$

where I_0 is the intensity entering DC2. The sensitivity of the interferometer is zero around $\Delta\phi = 0$. This problem is similar to that in the Mach–Zehnder interferometer. Thus, for maximum sensitivity the operating point should be shifted to the quadrature point where, with no rotation, the phase difference between the two interfering beams must be $\pi/2$. Assuming such a phase shift, equation (18.34) gets modified to

$$I = \frac{I_0}{2}\left[1 + \cos\left(\frac{\pi}{2} + \Delta\phi\right)\right]$$

$$= \frac{I_0}{2}(1 - \sin\Delta\phi)$$

which for small $\Delta\phi$ becomes

$$I \simeq \frac{I_0}{2}(1 - \Delta\phi) \tag{18.35}$$

If the detection is shot noise limited, then, as discussed in Section 18.2, the minimum detectable phase shift and the corresponding minimum detectable rotation rate are

$$\Delta\phi_{min} \simeq 2\left(\frac{e\Delta f}{\rho I_0}\right)^{1/2} = \frac{8\pi NA\Omega_{min}}{c\lambda_0}$$

Thus

$$\Omega_{min} = \frac{c\lambda_0}{4\pi NA}\left(\frac{e\Delta f}{\rho I_0}\right)^{1/2}$$

If we assume a loop radius of 5 cm and a total fiber length of 1 km, $\rho = 0.5$ A/W, $I_0 = 1$ mW, $\lambda_0 = 850$ nm, and $\Delta f = 1$ Hz, we obtain for the minimum measurable rotation rate as determined by shot noise

$$\Omega_{min} \simeq 1.5 \times 10^{-8}\ \text{rad/s}$$

$$= 0.003°/\text{hr}$$

In an actual gyroscope the total fiber length is limited by loss, the radius of the loop is limited by other constraints, and the optical power is the only free parameter.

The configuration shown in Figure 18.5 is called the minimum configuration in which the detection is done not at the output of DC2 but at the output of DC1. This is because the output in the other arm of DC2 is formed by the interference of two beams that have not followed perfectly identical paths. For example, the anticlockwise beam exits after two crossovers in DC2, whereas the clockwise beam never couples over to the other port. Thus, the output returning along the input fiber in DC2 is the reciprocal port and is used in sensing. The output from the free end of DC2 would be sensitive to even reciprocal perturbations such as temperature, vibrations, and so forth whereas the other output is free from this problem.

Fiber optic gyroscopes for various applications are available commercially. These applications include missile guidance, vehicle stabilization, land navigation, industrial robots, and so forth. The gyroscopes use polarization maintaining fibers and broadband light sources. With lowering costs, fiber optic gyroscopes should find a variety of applications [Burns (1994)].

Problems

18.1 In the discussion in Section 18.2, we assumed the source to be perfectly monochromatic. If the source is a laser with a spectral width $\Delta\lambda$, what is the maximum permissible difference in length between the two arms of the Mach–Zehnder interferometer in order to obtain a signal at the output?

18.2 Consider a fiber optic gyroscope with 500 m of fiber wound on a coil of radius 10 cm. What is the phase difference between the two outputs for a rotation rate corresponding to earth's rotation at 15°/hr? Assume $\lambda_0 = 633$ nm.

[ANSWER: $\Delta\phi \sim 250\,\mu$rad.]

18.3 If the frequency of the laser source in a Mach–Zehnder interferometric sensor changes by $\Delta\nu$, what is the corresponding effect on the two outputs of the interferometer?

18.4 What would be the effect of magnetic field on a fiber optic gyroscope with regard to the phase shifts?

19

Measurement methods in optical fibers: I

19.1 Introduction

Characterization of optical fibers is very important for a number of reasons. The users of optical fibers need the fiber characteristics to design the optical fiber communication system, whereas the manufacturers need them for optimizing their fabrication processes to obtain fibers with desired characteristics. The fiber characteristics are also necessary for the development and verification of various theoretical models for predicting various performance properties of the fiber. The two most important characteristics of an optical fiber are bandwidth (or pulse dispersion) and loss. In addition, one requires knowledge of various other parameters such as refractive index profile, core diameter, and so forth for predicting losses at joints. Table 19.1 lists the various characteristics of optical fibers along with their effect on system performance.

A large number of techniques have been developed for measuring various fiber characteristics. In this and the following chapter, we briefly discuss some of the standard techniques used in fiber characterization; for more details of the various techniques, readers may consult Pal, Thyagarajan, and Kumar (1988) and Thyagarajan, Pal, and Kumar (1988b).

In Section 19.2 we discuss some general experimental considerations relevant to fiber measurements, and in Section 19.3 we discuss various techniques for the measurement of refractive index profile, spectral attenuation and pulse dispersion, or bandwidth. In Chapter 20 we discuss measurement of characteristics specific to single-mode fibers only – namely, mode field diameter, cutoff wavelength, and birefringence.

19.2 General experimental considerations

19.2.1 Fiber end preparation

In most fiber measurements, one has to launch light into an optical fiber and observe the light output from the fiber. The quality of fiber cut at the input end will essentially determine the power coupled into the optical fiber from the source, which could be an incoherent source such as a tungsten halogen lamp or a laser. The quality of fiber cut at the output end is very critical, especially in experiments requiring measurement of radial power distribution. A number of

Table 19.1. *Optical fiber characteristics and their effect on system performance*

Fiber type	Measured parameter	Effect on system performance
Multimode	Attenuation	Repeater spacing
	Refractive index Profile	Bandwidth, number of modes, splice loss, source–fiber coupling
	Dispersion/bandwidth	Bit rate, repeater spacing
	Numerical aperture	Source–fiber coupling, splice loss
	Geometry	Splice loss, connector design
Single mode	Attenuation	Repeater spacing
	Refractive index profile	Dispersion characteristics
	Cutoff wavelength	Single-mode operating region, modal noise
	Chromatic dispersion [zero dispersion wavelength (ZDW) and slope at ZDW]	Bit rate, repeater spacing, optimum operating λ, WDM capability
	Mode field diameter	Splice loss, microbend loss, source–fiber coupling loss, waveguide dispersion
	Beat length	Birefringence, polarization mode dispersion.

Note: Adapted from Pal, Thyagarajan, and Kumar (1988).

manufacturers supply a fiber-cutting tool and most of them rely on scribing the bare fiber (after removing the plastic coating) placed under some tension along a curved surface and pulling to break the fiber. In the absence of a commercial tool, one could employ a simple tungsten carbide blade for scribing the fiber placed along a finger tip and then pulling it to obtain a good cut.

The quality of the fiber end cut could be assessed either by illuminating with a white light source and observing under a microscope or by coupling laser light (e.g., a He–Ne laser) and observing the radiation pattern on a screen. A circularly symmetric and uniform pattern essentially implies a good cut.

19.2.2 Cladding mode stripping

Since light can be total internally reflected at the cladding–air interface, light launched at the input end may also be guided along the cladding. It is very important in all measurement setups to remove the cladding light. This is accomplished by using what are referred to as cladding mode strippers. To remove the cladding light, one removes the plastic jacket covering the fiber over a length of about 50 mm at the input and output ends. This is then covered by a few drops of an index matching oil (like paraffin), which has a refractive index very close to, but slightly higher than, the cladding. This will then attenuate the light propagating in the cladding. Some fiber manufacturers use a buffer coating over the fiber, which itself serves as a cladding mode stripper.

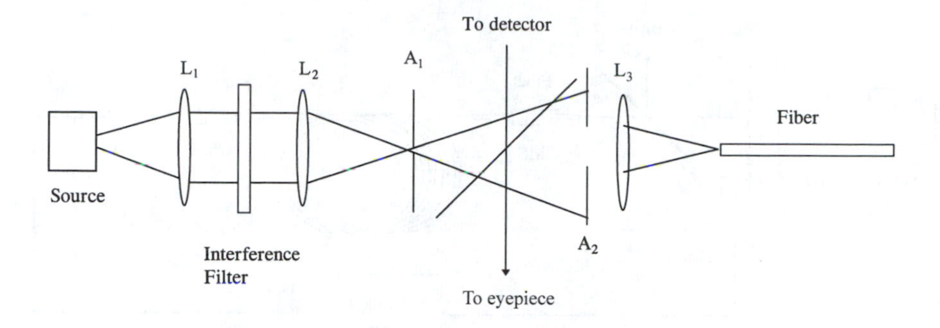

Fig. 19.1: A typical setup for launching light into optical fibers. The aperture A_1 essentially controls the size of the focused spot on the fiber end, and the aperture A_2 controls the NA of the input beam.

19.2.3 Launching light into fibers

Figure 19.1 shows a general launch optics setup that can be used to excite the fiber under different launching conditions. Lenses L_1 and L_2 form a condenser pair and help in imaging the incoherent source onto an aperture A_1. The beam emerging from A_1 falls on another aperture A_2 through a beam splitter and is finally focused by lens L_3 to the fiber tip. The size of the focused image on the fiber end can be controlled by varying the size of A_1, and the NA of the focused spot can be controlled by varying A_2. The beams emerging from the beam splitter can be used for detecting source power fluctuations as well as for viewing the input fiber end. For measuring wavelength-dependent characteristics, one can introduce interference filters between the lenses L_1 and L_2 as shown in Figure 19.1. In this space one could also introduce a mechanical chopper for phase-sensitive detection.

There are mainly two different input launch conditions that are used in the measurement of multimode fiber characteristics.

(a) **Overfilled launch:** In this case the apertures A_1 and A_2 are so adjusted that the launched spot at the fiber end has an NA greater than the fiber NA and the spot size is greater than the core diameter.

(b) **Limited phase-space (LPS) launch:** The apertures A_1 and A_2 are so adjusted that the focused spot has a dimension that is 70% of the core size and has an NA that is 70% of the NA of fiber.

19.2.4 Coupling from fiber to detector

Care must be excercised while coupling light from the fiber to the detector. Thus, one shold take care that the light emerging from the fiber illuminates about 70% of the detector area around the center. This would take care of any nonuniformities present in the detector surface and also would reduce the intensity level on the detector, thus reducing problems of saturation. While performing measurements on multimode fibers, if the detector area happens to be smaller than the illuminated spot, then one may have problems of modal noise.

19.3 Measurement of attenuation

Attenuation is one of the basic characteristics of the fiber and, along with dispersion, limits the repeater spacing. One of the most commonly used techniques is the cutback technique, and Figure 19.2 shows a typical experimental setup used

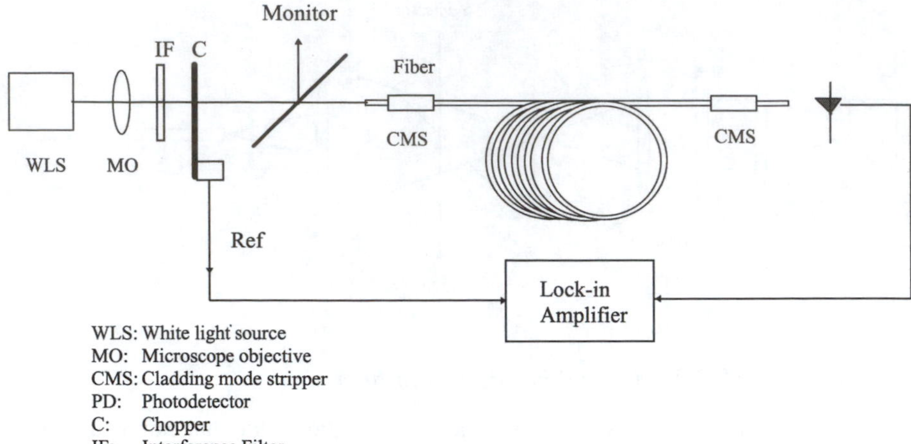

Fig. 19.2: A typical experimental setup for measurement of spectral attenuation in fibers.

in attenuation measurement. Light from a tungsten halogen lamp is coupled into the test optical fiber typically of length 1–2 km and the spectral variation of output power [say $P_l(\lambda)$] is first measured. Then the fiber is cut back, leaving typically 2 m of the fiber, and a repeat measurement of power [say $P_r(\lambda)$] at different wavelengths is performed. The attenuation at any λ is then given by

$$\alpha(\lambda) = \frac{10}{L} \log \frac{P_r(\lambda)}{P_l(\lambda)} \text{ dB/km} \tag{19.1}$$

where L is the length of the fiber cut in kilometers.

Although this procedure seems very simple, it is not trivial, especially for multimode fibers, because in a multimode fiber different mode groups suffer different attenuation rates. This is referred to as differential mode attenuation (DMA), which results in a measured attenuation that critically depends on the excitation conditions of the optical fiber. DMA also leads to a length dependence of the attenuation coefficient unless light propagates in a steady-state or equilibrium-mode distribution (EMD). Thus, it is very important to measure attenuation under steady-state mode distribution so that one can scale attenuation linearly with length for design in fiber communication systems.

In a long multimode optical fiber, EMD is established after propagation through a sufficiently long length ($\sim$ several hundred meters to a kilometer) of the fiber. In such a case the output mode distribution becomes independent of the launch conditions. To simulate an EMD one uses either of two approaches.

(a) To use an *overfilled launch* followed by a mode filter. The overfilled launch condition results in the excitation of all modes and the mode filter is used to generate an EMD. Figure 19.3(a) shows a typical mode filter consisting of five turns of the fiber wound on a mandrel of diameter 1–1.5 cm. The output after the mandrel is expected to be a steady-state mode distribution.

To determine whether the mode filter is generating the EMD, one measures the angle at which the intensity falls to 5% of the peak in the fiber far field with a long (1–2 km) fiber and at the output of the mode

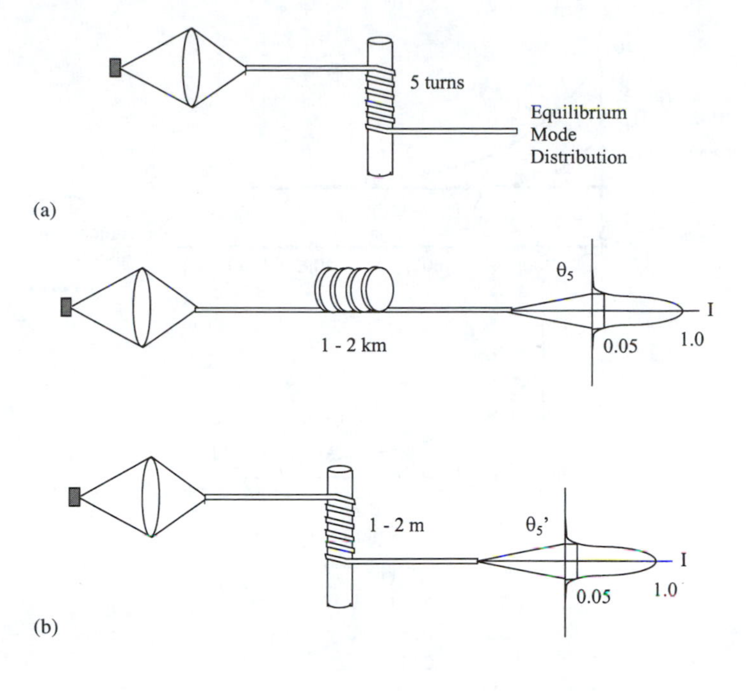

Fig. 19.3: (a) A typical mode filter for generating equilibrium-mode distribution in a short length of fiber. (b) For a mode filter $0.94\,\theta_5 < \theta_5' < \theta_5$. If $\theta_5' > \theta_5$, this implies incomplete mode filtering.

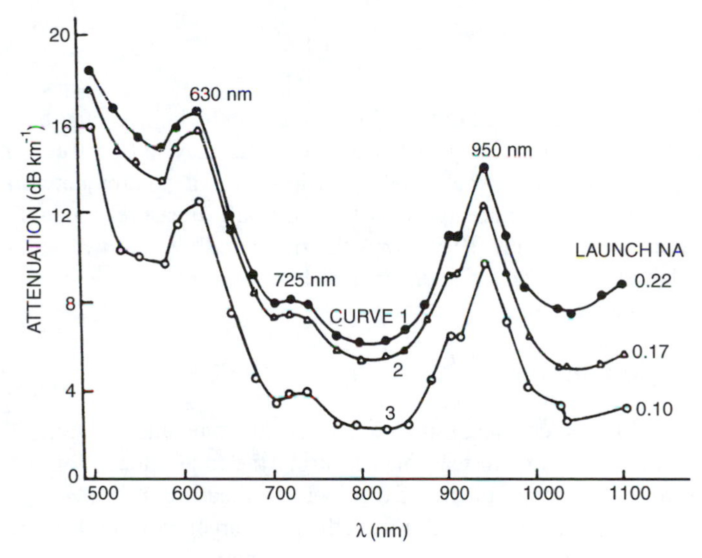

Fig. 19.4: A typical attenuation spectrum of a multimode step index fiber with a core diameter of 200 μm, NA = 0.17, and launch spot size of 130 μm. [After Gupta (1984).]

filter (see Figure 19.3(b)). If the former is termed θ_5 and the latter θ_5', then for a mode filter

$$0.94\theta_5 < \theta_5' < \theta_5 \tag{19.2}$$

(b) To use an *LPS launch* in which the launch NA is 0.7 of fiber NA and launch spot size is 0.7 of fiber core size.

Figure 19.4 shows a typical attenuation spectrum corresponding to a step index multimode fiber for different launch NAs. The dependence of attenuation on launch NA can be clearly seen.

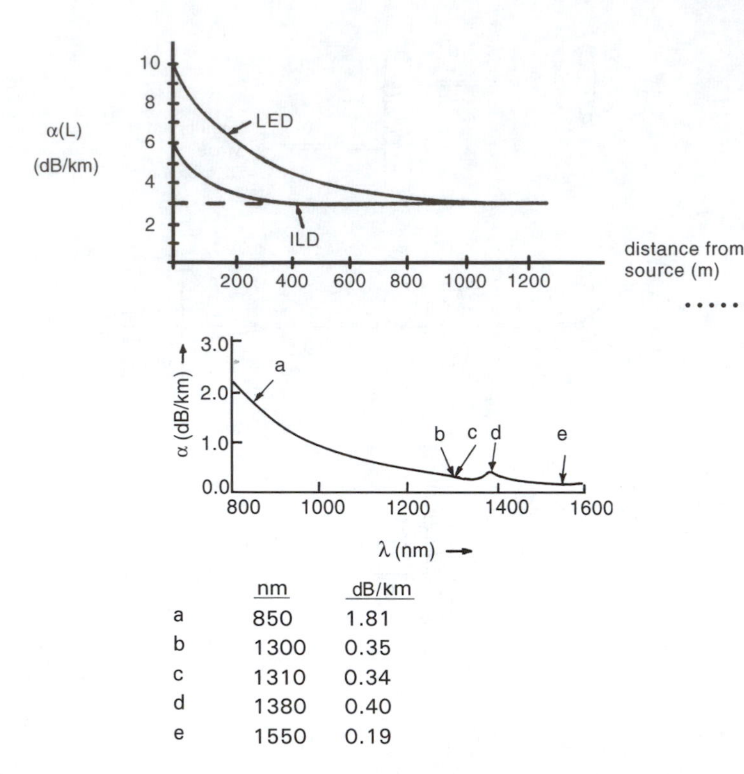

	nm	dB/km
a	850	1.81
b	1300	0.35
c	1310	0.34
d	1380	0.40
e	1550	0.19

Fig. 19.5: A typical measured variation of attenuation per unit length for various fiber lengths with LED and LD excitations. [After Kapron (1987).]

Fig. 19.6: Spectral attenuation of a commercial single-mode fiber from Corning Glass Works.

Although the attenuation obtained by the above technique may be used to estimate link losses over long distances, the same may not be true for short distances (for example, in multimode fiber links in a local area network). For such cases, it may be more appropriate to actually measure nonequilibrium attenuation values. Figure 19.5 shows the measured variation of attenuation per unit length for different fiber lengths. For such cases one may define an attenuation coefficient as

$$\alpha(L) = \alpha_0 + \alpha_t e^{-L/L_t} \tag{19.3}$$

where α_t and L_t are constants and α_0 is the steady-state attenuation.

In the above context, instead of measuring the total attenuation, one may excite different groups of modes of the fiber and measure the DMA.

Single-mode fibers do not suffer from the problem of multimode effects and, hence, it is much more straightforward to measure attenuation. One usually overfills the input (typically with spot sizes of 200 μm and NA of 0.2) to minimize sensitivity to input fiber end position. In addition, since cladding to core area is much larger in a single-mode fiber than in a multimode fiber, good cladding mode strippers should be used. Figure 19.6 shows the spectral loss of a typical commercial single-mode fiber from Corning Glass Works.

19.4 Measurement of refractive index profile

The bandwidth of a multimode optical fiber can be optimized by choosing a proper transverse refractive index profile (see Chapter 5). Even in single-mode fibers, one can tailor the dispersion characteristics by an appropriate choice of

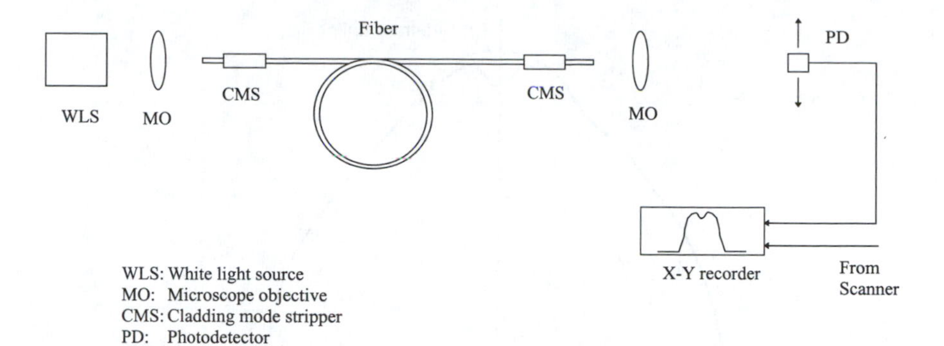

Fig. 19.7: Experimental setup for measurement of refractive index profile using the TNF method.

the transverse refractive index profile (see Chapter 10). Thus, measurement of refractive index profiles of optical fibers is quite important, especially in the optimization during fabrication. Although there are many techniques for the measurement of refractive index profile, here we discuss two of the standard techniques that are also the most widely used.

19.4.1 The transmitted near-field (TNF) method

This method is based on the fact that if all guided modes of a multimode fiber are excited equally, then the near-field power distribution $P(r)$ (i.e., the variation of optical power $P(r)$ with radial distance r from the axis) is given by

$$P(r) = P(0)\frac{n^2(r) - n^2(a)}{n^2(0) - n^2(a)} \tag{19.4}$$

where a is the core radius. For a profile described by

$$\begin{aligned} n^2(r) &= n^2(0)\left[1 - 2\Delta f\left(\frac{r}{a}\right)\right]; \quad r < a \\ &= n^2(0)(1 - 2\Delta); \quad r > a \end{aligned} \tag{19.5}$$

Equation (19.4) becomes

$$\frac{P(r)}{P(0)} = 1 - f\left(\frac{r}{a}\right) \tag{19.6}$$

Thus, the profile described by $f(\frac{r}{a})$ can be estimated by just measuring $P(r)$.

Figure 19.7 shows the experimental setup used. Light from an incoherent source (such as a tungsten halogen lamp or an LED) is used to excite a 1-m-long fiber under overfilled launched conditions (typically input spot of $\sim$70 μm and NA $\simeq$ 0.3 for a 50-μm core diameter, 0.2 NA fiber). Cladding modes are stripped by using a cladding mode stripper at either end. The output end of the fiber is magnified and imaged on a scanning photodetector that measures $P(r)$. Figure 19.8 shows a typical result obtained on a graded index multimode fiber.

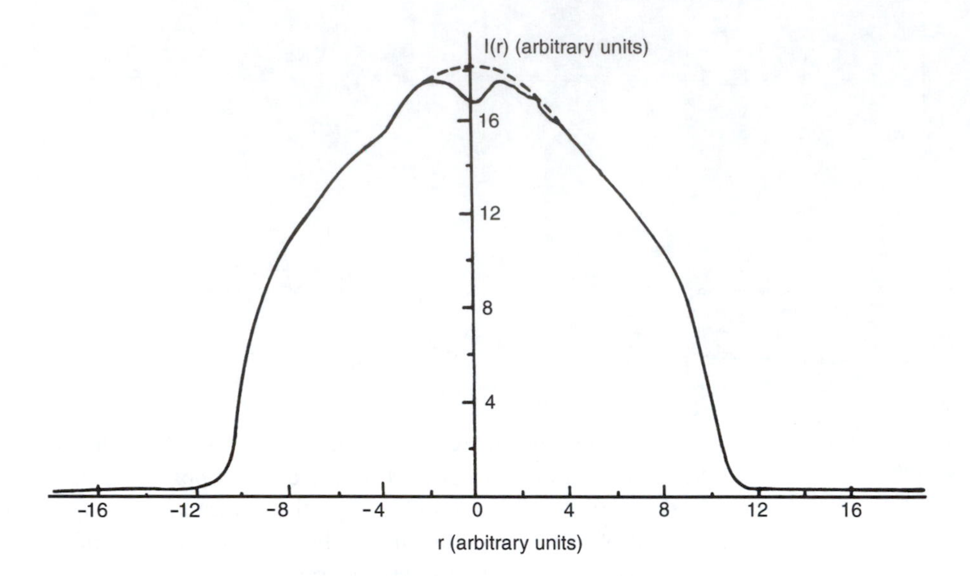

Fig. 19.8: A typical measured refractive index profile of a graded index multimode fiber using the TNF method. [After Dilip (1981).]

The measurement resolution is primarily limited by fiber NA. Figure 19.9 shows the output end of the fiber being imaged by a lens of focal length f on a screen. The effective lens diameter is

$$D \simeq 2f\theta \simeq 2f\mathrm{NA} \tag{19.7}$$

Thus, the limit of resolution is given by

$$\Delta r = 1.22\frac{\lambda f}{D} = 0.61\frac{\lambda}{\mathrm{NA}} \tag{19.8}$$

For $\lambda \sim 600$ nm, and NA ~ 0.2

$$\Delta r \simeq 1.8\,\mu\mathrm{m}$$

Also, since the local NA $\rightarrow 0$ near the core–cladding interface, the resolution drops as one nears the core–cladding boundary.

The above method is very simple to employ, but the estimated profile may suffer from errors due to propagation of leaky modes. For near parabolic index profiles, leaky mode corrections do not seem necessary. In addition, the proportionality between $P(r)$ and $n^2(r)$ is true only for equal mode excitation. In spite of the above, the TNF method is convenient for a quick estimate of the refractive index profile.

The TNF method in the above form is valid only for multimode fibers. If one replaces the incoherent source by a laser and measures the near-field power distribution of a single-mode fiber, one will essentially obtain the modal power distribution. If $P(r)$ can be measured very accurately, one may invert the wave equation to obtain $n^2(r)$. Thus, if $R(r)$ represents the radial distribution of modal field, then it satisfies the following scalar wave equation (see Section 8.3).

$$\frac{d^2R}{dr^2} + \frac{1}{r}\frac{dR}{dr} + k_0^2\left[n^2(r) - n_{eff}^2\right]R = 0 \tag{19.9}$$

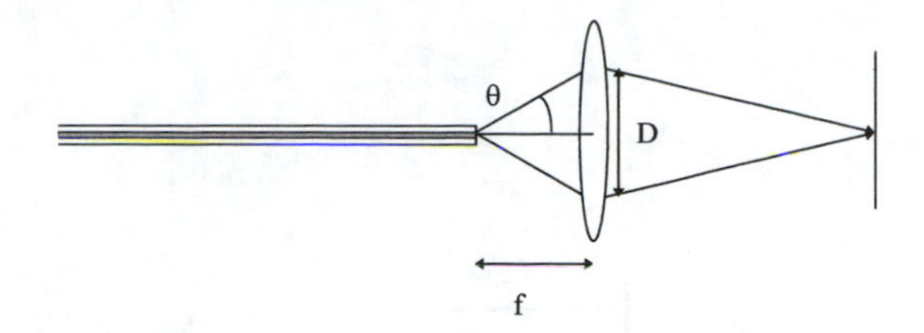

Fig. 19.9: Calculation of limit of resolution in the TNF method.

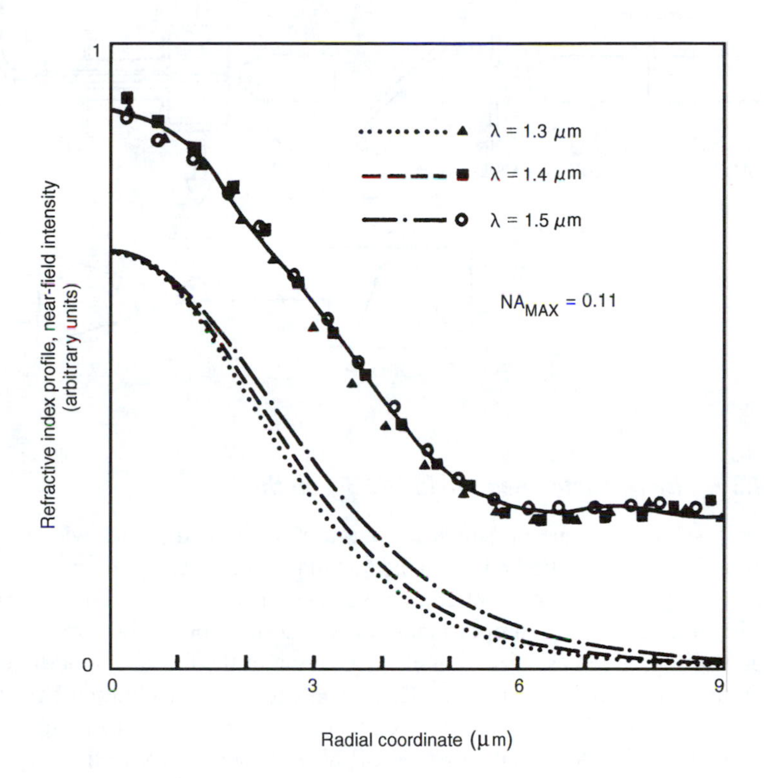

Fig. 19.10: Measured near-field intensity profiles at $\lambda = 1.3$, 1.4, and 1.5 μm and comparison of the refractive index profile calculated from the near field with that measured by the RNF method. [Coppa et al. (1984).]

where n_{eff} is the effective index of the mode ($= \beta/k_0$) and $k_0 = 2\pi/\lambda_0$ is the free space wavelength. The modal power distribution $P(r)$ is

$$P(r) = KR^2(r) \tag{19.10}$$

where K is a proportionality constant. In terms of $P(r)$, equation (19.9) can be modified to

$$n^2(r) = n_{eff}^2 - \left(\frac{\lambda_0}{4\pi P}\right)^2 \left[2P\frac{d^2P}{dr^2} - \left(\frac{dP}{dr}\right)^2 + \frac{2P}{r}\frac{dP}{dr}\right] \tag{19.11}$$

Thus, an accurate measurement of $P(r)$ can lead to an estimation of $n^2(r)$. Figure 19.10 shows typical results obtained with the above technique in comparison with the refracted near-field (RNF) method.

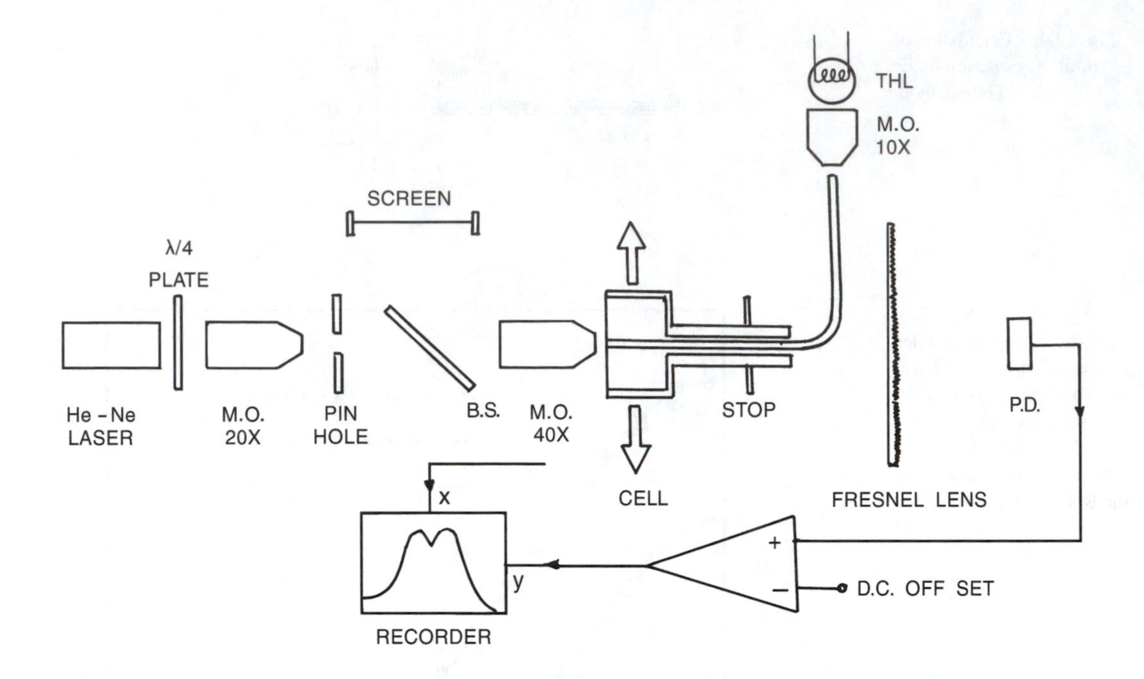

Fig. 19.11: Typical experimental setup to measure refractive index profile by the RNF method.

19.4.2 *The refracted near-field (RNF) method*

In the TNF method one measures the variation of guided power with radial position. The RNF method relies on the fact that if we focus a light beam with a high NA at any point in the fiber cross section, the power that is not coupled into the guided modes is proportional to $n^2(r)$ at that point. Thus, measuring the power not guided as a function of the position of the focused spot directly yields the refractive index profile. The advantage of this technique over the TNF technique lies in the fact that the effect of leaky modes can be completely eliminated and, in addition, the method can also be used to profile single-mode fibers.

Figure 19.11 shows a typical experimental setup. Light from an unpolarized (or circularly polarized) He–Ne laser is focused at the entrance face of an optical fiber immersed in a cell filled with a liquid of refractive index slightly higher than the cladding. The rays refracted by the fiber are collected by a large-diameter lens (like a Fresnel lens) and focused onto a detector. A stop is placed as shown in Figure 19.10 and subtends an angle larger than the NA of the fiber at the input end; this stop is to ensure that no leaky rays are allowed to reach the detector.

To understand the basis of the RNF method, consider a bare fiber of refractive index n immersed in a liquid of refractive index n_L (see Figure 19.12). Consider a ray of light incident at an angle θ as shown in Figure 19.12. Using Snell's law, we may write

$$\sin\theta = n \sin\theta_1 \tag{19.12}$$

$$n\cos\theta_1 = n_L \cos\theta_2 \tag{19.13}$$

$$n_L \sin\theta_2 = \sin\theta' \tag{19.14}$$

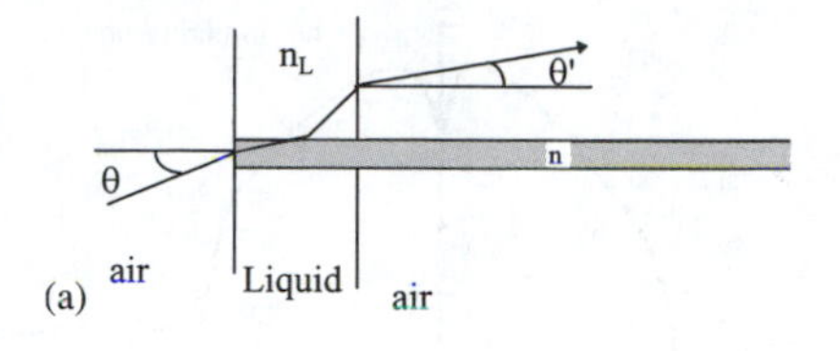

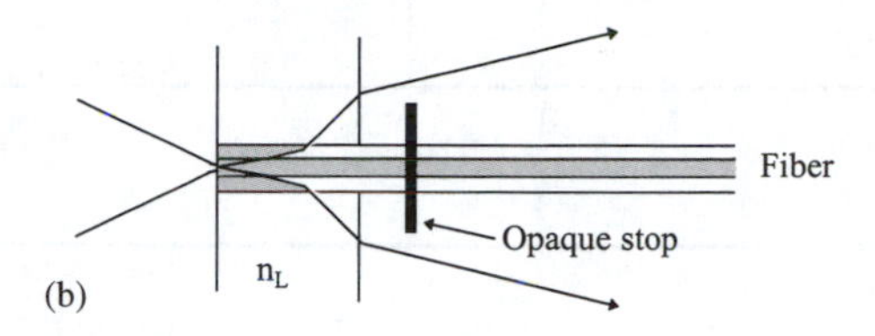

Fig. 19.12: (a) Path of a ray as it propagates through the fiber and refracts out. (b) An opaque stop blocks light rays up to an angle θ_5, which encompasses the leaky rays.

On eliminating θ_1 and θ_2 we obtain

$$n^2 = n_L^2 + \sin^2\theta - \sin^2\theta' \tag{19.15}$$

Thus, for a given θ, the angle θ' at which the ray emerges from the cell is related to n. For a fiber, n varies with the radial coordinate r; thus, a given cone of rays focused on the fiber end would emerge within a certain cone whose angular region is determined by the refractive index at the point of focus of the incident beam.

For a Lambertian source, the power emerging within a cone with vertex angle θ is

$$P(\theta') = A\sin^2\theta' \tag{19.16}$$

Hence, if a disc is placed behind the fiber (see Figure 19.12(b)) such that it stops all rays emerging at an angle less than θ_s, then the power crossing the stop is

$$P(\theta') = A(\sin^2\theta' - \sin^2\theta_s) \tag{19.17}$$

If the power collected for focusing on the liquid of refractive index n_L is P_0, then $\theta' = \theta$ (see equation (19.15)) and

$$P_0 = A(\sin^2\theta - \sin^2\theta_s) \tag{19.18}$$

Substituting from equations (19.17) and (19.18) in equation (19.15), we obtain

$$n^2(r) = n_L^2 + \frac{P(\theta') - P_0}{P_0}(\sin^2\theta - \sin^2\theta_s) \tag{19.19}$$

Thus, a measurement of $P(\theta')$ with r gives $n^2(r)$.

Figures 19.13(a) and 19.13(b) show typical refractive index profiles corresponding to a graded index multimode fiber and a single-mode fiber.

The advantages of the RNF method over the TNF method are

(a) No leaky mode correction is required.

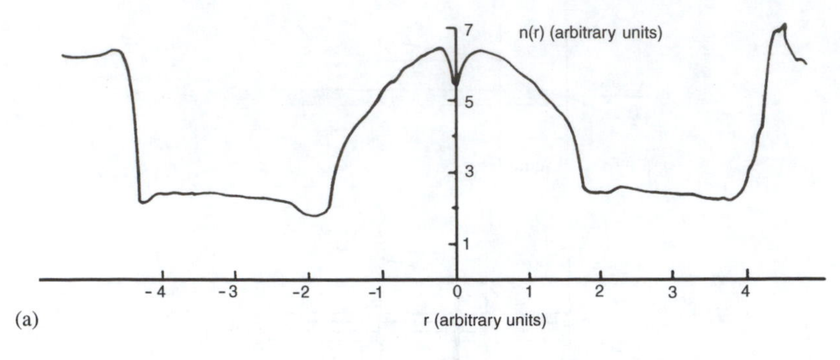

(a)

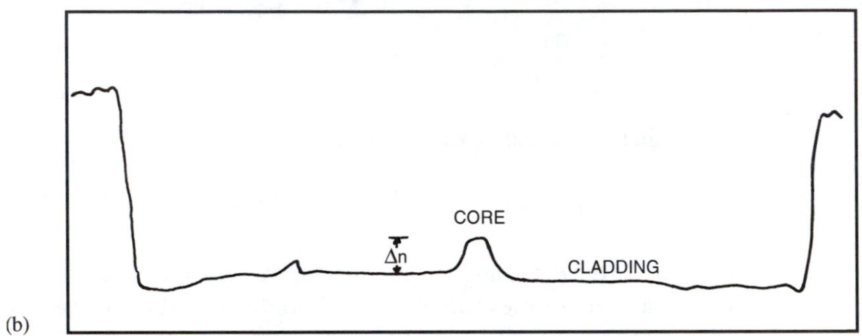

(b)

Fig. 19.13: Refractive index profiles of (a) the same multimode fiber as in Figure 19.8 and (b) a single-mode fiber using the RNF technique. [After Subramanyam (1983).]

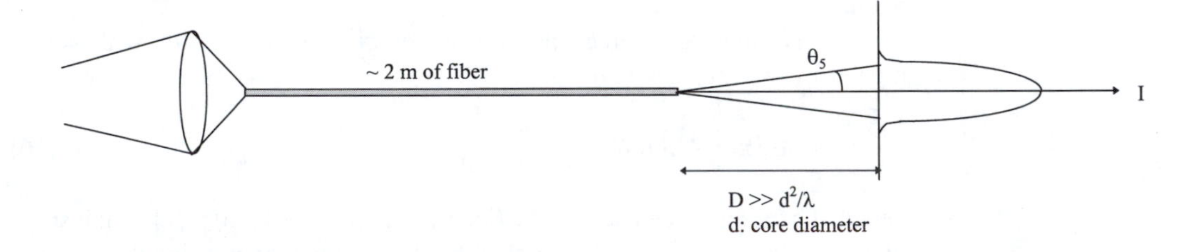

Fig. 19.14: Far-field technique for the measurement of NA of a multimode fiber.

(b) A coherent source such as a laser can be used.
(c) Very high resolutions in refractive index ($\sim 10^{-4}$) and spatial position ($\sim 0.4\ \mu$m) are possible.
(d) The method can be used to profile single-mode fibers also.

19.5 Measurement of NA

One can estimate the NA of a fiber from the measured refractive index profile

$$\mathrm{NA} = \left(n_1^2 - n_2^2\right)^{1/2} \tag{19.20}$$

where n_1 and n_2 represent the refractive index at the core center and of cladding, respectively.

A direct measurement of NA can also be performed by overfilling the input of a 2-m-long fiber and measuring the far field of the fiber (see Figure 19.14). If θ_5 represents the angle between the axis and 5% intensity point, then the NA is usually assumed to be given by

$$\mathrm{NA} = \sin\theta_5 \tag{19.21}$$

NA measured by equation (19.21) is generally less than that estimated from equation (19.20) due to DMA. In addition, NA is critically dependent on the excitation conditions and could also be length dependent because of the presence of leaky modes and DMA.

19.6 Measurement of pulse dispersion and bandwidth

It is well known that optical fibers may possess very high bandwidth (or very low pulse dispersion). Pulse dispersion essentially determines the information transmission capabilities of the optical fiber. We now discuss bandwidth measurements in multimode and single-mode fibers.

19.6.1 Multimode fibers

The main pulse broadening mechanism in multimode fibers is the intermodal broadening. In addition to this, another important broadening mechanism is material dispersion, which is approximately 90 ps/km·nm around 0.85-μm operating wavelength and less than 0.2 ps/km·nm around 1.3 μm (see Chapter 6). Normally encountered laser diodes around 0.85 μm have spectral width $\Delta\lambda_0$ of about 1 nm. Thus, the material dispersion over a 1-km fiber is

$$\Delta\tau = \frac{\lambda_0 L}{c}\left|\frac{d^2n}{d\lambda_0^2}\right|\Delta\lambda_0$$

$$\simeq 85\,\text{ps} \tag{19.22}$$

where we have used the fact that at 0.85 μm, $d^2n/d\lambda_0^2 \simeq 3 \times 10^{10}\,\text{m}^{-2}$. The dispersion given by equation (19.22) corresponds to a 3-dB bandwidth of about (see equation (19.27)).

$$f_{-3} \simeq \frac{0.44}{\Delta\tau} \simeq 5\,\text{GHz} \tag{19.23}$$

Thus, material dispersion becomes significant only for profiles having bandwidth in excess of about 1 GHz·km. Dispersion measurements around 850 nm with source spectral width of about 1 nm essentially give intermodal dispersion in fibers having a bandwidth of about 1 GHz·km. Since material dispersion is much lower around 1.3 μm, sources with much greater spectral width can be used for measurement of intermodal dispersion.

Figure 19.15 shows a typical experimental setup used for the measurement of pulse dispersion in multimode fibers. Pulses (of duration ~200–400 ps) are launched into the optical fiber after passing through a mode scrambler. The mode scrambler essentially mixes various modes to provide a well-defined launching condition irrespective of the source radiation pattern. The pulse coming out of the test fiber (typically 1–2 km) is then detected by an APD and is measured in a sampling oscilloscope. A delayed trigger from the laser is used to trigger the sampling oscilloscope. The broadened output pulse duration (FWHM) τ_0 is then measured. Then the fiber is cut back (as in the measurement of attenuation) and the pulse duration (FWHM) τ_i is again measured with the help of the

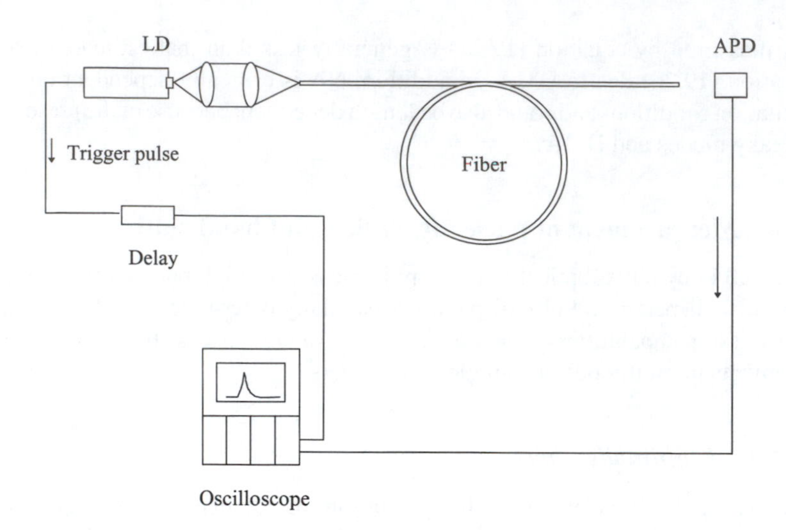

Fig. 19.15: A typical experimental setup for pulse dispersion measurement in the time domain in multimode fibers.

sampling oscilloscope. The impulse response is then approximately calculated as

$$\tau_f^2 \simeq \tau_0^2 - \tau_i^2 \tag{19.24}$$

The above approximation becomes an equality for Gaussian pulses.

Alternatively, one can first compute the Fourier transforms of the output and input pulses. If the variations of the corresponding amplitudes (in dB) with frequency are represented by $B_0(f)$ and $B_i(f)$, respectively, then one may estimate

$$B_F(f) = B_0(f) - B_i(f) \tag{19.25}$$

The -3-dB bandwidth is then the lowest f value for which $B_F(f) = -3$ dB. If the input and output pulses are approximately Gaussian, then by simple Fourier transform analysis one may show that (see Problem 19.3)

$$\text{3-dB Optical BW} \times \text{RMS impulse width} \simeq 187\ \text{MHz·ns} \tag{19.26}$$

$$\text{3-dB Optical BW} \times \text{FWHM impulse width} \simeq 440\ \text{MHz·ns} \tag{19.27}$$

The presence of a large number of modes in a multimode fiber leads to a dependence of the measured bandwidth on input launch conditions. The mode scrambler overfills the launch into the test fiber and, hence, may predict a bandwidth lower than that obtained with a small spot excitation.

In addition to the above method in the time domain, one may directly measure bandwidth in the frequency domain. In this domain, the source is sinusoidally amplitude modulated at various frequencies and the corresponding depth of modulation at the output (in dB) of a long (1–2 km) length of the fiber is measured. The same measurements are again taken after cutting back about 1–2 m of the fiber and one obtains $B_i(f)$ (in dB). The frequency response is given by equation (19.25) and the 3-dB bandwidth is the lowest f for which $B_F(f) = -3$ dB.

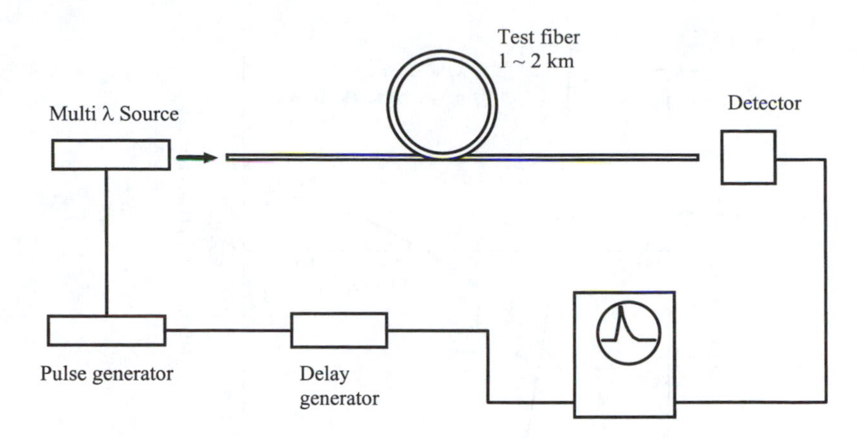

Fig. 19.16: Experimental setup for the measurement of pulse dispersion in a single-mode fiber using the pulse delay technique.

19.6.2 Single-mode fibers

Since dispersion is much smaller in single-mode fibers compared with multimode fibers, the conventional technique of measuring τ_0 and τ_i is not employed, and other techniques have been developed for measuring the dispersion in single-mode fibers.

Dispersion in single-mode fibers is specified in terms of the wavelength of zero dispersion (λ_z) and the dispersion slope S_0 at the zero dispersion wavelength (ZDW). Since the only dispersion mechanism in single-mode fibers is the chromatic dispersion – that is, variation of group velocity with wavelength – one can calculate the dispersion by first measuring the variation of time delay $\tau(\lambda)$ through a given length of the fiber as a function of λ and then obtaining $d\tau/d\lambda$. Figure 19.16 shows a simple arrangement for obtaining λ_z and S_0 using the pulse delay technique. Light from a multiwavelength source is coupled into the test fiber (~1–2 km) and the output is fed to a detector connected to an oscilloscope. A reference signal from the pulse generator is also connected to the oscilloscope through a delay generator. The variation of delay through the fiber is measured as a function of λ for both the long and the reference lengths. From these measurements one obtains the group delay $\tau(\lambda)$ per unit length. The dispersion coefficient is

$$D(\lambda) = \frac{d\tau}{d\lambda} \tag{19.28}$$

The multiple wavelength source could be a fiber Raman laser or a set of laser diodes at specified wavelengths. One could even use an LED in combination with interference filters to generate different wavelengths.

One usually fits the measured $\tau(\lambda)$ versus λ curve to an analytic function (such as Sellemeir fit or a polynomial) of the form

$$\tau(\lambda) = \tau_0 + \frac{S_0}{8}\left(\lambda - \frac{\lambda_z^2}{\lambda}\right)^2 \tag{19.29}$$

$$\tau(\lambda) = A\lambda^{-4} + B\lambda^{-2} + C + D\lambda^2 + E\lambda^4 \tag{19.30}$$

$$\tau(\lambda) = A + B(\lambda - D) + C(\lambda - D)^2 + E(\lambda - D)^3 \tag{19.31}$$

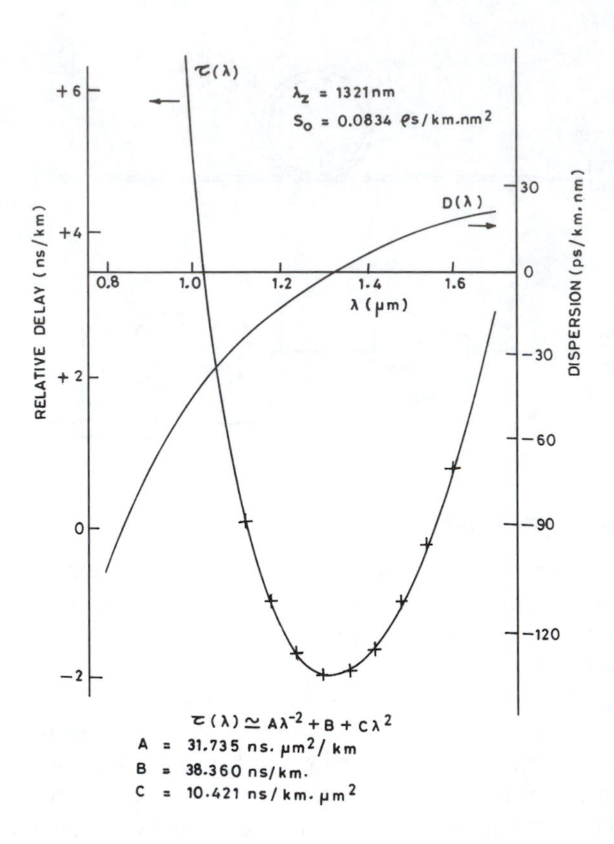

Fig. 19.17: Crosses represent typical measured relative delay at different wavelengths. $D(\lambda)$ crosses through zero at 1321 nm and $S_0 = 0.0834$ ps/km·nm^2.

where $A, \ldots, E$ are constants and in equation (19.29) λ_z is the wavelength of zero dispersion and S_0 is the slope at ZDW. Thus, from the fit one may easily obtain both λ_z and S_0.

Figure 19.17 shows a typical measured variation of $\tau(\lambda)$ with λ and the corresponding $D(\lambda)$. The fiber has $\lambda_0 = 1321$ nm and $S_0 = 0.0834$ ps/km·nm^2. The group delay $\tau(\lambda)$ can also be obtained by measuring the variation of the phase shift $\Phi(\lambda)$ suffered by a sinusoidally modulated source as a function of λ since (see Problem 6.7)

$$\Phi(\lambda) = 2\pi f \tau(\lambda) L \tag{19.32}$$

where f is the source modulation frequency and L is the length of the fiber. The relative variation of phase with λ can be measured by comparing the optical output (detected with an APD) and the electrical signal input by using a vector voltmeter. Again, measuring $\Phi(\lambda)$ for different λ essentially yields $\tau(\lambda)$ from which λ_z and S_0 can be obtained.

Dispersion can also be measured by using an interferometeric technique that requires only short lengths ($\sim$1 m) of fibers and, hence, can be used to pretest short fiber pieces before drawing a fiber from the perform or can also be used for measuring axial nonuniformity of dispersion characteristics of fibers. Figure 19.18 shows a typical experimental setup in which interference is formed between two beams, one passing through the test fiber of length L and the other through a reference arm length L_0. The reference beam can be given an additional variable delay via a movable mirror as shown in Figure 19.18. A source having spectral width $\Delta\lambda$ has coherence length $\lambda^2/\Delta\lambda$. Thus, as

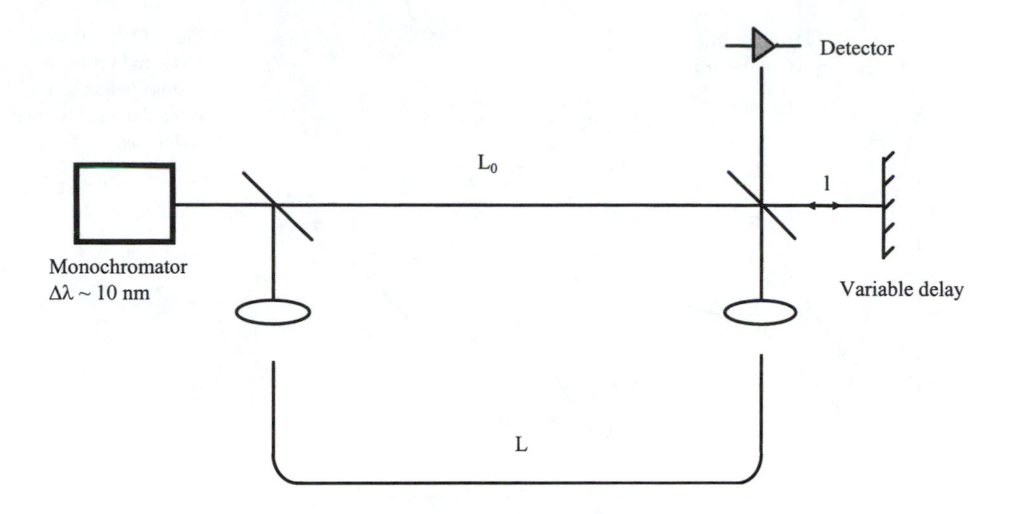

Fig. 19.18: A typical experimental setup for measurement of λ_z and S_0 using the interferometric technique.

the delay between the two beams is varied, maximum contrast will appear when

$$c\tau L = L_0 + 2l \tag{19.33}$$

where τ is the delay per unit length of the fiber and l is shown in Figure 19.18. For path differences much larger than $\lambda^2/\Delta\lambda$, the contrast in the interference pattern would be very poor. If one measures l for position of maximum contrast for different center wavelengths, one essentially obtains $\tau(\lambda)$. From the measured $\tau(\lambda)$, one can again obtain λ_z and S_0. It may be mentioned that for good resolution in $\tau(\lambda)$ and good signal level, $\Delta\lambda$ should be large. On the other hand, large $\Delta\lambda$ reduces the accuracy in $d\tau/d\lambda$. Figure 19.19 shows a typical measured variation of ZDW along a fiber and demonstrates a typical application of the interferometric technique.

Problems

19.1 If the measured variation of $P(r)$ is Gaussian – that is

$$P(r) = P_0 e^{-2r^2/w^2} \tag{19.34}$$

show that the corresponding refractive index profile is parabolic.

19.2 If, instead of using an incoherent source, a laser is used in the TNF measurement of multimode fibers, what would happen to the near-field profile?

19.3 If the input and output pulses are Gaussian, derive the relationships of equations (19.26) and (19.27).

Solution: If the input and output pulses are Gaussian, then the impulse response of the fiber is also Gaussian. Hence, for an impulse input, we write the output optical power variation as

$$P = P_0 \, e^{-t^2/\tau_f^2} \tag{19.35}$$

The electrical signal from the detector would be proportional to P and, hence, would have a time variation of the form e^{-t^2/τ_f^2}. Now, the FWHM of the pulse

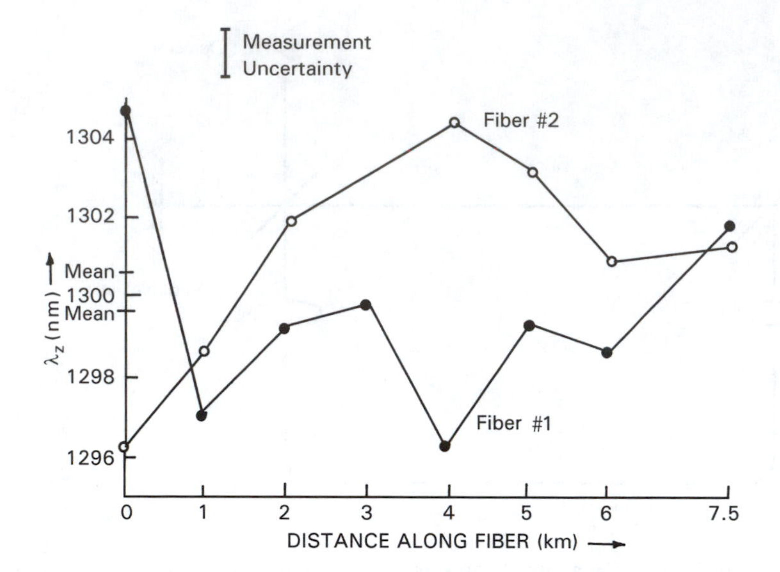

Fig. 19.19: Typical measured variation of λ_z at various points in a fiber using the interferometric technique.

represented by equation (19.35) can be obtained by putting

$$e^{-t^2/\tau_f^2} = \frac{1}{2}$$

implying

$$t = \pm\tau_f\sqrt{\ln 2} \tag{19.36}$$

Thus

$$\text{FWHM} = 2\tau_f\sqrt{\ln 2} \tag{19.37}$$

Also

$$\text{RMS width} = \left[\frac{\int_{-\infty}^{\infty} t^2 e^{-t^2/\tau_f^2}\,dt}{\int_{-\infty}^{\infty} e^{-t^2/\tau_f^2}\,dt}\right]^{1/2} = \frac{\tau_f}{\sqrt{2}} \tag{19.38}$$

To obtain the 3-dB optical bandwidth we first take a Fourier transform of equation (19.35) and obtain for the frequency response

$$\begin{aligned}\tilde{P}(f) &= \int_{-\infty}^{\infty} P(t)\,e^{-2\pi i f t}\,dt \\ &= P_0\int_{-\infty}^{\infty} e^{-t^2/\tau_f^2}e^{-2\pi i f t}\,dt \\ &= P_0\,e^{-\pi^2 f^2\tau_f^2}\end{aligned} \tag{19.39}$$

Hence, the 3-dB bandwidth f_3 is obtained by setting

$$\exp\left[-\pi^2 f_3^2\tau_f^2\right] = \frac{1}{2}$$

or

$$f_3 = \frac{\sqrt{\ln 2}}{\pi\tau_f} \tag{19.40}$$

Using equations (19.37), (19.38), and (19.40) we obtain

$$\text{3-dB optical BW} \times \text{RMS Impulse width} = \frac{f_3\tau_f}{\sqrt{2}} = \frac{1}{\pi}\sqrt{\frac{\ln 2}{2}}$$

$$\simeq 187\,\text{MHz·ns} \qquad (19.41)$$

$$\text{3-dB optical BW} \times \text{FWHM Impulse width} = f_3 2\tau_f\sqrt{\ln 2}$$

$$= \frac{2\ln 2}{\pi}$$

$$\simeq 440\,\text{MHz·ns} \qquad (19.42)$$

19.4 In the phase shift technique, if the maximum relative phase shift is to be less than 2π over the whole measured wavelength region, show that the maximum allowed modulation frequency is

$$f_{\max} = \frac{8\times 10^6}{S_0 L}\left(\lambda_i - \frac{\lambda_z^2}{\lambda_i}\right)^{-2}\ \text{MHz} \qquad (19.43)$$

where L is the length of the fiber (in km), λ_z (in nm) and S_0 (in ps/km/nm^2) are the ZDW and slope at ZDW, respectively, and λ_i is the source wavelength that minimizes $f_{\max}$.

Solution: The phase change of the propagating modulated wave is (see Problem 6.7)

$$\Phi(\lambda) = L\Omega\tau(\lambda) \qquad (19.44)$$

where

$$\tau(\lambda) = \tau_0 + \frac{S_0}{8}\left(\lambda - \frac{\lambda_z^2}{\lambda}\right)^2 \qquad (19.45)$$

Thus, if the difference in phase change between the ZDW (λ_z) and another source wavelength λ_i should be less than 2π we must have

$$L\Omega[\tau(\lambda_i) - \tau(\lambda_z)] < 2\pi$$

implying

$$L\Omega\frac{S_0}{8}\left(\lambda_i - \frac{\lambda_z^2}{\lambda_i}\right)^2 < 2\pi$$

which gives

$$f_{\max} = \frac{8\times 10^6}{S_0 L}\left(\lambda_i - \frac{\lambda_z^2}{\lambda_i}\right)^{-2}\ \text{MHz}$$

where $\Omega = 2\pi f$.

20

Measurement methods in optical fibers: II

20.1 Introduction

In the previous Chapter we discussed various techniques for the measurement of refractive index profile, spectral attenuation, and dispersion in both multimode and single-mode fibers. Some characteristics are relevant only to single-mode fibers; these are the cutoff wavelength, mode field diameter, birefringence, etc. In this chapter we discuss the principles and techniques for the measurement of these parameters in single-mode fibers. In Section 20.2 we discuss various cutoff wavelength measurements, in Section 20.4 mode field diameter measurements are discussed, and finally in Section 20.6 we discuss various methods for measurement of birefringence.

20.2 Cutoff wavelength

Cutoff wavelength of the LP_{11} mode forms one of the most important characteristics of a single-mode fiber as it essentially determines the wavelength above which the fiber behaves as a single-mode fiber. We have seen in Chapter 8 that for a step index fiber, the cutoff wavelength λ_c of the LP_{11} mode is given by

$$\lambda_c = \frac{2\pi a}{2.4048}\sqrt{n_1^2 - n_2^2} \tag{20.1}$$

where a is the fiber core radius, and n_1 and n_2 are the core and cladding refractive indices. For other refractive index profiles, one can estimate theoretically the cutoff wavelength. However, a direct measurement of λ_c essentially leads to a value also known as the effective cutoff wavelength, and this value is, in general, different from the theoretical value λ_c. This is essentially due to the fact that, as one approaches the cutoff wavelength from the lower wavelength side, the LP_{11} mode gets less and less confined to the core, and small deviations from straightness of the fiber in the experimental setup result in a large attenuation of the LP_{11} mode. The effective cutoff wavelength is usually defined as that wavelength at which the ratio of the total power to the fundamental mode power is 0.1 dB. The effective cutoff wavelength is slightly smaller than the theoretical value of the cutoff wavelength.

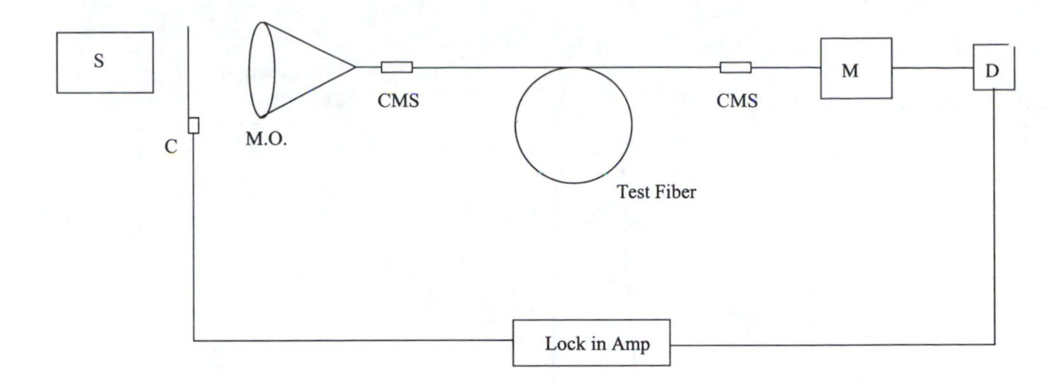

S: White light source
C: Chopper
MO: Microscope Objective
CMS: Cladding Mode Stripper
M: Monochromator
D: Detector

Fig. 20.1: Measurement setup for the determination of effective cutoff wavelength of a fiber using the bend reference technique. From the measured variations $P_s(\lambda)$ and $P_b(\lambda)$ one can obtain λ_{ce}.

It is very essential in a single-mode fiber link that, at the operating wavelength, the residual power in the LP_{11} mode be as low as possible so that no modal noise is introduced in the link. Thus, in this respect it is advantageous to have the cutoff wavelength much lower than the operating wavelength. At the same time, since the fundamental LP_{01} mode should also be tightly confined to the core (to reduce bending loss), one would like to operate as close as possible to the cutoff wavelength. Thus, knowledge of the cutoff wavelength is very important from the utilization point of view.

There are many techniques for the measurement of cutoff wavelength. Here we discuss the bend reference technique and multimode reference technique in detail.

20.2.1 Bend reference technique

Figure 20.1 shows the experimental setup used in the bend reference technique. The spectral variation of power output $P_s(\lambda)$ from a 2-m-long test fiber having a single circular loop of 140-mm radius is first measured. Remaining portions of the fiber should be substantially free of any external stresses and should not contain any bends of 140-mm radius or smaller. Then, without changing the input excitation condition, an additional loop of radius 30 mm is introduced, and again the spectral variation of output power $P_b(\lambda)$ is measured. Then, a plot of the following quantity

$$R_b(\lambda) = 10 \log \left[\frac{P_s(\lambda)}{P_b(\lambda)} \right] \tag{20.2}$$

is made as a function of λ. Figure 20.2 shows a typical plot for a fiber from York Technology (United Kingdom). The cutoff wavelength is defined to be that wavelength where the long-wavelength edge of the bend-induced loss is greater than the long-wavelength baseline by 0.1 dB. Thus, from the measured variation of $R_b(\lambda)$ we can obtain λ_{ce} by determining the point of intersection of a straight

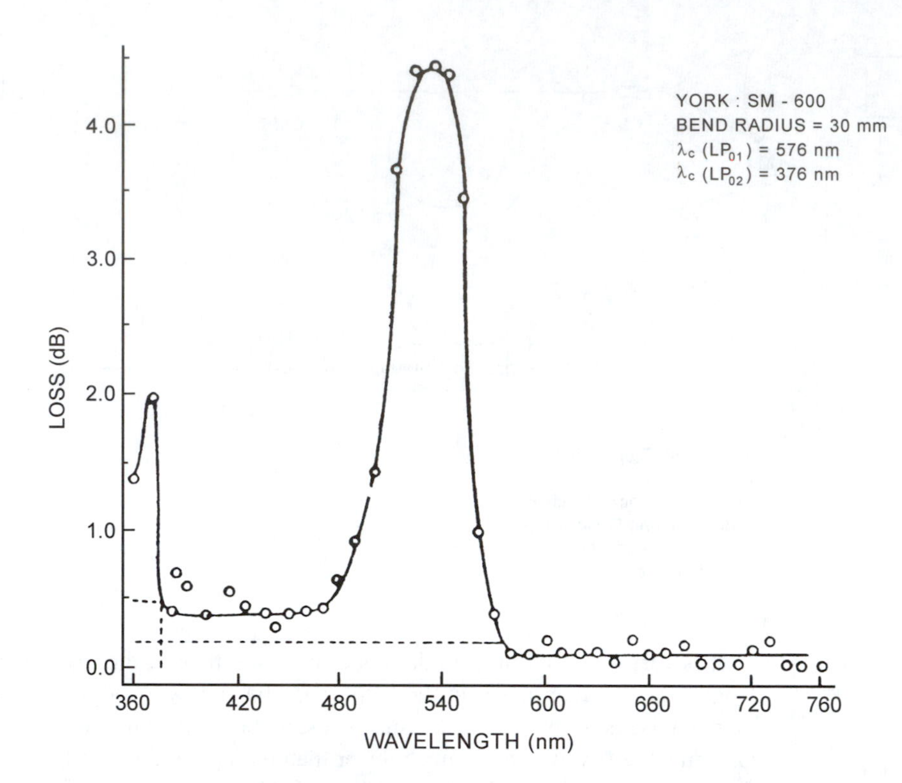

Fig. 20.2: Typical variation of $R_b(\lambda)$ given by equation (20.2) for a fiber from York Technology (United Kingdom). The peak on the right corresponds to the cutoff of LP_{11} mode and that on the left corresponds to the LP_{02} mode. [After Banerjee (1986).]

line parallel to the λ-axis and displaced by 0.1 dB from the long-wavelength baseline.

The above method of defining the effective cutoff wavelength essentially determines the wavelength at which the attenuation of the LP_{11} mode caused by the 140-mm bend is $\simeq$20 dB. To show this, we first recall that the LP_{01} mode is doubly degenerate (due to the two independent polarization directions) and the LP_{11} mode has four-fold degeneracy (due to $\sin\phi$, $\cos\phi$ dependence, and two orthogonal polarization states) (see Chapter 8). We assume that the input incoherent source excites all modes equally, say each with power $s(\lambda)$. Thus, the power in the fiber before the 140-mm loop is $6s(\lambda)$. Close to the cutoff of the LP_{11} mode, the LP_{01} mode is well confined and, thus, the 140-mm bend may be assumed to introduce negligible loss of the LP_{01} mode but would introduce some loss $\epsilon(\lambda)$ for the LP_{11} mode. Thus, we may write for the power after the 140-mm loop as (see Figure 20.3)

$$P_s(\lambda) = 2s(\lambda) + \epsilon(\lambda)\, s(\lambda) \tag{20.3}$$

where $\epsilon(\lambda)\, s(\lambda)$ is the power left over in the LP_{11} mode. When we introduce the 30-mm bend, even the remaining power in the LP_{11} is lost and we essentially measure the power in the LP_{01} mode, which is (see Figure 20.3)

$$P_b(\lambda) = 2s(\lambda) \tag{20.4}$$

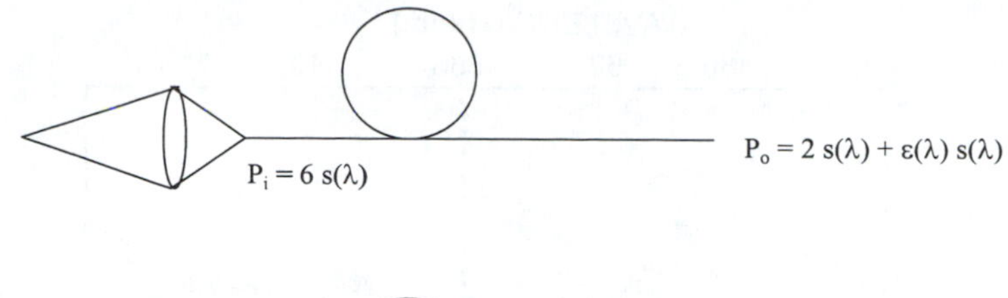

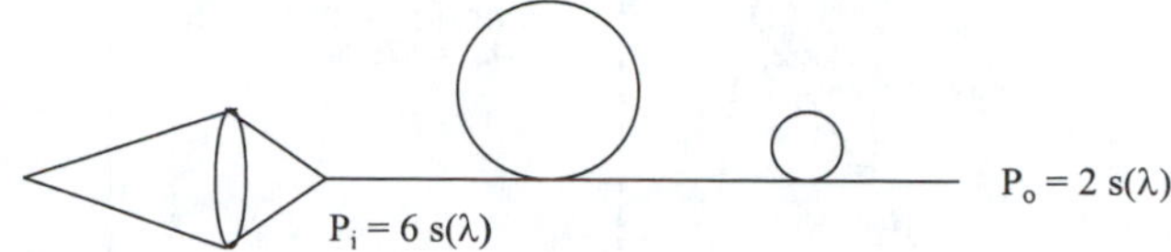

Fig. 20.3: Without the additional 30-mm bend, $P_s(\lambda) = 2s(\lambda) + \epsilon(\lambda)\, s(\lambda)$, where $s(\lambda)$ is the power into each of the modes of the fiber and $\epsilon(\lambda)$ is the loss in LP_{11} mode due to the 140-mm bend. After the 30-mm bend, one has $P_b(\lambda) = 2s(\lambda)$.

Thus

$$R_b(\lambda) = 10 \log\left[\frac{2 + \epsilon(\lambda)}{2}\right] \tag{20.5}$$

The wavelength $\lambda = \lambda_{ce}$ at which $R_b(\lambda) = 0.1$ dB corresponds to

$$\epsilon(\lambda) \simeq 4.7 \times 10^{-2} \tag{20.6}$$

Thus, the initial power $4s(\lambda)$ (in the LP_{11} modes) before the 140-mm bend reduces to $4.7 \times 10^{-2}\ s(\lambda)$ after the 140-mm bend. This corresponds to a bend-induced attenuation of

$$10 \log\left[\frac{\epsilon(\lambda)}{4}\right] \simeq 20\,\text{dB} \tag{20.7}$$

Thus, at the effective cutoff wavelength, the 140-mm bend induces an attenuation of approximately 20 dB in the LP_{11} mode.

As the wavelength reduces below λ_{ce}, the attenuation induced by the 140-mm bend decreases and, thus, $\epsilon(\lambda)$ increases; consequently, $R_b(\lambda)$ increases as λ decreases (see Figure 20.2). The maximum value of $R_b(\lambda)$ would correspond to a point where the 140-mm bend does not introduce any loss for the LP_{11} mode while the 30-mm bend removes the LP_{11} mode completely. Thus, at this point

$$P_s(\lambda) = 6s(\lambda)$$

$$P_b(\lambda) = 2s(\lambda)$$

and thus

$$R_b(\lambda) = 10\ \log(3) \simeq 4.8\,\text{dB} \tag{20.8}$$

In Figure 20.2 we see that the peak value of $R_b(\lambda)$ is almost 4.5 dB.

As the wavelength decreases further, even the LP_{11} mode gets tightly confined and, thus, both the 140-mm bend and the 30-mm bend do not introduce any attenuation of the LP_{11} mode; thus, $R_b(\lambda) \to 0$ dB. $R_b(\lambda)$ would exhibit other peaks corresponding to the cutoff of other modes (see Problem 20.1).

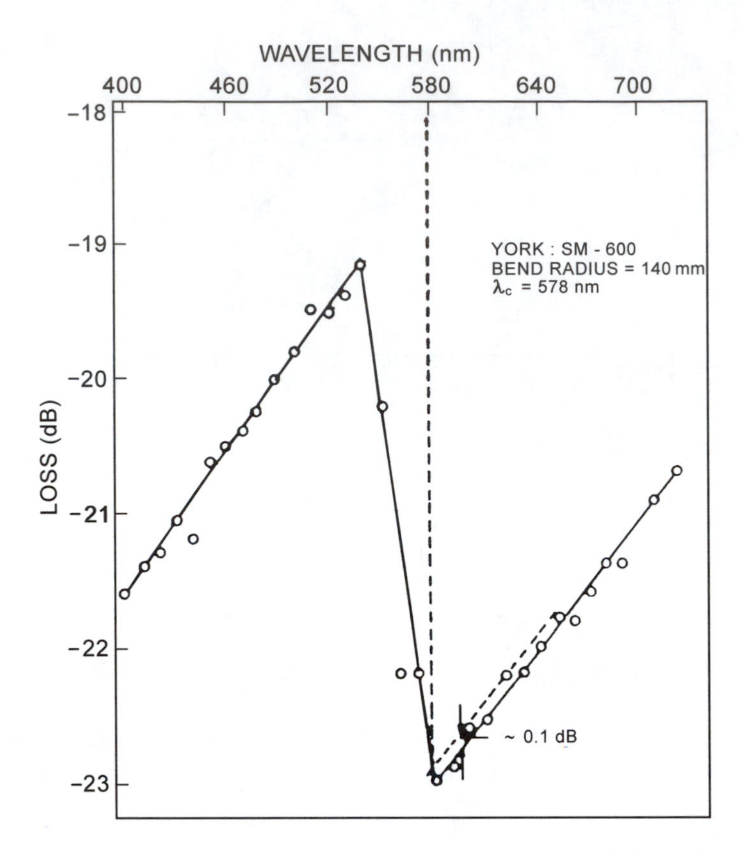

Fig. 20.4: Variation of loss $R_m(\lambda)$ given by equation (20.9) for the same fiber as in Figure 20.2 using the multimode reference technique. [After Banerjee (1986).]

20.2.2 *Multimode reference technique*

In the multimode reference technique one first measures $P_s(\lambda)$ as described earlier, and then the single-mode fiber is replaced by a multimode fiber and the spectral dependence of power $P_m(\lambda)$ is then measured. One then plots

$$R_m(\lambda) = 10 \log \left[\frac{P_s(\lambda)}{P_m(\lambda)} \right] \tag{20.9}$$

as a function of λ. Figure 20.4 shows the measured variation of $R_m(\lambda)$ with λ for the same single-mode fiber as in Figure 20.2. A straight line is fitted to the long-wavelength region of $R_m(\lambda)$ and shifted up by 0.1 dB (see Figure 20.4). The λ value at the point of intersection of this line with the $R_m(\lambda)$ curve gives the cutoff wavelength.

Table 20.1 gives a comparison of the effective cutoff wavelength measured by the two techniques of a single-mode fiber from York Technology (United Kingdom) and International Telephone and Telegraph (United States). For single-mode fibers operating at 1300 nm, the recommended cutoff wavelength is 1280 nm or lower.

20.3 Bending loss measurement

One can use the above experimental setup to measure bend-induced loss. To do this, one first measures the spectral variation of output power with λ for a fiber

Table 20.1. *Comparison of cutoff wavelengths obtained using the bend reference technique and the multimode reference technique*

	λ_{ce}(nm)	
Technique	York SM-600 fiber	ITT T1601 fiber
Bend reference	576 ± 4	596 ± 4
Multimode reference	578 ± 4	598 ± 4

with a bend radius greater than about 140 mm. Then, about 40–100 turns are given on a 75-mm-diameter mandrel to simulate splice housings and one again measures $P_b(\lambda)$. The bend-induced excess attenuation is

$$A\,(\text{dB}) = 10 \log\left[\frac{P_s(\lambda)}{P_b(\lambda)}\right] \tag{20.10}$$

Typical specifications for telecommunication-grade fibers are $A < 0.1$ dB at 1310 nm and <1 dB at 1550 nm for 100 turns on a mandrel of diameter 75 mm. The bend-induced attenuation depends critically on the mode field diameter (see Chapter 8).

20.4 Mode field diameter (MFD)

In single-mode fibers, it is the MFD rather than the core diameter that is an important parameter. The MFD essentially specifies the transverse extent of the fundamental modal field. For typical single-mode fibers, the modal field extends far into the cladding and, thus, MFD can be very different from core diameter. MFD can be used to estimate joint losses between two single-mode fibers, coupling efficiency, cutoff wavelength, backscattering characteristics, microbending and macrobending losses, and even waveguide dispersion.

There are various definitions for MFD. In the following we define three commonly used definitions.

(1) **Near-field rms MFD:** If $\psi^2(r)$ is the transverse modal intensity profile, then the near-field rms MFD is defined by

$$d_N = 2\left[\frac{2\int_0^\infty r^2\psi^2(r)\,r\,dr}{\int_0^\infty \psi^2(r)\,dr}\right]^{1/2} \tag{20.11}$$

The near-field spot size is defined as

$$w_r = \frac{1}{2}d_N \tag{20.12}$$

Joint losses for small angular tilts are proportional to d_N^2 and microbending loss is proportional to d_N^6 [see, e.g., Jeunhomme (1983)].

(2) **Petermann-2 spot size:** The Petermann-2 spot size is defined by the following equation

$$w_P = \left[\frac{2 \int_0^\infty \psi^2(r)\, r\, dr}{\int_0^\infty \left(\frac{d\psi}{dr}\right)^2 r\, dr} \right]^{1/2} \tag{20.13}$$

Losses across a joint for small transverse offset are proportional to λ/w_P^2 (see Problem 8.9) and the waveguide dispersion is proportional to $d/d\lambda(\lambda/w_P^2)$ (see Problem 8.12). The far-field rms MFD [see equation (8.105)] is given by

$$d_F = 2w_P \tag{20.14}$$

The above relation has been derived in Section 8.5.4.

20.5 Gaussian mode field diameter

In Chapter 8 we discussed the Gaussian approximation for the fundamental mode of single-mode fibers. We recall that for conventional fibers operating at 1300 nm, the fundamental mode can be accurately represented by a Gaussian function of the form

$$\psi_G(r) = A\, e^{-r^2/w_0^2} \tag{20.15}$$

where w_0^2 is usually known as the spot size. Using this Gaussian modal field, one can calculate the various splice losses, launching efficiency, and so forth with considerable ease (see Chapter 8). It is obvious that one may use a variety of methods for the determination of the Gaussian spot size, w_0 for a modal field having a near Gaussian field distribution. For example, one may assume the MFD to be the diameter of the mode at $1/e^2$ intensity points, which is given by

$$d_G = 2w_0 \tag{20.16}$$

However, this usually leads to a poor approximation. A widely used criterion is to choose the value of w_0 so that the Gaussian beam leads to the maximum launching efficiency of the actual fundamental mode. For example, for a step index fiber lying in the range $0.8 < V < 2.5$, the following empirical expression gives a value of w_0 that has an accuracy better than 1% [Marcuse (1977), Chapter 8]

$$\frac{w_0}{a} \approx 0.65 + \frac{1.619}{V^{3/2}} + \frac{2.879}{V^6} \tag{20.17}$$

where a is the core radius. Similar empirical relations for a graded index fiber are given in Marcuse (1978).

For CSF profiles such as single clad step index or graded index, the three spot sizes – namely, near-field, far-field rms, and Gaussian spot sizes are nearly equal, particularly near the cutoff wavelength; this is because the Gaussian approximation works well for such fibers. On the other hand, for specialized

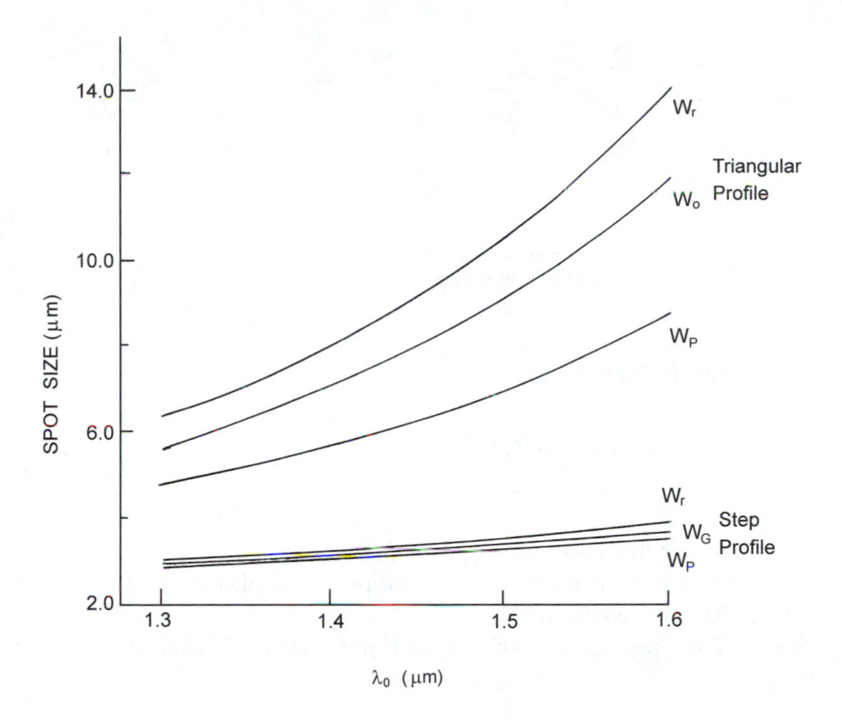

Fig. 20.5: Wavelength variation of the three spot sizes – namely, the near-field rms spot size w_r, the far-field rms spot size w_F, and the Gaussian spot size w_0 for a step index fiber and a triangular profile fiber corresponding to $\Delta = 0.75\%$ and $a = 2.3$ μm. Note that for a step index fiber the three spot sizes are very nearly equal, whereas for the triangular core fiber they are substantially different. This is due to the almost Gaussian nature of the field for a step index fiber. [After Ghatak and Sharma (1986).]

profiles such as the triangular core or multiple clad, the two spot sizes, w_r and w_P, may differ substantially and it is not appropriate to approximate them by w_0 [Das, Goyal, and Srivastava (1987)]. This point is illustrated in Figure 20.5, where we have plotted the variation of w_r, w_P, and w_0 as a function of λ_0 for a step index fiber and for a triangular profile fiber.

The two spot sizes, w_P and w_r, give a good estimate of some of the important characteristics of single-mode fibers and, hence, play a very important role in the design and characterization of single-mode fibers.

There are various techniques for the measurement of MFD. These include

(a) Near-field technique
(b) Far-field technique
(c) Transverse offset technique

20.5.1 Near-field technique

Figure 20.6 shows the experimental setup in which a laser beam is launched into the test fiber and the near field of the fiber is magnified by using a microscope objective on a scanning photodetector or a vidicon. Cladding mode strippers are used at either ends. The measured intensity pattern is

$$I(r) \propto \psi^2(r) \tag{20.18}$$

For near Gaussian fields one does a curve fitting to a Gaussian

$$G(r) = G_0\, e^{-2r^2/w_0^2} \tag{20.19}$$

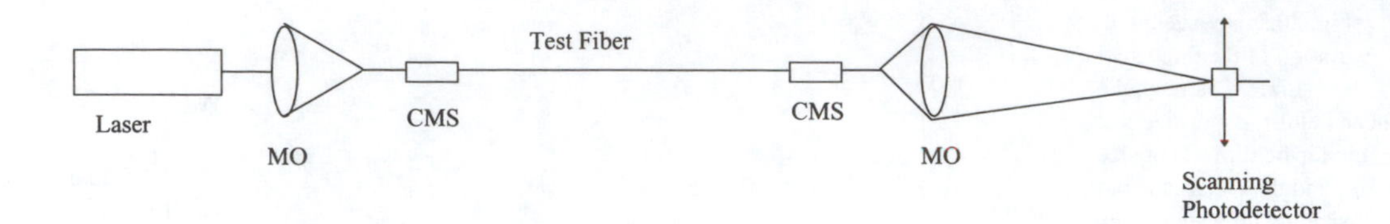

Fig. 20.6: A typical experimental setup for the measurement of the near-field intensity pattern of a single-mode fiber.

that is, maximize the overlap

$$\Gamma = \frac{\left[\int_0^\infty \sqrt{I(r)\,G(r)}\,r\,dr\right]^2}{\int_0^\infty I(r)\,dr\,\int_0^\infty G(r)\,r\,dr} \tag{20.20}$$

If w_0 is the value that maximizes Γ, then $d_N = 2w_0$.

For non-Gaussian modal fields one can directly calculate d_N or d_F by using equation (20.11) or equation (20.13).

Some of the problems associated with measuring d_N using this method are

(a) Difficulty in determining exact position of image plane
(b) The quality of the fiber end face, which must be good

20.5.2 *Far-field technique*

Figure 20.7 shows a typical experimental arrangement for measurement of d_F by using the far-field technique. Here the far-field intensity profile is measured with a photodetector. For Gaussian near field, the far-field intensity pattern may be written as (see Problem 20.2)

$$|\Psi(\theta)|^2 = |\Psi(0)|^2\, e^{-2\sin^2\theta/\sin^2\theta_f} \tag{20.21}$$

where

$$\sin\theta_f = \frac{\lambda}{\pi w_0} \tag{20.22}$$

For small θ and θ_f, equation (20.21) may be written as

$$|\Psi(\theta)|^2 \simeq |\Psi(0)|^2\, e^{-2\theta^2/\theta_f^2} \tag{20.23}$$

Thus, the measured far field can be fitted to equation (20.23) to obtain the optimum value of θ_f and using equation (20.22) we have

$$d_f = 2w_0 = \frac{2\lambda}{\pi\,\sin\theta_f} \tag{20.24}$$

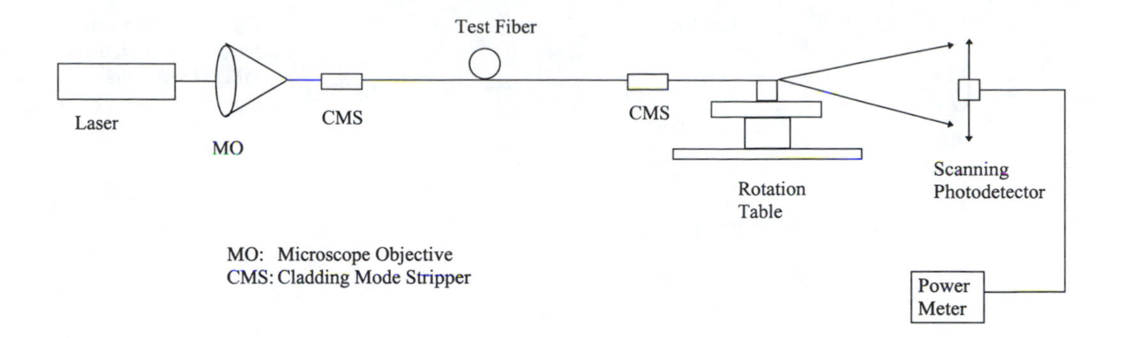

Fig. 20.7: Typical experimental setup for measurement of the far-field intensity pattern of a single-mode fiber.

From the measured far-field data, one can also obtain d_F by directly using equation (8.105).

20.5.3 Transverse offset technique

Figure 20.8 shows the transverse offset technique for the measurement of the MFD. In this, one measures the variation of power through a transverse offset between two pieces of the test fiber. Now, for a transverse misalignment of d the loss is given by (see equation (8.114))

$$\alpha \, \mathrm{dB} \approx 4.343\left(\frac{d}{w_P}\right)^2 \tag{20.25}$$

for $d \ll w_P$. Thus, by fitting a parabola to α (as a function of d) one can obtain w_P.

In these measurements one must note the following:

(a) The separation between the fiber ends should be less than about 5 μm.
(b) Overfilled launch conditions should be used to reduce sensitivity to input fiber position.
(c) Good cladding mode strippers must be used.
(d) For measurements close to the cutoff of LP_{11} mode, the LP_{11} mode should be filtered off by using a bend on either side of the joint.

This method can be used with ease for the measurement of variation of MFD with wavelength by using a white light source and a monochromator/interference filter at the launch end. Figure 20.9 shows a typical variation of measured spot size with wavelength.

We may mention here that for power law profiles – such as step index, parabolic index, and triangular core index – β has been found to be accurately described by the following empirical relation [Tewari, Thyagarajan, and Ghatak (1986)].

$$\beta^2 = \frac{4\pi^2}{\lambda_0^2}\left[n_2^2 + \delta\, e^{-\alpha\lambda_0^h}\right] \tag{20.26}$$

where α, $\delta[= (n_1^2 - n_2^2)]$, and h are fitting parameters. Using equations (20.26)

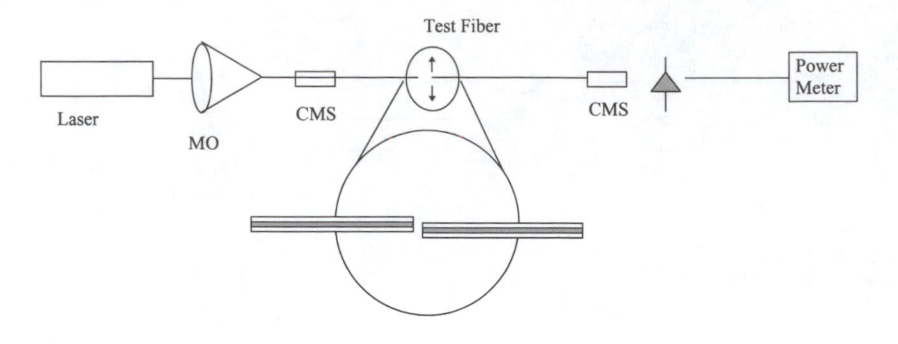

Fig. 20.8: Experimental setup for measurement of MFD by using the transverse offset technique.

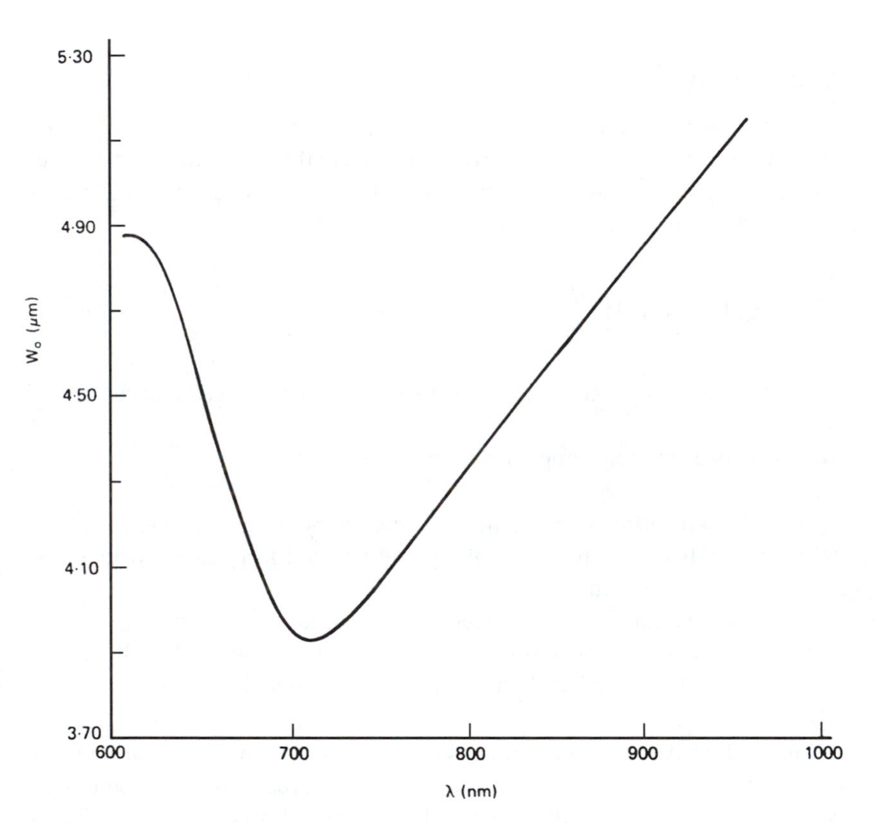

Fig. 20.9: A typical measured variation of mode field diameter with wavelength using the transverse offset technique. For this fiber, the cutoff wavelength corresponding to the LP_{11} mode is around 705 nm. [After Mahadevan (1985).]

and (8.131), we obtain

$$w_P^2 = \frac{e^{\alpha\lambda_0^h}}{\pi^2 \alpha h \lambda_0^{h-2} \delta} \tag{20.27}$$

Thus, if one measures the wavelength dependence of w_P and fits the experimental data using equation (20.27), one can obtain the parameters α, δ, and h, which on substitution in equation (20.26) gives the λ_0 variation of β in

Table 20.2. *Values of various parameters appearing in equation (20.26)*

Fiber		$q = \infty$		$q = 2$		$q = 1$	
		#1($a = 5.0$)	#2($a = 1.5$)	#3($a = 5.0$)	#4($a = 1.5$)	#5($a = 5.0$)	#6($a = 1.5$)
δ	Using (20.27)	0.01428	0.03716	0.01828	0.05559	0.02204	0.06069
	Exact	0.01313	0.04447	0.01973	0.07634	0.02723	0.10981
α	Using (20.27)	0.30421	0.72217	0.45288	0.92709	0.55582	0.90097
h	Using (20.27)	1.66157	2.08060	1.58386	1.95202	1.43889	1.97728
β (in μm^{-1}) at	Using (20.26)	6.04328	6.04193	6.04309	6.04230	6.04368	6.04221
$\lambda_0 = 1.5\,\mu$m	Exact	6.04197	6.04197	6.04197	6.04197	6.04197	6.04197

Note: The quantities a and λ_0 are measured in μm and β is measured in μm^{-1}; the value of α is determined accordingly. [Adapted from Tewari et al. (1986).]

the wavelength range of interest. The accuracy of this procedure is illustrated in Table 20.2, where exact values of β have been compared with the corresponding value obtained by fitting equation (20.27) in the exact computed λ_0 dependence of w_P^2. The two values of β agree very well. An important application of the λ_0 dependence of β obtained above is in the calculation of the crossover wavelength λ_x in a wavelength filter consisting of a pair of nonidentical single-mode fibers [Marcuse (1985)]. The crossover wavelength λ_x is the wavelength at which the propagation constants of the two fibers are identical. One can easily plot β as a function of λ_0 for both fibers and the point of intersection will correspond to λ_x. For example, using the fitted values of the parameters for fiber 1 and fiber 4 (see Table 20.2), the calculated value of λ_x comes out to be 1.46 μm, which is within 3% of the exact numerical value of 1.50 μm.

20.6 Birefringence measurements

A perfectly circular core single-mode fiber actually supports two modes that have almost orthogonal linear polarization states and have the same propagation constant. In practical single-mode fibers the core is nominally circular and with the presence of stress, twist, bend, and so forth, such fibers are slightly birefringent. This small difference in propagation constant leads to a coupling of power between the orthogonal polarization states even under the smallest external perturbation. Thus, for such fibers, input linearly polarized light becomes elliptically polarized over short distances and, in addition, the output state of polarization changes with changes in external perturbations. The random change in the state of polarization creates problems when the single-mode fiber is to be used in coherent communications, in fiber optic phase sensors, or in applications in which the fiber has to be coupled to polarization-sensitive devices such as integrated optic devices.

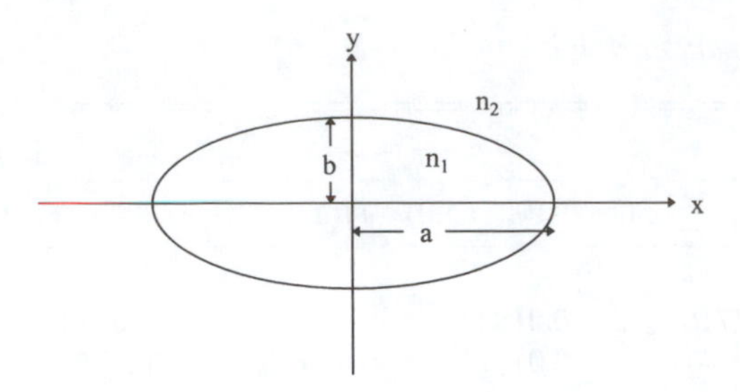

Fig. 20.10: Cross-section of an elliptic core fiber that is birefringent – the modes of the fiber are almost x-polarized and y-polarized and their propagation constants are different.

To overcome these problems, special fibers such as elliptic core fibers and stress-induced birefringent fibers have been developed. Not only do these fibers have applications in areas mentioned earlier, but very interesting optics experiments can be done with them; these experiments also enable us to determine the birefringence in the fiber.

For an elliptic core fiber (see Figure 20.10) the "modes" of the fiber are (approximately) x- and y-polarized – that is, if an x-polarized beam is incident it will propagate without any change in the SOP with a certain phase velocity ω/β_x. Similarly, a y-polarized beam will propagate as a y-polarized beam with velocity ω/β_y. The birefringence in a fiber is a measure of the difference in the effective indices of the two orthogonally polarized modes and is defined by

$$B = \delta n_e = \delta\left(\frac{\beta}{k_0}\right) = \frac{\lambda_0}{2\pi}\delta\beta \tag{20.28}$$

Let us consider a circularly polarized beam to be incident on the input face of the fiber at $z = 0$; then we must resolve the incident beam into x- and y-polarized beams propagating with slightly different phase velocities. Thus

$$\mathbf{E}(\mathbf{r}, z) = \psi(x, y)[\hat{\mathbf{x}}\cos(\omega t - \beta_x z) + \hat{\mathbf{y}}\sin(\omega t - \beta_y z)] \tag{20.29}$$

where $\psi(x, y)$ is the transverse field distribution of the fundamental mode. (It may be readily seen that if $\beta_x = \beta_y$, as is indeed true for circular core fibers, the beam will remain circularly polarized for all values of z.) Now, at $z = 0$

$$\begin{aligned} E_x &= \psi(x, y)\cos\omega t \\ E_y &= \psi(x, y)\sin\omega t \end{aligned} \tag{20.30}$$

which represents a right circularly polarized wave (see Figure 20.11). For

$$z = z_1 = \frac{\pi}{2(\beta_y - \beta_x)} \tag{20.31}$$

that is, for $\beta_y z_1 = \beta_x z_1 + \pi/2$

$$\begin{aligned} E_x &= \psi(x, y)\cos(\omega t - \phi_1) \\ E_y &= \psi(x, y)\sin(\omega t - \phi_1 - \pi/2) = -\psi(x, y)\cos(\omega t - \phi_1) \end{aligned} \tag{20.32}$$

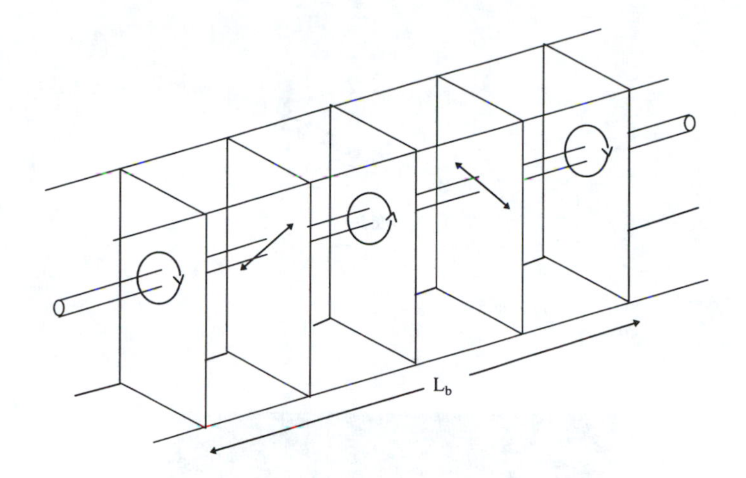

Fig. 20.11: A circularly polarized input periodically changes its state of polarization as it propagates along a birefringent fiber. The beat length L_b corresponds to the separation between two adjacent points along the fiber where the state of polarization repeats itself.

where $\phi_1 = \beta_x z_1$. The above equations represent a linearly polarized wave (see Figure 20.11); we assume the direction of the **E**–vector to be along the y' axis. Similarly, at

$$z = z_2 = \frac{\pi}{(\beta_y - \beta_x)} = 2\,z_1 \tag{20.33}$$

$$E_x = \psi(x, y)\,\cos(\omega t - \phi_2)$$

$$E_y = \psi(x, y)\,\sin(\omega t - \phi_2 - \pi) = -\psi(x, y)\,\sin(\omega t - \phi_2) \tag{20.34}$$

where

$$\phi_2 = \beta_x z_2$$

and the wave will be left circularly polarized (see Figure 20.11). At

$$z = z_3 = \frac{3\pi}{2(\beta_y - \beta_x)} = 3\,z_1 \tag{20.35}$$

we will have

$$E_x = \psi(x, y)\,\cos(\omega t - \phi_3)$$

$$E_y = \psi(x, y)\,\sin\left(\omega t - \phi_3 - \frac{3\pi}{2}\right) = \psi(x, y)\,\cos(\omega t - \phi_3)$$

where

$$\phi_3 = \beta_x z_3$$

Thus, the wave would again be linearly polarized, but now the direction of the oscillating electric field will be at right angles to the field at $z = z_1$. In a similar manner, we can easily continue to determine the SOP of the propagating beam.

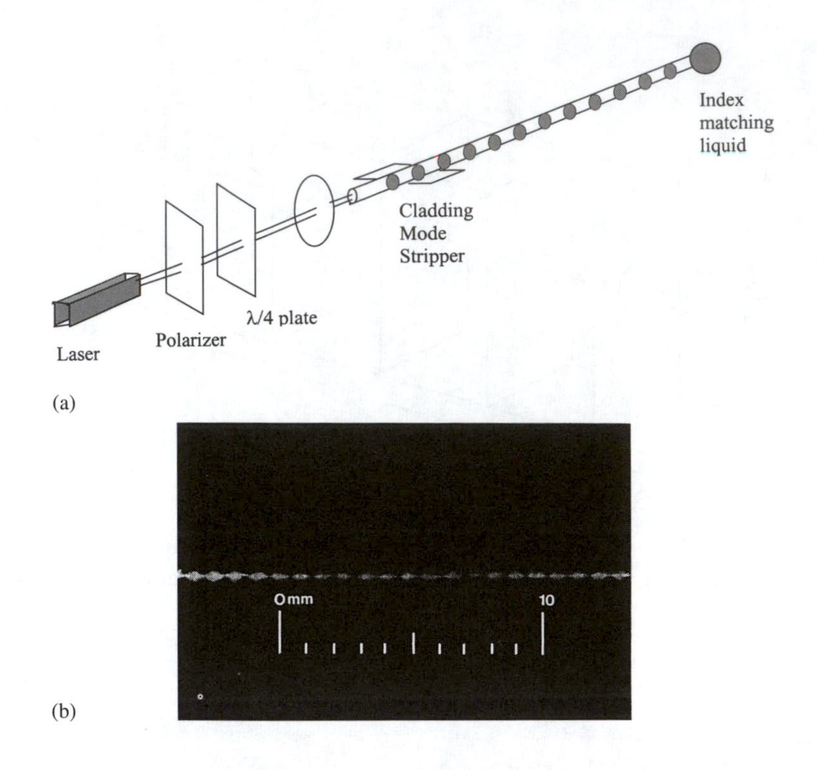

Fig. 20.12: (a) Experimental arrangement for measurement of beat length of a high birefringence fiber using the Rayleigh scattering technique. (b) Photograph showing the transverse beat pattern observed on the fiber. [Adapted from a photograph by Andrew Corporation, Orland Park, IL, USA; the authors came across this photograph in Jeunhomme (1983).]

Thus, at $z = 5z_1, 9z_1, 13z_1, \ldots$ the SOP will be the same as at $z = z_1$ and at $z = 7z_1, 11z_1, 15z_1, \ldots$ the SOP will be the same as at $z = 3z_1$. Similarly, at $z = 4z_1, 8z_1, 12z_1, \ldots$ the beam will be RCP and at $z = 2z_1, 6z_1, 10z_1, \ldots$ the beam will be left circularly polarized (LCP).

We have seen above that a given SOP repeats itself after a distance of $4z_1 = 2\pi/\delta\beta$; this length is referred to as the beat length of the fiber.

$$L_b = \frac{2\pi}{\delta\beta} \tag{20.36}$$

The beat length is a measure of the birefringence of the fiber; the smaller the beat length the larger is the corresponding birefringence.

The birefringence property of the fiber can be measured either by measuring B directly or by measuring the beat length L_b. We now discuss some techniques for measurement of L_b and also for direct measurement of $\delta\beta$.

20.6.1 Rayleigh scattering technique

Rayleigh scattering is a fundamental mechanism by which light is scattered out of the fiber. It is well known that for a linearly polarized light undergoing scattering, there is no scattered wave along the axis of the dipole – that is, along the direction of polarization – and the scattering is maximum in a plane perpendicular to the direction of polarization.

To use the above principle in the measurement of beat length, one uses an experimental setup as shown in Figure 20.12. Light from a linearly polarized laser (or an unpolarized laser with a polarizer in front) is first passed through

Table 20.3. *Comparison of beat lengths of an elliptic core fiber from Andrew Corporation (United States) using the Rayleigh scattering technique and the prism coupling technique. [After Thyagarajan, Shenoy and Ramadas (1986)]*

Technique	Beat length (mm)
Prism coupling	1.50 ± 0.01
Rayleigh scattering	1.52 ± 0.08

a $\lambda/4$ plate to convert it into a circularly polarized beam and is subsequently focused on the entrance face of the test fiber. Cladding modes are removed with a cladding mode stripper, and the fiber is kept almost straight with the other end dipped into an index matching fluid to avoid any reflections from the fiber end. Now, as discussed earlier, as the beam propagates through the fiber, its state of polarization oscillates between a circular and a linear polarization state (see Figure 20.11). If one observes the fiber transversely along a direction making an angle of 45° with the fiber eigen axis, then, as discussed earlier, whenever the state of polarization is linear and directed along the direction of observation, the intensity of the scattered light will be very small. Half–a–beat length later, the state of polarization again will be linear with its direction at right angles to the direction of observation. At this point, one would receive a maximum of scattered light. Thus, observing the fiber, one would see alternate bright and dark regions (see Figure 20.12) and the distance between two adjacent bright or dark bands would just be the beat length. Table 20.3 presents results obtained by using this technique on an elliptic core fiber measured with a He–Ne laser at 0.6328 μm.

As a numerical example, we consider an elliptic core fiber for which

$$2a = 2.14\ \mu\text{m}, \quad 2b = 0.85\ \mu\text{m}$$
$$n_1 = 1.535, \quad n_2 = 1.47$$

For such a fiber operating at $\lambda_0 = 0.6328\ \mu$m ($k_0 \simeq 9.929 \times 10^4\ \text{cm}^{-1}$),

$$\frac{\beta_x}{k_0} \simeq 1.506845 \quad \text{and} \quad \frac{\beta_y}{k_0} \simeq 1.507716$$

The quantity

$$L_b = \frac{2\pi}{\Delta\beta} = \frac{2\pi}{\beta_y - \beta_x} \simeq 0.727\ \text{mm}$$

is known as the beat length. Obviously, if we measure L_b we can calculate $\Delta\beta$.

The Rayleigh scattering technique has also been used to measure twist-induced birefringence in single-mode fibers [Graindorge et al. (1982)].

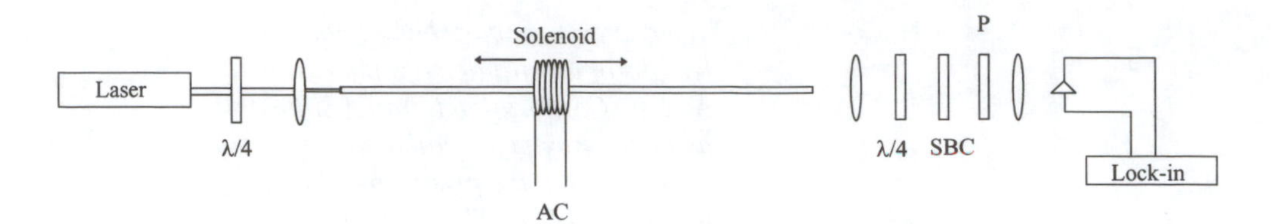

Fig. 20.13: Experimental setup for measurement of beat length of birefringent fibers using the magneto optic modulation techique. The solenoid is moved along the fiber length and the modulated signal is detected by a photodetector.

Although the Rayleigh scattering technique is a direct and accurate method to measure L_b and hence $\Delta\beta$, it has some limitations:

(a) The method is best suited to highly birefringent fibers with a beat length of only a few millimeters.
(b) Since Rayleigh scattering decreases as λ^{-4}, the method is convenient to use only in the visible region.

20.6.2 *Magneto optic modulation*

The magneto optic modulation technique is based on the Faraday effect, in which an application of a longitudinal magnetic field on a material induces circular birefringence in the medium. Consider again a circularly polarized input beam propagating through a birefringent fiber and periodically changing its state of polarization. If a small electromagnet providing a magnetic field along the length of the fiber is placed around the fiber, then the applied magnetic field will induce a local circular birefringence at the point of application of the magnetic field. The effect of this induced circular birefringence depends on the state of polarization of the light beam at that point. Thus, if the polarization is circular, there is no effect and maximum effect will be felt when the polarization is linear. Hence, if the electromagnet is scanned along the length of the fiber, and the modulation in the state of polarization at the output is detected, then one will observe a periodic variation in the depth of modulation. The distance between two consecutive points of maximum or minimum modulation will be half the beat length.

The experimental arrangement is shown in Figure 20.13. Light from a laser is first passed through a polarizer, $\lambda/4$ plate combination to achieve a circular polarization and is coupled into the optical fiber. A microscope objective collimates the light coming out of the fiber, which is then passed through a $\lambda/4$ plate followed by a Soleil Babinet compensator (SBC) and a linear analyzer. The $\lambda/4$ plate is adjusted so that in the absence of any magnetic field, the light coming out of the fiber is made linear. The SBC is then adjusted so that its slow axis is aligned with the linear state of polarization emerging from the $\lambda/4$ plate, and then the linear analyzer is aligned with its pass axis at 45° to the axes of the the SBC. Such an arrangement results in an analyzer that is most sensitive to the changes in the state of polarization. The phase retardation of

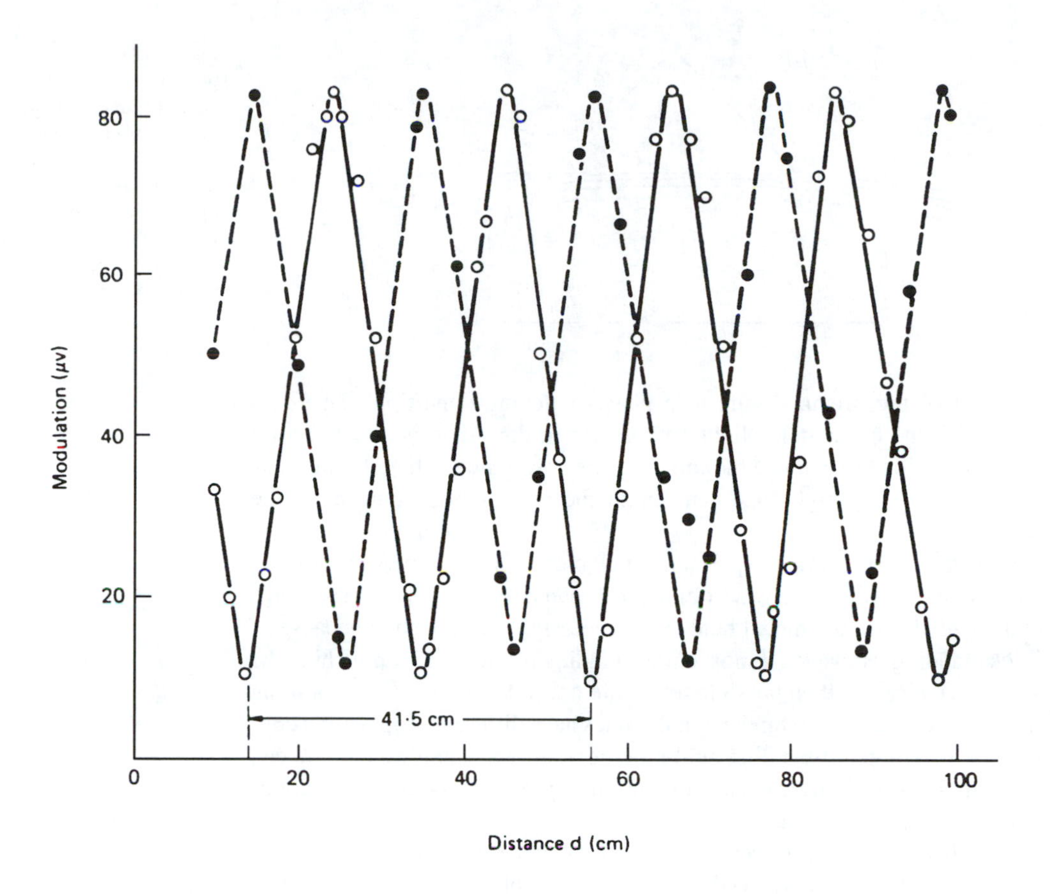

Fig. 20.14: Output signal as a function of the position of the solenoid along the fiber. The solid curve corresponds to an input linear SOP and the dashed line corresponds to an input circular SOP. The beat length of the fiber is 41.5 cm. [After Bhat (1984).]

the SBC is adjusted for every position of the solenoid to maximize the modulation signal. An alternating current across the electromagnet leads to an output signal varying periodically with time, which can be detected with a lock-in amplifier.

Figure 20.14 shows a typical variation of the amplitude of modulation signal as a function of the position of the solenoid for linearly and circularly polarized input. One does indeed observe a sinusoidal modulation and also a phase shift of half a period between linear and circular inputs (why?). For the fiber used, the beat length (which is twice the period) is 41.5 cm.

20.6.3 Prism-coupling technique

In the methods based on Rayleigh scattering and magneto optic modulation one estimates the modal birefringence by measuring the beat length. We now discuss another technique in which one directly measures the birefringence rather than the beat length. This method is the same as the conventional prism–film coupling technique used in integrated optics to couple into and out of optical waveguides. In this technique a prism whose refractive index is greater than the film is placed close to the waveguiding film. Since the waveguide becomes leaky in the presence of the prism, the light propagating in the waveguide

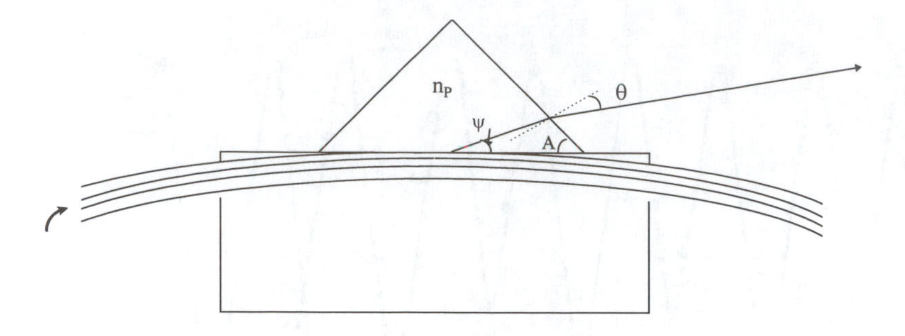

Fig. 20.15: A high-index prism is placed in close contact with the laterally polished fiber half block (see Section 17.3.1), and light coupled into the optical fiber couples out through the prism at specific angles corresponding to the phase-matching condition.

couples out of the prism. Owing to a phase-matching condition, the direction at which the out-coupled light emerges from the prism is characteristic of the propagation constant of the mode of the waveguide. Thus, by measuring the out-coupling angle, one can obtain the propagation constant of the mode.

For this method to work, the prism must be placed close to the waveguiding region so that the evanescent wave of the mode can interact with the prism. In a normal cladded fiber, the modal field is not accessible from outside because of the thick cladding. However, if most of the cladding is removed by polishing the fiber transversely, it is then possible to couple out light by the prism-coupling technique. Because the two fundamental modes have different propagation constants in a birefringent fiber, they will emerge at different angles and one can then obtain the modal birefringence by measuring the propagation constant of each individual polarization.

Figure 20.15 shows a transverse section of the experimental arrangement in which a high-index prism is placed on top of a side-polished single-mode fiber. The phase matching condition is

$$\beta = k_0 n_p \cos \psi \tag{20.37}$$

where n_p is the refractive index of the prism, β is the propagation constant of the mode, and ψ is shown in Figure 20.15. If A represents the angle of the prism, then the effective index can be calculated through the relation

$$\tilde{\beta} = \frac{\beta}{k_0} = n_p \sin\left[A + \sin^{-1}\left(\frac{\sin\theta}{n_p}\right)\right] \tag{20.38}$$

Thus, by measuring θ, and knowing A and n_p, one can obtain β.

Figure 20.16 shows the output from an elliptic core fiber polished with its major axis perpendicular to the polished surface. The two lines are orthogonally polarized and are separated by an angle $\sim 4'$. Table 20.3 shows a comparison of the results obtained from the prism-coupling method and the Rayleigh scattering method.

The above method is suitable only for high-birefringence fibers (with a beat length $<$ a few millimeters). Since polished fibers are used in the fabrication of fiber directional couplers, the above technique can indeed be used for a local measurement of birefringence of the polished fiber. This is important

(a)

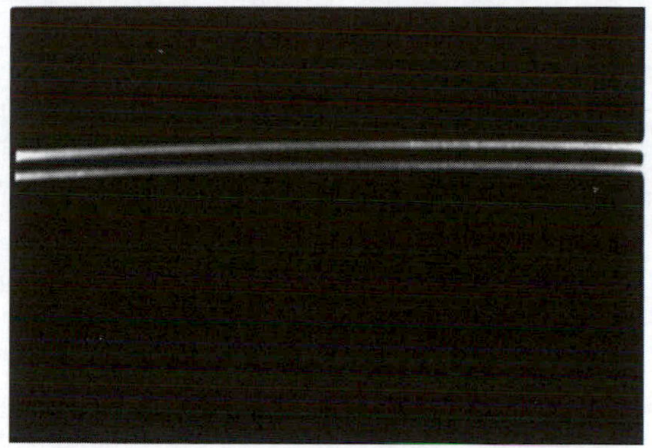

(b)

Fig. 20.16: Photograph showing the output coupled m lines through a prism from an elliptic core fiber from Andrew Corporation (United States) at a distance of (a) ~15 cm and (b) ~2 m from the polished half block. The two lines are separated by an angle of ~4′ of arc. [After Thyagarajan, Shenoy, and Ramadas (1986).]

because the polishing operation and the removal of cladding do change the birefringence.

Problems

20.1 For a step index fiber, if the cutoff wavelength of LP_{11} mode is 576 nm, estimate what should be the approximate cutoff wavelength of the next higher order LP_{02} mode. Compare with the corresponding value obtained from Figure 20.2.

Solution: The cutoff V values of LP_{11} and LP_{02} modes are 2.405 and 3.832, respectively. If we neglect variation of NA with λ, then the ratio of cutoff wavelengths of LP_{02} and LP_{11} modes should be

$$\frac{\lambda_{c02}}{\lambda_{c11}} = \frac{V_{c11}}{V_{c02}} = \frac{2.405}{3.832} \simeq 0.63$$

Thus, if λ_{c11} is 576 nm, λ_{c02} should be approximately 362 nm. From Figure 20.2 we get the cutoff wavelength of the LP_{02} mode to be 376 nm, which is not very far from the predicted value.

20.2 Show that for $\psi(r)$ given by equation (20.15) $d_N = d_F = 2w_0$.

Solution: for $\psi(r)$ given by equation (20.15),

$$d_N = 2\left[\frac{2\int_0^\infty r^3 e^{-2r^2/w_0^2} dr}{\int_0^\infty e^{-2r^2/w_0^2} r\, dr}\right]^{1/2}$$

$$= 2\left[2\frac{w_0^4/8}{w_0^2/4}\right]^{1/2} = 2w_0 \tag{20.39}$$

To calculate d_F we first calculate the far field corresponding to $\psi(r)$ given by equation (20.15). A Gaussian field distribution remains Gaussian as it propagates, and, for a near-field pattern given by equation (20.15), the far field is given by (see Chapter 2)

$$|\Psi(\theta)|^2 = |\Psi(0)|^2 e^{-2\sin^2\theta/\sin^2\theta_f} \tag{20.40}$$

where

$$\sin\theta_f = \frac{\lambda}{\pi w_0} \tag{20.41}$$

writing

$$q = \frac{2\pi}{\lambda}\sin\theta \tag{20.42}$$

we get

$$|\Psi(q)|^2 = |\Psi(0)|^2 e^{-q^2 w_0^2/2} \tag{20.43}$$

Substituting in equation (20.13) for $\Psi^2(q)$ from equation (20.43) and integrating, we obtain

$$d_F = 2w_0 \tag{20.44}$$

20.3 Using the expression for the modal field for a step index fiber, show that [Neumann (1988), pages 225, 226]

$$d_N = \frac{4a}{\sqrt{6}}\left[\frac{J_0(U)}{U\,J_1(U)} - \frac{1}{U^2} + \frac{1}{W^2} + \frac{1}{2}\right]^{1/2} \tag{20.45}$$

$$d_F = 2\sqrt{2}\frac{a J_1(U)}{W J_0(U)} \tag{20.46}$$

20.4 For a single-mode fiber operating at 1300 nm, the MFD is approximately 10 μm. For far-field measurements, at what distance from the fiber tip should the detector be placed?

Solution: For far field, we must have

$$z \gg \frac{d^2}{\lambda} \simeq 77\,\mu\text{m}$$

The detector is usually placed at a distance greater than 20 mm from the fiber tip.

20.5 If the expected MFD is d, what pinhole diameter in front of the detector would you use in the far-field measurement setup?

Solution: For a good resolution one has to use as small a pinhole size as possible, but that then reduces the power detected by the detector and, hence, a compromise has to be made. Thus, if one arbitrarily chooses ten resolvable points in the far field, if h is the pinhole diameter and z is the distance between pinhole and the fiber tip, then we may write

$$\frac{h}{z} \lesssim \frac{\lambda}{10d} \quad \text{or} \quad h \lesssim z\frac{\lambda}{10d}$$

Hence, if $d = 10\ \mu\text{m}$, $z = 50$ mm, $\lambda = 1.3\ \mu\text{m}$, then

$$h \lesssim 0.65\ \text{mm}$$

20.6 In Figure 20.10 explain why the measured MFD first reduces and then starts to increase at some wavelength when the wavelength is continuously decreased.

20.7 (a) In the Rayleigh scattering method, if one observes along an eigen polarization direction, would one see any beats? Explain. (b) If the incident beam is linearly polarized along an eigen polarization direction, can one see beats along any direction of observation? Explain. (c) What will happen if, instead of an input circular polarization, one chooses an elliptic polarization state?

20.8 Can the magneto optic method discussed in Section 20.5.2 be used to measure circular birefringence?

20.9 Consider a polished conventional single-mode fiber with a prism placed on top. (a) What will happen to the output coupled light as the wavelength is reduced? (b) Can this be used to measure the cutoff wavelength of the fiber?

21

Periodic interactions in waveguides

21.1 Introduction

Periodic waveguides in which either the thickness, refractive index, or any other waveguide characteristic varies periodically along the propagation direction are used in many fiber optic (see Chapter 17) and integrated optic devices. These include mode converters, polarization transformers, wavelength filters, input–output couplers, frequency shifters, second harmonic generators, etc. Periodic waveguides also form basic components in distributed feedback and distributed Bragg reflector lasers (see Section 11.5.2). Figure 21.1 shows two common types of periodic waveguides with a periodic modulation of either the refractive index or the thickness of the waveguide. The periodic refractive index modulation can be inbuilt in the waveguide (such as produced by UV light irradiation in an optical fiber; see Chapter 17) or can be externally induced by using electrooptic or acoustooptic effects.

One of the chief characteristics of periodic coupling is that the periodic perturbation couples power mainly among two modes that satisfy a quasi-phase-matching condition. According to this, if β_1 and β_2 are the propagation constants of two modes, then a periodic perturbation with period Λ will induce coupling between these modes if

$$\beta_1 - \beta_2 \simeq \pm K \tag{21.1}$$

where

$$K = \frac{2\pi}{\Lambda} \tag{21.2}$$

Because of the quasi-phase-matching condition, periodic coupling is very selective and this leads to its various applications in guided-wave optics. Figures 21.2(a) and 21.2(b) show the transverse intensity distributions for the first two TE modes of an unperturbed planar waveguide as they propagate through the waveguide. For example, the propagation of the fundamental and the first higher order TE mode is given by

$$\Psi_0(x, z) = \psi_0(x)\, e^{i(\omega t - \beta_0 z)} \tag{21.3}$$

$$\Psi_1(x, z) = \psi_1(x)\, e^{i(\omega t - \beta_1 z)} \tag{21.4}$$

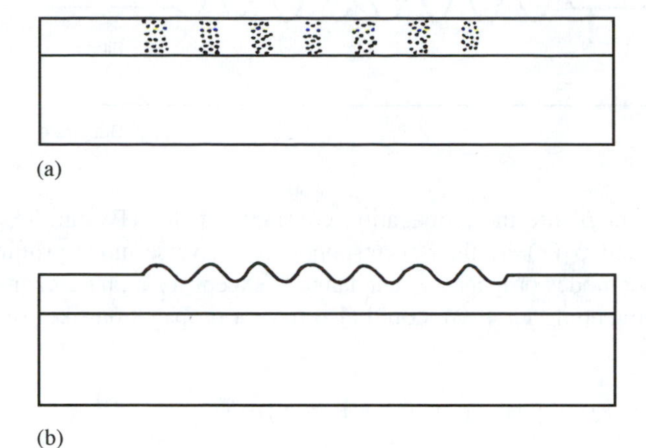

Fig. 21.1: Two common types of periodic waveguides in which (a) the refractive index or (b) the thickness of the waveguide varies periodically along the propagation direction.

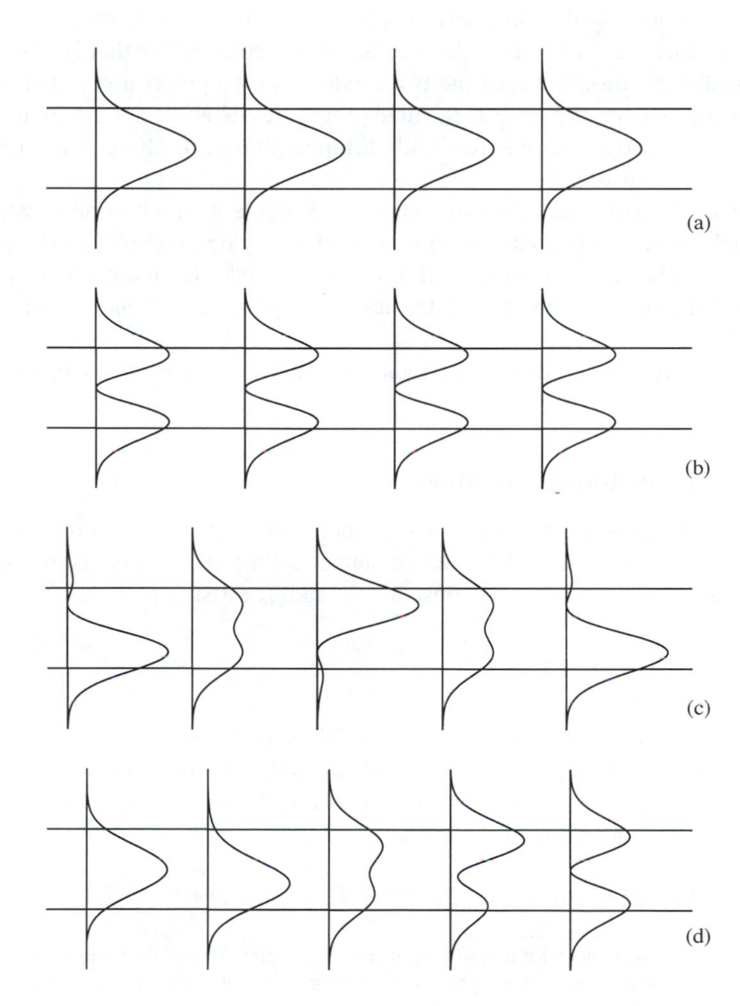

Fig. 21.2: (a) and (b) The transverse intensity pattern at different z values when a TE_0 and a TE_1 mode propagates in an unperturbed waveguide. Note that the transverse intensity pattern does not change as the modes propagate. (c) At the input $z = 0$ of an unperturbed waveguide we have excited a linear combination of TE_0 and TE_1 modes in phase and with equal powers. The figure shows the corresponding intensity patterns at different z values. The changing transverse intensity pattern is due to interference among the modes. (d) Propagation in a periodically perturbed waveguide. At the input ($z = 0$), TE_0 mode is excited. The period Λ of the perturbation is chosen to satisfy a quasi-phase-matching condition and length L is chosen so that complete conversion of power from TE_0 to TE_1 mode takes place.

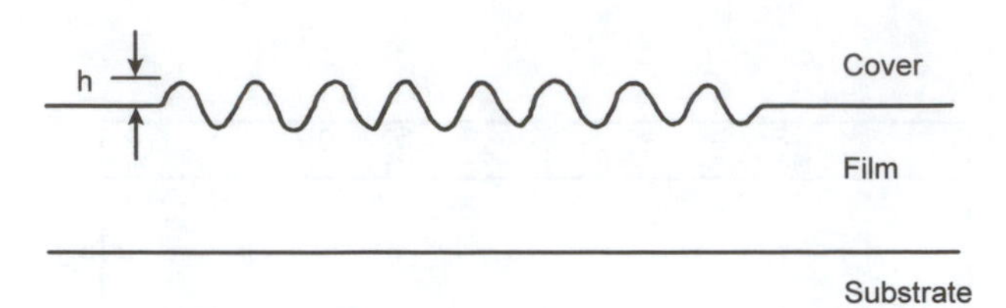

Fig. 21.3: A sinusoidal variation of film thickness in a planar waveguide.

where β_0 and β_1 are the propagation constants of the TE_0 and TE_1 modes and $\psi_0(x)$ and $\psi_1(x)$ are their corresponding transverse mode profiles. Each of the above modes propagates "unchanged" except for a phase change given by the exponential factor. We could also have a propagation like (see Figure 21.2c)

$$\Psi(x, z) = \left[A\psi_0(x)\, e^{-i\beta_0 z} + B\psi_1(x)\, e^{-i\beta_1 z}\right] e^{i\omega t} \tag{21.5}$$

which represents a linear combination of the two modes and A and B are constants; the corresponding intensity distribution for $A = B$ is plotted in Figure 21.2(c). Note that in this case there is no "mode coupling" – that is, there is no transfer of power between the two modes, and the power associated with each mode remains the same as the field propagates. However, at different values of z, the two modes interfere with different phases, resulting in a varying transverse intensity pattern.

Figure 21.2(d) shows the same planar waveguide with a section in which the thickness varies periodically with a period satisfying equation (21.1). Thus, power launched in the fundamental mode couples into the first excited mode, which then exits from the waveguide, thus showing the phenomenon of periodic coupling.

In this chapter we discuss periodic coupling and some of its important applications.[1]

21.2 Coupled-mode equations

As discussed in the previous section, in the presence of a periodic perturbation of the type shown in Figure 21.3, predominant coupling takes place only between those modes with propagation constants β_m and β_n satisfying

$$\beta_m - \beta_n \simeq \pm K = \pm\frac{2\pi}{\Lambda}$$

and coupling to other modes is negligible. Thus, if at $z = 0$, we launch power in the qth mode, then power gets coupled only to those modes for which $\Delta\beta \simeq \beta_m - \beta_q \simeq \pm K$. Assuming that power gets coupled only among two modes, we may write the total field at any z as

$$E(x, z) = A(z)E_1(x, y)\, e^{-i\beta_1 z} + B(z)E_2(x, y)\, e^{-i\beta_2 z} \tag{21.6}$$

[1] It is interesting to note that the problem of periodic coupling is very similar to the quantum mechanical coupling of two energy eigenstates by the application of a time-varying harmonic perturbation. Indeed, the coupled mode theory to be described here is very similar to the time-dependent perturbation theory used in quantum mechanics, and through experimental and theoretical studies on periodic waveguides one can very easily appreciate some of the effects related to harmonic perturbation in quantum mechanics.

where $E_1(x, y)$ and $E_2(x, y)$ are the modal field profiles of the two interacting modes and β_1 and β_2 are their corresponding propagation constants. The coupling between the two modes is described by the following coupled-mode equations (see Appendix F)

$$\frac{dA}{dz} = \kappa B e^{i\Gamma z} \tag{21.7}$$

$$\frac{dB}{dz} = -\kappa A e^{-i\Gamma z} \tag{21.8}$$

where

$$\Gamma = \beta_1 - \beta_2 - K \tag{21.9}$$

is the phase mismatch and κ is known as the coupling coefficient since it is responsible for coupling among the modes of the waveguide.

The value of κ depends on the waveguide parameters, wavelength of operation, and the extent of the periodic perturbation. For example, in a planar waveguide for a sinusoidal periodic perturbation of the type shown in Figure 21.3, the value of κ is given by

$$\kappa \simeq \frac{\pi}{\lambda_0} \frac{h}{\sqrt{d_1 d_2}} \left[\frac{\left(n_f^2 - n_{e1}^2\right)\left(n_f^2 - n_{e2}^2\right)}{n_{e1} n_{e2}} \right]^{1/2} \tag{21.10}$$

where h is the amplitude of the periodic thickness variation

$$d_1 = d + \frac{1}{k_0\sqrt{n_{e1}^2 - n_s^2}} + \frac{1}{k_0\sqrt{n_{e1}^2 - n_c^2}} \tag{21.11}$$

$$d_2 = d + \frac{1}{k_0\sqrt{n_{e2}^2 - n_s^2}} + \frac{1}{k_0\sqrt{n_{e2}^2 - n_c^2}} \tag{21.12}$$

are the effective waveguide thickness for the two modes and $n_{e1} = \beta_1/k_0$, $n_{e2} = \beta_2/k_0$ are the effective indices of the modes 1 and 2 that are coupled; n_f, n_s, and n_c correspond to the film, substrate, and cover refractive indices, respectively, and d is the film thickness.

We now solve the coupled-mode equations and discuss various aspects of the coupling process.

21.2.1 Solution under phase matching

We first consider the case $\Gamma = 0$ – that is, the periodic perturbation has a period

$$\Lambda = \frac{2\pi}{\beta_1 - \beta_2} = \frac{\lambda_0}{(n_{e1} - n_{e2})} \tag{21.13}$$

where $\beta_1 = 2\pi n_{e1}/\lambda_0$, $\beta_2 = 2\pi n_{e2}/\lambda_0$ and n_{e1} and n_{e2} are the effective indices of the modes. In such a case, equations (21.7) and (21.8) become

$$\frac{dA}{dz} = \kappa B \tag{21.14}$$

$$\frac{dB}{dz} = -\kappa A \tag{21.15}$$

Differentiating equation (21.15) with respect to z and using equation (21.14) we obtain

$$\frac{d^2 B}{dz^2} = \kappa^2 B \tag{21.16}$$

whose solution is

$$B(z) = b_1 \cos \kappa z + b_2 \sin \kappa z \tag{21.17}$$

Using equation (21.17) in equation (21.15), we have

$$A(z) = b_1 \sin \kappa z - b_2 \cos \kappa z \tag{21.18}$$

The constants b_1 and b_2 in equations (21.17) and (21.18) are determined through the initial conditions at $z = 0$. If we assume that at $z = 0$, mode E_1 is launched with unit power (see Figure 21.1(d)), then

$$A(z = 0) = 1, \quad B(z = 0) = 0 \tag{21.19}$$

Thus

$$b_1 = 0 \quad \text{and} \quad b_2 = -1$$

and we obtain

$$A(z) = \cos \kappa z \tag{21.20}$$

$$B(z) = -\sin \kappa z \tag{21.21}$$

Thus, the powers carried by modes E_1 and E_2 vary with z as

$$P_1 = |A(z)|^2 = \cos^2 \kappa z \tag{21.22}$$

$$P_2 = |B(z)|^2 = \sin^2 \kappa z \tag{21.23}$$

We notice from equations (21.22) and (21.23) that there is a periodic exchange of power among the two modes and at

$$z = L_c = \frac{\pi}{2\kappa} \tag{21.24}$$

all power from the input mode E_1 is transferred into mode E_2 (see Figure 21.1(d)). Figure 21.4 shows the periodic power exchange given by equations

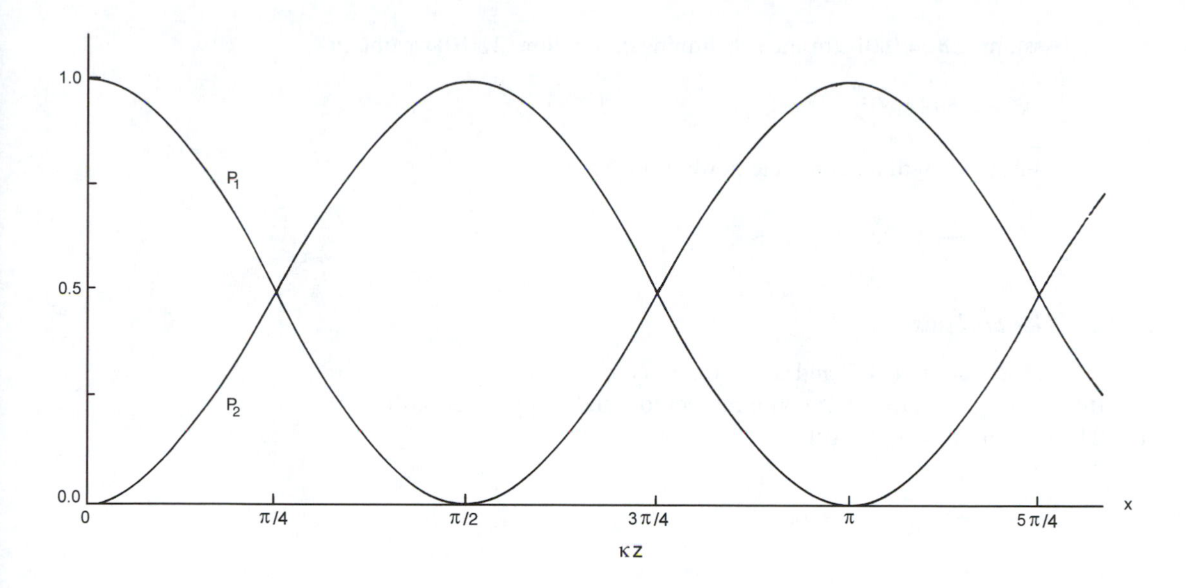

Fig. 21.4: Variation of power in the two modes satisfying the quasi-phase-matching condition (equation 21.1) with z. Note the periodic and complete exchange of power between the modes.

(21.22) and (21.23). We will see in Section 21.2.2 that such complete power conversion is possible only for $\Gamma = 0$ – that is, only if there is phase matching.

Example 21.1: We consider a planar waveguide with

$$n_f = 1.51, \quad n_s = 1.50, \quad n_c = 1.0, \quad d = 4\,\mu\text{m}$$

At $\lambda_0 = 0.6\,\mu$m such a waveguide supports two TE modes whose propagation constants can be determined from the transcendental equation for modes (see Chapter 7). The corresponding effective indices for the first two TE modes are

$$\beta_1/k_0 = n_{e1} = n_e(\text{TE}_0) = 1.50862$$

$$\beta_2/k_0 = n_{e2} = n_e(\text{TE}_1) = 1.5046$$

To efficiently couple power among these two modes, we must choose a periodic perturbation whose periodicity is given by (see equation (21.13))

$$\Lambda = \frac{2\pi}{\Delta\beta} = \frac{\lambda_0}{\Delta n_e}$$

$$= 149.3\,\mu\text{m}$$

where $\Delta n_e = n_{e1} - n_{e2}$. Such a periodic perturbation can be provided by having a periodic thickness variation as shown in Figure 21.1(b). Using equations (21.11) and (21.12) we obtain

$$d_1 = 4.678\,\mu\text{m}$$

$$d_2 = 4.897\,\mu\text{m}$$

Assuming $h = 0.01\,\mu$m and substituting in equation (21.10) we obtain

$$\kappa = 0.598\,\text{cm}^{-1}$$

Thus, the length for complete power transfer is

$$L_c = \frac{\pi}{2\kappa} \simeq 2.63\,\text{cm}$$

21.2.2 General case

We now solve equations (21.7) and (21.8) for the case when $\Gamma = \beta_1 - \beta_2 - K \neq 0$. Differentiating equation (21.8) with respect to z and using equations (21.7) and (21.8) to eliminate A, we get

$$\frac{d^2B}{dz^2} = -\kappa\frac{dA}{dz}e^{-i\Gamma z} + i\kappa\Gamma A e^{-i\Gamma z}$$

$$= -\kappa^2 B - i\Gamma\frac{dB}{dz}$$

or

$$\frac{d^2B}{dz^2} + i\Gamma\frac{dB}{dz} + \kappa^2 B = 0 \tag{21.25}$$

The general solution of equation (21.25) is

$$B(z) = e^{-i\Gamma z/2}[b_1 e^{i\gamma z} + b_2 e^{-i\gamma z}] \tag{21.26}$$

where

$$\gamma^2 = \kappa^2 + \frac{\Gamma^2}{4} \tag{21.27}$$

and b_1 and b_2 are constants. Substituting from equation (21.26) in equation (21.8), we obtain

$$A(z) = \frac{i}{\kappa}e^{i\Gamma z/2}\left[\left(\frac{\Gamma}{2} - \gamma\right)b_1 e^{i\gamma z} + \left(\frac{\Gamma}{2} + \gamma\right)b_2 e^{-i\gamma z}\right] \tag{21.28}$$

The constants b_1 and b_2 are determined by the initial conditions at $z = 0$. As an example, we assume that at $z = 0$, unit power is launched in mode 1 – that is

$$A(z = 0) = 1, \quad B(z = 0) = 0 \tag{21.29}$$

Hence, from equations (21.26) and (21.28) we have

$$b_1 + b_2 = 0 \tag{21.30}$$

$$\frac{i}{\kappa}\left[\left(\frac{\Gamma}{2} - \gamma\right)b_1 + \left(\frac{\Gamma}{2} + \gamma\right)b_2\right] = 1 \tag{21.31}$$

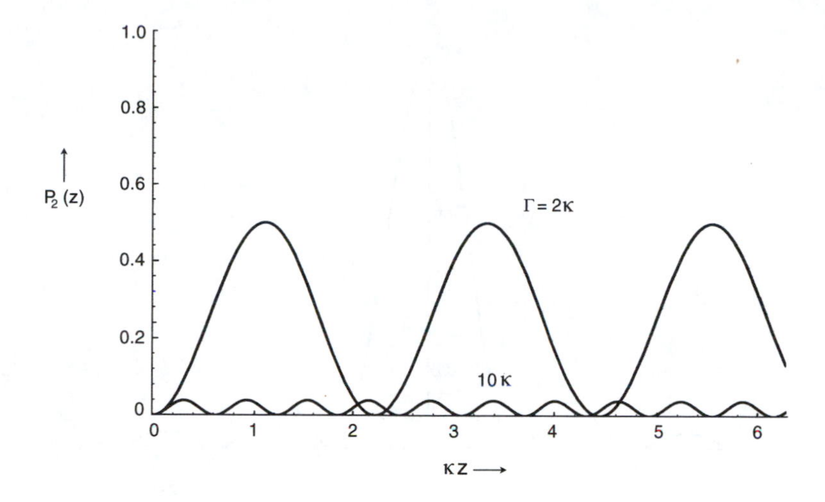

Fig. 21.5: Variation of $P_1(z)$ and $P_2(z)$ with z for $\Gamma = 2\kappa$ and $\Gamma = 10\kappa$ – that is, non-quasi-phase-matched condition when unit power is incident in mode 1 at $z = 0$. Note the incomplete power exchange. The corresponding variations for $\Gamma = 0$ are shown in Figure 21.4.

Solving equations (21.30) and (21.31) we have

$$b_1 = i\frac{\kappa}{2\gamma} = -b_2$$

Thus

$$B(z) = -\frac{\kappa}{\gamma} e^{-i\Gamma z/2} \sin \gamma z \tag{21.32}$$

$$A(z) = e^{i\Gamma z/2}\left[\cos \gamma z - i\frac{\Gamma}{2\gamma} \sin \gamma z\right] \tag{21.33}$$

Thus, the power in modes 1 and 2 at any value of z will be

$$P_1(z) = |A(z)|^2 = \cos^2 \gamma z + \frac{\Gamma^2}{4\gamma^2} \sin^2 \gamma z \tag{21.34}$$

$$P_2(z) = |B(z)|^2 = \frac{\kappa^2}{\gamma^2} \sin^2 \gamma z \tag{21.35}$$

Equations (21.34) and (21.35) describe the variation of power in the two modes with z. Figure 21.5 shows a plot of $P_2(z)$ with z for $\Gamma = 2\kappa$ and $\Gamma = 10\kappa$. The corresponding variation with $\Gamma = 0$ is shown in Figure 21.4. From equation (21.35) we notice that the maximum value of P_2 is κ^2/γ^2, which is always less than or equal to 1; it is equal to unity only when $\gamma = \kappa$ – that is, $\Gamma = 0$. Thus, complete power transfer is possible only if $\Gamma = 0$. Also for $\gamma \gg \kappa$, – that is, for $\Gamma \gg 2\kappa$, $P_2(\max) \ll 1$. Thus, strong exchange of power takes place only between modes for which $\frac{\Gamma}{2\kappa} \ll 1$. This justifies our assumption made at the beginning to consider coupling only between two modes, which almost satisfy the phase-matching condition given by equation (21.1). Figure 21.6 gives the variation of $P_2(z)$ with $\Gamma/2\kappa$ for $\kappa z = \pi/2$. The figure shows how the coupling of power decreases as we move away from the point $\Gamma = 0$. This characteristic is used in the design of wavelength filters (see Section 21.3).

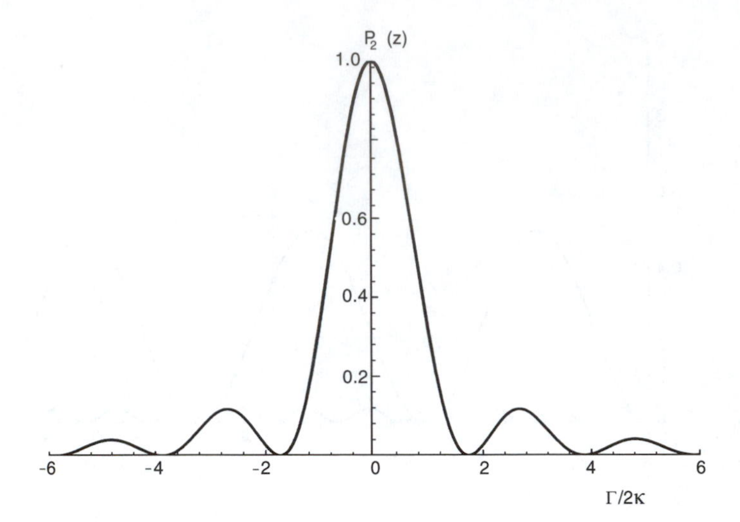

Fig. 21.6: Variation of $P_2(z)$ with Γ (the phase mismatch) at $\kappa z = \pi/2$, showing how the power coupling between the two modes decreases as Γ increases. For $\Gamma \gg 2\kappa$, P_2 is very small.

Example 21.2: We consider the same waveguide as in Example 21.1 and assume that the periodic perturbation has a period of (i) 100 μm and (ii) 148 μm. We will calculate the maximum power that can be transferred between the modes and obtain the distance at which P_2 becomes maximum.

(i) For a period of $\Lambda = 100\,\mu$m

$$K = \frac{2\pi}{\Lambda} = 0.02\pi\,\mu\text{m}^{-1}$$

and

$$\Gamma = \beta_1 - \beta_2 - K = \frac{2\pi}{\lambda_0}(n_{e1} - n_{e2}) - \frac{2\pi}{\Lambda}$$

$$= -2.07 \times 10^{-2}\,\mu\text{m}^{-1}$$

Thus

$$\gamma = \left(\kappa^2 + \frac{\Gamma^2}{4}\right)^{1/2} \simeq 1.035 \times 10^{-2}\,\mu\text{m}^{-1}$$

Since $\kappa = 0.598 \times 10^{-4}\,\mu\text{m}^{-1}$ (see Example 21.1), we see that in this example $\Gamma \gg \kappa$. The maximum power that gets transferred between the modes is

$$P_{2\max} = \frac{\kappa^2}{\gamma^2} \simeq 3.3 \times 10^{-5}$$

Hence, for this case the power transfer between the modes is very weak. The distance for maximum power transfer is

$$L = \frac{\pi}{2\gamma} \simeq 152\,\mu\text{m}$$

which is much smaller than the coupling length with $\Gamma = 0$.

(ii) For $\Lambda = 148\,\mu$m,

$$\Gamma = \beta_1 - \beta_2 - K = \frac{2\pi}{\lambda_0}(n_{e1} - n_{e2}) - \frac{2\pi}{\Lambda}$$

$$= -3.566 \times 10^{-4}\,\mu\text{m}^{-1}$$

Thus

$$\gamma = 1.89 \times 10^{-4}\,\mu\text{m}^{-1}$$

and maximum power transfer is given by

$$P_{2\max} = \frac{\kappa^2}{\gamma^2} \simeq 10^{-1}$$

which is much greater than the first case since the period is close to the phase-matching periodicity of 149.3 μm.

21.3 Application to a wavelength filter

We consider a waveguide device consisting of two unperturbed waveguides joined together by a periodically perturbed region as shown in Figure 21.7. If mode 1 is incident from the left, then the periodic perturbation will couple power from mode 1 to mode 2 depending on the value of Γ. Now the power coupled from mode 1 to mode 2 after propagation through length L of the periodic waveguide is (see equation (21.35))

$$P_2(L) = \frac{\kappa^2}{\kappa^2 + \Gamma^2/4} \sin^2\left[\left(\kappa^2 + \frac{\Gamma^2}{4}\right)^{1/2} L\right] \tag{21.36}$$

where

$$\Gamma = \beta_1 - \beta_2 - K$$

$$= \frac{2\pi}{\lambda_0}(n_{e1} - n_{e2}) - \frac{2\pi}{\Lambda} \tag{21.37}$$

It is obvious from equation (21.37) that Γ can be made equal to zero only at a particular wavelength $\lambda_0 = \lambda_c$ (say) and will be nonzero at other wavelengths. Thus, complete power transfer will be possible only at λ_c and the coupling of power from mode 1 to mode 2 will exhibit a bandpass characteristic. Thus, if mode 2 is filtered out at the output, then the device will act as a wavelength filter.

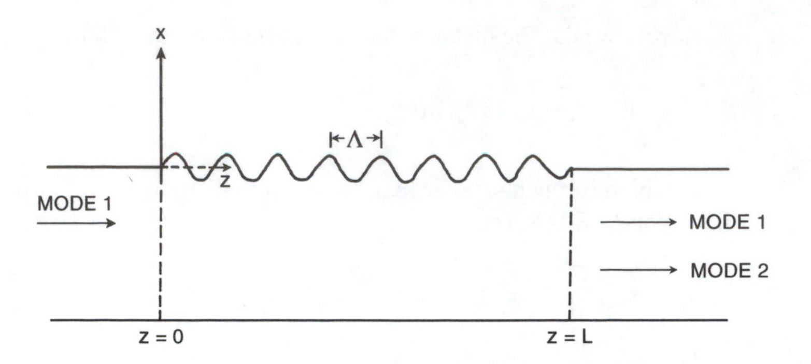

Fig. 21.7: Two unperturbed waveguides joined by a periodically perturbed waveguide of length L. The period of the perturbation is such that at a center wavelength $\lambda_0 = \lambda_c$, all incident power in mode 1 gets coupled to mode 2 after length L. Such a device acts as a wavelength filter.

Using equation (21.36) we have calculated the variation of the power in TE_1 mode at the output of a periodic waveguide when a TE_0 mode (with unit power) is incident at $z = 0$. The calculations correspond to the following values of various parameters.

$$n_f = 1.51, \quad n_s = 1.50 = n_c$$

$$d = 4\,\mu\text{m}, \quad h = 0.01\,\mu\text{m}$$

$$\lambda_0 = 0.6\,\mu\text{m}, \quad \Lambda = 149.3\,\mu\text{m}$$

The corresponding variation is shown in Figure 21.8. In these calculations we have neglected material dispersion – that is, assumed n_f and n_s to be independent of wavelength. The length of the periodically varying region is chosen so that at $\lambda_0 = 0.6\,\mu$m, complete conversion from TE_0 to TE_1 mode takes place. The period Λ is also chosen so that at $\lambda_0 = \lambda_c = 0.6\,\mu$m, the TE_0 and TE_1 modes satisfy the condition $\Gamma = \beta_1 - \beta_2 - K = 0$ – that is,

$$\frac{2\pi}{\lambda_c}[n_{e1}(\lambda_c) - n_{e2}(\lambda_c)] = \frac{2\pi}{\Lambda}$$

or

$$\Lambda = \frac{\lambda_c}{n_{e1}(\lambda_c) - n_{e2}(\lambda_c)} \tag{21.38}$$

The bandpass characteristic of the device is obvious from Figure 21.8.

We shall now obtain an approximate expression for the bandwidth of the filter. We choose the length L of the periodic waveguide equal to one coupling length at λ_c. Thus

$$L = \frac{\pi}{2\kappa(\lambda_c)} \tag{21.39}$$

and $P_2(z = L_c) = 1$. For any other value of λ_0,

$$P_2(z = L_c, \lambda_0) = \frac{\kappa^2(\lambda_0)}{\kappa^2(\lambda_0) + \frac{\Gamma^2(\lambda_0)}{4}} \sin^2\left[\frac{\sqrt{\kappa^2(\lambda_0) + \frac{\Gamma^2(\lambda_0)}{4}}\,\pi}{2\kappa(\lambda_c)}\right]$$

Neglecting the wavelength dependence of κ, the coupled power drops to 0.5

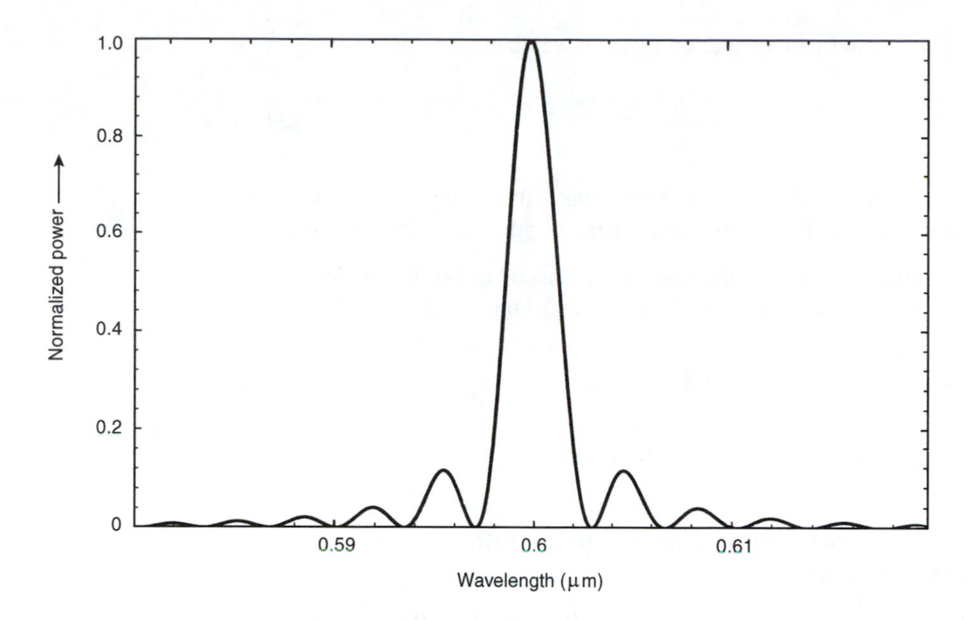

Fig. 21.8: Calculated wavelength variation of power in TE_1 mode at the output of a periodic waveguide with $n_f = 1.51, n_s = 1.50, n_c = 1.50, d = 4\ \mu$m and having a periodic perturbation of period $\Lambda = 149.3\ \mu$m. The bandwidth of the filter is approximately 3 nm.

for a wavelength satisfying

$$\left[1 + \frac{\Gamma^2(\lambda_0)}{4\kappa^2(\lambda_0)}\right]^{-1} \sin^2\left[\sqrt{1 + \frac{\Gamma^2(\lambda_0)}{4\kappa^2(\lambda_0)}}\frac{\pi}{2}\right] = 0.5 \tag{21.40}$$

whose solution is

$$\frac{\Gamma(\lambda_0)}{2\kappa(\lambda_0)} \simeq \pm 0.8$$

or

$$2\pi\left[\frac{n_{e1}(\lambda_0) - n_{e2}(\lambda_0)}{\lambda_0} - \frac{1}{\Lambda}\right] \simeq \pm 0.8\frac{\pi}{L} \tag{21.41}$$

where

$$\lambda_0 = \lambda_c \pm \frac{\Delta\lambda}{2} \tag{21.42}$$

are the wavelengths where $P_2 = 0.5$ and $\Delta\lambda$ would represent the FWHM of the filter. For $\Delta\lambda \ll \lambda_c$, we have

$$\begin{aligned}\frac{\Delta n_e(\lambda_0)}{\lambda_0} &\simeq \frac{\Delta n_e(\lambda_c)}{\left(\lambda_c \mp \frac{\Delta\lambda}{2}\right)} \\ &\simeq \frac{\Delta n_e(\lambda_c)}{\lambda_c} \pm \Delta n_e(\lambda_c)\frac{\Delta\lambda}{2\lambda_c}\end{aligned} \tag{21.43}$$

where we have assumed $\Delta n_e(\lambda_0) \simeq \Delta n_e(\lambda_c)$. Substituting equation (21.43) in equation (21.41) and using equation (21.38), we obtain

$$\frac{\Delta\lambda}{\lambda_c} \simeq \frac{0.8\lambda_c}{\Delta n_e(\lambda_c)L} = 0.8\frac{\Lambda}{L} \tag{21.44}$$

Thus, the fractional bandwidth of the wavelength filter is approximately equal to the ratio of the period of the perturbation to the length of the periodic waveguide.

Example 21.3: For the waveguide considered in Example 21.1, we have $\Lambda = 149.3\,\mu\text{m}$ and $L = 2.63$ cm. Thus

$$\Delta\lambda = 0.8\frac{\Lambda}{L}\lambda_c \simeq 27\,\text{Å}$$

where we have used $\lambda_c = 0.6\,\mu\text{m}$.

21.4 Coupling between orthogonal polarization in a birefringent fiber

In Chapter 20 we learned about birefringent fibers in which the x-polarized and y-polarized modes travel with different velocities. If n_x and n_y represent the effective indices of the x- and y-polarized LP_{01} modes, then the birefringence of the fiber is defined by

$$B = |n_x - n_y| \tag{21.45}$$

and the corresponding beat length is

$$L_b = \frac{\lambda_0}{B} \tag{21.46}$$

High-birefringence fibers have typical beat lengths of 2 mm at 1300 nm corresponding to a birefringence of 6.5×10^{-4}.

As described in Chapter 20, one of the primary advantages of high-birefringence fibers is its polarization maintaining capability. Thus, if at the input, light is launched in the x polarization, under normal perturbations of the fiber the light does not couple to the y-polarization state. This is primarily because to couple power from x-polarized mode to y-polarized mode, we need a periodic perturbation with a period given by (see equation (21.13)).

$$\Lambda = \frac{\lambda_0}{(n_x - n_y)} = L_b \tag{21.47}$$

Since $L_b \sim 2$ mm, under normal perturbations such as bending, twisting, etc., the amplitude of perturbations corresponding to such small spatial period (or such large spatial frequencies) would be negligible. One could, however, purposely create perturbations with period $\sim L_b$ and hence induce coupling among the x- and y-polarized modes. Thus, if a birefringent fiber is periodically stressed by placing the fiber between a pair of corrugated plates of period Λ as shown in Figure 21.9(a), one can induce a strong coupling of power between the modes. For coupling of power, it is necessary to have a

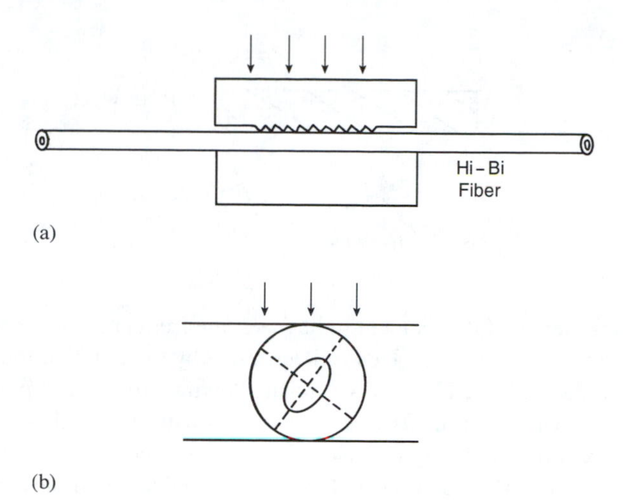

Fig. 21.9: (a) If a birefringent fiber is stressed periodically by placing it under a corrugated plate having corrugations with a period given by equation (21.47), then there is strong coupling of power between the two polarizations. (b) The cross section showing the direction of applied stress with respect to the eigen axes.

finite coupling coefficient. This can be met by orienting the fiber birefringence axes so that the stress is applied at 45° to the eigen axes (see Figure 21.9(b)).

By measuring the changes in output power in the x and y polarization states as a function of pressure, one can use this principle to construct pressure sensors. In the following section we show how this coupling can be used for making in-line fiber frequency shifters.

21.5 In-line fiber frequency shifters

In-line fiber frequency shifters are required in many applications such as in optical heterodyning in fiber optic sensors or in coherent communications, etc. One of the most interesting in-line fiber frequency shifters uses traveling acoustic waves to couple light between the two orthogonal polarization modes. Unlike the earlier cases of coupling by periodic gratings, which were all fixed (i.e., not moving), the periodic grating caused because of the periodic strain accompanying a traveling acoustic wave is a propagating grating. The consequence of coupling due to a propagating periodic grating is to induce a frequency shift [see, e.g., Ghatak and Thyagarajan (1989)]. If n_x and n_y are again the effective indices of the x- and y-polarized LP_{01} modes of the fiber, then the required period of the grating is given by

$$\Lambda = \frac{\lambda_0}{|n_x - n_y|} \tag{21.48}$$

If the acoustic wave is traveling in the same direction as the propagating light waves, then

$$\Lambda = \frac{v_a}{f} \tag{21.49}$$

where v_a and f are the velocity and frequency of the acoustic wave. Thus

$$f = \frac{v_a|n_x - n_y|}{\lambda_0} = \frac{v_a}{L_b} \tag{21.50}$$

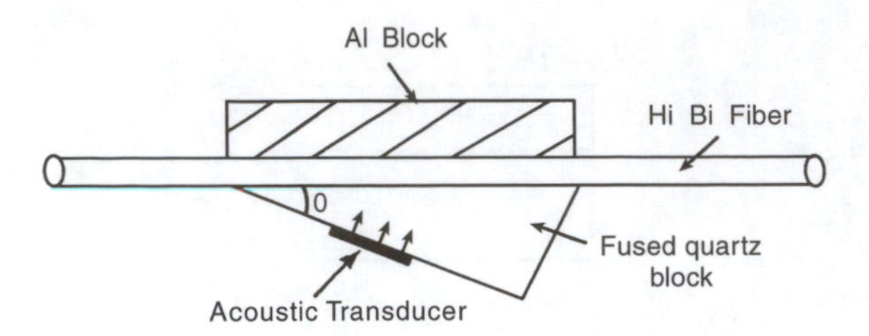

Fig. 21.10: A fiber optic frequency shifter in which acoustic waves launched into the quartz wedge induce periodic traveling refractive index perturbation in the fiber through the strain optic effect. The perturbation induces a coupling between the two eigenmodes, and the coupled lightwave is frequency shifted due to the traveling periodic perturbation. [Adapted from Risk et al. (1984).]

If ν_i is the frequency of the incident light wave, the frequency of the diffracted wave will be $\nu_i + f$ or $\nu_i - f$. This is determined by the polarization states of the input and the diffracted light wave and also by the direction of propagation of the acoustic wave. Let us assume that the acoustic wave also propagates along the direction of the light wave. If $n_x > n_y$ and incident light is x-polarized, then the diffracted wave will be y-polarized and will have frequency $\nu_i - f$. On the other hand, if the incident light is y-polarized, the diffracted wave will be x-polarized and will have frequency $\nu_i + f$. One of the techniques to launch an acoustic wave along the fiber is to use an arrangement shown in Figure 21.10 [Risk et al. (1984)]. It consists of a wedge of fused quartz with a piezoelectric transducer to excite acoustic waves, and the fiber is sandwiched between the quartz block and an aluminum base for mechanical contact. In such a case the acoustic wave vector makes angle $(\frac{\pi}{2} - \theta)$ with the fiber axis and the corresponding periodic perturbation on the fiber has a period

$$\Lambda = \frac{v_a}{f \sin\theta} \tag{21.51}$$

For this case the required acoustic frequency is

$$f = \frac{v_a}{L \sin\theta} \tag{21.52}$$

Experimental results have been demonstrated for in-line fiber frequency shifts of 15 MHz [Risk et al. (1984)].

Example 21.4: In silica $v_a = 5.96$ km/s. Thus, for $L_b = 2$ mm the acoustic frequency required to couple between the two polarizations is (for $\theta = 0$)

$$f = \frac{v_a}{L_b} \simeq 2.98\,\text{MHz}$$

Example 21.5: Figure 21.10 shows a schematic of the arrangement used for frequency shifting by using bulk acoustic waves. If the beat length of the fiber is 1.7 mm at 632.8 nm, and if $\theta = 13.5°$, the corresponding acoustic frequency required to induce coupling is

$$f_a = \frac{5.96 \times 10^3}{1.7 \times 10^{-3} \times \sin 13.5} \simeq 15\,\text{MHz}$$

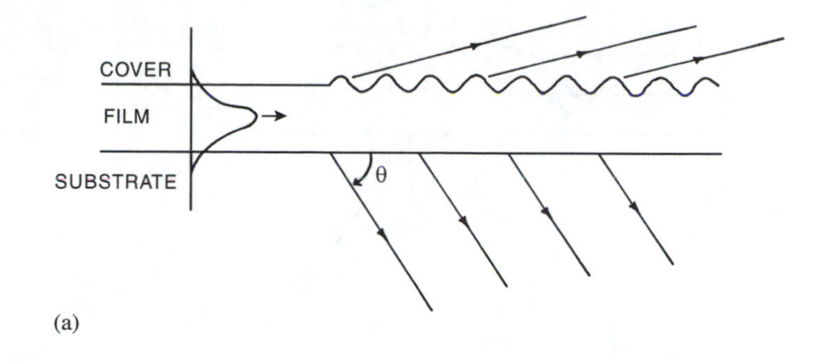

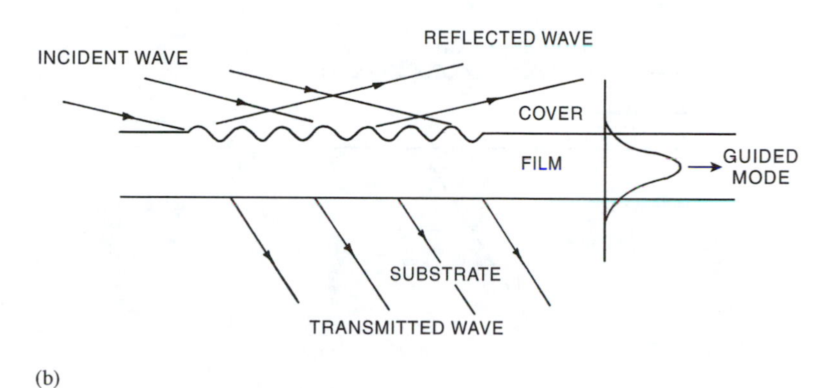

Fig. 21.11: (a) A grating output coupler in which coupling between a guided mode and radiation modes propagating in the substrate and cover is induced by a periodic perturbation. (b) A grating input coupler in which an incident beam is coupled to a guided mode by a periodic perturbation in the waveguide.

If the interaction length L is 1.03 cm, the corresponding bandwidth is

$$\frac{\Delta f_a}{f_a} \simeq \frac{\Lambda}{L} = \frac{L_b}{L} = \frac{1.7 \times 10^{-3}}{1.03 \times 10^{-2}} \simeq 0.17$$

consistent with the measured values [Risk et al. (1984)].

21.6 Grating input-output couplers

In the previous sections we have restricted our consideration to coupling of power among two guided modes of the given waveguide due to a periodic perturbation. Periodic waveguides can also be used to couple power from a guided mode to radiation modes (waveguide output coupler), which carry power out of the waveguide (see Figure 21.11(a)), or to couple power from an incident propagating beam into a guided mode of the waveguide (input coupler) (see Figure 21.11(b)). Such a coupler, which is known as a grating coupler, is very similar to a prism–film coupler, wherein one uses the phenomenon of frustrated total internal reflection.

Just like the requirement of a quasi-phase-matching condition for efficient coupling of power among two guided modes (see equation (21.1)), even for this case one has to satisfy a quasi-phase-matching condition. To illustrate this, we consider a planar waveguide consisting of a film, substrate, and cover of refractive indices n_f, n_s, and n_c (with $n_f > n_s \geq n_c$), which is assumed to

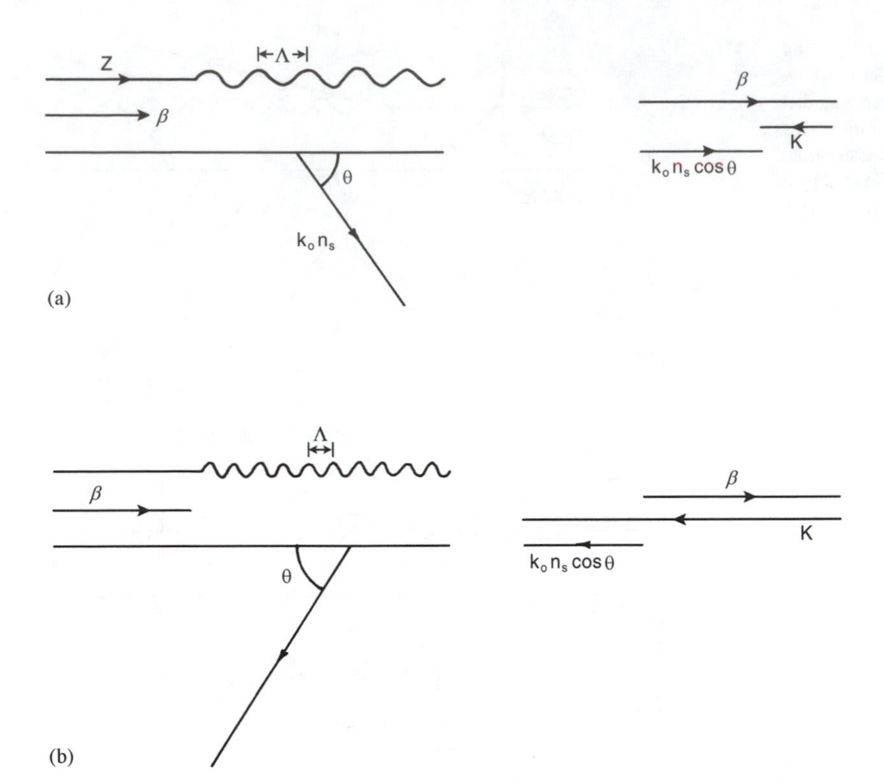

Fig. 21.12: Quasi-phase-matching condition implies that the radiation coming into the substrate (or cover) appears at a certain well-defined direction (as given by equation (21.55)). In (a) the period of the grating is such as to induce coupling to a forward propagating wave in the substrate and in (b) to a backward propagating wave in the substrate.

support a guided mode with the propagation constant satisfying

$$k_0 n_c \leq k_0 n_s < \beta < k_0 n_f \tag{21.53}$$

Let a sinusoidal perturbation of spatial period $\Lambda(= 2\pi/K)$ be introduced on the waveguide. If the periodic perturbation is to couple power from the guided mode to a wave propagating into the substrate and making angle θ with the z-direction (see Figure 21.11(a)), then we must have

$$\beta - K = k_0 n_s \cos\theta \tag{21.54}$$

or

$$\theta = \cos^{-1}\left(\frac{\beta - K}{k_0 n_s}\right) \tag{21.55}$$

Note that if $|\beta - K|$ is greater than $k_0 n_s$, such a coupling is not possible since then θ becomes imaginary. Also, if $\beta - K$ is positive, then coupling takes place to a wave propagating in the substrate in the forward direction (see Figure 21.12(a)); on the other hand, if $\beta - K$ is negative, then coupling takes place to a wave propagating in the backward z-direction (see Figure 21.12(b)). Similar considerations are applicable to coupling to a wave propagating in the cover.

In an identical fashion one can also discuss coupling of power from a propagating beam into a guided mode. One obvious fact is that, whereas in an output coupler the coupled beam automatically chooses the propagation direction so as to satisfy the quasi-phase-matching condition, in the case of the input coupler,

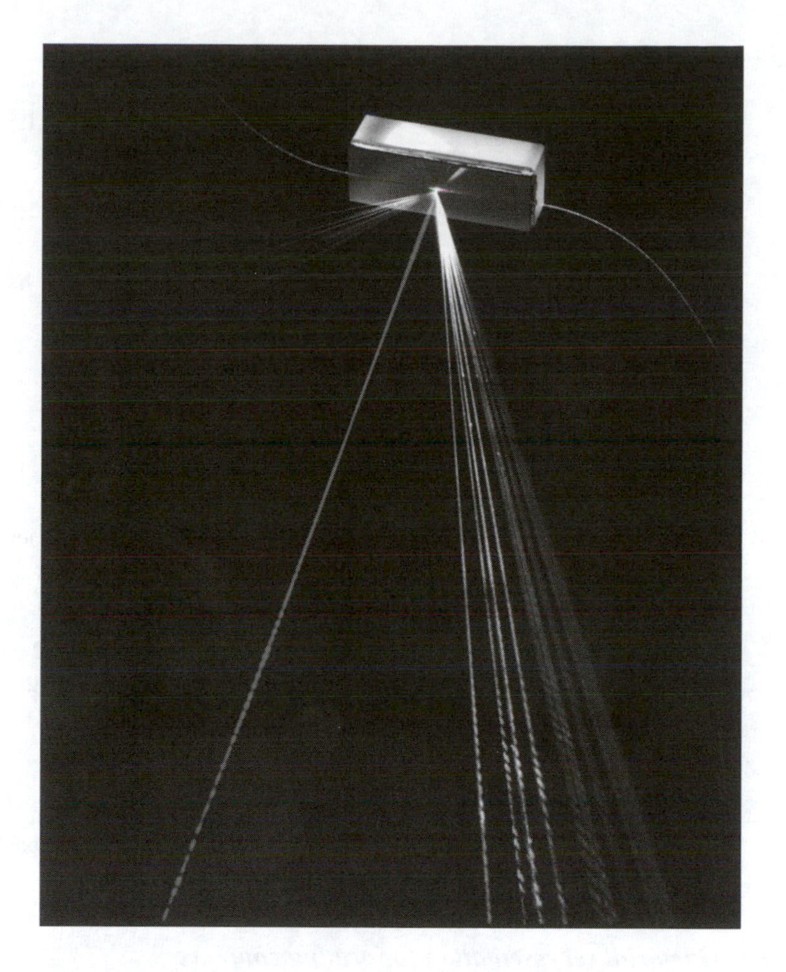

Fig. 21.13: Simultaneous guided to radiation-mode coupling by a number of guided, discrete, visible laser wavelengths incident on a fiber grating structure. [After Rowe, Bennion, and Reid (1987). Photograph courtesy Professor I. Bennion.]

the angle of incidence should be such as to satisfy the quasi-phase-matching condition.

Figure 21.13 shows simultaneous guided to radiation-mode coupling by a number of guided, discrete, visible laser wavelengths incident on a fiber grating structure.

Grating couplers can be fabricated on a waveguide either by etching the waveguide surface or by depositing a film on the waveguide, whose thickness is then periodically modulated. One can achieve efficiences of greater than 50%.

Example 21.6: Consider a waveguide with the values of various parameters as given in Example 21.1. Assuming a sinusoidally periodic thickness variation with period $\Lambda = 6\,\mu$m, we will now obtain the angles at which the radiation coupled out from the TE_0 and TE_1 modes will propagate in the substrate.

If θ_{s0} and θ_{s1} are the angles with the z-axis made by the waves coupled out from TE_0 and TE_1 modes, then

$$\beta(TE_0) - \frac{2\pi}{\Lambda} = k_0 n_s \cos\theta_{s0} \tag{21.56}$$

$$\beta(TE_1) - \frac{2\pi}{\Lambda} = k_0 n_s \cos\theta_{s1} \tag{21.57}$$

or in terms of effective indices

$$n_{e1} - \frac{\lambda_0}{\Lambda} = n_s \cos\theta_{s0} \tag{21.58}$$

$$n_{e2} - \frac{\lambda_0}{\Lambda} = n_s \cos\theta_{s1} \tag{21.59}$$

Using the values of n_{e1} and n_{e2} as obtained in Example 21.1, we have

$$\theta_{s0} = \cos^{-1}\left[\frac{1}{n_s}\left(n_{e1} - \frac{\lambda_0}{\Lambda}\right)\right] \simeq 20.1^\circ$$

$$\theta_{s1} = \cos^{-1}\left[\frac{1}{n_s}\left(n_{e2} - \frac{\lambda_0}{\Lambda}\right)\right] \simeq 20.5^\circ$$

The corresponding angles in the cover would be given by replacing n_s by n_c. Since both $(n_{e1} - \lambda_0/\Lambda)$ and $(n_{e2} - \lambda_0/\Lambda)$ are greater than unity, no coupling can occur to a wave propagating in the cover.

Example 21.7: In the above example if we had chosen $\Lambda = 0.2\,\mu$m, then the radiation would appear in the substrate at angles given by

$$\theta_{s0} = 173.9^\circ, \quad \theta_{s1} = 175.5^\circ$$

that is, the coupling takes place to a wave propagating along the backward direction as shown in Figure 21.12(b).

21.6.1 Pictorial representation for grating couplers

The phase-matching condition imposed by the equation (see equation (21.54))

$$\beta - K = k_0 n_s \cos\theta \tag{21.60}$$

can be pictorially represented by a wave vector diagram as shown in Figure 21.14(a), in which the horizontal axis represents the z component of the propagation vector of the propagating waves. The upper semicircle is of radius $k_0 n_c$ (corresponding to the cover), and the lower semicircle is that of radius $k_0 n_s$ (corresponding to the substrate). A guided mode can have the β value so that either

$$k_0 n_s < \beta < k_0 n_f \tag{21.61}$$

or

$$-k_0 n_f < \beta < -k_0 n_s \tag{21.62}$$

the latter corresponding to a wave propagating in the $-z$-direction. Any wave propagating in the cover will have its $\boldsymbol{k}$ of magnitude $k_0 n_c$ and any arbitrary direction. Thus, waves emanating from the waveguide and propagating in the

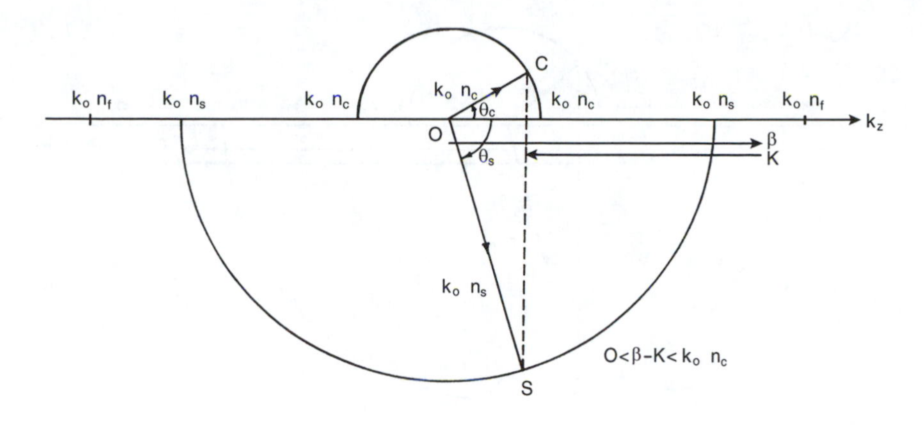

(a)

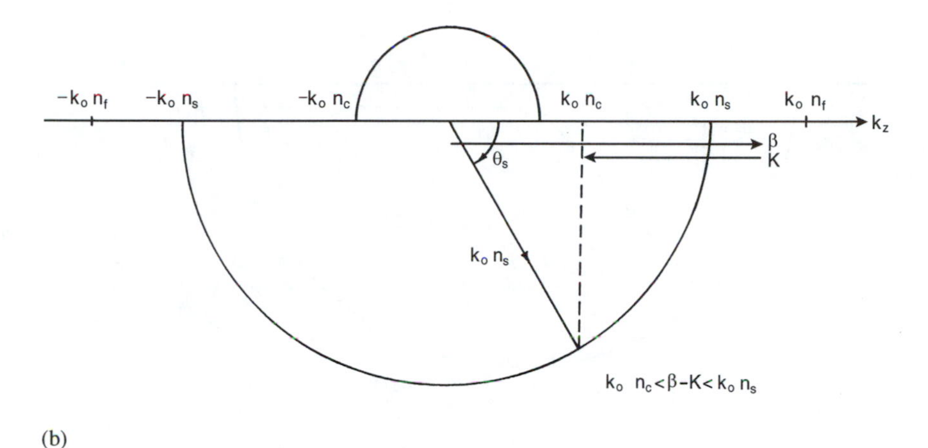

(b)

Fig. 21.14: The wave vector diagram corresponding to a grating coupler. The β value of the propagating mode lies between $k_0 n_f$ and $k_0 n_s$; K represents the wave vector of the periodic perturbation. The vectors $\overrightarrow{OS}$ and $\overrightarrow{OC}$ give the directions of the propagating waves in the substrate and cover that satisfy the phase-matching condition. (a) Coupling to both cover and substrate waves, (b) coupling to only substrate waves.

cover will be represented by a vector starting from the point O and ending on the upper semicircle. Similarly, any wave emanating from the waveguide and propagating in the substrate will be represented by a vector starting from O and ending on the lower semicircle.

Because the periodic structure has a z dependence of the form $\sin Kz$, we can also represent the periodic structure by a vector of length K pointing to the right or to the left along the β axis in Figure 21.14(a).

Now, the phase-matching condition as represented by equation (21.60) can be interpreted to imply that the z component of the propagation vector of the wave in the substrate must be equal to $\beta - K$. This can be pictorially represented as shown in Figure 21.14(a), where the vector $\overrightarrow{OS}$ gives the direction of propagation of the substrate radiation that satisfies the phase-matching condition. Note that one can also satisfy a phase-matching condition for a wave propagating in the cover, and the corresponding direction is given by the vector $\overrightarrow{OC}$. Thus, as is obvious from Figure 21.14, if the spatial period of the periodic structure is such that

$$-k_0 n_c < \beta - K < k_0 n_c \tag{21.63}$$

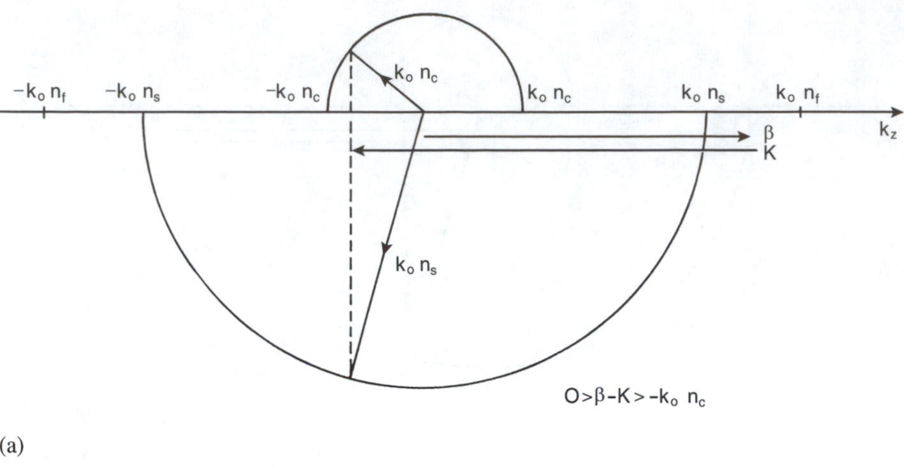

(a)

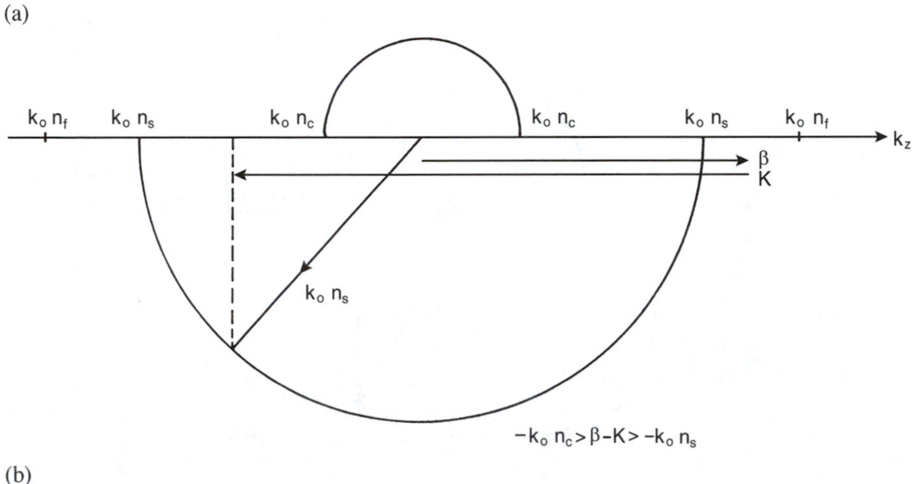

(b)

Fig. 21.15: Wave vector diagrams corresponding to coupling to backward propagating waves.

then coupling will take place to radiation propagating in both cover and substrate (we are assuming $n_c < n_s$). On the other hand, if

$$k_0 n_c < \beta - K < k_0 n_s \tag{21.64}$$

then coupling takes place only to the forward propagating wave in the substrate (see Figure 21.14(b)) since phase matching cannot be satisfied in the cover. Figures 21.15(a) and (b) show wave vector diagrams corresponding to

$$0 > \beta - K > -k_0 n_c$$

and

$$-k_0 n_c > \beta - K > -k_0 n_s$$

Note that in the above cases the radiation in the substrate (and in the cover) propagates in the backward direction. Also note that depending on the K value, coupling may take place to radiation propagating only in the substrate or in both substrate and cover (remember, we are assuming $n_c \leq n_s$).

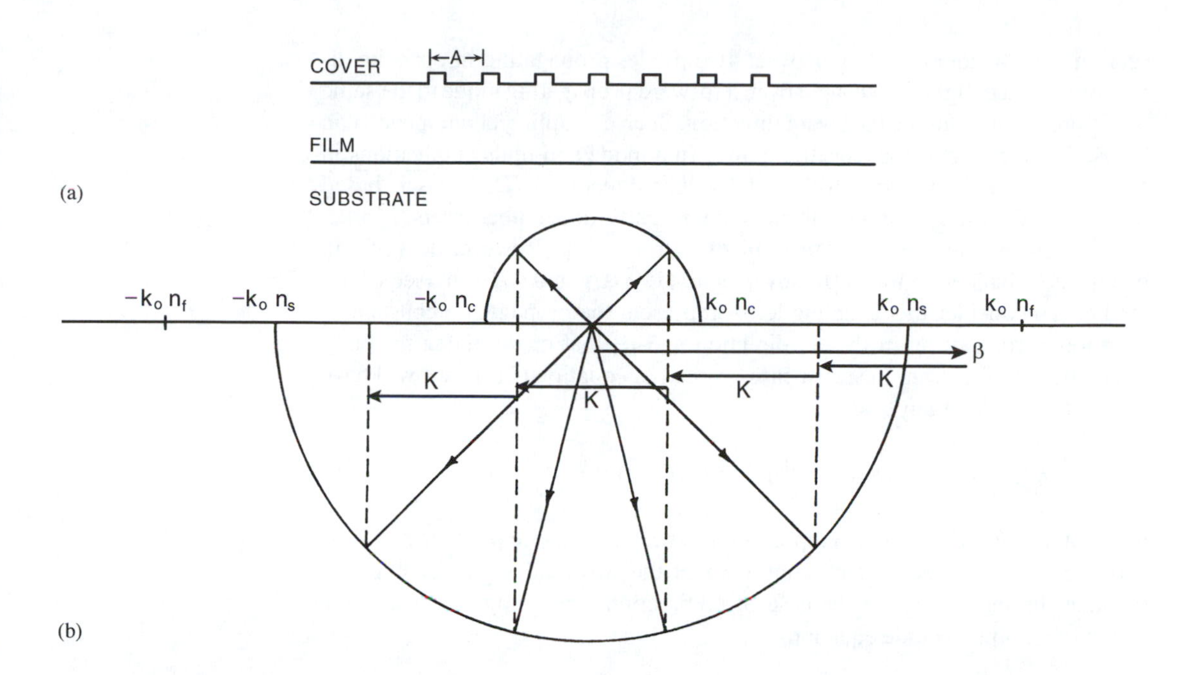

Fig. 21.16: (a) A periodic structure with a nonsinusoidal thickness variation. (b) Wave vector diagram showing multiple-beam coupling.

In the above discussion we have assumed that the periodic perturbation is perfectly sinusoidal. If, for example, the thickness variations are of the form shown in Figure 21.16(a), then the periodic perturbation can be described by a Fourier series expansion, which would thus contain terms such as $\sin Kz$, $\sin 2Kz$, $\sin 3Kz$, and so forth, where $K = 2\pi/\Lambda$, Λ being the period (see Figure 21.16(a)). In such a case, one could indeed have multiple-order beam coupling, and the phase-matching condition (equation (21.60)), gets modified to

$$k_0 n_s \sin\theta_n = \beta_0 \pm pK; p = 1, 2, 3 \ldots \tag{21.65}$$

This is shown in Figure 21.16(b), where one can see multiple beam coupling. Note that if $-k_0 n_c > \beta - K > -k_0 n_s$, then even for a nonsinusoidal perturbation one will have only a single-beam coupling into the substrate. Such single-beam coupling is desirable to achieve high coupling efficiency.

Typical grating couplers have a length of a few millimeters and one can achieve a maximum launching efficiency of 81% by using grating input couplers. Some of the major advantages of using grating couplers are as follows:

(1) Guided waves can be excited by a large width input beam and positioning is not very critical as in end-fire coupling.
(2) One can selectively excite various guided modes.
(3) Gratings are compact, stable, and integrable with the waveguide.

21.7 Contradirectional coupling

In Section 21.2 we discussed coupling between two modes propagating along the same direction. This leads to what is called codirectional coupling. In this section we consider coupling between two modes propagating in opposite directions. Hence, light may couple from a forward propagating mode to the same mode propagating in the backward direction. Such a coupling phenomenon can be used as a reflector for reflecting power in a mode and finds applications in distributed Bragg reflector (DBR) and distributed feedback (DFB) lasers based on semiconductors, as Bragg reflectors in rare-earth-doped fiber lasers [Morkel (1993)], in external fiber cavity semiconductor lasers [Rowe et al. (1987)], bandpass or band stop filters [Kashyap et al. (1993a)], and so forth. (see Chapter 17). To consider this coupling let β_1 represent the propagation constant of the mode propagating in the $+z$-direction and β_2 that of the mode traveling along the $-z$-direction. In such a case, instead of equation (21.3), we will have for the total field at any z as

$$E(x,z) = A(z)E_1(x)\,e^{-i\beta_1 z} + B(z)E_2(x)\,e^{i\beta_2 z} \tag{21.66}$$

where again $E_1(x)$ and $E_2(x)$ represent the transverse mode patterns and $A(z)$ and $B(z)$ are the z-dependent amplitudes of the two modes. Proceeding in a manner similar to that for the case of codirectional coupling, we obtain the following coupled-mode equations

$$\frac{dA}{dz} = \kappa B e^{i\Gamma z} \tag{21.67}$$

$$\frac{dB}{dz} = \kappa A e^{-i\Gamma z} \tag{21.68}$$

where now

$$\Gamma = \beta_1 + \beta_2 - K \tag{21.69}$$

For the present case, for phase-matching we require $\Gamma = 0$ implying

$$\beta_1 + \beta_2 = K \tag{21.70}$$

If the coupling is between two identical modes traveling in opposite directions, then $\beta_1 = \beta_2 = (2\pi/\lambda_0)n_{eff}$, where n_{eff} is the effective index of the mode. If we let $K = 2\pi/\Lambda$, then equation (21.70) implies

$$\Lambda = \frac{\lambda_0}{2n_{eff}} \tag{21.71}$$

Comparing with equation (21.13) we note that the periodicity required here is much smaller than in the case of codirectional coupling. We will solve equations (21.67) and (21.68) when the modes are phase-matched – that is, $\Gamma = 0$. We differentiate equation (21.68) with respect to z and obtain (using equation (21.67))

$$\frac{d^2B}{dz^2} = \kappa^2 B$$

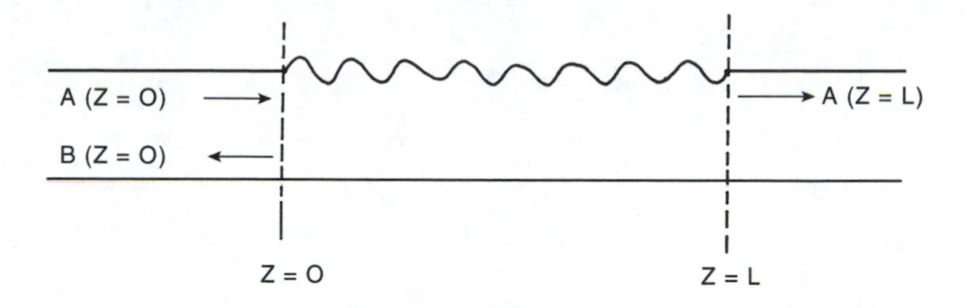

Fig. 21.17: If the period of perturbation is so chosen to quasi-phase-match a forward and a backward propagating mode (corresponding to contradirectional coupling), then the periodic region acts as a reflector. The corresponding period is given by equation (21.63).

whose solution is

$$B(z) = b_1 e^{\kappa z} + b_2 e^{-\kappa z} \tag{21.72}$$

Note that unlike the codirectional case, here the solutions are not oscillatory. Substituting in equation (21.68) we obtain

$$A(z) = [b_1 e^{\kappa z} - b_2 e^{-\kappa z}] \tag{21.73}$$

We now assume that unit power is incident in mode A on a periodic waveguide of length L – that is, $A(z = 0) = 1$ (see Figure 21.17). Since there is no back-coupled wave beyond $z = L$, we must have

$$B(z = L) = 0 \tag{21.74}$$

Thus

$$b_1 e^{\kappa L} + b_2 e^{-\kappa L} = 0; \quad (b_1 - b_2) = 1$$

which give us

$$b_1 = \frac{e^{-\kappa L}}{2\cosh \kappa L}; \quad b_2 = \frac{-e^{+\kappa L}}{2\cosh \kappa L} \tag{21.75}$$

Hence

$$B(z) = \frac{\sinh \kappa(z - L)}{\cosh \kappa L} \tag{21.76}$$

and

$$A(z) = \frac{\cosh \kappa(z - L)}{\cosh \kappa L} \tag{21.77}$$

Note that in the present case

$$|A(z)|^2 - |B(z)|^2 = (\cosh^2 \kappa L)^{-1} = \text{constant} \tag{21.78}$$

which is the equation for energy conservation since the two waves are now traveling along opposite directions. Figure 21.18 shows the variation of power carried by the two modes as a function of z.

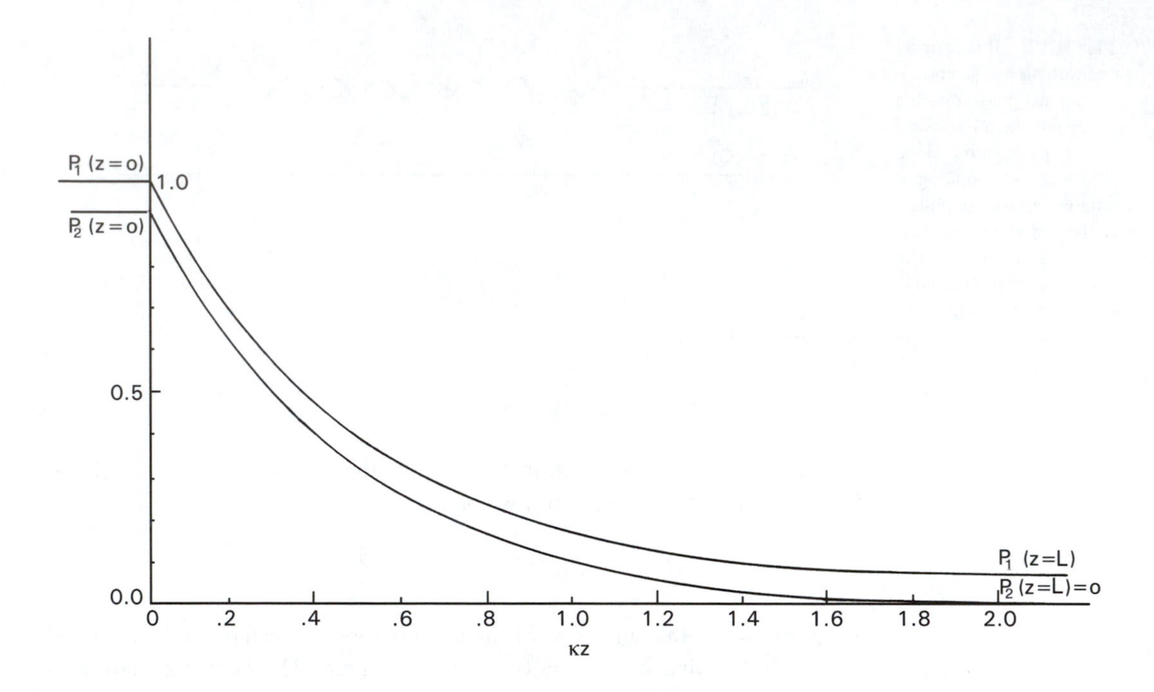

Fig. 21.18: z variation of power carried by the two modes undergoing contradirectional coupling.

The mode $B(z)$ corresponds to the reflected mode and, hence, we define a reflection coefficient from the periodic structure as

$$r = \frac{B(z=0)}{A(z=0)} = -\tanh \kappa L \tag{21.79}$$

The energy reflection coefficient is given by

$$R = |r|^2 = \tanh^2 \kappa L \tag{21.80}$$

For a medium of refractive index n having a periodic refractive index grating given by

$$n(z) = n_0 + \Delta n \sin Kz \tag{21.81}$$

the coupling coefficient is [see, e.g., Ghatak and Thyagarajan (1989)]

$$\kappa = \frac{\pi \Delta n}{\lambda_0} \tag{21.82}$$

If a similar expression is assumed for an optical fiber with a refractive index grating in the core given by equation (21.81), then the reflectivity of a fiber grating of length L is

$$R = \tanh^2 \left(\frac{\pi \Delta n L}{\lambda_0} \right) \tag{21.83}$$

It has recently been demonstrated that refractive index gratings can indeed be written directly into the core of a single-mode fiber by irradiation with UV light (see Chapter 17). If we wish to fabricate a reflector centered around 1550 nm,

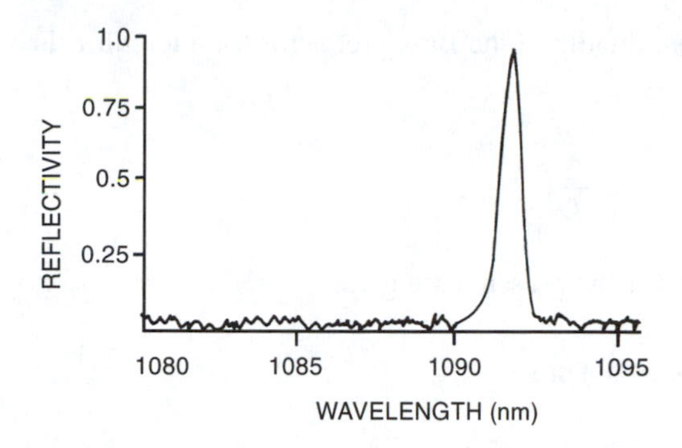

Fig. 21.19: Reflectivity of a Bragg fiber grating as a function of wavelength. [After Rowe et al. (1987).]

the required grating period is

$$\Lambda = \frac{\lambda_0}{2n_{eff}} \simeq \frac{1550}{2 \times 1.46} = 531 \text{ nm}$$

Typical UV written gratings have $\Delta n \simeq 0.4 \times 10^{-3}$. Hence, if the grating length is 2 mm, the reflectivity is given by

$$R = \tanh^2\left(\frac{\pi \times 0.4 \times 10^{-3} \times 2 \times 10^{-3}}{1.55 \times 10^{-3}}\right) = 0.85$$

The corresponding bandwidth of the reflector is given by [see, e.g., Ghatak and Thyagarajan (1989)]

$$\Delta\lambda_0 = \frac{\lambda_B^2}{\pi n_{eff} L}(\kappa^2 L^2 + \pi^2)^{1/2}$$

$$\simeq 0.8 \text{ nm} \tag{21.84}$$

Example 21.8: Figure 21.19 shows the reflectivity of a Bragg fiber grating as a function of wavelength. The center wavelength corresponds to 1092 nm and the measured bandwidth (FWHM) of 0.8 nm. The length of interaction was 1 mm. Since peak reflectivity is $R = 0.98$, we have

$$R = \tanh^2 \kappa L = 0.98$$

or

$$\kappa = \frac{1}{2L} \ln\left(\frac{1 + \sqrt{R}}{1 - \sqrt{R}}\right)$$

$$= 2.64 \text{ mm}^{-1}$$

Assuming an effective index of 1.46 for the mode, we obtain for the required period Λ

$$\Lambda = \frac{\lambda_0}{2n_{eff}} = \frac{1.092}{2 \times 1.46} \simeq 0.37 \,\mu\text{m}$$

The bandwidth of the Bragg reflector for a length L is approximately given by

$$\Delta\lambda \simeq \lambda_0 \frac{\Lambda}{L}$$

which for the present case gives

$$\Delta\lambda \simeq 0.4\,\text{nm}$$

Example 21.9: Consider a Bragg reflector formed on a side-polished fiber or on a D-fiber for operation around $\lambda_0 = 1.55\,\mu\text{m}$ [Ragdale, Reid, and Bennion (1989)]. If we assume $n_{eff} \simeq 1.46$, the required periodicity comes out to be

$$\Lambda = \frac{\lambda_0}{2n_{eff}} = \frac{1550}{2 \times 1.46} \simeq 531\,\text{nm}$$

Gratings with interaction lengths of $L = 1$ cm with peak reflectivity of 0.85 have been made. This implies a coupling coefficient κ of

$$\kappa = \frac{1}{L}\tanh^{-1}\left(\sqrt{R}\right)$$

$$= 1.589\,\text{m}^{-1}$$

The bandwidth of the Bragg reflector is approximately given by (see equation (21.84))

$$\Delta\lambda \simeq 0.16\,\text{nm}$$

Problems

21.1 For the waveguide as in Example 21.1, plot the variation of $P_{2\max}$ with Λ.

21.2 Consider a symmetric planar waveguide with

$$n_f = 1.50, \quad n_s = n_c = 1.48 \quad \text{and} \quad d = 6.2\,\mu\text{m}$$

operating at $\lambda_0 = 1.0\,\mu\text{m}$.

(a) Show that the waveguide supports three TE modes – namely, TE_0, TE_1, and TE_2. Obtain their effective indices.

(b) Calculate the perturbation period required for coupling of power among TE_0 and TE_1 modes, TE_0 and TE_2 modes, and TE_1 and TE_2 modes.

(c) For a thickness variation with $h = 0.01\,\mu\text{m}$, calculate the coupling coefficients between $TE_0 \leftrightarrow TE_1$ modes, $TE_0 \leftrightarrow TE_2$ modes, and $TE_1 \leftrightarrow TE_2$ modes.

(d) Assuming a thickness variation with a period corresponding to phase matching between $TE_0 \leftrightarrow TE_1$ modes, show that the coupling between TE_0 and TE_2 modes and between TE_1 and TE_2 modes is very weak.

[ANSWERS:

(a) $n_{e0} = 1.49792, \quad n_{e1} = 1.49187, \quad n_{e2} = 1.4829$
(b) $\Lambda_{01} = 165.29\,\mu\text{m}, \quad \Lambda_{02} = 66.58\,\mu\text{m}, \quad \Lambda_{12} = 111.48\,\mu\text{m}$
(c) $\kappa_{01} = 4.77 \times 10^{-4}\,\mu\text{m}^{-1}, \quad \kappa_{02} = 5.13 \times 10^{-4}\,\mu\text{m}^{-1},$
$\kappa_{12} = 11.9 \times 10^{-4}\,\mu\text{m}^{-1}$
(d) $\Lambda = \Lambda_{01}, \quad \Delta\beta(0 \leftrightarrow 1) - K = 0, \quad \Delta\beta(0 \leftrightarrow 2) - K = 0.0564 \gg \kappa_{02}$
$= \Delta\beta(1 \leftrightarrow 2) - K = 0.0184 \gg \kappa_{12}$.]

21.3 Expand $\Delta n_e(\lambda_0)$ around $\lambda_0 = \lambda_c$ and retain terms up to $\Delta\lambda$ and show that a more accurate expression for FWHM of the filter is

$$\frac{\Delta\lambda}{\lambda_c} \simeq 0.8\frac{\Lambda}{L}\frac{\Delta n_e(\lambda_c)}{\left[\Delta n_e(\lambda_c) - \lambda_c \frac{d\Delta n_e}{d\lambda_0}\Big|_{\lambda_c}\right]} \tag{21.85}$$

22

The ray equation in cartesian coordinates and its solutions

22.1 Introduction

In this chapter we obtain rigorously correct solutions of the ray equation for a square law medium characterized by the following refractive index variation

$$n^2(x, y) = n_1^2\left[1 - 2\Delta\left(\frac{x^2 + y^2}{a^2}\right)\right] \tag{22.1}$$

Fibers possessing such parabolic or near-parabolic refractive index variation have very low pulse dispersion and, hence, have extremely high information-carrying capacities (see Chapter 5). The ray equation itself will be derived in the next section.

22.2 The optical Lagrangian and the ray equation

In geometrical optics we have the Fermat's principle, which determines the path of rays. According to this principle, a ray from point A to point B (see Figure 22.1) will be such that the time taken by the ray is an extremum. Now the time taken by the ray from point A to point B along path ACB (see Figure 22.1) is given by

$$\tau = \int_{\mathrm{A}}^{\mathrm{B}} \frac{ds}{v} = \frac{1}{c}\int_{\mathrm{A}}^{\mathrm{B}} n\, ds \tag{22.2}$$

where ds represents the arc length along the path of the ray. For an allowed ray path, τ should be an extremum – that is,

$$\delta\tau = \frac{1}{c}\,\delta\int_{\mathrm{A}}^{\mathrm{B}} n\, ds = 0$$

or simply

$$\delta\int_{\mathrm{A}}^{\mathrm{B}} n\, ds = 0 \tag{22.3}$$

Now, the arc length ds along the path of the ray is given by

$$(ds)^2 = (dx)^2 + (dy)^2 + (dz)^2 \tag{22.4}$$

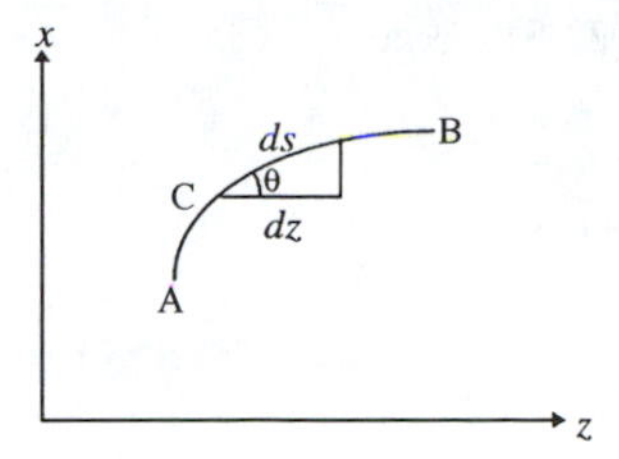

Fig. 22.1: According to Fermat's principle, the ray will take that path from A to B so that the time taken by the ray is an extremum.

which can be written in the form

$$ds = \sqrt{1 + \dot{x}^2 + \dot{y}^2}\, dz \tag{22.5}$$

where

$$\dot{x} = \frac{dx}{dz} \quad \text{and} \quad \dot{y} = \frac{dy}{dz} \tag{22.6}$$

Thus, equation (22.3) becomes

$$\delta \int_A^B n(x, y, z)\sqrt{1 + \dot{x}^2 + \dot{y}^2} dz = 0 \tag{22.7}$$

or

$$\delta \int L\, dz = 0 \tag{22.8}$$

where

$$L = n(x, y, z)\sqrt{1 + \dot{x}^2 + \dot{y}^2} \tag{22.9}$$

represents the optical Lagrangian. Since equation (22.8) is of a form identical to the Hamilton's principle in classical mechanics, we can immediately write the corresponding Lagrange's equations of motion (see Problem 22.4)

$$\frac{d}{dz}\left(\frac{\partial L}{\partial \dot{x}}\right) = \frac{\partial L}{\partial x}$$

or

$$\frac{d}{dz}\left(\frac{n\dot{x}}{\sqrt{1 + \dot{x}^2 + \dot{y}^2}}\right) = \sqrt{1 + \dot{x}^2 + \dot{y}^2}\, \frac{\partial n}{\partial x} \tag{22.10}$$

Similarly

$$\frac{d}{dz}\left(\frac{n\dot{y}}{\sqrt{1 + \dot{x}^2 + \dot{y}^2}}\right) = \sqrt{1 + \dot{x}^2 + \dot{y}^2}\, \frac{\partial n}{\partial y} \tag{22.11}$$

Now, equation (22.10) may be written in the form

$$\frac{1}{\sqrt{1 + \dot{x}^2 + \dot{y}^2}} \frac{d}{dz}\left(n \frac{1}{\sqrt{1 + \dot{x}^2 + \dot{y}^2}} \frac{dx}{dz}\right) = \frac{\partial n}{\partial x} \tag{22.12}$$

If we now use equation (22.5) we get

$$\frac{d}{ds}\left(n\frac{dx}{ds}\right) = \frac{\partial n}{\partial x} \tag{22.13}$$

Similarly

$$\frac{d}{ds}\left(n\frac{dy}{ds}\right) = \frac{\partial n}{\partial y} \tag{22.14}$$

and[1]

$$\frac{d}{ds}\left(n\frac{dz}{ds}\right) = \frac{\partial n}{\partial z} \tag{22.15}$$

The above three equations can be combined into the following vector form

$$\frac{d}{ds}\left(n\frac{d\mathbf{r}}{ds}\right) = \nabla n \tag{22.16}$$

which is known as the ray equation. The above equation can also be derived from Maxwell's equations [see, e.g., Born and Wolf (1975)]; however, the algebra is much more involved.

22.3 The ray invariant $\widetilde{\beta}$ for a waveguide

We next consider a waveguide with a z-independent refractive index profile given by

$$n = n(x, y) \tag{22.17}$$

Thus, equation (22.15) becomes

$$\frac{d}{ds}\left(n\,\frac{dz}{ds}\right) = \frac{\partial n}{\partial z} = 0 \tag{22.18}$$

implying

$$n\,\frac{dz}{ds} = \tilde{\beta} \text{ (an invariant of the ray path)} \tag{22.19}$$

If θ is the angle that the ray makes with the z-axis (see Figure 22.1), then

$$\frac{dz}{ds} = \cos\theta$$

[1] Equation (22.15) follows from the fact that one could equally well have written $ds = [1+\dot{y}^2+\dot{z}^2]^{1/2}\,dx$ with dots now representing differentiation with respect to the x coordinate. Alternatively, equation (22.15) can be derived by using equations (22.13) and (22.14) [see, e.g., Ghatak and Thyagarajan (1978), Section 1.3].

and

$$n(x, y) \cos\theta(x, y) = \tilde{\beta} \tag{22.20}$$

The above equation implies that as the ray propagates through the waveguide, it will bend in such a way that the product $n \cos\theta$ will remain unchanged. We rewrite equation (22.19) and use equation (22.5) to get

$$\tilde{\beta} = n\,\frac{dz}{ds} = \frac{n(x, y)}{\sqrt{1 + \dot{x}^2 + \dot{y}^2}} \tag{22.21}$$

If we use the above equation in equation (22.10) we get

$$\frac{d}{dz}\left(\tilde{\beta}\,\frac{dx}{dz}\right) = \frac{n(x, y)}{\tilde{\beta}}\,\frac{\partial n}{\partial x}$$

Thus

$$\frac{d^2x}{dz^2} = \frac{1}{2\tilde{\beta}^2}\,\frac{\partial n^2}{\partial x} \tag{22.22}$$

Similarly

$$\frac{d^2y}{dz^2} = \frac{1}{2\tilde{\beta}^2}\,\frac{\partial n^2}{\partial y} \tag{22.23}$$

The above equations represent rigorously correct ray equations for media with n^2 independent of the z coordinate.

22.4 Exact solutions for a parabolic index fiber

We consider a parabolic index fiber characterized by the following refractive index variation

$$n^2 = \begin{cases} n_1^2\left(1 - 2\Delta\frac{r^2}{a^2}\right) = n_1^2\left(1 - 2\Delta\frac{x^2+y^2}{a^2}\right); & 0 < r < a \text{ core} \\ n_2^2 = n_1^2\,(1 - 2\Delta); \quad r > a \text{ cladding} \end{cases} \tag{22.24}$$

where n_1 and n_2 represent the core and cladding refractive indices, respectively, a being the core radius, and $r = a$ represents the core–cladding interface. In the core of the fiber, the refractive index is given by equation (22.24), and equation (22.22) takes the form

$$\frac{d^2x}{dz^2} + \Gamma^2 x(z) = 0 \tag{22.25}$$

where

$$\Gamma = \frac{n_1\sqrt{2\Delta}}{a\tilde{\beta}} \tag{22.26}$$

Thus

$$x(z) = A \sin \Gamma z + B \cos \Gamma z \tag{22.27}$$

and similarly

$$y(z) = C \sin \Gamma z + D \cos \Gamma z \tag{22.28}$$

which represent the rigorously correct ray paths inside the core of a parabolic index fiber. The constants A, B, C, and D are determined from the initial launching conditions of the ray; this is indeed very similar to the two-dimensional harmonic oscillator problem where the trajectory of the oscillator is determined from the initial conditions. We next consider some special launching conditions:

22.4.1 *Ray launched in the x–z plane (an example of a meridional ray)*

We first consider a ray launched in the x–z plane on the z-axis making an angle θ_1 with the z-axis. Thus, the launching conditions on the plane $z = 0$ are

$$y(z = 0) = 0 \Rightarrow D = 0$$

$$\frac{dy}{dz}(z = 0) = 0 \Rightarrow C = 0$$

$$x(z = 0) = 0 \Rightarrow B = 0$$

$$\left.\frac{dx}{dz}\right|_{z=0} = \tan \theta_1$$

The last equation implying

$$A = \frac{1}{\Gamma} \tan \theta_1 = \frac{a\tilde{\beta}}{n_1\sqrt{2\Delta}} \tan \theta_1$$

or

$$A = \frac{a \sin \theta_1}{\sqrt{2\Delta}} \tag{22.29}$$

where in the last step we have used the fact that $\tilde{\beta} = n_1 \cos \theta_1$ (see equation (22.20)). Thus, the ray paths are

$$\left.\begin{aligned} x(z) &= \frac{a \sin \theta_1}{\sqrt{2\Delta}} \sin\left(\frac{\sqrt{2\Delta}}{a \cos \theta_1} z\right) \\ y(z) &= 0 \end{aligned}\right\} \text{rigorously correct meridional ray paths in a parabolic index fiber} \tag{22.30}$$

The above equations describe the meridional rays confined in the x–z plane; we point out that the meridional rays are defined so that they are confined in a plane and intersect the z-axis. Obviously, there would be meridional rays confined in the y–z plane – indeed, in any plane containing the z-axis. The ray paths given by equation (22.30) are exactly the same as given in Section 4.2 and the discussion there is equally applicable here. Once again, for paraxial rays, the periodic length

$$z_p = \frac{2\pi a \cos\theta_1}{\sqrt{2\Delta}} \approx \frac{2\pi a}{\sqrt{2\Delta}} \tag{22.31}$$

is independent of θ_1 and therefore in this approximation all rays take the same amount of time. Furthermore, for bound rays we must have

$$n_2 < \tilde{\beta} < n_1 \tag{22.32}$$

a result that is also valid for nonmeridional rays.

22.4.2 Helical ray

In general, equations (22.27) and (22.28) describe what are known as skew rays, which, in general, do not remain confined to a plane. We consider an extreme form of skew rays in which the ray is launched on the x-axis (at $x = a'$) in the y–z plane (making angle θ' with the z-axis). Thus

$$x|_{z=0} = a' \Rightarrow B = a'$$

and

$$\left.\frac{dx}{dz}\right|_{z=0} = 0 \Rightarrow A = 0$$

Thus

$$x(z) = a' \cos\Gamma z \tag{22.33}$$

Further,

$$y|_{z=0} = 0 \Rightarrow D = 0$$

and

$$\left.\frac{dy}{dz}\right|_{z=0} = \tan\theta' \Rightarrow C = \frac{\tan\theta'}{\Gamma} = \frac{a\tan\theta'}{n_1\sqrt{2\Delta}}\tilde{\beta}$$

If at the launching point, $n = n'$, then $\tilde{\beta} = n'\cos\theta'$ and

$$C = \frac{an'\sin\theta'}{n_1\sqrt{2\Delta}}$$

Thus, the ray path is given by

$$\left.\begin{aligned} x(z) &= a'\cos\Gamma z \\ y(z) &= \frac{an'\sin\theta'}{n_1\sqrt{2\Delta}}\sin\Gamma z \end{aligned}\right\} \tag{22.34}$$

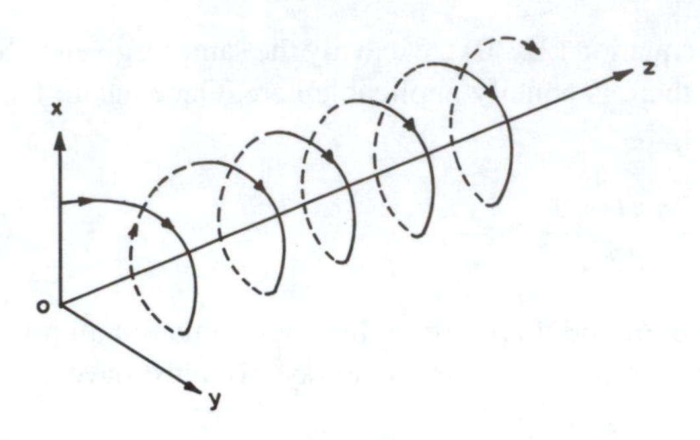

Fig. 22.2: Propagation of a helical ray in a parabolic index fiber.

If

$$\sin\theta' = \frac{a' n_1 \sqrt{2\Delta}}{a n'} = \frac{\sqrt{n_1^2 - n'^2}}{n'} \tag{22.35}$$

then

$$\left.\begin{aligned} x(z) &= a' \cos \Gamma z \\ y(z) &= a' \sin \Gamma z \end{aligned}\right\} \tag{22.36}$$

and

$$x^2(z) + y^2(z) = a'^2 \quad \text{independent of } z \tag{22.37}$$

that is, the ray spirals around the z-axis as a helix at a constant distance a' from it (see Figure 22.2). Such a ray is an extreme form of a skew ray and is known as a helical ray.

Problems

22.1 Consider a typical parabolic index fiber with

$$n_1 = 1.47, \quad \Delta = 0.01, \quad a = 50\,\mu\text{m}$$

Consider bound meridional rays launched on the z-axis making angle θ_1 with the z-axis. Calculate the minimum and maximum values of $\tilde{\beta}$, θ_1, and z_p.

[ANSWER: $1.4553 < \tilde{\beta} < 1.47$.]

22.2 For the fiber described in Problem 22.1, calculate the launching angle θ' for the ray to be helical if the ray is launched at $x = 20\,\mu$m and at $x = 40\,\mu$m.

22.3 Obtain the ray paths in an elliptic parabolic index fiber characterized by the following refractive index distribution.

$$\left.\begin{aligned} n^2 &= n_1^2\left[1 - 2\Delta\left(\tfrac{x^2}{a^2} + \tfrac{y^2}{b^2}\right)\right] \quad \text{for } \tfrac{x^2}{a^2} + \tfrac{y^2}{b^2} < 1 \\ &= n_1^2(1 - 2\Delta) = n_2^2 \quad \text{for } \tfrac{x^2}{a^2} + \tfrac{y^2}{b^2} > 1 \end{aligned}\right\} \tag{22.38}$$

Solution: The ray paths inside the core of the fiber are given by

$$\left.\begin{aligned} x(z) &= A \cos \Gamma_x z + B \sin \Gamma_x z \\ y(z) &= C \cos \Gamma_y z + D \sin \Gamma_y z \end{aligned}\right\} \tag{22.39}$$

where

$$\Gamma_x = \frac{n_1\sqrt{2\Delta}}{a\tilde{\beta}}, \quad \Gamma_y = \frac{n_1\sqrt{2\Delta}}{b\tilde{\beta}} \tag{22.40}$$

Thus, the periodic lengths along x and y would be different.

22.4 Derive Lagrange's equation from Fermat's principle. [Goldstein (1950) Section 2.3.]

Solution: We consider only the case in which we assume the Lagrangian to be independent of the y coordinate. Fermat's principle tells us that

$$0 = \delta \int_{z_1}^{z_2} L[x(z), \dot{x}(z), z]\, dz = \int_{z_1}^{z_2} \delta L[x(z), \dot{x}(z), z]\, dz$$

$$= \int_{z_1}^{z_2} \left[\frac{\partial L}{\partial x}\,\delta x + \frac{\partial L}{\partial \dot{x}}\,\delta \dot{x}\right] dz \tag{22.41}$$

where δL represents the change in the Lagrangian as we go from the actual ray path to a nearby path having the same endpoints (i.e., $\delta x|_{z_1} = \delta x|_{z_2} = 0$). Now

$$\int_{z_1}^{z_2} \frac{\partial L}{\partial \dot{x}}\delta \dot{x}\, dz = \int_{z_1}^{z_2} \frac{\partial L}{\partial \dot{x}}\frac{d}{dz}\delta x\, dz$$

$$= \frac{\partial L}{\partial \dot{x}}\,\delta x\bigg|_{z_1}^{z_2} - \int_{z_1}^{z_2} \frac{d}{dz}\left(\frac{\partial L}{\partial \dot{x}}\right)\delta x\, dz$$

$$= \int_{z_1}^{z_2} \frac{d}{dz}\left(\frac{\partial L}{\partial \dot{x}}\right)\delta x\, dz$$

Thus, equation (22.41) becomes

$$\int_{z_1}^{z_2} \left[\frac{\partial L}{\partial x} - \frac{d}{dz}\left(\frac{\partial L}{\partial \dot{x}}\right)\right]\delta x\, dz = 0 \tag{22.42}$$

Although $\delta x(z)$ is an infinitesimal quantity, it is an arbitrary function of z; thus, the integrand of equation (22.42) must vanish, which gives us

$$\frac{d}{dz}\left(\frac{\partial L}{\partial \dot{x}}\right) = \frac{\partial L}{\partial x} \tag{22.43}$$

The above considerations can easily be generalized to the Lagrangian depending on the y coordinate also.

23

Ray paths and their classification in optical fibers

23.1 Introduction

In most optical fibers, the refractive index depends only on the distance from the axis – that is,

$$n = n(r) \tag{23.1}$$

which are usually referred to as cylindrically symmetric media. In Section 23.2 we use the Lagrangian formalism to derive the equations determining the ray paths in a cylindrically symmetric medium; these equations were used in Section 5.3 to calculate pulse dispersion for a general class of graded index optical fibers. Such calculations are of extreme importance in obtaining the optimum profile that gives rise to minimum pulse dispersion (see Section 5.2).

In Section 23.3 we obtain solutions of the ray equation for a parabolic index fiber. We compare these solutions with the ones obtained in the previous chapter, where we solved the ray equation in Cartesian coordinates.

In Section 23.4 we use the equations determining the ray paths to classify different types of rays that are excited in the fiber. We show that the rays that are excited in the fiber can be classified into

(1) bound rays
(2) refracting leaky rays
(3) tunneling leaky rays

Bound rays remain guided in the fiber. Refracting leaky rays leak out of the fiber in a very short distance. On the other hand, tunneling leaky rays leak out gradually from the core of the fiber; the attenuation distance of tunneling leaky rays can vary from a few millimeters to kilometers! Thus, a study of excitation of different types of rays is of importance in the understanding of the propagation of an optical beam through the fiber.

The analysis given in this chapter is based on papers by Ankiewicz and Pask (1977) and Ghatak and Sauter (1989).

23.2 Ray equation for cylindrically symmetric refractive index profiles

In the previous chapter we derived the ray equation

$$\frac{d}{ds}\left[n\frac{d\mathbf{r}}{ds}\right] = \nabla n \tag{23.2}$$

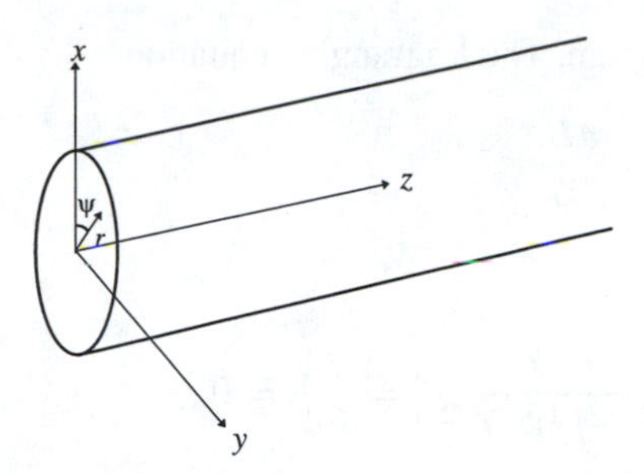

Fig. 23.1: The cylindrical system of coordinates (r, ψ, z). The z-axis represents the axis of the fiber.

where ds represents the arc length along the ray path. If we take the z component of the above equation we get

$$\frac{d}{ds}\left[n\frac{dz}{ds}\right] = \frac{\partial n}{\partial z} = 0 \tag{23.3}$$

where we have used equation (23.1). Thus, we obtain the invariant

$$\tilde{\beta} = n\,\frac{dz}{ds} = n(r)\cos\theta \tag{23.4}$$

where θ is the angle that the ray makes with the z-axis (cf. equation (4.3)).

Now, in a cylindrically symmetric medium (i.e., $n = n(r)$), it is obviously more convenient to use the cylindrical system of coordinates (r, ψ, z) (see Figure 23.1) for which the arc length along the path of the ray is given by

$$\begin{aligned} ds &= [(dz)^2 + (dr)^2 + (r\,d\psi)^2]^{1/2} \\ &= [1 + \dot{r}^2 + r^2\dot{\psi}^2]^{1/2}\,dz \end{aligned} \tag{23.5}$$

where dots represent differentiation with respect to the z coordinate.

$$\dot{r} \equiv \frac{dr}{dz}; \quad \dot{\psi} \equiv \frac{d\psi}{dz} \tag{23.6}$$

Thus

$$\tilde{\beta} = n\,\frac{dz}{ds} = \frac{n(r)}{\sqrt{1 + \dot{r}^2 + r^2\dot{\psi}^2}} \quad \text{ray invariant} \tag{23.7}$$

Furthermore, Fermat's principle gives us

$$0 = \delta\int n\,ds = \delta\int n(r)[1 + \dot{r}^2 + r^2\dot{\psi}^2]^{1/2}\,dz$$

implying

$$L = L_{opt} = n(r)[1 + \dot{r}^2 + r^2\dot{\psi}^2]^{1/2} \tag{23.8}$$

as the optical Lagrangian. The Lagrangian equation

$$\frac{d}{dz}\left(\frac{\partial L}{\partial \dot{\psi}}\right) = \frac{\partial L}{\partial \psi}$$

gives

$$\frac{d}{dz}\left[\frac{n(r)\,r^2\dot{\psi}}{(1+\dot{r}^2+r^2\dot{\psi}^2)^{1/2}}\right] = \frac{\partial L}{\partial \psi} = 0$$

or

$$\frac{d}{dz}[\tilde{\beta} r^2 \dot{\psi}] = 0 \tag{23.9}$$

where we have used equation (23.7). Thus, we obtain another invariant of the ray path

$$\tilde{\beta}\,\frac{r^2}{a}\,\frac{d\psi}{dz} = \tilde{l} \quad \text{ray invariant} \tag{23.10}$$

In the above equation the parameter a represents the core radius, which has been introduced so that the invariant $\tilde{l}$ is dimensionless. Physically, the invariants $\tilde{\beta}$ and $\tilde{l}$ are manifestations of the translational and rotational invariance of the refractive index profile. The parameter $\tilde{l}$ is usually referred to as the *skewness parameter*; obviously, for meridional rays, the angle ψ remains unchanged and $\tilde{l} = 0$.

Now, from equation (23.7) we have

$$\left[\frac{n(r)}{\tilde{\beta}}\right]^2 = 1 + \dot{r}^2 + r^2\dot{\psi}^2$$

$$= 1 + \dot{r}^2 + r^2\left(\frac{\tilde{l}a}{\tilde{\beta}r^2}\right)^2$$

Simple rearrangement gives us

$$\dot{r} = \frac{dr}{dz} = \pm\frac{1}{\tilde{\beta}}[f(r)]^{1/2} \tag{23.11}$$

where

$$f(r) \equiv n^2(r) - \frac{\tilde{\beta}^2}{(r/a)^2} - \tilde{l}^2 \tag{23.12}$$

The evaluation of the integral

$$\tilde{\beta}\int \frac{dr}{[f(r)]^{1/2}} = \pm \int dz \tag{23.13}$$

would give us the r coordinate of the ray as a function of z as it propagates through the fiber. Once $r(z)$ is known, the integration of the equation

$$\dot{\psi} = \frac{d\psi}{dz} = \frac{a\tilde{l}}{\tilde{\beta}r^2(z)} \tag{23.14}$$

would determine the azimuthal coordinate ψ as a function of z. Obviously, $f(r)$ should be positive; this condition will be used later for classification of different types of rays.

23.3 Exact ray paths in a parabolic index fiber

In Section 22.4, we obtained ray paths in a parabolic index fiber by using the Cartesian system of coordinates. We now use equation (23.13) to determine the exact ray paths in a parabolic index fiber for which the refractive index variation (inside the core) is given by

$$n^2(r) = n_1^2\left[1 - 2\Delta\left(\frac{r}{a}\right)^2\right] = n_1^2 - g^2r^2 \tag{23.15}$$

where

$$g^2 = n_1^2\frac{2\Delta}{a^2} \tag{23.16}$$

Typical parameters for a parabolic index fiber are

$$n_1 \simeq 1.45, \quad \Delta \simeq 0.01, \quad a \simeq 25\,\mu\text{m}$$

implying

$$g \simeq 8.2 \times 10^3\,\text{m}^{-1}$$

Substituting the parabolic refractive index variation in equation (23.12), equation (23.13) takes the form

$$\tilde{\beta}\int \frac{dr}{\left[n_1^2 - g^2r^2 - \frac{\tilde{l}^2}{(r/a)^2} - \tilde{\beta}^2\right]^{1/2}} = \pm\int dz \tag{23.17}$$

Carrying out the integration we obtain (see Problem 23.3)

$$r^2(z) = \gamma + \alpha\cos\frac{2g}{\tilde{\beta}}(z - z_1) \tag{23.18}$$

which represents exact ray paths in a parabolic index fiber. In equation (23.18) z_1 is the constant of integration,

$$\gamma = \frac{n_1^2 - \tilde{\beta}^2}{2g^2} = \frac{1}{2}\left[1 - \frac{\tilde{\beta}^2}{n_1^2}\right]\frac{a^2}{2\Delta} \tag{23.19}$$

and

$$\alpha^2 = \gamma^2 - \frac{\tilde{l}^2a^2}{g^2} = \left(\frac{a^2}{2\Delta}\right)^2\left[\frac{1}{4}\left(1 - \frac{\tilde{\beta}^2}{n_1^2}\right)^2 - \tilde{l}^2\frac{2\Delta}{n_1^2}\right] \tag{23.20}$$

Since

$$-1 < \cos\frac{2g}{\tilde{\beta}}(z - z_1) < 1$$

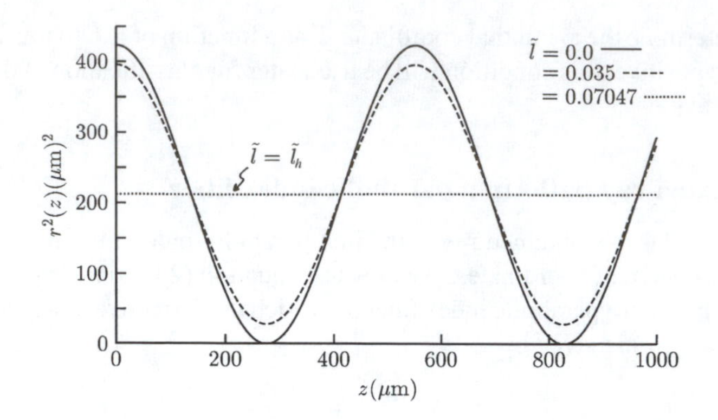

Fig. 23.2: Variation of r^2 with z for different values of $\tilde{l}$. The values of various parameters are $n_1 = 1.45$, $\Delta = 0.01$, and $a = 25\ \mu m$.

we have

$$\gamma - \alpha < r^2(z) < \gamma + \alpha \tag{23.21}$$

as shown in Figure 23.2. The circles

$$r(z) = \sqrt{\gamma - \alpha} = r_1 \tag{23.22}$$

and

$$r(z) = \sqrt{\gamma + \alpha} = r_2 \tag{23.23}$$

represent the inner and outer caustics where $f(r)$ (and hence dr/dz) vanishes (see Figure 23.3). For $\tilde{l} = 0$, $\alpha = \gamma$ and the radius of the inner caustic becomes zero and the ray intersects the axis (see also Figure 5.9(a) and (b)). To obtain $r(\psi)$ we divide equation (23.10) by equation (23.11) to obtain

$$\tilde{\beta}\frac{r^2}{a}\frac{d\psi}{dr} = \frac{\tilde{l}\tilde{\beta}}{[f(r)]^{1/2}}$$

or

$$\int d\psi = \pm a\tilde{l}\int \frac{dr}{r^2\left[n^2(r) - \frac{\tilde{l}^2}{(r/a)^2} - \tilde{\beta}^2\right]^{1/2}} \tag{23.24}$$

If we now substitute equation (23.15) for $n^2(r)$ and carry out the integration (see Problem 23.4) we get

$$\frac{1}{r^2} = \nu + \mu\cos 2(\psi - \psi_1) \tag{23.25}$$

where ψ_1 is a constant of integration,

$$\nu = \frac{n_1^2 - \tilde{\beta}^2}{2\tilde{l}^2 a^2} = \frac{\gamma}{l^2}; \quad l = \frac{\tilde{l}a}{g} \tag{23.26}$$

and

$$\mu = \left[\nu^2 - \frac{g^2}{\tilde{l}^2 a^2}\right]^{1/2} = \left[\frac{\gamma^2}{l^4} - \frac{1}{l^2}\right]^{1/2} \tag{23.27}$$

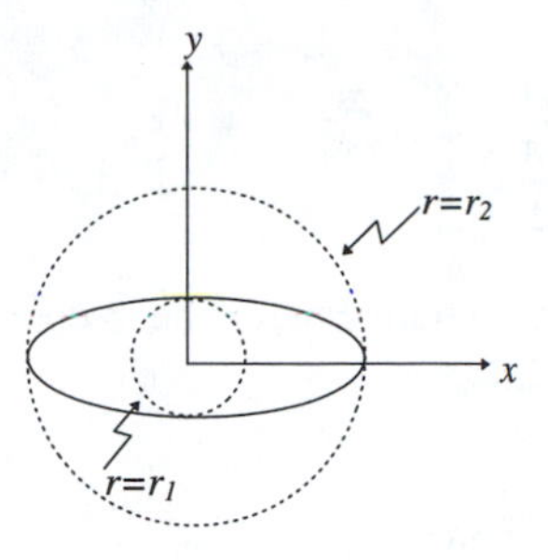

Fig. 23.3: The projection of a ray on the x–y plane is an ellipse. The two circles ($r = r_1$ and $r = r_2$) represent the inner and outer caustics. In general, the axes of the ellipse will make an angle with the x- and y-axes. For meridional rays, the ellipse will become a straight line, and for helical rays $r_1 = r_2$.

Using equations (23.18) and (23.25) we can calculate r and ψ as functions of z. The projection of the ray path in a plane perpendicular to the z-direction is shown in Figure 23.3 and, as can be seen, the projection is a closed ellipse (see also Section 22.4 and Problem 23.4). In Figures 5.9(a) and 5.9(b) we have shown typical variations of $r(z)$ and also the projection of the ray path on the transverse plane for $\tilde{l} > 0$ and $\tilde{l} = 0$, respectively.

We next consider some special cases.

23.3.1 Meridional rays ($\tilde{l} = 0$)

For meridional rays, $\tilde{l} = 0$ and

$$\alpha = \gamma = \frac{1}{2}\left(1 - \frac{\tilde{\beta}^2}{n_1^2}\right)\frac{a^2}{2\Delta}$$

Thus, equation (23.18) can be written in the form

$$\begin{aligned} r^2(z) &= \gamma\left[1 + \cos\frac{2g}{\tilde{\beta}}(z - z_1)\right] \\ &= 2\gamma\cos^2\left[\frac{g}{\tilde{\beta}}(z - z_1)\right] \end{aligned} \tag{23.28}$$

or

$$\begin{aligned} r(z) &= \pm\sqrt{2\gamma}\,\cos\left[\frac{g}{\tilde{\beta}}(z - z_1)\right] \\ &= \pm\left[\left(1 - \frac{\tilde{\beta}^2}{n_1^2}\right)\frac{a^2}{2\Delta}\right]^{1/2}\cos\left(\frac{n_1\sqrt{2\Delta}}{a\tilde{\beta}}\,z - \phi_1\right) \end{aligned} \tag{23.29}$$

Since r should always be positive, we must choose the + sign where the cosine function is positive and the − sign when the cosine function is negative (see Figure 5.9(b)). If we put $\tilde{\beta} = n_1\cos\theta_1$ and $\phi_1 = \pi/2$, the above equation is identical in form to the ray path given by equation (22.30). Further, when $\tilde{l} = 0$, we must have

$$\cos 2(\psi - \psi_1) = 0$$

or

$$\psi - \psi_1 = \frac{\pi}{2}, \frac{3\pi}{2}$$

implying that ψ should not change with the z coordinate except for abrupt changes from $\pi/2$ to $3\pi/2$.

23.3.2 Helical rays ($\tilde{l} = \tilde{l}_h$)

From equation (23.20) one can readily see that the maximum value of $\tilde{l}$ is given by

$$\tilde{l} = \tilde{l}_h = \frac{\gamma g}{a} = \frac{n_1^2 - \tilde{\beta}^2}{2n_1\sqrt{2\Delta}} \tag{23.30}$$

and $\alpha = 0$ so that

$$r^2(z) = \gamma = \frac{1}{2}\left(1 - \frac{\tilde{\beta}^2}{n_1^2}\right)\frac{a^2}{2\Delta} \tag{23.31}$$

implying that the ray will always be at a constant distance from the axis; this is the helical ray, which we discussed in Section 22.4.2.

23.4 Classification of rays

We now discuss the different types of rays that can propagate through an optical fiber. Our starting point will be the equation determining the ray path (see equation (23.11)).

$$\tilde{\beta}\int \frac{dr}{[f(r)]^{1/2}} = \int dz \tag{23.32}$$

where

$$f(r) = n^2(r) - \tilde{\beta}^2 - \frac{\tilde{l}^2}{(r/a)^2} \tag{23.33}$$

Obviously, for a ray path to be allowed, $f(r)$ should be positive. The forbidden regions correspond to $f(r)$ being negative.

We assume the refractive index variation to be of the form

$$\begin{aligned} n^2(r) &= n_1^2\left[1 - 2\Delta\left(\frac{r}{a}\right)^q\right], \quad 0 < r < a \\ &= n_2^2 = n_1^2(1 - 2\Delta), \quad r > a \end{aligned} \tag{23.34}$$

which is known as the power law profile (see Figure 5.1). Most multimode fibers can be approximately characterized by a power law profile. We first consider the case $\tilde{l} = 0$, which will be followed by the case $\tilde{l} > 0$.

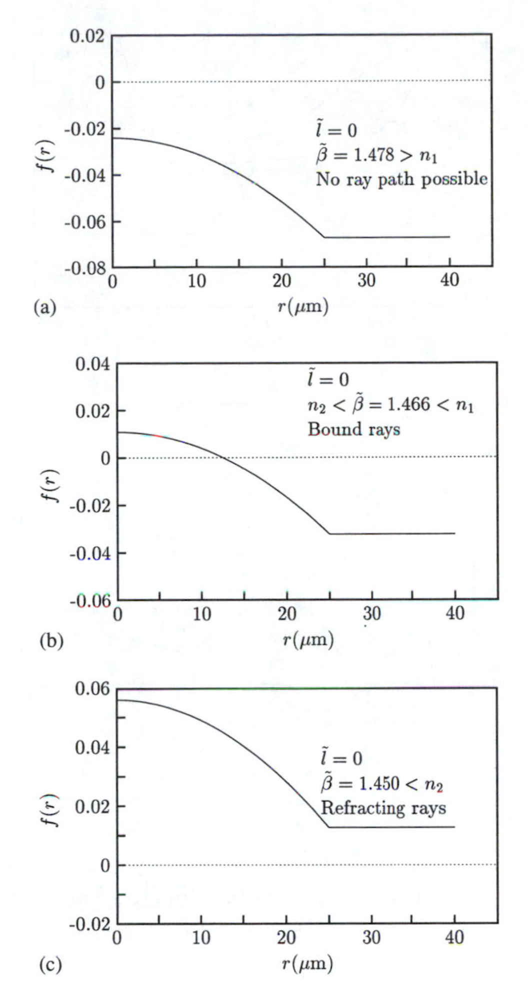

Fig. 23.4: Variation of $f(r)$ for meridional rays ($\tilde{l} = 0$). No ray paths are possible for $\tilde{\beta} > n_1$ (see (a)). Bound rays and refracting rays correspond to $n_2 < \tilde{\beta} < n_1$ and $0 < \tilde{\beta} < n_2$, respectively (see (b) and (c)).

23.4.1 $\tilde{l} = 0$ meridional rays

From equation (23.33) we have

$$f(r) = n^2(r) - \tilde{\beta}^2 \quad \text{for } \tilde{l} = 0 \tag{23.35}$$

The $f(r)$ variations for $\tilde{\beta}^2 > n_1^2$, $n_2^2 < \tilde{\beta}^2 < n_1^2$, and $\tilde{\beta}^2 < n_2^2$ are shown in Figures 23.4(a), 23.4(b), and 23.4(c), respectively. Although the figures correspond to the parabolic index profile, the general behavior will be the same for an arbitrary power law variation. As can be seen, for $\tilde{\beta}^2 > n_1^2$, $f(r)$ is everywhere negative and no ray path is possible. This is also obvious from the fact that

$$\tilde{\beta} = n(r) \cos\theta(r)$$

and since the maximum values of $n(r)$ and $\cos\theta$ are n_1 and 1, $\tilde{\beta}$ can never be greater than n_1 – that is,

$$\tilde{\beta} \ngtr n_1 \tag{23.36}$$

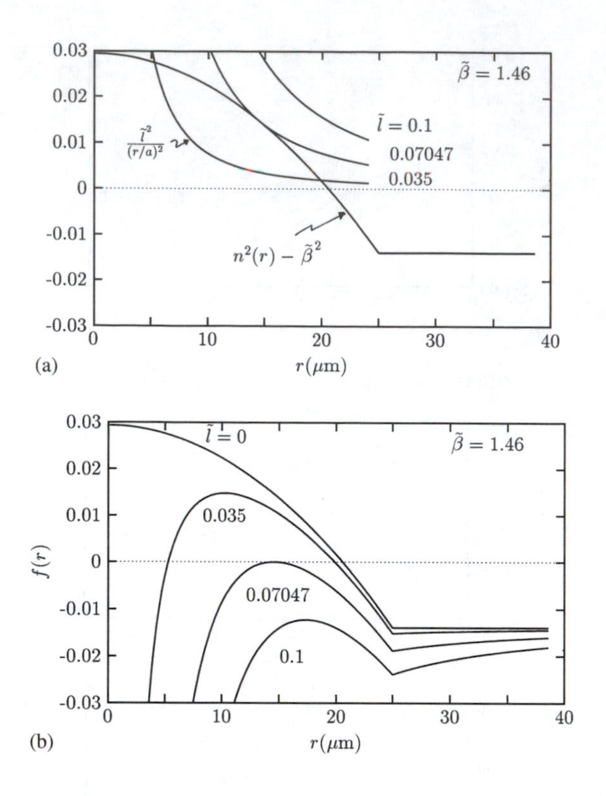

Fig. 23.5: (a) Variations of $n^2(r) - \tilde{\beta}^2$ and $\tilde{l}^2/(r/a)^2$ corresponding to a $q = 2$ fiber for bound rays ($n_2 < \tilde{\beta} < n_1$). (b) The corresponding $f(r)$ variations; the values of various parameters are $n_1 = 1.47$, $\Delta = 0.01$, $a = 25\ \mu$m, and $\tilde{\beta} = 1.46$. The value of $\tilde{l}_h = 0.07047$.

When

$$n_2^2 < \tilde{\beta}^2 < n_1^2$$

$f(r)$ is positive in the domain $0 < r < r_2$ (see Figure 23.4(b)) and $r = r_2$ is so that

$$n^2(r_2) = \tilde{\beta}^2 \tag{23.37}$$

These are the *bound meridional rays* and $r = r_2$ is known as the turning point, which is given by

$$r_2 = a\left[\frac{n_1^2 - \tilde{\beta}^2}{2\Delta n_1^2}\right]^{1/q} \tag{23.38}$$

For a step index fiber, ($q = \infty$) and $r_2 = a$; the rays undergo total internal reflection at the core–cladding interface.

Finally, for $\tilde{\beta}^2 < n_2^2$, $f(r)$ is everywhere positive (see Figure 23.4(c)) and the ray just refracts away at the core–cladding interface.

23.4.2 $\tilde{l} > 0$ skew rays

We next consider the case $\tilde{l} > 0$, which corresponds to skew rays where the rays are not confined to a single plane. Once again, for $\tilde{\beta}^2 > n_1^2$, $f(r)$ will be everywhere negative, implying that no ray paths are possible.

For $n_2^2 < \tilde{\beta}^2 < n_1^2$, the variations of $n^2(r) - \tilde{\beta}^2$ and $\tilde{l}^2/(r/a)^2$ for $q = 2$ are shown in Figure 23.5(a). The points of intersection of the two curves determine

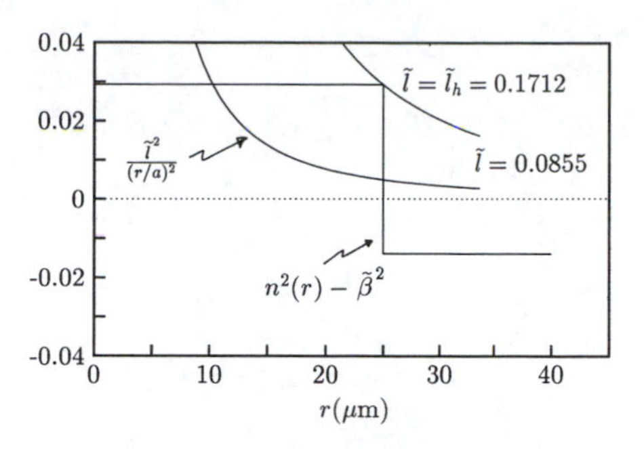

Fig. 23.6: The step function shows the variation of $n^2(r) - \tilde{\beta}^2$ with r for a step index fiber; the values of various parameters are $n_1 = 1.47$, $\Delta = 0.001$, $a = 25\,\mu$m, and $\tilde{\beta} = 1.46$. The two curves represent the variation of $\tilde{l}^2/(r/a)^2$ for $\tilde{l} = \tilde{l}_h = 0.17117$ and $\tilde{l} = \frac{1}{2}\tilde{l}_h = 0.0855$.

the domain of the ray path. When the value of $\tilde{l}$ is such that the two curves touch at one point, we have a helical ray ($\tilde{l} = \tilde{l}_h$); the value of $\tilde{l}_h$ will depend on the value of $\tilde{\beta}$. In Figure 23.5(b) we have plotted the corresponding variation of $f(r)$. Figure 23.5 corresponds to $q = 2$ with

$$n_1 = 1.47, \quad a = 25\,\mu\text{m}, \quad \Delta = 0.01 \tag{23.39}$$

and

$$\tilde{\beta} = 1.46 \tag{23.40}$$

Thus, from equation (23.30) we find

$$\tilde{l}_h = 0.07047 \tag{23.41}$$

For $\tilde{l} < \tilde{l}_h$, we will have two values of r (equal to r_1 and r_2), where $f(r) = 0$; the points r_1 and r_2 at which $f(r) = 0$ (inside the core) are known as the turning points, and the cylindrical surfaces $r = r_1$ and $r = r_2$ are known as the inner and outer caustic, respectively. For $q = 2$, the projection of a bound skew ray path on the x–y plane is an ellipse as shown in Figure 23.3.

For a step index fiber ($q = \infty$) $r_2 = a$ – that is, the outer caustic is at the core–cladding interface. Figure 23.6 corresponds to a step index fiber ($q = \infty$) with the same values of n_1, a, Δ, and $\tilde{\beta}$ as given by equations (23.39) and (23.40). Obviously,

$$\tilde{l}_h = \left(n_1^2 - \tilde{\beta}^2\right)^{1/2} = 0.17117 \tag{23.42}$$

and the helical ray slides along the core-cladding interface. For $\tilde{l} > \tilde{l}_h$, $f(r)$ will be negative everywhere and there will be no ray path possible.

For $q \neq 2$ and $\neq \infty$, the projection of the bound skew ray on the x–y plane is qualitatively shown in Figure 23.7; for meridional rays ($\tilde{l} = 0$), the curve will degenerate into a straight line.

We finally consider the case when $\tilde{\beta}^2 < n_2^2$. Corresponding to different values of $\tilde{l}$, four different situations may arise. For small values of $\tilde{l}$, there is only one point of intersection between the two curves and we have a refracting (skew) ray as shown in Figure 23.8(a). As we increase the value of $\tilde{l}$, we obtain three values of r where $f(r) = 0$ and we have the tunneling leaky rays, where we have a guided ray inside the core of the fiber, and since the ray path is also

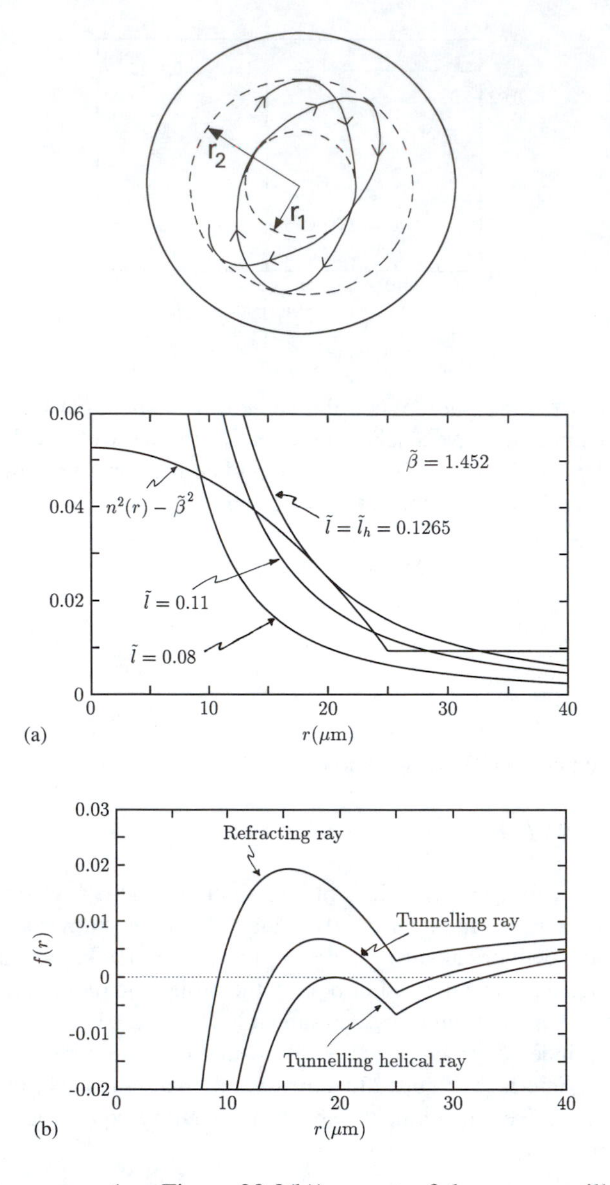

Fig. 23.7: Projection of the ray path onto the fiber cross section for $q \neq 2$, $\tilde{l} \neq 0$. For $\tilde{l} = 0$, the ray is meridional with $r_1 = 0$ and the projection is a straight line. For a parabolic index fiber ($q = 2$), the projection is a closed ellipse. [Adapted from Ankiewicz and Pask (1977).]

Fig. 23.8: (a) For $\tilde{\beta}\,(=1.452) < n_2$, variations of $n^2(r) - \tilde{\beta}^2$ and $\tilde{l}^2/(r/a)^2$ for $\tilde{l} = \tilde{l}_h = 0.1265$ (tunneling helical ray), $\tilde{l} = 0.11$ (tunneling skew ray), and $\tilde{l} = 0.08$ (refracting ray) in a parabolic index fiber ($n_1 = 1.47$, $\Delta = 0.01$, $a = 25\,\mu$m). (b) The corresponding variations of $f(r)\,[=n^2(r) - \tilde{\beta}^2 - \tilde{l}^2/(r/a)^2]$.

allowed for $r > r_3$ (see Figure 23.8(b)), a part of the energy will tunnel out at the outer surface caustic. This tunneling of power is entirely a wave optic phenomenon, with the tunneling probability at each caustic approximately given by the following expression:

$$T \simeq \exp\left[-2k_0 \int_{r_2}^{r_3} \left(n^2(r) - \tilde{\beta}^2 - \frac{\tilde{l}^2 a^2}{r^2}\right)^{1/2} dr\right] \tag{23.43}$$

where $k_0 = 2\pi/\lambda_0$ and $r = r_2$ represents the outer caustic.

As we further increase the value of $\tilde{l}$, we get a tunneling helical ray ($\tilde{l} = \tilde{l}_h$). For $\tilde{l} > \tilde{l}_h$, $f(r)$ will not be positive inside the core and no ray path (inside the core) will be possible.

Problems

23.1 (a) Solve equation (23.11) for a step index optical fiber to obtain

$$r^2(z) = \frac{\tilde{l}^2 a^2}{n_1^2 - \tilde{\beta}^2} + \frac{n_1^2 - \tilde{\beta}^2}{\tilde{\beta}^2}(z-z_1)^2; \quad 0 < r < a \tag{23.44}$$

where we have assumed that $r^2(z)$ is a minimum at $z = z_1$ and

$$r^2\Big|_{min} = \frac{\tilde{l}^2 a^2}{n_1^2 - \tilde{\beta}^2} \tag{23.45}$$

Show also that the ray will hit the core–cladding interface at

$$z = z_1 \pm z_0 \tag{23.46}$$

where

$$z_0 = \left[\frac{\tilde{\beta}^2}{n_1^2 - \tilde{\beta}^2}\left(1 - \frac{\tilde{l}^2}{n_1^2 - \tilde{\beta}^2}\right)\right]^{1/2} a \tag{23.47}$$

(b) Solve equation (23.14) to obtain

$$r\cos(\psi - \psi') = \pm\frac{\tilde{l}\,a}{\left(n_1^2 - \tilde{\beta}^2\right)^{1/2}} \tag{23.48}$$

or

$$x\cos\psi' + y\sin\psi' = \pm\frac{\tilde{l}\,a}{\left(n_1^2 - \tilde{\beta}^2\right)^{1/2}} \tag{23.49}$$

showing that the projection of the ray on the x–y plane is a straight line.

23.2 Consider again a step index fiber.

(a) Show that no ray paths are possible for $\tilde{\beta} > n_1$.

(b) For $n_2 < \tilde{\beta} < n_1$ show that guided skew rays correspond to $0 < \tilde{l} < \tilde{l}_h$ where

$$\tilde{l} = \tilde{l}_h = \left(n_1^2 - \tilde{\beta}^2\right)^{1/2} \tag{23.50}$$

For the helical ray ($\tilde{l} = \tilde{l}_h$), show that

$$r(z) = a$$

and

$$\psi(z) = \pm\frac{\tilde{l}}{\tilde{\beta}}\,z + \text{constant} \tag{23.51}$$

(c) For $\tilde{\beta} < n_2$, show that we will have tunneling leaky rays for $\tilde{l}_1 < \tilde{l} < \tilde{l}_h$, where

$$\tilde{l}_1 = \left(n_2^2 - \tilde{\beta}^2\right)^{1/2} \tag{23.52}$$

23.3 Carry out the integration in equation (23.17) and derive equation (23.18)

Solution:

$$z - z_1 = \pm\tilde{\beta} \int \frac{dr}{\left[n_2^2 - g^2 r^2 - \frac{\tilde{l}^2 a^2}{r^2} - \tilde{\beta}^2\right]^{1/2}}$$

Making the substitution $\zeta = r^2$, we readily obtain

$$z - z_1 = \pm\frac{1}{2g}\tilde{\beta} \int \frac{d\zeta}{[\alpha^2 - (\zeta - \gamma)^2]^{1/2}}$$

$$= \mp\frac{1}{2g}\tilde{\beta} \cos^{-1}\left(\frac{\zeta - \gamma}{\alpha}\right)$$

where α and γ are defined through equations (23.19) and (23.20). From the above equation one readily obtains equation (23.18).

23.4 (a) For $n^2(r) = n_1^2 - g^2 r^2$, carry out the integration in equation (23.24) to derive equation (23.25) and
(b) show that it represents an ellipse.

Solution:

(a) If we use the transformation

$$\xi = \frac{1}{r^2} \Rightarrow \frac{dr}{r^2} = -\frac{1}{2\sqrt{\xi}}\, d\xi \tag{23.53}$$

in equation (23.24) we obtain

$$\psi - \psi_1 = \mp\frac{1}{2}\tilde{l}a \int \frac{d\xi}{\sqrt{\xi f(\xi)}} \tag{23.54}$$

where ψ_1 is the constant of integration and

$$\xi f(\xi) = \xi\left[n_1^2 - g^2 r^2 - \tilde{\beta}^2 - \frac{\tilde{l}^2 a^2}{r^2}\right]$$

$$= \xi\left[\left(n_1^2 - \tilde{\beta}^2\right) - \frac{g^2}{\xi} - \tilde{l}^2 a^2 \xi\right]$$

$$= \tilde{l}^2 a^2 [\mu^2 - (\xi - \nu)^2]$$

and ν and μ have been defined through equations (23.26) and (23.27). Thus

$$2(\psi - \psi_1) = \mp \int \frac{d\xi}{[\mu^2 - (\xi - \nu)^2]^{1/2}} = \pm\cos^{-1}\left(\frac{\xi - \nu}{\mu}\right)$$

or

$$\xi = \frac{1}{r^2} = \nu + \mu \cos[2(\psi - \psi_1)] \tag{23.55}$$

(b) Without any loss of generally we may assume $\psi_1 = \pi/2$ (which essentially represents a reorientation of the axes). Elementary manipulations give us

$$\frac{1}{r^2} = \nu(\cos^2\psi + \sin^2\psi) - \mu(\cos^2\psi - \sin^2\psi) \tag{23.56}$$

or

$$\frac{x^2}{r_2^2} + \frac{y^2}{r_1^2} = 1 \tag{23.57}$$

where

$$r_2 = \frac{1}{\sqrt{\nu - \mu}} \quad \text{and} \quad r_1 = \frac{1}{\sqrt{\nu + \mu}} \tag{23.58}$$

Equation (23.57) represents an ellipse with r_1 and r_2 representing the minor and major axes (see Figure 23.3).

23.5 Show that

$$r_1 = \frac{1}{\sqrt{\nu + \mu}} = \sqrt{\gamma - \alpha} \quad \text{and} \quad r_2 = \frac{1}{\sqrt{\nu - \mu}} = \sqrt{\gamma + \alpha} \tag{23.59}$$

which represent the radii of inner and outer caustic, respectively.

23.6 Show that in the limit of $\tilde{\beta} \to 0$, the ellipse degenerates to a straight line.

Solution:

$$\frac{1}{\nu \pm \mu} = \left[\frac{\gamma}{l^2} \pm \sqrt{\frac{\gamma^2}{l^4} - \frac{1}{l^2}}\right]^{-1}$$

$$= \frac{l^2}{\gamma}\left[1 \pm \sqrt{1 - \frac{l^2}{\gamma^2}}\right]^{-1}$$

Thus, as $\tilde{l} \to 0$

$$r_1^2 = \frac{1}{\nu + \mu} \simeq \frac{l^2}{2\gamma} \to 0$$

and

$$r_2^2 = \frac{1}{\nu - \mu} \simeq 2\gamma = \frac{n_1^2 - \tilde{\beta}^2}{g^2}$$

which is consistent with equation (23.29).

23.7 Using the Cartesian system of coordinates, show that the invariant $\tilde{l}$ is given by

$$\tilde{l} = \frac{\tilde{\beta}}{a}\left[x(z)\frac{dy}{dz} - y(z)\frac{dx}{dz}\right] \tag{23.60}$$

Solution: Since $\tan\psi = y/x$, we have

$$\sec^2\psi \frac{d\psi}{dz} = \frac{1}{x}\frac{dy}{dz} - \frac{y}{x^2}\frac{dx}{dz}$$

But

$$\sec^2\psi = 1 + \tan^2\psi = 1 + \frac{y^2}{x^2} = \frac{r^2}{x^2}$$

Thus

$$r^2\frac{d\psi}{dz} = x\frac{dy}{dz} - y\frac{dx}{dz}$$

If we now use equation (23.10) we obtain equation (23.60). We may point out that if we replace z by the time coordinate, then the invariant $\tilde{l}$ is directly related to the z component of angular momentum in classical mechanics.

23.8 In Section 22.4 we showed that the general ray path in a square law medium can be written in the form

$$x(z) = A\sin(\Gamma z + \theta_1)$$

$$y(z) = B\sin(\Gamma z + \theta_2)$$

where

$$\Gamma = \frac{n_1\sqrt{2\Delta}}{\tilde{\beta}a} = \frac{g}{\tilde{\beta}}$$

Using the results of the previous problem, show that

$$\tilde{l} = n_1\sqrt{2\Delta}\,\frac{AB}{a^2}\,\sin(\theta_1 - \theta_2) \tag{23.61}$$

23.9 For a helical ray we may write

$$B = A, \quad \theta_1 - \theta_2 = \frac{\pi}{2}$$

so that $x^2 + y^2$ is independent of z. Thus, for the helical ray

$$\tilde{l} = n_1\sqrt{2\Delta}\,\frac{A^2}{a^2} \tag{23.62}$$

For $a = 25\,\mu$m, $n_1 = 1.45$, $\Delta = 0.01$ calculate the values of $\tilde{l}$ for helical rays launched at distances of 10, 15, and 20 μm from the axis.

[ANSWER: $\simeq$ 0.0328, 0.0738, 0.1312.]

Note that the axial ray ($A = 0$) is a helical ray with $\tilde{l} = 0$.

23.10 Consider a parabolic core profile ($q = 2$) with $n_1 = 1.5$, $\Delta = 0.01$, and $a = 40\,\mu$m. Assume $\tilde{l} = 0$ and plot $f(r)$ for $\tilde{\beta} = 1.6$, 1.49, and 1.46. Show that no ray path is possible for $\tilde{\beta} = 1.6$ and that $\tilde{\beta} = 1.49$ and $\tilde{\beta} = 1.46$ correspond to a bound ray and a refracting ray, respectively.

23.11 (a) For a skew ray inside a step index fiber, show that the radii of the inner and outer caustics are given by

$$r = r_1 = \left(\frac{\tilde{l}}{\tilde{l}_h}\right)a \quad \text{and} \quad r = r_2 = a \tag{23.63}$$

where

$$\tilde{l}_h = \sqrt{n_1^2 - \tilde{\beta}^2} \tag{23.64}$$

represents the value of $\tilde{l}$ corresponding to the helical ray.

(b) Consider a step index fiber with $n_1 = 1.5$, $\Delta = 0.01$, and $a = 30\,\mu$m. Calculate the cladding refractive index and r_1 and r_2 for $\tilde{\beta} = 1.49$ and $\tilde{l} = 0.06$. Repeat the calculations for $\tilde{l} = 0.2$.

[ANSWER: $n_2 = 1.4849$, $r_1 \simeq 10.4\,\mu$m, $r_2 \simeq 30\,\mu$m. For $\tilde{l} = 0.2$, ray paths cannot exist in the core.]

23.12 Determine the conditions for launching of a helical ray in a parabolic index fiber and show that if the helical ray is launched at $r = r_h < a/\sqrt{2}$, we will have a *bound helical ray*, and if the helical ray is launched at $r = r_h > a/\sqrt{2}$, we will have a *tunneling helical ray*.

Solution: Inside the core of a parabolic index fiber we have

$$f(r) = n_1^2\left[1 - 2\Delta\left(\frac{r}{a}\right)^2\right] - \tilde{\beta}^2 - \frac{\tilde{l}^2}{(r/a)^2} \quad \text{for } r < a$$

or

$$f(\rho) = \left(n_1^2 - \tilde{\beta}^2\right) - (2\Delta)n_1^2\rho^2 - \frac{\tilde{l}^2}{\rho^2} \quad \text{for } \rho < 1$$

where $\rho = r/a$. The value of ρ ($=\rho_0$) at which $f(\rho)$ attains its maximum value will satisfy the equation

$$\frac{df}{d\rho} = 0 = -(4\Delta)n_1^2\rho + \frac{2\tilde{l}^2}{\rho^3}$$

or

$$\rho_0 = \left[\frac{\tilde{l}^2}{2n_1^2\Delta}\right]^{1/4} \tag{23.65}$$

Thus

$$f(\rho_0) = \left(n_1^2 - \tilde{\beta}^2\right) - 2\tilde{l}\, n_1\sqrt{2\Delta}$$

Obviously

$$\tilde{l} < \tilde{l}_h = \frac{n_1^2 - \tilde{\beta}^2}{2n_1\sqrt{2\Delta}} \quad (q = 2) \tag{23.66}$$

The distance at which the helical ray occurs is given by

$$\frac{r_h}{a} = \rho_h = \left[\frac{\tilde{l}_h^2}{2n_1^2\Delta}\right]^{1/4} \tag{23.67}$$

or

$$r_h = \frac{a}{\sqrt{4\Delta}}\left(\frac{n_1^2 - \tilde{\beta}^2}{n_1^2}\right)^{1/2} \quad (q = 2) \tag{23.68}$$

Thus, for a ray to be helical at a distance of r_h from the axis, the value of $\tilde{\beta}$ should be given by

$$\tilde{\beta} = n_1\left[1 - 4\Delta\left(\frac{r_h}{a}\right)^2\right]^{1/2} \tag{23.69}$$

From the above equation it readily follows that

$$r_h < \frac{a}{\sqrt{2}} \Rightarrow n_2 < \tilde{\beta} < n_1$$

and we have a bound helical ray. On the other hand

$$r_h > \frac{a}{\sqrt{2}} \Rightarrow \tilde{\beta} < n_2$$

and we have a tunneling helical ray.

23.13 Consider a parabolic index fiber ($q = 2$) with $n_1 = 1.47$, $\Delta = 0.01$, and $a = 25\,\mu$m. Determine the value of $\tilde{l}_h$ for $\tilde{\beta} = 1.46$ and 1.44 which would correspond, respectively, to a bound helical ray and a skew helical ray.

[ANSWER: 0.07047, 0.21.]

23.14 (a) For a skew ray inside a parabolic index fiber, show that the radii of the inner and outer caustics are given by

$$r = r_1 = \left[\frac{1}{n_1^2(2\Delta)}\right]^{1/4}\left[\tilde{l}_h - \sqrt{\tilde{l}_h^2 - \tilde{l}^2}\right]^{1/2} a \tag{23.70}$$

and

$$r = r_2 = \left[\frac{1}{n_1^2(2\Delta)}\right]^{1/4}\left[\tilde{l}_h + \sqrt{\tilde{l}_h^2 - \tilde{l}^2}\right]^{1/2} a \tag{23.71}$$

where $\tilde{l}_h$ is given by equation (23.30).

(b) Consider a parabolic index fiber with $n_1 = 1.47$, $\Delta = 0.01$, and $a = 25\,\mu$m. Calculate r_1 and r_2 for $\tilde{\beta} = 1.46$ and $\tilde{l} = 0.035$.

[ANSWER: $r_1 \simeq 20\,\mu$m, $r_2 \simeq 5.29\,\mu$m.]

23.15 Consider a power law profile with $\tilde{\beta} < n_2$. Show that if $\tilde{l}^2 < n_2^2 - \tilde{\beta}^2$, $f(a) > 0$ and we have a refractive ray. On the other hand, if $\tilde{l}^2 > n_2^2 - \tilde{\beta}^2$, $f(a) < 0$ and we have a tunneling ray.

23.16 In Section 22.4.2, using Cartesian coordinates, we derived the condition for a helical ray in a parabolic index fiber. For a ray launched in the y–z plane at $x = r_h$, $y = 0$, a helical ray is described by the equations

$$r\cos\psi = x = r_h \cos\Gamma z \tag{23.72}$$

and

$$r\sin\psi = y = r_h \sin\Gamma z \tag{23.73}$$

where

$$\Gamma = \frac{n_1\sqrt{2\Delta}}{a\tilde{\beta}} \tag{23.74}$$

Show that the above equations are consistent with the results derived in Problem 23.12.

Solution: At $z = 0$, if the ray makes angle θ' with the z-axis, then

$$\tan\theta' = \left.\frac{dy}{dz}\right|_{z=0} = r_h\Gamma$$

Now

$$\tilde{\beta} = n(r = r_h)\cos\theta'$$

implying

$$\sec^2\theta' = \frac{n^2(r = r_h)}{\tilde{\beta}^2} = \frac{n_1^2}{\tilde{\beta}^2}\left[1 - 2\Delta\left(\frac{r_h}{a}\right)^2\right]$$

or

$$1 + r_h^2\Gamma^2 = \frac{n_1^2}{\tilde{\beta}^2}\left[1 - 2\Delta\left(\frac{r_h}{a}\right)^2\right]$$

Thus

$$r_h = \left[\frac{n_1^2 - \tilde{\beta}^2}{n_1^2}\right]^{1/2} \frac{a}{\sqrt{4\Delta}} \Rightarrow \tilde{\beta} = n_1\left[1 - 4\Delta\left(\frac{r_h}{a}\right)^2\right]^{1/2}$$

where we have used equation (23.74). The above equation is consistent with equation (23.68). From equations (23.72) and (23.73) we readily get

$$\psi(z) = \Gamma z + 2m\pi \tag{23.75}$$

or

$$\frac{d\psi}{dz} = \Gamma$$

If we now use equations (23.10) and (23.74) we obtain for the helical ray

$$\frac{a\tilde{l}_h}{\tilde{\beta}r_h^2} = \Gamma = \frac{n_1\sqrt{2\Delta}}{a\tilde{\beta}}$$

or

$$\tilde{l}_h = \frac{n_1^2 - \tilde{\beta}^2}{2n_1\sqrt{2\Delta}} \tag{23.76}$$

23.17 Consider a parabolic index fiber with $n_1 = 1.46$, $a = 25\,\mu$m, and $\Delta = 0.01$. At what angle should a ray be launched (at $x = a/2$, $y = 0$) so that the ray is helical? Will the ray be bound or tunneling?

Solution:

$$n_1 = n_2(1 - 2\Delta)^{-1/2} \simeq 1.4748$$

Thus

$$\tilde{\beta} = n_1\left[1 - 4\Delta\left(\frac{a/2}{a}\right)^2\right]^{1/2} \simeq 1.4674$$

$$n' = n_1\left[1 - 2\Delta\left(\frac{1}{2}\right)^2\right]^{1/2} \simeq 1.4711$$

$$\Rightarrow \theta' = \cos^{-1}(\tilde{\beta}/n') \simeq 4.06°$$

Thus, the ray should be launched in the y–z plane making an angle of about 4° to the z-axis.

23.18 In the Lagrangian formulation, choose r as the independent co-ordinate and derive equations (23.7) and (23.10).

Solution:

$$ds = [(dr)^2 + (r d\psi)^2 + (dz)^2]^{1/2}$$

$$= [1 + r^2\dot{\psi}^2 + \dot{z}^2]^{1/2}\, dr$$

where dots now represent differentiation with respect to the r coordinate. The optical Lagrangian is thus given by

$$L = n(r)[1 + r^2\dot{\psi}^2 + \dot{z}^2]^{1/2} \tag{23.77}$$

and the Lagrange's equations give us

$$\frac{d}{dr}\left(\frac{\partial L}{\partial \dot{z}}\right) = \frac{\partial L}{\partial z} = 0 \Rightarrow n\frac{dz}{ds} = \tilde{\beta} \tag{23.78}$$

and

$$\frac{d}{dr}\left(\frac{\partial L}{\partial \dot{\psi}}\right) = \frac{\partial L}{\partial \psi} = 0$$

implying

$$\frac{\partial L}{\partial \dot{\psi}} = \frac{nr^2\dot{\psi}}{\sqrt{1 + r^2\dot{\psi}^2 + \dot{z}^2}} = \text{an invariant of the ray path}$$

or

$$nr^2\frac{d\psi}{ds} = nr^2\frac{d\psi}{dz}\frac{dz}{ds} = \text{an invariant of the ray path}$$

If we now use equation (23.78), we get

$$\tilde{\beta}r^2\frac{d\psi}{dz} = \text{an invariant of the ray path}$$

24

Leaky modes

24.1 Introduction

In any optical waveguide such as a planar waveguide or an optical fiber, guidance of light takes place through the phenomenon of total internal reflection. For this to happen, the guiding region has to have a refractive index larger than the surrounding regions. In a class of waveguides called leaky waveguides, the low index surrounding region has a finite thickness comparable to the penetration depth of the guided field, and beyond this distance the medium has an index equal to or greater than that of the guiding region. In such a case, the waves do not undergo total internal reflection and, thus, the reflection coefficient is less than unity. Such a phenomenon is known as frustrated total internal reflection (FTIR). Hence, in such waveguides, there are no perfectly guided modes. On the other hand, such waveguides have leaky modes that are characterized by a finite loss coefficient. Such leaky modes find applications in the realization of many devices such as in-line fiber polarizers (see Section 17.5.1). We also see in Section 24.5 that a bent waveguide is a leaky waveguide, and the loss due to bending can be understood as due to a leakage mechanism.

In Section 24.2 we discuss the leakage loss calculations using the ray picture, and in Section 24.3 we discuss the concept of quasimodes. In Section 24.4 we discuss the matrix method to numerically obtain leakage loss. Finally, in Section 24.5 we discuss the bending loss in optical waveguides.

24.2 Leakage loss calculations: approximate theory

We first consider the transmission of a (y-polarized) plane wave propagating in a medium of refractive index n_1 and incident on a layer of refractive index n_2 and of thickness h (see Figure 24.1). The electric field (assumed to be in the y-direction) is given by

$$\left.\begin{array}{ll} \mathcal{E}_1=\left[E_1^+ e^{-i(k_1\cos\phi_1)x}+E_1^- e^{+i(k_1\cos\phi_1)x}\right]e^{i(\omega t-\beta z)} & x<0 \\ \mathcal{E}_2=\left[E_2^+ e^{-i(k_2\cos\phi_2)x}+E_2^- e^{+i(k_2\cos\phi_2)x}\right]e^{i(\omega t-\beta z)} & 0<x<h \\ \mathcal{E}_3=\left[E_3^+ e^{-i(k_1\cos\phi_1)(x-h)}\right]e^{i(\omega t-\beta z)} & x>h \end{array}\right\} \tag{24.1}$$

where

$$\left.\begin{array}{l} k_1=n_1k_0, \quad k_2=n_2k_0, \quad k_0=\frac{\omega}{c} \\ \beta=k_1\sin\phi_1=k_2\sin\phi_2 \end{array}\right\} \tag{24.2}$$

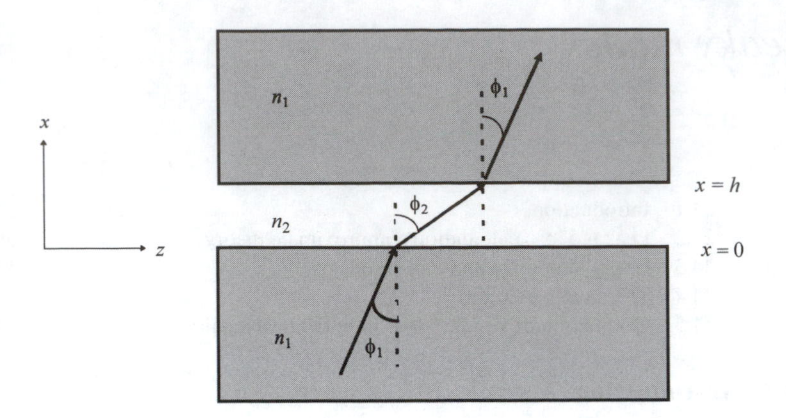

Fig. 24.1: A medium of refractive index n_2 (of thickness h) sandwiched between two media of refractive index n_1.

giving

$$\frac{\beta}{k_0} = n_1 \sin\phi_1 = n_2 \sin\phi_2 \tag{24.3}$$

In equation (24.1), E_i^+ and E_i^- ($i = 1, 2, 3$) represent the amplitudes of the plane waves propagating in the $+x$ and $-x$ directions, respectively. When ϕ_1 is greater than the critical angle, $\sin\phi_2 > 1$ and $\cos\phi_2$ will be imaginary.

$$k_2 \cos\phi_2 = k_0\sqrt{n_2^2 - n_2^2 \sin^2\phi_2} = i\gamma \tag{24.4}$$

where

$$\gamma = \sqrt{\beta^2 - k_0^2 n_2^2} \tag{24.5}$$

The fields can therefore be written as

$$\left.\begin{aligned} \mathcal{E}_1 &= \left[E_1^+ e^{-i\kappa x} + E_1^- e^{+i\kappa x}\right] e^{i(\omega t - \beta z)} && x < 0 \\ \mathcal{E}_2 &= \left[E_2^+ e^{\gamma x} + E_2^- e^{-\gamma x}\right] e^{i(\omega t - \beta z)} && 0 < x < h \\ \mathcal{E}_3 &= E_3^+ e^{-i\kappa(x-h)}\, e^{i(\omega t - \beta z)} && x > h \end{aligned}\right\} \tag{24.6}$$

where

$$\kappa = k_1 \cos\phi_1 = \sqrt{k_0^2 n_1^2 - \beta^2} \tag{24.7}$$

Continuity of $\mathcal{E}_y$ and $\partial\mathcal{E}_y/\partial x$ at $x = 0$ and at $x = h$ readily yield

$$\begin{pmatrix} E_1^+ \\ E_1^- \end{pmatrix} = \frac{1}{2}\begin{pmatrix} \left(1 + \frac{i\gamma}{\kappa}\right) & \left(1 - \frac{i\gamma}{\kappa}\right) \\ \left(1 - \frac{i\gamma}{\kappa}\right) & \left(1 + \frac{i\gamma}{\kappa}\right) \end{pmatrix}\begin{pmatrix} E_2^+ \\ E_2^- \end{pmatrix} \tag{24.8}$$

and

$$\begin{pmatrix} E_2^+ \\ E_2^- \end{pmatrix} = \frac{1}{2}\begin{pmatrix} \left(1 - \frac{i\kappa}{\gamma}\right)e^{-\gamma h} & 0 \\ \left(1 + \frac{i\kappa}{\gamma}\right)e^{\gamma h} & 0 \end{pmatrix}\begin{pmatrix} E_3^+ \\ 0 \end{pmatrix} \tag{24.9}$$

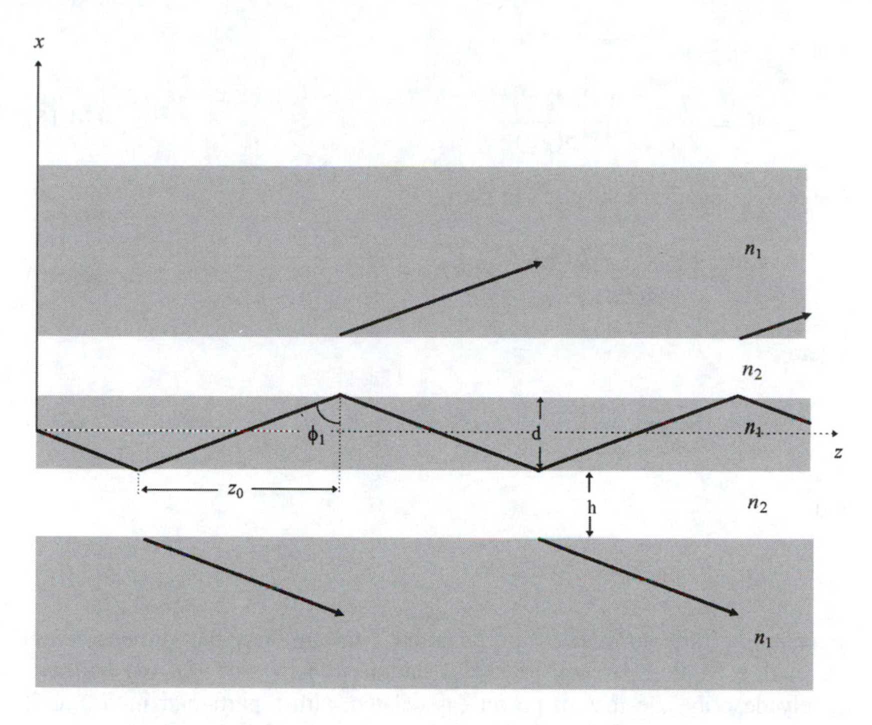

Fig. 24.2: A simple leaky structure.

Elementary calculations would give us the following expression for the tunneling probability

$$T = \left| \frac{E_3^+}{E_1^+} \right|^2 \approx 16 \frac{\gamma^2 \kappa^2}{\delta^4} \, e^{-2\gamma h} \tag{24.10}$$

where

$$\delta^2 = \kappa^2 + \gamma^2 = k_0^2 \left(n_1^2 - n_2^2 \right) \tag{24.11}$$

and we have assumed $e^{\gamma h} \gg 1$.

We next consider a leaky waveguide shown in Figure 24.2. A beam will lose power by tunneling whenever it undergoes FTIR and this will happen after traversing a distance z_0 along the waveguide

$$z_0 = d \tan \phi_1 = \frac{\beta d}{\kappa} \tag{24.12}$$

where d is the thickness of the guiding layer (see Figure 24.2). Thus

$$\frac{dP}{dz} \approx -\frac{PT}{z_0} \tag{24.13}$$

giving

$$P = P_0 \, e^{-2\Gamma z} \tag{24.14}$$

where

$$\Gamma = \frac{T}{2z_0} = \frac{8\gamma^3\kappa^3 e^{-2\gamma h}}{\delta^4 \beta(\gamma d)} \tag{24.15}$$

The above equation can be put in the form

$$\Gamma = \frac{8b^{3/2}(1-b)^{3/2}V^2}{d(\beta d)(\gamma d)}\, e^{-2\gamma h} \tag{24.16}$$

where

$$b = \frac{\beta^2/k_0^2 - n_2^2}{n_1^2 - n_2^2} \tag{24.17}$$

and

$$V = k_0 d\sqrt{n_1^2 - n_2^2} \tag{24.18}$$

representing the normalized propagation constant and the dimensionless waveguide parameter, respectively. Equations (24.14) and (24.16) approximately describe the loss of power (associated with a particular mode) as it propagates through the (leaky) waveguide.

24.3 Leaky structures and quasimodes

In the previous section we gave a simple ray picture for understanding the power loss in a leaky structure. For a more rigorous analysis, we must understand the concept of quasimodes, which is discussed in this section; this will be followed by an analysis of leakage loss calculations.

24.3.1 *Quasimodes*

We first consider the guiding structure shown in Figure 24.3. We consider only the antisymmetric modes; the symmetric modes can be considered in an identical manner. For TE modes, we write

$$\mathcal{E}_y(x, z, t) = \psi_g(x)\, e^{i(\omega t - \beta_g z)} \tag{24.19}$$

where

$$\left.\begin{aligned} \psi_g(x) &= A_g \sin \kappa_g x && 0 < |x| < d/2 \\ &= \tfrac{x}{|x|} A_g \sin(\kappa_g d/2)\, e^{-\gamma_g(|x| - d/2)} && |x| > d/2 \end{aligned}\right\} \tag{24.20}$$

with

$$\kappa_g = \sqrt{k_0^2 n_1^2 - \beta_g^2}; \quad \gamma_g = \sqrt{\beta_g^2 - k_0^2 n_2^2} \tag{24.21}$$

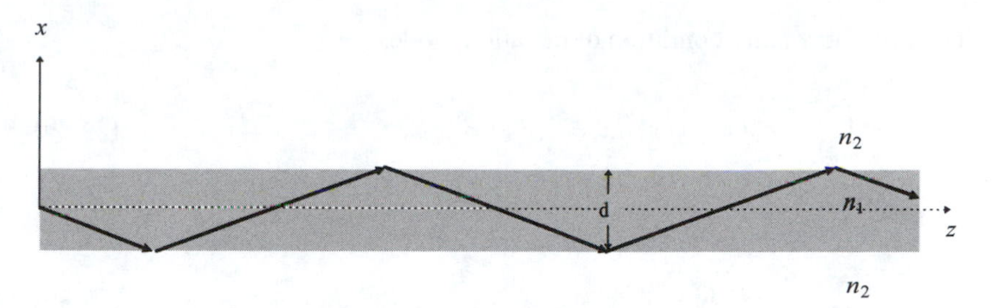

Fig. 24.3: A simple planar waveguide.

and β_g represents the propagation constant, which is determined from the transcendental equation (see Section 7.3)

$$L = \sin\left(\kappa_g \frac{d}{2}\right) + \frac{\kappa_g}{\gamma_g} \cos\left(\kappa_g \frac{d}{2}\right) = 0 \tag{24.22}$$

Obviously, $\psi_g(-x) = -\psi_g(x)$. The normalization condition

$$\int_{-\infty}^{+\infty} [\psi_g(x)]^2 \, dx = 1 \tag{24.23}$$

readily gives

$$A_g = \left(\frac{\gamma_g}{1 + \gamma_g d/2}\right)^{1/2} \tag{24.24}$$

We next consider the leaky structure shown in Figure 24.2. Obviously, there are no guided modes and all values of $\beta^2 < k_0^2 n_1^2$ are allowed; these form the continuum radiation modes of the system. The solutions for the (antisymmetric) TE modes are given by

$$\left.\begin{aligned} \psi_\beta(x) &= A \sin \kappa x && 0 < |x| < d/2 \\ &= B e^{+\gamma(x-\frac{1}{2}d)} + C e^{-\gamma(x-\frac{1}{2}d)} && \tfrac{d}{2} < x < \tfrac{d}{2} + h \\ &= D_+ \, e^{i\kappa(x-\frac{1}{2}d-h)} + D_- \, e^{-i\kappa(x-\frac{1}{2}d-h)} && x > \tfrac{d}{2} + h \end{aligned}\right\} \tag{24.25}$$

with κ and γ defined through equations (24.7) and (24.5). Continuity of $\psi_\beta(x)$ and its derivative at $x = d/2$ and at $x = d/2 + h$ give us

$$B = \frac{1}{2} A \left[\sin(\kappa d/2) + \frac{\kappa}{\gamma} \cos(\kappa d/2)\right] \tag{24.26}$$

$$C = \frac{1}{2} A \left[\sin(\kappa d/2) - \frac{\kappa}{\gamma} \cos(\kappa d/2)\right] \tag{24.27}$$

$$D_-^* = D_+ = \frac{1}{2} B \left[1 + \frac{\gamma}{i\kappa}\right] e^{\gamma h} + \frac{1}{2} C \left(1 - \frac{\gamma}{i\kappa}\right) e^{-\gamma h} \tag{24.28}$$

The orthonormality condition of radiation modes

$$\int_{-\infty}^{+\infty} \psi_{\beta'}^*(x)\psi_\beta(x)\, dx = \delta(\beta - \beta') \tag{24.29}$$

gives us

$$|D_\pm| = \sqrt{\frac{\beta}{4\pi\kappa}} \tag{24.30}$$

Now, if we assume $\gamma h \gg 1$, then the exponentially decaying term in equation (24.28) can be neglected, and in this approximation

$$D_-^* = D_+ \approx \frac{1}{4} LA\left(1 + \frac{\gamma}{i\kappa}\right) e^{\gamma h} \tag{24.31}$$

where

$$L = \sin(\kappa d/2) + \frac{\kappa}{\gamma} \cos(\kappa d/2) \tag{24.32}$$

It is readily seen that since $\gamma h \gg 1$, unless $L \approx 0$,

$$\left|\frac{D_\pm}{A}\right| \gg 1 \tag{24.33}$$

implying that the amplitude of the field is large in the region $x > (\frac{d}{2} + h)$. However, for β values such that $L = 0$ (i.e., when the condition for guided modes are satisfied – see equation (24.22)), then $B = 0$ and we have only the exponentially decaying solution in the region $\frac{d}{2} < x < h + \frac{d}{2}$ (see equation (24.25)) and the oscillatory field in the region $x > (\frac{d}{2} + h)$ is very weak. Thus, when

$$L = 0 = B$$

the field has the properties of a guided mode in regions I and II becoming oscillatory in region III. We refer to the modes corresponding to $B = 0$ as the *quasimodes*.

24.3.2 Leakage of power

We next consider the incidence of a guided mode of the structure shown in Figure 24.3 on a leaky structure as shown in Figure 24.4. Thus

$$\psi(x, z = 0) = \psi_g(x) \tag{24.34}$$

Such an incident field would excite a packet of radiation modes and we may write

$$\psi(x, z = 0) = \psi_g(x) = \int \phi(\beta)\psi_\beta(x)\, d\beta \tag{24.35}$$

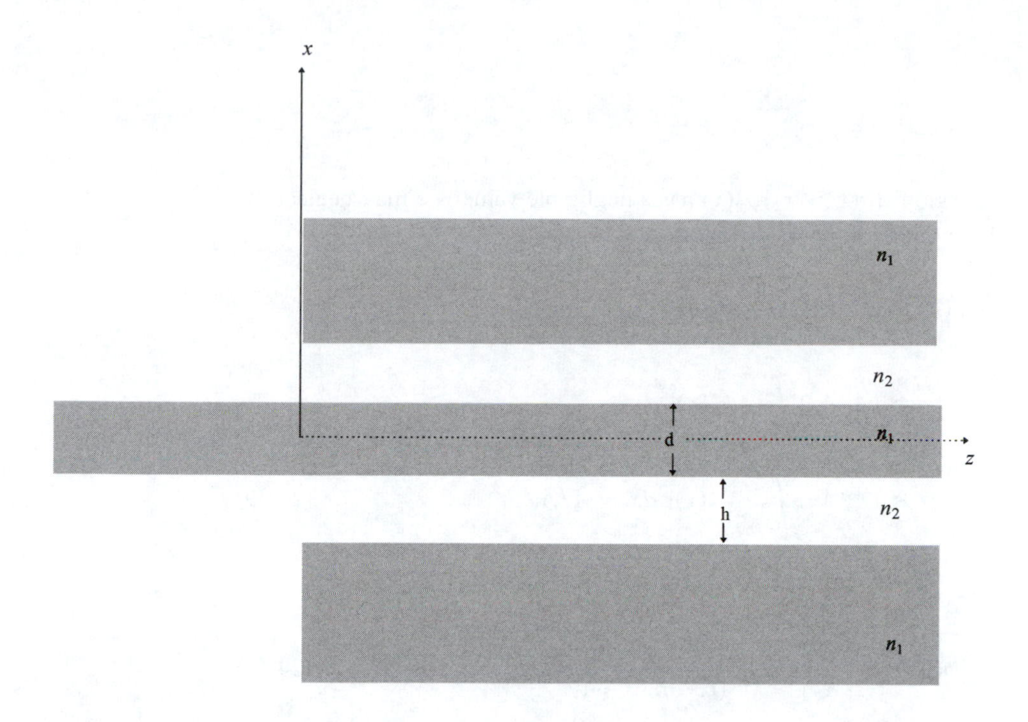

Fig. 24.4: A guiding structure followed by a leaky structure.

For $z > 0$, the field will be given by

$$\psi(x, z) = \int \phi(\beta)\psi_\beta(x)\, e^{-i\beta z}\, d\beta \tag{24.36}$$

To determine $\phi(\beta)$, we multiply equation (24.35) by $\psi^*_{\beta'}(x)$ and integrate over x to obtain

$$\int \psi^*_{\beta'}(x)\psi_g(x)\, dx = \int d\beta\, \phi(\beta) \int dx\, \psi^*_{\beta'}(x)\psi_\beta(x)$$

$$= \int d\beta\, \phi(\beta)\, \delta(\beta - \beta') = \phi(\beta')$$

where use has been made of equation (24.29). Thus

$$\phi(\beta) = \int_0^\infty \psi^*_\beta(x)\psi_g(x)\, dx \tag{24.37}$$

As discussed earlier, it is only around $\beta \approx \beta_g$ that $\psi_\beta(x)$ has almost the same spatial dependence as $\psi_g(x)$ in regions I and II (see Figure 24.2) and therefore in equation (24.37) $\phi(\beta)$ will be appreciable only around $\beta \approx \beta_g$; this we will explicitly find later. Thus, we may write

$$\psi_\beta(x) \approx (A/A_g)\psi_g(x) \text{ regions I and II} \tag{24.38}$$

For convenience we write

$$x_1 = \frac{1}{2} d$$

and

$$x_2 = \frac{1}{2}d + h$$

Since in region III ($x > x_2$)$\psi_g(x)$ has a negligible value, we may neglect the contribution from region III to the integral in equation (24.37) to write

$$\begin{aligned}\phi(\beta) &\approx \int_0^{x_2} \psi_\beta^*(x)\psi_g(x)\,dx \\ &\approx (A/A_g)\int_0^{x_2} |\psi_g(x)|^2 dx \\ &\approx (A/A_g)\int_0^{\infty} |\psi_g(x)|^2 dx = A/A_g \end{aligned} \qquad (24.39)$$

where $x_2 = h + \frac{d}{2}$. In Appendix G we show that around $\beta \approx \beta_g$

$$|\phi(\beta)|^2 = \left|\frac{A}{A_g}\right|^2 = \frac{1}{\pi}\frac{\Gamma}{(\beta - \beta_g')^2 + \Gamma^2} \qquad (24.40)$$

where

$$\Gamma = \frac{8\gamma_g^3\kappa_g^3 e^{-2\gamma_g h}}{\delta^4\beta_g(2 + \gamma_g d)} \qquad (24.41)$$

$$\beta_g' = \beta_g + \Delta\beta \qquad (24.42)$$

$$\Delta\beta = -\frac{\Gamma\left(\kappa_g^2 - \gamma_g^2\right)}{2\gamma_g\kappa_g} \qquad (24.43)$$

The fractional power $W(z)$ that remains inside the core at z is approximately given by

$$W(z) \approx \left|\int_0^{\infty} \psi^*(x, 0)\psi(x, z)\,dx\right|^2 \qquad (24.44)$$

which can be evaluated to give (see Appendix G)

$$W(z) = e^{-2\Gamma z} \qquad (24.45)$$

The above equation shows how the power inside the core "leaks" into region III. Equation (24.41) gives an analytical expression for the attenuation coefficient of the quasimodes. Note that as $x_2 \to \infty$, $\Gamma \to 0$ and there is no leakage of power.

Equation (24.41) may be compared with the approximate expression given by equation (24.15). If we replace d by $(d + 2/\gamma_f)$ in equation (24.15), we get equation (24.41). Since $1/\gamma_f$ represents the penetration depth of the mode field in the lower refractive index region, $d + 2/\gamma_f$ can be interpreted as the effective waveguide thickness. The approximate ray analysis will be accurate when $\gamma d \gg 1$. We may rewrite equation (24.41) as (cf. equation (24.16))

$$\Gamma = \frac{8b^{3/2}(1-b)^{3/2}V^2}{d(\beta_g d)(2+\gamma_g d)} e^{-2\gamma_g h} \tag{24.46}$$

Now, in Appendix G we also show that

$$\left|\frac{A}{D_\pm}\right|^2 \approx \frac{8\gamma_g \kappa_g \Gamma}{\beta_g(1+\gamma_g x_1)} \frac{1}{(\beta-\beta_g')^2+\Gamma^2} \tag{24.47}$$

Thus, if we are able to calculate $|A/D_\pm|^2$ as a function of β (i.e., as a function of the angle of incidence in Figure 24.2), we would get a series of Lorentzians, each Lorentzian corresponding to a quasimode of the structure. By fitting each peak to a Lorentzian, we would be able to get β_g' and Γ.

In the next section we develop a matrix method to calculate $|A/D_\pm|^2$ as a function of β, from which we will be able to get the propagation characteristics (including leakage/absorption losses) even in graded/absorbing structures.

24.4 The matrix method

In this section we develop the matrix method for determining the propagation characteristics of a leaky (as well as a guiding) structure. The matrix method can be used to solve the general scalar wave equation

$$\frac{d^2\psi}{dx^2} + \kappa^2(x)\psi(x) = 0 \tag{24.48}$$

For example, for the TE modes in a slab waveguide, $\psi(x)$ would represent $E_y(x)$ and

$$\kappa^2(x) = k_0^2 n^2(x) - \beta^2 \tag{24.49}$$

(see equation (7.23)). To solve equation (24.48), we consider an arbitrary variation of $\kappa^2(x)$, which we replace by discrete steps as shown in Figure 24.5. The solution in the jth region is given by

$$\psi_j(x) = A_j e^{i\kappa_j(x-\Delta_j)} + B_j e^{-i\kappa_j(x-\Delta_j)} \tag{24.50}$$

where κ_j is the value of $\kappa(x)$ at the middle of each region with

$$\left.\begin{array}{l}\Delta_1 = 0; \quad \Delta_2 = 0; \quad \Delta_3 = D_2 \\ \Delta_4 = D_2 + D_3; \quad \Delta_5 = D_2 + D_3 + D_4; \quad \text{and so forth}\end{array}\right\} \tag{24.51}$$

Obviously, if κ^2 is a negative quantity, κ will be imaginary, and the first and the second term on the RHS of equation (24.50) represent exponentially decaying and amplifying solutions, respectively. Now, if $\kappa^2(x)$ does not have an infinite

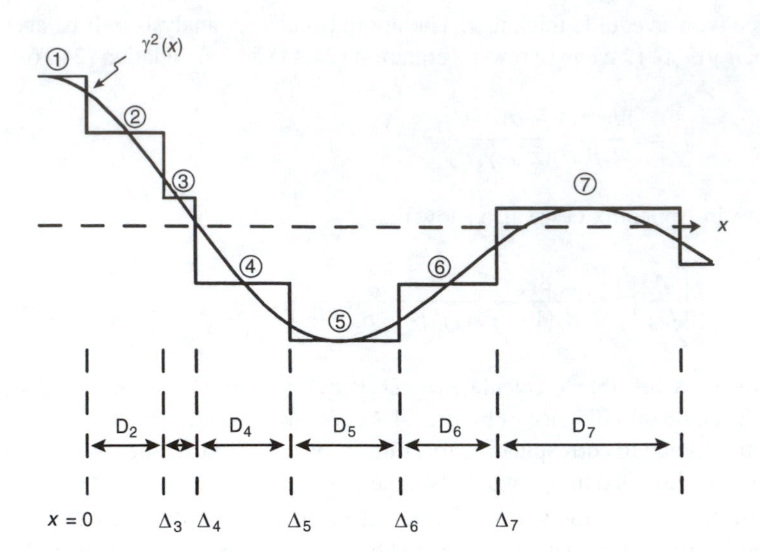

Fig. 24.5: An arbitrary variation of $\kappa^2(x)$ is replaced by a large number of steps.

discontinuity at any point, then $\psi(x)$ and $d\psi/dx$ will be continuous everywhere. The continuity conditions at the interface of the jth and $(j+1)$th regions will give

$$\begin{pmatrix} A_j \\ B_j \end{pmatrix} = \begin{pmatrix} S_{11}^{(j)} & S_{12}^{(j)} \\ S_{21}^{(j)} & S_{22}^{(j)} \end{pmatrix} \begin{pmatrix} A_{j+1} \\ B_{j+1} \end{pmatrix} \tag{24.52}$$

where

$$\begin{aligned} S_{11}^{(j)} &= \frac{1}{2}\left(1 + \frac{\kappa_{j+1}}{\kappa_j}\right) e^{-i\kappa_j D_j}; \quad S_{12}^{(j)} = \frac{1}{2}\left(1 - \frac{\kappa_{j+1}}{\kappa_j}\right) e^{-i\kappa_j D_j}; \\ S_{21}^{(j)} &= \frac{1}{2}\left(1 - \frac{\kappa_{j+1}}{\kappa_j}\right) e^{i\kappa_j D_j}; \quad S_{22}^{(j)} = \frac{1}{2}\left(1 - \frac{\kappa_{j+1}}{\kappa_j}\right) e^{i\kappa_j D_j}; \end{aligned} \tag{24.53}$$

With the help of the above equations one can determine all A_j and B_j in terms of only two unknowns (say, A_m and B_m). Thus

$$\begin{pmatrix} A_1 \\ B_1 \end{pmatrix} = S^{(1)} S^{(2)} \ldots\ldots\ldots S^{(m-1)} \begin{pmatrix} A_m \\ B_m \end{pmatrix} = G \begin{pmatrix} A_m \\ B_m \end{pmatrix} \tag{24.54}$$

where

$$G = S^{(1)} S^{(2)} \ldots\ldots\ldots S^{(m-1)} \tag{24.55}$$

If κ^2 is negative in the right-most or left-most region, then in a physical problem the coefficient of the exponentially amplifying solution in that region should be zero.

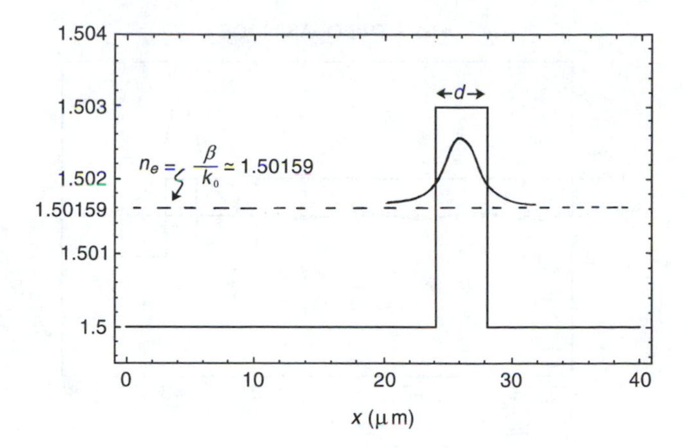

Fig. 24.6: A symmetric waveguide characterized by equation (24.58) with $n_1 = 1.503, n_2 = 1.500$, $d = 4\,\mu$m, and $d_1 = 24\,\mu$m.

Assuming that κ^2 is negative in the right-most region (say, mth), we may write

$$\begin{pmatrix} A_1 \\ B_1 \end{pmatrix} = \begin{pmatrix} G_{11} & G_{12} \\ G_{21} & G_{22} \end{pmatrix} \begin{pmatrix} A_m \\ 0 \end{pmatrix} \tag{24.56}$$

and

$$\left| \frac{A_m}{A_1} \right|^2 = \frac{1}{|G_{11}|^2} \tag{24.57}$$

Thus, $\psi(x)$ is known in the entire region of x apart from a multiplying constant. The only approximation in the above procedure is that $\kappa^2(x)$ has been assumed to be constant ($=\kappa_j^2$) in each region. By choosing sufficiently small values of D_j, one can obtain extremely accurate variation of $\psi(x)$. If equation (24.48) is an eigenvalue equation so that $\kappa^2(x)$ contains an eigenvalue parameter, then one may use a method similar to that described in the following simple example.

Example 24.1: We first consider a (symmetric) waveguide characterized by (see Figure 24.6)

$$n(x) = \begin{cases} n_1, & d_1 < x < d_1 + d \\ n_2, & x < d_1 \quad \& \quad x > d_1 + d \end{cases} \tag{24.58}$$

Actually, the value of d_1 can be arbitrary; it just shifts the origin. If we assume $n_1 = 1.503$, $n_2 = 1.500$, $d = 4\,\mu$m, and $\lambda_0 = 1\,\mu$m, then elementary algebra shows that the waveguide will support one TE mode with $\beta/k_0 = 1.50159$ (see Example 7.1); the corresponding value of $b \simeq 0.531223$.

To solve this problem by the matrix method we first note that, corresponding to a guided mode, both κ_1^2 and κ_3^2 should be negative. Considering region III we therefore must have

$$B_3 = 0$$

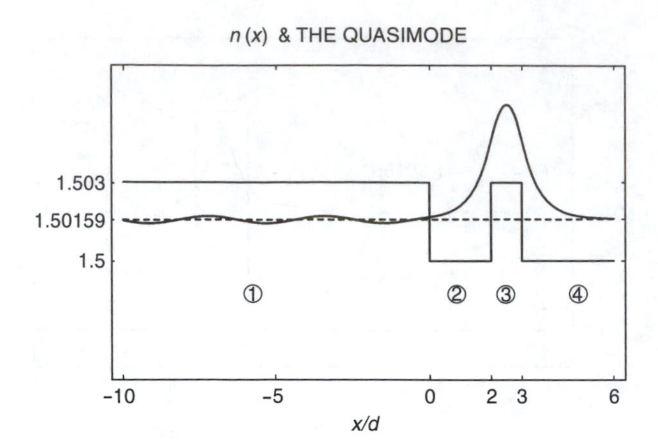

Fig. 24.7: The refractive index distribution of the leaky structure given by equation (24.59) and the field distribution of its quasimode.

so that there is only an exponentially decaying solution in this region ($x > d_1 + d$). One can now determine A_1 with the help of equation (24.56) as a function of b, which we vary from 0 to 1. Since κ_1^2 is also negative, A_1 represents the coefficient of the exponentially amplifying term in region I, and therefore we must have

$$A_1 = 0$$

One then finds out the value of b so that the above equation is satisfied. The value of b so obtained is 0.531223 ($\beta/k_0 = 1.50159$).

Example 24.2: We next consider the waveguide of Example 24.1 and make it *leaky* by introducing an infinitely extended layer of refractive index n_1(=1.503). Thus

$$n(x) = \begin{cases} n_1, & x < d_1 - d_2 \quad \& \quad d_1 < x < d_1 + d \\ n_2, & d_1 - d_2 < x < d_1 \quad \& \quad x > d_1 + d \end{cases} \tag{24.59}$$

Figure 24.7 shows the above refractive index distribution with $d_1/d = 2$ and $d_2/d = 2$. The above structure is no more a guiding structure and, as such, *no guided mode will exist*. All values of β (and therefore b) are allowed and the modes are called the radiation modes. As discussed earlier by taking

$$B_4 = 0$$

one can determine the field in the entire region apart from a multiplying constant. Figures 24.7 and 24.8 represent the radiation modes for $\beta/k_0 = 1.50159$ (which is the quasimode) and for $\beta/k_0 = 1.50156$, respectively. We see from the two figures that the field pattern for the quasimode is almost the same as that of the guided mode (shown in Figure 24.6) (except for the weak oscillatory behavior in the leaky region); however, if we go slightly away from the quasimode, the nature of the field pattern changes drastically.

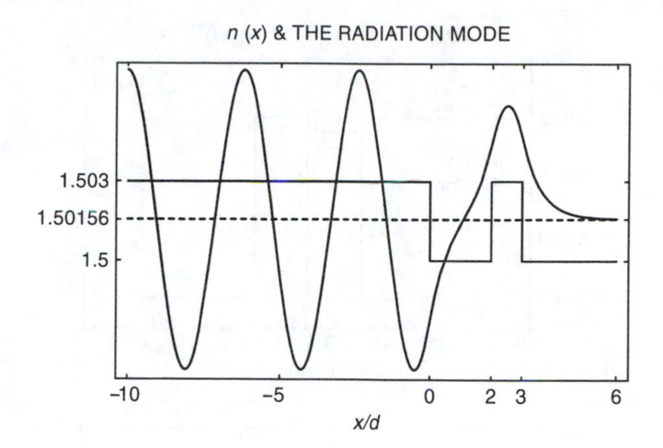

Fig. 24.8: A typical radiation mode of the leaky structure given by equation (24.59) with $\beta/k_0 = 1.50156$, which is only very slightly different from that of the quasimode (=1.50159).

Following the analysis given in Section 24.3.2, if we plot $|A_4/A_1|^2$ as a function of β (or b), we will obtain a sharply peaked Lorentzian very close to β of the guided structure. Since the structure is leaky, there will be a loss of power from the core, which is given by

$$P(z) = P_0 \, e^{-2\Gamma z} \tag{24.60}$$

where 2Γ is the FWHM of the $|A_4/A_1|^2$ versus β curve. If $|A_4/A_1|^2$ is plotted as a function of b, then the quantity Γ is directly related to the FWHM of the corresponding Lorentzian, $2\Delta b$.

$$\Gamma = \Delta b \frac{n_1^2 - n_2^2}{2(\beta/k_0)} \frac{2\pi}{\lambda_0} \tag{24.61}$$

As we increase the value of d_2, the Lorentzians become more sharply peaked and the peaks occur at values of b closer to the value of b for the corresponding guided structure. As $d_2 \to \infty$, $\Gamma \to 0$ and the peak occurs at the value of b for the guided mode. We explain the above concept through Example 24.3.

Example 24.3: We consider the leaky waveguide represented by equation (24.58) with the following values of various parameters (see Figure 24.9(a))

$$\begin{aligned} &n_1 = 1.503; \quad n_2 = 1.500; \quad d = 4\,\mu\text{m} \\ &d_2 = 4\,\mu\text{m} \quad \text{and} \quad \lambda_0 = 1\,\mu\text{m} \end{aligned} \tag{24.62}$$

As discussed earlier, we plot $|A_4/A_1|^2$ as a function of b, which is shown in Figure 24.9(b). The shape of the curve is Lorentzian and the peak occurs at $b = b_r = 0.531776$. This may be compared with b_{exact} (=0.531223), the value of b for the corresponding guided mode. The value of Γ comes out to be $\cong$1.528 cm^{-1}, which means that half of the power from the core leaks into the cladding in distance $z_0 \cong$ 0.227 cm.

If we now increase the value of d_2 to 12 μm and repeat the above calculations, we will obtain the $|A_4/A_1|^2$ versus b curve as shown

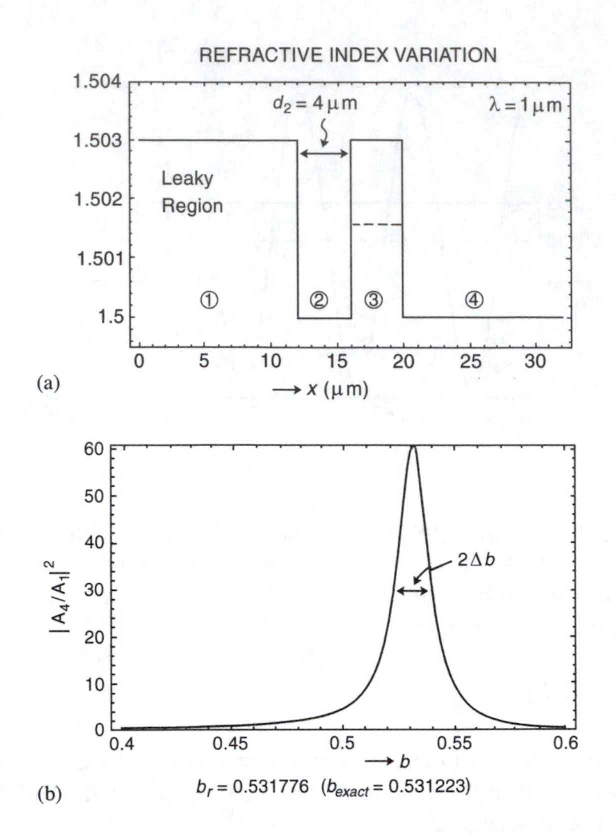

Fig. 24.9: The lower part of the figure shows the variation of $|A_4/A_1|^2$ with b corresponding to the leaky waveguide whose refractive index variation is shown in (a).

in Figure 24.10. The peak now occurs at $b = b_r \cong 0.531223$ – that is, almost at $b = b_{exact}$ – and Γ comes out to be $\cong 0.00149$ cm^{-1}, which gives $z_0 \cong 232.6$ cm. This shows that increasing the separation between the leaky layer and the core results in the decrease of the value of Γ and, hence, of the leakage loss.

Example 24.4: We repeat the above example with $\lambda_0 = 0.5\ \mu$m and $d_2 = 4\ \mu$m (see Figure 24.11). Simple calculations will show that the corresponding guiding structure ($d_2 = \infty$) will support two guided modes with $b_{exact} = 0.789584$ and 0.238762 for the first and second modes, respectively. The $|A_4/A_1|^2$ versus b curve for the leaky structure ($d_2 = 4\ \mu$m) has been shown in Figure 24.12. It may be seen that the peak corresponding to the first quasimode is much narrower than the peak for the second quasimode. This is expected because the first mode will be more tightly bound to the core than the higher order mode. Thus, Γ and, hence, the leakage loss for the second quasimode will be much higher than the first mode.

Example 24.5: We next consider an example of an inhomogeneous refractive index profile. Let the refractive index profile be given by (see Figure 24.13(a))

$$n^2(x) = \begin{cases} 1.0 & x < 0 \\ n_2^2 + \left(n_1^2 - n_2^2\right) e^{-x/d} & x > 0 \end{cases} \tag{24.63}$$

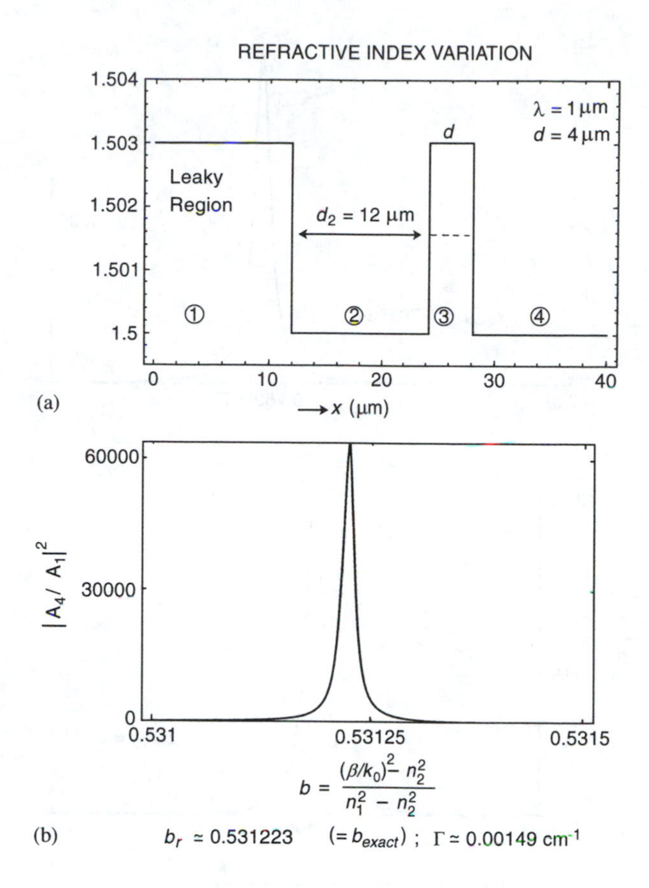

Fig. 24.10: The lower part of the figure shows the variation of $|A_4/A_1|^2$ with b corresponding to the leaky waveguide whose refractive index variation is shown in (a). [After Ghatak and Goyal (1994)].

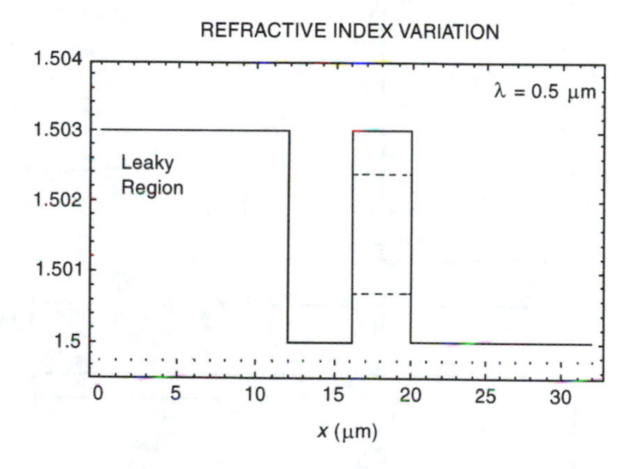

Fig. 24.11: A leaky structure with two quasimodes. The dashed horizontal lines represent the propagation constants of the corresponding guiding structure.

To apply the matrix method, the inhomogeneous region is replaced by a number of slabs of homogeneous refractive indices as shown in Figure 24.13(b). We also make it leaky by introducing an infinitely extended layer of refractive index n_1 beyond a certain value of x, which we choose as $x = 2d$. Following the procedure of previous examples, we calculate $|A_7/A_1|^2$ as a function of b, which would nearly be a Lorentzian; the FWHM of the Lorentzian would give the leakage loss. If one puts the leaky layer at a larger distance from the core, the peak will come closer to the value of b corresponding to the guided

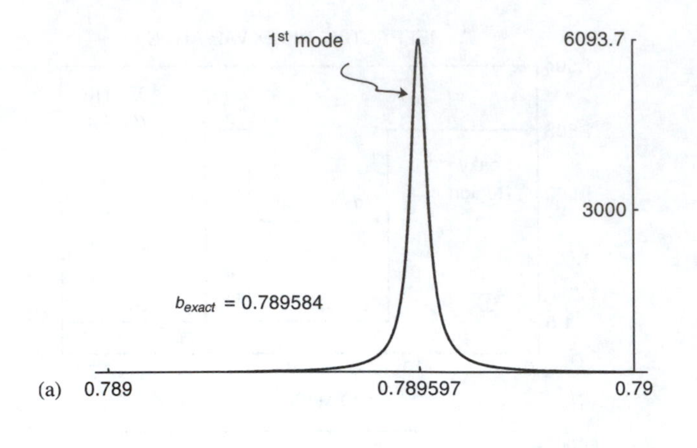

Fig. 24.12: In (a) and (b) we have plotted $|A_4/A_1|^2$ versus b for the two quasimodes of the waveguide shown in Figure 24.11. Note the difference in the scales of the two figures. [After Ghatak and Goyal (1994)].

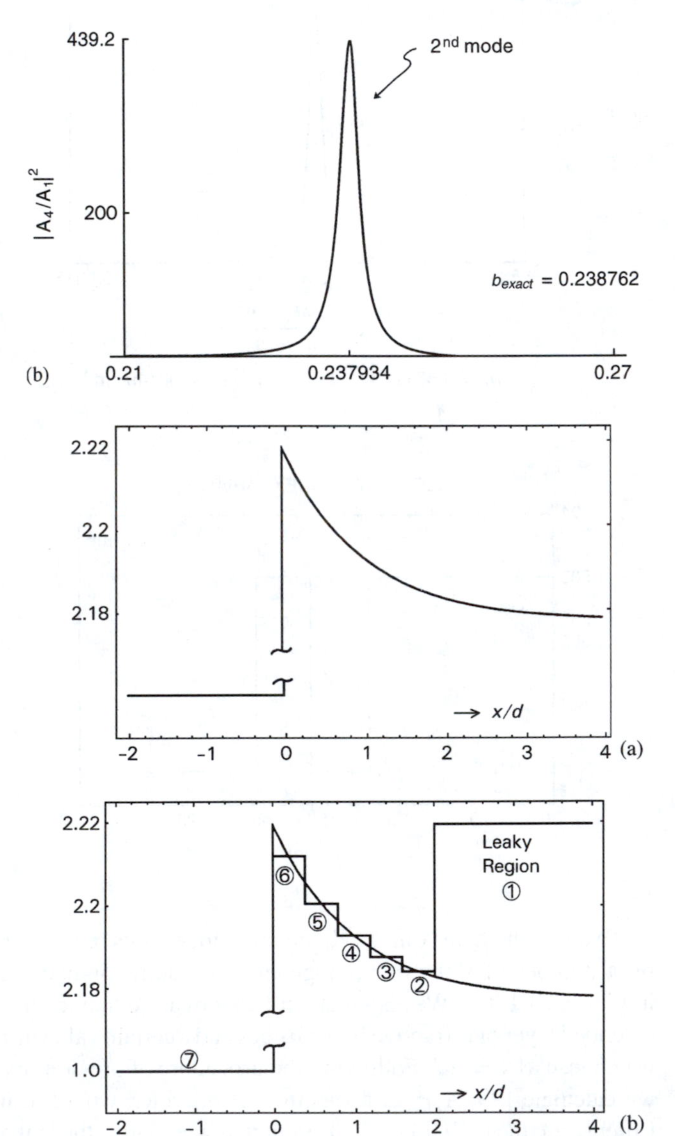

Fig. 24.13: The exponential profile.

mode. Increasing the number of steps will improve the accuracy of calculations.

24.4.1 The general procedure

What we describe now is a general procedure applicable to structures that could be either leaky or nonleaky with or without absorption.

(1) If $\kappa^2(x)$ in equation (24.48) is completely known – that is, it does not contain any eigenvalue parameter – then the procedure described in Section 24.4 is followed and one can determine $\psi(x)$ in the entire region apart from a multiplying constant. If, however, an eigenvalue parameter exists in $\kappa^2(x)$ (see, e.g., equation (24.49)), we proceed as follows.

(2) If the structure is nonleaky, then we artificially introduce an infinite *leaky* layer (see Figures 24.7 and 24.13(b)) of refractive index equal to the largest value of the real part of the refractive index in the guiding region.

(3) Following the procedure discussed earlier, one plots $|A_m/A_1|^2$ as a function of β, where m refers either to a layer in the guiding region or to the region where there exists only an exponentially decaying solution. The plot will be nearly a Lorentzian peaked at $\beta = \beta_r$, the real part of the propagation constant of the quasimode of the structure. The FWHM of the Lorentizian will give 2Γ.

(4) If the given structure to be analyzed is leaky as well as absorbing, then its propagation constant β will be given by

$$\beta = \beta_r - j\Gamma \tag{24.64}$$

The accuracy of calculations may be increased by increasing the number of steps in the inhomogeneous region of the refractive index.

The following particular cases may arise.

(i) If the structure to be analyzed is nonabsorbing but leaky, then β_r represents the propagation constant of the quasimode and power from the core leaks into cladding as given by equation (24.60).

(ii) If the structure is nonleaky, then by increasing the separation between the guiding layer and the artificially introduced leaky layer, the position of the peak and the width of the Lorentzian will approach to certain fixed values. If the structure is nonabsorbing, Γ will converge to 0 and β_r will tend to the propagation constant of the given guided structure. Thus, there will be no loss of power as the mode propagates. On the other hand, if the structure is absorbing (i.e., if $n(x)$ is complex), then Γ will converge to a finite value [see, e.g., Ghatak, Thyagarajan, and Shenoy (1987) where an air–polymer–metal waveguide has been analyzed]. In this case also the loss of power will again be given by equation (24.60), but now the energy does not leak from the core; it is being absorbed in the structure as it propagates in the z-direction.

We should mention here that even for an optical fiber characterized by the (cylindrically symmetric) refractive index variation $n(r)$, the radial part of the wave equation (see equation (8.11)).

$$\frac{1}{r}\frac{d}{dr}\left[r\frac{dR}{dr}\right] + \left[k_0^2 n^2(r) - \beta^2 - \frac{l^2}{r^2}\right]R(r) = 0 \tag{24.65}$$

can be transformed to

$$\frac{d^2u}{dr^2} + \kappa^2(r)\,u(r) = 0 \tag{24.66}$$

where

$$u(r) = \sqrt{r}\,R(r) \tag{24.67}$$

and

$$\kappa^2(r) = k_0^2 n^2(r) - \frac{l^2 - \frac{1}{4}}{r^2} \tag{24.68}$$

Equation (24.66) is of the same form as equation (24.48) and the matrix method can easily be applied for the analysis of leaky as well as nonleaky structures [Shenoy, Thyagarajan, and Ghatak (1988)]. The matrix method has also been applied to absorbing waveguides [Ramadas et al. (1989a)], to study whispering gallery modes [Goyal, Gallawa and Ghatak (1990)], nonlinear waveguides [Ramadas et al. (1989b)], ARROW waveguides [Pal (1993)], DIC fibers, and also quantum well structures [Ghatak, Thyagarajan, and Shenoy (1988), Ghatak, Goyal, and Gallawa (1990)]. The technique, for example, when applied to quantum well structures, not only yields the energy eigenvalues and the wave functions but also makes an accurate prediction of lifetimes of quasi-bound states.

24.5 Calculation of bending loss in optical waveguides

Bending loss calculation is a topic of great interest in optical waveguides. We show that a bent waveguide is essentially a leaky structure and therefore the theory developed in the previous section can be used to calculate the bending loss. We follow the analysis given by Thyagarajan et al. (1987a, 1987b).

We consider a planar waveguide that is bent along the arc of a circle of radius ρ as shown in Figure 24.14. We assume the validity of the scalar wave equation

$$\nabla^2\psi + k_0^2 n^2 \psi = 0 \tag{24.69}$$

We use a cylindrical system of coordinates (r, ϕ, z) whose origin is at the center of the arc. The refractive index depends only on r and, since the waveguide is of infinite extent in the z direction, we may neglect the z dependence of the fields. Thus, we assume a solution of the form

$$\psi = R(r)\,e^{-i\beta\rho\phi} \tag{24.70}$$

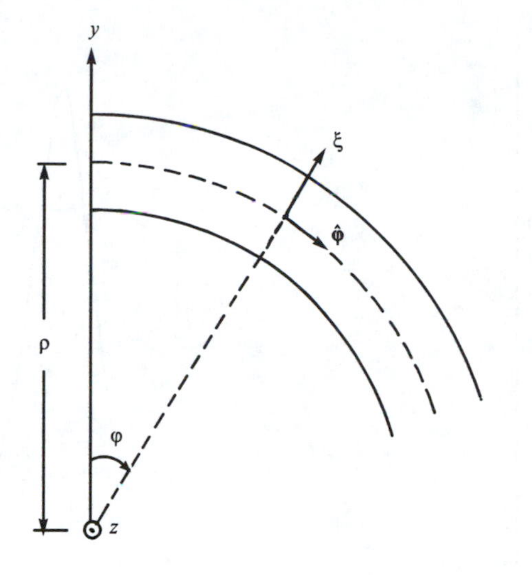

Fig. 24.14: The coordinate system for a bent planar waveguide.

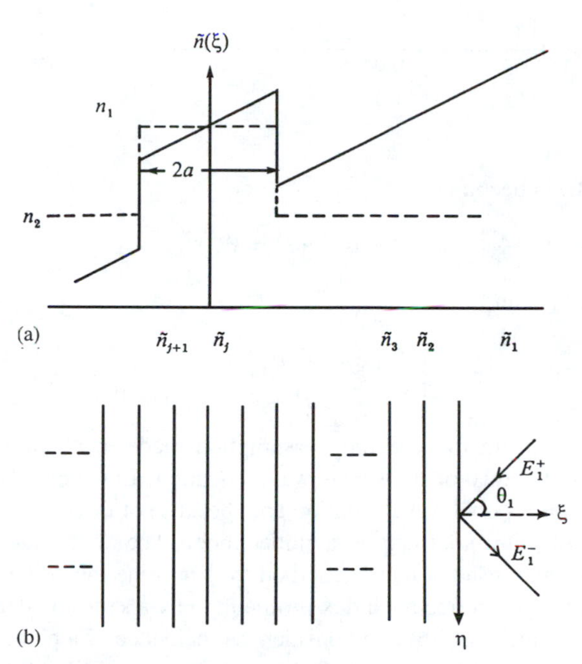

Fig. 24.15: (a) The solid line gives the effective refractive index profile corresponding to a bent slab waveguide. The dashed line gives the corresponding profile when the waveguide is straight. (b) The effective profile replaced by a large number of homogeneous layers.

and substitute in equation (24.69) to obtain

$$\frac{1}{r}\frac{d}{dr}\left(r\frac{dR}{dr}\right) + \left[k_0^2 n^2(r) - \frac{\beta^2\rho^2}{r^2}\right]R(r) = 0 \tag{24.71}$$

Obviously, in the limit of $\rho \to \infty$, β will correspond to the propagation constant of the straight guide. If we write

$$R(r) = u(r)/\sqrt{r} \tag{24.72}$$

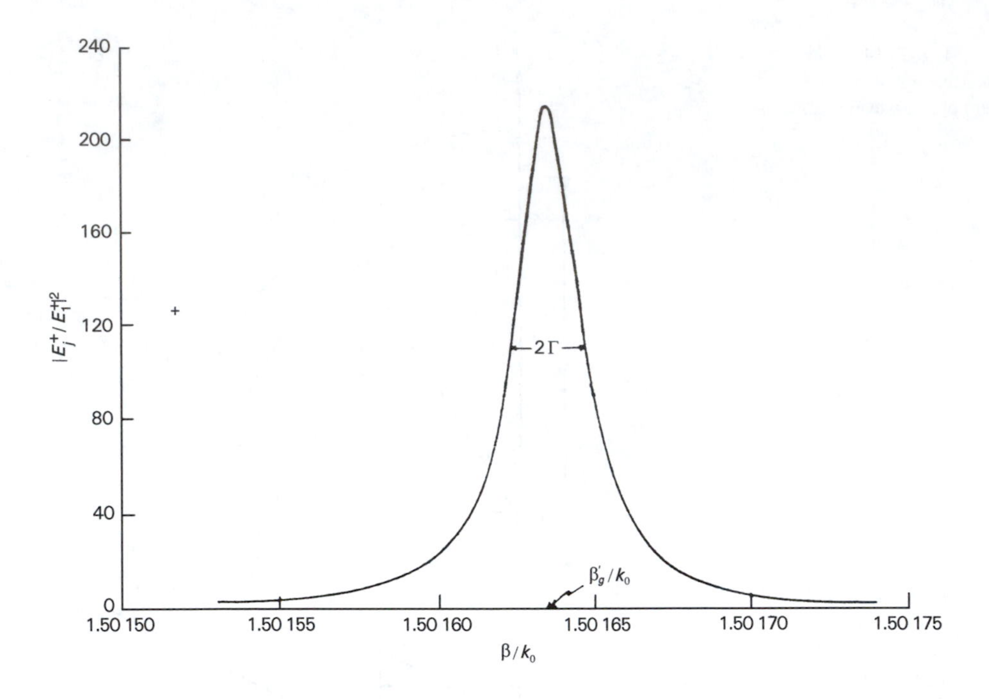

Fig. 24.16: Typical variation of $|E_j^+/E_1^+|^2$ as a function of β for $\rho = 1$ cm corresponding to the waveguide defined by equation (24.74). The peak corresponds to β_g' and the FWHM, 2Γ gives the loss coefficient of the bent waveguide; j corresponds to a layer inside the core of the waveguide.

equation (24.71) becomes

$$d^2u/d\xi^2 + \left[k_0^2\tilde{n}^2(\xi) - \beta^2\right]u(\xi) = 0 \tag{24.73}$$

where $\xi = r - \rho$ and

$$\tilde{n}^2(\xi) = n^2(r) + \left\{\frac{\beta^2}{k_0^2}\left[1 - \frac{\rho^2}{(\rho+\xi)^2}\right] + \frac{1}{4k_0^2(\rho+\xi)^2}\right\} \tag{24.74}$$

We should point out that the only assumption made in obtaining equation (24.73) is the validity of the scalar wave equation. Equation (24.73) represents a one-dimensional wave equation and, because of the form of $\tilde{n}^2(\xi)$ (see Figure 24.15(a)), it cannot support a guided mode. The waveguide is therefore leaky and we may use the matrix method to determine the bending loss. We replace the effective refractive index profile $\tilde{n}(\xi)$ by a series of step variations as shown in Figure 24.15(b) and consider the incidence of a plane wave from medium 1 (as shown in Figure 24.15(b)), which is given by $E_1^+ e^{i(k_{1\xi}\xi - \beta\eta)}$, where $k_{1\xi} = k_0\tilde{n}_1\cos\theta$, $\beta = k_0\tilde{n}_1\sin\theta$, and $\eta = \rho\phi$. We next calculate the quantity $|E_j^+/E_1^+|^2$ as a function of β; here E_j^+ is the amplitude of the downward traveling plane wave in the jth medium which, in this case, is taken as one of the layers lying inside the core of the waveguide. The quantity $|E_j^+/E_1^+|^2$ will be sharply peaked around each quasimode and will be a Lorentzian function given by

$$\left|\frac{E_j^+}{E_1^+}\right|^2 \sim \frac{1}{\left(\beta - \beta_g'\right)^2 + \Gamma^2} \tag{24.75}$$

where β_g' is the propagation constant of the quasimode and 2Γ (which represents the FWHM of the Lorentzian) represents the leakage loss of the mode – that is,

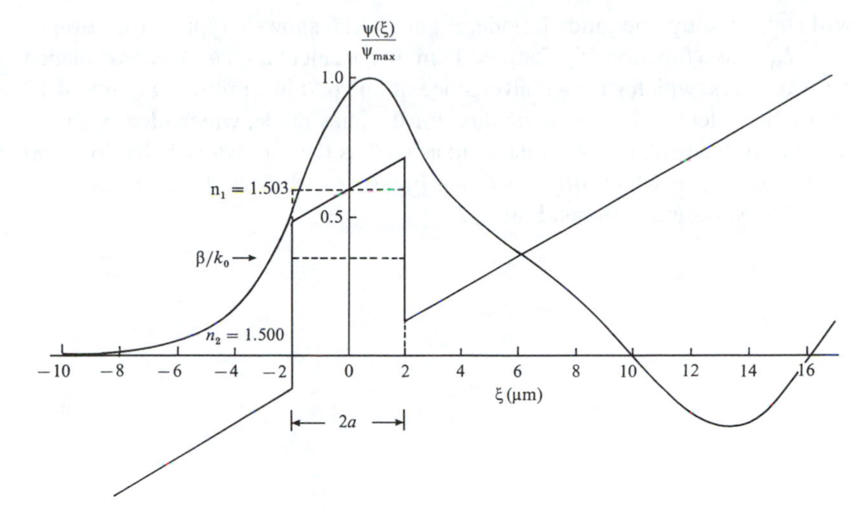

Fig. 24.17: The effective refractive index profile of the planar waveguide defined by equation (24.75) and the corresponding field distribution of the quasimode for a bend radius of $\rho = 0.5$ cm.

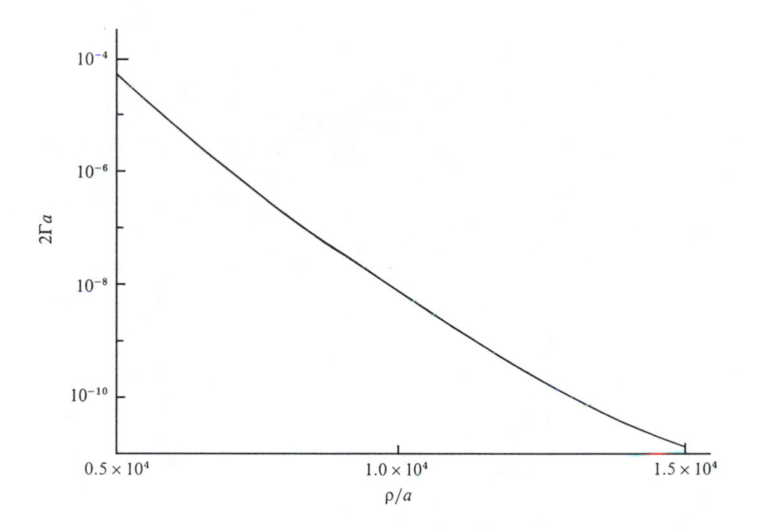

Fig. 24.18: Variation of bend loss with radius of curvature for a planar waveguide defined by equation (24.76). [Adapted from Thyagarajan et al. (1987b).]

the power inside the core would decrease as $P(\phi) = P(0)e^{-2\Gamma\rho\phi}$. Since β^2 also appears in the expression for $\tilde{n}^2(\xi)$, one can use the following iterative method.

We first substitute the value of β corresponding to the straight waveguide in equation (24.73), calculate β_g' and Γ using equation (24.75), then replace β in equation (24.74) by the obtained value of β_g', and iterate until the value of β_g' converges. The iterated values of β_g' and Γ will give us the propagation constant and loss of the quasimode of the bent waveguide.

As an example, we consider a step index slab waveguide with [Thyagarajan, et al. (1987b)]

$$n(x) = \begin{cases} n_1 = 1.503; & |x| < d/2 \\ n_2 = 1.500; & |x| > d/2 \end{cases} \tag{24.76}$$

$$d = 4\,\mu\text{m} \quad \text{and} \quad \lambda_0 = 1\,\mu\text{m}$$

so that $V = (2\pi/\lambda_0)d(n_1^2 - n_2^2)^{1/2} \approx 2.385$ and therefore the straight waveguide will support only one guided mode. Figure 24.16 shows a typical variation of $|E_j^+/E_1^+|^2$ as a function of β for $\rho = 1$ cm. In the calculation $\tilde{n}^2(\xi)$ was replaced by 100 layers, which gave a convergence of one part in a million. Figure 24.17 shows the calculated field distribution for the quasimode, which clearly shows the shift in the peak of the modal field as well as the oscillatory behavior from the value of ξ at which $\tilde{n}(\xi) > \beta/k_0$. Figure 24.18 gives the corresponding bend loss variation with bend radius.

Appendix A

Solution of the scalar wave equation for an infinite square law medium

For an infinitely extended square law medium

$$n^2(x) = n_1^2[1 - 2\Delta(x/a)^2] \tag{A.1}$$

the scalar wave equation

$$d^2\psi/dx^2 + \left[k_0^2 n^2(x) - \beta^2\right] \psi(x) = 0 \tag{A.2}$$

can be written in the form

$$d^2\psi/d\xi^2 + [\Lambda - \xi^2]\,\psi(\xi) = 0 \tag{A.3}$$

where

$$\xi = \gamma x, \quad \gamma = \left[k_0^2 n_1^2 (2\Delta)/a^2\right]^{1/4} \tag{A.4}$$

and

$$\Lambda = \frac{k_0^2 n_1^2 - \beta^2}{\gamma^2} = \frac{k_0^2 n_1^2 - \beta^2}{(k_0 n_1/a)(2\Delta)^{1/2}} \tag{A.5}$$

Equation (A.3) is the same as one obtains in the linear harmonic oscillator problem in quantum mechanics [see, e.g., Ghatak (1996), Chapter 7]. To solve equation (A.3) we write

$$\psi(\xi) = e^{-\xi^2/2}\, u(\xi) \tag{A.6}$$

Elementary manipulations give us

$$\frac{d^2u}{d\xi^2} - 2\xi\frac{du}{d\xi} + (\Lambda - 1)\, u(\xi) = 0 \tag{A.7}$$

We solve the above equation by the power series method.

$$u(\xi) = \sum_r a_r \xi^{r+s} = \xi^s[a_0 + a_1\xi + a_2\xi^2 + \cdots] \tag{A.8}$$

Substituting in equation (A.7) we get

$$s(s-1)\, a_0 = 0 \tag{A.9}$$

$$s(s+1)\, a_1 = 0 \tag{A.10}$$

and

$$\frac{a_{r+2}}{a_r} = \frac{2r + 2s + 1 - \Lambda}{(r+s+2)(r+s+1)} \tag{A.11}$$

According to one of the theorems in the theory of differential equations, since the root $s = 0$ (of equation (A.10)) makes a_1 indeterminate, it should determine both the solutions. Further, in the limit of $r \to \infty$; a_{r+2}/a_r tends to $2/r$ and therefore it behaves (for large r) as e^{ξ^2}. Thus, for $u \to 0$ as $x \to \pm\infty$ (the condition for a mode to be guided), the infinite series in equation (A.8) should become a polynomial and therefore we must have

$$\Lambda = 2m + 1; \quad m = 0, 1, 2, \ldots \tag{A.12}$$

For $m = 0, 2, 4, \ldots$ the series involving even powers of ξ becomes a polynomial and we must set $a_1 = 0$. Similarly, for $m = 1, 3, 5, \ldots$ the series involving odd powers of ξ becomes a polynomial and we must set $a_0 = 0$. These polynomials are the Hermite polynomials (see equation (7.122)) and because of equation (A.6) the field patterns are the Hermite–Gauss functions. If we normalize these functions we obtain equations (7.118)–(7.122). Substituting equation (A.12) in equation (A.5) we obtain

$$\beta_m^2 = k_0^2 n_1^2 - \frac{k_0 n_1}{a}(2\Delta)^{1/2}(2m + 1); \quad m = 0, 1, 2, \ldots \tag{A.13}$$

which represents the allowed values of the propagation constant.

Appendix B
The far-field pattern

For a near-field pattern $\psi(x, y)$, the far-field pattern is given by [see, e.g., Ghatak and Thyagarajan (1989), Chapter 5]

$$u = C \iint_{-\infty}^{+\infty} \psi(\xi, \eta)\, e^{ik_0(l\xi + m\eta)}\, d\xi\, d\eta \tag{B.1}$$

where

$$l = \frac{x}{r'} \quad \text{and} \quad m = \frac{y}{r'}$$

represent the x- and y-direction cosines of the observation direction and k_0 is the free space wave number. Since the fundamental mode distribution of a circular core optical fiber is cylindrically symmetric, the far-field pattern will also be cylindrically symmetric and we may calculate the field distribution along the x-axis for which $m = 0$ and $l = \sin\theta$. Thus

$$u = C \iint_{-\infty}^{+\infty} \psi(\xi, \eta)\, e^{ik_0 l\xi}\, d\xi\, d\eta \tag{B.2}$$

Since ψ depends only on the cylindrical r coordinate, we change over to the cylindrical system of coordinates.

$$\xi = r\cos\phi, \qquad \eta = r\sin\phi \tag{B.3}$$

and obtain

$$u = C \int_0^{\infty} r\, dr \int_0^{2\pi} d\phi\, \psi(r)\, e^{ik_0 r\sin\theta\cos\phi} \tag{B.4}$$

Now

$$J_0(\zeta) = \frac{1}{2\pi} \int_0^{2\pi} e^{i\zeta\cos\phi}\, d\phi \tag{B.5}$$

and equation (B.4) becomes

$$u(\theta) = 2\pi C \int_0^{\infty} \psi(r) J_0(k_0 r\sin\theta) r\, dr \tag{B.6}$$

which represents the far-field pattern. The above equation can also be written in the form

$$u(q) = 2\pi C \int_0^\infty \psi(r) J_0(qr) r\, dr \tag{B.7}$$

where

$$q = k_0 \sin\theta = \frac{2\pi}{\lambda_0} \sin\theta \tag{B.8}$$

The field pattern for a step index fiber is given by

$$\psi(r) = \begin{cases} \frac{J_0\left(\frac{Ur}{a}\right)}{J_0(U)}; & 0 < r < a \\ \frac{K_0\left(\frac{Wr}{a}\right)}{K_0(W)}; & r > a \end{cases} \tag{B.9}$$

Substituting from equation (B.9) into equation (B.6) we get

$$u(\theta) = \frac{2\pi C}{J_0(U)} \left[\int_0^a J_0\left(\frac{Ur}{a}\right) J_0(qr) r\, dr + \frac{J_0(U)}{K_0(W)} \int_a^\infty K_0\left(\frac{Wr}{a}\right) J_0(qr) r\, dr \right] \tag{B.10}$$

or

$$u(\theta) = \frac{2\pi C a^2}{J_0(U)} \left[\int_0^1 J_0(U\zeta) J_0(\alpha\zeta) \zeta\, d\zeta + \frac{J_0(U)}{K_0(W)} \int_1^\infty K_0(W\zeta) J_0(\alpha\zeta) \zeta\, d\zeta \right] \tag{B.11}$$

where

$$\alpha = qa = k_0 a \sin\theta \quad \text{and} \quad \zeta = \frac{r}{a}$$

The field amplitude along the axis – that is, $\theta = 0$ – is

$$u(0) = \frac{2\pi C a^2}{J_0(U)} \left[\int_0^1 J_0(U\zeta) \zeta\, d\zeta + \frac{J_0(U)}{K_0(W)} \int_1^\infty K_0(W\zeta) \zeta\, d\zeta \right] \tag{B.12}$$

Using the relations

$$\begin{aligned} \frac{d}{dx}[x J_1(x)] &= x J_0(x) \\ \frac{d}{dx}[x K_1(x)] &= -x K_0(x) \end{aligned} \tag{B.13}$$

the integrals in equation (B.12) can be evaluated to give

$$u(0) = \frac{2\pi C a^2}{J_0(U)}\left[\frac{J_1(U)}{U} + \frac{J_0(U)K_1(W)}{W K_0(W)}\right]$$

$$= \frac{2\pi C a^2}{J_0(U)}\frac{J_1(U)}{U W^2}(U^2 + W^2) \tag{B.14}$$

where we have used the relation (see equation (8.34))

$$U\frac{J_1(U)}{J_0(U)} = W\frac{K_1(W)}{K_0(W)} \tag{B.15}$$

which determines the propagation constant. Hence, the relative far-field pattern is given by

$$\tilde{u}(\theta) = \frac{u(\theta)}{u(0)}$$

$$= \frac{\left[\int_0^1 J_0(U\zeta)J_0(\alpha\zeta)\zeta\, d\zeta + \frac{J_0(U)}{K_0(W)}\int_1^\infty K_0(W\zeta)J_0(\alpha\zeta)\zeta\, d\zeta\right]}{\frac{J_1(U)}{U W^2}(U^2 + W^2)} \tag{B.16}$$

The integrals appearing in equation (B.15) can be solved by using the following standard integrals

$$\int z J_p(\alpha z)J_p(\beta z)\, dz = \frac{z}{\alpha^2 - \beta^2}[\beta J_p(\alpha z)J_{p-1}(\beta z) - \alpha J_{p-1}(\alpha z)\, J_p(\beta z)] \tag{B.17}$$

$$\int z J_p(\alpha z)K_p(\beta z)\, dz = \frac{z}{\alpha^2 + \beta^2}[\alpha K_p(\beta z)J_{p+1}(\alpha z) - \beta J_p(\alpha z)K_{p+1}(\beta z)] \tag{B.18}$$

$$\int z J_0^2(\alpha z)\, dz = \frac{z^2}{2}\left[J_0^2(\alpha z) + J_1^2(\alpha z)\right] \tag{B.19}$$

Thus

$$\int_0^1 J_0^2(U\zeta)J_0(\alpha\zeta)\zeta\, d\zeta = \frac{\alpha J_0(U)J_{-1}(\alpha) - U J_{-1}(U)J_0(\alpha)}{U^2 - \alpha^2}$$

$$= \frac{U J_1(U)J_0(\alpha) - \alpha J_0(U)J_1(\alpha)}{U^2 - \alpha^2} \tag{B.20}$$

For $U = \alpha$ we have

$$\int_0^1 J_0^2(U\zeta)\zeta \, d\zeta = \frac{J_0^2(U) + J_1^2(U)}{2} \tag{B.21}$$

Similarly

$$\begin{aligned}
&\int_1^\infty K_0(W\zeta)J_0(\alpha\zeta)\zeta \, d\zeta \\
&\quad = \frac{z}{W^2+\alpha^2}[\alpha K_0(W\zeta)J_1(\alpha\zeta) - W J_0(\alpha\zeta)K_1(W\zeta)]\Big|_1^\infty \\
&\quad = \frac{W J_0(\alpha)K_1(W) - \alpha K_0(W)J_1(\alpha)}{W^2+\alpha^2}
\end{aligned} \tag{B.22}$$

Hence

$$\begin{aligned}
&\int_0^1 J_0(U\zeta)J_0(\alpha\zeta)\zeta \, d\zeta + \frac{J_0(U)}{K_0(W)}\int_1^\infty K_0(W\zeta)J_0(\alpha\zeta)\zeta \, d\zeta \\
&\quad = \frac{(W^2+U^2)U J_1(U)}{(U^2-\alpha^2)(W^2+\alpha^2)}\left[J_0(\alpha) - \alpha J_1(\alpha)\frac{J_0(U)}{U J_1(U)}\right]
\end{aligned} \tag{B.23}$$

where we have used the eigenvalue equation

$$U\frac{J_1(U)}{J_0(U)} = W\frac{K_1(W)}{K_0(W)} \tag{B.24}$$

Hence, we obtain for $\alpha \neq U$

$$I(\theta) = \left\{\frac{U^2W^2}{(U^2-\alpha^2)(W^2+\alpha^2)}\left[J_0(\alpha) - \alpha J_1(\alpha)\frac{J_0(U)}{U J_1(U)}\right]\right\}^2 \tag{B.25}$$

For $\alpha = U$, the first integral in the numerator of equation (B.16) is given by equation (B.19) and the second integral becomes zero on using equation (B.24). Thus

$$I(\theta) = \left\{\frac{U^2W^2}{2V^2}\frac{1}{U J_1(U)}\left[J_0^2(U) + J_1^2(U)\right]\right\}^2 \quad \text{for } \alpha = U \tag{B.26}$$

where

$$V^2 = U^2 + W^2 = k_0^2 a^2 \left(n_1^2 - n_2^2\right) \tag{B.27}$$

Appendix C
WKB analysis of multimode fibers

The WKB method is applicable to multimode fibers that have profiles so that the wave equation is separable to one-dimensional equations and when the variation of the refractive index is small in distances $\sim\lambda$. In this appendix we use the WKB method to study the propagation characteristics of multimode fibers and determine the time taken by the various modes to propagate through a certain distance of the fiber. We follow the analysis of Gloge and Marcatili (1973).

C1.1 The propagation constants

We start with the scalar wave equation (see equation (7.23))

$$d^2\psi/dx^2 + \left[k_0^2 n^2(x) - \beta^2\right]\psi(x) = 0 \tag{C.1}$$

In the WKB approximation, the propagation constants β_m are determined from the relation

$$\int_{x_1}^{x_2} \left[k_0^2 n^2(x) - \beta_m^2\right]^{1/2} dx = \left(m + \frac{1}{2}\right)\pi; \quad m = 0, 1, 2, \ldots \tag{C.2}$$

Now, for a cylindrically symmetric profile (i.e., for n depending on the cylindrical coordinate r only), the propagation constants are determined by solving the radial part of the scalar wave equation (see equation (8.11)), which, on making the transformation

$$r = e^x \tag{C.3}$$

takes the form

$$d^2R/dx^2 + \left(\left\{\left[n^2(x)k_0^2 - \beta^2\right]\right\} e^{2x} - l^2\right) R = 0 \tag{C.4}$$

Observe that even though r goes from 0 to ∞, x goes from $-\infty$ to $+\infty$. Equation (C.4) resembles the one-dimensional wave equation, and the quantization condition is

$$\int_{x_1}^{x_2} \left\{\left[n^2(x)k_0^2 - \beta^2\right] e^{2x} - l^2\right\}^{1/2} dx = \left(m + \frac{1}{2}\right)\pi \tag{C.5}$$

where x_1 and x_2 are the turning points where the quantity in brace-style brackets vanishes. Using the transformation in equation (C.3), we can write equation (C.5) as

$$\int_{r_1}^{r_2} \left[n^2(r)k_0^2 - \beta^2 - l^2/r^2\right]^{1/2} dr \approx m\pi \tag{C.6}$$

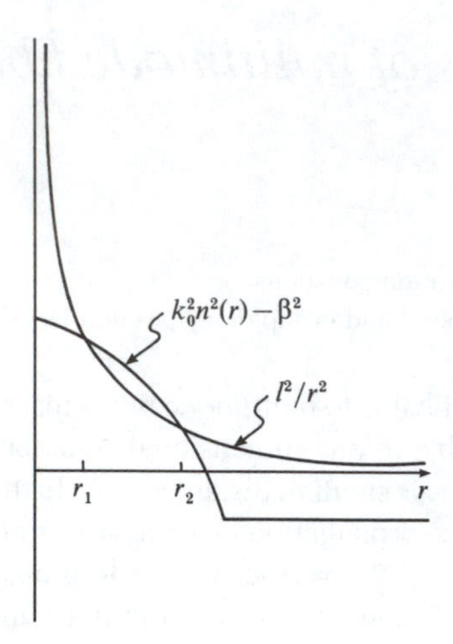

Fig. C.1: Qualitative plots of $k_0^2n^2(r) - \beta^2$ and l^2/r^2 versus r. The points of intersection give the turning points.

where r_1 and r_2 are the turning points and we have assumed $m \gg 1$ so that on the RHS we replace $m + \frac{1}{2}$ by m; only then is it possible to obtain an analytical expression for the propagation constant for a power law profile. Further, the assumption of $l \gg 1$ and $m \gg 1$ implies that we are dealing with highly multimoded fibers.

For a smooth profile of the type shown in Figure 5.1, the discrete values of β will lie between $k_0 n_1$ and $k_0 n_2$ – that is

$$k_0 n_1 > \beta > k_0 n_2 \tag{C.7}$$

In Figure C.1, we have given qualitative plots of $k_0^2 n^2(r) - \beta^2$ and l^2/r^2; the points of intersection are the turning points r_1 and r_2. Now, for a given value of l, the number of modes will be equal to the maximum value of m (which we denote by m') and the value m' will correspond to the minimum value of β^2.

$$m'(l) = (1/\pi) \int_{r_1}^{r_2} \left[k_0^2 n^2(r) - \beta_{min}^2 - l^2/r^2\right]^{1/2} dr \tag{C.8}$$

Obviously,

$$\beta_{min} \approx k_0 n_2 \tag{C.9}$$

The number of modes whose propagation constants are greater than a certain value of β (say, equal to β') will be the value of m *corresponding to* β'. Thus, if the number of modes is designated as $m'(l, \beta')$, then

$$m'(l, \beta') \approx (1/\pi) \int_{r_1}^{r_2} \left[k_0^2 n^2(r) - \beta'^2 - l^2/r^2\right]^{1/2} dr \tag{C.10}$$

The *total* number of modes (whose propagation constants are greater than β') will be given by

$$\nu(\beta') = 2m'(l=0, \beta') + 4m'(l=1, \beta') + 4m'(l=2, \beta') + \cdots + 4m'(l_{max}, \beta') \tag{C.11}$$

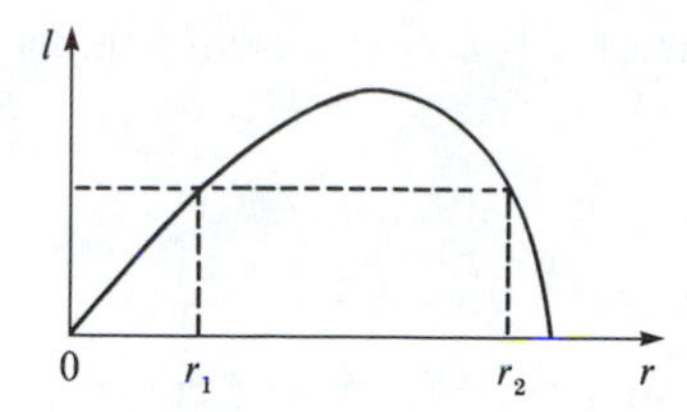

Fig. C.2: The domain of integration appearing in equation (C.12).

where l_{max} denotes the maximum value of l corresponding to a given value of β'. In writing the above equation, we have used the fact that the $l = 0$ mode is two-fold degenerate and $l \geq 1$ modes are four-fold degenerate. Replacing the sum in equation (C.11) by an integral we get

$$\nu(\beta') \approx \frac{4}{\pi}\int_0^{l_{max}} m'(l,\beta')\,dl$$

$$\approx \frac{4}{\pi}\int_0^{l_{max}}\int_{r_1}^{r_2}\left[k_0^2n^2(r) - \beta'^2 - l^2/r^2\right]^{1/2} dr\,dl \tag{C.12}$$

To evaluate the above integral we interchange the order of integration. The domain of integration is shown in Figure C.2. Obviously, for a given value of r, l goes from 0 to $r(k_0^2n^2 - \beta^2)^{1/2}$; further, the value of r goes from 0 to r_{max} where

$$k_0^2n^2(r_{max}) - \beta^2 = 0 \tag{C.13}$$

Thus

$$\nu(\beta) \approx \frac{4}{\pi}\int_0^{r_{max}}\int_0^{r\left[k_0^2n^2(r)-\beta^2\right]^{1/2}}\left[k_0^2n^2(r) - \beta^2 - l^2/r^2\right]^{1/2} dl\,dr$$

The integration over l is very easy to carry out and gives

$$\nu(\beta) \approx \frac{4}{\pi}\int_0^{r_{max}}\left[k_0^2n^2(r) - \beta^2\right] r\,dr \tag{C.14}$$

We now consider the power law profile (see equation (9.45)), so that

$$\nu(\beta) \approx \int_0^{r_{max}}\left[\left(k_0^2n_1^2 - \beta^2\right) - k_0^2n_1^2 2\Delta(r/a)^q\right] r\,dr$$

with

$$r_{max} = a\left[\frac{k_0^2n_1^2 - \beta^2}{2\Delta k_0^2n_1^2}\right]^{1/q} \tag{C.15}$$

On evaluating the integral we obtain

$$\nu(\beta) = a^2k_0^2n_1^2\Delta\frac{q}{q+2}\left(\frac{k_0^2n_1^2 - \beta^2}{2\Delta k_0^2n_1^2}\right)^{(q+2)/q} \tag{C.16}$$

Since the minimum value of β is $k_0 n_2$, the *total* number of guided modes will approximately be

$$N = a^2 k_0^2 n_1^2 \Delta \frac{q}{q+2} = \frac{1}{2} \frac{q}{q+2} V^2 \tag{C.17}$$

where we have used equation (9.48) and the fact that

$$\frac{k_0^2 n_1^2 - \beta_{min}^2}{2k_0^2 n_1^2 \Delta} = \frac{k_0^2 (n_1^2 - n_2^2)}{2k_0^2 n_1^2 \Delta} = 1$$

For a typical multimode graded index fiber we have $q \approx 2$, $V \approx 50$, and the total number of guided modes will be approximately 600. Equation (C.17) also tells us that for a given value of the waveguide parameter V, the total number of guided modes in a step index ($q = \infty$) fiber is twice as many as in a parabolic index ($q = 2$) fiber.

Now, if we label the propagation constants as $\beta_1, \beta_2, \ldots (\beta_1$ corresponding to the maximum value of β), then equation (C.16) gives us

$$\beta_\nu = k \left[1 - 2\Delta \left(\frac{q+2}{q} \frac{\nu}{a^2 k^2 \Delta} \right)^{q/(q+2)} \right]^{\frac{1}{2}}$$

$$= k \left[1 - 2\Delta \left(\frac{\nu}{N} \right)^{q/(q+2)} \right]^{\frac{1}{2}} \tag{C.18}$$

where $k \equiv k_0 n_1$. We should mention here that the label ν stands for the composite pair (l, m). The above equation can be rewritten in the form

$$\beta_\nu^2 = k^2 - \Gamma \Delta^{2/(q+2)} k^{4/(q+2)} = k^2 - 2k^2 \delta \tag{C.19}$$

where

$$\Gamma = 2 \left[\frac{q+2}{q} \frac{\nu}{a^2} \right]^{q/(q+2)}$$

$$\delta = \left(\frac{\nu}{N} \right)^{q/(q+2)} = \frac{1}{2k^2} \left(\Gamma \Delta^{2/(q+2)} k^{4/(q+2)} \right) \tag{C.20}$$

Since $\nu < N$, the value of δ lies between 0 and Δ

$$0 < \delta < \Delta \tag{C.21}$$

C1.2 Group velocity and group delay per unit length

To evaluate the group velocity, we evaluate $d\beta/dk$.

$$2\beta_\nu \frac{d\beta_\nu}{dk} = 2k - \frac{4}{(q+2)k} \left(\Gamma \Delta^{2/(q+2)} k^{4/(q+2)} \right)$$

$$- \frac{2}{q+2} \frac{1}{\Delta} \frac{d\Delta}{dk} \left(\Gamma \Delta^{2/(q+2)} k^{4/(q+2)} \right)$$

or

$$\frac{d\beta_\nu}{dk} = (1-2\delta)^{-1/2}\left(1 - \frac{4}{q+2}\delta - \frac{\epsilon}{q+2}\delta\right) \tag{C.22}$$

where

$$\epsilon = \frac{2k}{\Delta}\frac{d\Delta}{dk} = -\frac{2n_1}{N_1}\left(\frac{\lambda_0\Delta'}{\Delta}\right) \tag{C.23}$$

$$N_1 = n_1 - \lambda_0 dn_1/d\lambda_0 = n_1 - \lambda_0 n_1' \tag{C.24}$$

and primes denote differentiation with respect to the free spece wavelength λ_0. In writing the last step of equation (C.23), we have made use of the relation

$$\frac{dk}{d\lambda_0} = \frac{d}{d\lambda_0}\left(n_1\frac{2\pi}{\lambda_0}\right) = -\frac{2\pi}{\lambda_0^2}[n_1 - \lambda_0 n_1'] = -\frac{2\pi N_1}{\lambda_0^2} \tag{C.25}$$

Thus

$$\frac{k}{\Delta}\frac{d\Delta}{dk} = n_1\frac{2\pi}{\lambda_0}\frac{1}{\Delta}\left(\frac{d\Delta}{d\lambda_0}\right)\left(\frac{dk}{d\lambda_0}\right)^{-1} = -\frac{n_1}{N_1}\left(\frac{\lambda_0\Delta'}{\Delta}\right)$$

Making the binomial expansion in equation (C.22) and retaining terms up to $O(\Delta^2)$ we get

$$\frac{d\beta_\nu}{dk} \approx \left[1 - \frac{q-2-\epsilon}{q+2}\delta + \frac{3q-2-2\epsilon}{2(q+2)}\delta^2\right] + O(\delta^3) \tag{C.26}$$

Now the group delay per unit length (which is the inverse of the group velocity) is given by

$$\begin{aligned}\tau_\nu &= \frac{1}{v_\nu} = \frac{d\beta_\nu}{d\omega} = -\frac{\lambda_0^2}{2\pi c}\frac{d\beta_\nu}{d\lambda_0}\\ &= -\frac{\lambda_0^2}{2\pi c}\frac{d\beta_\nu}{dk}\frac{dk}{d\lambda_0} = \frac{N_1}{c}\frac{d\beta_\nu}{dk}\end{aligned} \tag{C.27}$$

Thus, the time taken for the νth mode to propagate through distance z of the fiber will be given by

$$t_\nu = \frac{z}{v_\nu} = \frac{N_1 z}{c}\left[1 + \frac{q-2-\epsilon}{q+2}\delta + \frac{3q-2-2\epsilon}{2(q+2)}\delta^2\right] + O(\delta^3) \tag{C.28}$$

The above equation is in a form identical to the one given by Olshansky and Keck (1976).

Appendix D
Gaussian envelope approximation

For a step index fiber, the modal intensity patterns are analytically known (see Chapter 8) and for the LP_{01} mode are given by

$$f(r) = A^2 J_0^2\left(\frac{Ur}{a}\right); \quad r < a$$

$$= A^2 \frac{J_0^2(U)}{K_0^2(W)} K_0^2\left(\frac{Wr}{a}\right); \quad r > a \tag{D.1}$$

where A is a normalization constant so that

$$2\pi \int_0^\infty f(r) r\, dr = 1 \tag{D.2}$$

Here

$$U = a\left(k_0^2 n_1^2 - \beta^2\right)^{1/2} \tag{D.3}$$

$$W = a\left(\beta^2 - k_0^2 n_2^2\right)^{1/2}$$

$$k_0 = \frac{2\pi}{\lambda_0} \tag{D.4}$$

where λ_0 is the free space wavelength, β is the propagation constant of the mode, and n_1 and n_2 are the core and cladding indices. We also have

$$U^2 + W^2 = V^2 \tag{D.5}$$

For a given V, W can be obtained with reasonable accuracy by using the following empirical relationship (see Chapter 8)

$$W = 1.1428V - 0.996 \tag{D.6}$$

which is valid for V lying in the range $1.5 < V < 2.5$. Also

$$U = (V^2 - W^2)^{1/2} \tag{D.7}$$

Using the relationship satisfied by Bessel functions, one can show that (see Problem D.1)

$$A^2 = \frac{1}{\pi a^2} \cdot \frac{U^2}{V^2 J_0^2(U)} \cdot \frac{K_0^2(W)}{K_1^2(W)} \tag{D.8}$$

Since the peak intensity value at $r = 0$ is A^2, we define the mode radius Ω by the relation

$$\pi \Omega^2 A^2 = 1 \tag{D.9}$$

Using equation (D.8), we get

$$\Omega = a J_0(U) \frac{V}{U} \cdot \frac{K_1(W)}{K_0(W)} \tag{D.10}$$

Since the mode intensity patterns at the pump and signal wavelengths resemble a Gaussian distribution (see Section 14.5), we approximate $f_p(r)$ and $f_s(r)$ by Gaussian distributions with spot sizes Ω_p and Ω_s, respectively

$$f_p(r) = \frac{1}{\pi \Omega_p^2} e^{-r^2/\Omega_p^2} \tag{D.11}$$

$$f_s(r) = \frac{1}{\pi \Omega_s^2} e^{-r^2/\Omega_s^2} \tag{D.12}$$

where Ω_p and Ω_s correspond to the pump and signal wavelengths, respectively, and can be obtained from equation (D.10).

Problem

D.1 Show that for $f(r)$ given by equation (D.1), to satisfy the normalization condition (equation (D.2)), A^2 should be given by equation (D.8).

Solution: Substituting for $f(r)$ from equation (D.1) in equation (D.2) we obtain

$$2\pi A^2 \left[\int_0^a J_0^2\left(\frac{Ur}{a}\right) r\, dr + \frac{J_0^2(U)}{K_0^2(W)} \int_a^\infty K_0^2\left(\frac{Wr}{a}\right) r\, dr \right] = 1$$

Substituting $\zeta = Ur/a$, $\xi = Wr/a$, we obtain

$$2\pi A^2 a^2 \left[\frac{1}{U^2} \int_0^u J_0^2(\zeta)\zeta\, d\zeta + \frac{J_0^2(U)}{K_0^2(W)} \frac{1}{W^2} \int_W^\infty K_0^2(\xi)\xi\, d\xi \right] = 1$$

Using the following identities

$$\int_0^u J_0^2(\zeta)\zeta\, d\zeta = \frac{1}{2}\left[U^2 J_1^2(U) + U^2 J_0^2(U)\right]$$

$$= \frac{U^2}{2}\left(J_0^2(U) + J_1^2(U)\right)$$

$$\int_W^\infty K_0^2(\xi)\xi\, d\xi = \frac{1}{2}\, \xi^2 K_1^2(\xi) + \xi^2 K_0^2(\xi)\Big|_W^\infty$$

$$-\frac{w^2}{2}\left[K_0^2(W) + K_1^2(W)\right]$$

we obtain

$$\pi A_p^2 a^2 \left[J_0^2(U) + J_1^2(U) - J_0^2(U) + \frac{K_1^2(W)}{K_0^2(W)} J_0^2(U) \right] = 1$$

or

$$\pi A_p^2 a^2 J_0^2(U) \left[\frac{J_1^2(U)}{J_0^2(U)} + \frac{K_1^2(W)}{K_0^2(W)} \right] = 1$$

Using the eigenvalue equation for LP_{01} mode (see Chapter 8), the above equation reduces to

$$A_p^2 = \frac{U^2 K_0^2(W)}{\pi a^2 V^2 K_1^2(W) J_0^2(U)}$$

Appendix E

Coupled-mode equations

In this appendix we derive the coupled-mode equations, which describe the variation in the amplitude of the waves propagating in each individual waveguide of a directional coupler. Let $n_1(x, y)$ and $n_2(x, y)$ represent the refractive index variation in the transverse plane of waveguide 1 in the absence of waveguide 2 and that of waveguide 2 in the absence of waveguide 1. Let $n(x, y)$ represent the refractive index variation of the directional coupler consisting of the waveguides 1 and 2. For example, for a directional coupler consisting of two step index planar waveguides, $n_1(x)$, $n_2(x)$, and $n(x)$ are shown in Figure E.1.

If β_1 and β_2 represent the propagation constants of the modes of waveguides 1 and 2 in the absence of the other then we may write

$$\nabla_t^2 \psi_1 + \left[k_0^2 n_1^2(x, y) - \beta_1^2\right]\psi_1 = 0 \tag{E.1}$$

$$\nabla_t^2 \psi_2 + \left[k_0^2 n_2^2(x, y) - \beta_2^2\right]\psi_2 = 0 \tag{E.2}$$

where

$$\nabla_t^2 = \nabla^2 - \frac{\partial^2}{\partial z^2} = \frac{\partial^2}{\partial x^2} + \frac{\partial^2}{\partial y^2} \tag{E.3}$$

and $\psi_1(x, y)$ and $\psi_2(x, y)$ represent the transverse mode field patterns of waveguides 1 and 2, respectively, in the absence of the other.

If $\Psi(x, y, z)$ represents the total field of the directional coupler structure, then we have

$$\nabla_t^2 \Psi + \frac{\partial^2 \Psi}{\partial z^2} + k_0^2 n^2(x, y)\Psi = 0 \tag{E.4}$$

We now approximate Ψ as follows

$$\Psi(x, y, z) = A(z)\psi_1(x, y)\, e^{-i\beta_1 z} + B(z)\psi_2(x, y)\, e^{-i\beta_2 z} \tag{E.5}$$

which is valid when the two waveguides are not very strongly interacting. In equation (E.5) we have written the total field as a superposition of the fields in the first and second waveguides with amplitudes $A(z)$ and $B(z)$, which are functions of z. For infinite separation between the two waveguides, obviously the waveguides are noninteracting, and A and B would then be independent of z. The coupling between the two waveguides leads to z-dependent amplitudes. Substituting for Ψ in equation (E.4) we obtain

$$\begin{aligned} &Ae^{-i\beta_1 z}\left(\nabla_t^2 \psi_1 - \beta_1^2 \psi_1 + k_0^2 n^2 \psi_1\right) + Be^{-i\beta_2 z}\left(\nabla_t^2 \psi_2 - \beta_2^2 \psi_2 + k_0^2 n^2 \psi_2\right) \\ &\quad - 2i\beta_1 (dA/dz)\psi_1 e^{-i\beta_1 z} - 2i\beta_2 (dB/dz)\psi_2 e^{-i\beta_2 z} = 0 \end{aligned} \tag{E.6}$$

where we have neglected terms proportional to d^2A/dz^2 and d^2B/dz^2, which is

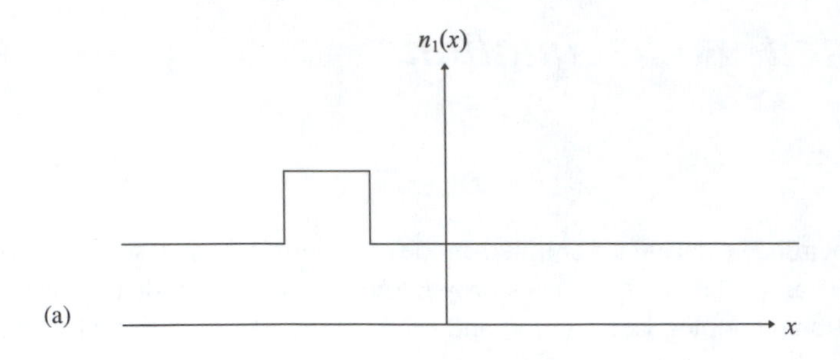

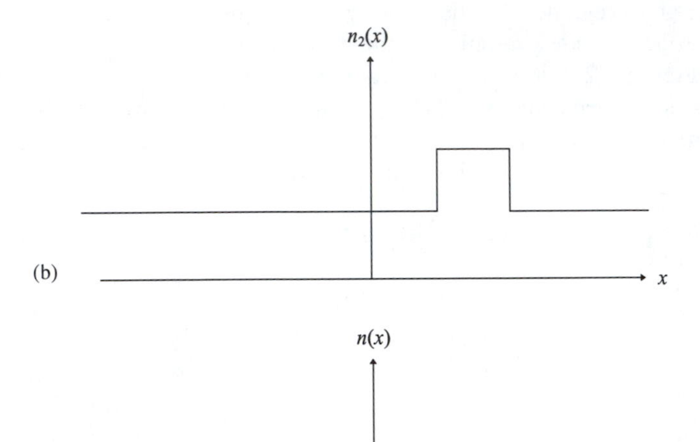

Fig. E.1: (a) and (b) The refractive index profiles corresponding to two isolated step index planar waveguides. (c) The refractive index profile corresponding to a directional coupler formed by the two waveguides.

justified when $A(z)$ and $B(z)$ are slowly varying functions of z. Using equations (E.1) and (E.2), equation (E.6) becomes

$$k_0^2 \Delta n_1^2 A \psi_1 + k_0^2 \Delta n_2^2 B \psi_2 \, e^{i\Delta\beta z} - 2i\beta_1 (dB/dz)\psi_1$$
$$- 2i\beta_2 (dA/dz)\psi_2 \, e^{i\Delta\beta z} = 0 \tag{E.7}$$

where

$$\Delta n_1^2 = n^2(x, y) - n_1^2(x, y) \tag{E.8}$$

$$\Delta n_2^2 = n^2(x, y) - n_2^2(x, y) \tag{E.9}$$

$$\Delta\beta = \beta_1 - \beta_2 \tag{E.10}$$

Multiplying equation (E.7) by ψ_1^* and integrating over the whole cross section, we obtain

$$dA/dz = -i\kappa_{11} A(z) - i\kappa_{12} B e^{i\Delta\beta z} \tag{E.11}$$

where

$$\kappa_{11} = \frac{k_0^2}{2\beta_1} \frac{\iint_{-\infty}^{\infty} \psi_1^* \Delta n_1^2 \psi_1 \, dx \, dy}{\iint_{-\infty}^{\infty} \psi_1^* \psi_1 \, dx \, dy} \tag{E.12}$$

$$\kappa_{12} = \frac{k_0^2}{2\beta_1} \frac{\iint_{-\infty}^{\infty} \psi_1^* \Delta n_2^2 \psi_2 \, dx \, dy}{\iint_{-\infty}^{\infty} \psi_1^* \psi_1 \, dx \, dy} \tag{E.13}$$

In writing equation (E.11) we have neglected the overlap integral of the modes – that is, we assume

$$\iint_{-\infty}^{\infty} \psi_1^* \psi_2 \, dx \, dy \ll \iint_{-\infty}^{\infty} \psi_1^* \psi_1 \, dx \, dy \tag{E.14}$$

which is valid for weak coupling between the waveguides.

Similarly, if we multiply equation (E.7) by ψ_2^* and integrate, we obtain

$$dB/dz = -i\kappa_{22} B - i\kappa_{21} A \, e^{-i\Delta\beta z} \tag{E.15}$$

where

$$\kappa_{22} = \frac{k_0^2}{2\beta_2} \frac{\iint_{-\infty}^{\infty} \psi_2^* \Delta n_2^2 \psi_2 \, dx \, dy}{\iint_{-\infty}^{\infty} \psi_2^* \psi_2 \, dx \, dy} \tag{E.16}$$

$$\kappa_{21} = \frac{k_0^2}{2\beta_2} \frac{\iint_{-\infty}^{\infty} \psi_2^* \Delta n_1^2 \psi_1 \, dx \, dy}{\iint_{-\infty}^{\infty} \psi_2^* \psi_2 \, dx \, dy} \tag{E.17}$$

We can write equations (E.11) and (E.15) in a different form if we define

$$a(z) = A(z) \, e^{-i\beta_1 z} \tag{E.18}$$

$$b(z) = B(z) \, e^{-i\beta_2 z} \tag{E.19}$$

Substituting from equations (E.18) and (E.19) in equations (E.11) and (E.15), we have

$$da/dz = -i(\beta_1 + \kappa_{11}) a - i\kappa_{12} b \tag{E.20}$$

$$db/dz = -i(\beta_2 + \kappa_{22}) b - i\kappa_{21} a \tag{E.21}$$

The above two equations represent the coupled-mode equations. It follows from equations (E.20) and (E.21) that κ_{11} and κ_{22} represent the corrections to the propagation constants of each individual waveguide mode due to the presence of the other waveguide. These correction factors are normally neglected in the analysis, although one can very easily incorporate them. Thus, the coupled equations may be written as

$$da/dz = -i\beta_1 a - i\kappa_{11} b \tag{E.22}$$

$$db/dz = -i\beta_2 b - i\kappa_{21} a \tag{E.23}$$

These are the coupled-mode equations that have been used in Chapter 14.

The quantities κ_{11} and κ_{22} in equations (E.12) and (E.16) represent corrections to the propagation constants of each individual waveguide mode due to the presence of the other waveguide and are normally neglected in the analysis, although they can be incorporated easily. Thus, we write equations (E.11) and (E.15) as

$$\frac{dA}{dz} = -i\kappa_{12} B e^{i\Delta\beta z} \tag{E.24}$$

$$\frac{dB}{dz} = -i\kappa_{21} A e^{-i\Delta\beta z} \tag{E.25}$$

Differentiating equation (E.24) with respect to z and eliminating B-dependent terms, we obtain

$$\frac{d^2 A}{dz^2} - i\Delta\beta\frac{dA}{dz} + \kappa^2 A = 0 \tag{E.26}$$

whose general solution is

$$A(z) = e^{i\Delta\beta z/2}(a_1 e^{i\gamma z} + a_2 e^{-i\gamma z}) \tag{E.27}$$

where

$$\gamma^2 = \kappa^2 + \Delta\beta^2/4, \quad \kappa^2 = \sqrt{\kappa_{12}\kappa_{21}} \tag{E.28}$$

Substituting the value of $A(z)$ in equation (E.24), we obtain

$$B(z) = -\frac{1}{\kappa_{12}} e^{-i\Delta\beta z/2}\left[\left(\frac{\Delta\beta}{2} + \gamma\right) a_1 e^{i\gamma z} + \left(\frac{\Delta\beta}{2} - \gamma\right) a_2 e^{-i\gamma z}\right] \tag{E.29}$$

If at $z = 0$, power is incident only in waveguide 1, then

$$A(0) = A_0; \quad B(0) = 0 \tag{E.30}$$

Using these initial conditions in equations (E.27) and (E.29), the constants a_1 and a_2 can be determined.

Finally, the power in each waveguide at any value of z is given by

$$\frac{P_1(z)}{P_1(0)} = 1 - \frac{\kappa^2}{\gamma^2}\sin^2\gamma z \tag{E.31}$$

$$\frac{P_2(z)}{P_1(0)} = \frac{\kappa^2}{\gamma^2}\sin^2\gamma z \tag{E.32}$$

where $P_1(0)$ is the power launched into waveguide 1. In obtaining the final equations, we have assumed $\kappa_{12} \simeq \kappa_{21}$.

Appendix F

Derivation of coupled-mode equation for periodic coupling

Let us consider an optical fiber with a refractive index profile $n^2(x, y)$ in which there is a periodic z-dependent perturbation given by $\Delta n^2(x, y, z)$. This perturbation could correspond to periodic index variation as in a fiber Bragg grating (see Chapter 17) or it could be a periodic stress or a periodic undulation of the fiber axis.

If $\psi_1(x, y)$ and $\psi_2(x, y)$ are two modes of the fiber, then the periodic perturbation can, under certain conditions (to be derived), couple power among the modes. Thus, we write for the total field at any value of z as

$$\psi(x, y, z) = A(z)\psi_1(x, y)\, e^{-i\beta_1 z} + B(z)\psi_2(x, y)\, e^{-i\beta_2 z} \tag{F.1}$$

Here β_1 and β_2 are the absence of perturbation and $A(z)$ and $B(z)$ are their corresponding amplitudes. In the absence of the perturbation A and B would be constants; the perturbation couples power among the modes and, hence, A and B are z-dependent.

Since ψ_1 and ψ_2 are the modes of the fiber in the absence of any perturbation, they satisfy the following equations.

$$\nabla_t^2 \psi_1 + \left(k_0^2 n^2(x, y) - \beta_1^2\right)\psi_1 = 0 \tag{F.2}$$

$$\nabla_t^2 \psi_2 + \left(k_0^2 n^2(x, y) - \beta_2^2\right)\psi_2 = 0 \tag{F.3}$$

where

$$\nabla_t^2 = \nabla^2 - \frac{\partial^2}{\partial z^2} \tag{F.4}$$

They also satisfy the orthogonality condition

$$\int_{-\infty}^{\infty}\int_{-\infty}^{\infty} \psi_1^*(x, y)\psi_2(x, y)\, dx\, dy = 0 \tag{F.5}$$

The wave equation to be satisfied by $\psi(x, y, z)$ is

$$\nabla_t^2 \psi + \frac{\partial^2 \psi}{\partial z^2} + k_0^2[n^2(x, y) + \Delta n^2(x, y, z)]\psi = 0 \tag{F.6}$$

Substituting for ψ from equation (F.1), neglecting the second derivatives of $A(z)$ and $B(z)$ with z (also referred to as the slowly varying envelope approximation), and using equations (F.2) and (F.3), we obtain

$$\begin{aligned} &-2i\beta_1 \frac{dA}{dz}\psi_1 - 2i\beta_2 \frac{dB}{dz}\psi_2\, e^{i\Delta\beta z} \\ &\qquad + k_0^2 \Delta n^2(x, y, z)[A\psi_1 + B\psi_2\, e^{i\Delta\beta z}] = 0 \end{aligned} \tag{F.7}$$

where

$$\Delta\beta = \beta_1 - \beta_2 \tag{F.8}$$

Multiplying equation (F.7) by ψ_1^* and integrating we get after some simplifications

$$\frac{dA}{dz} = -ic_{11}A - ic_{12}Be^{i\Delta\beta z} \tag{F.9}$$

where we have used the orthogonality condition (see equation (F.5)) and

$$c_{11}(z) = \frac{k_0^2}{2\beta_1}\frac{\iint \psi_1^*\Delta n^2\psi_1\,dx\,dy}{\iint \psi_1^*\psi_1\,dx\,dy} \tag{F.10}$$

$$c_{12}(z) = \frac{k_0^2}{2\beta_1}\frac{\iint \psi_1^*\Delta n^2\psi_2\,dx\,dy}{\iint \psi_1^*\psi_1\,dx\,dy} \tag{F.11}$$

Similarly, multiplying equation (F.7) by ψ_2^* and integrating, we get

$$\frac{dB}{dz} = -ic_{22}B - ic_{21}Ae^{-i\Delta\beta z} \tag{F.12}$$

where

$$c_{22}(z) = \frac{k_0^2}{2\beta_2}\frac{\iint \psi_2^*\Delta n^2\psi_2\,dx\,dy}{\iint \psi_2^*\psi_2\,dx\,dy} \tag{F.13}$$

$$c_{21}(z) = \frac{k_0^2}{2\beta_2}\frac{\iint \psi_2^*\Delta n^2\psi_1\,dx\,dy}{\iint \psi_2^*\psi_2\,dx\,dy} \tag{F.14}$$

Equations (F.9) and (F.12) together represent the coupled-mode equations and describe the z-dependence of A and B.

In the presence of a periodic z-dependent perturbation, we may write

$$\Delta n^2(x, y, z) = \Delta n^2(x, y)\sin Kz \tag{F.15}$$

where $K = 2\pi/\Lambda$, and Λ represents the spatial period of the perturbation. Equation (F.15) very well represents the refractive index perturbation in a Bragg fiber (see Chapter 17). For $\Delta n^2(x, y, z)$ given by equation (F.15), we have

$$\begin{aligned} c_{12} &= \frac{k_0^2}{2\beta_1}\frac{\iint \psi_1^*\Delta n^2(x, y)\psi_2\,dx\,dy}{\iint \psi_1^*\psi_1\,dx\,dy}\sin Kz \\ &= 2\kappa_{12}\sin Kz \end{aligned} \tag{F.16}$$

where

$$\kappa_{12} = \frac{k_0^2}{4\beta_1}\frac{\iint \psi_1^*\Delta n^2(x, y)\psi_2\,dx\,dy}{\iint \psi_1^*\psi_1\,dx\,dy} \tag{F.17}$$

is z-independent. Similarly, we have

$$\begin{aligned} c_{11} &= 2\kappa_{11}\sin Kz, \quad c_{22} = 2\kappa_{22}\sin Kz \\ c_{21} &= 2\kappa_{21}\sin Kz \end{aligned} \tag{F.18}$$

with κ_{11}, κ_{22}, and κ_{21} being z-independent and given by equations of the type equation (F.17).

Substituting from equations (F.16) and (F.18) in equation (F.9) we have

$$\frac{dA}{dz} = -2i\kappa_{11}A\sin Kz - \kappa_{12}Be^{i(\Delta\beta+K)z} + \kappa_{12}Be^{i(\Delta\beta-K)z} \tag{F.19}$$

For weak perturbations, κ_{12} and κ_{21} are small and, hence, the typical length scale over which the mode amplitudes change appreciably $\sim 1/\kappa_{12} \simeq 1/\kappa_{21}$, which is large. If we integrate equation (F.19) over distance L, small compared with the distance over which A and B change appreciably, then we obtain

$$\begin{aligned} &A\left(z+\frac{L}{2}\right) - A\left(z-\frac{L}{2}\right) \\ &\quad = +4i\kappa_{11}A\cos Kz\frac{\sin KL/2}{K} \\ &\quad\quad -2i\,\kappa_{12}Be^{i(\Delta\beta+K)z}\left\{\frac{\sin(\Delta\beta+K)L/2}{(\Delta\beta+K)}\right\} \\ &\quad\quad +2i\kappa_{12}Be^{i(\Delta\beta-K)z}\,\frac{\sin(\Delta\beta-K)L/2}{(\Delta\beta-K)} \end{aligned} \tag{F.20}$$

Now,

$$\Delta\beta = \frac{2\pi}{\lambda_0}\Delta n_{eff} \tag{F.21}$$

where Δn_{eff} is the effective index difference between the modes. As a typical value, Δn_{eff} is approximately the index difference between core and cladding, which if assumed to be about 0.005, gives for $\lambda_0 = 1\ \mu$m

$$\Delta\beta \simeq 3\times 10^4\ \text{m}^{-1}$$

Thus, if K is chosen so that $K \simeq \Delta\beta$ and $L \simeq 2\times 10^{-3}$ m, then

$$\left|\frac{\sin(\Delta\beta-K)L/2}{(\Delta\beta-K)}\right| \simeq \frac{L}{2} = 10^{-3}\ \text{m}$$

$$\left|\frac{\sin(\Delta\beta+K)L/2}{(\Delta\beta+K)}\right| \le \frac{1}{(\Delta\beta+K)} \simeq \frac{1}{2\Delta\beta} \simeq 1.7\times 10^{-5}\ \text{m}$$

$$\left|\frac{\sin KL/2}{K}\right| \le \frac{1}{K} \simeq \frac{1}{\Delta\beta} \simeq 3\times 10^{-5}\ \text{m}$$

Hence, we note that for $K \simeq \Delta\beta$, the contributions from the first and the second terms in the RHS of equation (F.20) are negligible compared with the last term and, hence, can be neglected. The second term would have made significant contribution if $\Delta\beta = \beta_1 - \beta_2 = -K$ – that is, if $\beta_2 \simeq \beta_1 + K$.

Thus, in the presence of a periodic perturbation, coupling takes place mainly among modes for which $\Delta\beta$ is close to either K or to $-K$. This justifies the two-term expansion of equation (F.1). The approximation retaining either $e^{i(\Delta\beta+K)z}$ or $e^{i(\Delta\beta-K)z}$ term in equation (F.19) is called the synchronous approximation and corresponds to the rotating wave approximation used in time-dependent perturbation theory in quantum mechanics.

Thus, if we choose

$$K = \frac{2\pi}{\Lambda} \simeq \Delta\beta = \beta_1 - \beta_2 \tag{F.22}$$

then equation (F.19) can be approximated by

$$\frac{dA}{dz} = \kappa_{12} B e^{i\Gamma z} \tag{F.23}$$

where $\Gamma = \Delta\beta - K$. Similarly, equation (F.12) leads to

$$\frac{dB}{dz} = -\kappa_{21} A e^{-i\Gamma z} \tag{F.24}$$

If the modes ψ_1 and ψ_2 are normalized to carry unit power, then under the weakly guiding approximation we may write

$$\frac{\beta_1}{2\omega\mu_0} \iint \psi_1^* \psi_1 \, dx \, dy = 1 \tag{F.25}$$

and a similar equation for ψ_2, then $|A|^2$ and $|B|^2$ would directly give the power carried by both modes. Using equation (F.25) in equation (F.17) we obtain

$$\kappa_{12} = \frac{\omega\epsilon_0}{8} \iint \psi_1^* \Delta n^2(x, y) \psi_2 \, dx \, dy \tag{F.26}$$

By using the orthonormality relation for ψ_2, one can show that

$$\kappa_{21} = \kappa_{12} = \kappa \tag{F.27}$$

Thus, the two coupled equations become

$$\frac{dA}{dz} = \kappa B e^{i\Gamma z} \tag{F.28}$$

$$\frac{dB}{dz} = -\kappa A e^{-i\Gamma z} \tag{F.29}$$

with κ given by equation (F.26).

Equations (F.28) and (F.29) describe coupling among two modes propagating along the same direction – that is, with β_1 and β_2 both positive or both negative.

This is referred to as codirectional coupling. Another very important coupling is referred to as contradirectional coupling, wherein coupling takes place between two modes propagating in opposite directions. In such a case, instead of equation (F.1), we have

$$\psi(x, y, z) = A(z)\psi_1(x, y)\, e^{-i\beta_1 z} + B(z)\psi_2(x, y)\, e^{i\beta_2 z} \tag{F.30}$$

Thus, the mode with propagation constant β_1 is propagating in the $+z$-direction and that with propagation constant β_2 is propagating in $-z$-directions. If they correspond to the same mode, then $\beta_1 = \beta_2 = \beta$. Following a procedure identical to the one described earlier, we obtain the following two coupled-mode equations

$$\frac{dA}{dz} = \kappa B e^{i\Gamma z} \tag{F.31}$$

$$\frac{dB}{dz} = \kappa A e^{-i\Gamma z} \tag{F.32}$$

with

$$\Gamma = \beta_1 + \beta_2 - K \tag{F.33}$$

Note that the signs on the RHS of equations (F.31) and (F.32) are the same, which is opposite the case of codirectional coupling. Because of this, the solutions for the contradirectional case are not oscillatory.

Equations (F.31) and (F.32) have been solved in Section 21.7 for the case $\Gamma = 0$.

Example F.1: For contradirectional coupling between two LP_{01} modes propagating in opposite directions, $\psi_1 = \psi_2$ and equation (F.26) becomes

$$\kappa = \frac{\omega\epsilon_0}{8} \iint \Delta n^2(x, y)|\psi|^2\, dx\, dy \tag{F.34}$$

If we use the Gaussian approximation for ψ (see Chapter 8), then

$$\psi = \frac{2}{w_0}\sqrt{\frac{\omega\mu_0}{\pi\beta}}\, e^{-r^2/w_0^2} \tag{F.35}$$

where the multiplying constant will satisfy equation (F.25).

For fiber Bragg grating, we may assume that the periodic refractive index perturbation is uniform inside the core and zero outside – that is,

$$\begin{aligned} \Delta n^2(x, y) &= \Delta n^2; \quad x^2 + y^2 < a^2 \\ &= 0; \quad \text{otherwise} \end{aligned} \tag{F.36}$$

and equation (F.34) becomes

$$\kappa = \frac{\omega\epsilon_0}{8}\Delta n^2 \frac{4}{w_0^2}\frac{\omega\mu_0}{\pi\beta}\int_0^a r\,dr\int_0^{2\pi} d\phi e^{-2r^2/w_0^2}$$

$$= \frac{k_0 \Delta n^2}{4n_{eff}}\left(1 - e^{-2a^2/w_0^2}\right) \tag{F.37}$$

where we have used the relationship $\beta = k_0 n_{eff}$. If we write $\Delta n^2 \simeq 2n\Delta n$ and assume $n \simeq n_{eff}$, we obtain

$$\kappa = \frac{\pi \Delta n}{\lambda_0}\left(1 - e^{-2a^2/w_0^2}\right) \tag{F.38}$$

which is the same relation as equation (17.44) with I given by $(1 - e^{-2a^2/\omega_0^2})$ under the Gaussian approximation.

One can, in general, evaluate equation (F.26) for κ for any given perturbation $\Delta n^2(x, y)$ and modal field profiles $\psi_1(x, y)$ and $\psi_2(x, y)$.

Appendix G

Leakage loss in optical waveguides

The fractional power that remains inside the core at z is approximately given by the overlap integral

$$W(z) \approx \left| \int_0^\infty \psi^*(x,0)\psi(x,z)\,dx \right|^2 \tag{G.1}$$

Now

$$\psi(x,z) = \int d\beta \phi(\beta)\psi_\beta(x) e^{i\beta z} \tag{G.2}$$

(see equation (24.36)). Thus

$$\begin{aligned} W(z) &\approx \left| \int dx \left[\int d\beta \phi^*(\beta)\psi_\beta^*(x) \right] \left[\int d\beta' \phi(\beta')\psi_{\beta'}(x)\, e^{i\beta' z} \right] \right|^2 \\ &\approx \left| \int d\beta \phi^*(\beta) \int d\beta' \phi(\beta') e^{i\beta' z} \int dx \psi_{\beta'}^*(x)\psi_\beta(x) \right|^2 \end{aligned} \tag{G.3}$$

The last integral is $\delta(\beta - \beta')$. Since

$$\int d\beta' \phi(\beta')\, e^{i\beta' z} \delta(\beta - \beta') = \phi(\beta)\, e^{i\beta z} \tag{G.4}$$

we obtain

$$W(z) = \left| \int |\phi(\beta)|^2\, e^{i\beta z}\, d\beta \right|^2 \tag{G.5}$$

To evaluate the above integral we must evaluate $|\phi(\beta)|^2$, which is given by (see equation (24.39))

$$|\phi(\beta)|^2 \approx |A/A_g|^2 \tag{G.6}$$

We know A_g (see equation (24.24)). To evaluate A, we express it in terms of D and then use equation (24.30). Since the wave packet is a superposition of radiation modes around the "quasimode," $|\phi(\beta)|^2$ is very sharply peaked around $\beta \approx \beta_g$ (see also Figures 24.9 and 24.10) and therefore all calculations will be carried out near $\beta = \beta_g$.

We begin with the calculation of C around $\beta = \beta_g$.

$$\begin{aligned} C &= \frac{1}{2} A [\sin \kappa x_1 - (\kappa/\gamma) \cos \kappa x_1]_{\substack{\kappa \approx \kappa_g \\ \beta \approx \beta_g}} \quad \text{(see equation (24.27))} \\ &\approx \frac{1}{2} A [\sin \kappa_g x_1 - (\kappa_g/\gamma_g) \cos \kappa_g x_1] \\ &\approx \frac{1}{2} A \left[\frac{\kappa_g}{\delta} + \frac{\kappa_g}{\gamma_g} \frac{\gamma_g}{\delta} \right] \end{aligned}$$

where $x_1 = d/2$ and subscript g refers to the "guided mode" and use has been made of the relations

$$\sin \kappa_g x_1 = \kappa_g/\delta, \quad \cos \kappa_g x_1 = -\gamma_g/\delta \tag{G.7}$$

$$\delta^2 = \alpha^2 + \gamma^2 = k_0^2(n_1^2 - n_2^2) \tag{G.8}$$

Thus

$$C \approx (A/\delta)\alpha_g \tag{G.9}$$

Since

$$B|_{\beta=\beta_g} = 0 \tag{G.10}$$

we must make a Taylor series expansion of B around $\beta = \beta_g$.

$$B \approx \left.\frac{dB}{d\beta}\right|_{\beta=\beta_g}(\beta - \beta_g) = \left[\frac{dB}{d\alpha}\frac{d\alpha}{d\beta}\right]_{\substack{\kappa=\kappa_g\\ \beta=\beta_g}}(\beta - \beta_g)$$

$$\approx \frac{1}{2}A\left[\left(x_1 \cos\kappa x_1 - \frac{\kappa}{\gamma}x_1 \sin\alpha x_1\right.\right.$$

$$\left.\left. + \frac{1}{\gamma}\cos\kappa x_1 - \frac{\kappa}{\gamma^2}\frac{d\gamma}{d\kappa}\cos\kappa x_1\right)\frac{d\kappa}{d\beta}\right]_{\substack{\kappa=\kappa_g\\ \beta=\beta_g}}(\beta - \beta_g)$$

where we have used equation (24.26). If we use equation (G.8) and the relation

$$\kappa^2 = k_0^2 n_1^2 - \beta^2 \tag{G.11}$$

we obtain

$$\frac{d\gamma}{d\kappa} = -\frac{\kappa}{\gamma} \quad \text{and} \quad \frac{d\kappa}{d\beta} = -\frac{\beta}{\kappa} \tag{G.12}$$

On substitution in the expression for B and simplifying, we obtain

$$B \approx \frac{1}{2}A\frac{\beta_g\delta}{\gamma_g^2\kappa_g}(1 + \gamma_g x_1)(\beta - \beta_g) \tag{G.13}$$

Substituting for B and C in equation (24.28), we get

$$|D_\pm| \approx \frac{1}{4}|A|\frac{\delta\beta_g}{\gamma_g^2\kappa_g}(1 + \gamma_g x_1)\left|1 + \frac{\gamma_g}{i\kappa_g}\right| e^{\gamma_g(x_2 - x_1)}$$

$$\times \left|(\beta - \beta_g) + \frac{2\gamma_g^2\kappa_g^2}{\beta_g\delta^2(1 + \gamma_g x_1)}\frac{1 - (\gamma_g/i\kappa_g)}{1 + (\gamma_g/i\kappa_g)}e^{2\gamma_g(x_2 - x_1)}\right|$$

$$\approx \frac{1}{4}|A|^2\frac{\beta_g\delta^2}{\gamma_g^2\kappa_g^2}(1 + \gamma_g x_1)\, e^{\gamma_g(x_2 - x_1)}[(\beta - \beta_g')^2 + \Gamma^2]^{1/2} \tag{G.14}$$

where Γ and β_g' are given by equations (24.41) and (24.42). Equating the RHS of the above equation to the RHS of equation (24.30) we get

$$|A| \approx \frac{4\gamma_g^2 \kappa_g^2 e^{-\gamma_g(x_2-x_1)}}{\delta^2(2\pi\kappa_g\beta_g)^{1/2}(1+\gamma_g x_1)[(\beta-\beta_g')^2+\Gamma^2]^{1/2}} \tag{G.15}$$

Using equation (24.24) and simplifying, we get

$$|\phi(\beta)|^2 = \left|\frac{A}{A_g}\right|^2 \approx \frac{\Gamma}{\pi}\frac{1}{(\beta-\beta_g')^2+\Gamma^2} \tag{G.16}$$

Thus, equation (G.5) becomes

$$W(z) \approx \left|\frac{\Gamma}{\pi}\int \frac{e^{i\beta z}d\beta}{(\beta-\beta_g')^2+\Gamma^2}\right|^2 \tag{G.17}$$

We may evaluate the integral from $-\infty$ to $+\infty$ since most of the contribution will come from the region near resonance ($\beta \approx \beta_g$). We introduce the variable

$$\xi = \beta - \beta_g' \tag{G.18}$$

to write

$$W(z) \approx \left|\frac{\Gamma}{\pi}\int_{-\infty}^{+\infty} \frac{e^{i\xi z}}{(\xi+i\Gamma)(\xi-i\Gamma)}\right|^2 \tag{G.19}$$

For $z > 0$, the integral may be evaluated by using complex variable techniques and Jordan's lemma. In the complex ξ plane we choose a contour that consists of the real axis and a semicircle in the upper half plane where the integral vanishes. There is a simple pole within the contour at

$$\xi = i\Gamma$$

so that

$$W(z) \approx \left|\frac{\Gamma}{2\pi}2\pi i\frac{e^{-\Gamma z}}{2i\Gamma}\right|^2 = e^{-2\Gamma z} \tag{G.20}$$

which shows how the power inside the "core" decays exponentially. Further, use of equations (G.14) and (24.41) gives us equation (24.47).

References

Adams M.J. (1981). *An Introduction to Optical Waveguides*. John Wiley, Chichester.

Agrawal G.P. (1989). *Nonlinear Fiber Optics*. Academic Press, New York.

Agrawal G.P. (1993). *Fiber Optic Communication Systems*. John Wiley, Singapore.

Agrawal G.P. and Dutta N. (1993). *Semiconductor Laser*, 2nd ed. Van Nostrand Reinhold, New York.

Allen D., Bennet S.M., Brunner J., and Dyott R.D. (1994). A low cost fiber optic gyro for land navigation. *SPIE 2292: Fiber Optic and Laser Sensors* XII, 203.

Ankiewicz A. and Pask C. (1977). Geometric optics approach to light acceptance and propagation in graded index fibers. *Opt. Quant. Electron.* 9, 87.

Archambault J.L., Reekie L., and Russel P.S.J. (1993). 100% reflectivity Bragg reflector produced in optical fibers by single excimer laser pulses. *Electron. Letts.* 29, 453.

Askins C.G., Tsai T.E., Williams G.M., Putnam M.A., Bashkanshy M.A., and Friebele E.J. (1992). Fiber Bragg reflectors prepared by a single excimer pulse. *Opt. Letts.* 17, 833.

Banerjee D. (1986). Measurement of cutoff wavelength of single mode fibers: Comparison of various standard techniques. Master's thesis (Opto electronics and Optical Communication), Indian Institute of Technology, Delhi.

Bergano N.S. and Davidson C.R. (1996). Wavelength division multiplexing in long-haul transmission systems. *J. Lightwave Tech.* 14, 1299.

Bergh R.A., Lefevre H.C., and Shaw H.J. (1980). Single mode fiber optics polarizer. *Opt. Letts.* 5, 479.

Bhat A.K. (1984). Measurement of beat length of single mode fibers using magneto optic modulation. Master's thesis (Applied Optics), Indian Institute of Technology, Delhi.

Bilodeau F., Hill K.O., Malo B., Johnson D.C., and Skinner I.M. (1991). Efficient narrow band $LP_{01} \rightarrow LP_{02}$ mode converters fabricated in photosensitive fiber: Spectral response. *Electron. Letts.* 27, 682.

Bird D.M., Armitage J.R., Kashyap R., Fatah R.M.A., and Cameron K.H. (1991). Narrow line semiconductor laser using fiber grating. *Electron. Letts.* 27, 1115.

Blank L.C., Bickers L., and Walker S.D. (1985). Long span optical transmission experiments at 34 and 140 Mbit/s. *J. Lightwave Technol.* LT-3, 1017.

Born M. and Wolf E. (1975). Principles of Optics, Pergamon Press, Oxford.

Burns W.K. Ed. (1994). *Optical Fibers Rotation Sensing*. Academic Press, Boston.

Campbell J.C. (1989). Photodetectors for long wavelength lightwave systems. *Optoelectronics Technology and Lightwave Communications Systems*. Ed. C. Lin. Van Nostrand Reinhold, New York, chapter 14.

Cattermole K.W. (1969). *Principles of Pulse Code Modulation*. Iliffe Books, London.

Chen Y.K., Guo W.Y., Chi S., and Way W.I. (1995). Demonstration of in-service supervisory repeaterless bidirectional wavelength division multiplexing transmission system. *IEEE Photon. Tech. Letts.* 7, 1084.

Chraplyvy A.R., Gnauk A.H., Tkach R.W., Zyskind J.L., Sulhoff J.W., Lucero A.J., Sun Y., Jopson R.M., Forgieri F., Derosier R.M., Wolf C., and McCormick A.R. (1996). 1 Tb/s transmission experiment. *IEEE Photon. Tech. Letts.* 8, 1264.

Chraplyvy A.R., Nagel J.A., and Tkach R.W. (1992). Equalization in amplified WDM lightwave transmission systems. *IEEE Photon. Tech. Letts.* 4, 920.

Christiansen B., Jacobsen G., Mark J., and Mito I. (1994). 4 Gb/s soliton communication on standard non dispersion-shifted fiber. *IEEE Photon. Tech. Letts.* 6, 101.

Coppa G., Costa B., Vita P.D., and Rossi U. (1984). Characterization techniques for monomode fibers and cables. *CSELT Rapporte tecnic* XII, 257.

Creaney S., Johnstone W., and McCallion K. (1996). Continuous fiber modulator with high bandwidth coplanar strip electrodes. *Photon. Tech. Letts*. 8.

Croft T.D., Ritter J.E., and Bhagavatula V.A. (1985). Low loss dispersion shifted single mode fiber manufactured by the OVD process. *J. Lightwave Tech*. LT-3, 391.

Csencsits R.C., Lemaire P.J., Reed W.A., Shenk D.S., and Walker K.L. (1984). Fabrication of low loss single-mode fibers. *Tech. Dig. Conf. Opt. Fiber Commun*. TV-13.

Dakin J. and Culshaw B. (1988). *Optical Fiber Sensors: Principles and Components*, vols. I and II. Artech House, Boston.

Das U.K., Goyal I.C., and Srivastava R. (1987). Mode field radius of dispersion flattened single mode fibers. *Opt. Commun*. 61, 16.

Desurvire E. (1990). Analysis of noise figure spectral distribution in erbium doped fiber amplifiers pumped near 980 and 1480 nm. *Appl. Opt*. 29, 3118.

Desurvire E. (1992). Basic physics of erbium doped fiber amplifiers. *Guided Wave Nonlinear Optics*. Eds. D.B. Ostrowsky and R. Reinisch. Kluwer Academic, Dordrecht, Netherlands. 553.

Desurvire E. (1994). *Erbium Doped Fiber Amplifiers*. John Wiley, New York.

Desurvire E. and Simpson J.R. (1990). Evaluation of $^4I_{15/2}$ and $^4I_{13/2}$ Stark-level energies in erbium-doped aluminosilicate glass fibers. *Opt. Letts*. 15, 547.

Digonnet M.F. and Shaw H.J. (1982). Analysis of a tunable single mode fiber directional coupler. *IEEE J. Quant. Electron*. QE-18, 746.

Dilip K. (1981). Experimental studies on pulse dispersion in optical fibers. Master's thesis (Applied Optics), Indian Institute of Technology, Delhi.

Dong L., Reekie L., Cruz J.L., and Payne D.N. (1996). Grating formation in a phosphorous doped-germenosilicate fiber. Paper presented at Optical Fiber Communications '96, San Jose, CA. February 25–March 1, 1996, paper TuOZ.

Dyott R.B., Bello J., and Handreak V.A. (1987). Indium coated D shaped fiber polarizer. *Opt. Letts*. 12, 287.

Eisenmann M. and Weidel E. (1988). Single mode fused biconical couplers for wavelength division multiplexing with channel spacing between 100 and 300 nm. *J. Lightwave Tech*. 6, 113.

Forrestier D.S., Hill A.M., Lobett R.A., Wyatt R., and Carter S.F. (1991). 38.81 Gb/s 43.8 million-way WDM broadcast network with 527 km range. *Electron. Letts*. 27, 2051.

French W.G., Jaeger R.E., MacChesney J.B., Nagel S.R., Nassau K., and Pearson A.D. (1979). Fiber preform preparation. *Optical Fiber Telecommunications*. Eds. S.E. Miller and A.G. Chynoweth. Academic Press, New York.

Freude W., Ghatak A., and Grau G. (1997). Ray and wave optics in a graded index slab waveguide. Unpublished.

Ghatak A.K. (1985). Leaky modes in optical waveguides. *Opt. Quant. Electron*. 17, 311.

Ghatak A.K. (1992). *Optics*, 2nd ed. Tata McGraw Hill, New Delhi.

Ghatak A.K. (1996). *Introduction to Quantum Mechanics*. Macmillan India, New Delhi.

Ghatak A.K. and Goyal I.C. (1994). Modelling of leaky and absorbing optical waveguides and quantum well structures. In *Linear and Integrated Optics*, SPIE vol. 2212, 238.

Ghatak A.K., Goyal I.C., and Chua S.J. (1995). *Mathematical Physics*. Macmillan India, New Delhi.

Ghatak A.K., Goyal I.C., and Gallawa R.L. (1990). Mean lifetime calculations of quantum well structures: A rigorous analysis. *IEEE J. Quant. Electron*. QE-26, 305.

Ghatak A.K. and Sauter E.G. (1989). The harmonic oscillator problem and the parabolic index optical waveguide. I: Classical and ray optic analysis. *Eur. J. Phys*. 10, 136.

Ghatak A.K. and Sharma A. (1986). Single mode fiber characteristics. *J. Inst. Electron. Telecom. Engrs*. 32, 213.

Ghatak A.K. and Thyagarajan K. (1978). *Contemporary Optics*. Plenum Press, New York.

Ghatak A.K. and Thyagarajan K. (1989). *Optical Electronics*. Cambridge University Press, Cambridge (reprinted: Foundation Books, India, 1991).

Ghatak A.K., Thyagarajan K., and Shenoy M.R. (1987). Numerical analysis of planar optical waveguides using matrix approach. *J. Lightwave Technol.* LT-5, 660.

Ghatak A.K., Thyagarajan K., and Shenoy M.R. (1988). A novel technique for solving the one-dimensional Schrodinger equation using matrix approach: Application to quantum well structures. *IEEE J. Quant. Electron.* QE-24, 1524.

Giallorenzi T.G., Bucaro J.A., Dandridge A., Sigel G.H., Cole H.J., Rashleigh S.C., and Priest R.G. (1982). Optical fiber sensor technology. *IEEE Trans. Microwave Theory and Techniques*. MTT-30, 472.

Gloge D. (1971). Weakly guiding fibers. *Appl. Opt.* 10, 2252.

Gloge D. and Marcatili E.A.J. (1973). Multimode theory of graded-core fibers. *Bell. Syst. Tech. J.* 52, 1563.

Goldstein H. (1950). Classical Mechanics, Addison Wesley, Reading, Massachusetts.

Goyal I.C., Gallawa R.L., and Ghatak A.K. (1990). Bent planar waveguides and whispering gallery modes: A new method of analysis. *J. Lightwave Tech.* LT-8, 768.

Graindorge P., Thyagarajan K., Arditty H., and Papuchon M. (1982). Scattered light measurement of the circular birefringence in a twisted single mode fiber. *Opt. Commun.* 41, 164.

Grattan K.T.V. and Meggitt B.T. (1995). *Optical Fiber Sensor Technology*. Chapman and Hall, London.

Gupta A.K. (1984). Loss measurements in optical fibers. Master's thesis (Opto electronics and Optical Communication), Indian Institute of Technology, Delhi.

Halliday D., Resnick R., and Walker J. (1993). Fundamentals of Physics, John Wiley, New York.

Hawtof D.W., Berkey G.E., and Antos A.J. (1996). High figure of merit dispersion compensating fiber. Paper presented at Optical Fiber Communications '96, San Jose, CA, February 25–March 1, 1996, paper PD-6.

Henry P.S. (1985). Introduction to lightwave transmission. *IEEE Communications Magazine*, 23, 12.

Hill K.O., Bilodeas F., Malo B., and Johnson D.C. (1991). Birefringent photosensitivity in monomode optical fiber: Application to external writing of rocking filters. *Electron. Letts.* 27, 1548.

Hoss R.J. and Lacy E.A. (1993). *Fiber Optics*, 2nd ed. Prentice Hall, New Jersey.

Hussey C.D. and Martinez F. (1985). Approximate analytic forms for the propagation characteristics of single mode optical fibers. *Electron. Letts.* 21, 1103.

Irving J. and Mullineux I. (1959). *Mathematics in Physics and Engineering*. Academic Press, New York.

Iwatsuki K., Suzuki K., Nishi S., and Saruwatari M. (1993). 80 Gb/s optical soliton transmission over 80 km with time/polarization division multiplexing. *Photon. Tech. Letts.* 5, 245.

Jeunhomme L.B. (1983). *Single Mode Fiber Optics*. Marcel Dekker, New York.

Johnstone W., Fawcett G., and Yuiv L.W.K. (1994). In line fiber optic refractometry using index sensitive resonance positions in single mode fiber to planar polymer waveguide couplers. *IEE Proc. Optoelectron.* 141, 299.

Johnstone W., Murray S., Thursby G., Gill M., McDonach A., Moodie D., and Culshaw B. (1991). Fiber optic modulators using active multimode waveguide overlays. *Electron. Letts.* 27, 894.

Johnstone W., Stewart G., Culshaw B., and Hart T. (1988). Fiber optic polarizers and polarizing couplers. *Electron. Letts.* 24, 866.

Johnstone W., Stewart G., Hart T., and Culshaw B. (1990). Surface plasmon polaritons in thin metal films and their role in fiber optic polarizing devices. *J. Lightwave Tech.* 8, 538.

Johnstone W., Thursby G., Moodie D., and McCallion K. (1992). Fiber optic refractometer that utilizes multimode waveguide overlay devices. *Opt. Letts.* 17, 1538.

Juma S. (1996). Bragg gratings boost data transmission rates. *Laser Focus World.* November, 55.

Kao C.K. and Hockam G.A. (1966). Dielectric fiber surface waveguides for optical frequencies. *IEE Proc.* 133, 1151.

Kapron F.P. (1987). Fiber measurements: Factory and field. Paper presented at Optical Fiber Communications/Integrated Optics and Optical Communications, Boulder, Co, Minitutorial, Jan. 22, 1987.

Kashyap R., Armitage J.R., Campbell R.C., Williams D.L., Maxwell G.D., Ainslie B.J., and Millar C.A. (1993a). Light sensitive optical fibers and planar waveguides. *BT Technol. J.* 11.

Kashyap R., Armitage J.R., Wyatt R., Davey S.T., and Williams D.L. (1990). All fiber narrow band reflection gratings at 1550 nm. *Electron. Letts.* 26, 730.

Kashyap R., Wyatt R., and Mckee P.F. (1993b). Wavelength flattened saturated erbium amplifier using multiple side-tap Bragg gratings. *Electron. Letts.* 29, 1025.

Keiser G. (1991). *Optical Fiber Communications.* McGraw Hill, New York.

Kimura T. (1986). Basic concepts of the optical waveguide. *Optical Fiber Transmission.* Ed. K. Noda. North Holland Studies in Telecommunications, 6, North Holland, Amsterdam.

Kressel H. and Butler J.K. (1977). *Semiconductor Lasers and Heterojunction LEDs.* Academic Press, New York.

Laming R.I. and Payne D.N. (1990). Noise characteristics of erbium doped fiber amplifier pumped at 980 nm. *IEEE Photon. Tech. Letts.* 2, 418.

Lathi B.P. (1989). *Modern Digital and Analog Communication Systems*, 2nd ed. Oxford University Press, Delhi.

Lefevre H.C. (1980). Single mode fiber fractional wave devices and polarization controllers. *Electron. Letts.* 16, 778.

Lemaire P.J., Atkins R.M., Mizrahi V., and Reed W.A. (1993). High pressure H_2 loading as a technique for achieving ultrahigh photosensitivity and thermal sensitivity in GeO_2 doped optical fibers. *Electron. Letts.* 29, 1191.

Lin C. (1989). Optical communications: Single mode optical fiber transmission systems. *Optoelectronic Technology and Lightwave Communications Systems*, Ed. C. Lin. Van Nostrand Reinhold, New York.

Mahadevan T.V. (1985). Determination of equivalent step index parameters of single mode fibers. Master's thesis (Physics), Indian Institute of Technology, Delhi.

Marcuse D. (1977). Loss analysis of single mode fiber splices. *Bell Syst. Tech. J.* 56, 703.

Marcuse D. (1978). Gaussian approximation of the fundamental modes of graded index fibers. *J. Opt. Soc. Am.* 68, 103.

Marcuse D. (1979). Interdependence of waveguide and material dispersion. *Appl. Opt.* 18, 2930.

Marcuse D. (1985). Directional coupler filter using dissimilar optical fibers. *Electron. Letts.* 21, 726.

McCallion K., Creaney S., Madden I., and Johnstone W. (1995). A tunable fiber optic bandpass filter based on polished fiber to planar waveguide coupling techniques. *Opt. Fiber Tech.* 1, 271.

McCallion K., Johnstone W., and Fawcett G. (1994). Tunable in-line fiber optic bandpass filter. *Opt. Letts.* 19, 542.

McMohan G.W. and Cielo P.G. (1979). Fiber optic hydrophone sensitivity for different sensor configuration. *Appl. Opt.* 18, 3720.

Melchoir H., Hartman A.R., Shinke D., and Seidel T.E. (1978). Planar epitaxial silicon photodiode. *Bell Syst. Tech. J.* 57, 1791.

Meltz G., Morey W.N., and Glenn W.H. (1989). Formation of Bragg gratings in optical fibers by transverse holographic method. *Opt. Letts.* 14, 823.

Mendez A.C. and Sunak H.R.D. (1986). Chromatic dispersion characteristics of single mode mid–infrared fibers. *Proc. SPIE.* 722, 111.

Midwinter J.E. (1994). The threat of optical communications. *Electron. Comm. Eng. J.* February, 33.

Miya T., Terunama Y., Hosaka T., and Miyashita T. (1979). An ultimate low loss single mode fiber at 1.55 μm. *Electron. Letts.* 15, 106.

Miyajima Y., Komukai T., Sugawa T., and Yamamoto T. (1994). Rare earth doped fluoride fiber amplifiers and fiber lasers. *Opt. Fiber Tech.* 1, 35.

Morey W.W., Ball G.A., and Meltz G. (1994). Photoinduced Bragg gratings in optical fibers. *Opt. Photon. News.* February, 8.

Morishita K. (1989). Optical fiber devices using dispersive materials. *J. Lightwave Tech.* 7, 198.

Moriyama T., Fukuda O., Sanada K., Inada K., Edahvio T., and Chida K. (1980). Ultimately low OH content V.A.D. optical fibers. *Appl. Letts.* 16, 698.

Morkel P.R. (1993). Narrow line width and tunable fiber lasers. *Rare Earth Doped Fiber Lasers and Amplifiers*. Ed. M.J.F. Digonnet. Marcel Dekker, New York.

Mortimore D.B. (1985). Wavelength flattened fused couplers. *Electron. Letts.* 21, 742.

Nagel S. (1989). Optical fiber: The expanding medium. *IEEE Circuits Devices Magaz.* March, 36.

Nakazawa M. (1994). Soliton transmission in telecommunication networks. *IEEE Commun. Magaz.* March, 34.

Neumann E.G. (1988). *Single Mode Fibers*. Springer Verlag, Berlin.

Nikolaus B. and Grischowsky D. (1983). 90 fsec tunable optical pulses obtained by two-stage pulse compression. *Appl. Phys. Letts.* 43, 228.

Ohashi M. and Tsubshawa M. (1991). Optimum parameter design of Er^{3+} doped fiber for optical amplifiers. *Photon. Tech. Letts.* 3, 121.

Ohasi N., Kuwaki N., and Tanaka C. (1986). Characteristics of bend optimized convex index dispersion shifted fiber. Paper presented at First Optoelectronics Conference, Tokyo, paper PDP-22.

Olshansky R. and Keck D.B. (1976). Pulse broadening in graded index optical fibers. *Appl. Opt.* 15, 483.

Onaka H., Miyata H., Ishikawa G., Otsuka K., Ooi H., Kai Y., Kinoshita S., Seino M., Nishimoto H., and Chikama T. (1996). 1.1 Tb/s WDM transmission over a 150 km 1.3 μm zero dispersion single mode fiber. Paper presented at OFC Conference, San Jose, CA, paper PD19-1.

Ouellette F. (1987). Dispersion cancellation using linearly chirped Bragg grating filters in optical waveguides. *Opt. Letts.* 12, 847.

Paek U.C., Peterson G.E., and Carnevale A. (1981). Dispersionless single mode light guides with α index profiles. *Bell Syst. Tech. J.* 60, 583.

Pal B.P. (1979). Optical communication fiber waveguide fabrication: A review. *Fiber Integr. Opt.* 2, 195.

Pal B.P. (1993). Guided wave optics on silicon: Physics, technology and status. Progress in Optics. Ed. E. Wolf. vol. XXXII, 1.

Pal B.P. (1995). Optical fibers for lightwave communication: Design issues. *Trends in Fiber Optics and Optical Communications*. Eds. A.K. Ghatak, B. Culshaw, V. Nagarajan, and B.D. Khurana. Viva Books, New Delhi, 64.

Pal B.P., Thyagarajan K., and Kumar A. (1988). Characterization of optical fibers for telecommunication and sensors. Part I: Multimode fibers. *Int. J. Optoelectron.* 3, 45.

Palais J.C. (1988). *Fiber Optic Communications*. Prentice Hall, Englewood Cliffs, New Jersey.

Pask C. (1984). Physical interpretation of Petermann's strange spot size for single mode fibers. *Electron. Letts.* 20, 144.

Patrick H.J., Williams G.M., Kersey A.D., Pedrazzani J.R., and Vengsarkar A.M. (1996). Hybrid fiber Bragg grating/long period fiber grating sensor for strain/temperature discrimination. *IEEE Photon. Tech. Letts.* 8, 1223.

Pearson A.D., Cohen L.G., Reed W.A., Krause J.T., Sigety E.A., Dimarcello F.V., and Richardson A.G. (1984). Optical transmission in dispersion-shifted single mode spliced fibers and cables. *J. Lightwave Tech.* LT-2, 346.

Pedersen B. (1994). Small-signal erbium doped fiber amplifiers pumped at 980 nm: a design study. *Opt. Quant. Electron.* 26, S273.

Petermann K. (1983). Constraints for fundamental mode spot size for broad band dispersion compensated single mode fibers, *Electron. Letts.* 19, 712.

Poole C.D., Wlesenffold J.M., Digiovanni D.J., and Vengsarkar A.M. (1994). Optical fiber based dispersion compensation using higher order modes near cut off. *J. Lightwave Tech.* 12, 1746.

Ragdale C.M., Reid D.C.J., and Bennion I. (1989). Fiber grating devices. *Proc. SPIE* 1171, 148.

Raizada G. and Pal B.P. (1996). Refractometers and tunable components based on side–polished fibers with multimode overlay waveguides: Role of the superstrate. *Opt. Letts.* 21, 399.

Ramadas M.R., Garmire E., Ghatak A.K., Thyagarajan K., and Shenoy M.R. (1989a). Analysis of absorbing and leaky planar waveguides: A novel method. *Opt. Letts.* 14, 376.

Ramadas M.R., Varshney R.K., Thyagarajan K., and Ghatak A.K. (1989b). A matrix approach to study the propagation characteristics of a general nonlinear planar waveguide. *J. Lightwave Tech.* LT-7, 1901.

Risk W.P., Youngquist R.C., Kino G.S., and Shaw H.J. (1984). Acousto optic frequency shifting in birefringent fiber. *Opt. Letts.* 9, 309.

Rose A.H., Ren Z.B., and Day G.W. (1996). Twisting and annealing optical fiber for current sensors. *J. Lightwave Tech.* 14, 2492.

Rowe C.J., Bennion I., and Reid D.C.J. (1987). High reflectivity surface relief gratings in single mode optical fibers. *Proc. IEE.* 134, 197.

Roy S., Tewari R., and Thyagarajan K. (1991). Accurate empirical relations for characterizing a single mode matched clad fiber from its far field pattern. *J. Opt. Commun.* 12, 26.

Saifi M.A., Lang S.J., Cohe L.G., and Stone J. (1982). Triangular profile single mode fiber. *Opt. Letts.* 7, 43.

Saleh B.E.A. and Teich M.C. (1991). *Fundamentals of Photonics*. John Wiley, New York.

Schwartz M.I. (1984). Optical fiber transmission: From conception to prominence in 20 years. *IEEE Commun. Magaz.* 22, 38.

Shang H.T., Lenachan T.A., Glodis P.F., and Kalish D. (1985). Dispersion-shifted depressed-clad triangular-profile single-mode fiber. Paper presented at OFC'85, Conference on Optical Fiber Communication, San Jose, CA, 92.

Sharma A. (1995). Characteristics of single mode optical fibers. *Trends in Fiber Optics and Optical Communications*. Eds. A.K. Ghatak, B. Culshaw, V. Nagarajan, and B.D. Khurana. Viva Books, New Delhi, 35.

Shenoy M.R., Thyagarajan K., and Ghatak A.K. (1988). Numerical analysis of optical fibers using matrix approach. *J. Lightwave Tech.* LT-6, 1285.

Shimojoh N., Naito T., Deguchi H., Terahara T., and Chikama T. (1996). New gain equilization scheme in WDM optical amplifier repeatered transmission systems. Paper presented at First OECC Conference, Tokyo, paper 120.

Snyder A.W. (1972). Coupled mode theory for optical fibers. *J. Opt. Soc. Am.* 62, 1267.

Snyder A.W. and Love J.D. (1983). *Optical Waveguide Theory*. Chapman and Hall, London.

Sodha M.S. and Ghatak A.K. (1977). *Inhomogeneous Optical Waveguides*. Plenum Press, New York.

Strasser T.A. (1996). Photosensitivity in phosphorous fibers. Paper presented at OFC'96 Conference, San Jose, CA. paper TuO1.

Subramanyam T.V.B. (1983). Refracted near field scanning technique for refractive index profiling of optical fibers. Master's thesis (Applied Optics), Indian Institute of Technology, Delhi.

Taga H., Suzuki M., Edagawa N., Hideaki T., Yoshida T., Yamamoto S., Akiba S., and

Wakabayashi H. (1994). Multi thousand kilometer optical soliton data transmission experiments at 5 Gb/s using an electroabsorption modulater pulse generator. *IEEE Photon. Tech. Letts.* 12, 231.

Tewari R., Pal B.P., and Das U.K. (1992). Dispersion shifted dual-shape core fibers: Optimization based on spot size definitions. *J. Lightwave Technol.* 10, 1.

Tewari R. and Thyagarajan K. (1986). Analysis of tunable single mode fiber directional couplers using simple and accurate relations. *J. Lightwave Technol.* 4, 386.

Tewari R., Thyagarajan K., and Ghatak A.K. (1986). Novel method for characterization of single mode fibers and prediction of cross over wavelength and band pass in nonidentical fibers directional couplers. *Electron. Letts.* 22, 792.

Thyagarajan K., Bourbin Y., Enard A. Vatoux S., and Papuchon M. (1985). Experimental demonstration of TM-mode attenuation resonance in planar metal clad optical waveguides. *Opt. Letts.* 10, 288.

Thyagarajan K., Diggavi S., and Ghatak A.K. (1987a). Analytical investigations of leaky and absorbing planar structures. *Opt. Quant. Electron.* 19, 131.

Thyagarajan K., Diggavi S., and Ghatak A.K. (1988a). Design and analysis of a novel polarization splitting directional coupler. *Electron. Letts.* 24, 869.

Thyagarajan K., Diggavi S., Ghatak A.K., Johnstone W., Stewart G., and Culshaw B. (1990). Thin metal clad waveguide polarizers: Analysis and comparison with experiment. *Opt. Letts.* 15, 1041.

Thyagarajan K., Pal B.P., and Kumar A. (1988b). Characterization of optical fibers for telecommunication and sensors: Part II: Single mode fibers. *Int. J. Optoelectron.* 3, 153.

Thyagarajan K., Shenoy M.R., and Ghatak A.K. (1987b). Accurate numerical method for the calculation of bending loss in optical waveguides using a matrix approach. *Opt. Letts.* 12, 296 (erratum: *Opt. Letts.* March, 1989).

Thyagarajan K., Shenoy M.R., and Ramadas M.R. (1986). Prism coupling technique: A method for measurement of propagation constant and beat length in single mode fibers. *Electron. Letts.* 22, 832.

Thyagarajan K., Varshney R.K., Palai P., Ghatak A.K., and Goyal I.C. (1996). A novel design of a dispersion compensating fiber. *Photon. Tech. Letts.* 8, 1510.

Ulrich R., Rashleigh S.C., and Eikhoff W. (1980). Bending induced birefringence in single mode fibers. *Opt. Letts.* 5, 273.

Ulrich R. and Simon A. (1979). Polarization optics of twisted single mode fibers. *Appl. Opt.* 18, 2241.

Vengsarkar A.M., Lemaire P.J., Judkins J.B., Bhatia V., Erdogan T., and Sipe J.E. (1996a). Long period fiber gratings as band rejection filters. *J. Lightwave Technol.* 14, 58.

Vengsarkar A.M., Pedrazzani J.R., Judkins J.B., Lemaire P.E., Bergans N.S., and Davidson C.R. (1996b). Long period fiber-grating based gain equalisers. *Opt. Letts.* 21, 336.

Wang Q.Z., Lin Q.D., Lin D.H., and Alfano R. (1994). High resolution spectra of self phase modulation in optical fibers. *J. Opt. Soc. Am.* B-11, 1084.

Yariv A. (1989). *Quantum Electronics*, 3rd ed. John Wiley, New York.

Yariv A. (1991). *Optical Electronics*. Holt, Reinhart and Winston, New York.

Yoshida S., Kuwano S., and Iwashita K. (1995). Gain flattened EDFA with high Al concentration for multistage repeatered WDM transmission experiments. *Electron. Letts.* 31, 1765.

Yoshida S., Kuwano S., Takachio N., and Iwashita K. (1996). 10 Gbit/s × 10-channel WDM transmission experiment over 1200 km with repeater spacing of 100 km without gain equalization or pre-emphasis. Paper presented at Optical Fiber Communication 96 (OFC'96), San Jose, CA, USA, TuD6.

Zervas M.N. and Laming R.I. (1995). Twin core fiber erbium doped channel equalizer. *J. Lightwave Technol.* 13, 721.

Zheng W., Hulten O., and Rylander R. (1994). Erbium doped fiber splicing and splice loss estimation. *J. Lightwave Technol.* 12, 430.

Index